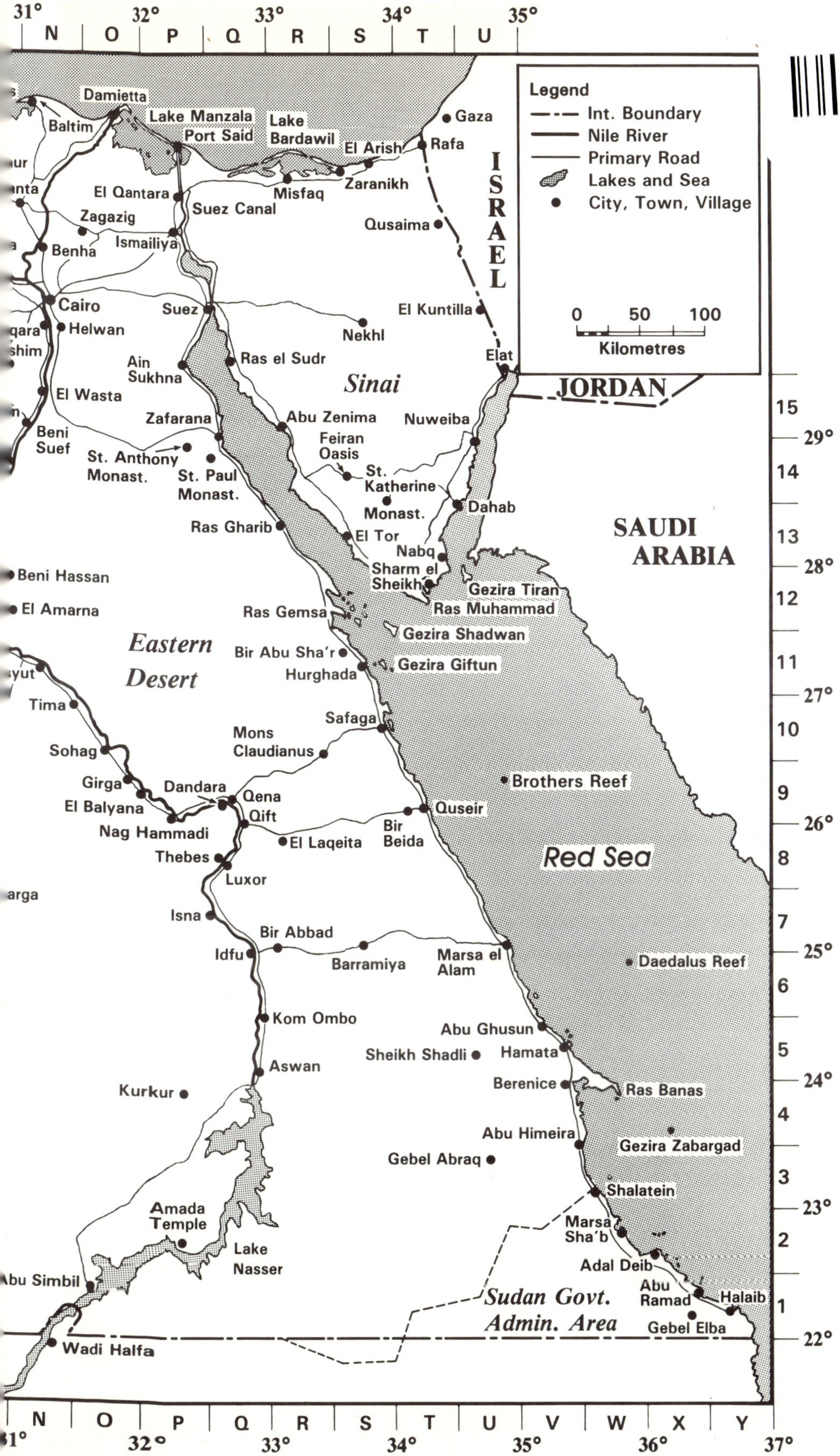

Legend
Int. Boundary
Nile River
Primary Road
Lakes and Sea
City, Town, Village
0 50 100
Kilometres
31° 32° 33° 34° 35°
N O P Q R S T U
N O P Q R S T U V W X Y
31° 32° 33° 34° 35° 36° 37°
15 14 13 12 11 10 9 8 7 6 5 4 3 2 1
29° 28° 27° 26° 25° 24° 23° 22°
ISRAEL
JORDAN
SAUDI ARABIA
Sinai
Eastern Desert
Red Sea
Sudan Govt. Admin. Area
Damietta
Baltim
Lake Manzala
Port Said
Lake Bardawil
El Arish
Gaza
Rafa
Zaranikh
Misfaq
El Qantara
Suez Canal
Zagazig
Ismailiya
Benha
Qusaima
Cairo
Helwan
Suez
Nekhl
El Kuntilla
Elat
Ain Sukhna
Ras el Sudr
El Wasta
Zafarana
Abu Zenima
Nuweiba
Beni Suef
St. Anthony Monast.
St. Paul Monast.
Feiran Oasis
St. Katherine Monast.
Dahab
Ras Gharib
El Tor
Nabq
Sharm el Sheikh
Gezira Tiran
Ras Muhammad
Beni Hassan
El Amarna
Ras Gemsa
Gezira Shadwan
Bir Abu Sha'r
Hurghada
Gezira Giftun
Tima
Safaga
Mons Claudianus
Sohag
Girga
El Balyana
Dandara
Qena
Qift
Nag Hammadi
Brothers Reef
Quseir
Bir Beida
El Laqeita
Thebes
Luxor
Isna
Bir Abbad
Idfu
Barramiya
Marsa el Alam
Daedalus Reef
Kom Ombo
Abu Ghusun
Hamata
Sheikh Shadli
Aswan
Berenice
Ras Banas
Kurkur
Abu Himeira
Gezira Zabargad
Gebel Abraq
Shalatein
Amada Temple
Lake Nasser
Marsa Sha'b
Adal Deib
Abu Ramad
Halaib
Gebel Elba
Abu Simbil
Wadi Halfa

AF684302

THE BIRDS OF EGYPT

THE BIRDS OF EGYPT

EDITORS
STEVEN M. GOODMAN
AND
PETER L. MEININGER

SHERIF M. BAHA EL DIN
JOSEPH J. HOBBS
WIM C. MULLIÉ

Paintings by Sherif M. Baha el Din

Graphics by Edward Khounganian and Peter L. Meininger

Oxford New York
OXFORD UNIVERSITY PRESS
1989

Oxford University Press, Walton Street, Oxford OX2 6DP
Oxford New York Toronto
Delhi Bombay Calcutta Madras Karachi
Petaling Jaya Singapore Hong Kong Tokyo
Nairobi Dar es Salaam Cape Town
Melbourne Auckland
and associated companies in
Berlin Ibadan

Oxford is a trade mark of Oxford University Press

Published in the United States
by Oxford University Press, New York

© *Steven M. Goodman, Peter L. Meininger, Sherif M. Baha el Din,*
Joseph J. Hobbs, Wim C. Mullié, 1989

All rights reserved. No part of this publication may be reproduced, stored in a retrieval system, or transmitted, in any form or by any means, electronic, mechanical, photocopying, recording, or otherwise, without the prior permission of Oxford University Press

British Library Cataloguing in Publication Data
The Birds of Egypt.
1. East & North-east Africa. Birds—Field guides
I. Goodman, Steven M. II. Meininger, Peter L.
598.2967
ISBN 0–19–857644–7

Library of Congress Cataloging in Publication Data
The Birds of Egypt / edited by Steven M. Goodman and Peter L. Meininger; paintings by Sherif M. Baha el Din; graphics by Edward Khounganian and Peter L. Meininger.
p. cm. Bibliography: p. Includes indexes.
1. Birds—Egypt. 2. Ornithology—Egypt—History. I. Goodman, Steven M. II. Meininger, Peter L.
QL692.E3B57 1989 598.2962—dc19 88–22546
ISBN 0–19–857644–7

Set by Wyvern Typesetting Ltd
Printed in Great Britain by
The Alden Press, Oxford

For Dorothé Schmiermann (1953–1985)

CONTENTS

FIGURES

TABLES

PREFACE

The details of the history of Egyptian ornithology are staggering and perhaps unparalleled anywhere else in the world. As early as the Predynastic period (4500–3000 BC) ancient peoples etched and painted rocks with depictions of birds. The relief wall sculpture of Old Kingdom mastabas (2649–2150 BC), such as those at Saqqara, clearly depict the myriads of birds occurring in the Nile Delta and Valley. There were also ancient bird fanciers. In the Dynasty XI (2040–1991 BC) tomb of Baket, at Beni Hassan, the walls are decorated with a partial catalogue of the local avifauna, and the detailed writings of Herodotus (fifth century BC), a Greek traveller and historian, clearly show his interest in documenting the birds of the country. This tradition continued for centuries. Early Islamic treatises on zoology, such as those of Ad-Damiri (*c.* AD 1341–1407), present a unique cultural perspective on understanding the organization of the natural world. As early modern explorers collected, described, and catalogued the local avifauna, Egypt emerged as one of the ornithologically best-known countries of Africa for that period. The nineteenth- and early twentieth-century books and papers by workers such as Adams, Gurney, Nicoll, and Shelley, to name a few, were paramount in forming the basis of our current knowledge of the local bird life. In 1930 Meinertzhagen summarized the information gathered to that time on Egyptian birds.

It is important to point out that the Egypt of most early modern ornithologists was essentially the Nile Delta and Valley, a fraction of the total land area of the country. Recent exploration of the vast regions on either side of the Nile, which were thought to be barren wastelands by many early natural historians, has revealed that portions are zoologically fascinating. In the past decade or so there has been a renewed interest in the birds of the country. Many interesting research projects have recently been implemented and completed, for example, the strategies used by trans-Saharan migrants to counter the extreme physiological stress of the crossing, the numbers and migrational patterns of raptors passing through the country, the importance of Egyptian wetlands for migrating and wintering Palearctic water-birds, the numbers and migrational patterns of water-birds passing in autumn along the north coast of Sinai, sea-birds breeding on Red Sea islands, and the distributional patterns of resident birds in remote desert regions.

It is our hope that this book brings together the many details on Egyptian birds scattered in the literature, museums, and unpublished records, and sets a clear picture of what is known about the avifauna of this fascinating country and what still needs to be done.

Ann Arbor, Michigan, U.S.A. S.M.G.
Middelburg, The Netherlands P.L.M.
1987

ACKNOWLEDGEMENTS

The culmination of a project such as this is never the sole effort of the authors. Without the gracious help of many people, agencies, and institutions this book would never have been completed. We send out a hearty thank you to the following individuals for submitting unpublished material, answering queries, helping with logistics and permits, and companionship in the field: Mohammed Abdel Rahim, M. Alibone, Peter L. Ames, Mohammed Amer, Lars N. Andersen, Ulrik Andersen, Xanja Andriesse, Gamil A. M. Atta, Ofer Bahat, George F. Barrowclough, Peter Basterfield, A. R. Bayoumi, Mark Beaman, Willem Beelaerts van Blokland, G. Belgraver, Amir Ben Dov, Carol A. Bennett, Brigitte Biebach, Herbert Biebach, Rob G. Bijlsma, Pieter Bison, Paul W. Boeren, Greetje Boerma, Johan Bogaert, John Borg, Celia W. Bosman, Len E. Bouldin, Chris G. R. Bowden, David Bradford, Jan Jaap Brinkman, H. G. Brownlow, Bertel Bruun, Erik Buchwald, Marc A. Buise, John Bull, Hidde Bult, Heinz Burgstaller, Lars Byrén, Christian Callsen, Steen Christensen, Victor Cilia, A. J. Cnoop Koopmans, Peter Colston, J. C. Cowley, L. P. Critchley, Abdel Halim A. Darrag, Stanley Davies, Gustaaf De Bièvre, Douwe de Boer, H. Deetjen, P. de Grissac, Nico A. de Haan, René Dekker, Piet den Besten, Frank E. de Roder, Mohammed M. Dessouki, Ben C. Dielissen, Chris Durdin, Enno Ebels, Abdel Gawad A. el Attar, Saad el Badrawi, Yehia M. el Din, Sayed M. el Gamei, Nabil el Hadidi, Samir el Wer, Hans Esselink, René D. Etchécopar, Derek J. Evans, Peter J. Ewins, David Ferguson, Robert Fernea, David J. Fisher, Ted Flaxman, Mustafa M. Fouda, Jack Frazier, Wolfgang Friedrich, Ron Frumkin, C. Hilary Fry, Zjemma M. Gamiss, Charles Gauci, Charles J. Gent, John D. Gerhart, Samir I. Ghabbour, Roberta Gilchrist, Mark Gilston, Paul Gnodde, Friedrich Goethe, Arthur E. Goodall, Derek Goodwin, Andrew Grieve, Leny G. Groen, Steffen Gysel, Ben J. M. Haase, Hassan Hafez, M. Hafez, Jürgen Haffer, Yousef Halim, Tine Harth, C. Vance Haynes, Georg Heine, Hermann Heinzel, Ibrahim Helmy, Mogens Henriksen, Erik Hirshfeld, Ofer Hochberg, Peter L. Hogg, Alan Hollett, Phil A. D. Hollom, John and Anne Homfray, Ted Hoogendoorn, Harry Hoogstraal, Oz Horin, Kenneth O. Horner, Lajos Horváth, Stanley Howe, Rob Hume, A. G. Hurrell, Paul Isenmann, Mounir M. Ishak, Eberhart Jany, Lukas and Susi Jenni, Michael C. Jennings, G. C. Johnson, Mohammed Kassas, Stuart Keith, Ahmed Khafagy, Edward Khounganian, Ragnar Kinzelbach, Per Klaesson, Eduard Koopman, Erling Krabbe, Bernard Krebs, Henry C. Kyllingstad, F. R. Lambert, Wesley Lanyon, Daphna Lavee, Henning Lege, Yossi Leshem, Margret and Otto Lewertoff, Wim Loode, Oluf Lou, Thomas Lovejoy, Mary C. McKitrick, Steve C. Madge, Mustafa A. Mahmoud, D. J. Manns, R. P. Martins, Chris O. Maser,

Lawrence Mason, Victor I. Massad, Gerry Matthews, Mary Megallie, Dick Meijer, Hans Meilstrup, Heinrich Mendelssohn, David Merrie, John Miles, David Mindell, Xavier Misonne, Richard Mitreiter, Gerhard Moll, Mervat Morcos, Nigel Muddiman, Eric Muller, Morten Müller, R. Jon Mycock, William Newmark, Steve F. Newton, E. Max Nicholson, Roger K. Norman, Fergus O'Gorman, Jan Ohlsson, Klaus M. Olsen, Ernst Oosterveld, Bedros Orchanian, Dale J. Osborn, Yoel Paran, Alan Parker, Donald Parr, Uzi Paz, Ib Petersen, Stefan Pihl, D.J. Pitman, Theo Postma, L.R. Prince, Dennis Puleston, Steven Quinn, G.W. Rayner, Adri Remeeus, Jürgen Resch, Ahmed Riad, Bernd Riedel, Hans Riehmann, Giles Roche, Roger Rödel, Karl Roth, Jaques Ruinaard, Hesham Sabry, Uriel Safriel, Stephanie Sagebiel, Saleh Salei, Aziz A. Salam, Wahid Salama, Magdy A. Saleh, Mustafa A. Saleh, Steve Santer, Eric Schaumberg, Dieter Schmidl, Dorothé Schmiermann, Annemarie Schramm, Karl-L. Schuchmann, Holger Schüle, Peter Schüle, Renske Schulting, David E. Sergeant, Tsila Shariv, Hadoram Shirihai, Lester Short, Manfred Siering, Kenneth E. L. Simmons, Ian Smart, Michael Smart, Chris Smeenk, Uffe Gjøl Sørensen, Bernard Spaans, K.G. Spencer, Gerard Steinhaus, Martin Stelzer, Mary Stewart, Robert W. Storer, Richard Stouthamer, Shalom Suaretz, Joe Sultana, Werner Suter, Ruben S.E. Swanquist, Hans Swegen, M. Talat Hashem, John Tennent, Sherif Tewfik, Elke Thielcke, C. Thomas, Derek K. Thomas, Robert Thoor, Ole Thorup, Hassan Touhamy, H.R. Toxopeus, Glen A. Tyler, Magnus Ullmann, Cees Valkenburg, Arnoud van den Berg, Nick van der Ham, Jan van der Kamp, Jan van der Laan, Anja van der Niet, Bert van der Struik, Rudy van Diggelen, Klaas van Dijk, Irene van Nimwegen, Ben van Os, Bob van Poelgeest, Jan van Tussenbroek, Jan Visser, Rudolf Völker, François Vuilleumier, Sudhir Vyas, Mohammed Wafaei, Pim Waldeck, John W. Wall, Terry A. Walsh, Arend Wassink, George E. Watson, Richard Weaving, Geoff Welch, S. M. Whitehouse, Martin Williams, Jörg Wittenberg, Hans Wohlmuth, Hans Wulffsberg, Raof Yhya, Yoram Yom-Tov, Ahmed Zahed, and Seif A. Zaher.

For allowing us access to or providing information on specimens under their care we are grateful to: Dr Kamil Wassif, Ain Shams University; Dr Mustafa A. Saleh, Al Azhar University Collection; Dr Lester L. Short and Ms Mary Lecroy, American Museum of Natural History; Messrs Graham S. Cowles and Ian C.J. Galbraith, British Museum (Natural History); Dr David M. Niles, Delaware Museum of Natural History; Mr Mohammed A. Amin, Entomological Society of Egypt; Mr Badr Abdel Moghany, Farafra Museum of Fine Arts; Drs John W. Fitzpatrick and David E. Willard, Field Museum of Natural History; Dr Dieter S. Peters, Forschungsinstitut Senckenberg; Mr Hassan Galal el Din (private collection); Drs Mohammed Amer and Mervat Morcos, Giza Zoological Museum; Dr Uriel Safriel, Zoological Museum, Hebrew University of Jerusalem; Mr Haim Hovel (private collection); Dr Abdel Hameid Darder, Hurghada Marine Biological Station; Dr Pierre Devillers, Institut Royal des Sciences Naturelles de Belgique; Dr Jan Wattel and Mr C.S. Roselaar, Instituut voor Taxonomische Zoologie, Zoologisch Museum, Universiteit van Amsterdam; Dr Raymond A. Paynter, Jr, Museum of Comparative Zoology,

Harvard; Dr George E. Watson, National Museum of Natural History; Dr Karl Edelstam, Naturhistoriska Riksmuseets; Dr G.F. Mees, Rijksmuseum van Natuurlijke Historie; Dr Jon C. Barlow, Royal Ontario Museum; Dr Heinrich Mendelssohn and Ms Tsila Shariv, Tel-Aviv University Museum; Drs Robert B. Payne and Robert W. Storer, University of Michigan Museum of Zoology; Dr Charles G. Sibley and Ms Eleanor H. Stickney, Yale Peabody Museum; Dr Jon Fjeldså, Zoological Museum, University of Copenhagen; Dr Karl-L. Schuchmann, Zoologisches Forschungsinstitut und Museum Alexander Koenig; and Dr K. Joysey, University Museum of Zoology, Cambridge, UK.

For providing details or answering queries about bird-ringing we are grateful to B. Gal, Bird Ringing Centre, Institute for Nature Conservation Research, Tel-Aviv University, Tel-Aviv, Israel; F. Roux, Centre de Recherches sur la Biologie des Populations d'Oiseaux, Museum National d'Histoire Naturelle, Paris, France; B. Behrouzi, Department of the Environment, Tehran, Iran; R.D. Wassenaar and A.C. Perdeck, Euring Data Bank and Vogeltrekstation, Institute for Ecological Research, Heteren, The Netherlands; M. Morcos and M. Abdel Rahim, Giza Zoological Museum and Egyptian Wildlife Service, Cairo, Egypt; G.I. Handrinos, Hellenic Ornithological Society, Ringing Scheme, Zographos, Greece; W. Token, Institut für Vogelforschung 'Vogelwarte Helgoland', Wilhelmshaven-Rüstersiel, Federal Republic of Germany; T. Gaultier, Institut National de Recherche Scientifique et Technique, Centre de Biologie et de Ressources Génétique, Unité d'Ornithologie, Hammam-Lif, Tunisia; M. Spagnesi, Instituto Nazionale di Biologia della Selvaggina, Ozzano Emilia, Italy; R. Patapavičius, Lithuanian Ringing Centre, T. Ivanausko Zoologijos Muziejus, Kaunas, Lithuania; R. Holyński, Magyar Madartani Egyesület, Budapest, Hungary; J. Sultana, Malta Ornithological Society, Valletta, Malta; S. Ashkenazi, Nature Reserves Authority, Jerusalem, Israel; R. Staav, Naturhistoriska Riksmuseet, Ringmärkningscentralen, Stockholm, Sweden; J. Gromadzka, Ornithological Station, The Polish Academy of Sciences, Institute of Zoology, Gdańsk, Poland; D. Šere, Ringing Centre, Kustodiat za Ornitologijo, Prirodoslovni muzej Slovenije, Ljubljana, Yugoslavia; O.W. Røstad, Ringing Department, Zoologisk Museum, Universitetet i Oslo, Norway; A. Vancoppenolle and A. Devos, Royale Fédération Colombophile Belge, Brussels, Belgium; T.B. Oatley, South African Bird-ringing Unit, University of Cape Town, Rondebosch, Cape, South Africa; S.P. Kharitonov, USSR Academy of Sciences, Institute of Evolutionary Animal Morphology and Ecology, The Centre of Ringing and Marking of Birds and Terrestrial Mammals, Moscow, USSR; R. Schmidt, Vogelwarte Hiddensee, Ernst-Moritz-Arndt Universität, Greifswald, German Democratic Republic; and G. Sirotic, Zavod za Ornitologiju, Jugoslavenska Akademija Zanosti i Umjetnosti, Zagreb, Yugoslavia.

Over the course of more than a decade of field research in Egypt we have been financially supported at various times by: British Petroleum Ltd, Egypt Branch; Cairo University; Egyptian Academy of Scientific Research and Technology; Frank M. Chapman Fund, American Museum of Natural History; The Fulbright Program;

Holy Land Conservation Fund; Horace H. Rackham School of Graduate Studies, University of Michigan; International Union for the Conservation of Nature and Natural Resources; International Waterfowl Research Bureau; National Geographic Society; The Netherlands Foundation for International Bird Preservation; The Netherlands Foundation for International Nature Conservation; Ornithological Society of the Middle East; Prins Bernhard Fonds; The Rolex Awards for Enterprise; Suez Canal University; University of Michigan Museum of Zoology; US Fish and Wildlife Service; and World Wildlife Fund. Further financial support during the writing phase of this project was received from the Ornithological Society of the Middle East and Shell Winning NV.

Our colleagues of the Egyptian Wildlife Service provided help in numerous ways. Dr Uzi Paz kindly allowed us to examine and cite his 1987 book when it was still in the galley proof stage. Mr Hadoram Shirihai, International Birdwatching Centre, Elat, provided unpublished information on his extensive experience with resident and migratory birds occurring in the area. Dr Edna Coffin and Anath Yahiv helped to translate some works from the Hebrew. We are indebted to Mr Edward Khounganian for graciously offering his skills in drafting the maps used in this book.

We would like to give special respect to the late Dr Harry Hoogstraal whose never-ending insight and help were a source of inspiration. He played a paramount role in the growth and development of modern zoological studies in Egypt and his death in early 1986 truly marked the end of an era.

Sherif M. Baha el Din would like to pay special tribute to his family, who throughout the years have been a constant source of encouragement, tolerance, and support. Bertel Bruun and David Ferguson also provided invaluable advice and generous help. The comradeship of Mustafa Fouda is deeply appreciated.

Steven M. Goodman would like personally to make the following acknowledgements. Several individuals at the University of Michigan helped me to see this work through: my mentor, Dr Robert W. Storer, has been a constant source of encouragement throughout the years in both the field and museum; Dr Philip Myers helped in numerous ways, including funds to subsidize text-editing costs; Mrs Janet Hinshaw provided untold aid with reference materials; and Dr Robert B. Payne generously provided working space and supplies. During the writing phase Michael W. Nachman and Richard O. Prum were both a source of encouragement and housing. The Horace H. Rackham School of Graduate Studies, University of Michigan, provided partial financial support during the completion of the book.

Joseph J. Hobbs extends his appreciation to Leo Tregenza, Robin W. Doughty, Ian R. Manners, Saleh Ali Suwaylim, and his parents Col. and Mrs Gregory J. Hobbs.

Peter L. Meininger would like to acknowledge his gratitude to Greetje Boerma for her patience and understanding during the writing of this book, and to his parents Mrs C. J. and Mr L. J. Meininger for their encouragement throughout the years.

Wim C. Mullié would like to add the following personal acknowledgements. Dorothé Schmiermann was my constant source of inspiration. Her tolerance was more than I could ever have asked for. Johan Dekker, Ursula Strik, and my parents Mrs M. and Mr D. Mullié were of crucial help in continuing the project after Dorothé's death. Their encouragement and support helped me through this difficult period in my life. The hospitality in Egypt of Mrs N. Nassar, Mr Bahi Ahmed, and David and Heddy Merrie is deeply appreciated.

The editors would like to acknowledge the help and co-operation of all the authors. We are grateful that during the evolution of this project the relationships of five colleagues, from three continents and with different backgrounds, merged into friendships.

Part 1

1. INTRODUCTION

Although Egypt has a long history of ornithological research until recently little information was available on the non-Nilotic portions. In a 1969 paper Reginald E. Moreau stated, 'notwithstanding the imposing two volumes of Meinertzhagen (1930), Egypt south of Cairo and on both sides of the Nile Valley remains one of the worst-documented parts of Africa . . .'. The past ten years have seen a surge of interest in Egyptian ornithology and substantial areas of the country, hitherto unknown, have been surveyed. The purpose of this book is to synthesize the new information on the birds of the country and to some extent re-evaluate the material available to Meinertzhagen (1930).

BRIEF HISTORY OF POST-1930 EGYPTIAN ORNITHOLOGY

After the death of Michael Nicoll in 1925, a partially composed manuscript that he had been writing on Egyptian birds was turned over to Richard Meinertzhagen. This manuscript clearly formed an integral portion of Meinertzhagen's 1930 book. Much of the late eighteenth-, nineteenth-, and early twentieth-century literature and work on Egyptian ornithology was summarized by Meinertzhagen, and the intent of this section is to briefly review work conducted in the country since then.

Alexander Koenig, a contemporary of Meinertzhagen, published several substantial monographs on the birds of the country and the Nile between 1907 and 1936. The majority of his papers appeared in the *Journal für Ornithologie*. Ahmed Al Hussaini was an important post-1930 figure in Egyptian ornithology. In the mid- to late 1930s he explored areas of the Western Desert and the Red Sea coast and published several papers on bird observations. His works also include a 1940 book in Arabic on the birds of Egypt, which was revised in 1954. Abdallah El Negumi surveyed remote desert areas, particularly the Eastern Desert, in the late 1940s and 1950s and made bird collections which are housed in the Giza Zoological Museum. In 1950 El Negumi, together with several colleagues, published a book that remains the standard work in Arabic on the birds of the country.

The British influence on twentieth-century Egyptian ornithology was significant. Until 1936 Egypt remained under British military occupation and numerous armed forces and civil personnel made important ornithological contributions. One individual at the forefront was Reginald E. Moreau, whose publications dealing with Egyptian ornithology span 44 years, from 1926 to 1970. At least from June 1944 to February 1945 R. H. Greaves compiled monthly bird observation notes which were distributed by educational authorities to local naturalists. Between the late 1940s and the mid-1950s the British army occupied various positions along the Suez Canal. From 1949 to at least 1951 the cyclostyled *Bulletin of the Fayid Ornithological Club* was distributed to members, mostly British military personnel stationed along the Suez Canal. At least 16 issues were circulated. H. G. Brownlow edited numbers 1–12 and L. P. Critchley numbers 13–16. These bulletins contain information on ringing projects, notable bird observations of members, and general bird notes. Several detailed natural history studies of local breeding birds were made by this group of people (e.g. Simmons 1952, 1954*a*) and a ringing station was operated at Fayid (see p. 7).

It was during this period that the magnitude of raptor migration along the Suez Canal was realized (Greaves 1945*a*; E. G. M. Goodwin 1947*a*). In 1945 a leaflet was published in the *Bulletin of the Zoological Society of Egypt* illustrating the flight profile and markings of several raptors to at least in part spur interest in raptor counts. That same year E. G. M. Goodwin (1947*b*) conducted one of the earliest radar studies on birds anywhere in the world; he measured the altitude and speed raptors pass at along the Suez Canal. More thorough raptor migration counts were subsequently performed by D. Goodwin (1949) and Simmons (1951).

Starting in the mid-1950s Harry Hoogstraal and his colleagues of the United States Naval Medical Research Unit Number Three (NAMRU-3), Cairo, commenced studies in Egypt on the relationships of ticks and their vertebrate hosts with the transmission of arboviruses. Birds were an important group in this research. Through netting, collecting, and visiting wild-bird markets a considerable amount of new information was obtained. Some of the earlier NAMRU-3 bird migration studies were summarized by Hoogstraal *et al.* (1961, 1963, 1964) and Hoogstraal and Kaiser (1961). The majority of NAMRU-3 bird material is deposited in the Field Museum of Natural History, Chicago. It should be noted that for several decades Hoogstraal and other researchers studying Egyptian zoology obtained specimens from the commercial hunters and trappers of Abu Rauwash. The men of this village travel widely throughout the country for their animals. Some birds procured from them simply bear the collection locality 'Egypt: Giza Governorate, Abu Rauwash'. The possibility that some specimens were taken elsewhere in the country and then brought to the village cannot be eliminated. Thus, caution should be exercised with specimens from this locality.

In 1957 an expedition from the Natural History Museum, Budapest, visited the country and obtained a considerable amount of new information. The general results were published by Horváth (1959). Some of the material collected on this trip included the type series of *Ammomanes deserti borosi* (Horváth 1958), the most recently named endemic bird form from the country.

In three field seasons between 1962 and 1965 the Yale University Prehistoric Expedition to Nubia worked in the portion of the Nile Valley now inundated by Lake Nasser. The field party contained several zoologists, including Thomas Lovejoy, Chris

Maser, and Charles Reed. The data and materials they gathered are particularly important because they represent the last information on the birds of this area before it was completely flooded by Lake Nasser. The specimens obtained by this group are deposited in the Yale Peabody Museum of Natural History, New Haven.

From 1966 to 1973 the Palearctic Migratory Bird Survey, under the direction of George E. Watson and in collaboration with NAMRU-3 and the Yale Arbovirus Research Unit, conducted a serological study of migratory birds and their ecto-parasites in north-eastern Africa. The autumn field seasons of 1966 and 1968–72 and spring seasons of 1967 and 1969–73 were held in Egypt, primarily along the north coast and in the Nile Delta. Information on some of the ornithological data gathered was summarized and analysed by Watson (1971) and Horner (1977). Voucher specimens were deposited in the National Museum of Natural History, Washington, DC.

In the early 1970s Daphna Lavee and Uriel Safriel of the Hebrew University, Jerusalem, studied bird migration in the mountains of southern Sinai. Their main research site was the gardens of the St Katherine Monastery, principally during the autumn. This work concentrated on the species passing through or wintering in the region and the importance of isolated oases as 'refuelling' stations for migrants (Lavee and Safriel 1974). Voucher specimens were collected and deposited in the Zoological Museum, Hebrew University of Jerusalem.

Perhaps no area of the country still remains as poorly known as the vast expanses of the Western Desert, particularly the more remote and inaccessible portions. Between 20 October 1968 and 15 January 1969 a Belgian expedition surveyed the vertebrate life of the Gebel Uweinat and Gilf Kebir areas. The ornithological results, which were published by Misonne (1974), form to a great extent our current knowledge of the birds occurring at these two isolated massifs. A number of groups have recently visited the more accessible portions of the Western Desert and this information has been summarized by Goodman *et al.* (1986). During four migratory seasons between 1980 and 1985 Herbert Biebach and his colleagues of the Max-Planck-Institut für Verhaltensphysiologie studied bird migration in Bahariya. This work has been particularly useful in elucidating the species composition, timing, and physiological and behavioural strategies used by birds crossing the Egyptian portion of the Sahara (Biebach 1985; Biebach *et al.* 1986).

We have widely used bird records collected by organized bird-watching tours to the country. The majority of these groups confine their travels to the Nile Delta and Valley. Two regular tours of particular note are the Nile cruises of the American Museum of Natural History and the Swan Hellenic/Royal Society for the Protection of Birds. The former group has visited the country almost every year since 1976 and records up until 1979 were summarized by Short and Horne (1981).

An important source of information on the migratory movements of birds passing through Egypt, particularly water-birds, has been the systematic counts conducted along the north coast of Sinai. The periods of these surveys include: 13 August–29

September 1973 and 15 August–7 October 1978 (Yoel Paran and Uzi Paz, unpublished), 11–20 October 1979 (Petersen and Sørensen 1980), 7–30 September 1980 (Petersen and Sørensen 1981*a*), 16 August–24 September 1981 (Petersen and Sørensen 1981*b*), and 2–24 September 1982 (Baha el Din and Salama 1984). During the winters of 1978–9 and 1979–80 surveys of water-birds using Egyptian wetlands clearly documented the importance of these areas, particularly for Palearctic migrants (Meininger and Mullié 1981*b*). In the spring of 1983 a group visited the Nile Delta lakes to document the local breeding birds (Meininger *et al.* 1986).

Other recent ornithological research projects in the country include raptor counts at Suez in autumn 1981 from 4 September to 5 November (Bijlsma 1983), spring 1982 from 23 February to 16 May (Wimpfheimer *et al.* 1983), and autumn 1984 from 5 September to 26 November (Rudy van Diggelen *et al.* unpublished); and inventories of sea-birds breeding on the Red Sea islands (Jennings *et al.* 1985; Goodman and Storer 1987). The Gebel Elba region in the extreme south-eastern corner of the Eastern Desert contains one of the most diverse and interesting biotas of the country. In the past five years this area has been visited several times for ornithological surveys (Goodman 1985; Goodman and Atta 1987; Goodman unpublished; Meininger unpublished).

The foundation of the Egyptian Wildlife Service (EWS) in 1979 heralded a new era in zoological field studies and national conservation practices. EWS rangers regularly visit remote areas of the country and have added to our knowledge of local avifaunas. Further, this agency has facilitated numerous ornithological field studies conducted by foreigners. Perhaps the most important recent development is the creation in 1982 of the Ornithological Society of Egypt (OSE). Some of the prime goals of the OSE are to promote ornithological studies in Egypt and national interest in educating local people about birds and conservation. The organ of the society, the *Courser*, provides an excellent medium for presenting this information.

In 1979 the Birds of Egypt Project was launched by P. L. Meininger and W. C. Mullié. In subsequent years the other three authors of the present book joined this working group. Requests for information on Egyptian birds were published in several European and North American ornithological journals and newsletters. By mid-May 1988 over 250 responses had been received from various institutions and people with unpublished information on the birds of the country.

This book is a collaborative effort by all the authors. Several points should, however, be mentioned about particular aspects of its production. The chapter on geography was principally written by Joseph J. Hobbs. Wim C. Mullié drafted the sections on conservation, hunting, and responses of birds to changing environment. Steven M. Goodman was principally responsible for the examination of museum specimens in various collections and the deposition of birds in the University of Michigan Museum of Zoology.

RINGING

Ringing operations in Egypt commenced in 1937 and the vast majority of birds handled since then have been fitted with Giza Zoological Museum rings. The first relatively intensive programme was conducted between 1937 and 1948 by ringers associated with the Giza Zoological Museum and the Zoological Society of Egypt. Mackintosh (1941) reported on the species and numbers handled up until 1940. The most commonly ringed birds during this period were Quail—4524 between 1937 and 1940; various ducks—8298 between 1937 and 1943 (predominantly Pintail); and White Wagtail—2232 between 1940 and 1941.

The next period of concentrated effort was from the spring seasons of 1949 to 1954, when British military personnel intermittently operated a Heligoland trap at Fayid, along the Great Bitter Lake. The ringing scheme and a portion of the information gathered at this site were reported by Brownlow (1952, 1960) and Simmons (1954*b*). The data from the last few years of this project were lost.

Between 1966 and 1973 field-workers of the Palearctic Migratory Bird Survey (see p. 5) ringed over 76 000 birds, mostly along the North Coast near Bahig. Information on the ringing programme was published by Hubbard (1967*a*, 1967*b*) and Hubbard and Horner (1969). Watson (1971) summarized the work conducted until the autumn of 1970, at which point over 47 000 birds of 75 species had been ringed, and tabulated post-Mackintosh (1941) recoveries in the Giza Zoological Museum Ringing Files (GZF) and some published literature.

Since Watson's 1971 report a considerable number of foreign ringed birds have been recovered or recaptured in Egypt. A large percentage of this information has been gleaned from the GZF and the Euring Data Bank. Most of the balance is from reports by various Eurasian and African national ringing centres. The vast majority of this information is summarized, along with previously published records, in the species accounts (Ch. 6). We have made a serious attempt to double-check many of these recoveries through correspondence with the ringing centres and/or by checking the literature. Information is outstanding for several hundred foreign birds recovered in Egypt and reported to the original ringing centre.

Approximately 100 000 birds have been ringed in Egypt to date. By May 1988 a total of 1337 recoveries of at least 90 species have been checked. Of these, 273 are birds ringed in Egypt and recovered within the country or abroad and 1064 are birds ringed abroad and recovered in Egypt. Of the recoveries, 76 per cent are of 10 species and much of this information is mapped under respective species accounts in Chapter 6. The 10 species include (number of recoveries in parentheses): White Stork (299), Slender-billed Gull (212), Quail (98), Pintail (57), Caspian Tern (76), White Wagtail (69), Lesser Black-backed Gull (52), Red-backed Shrike (60), Lesser Whitethroat (51), and Willow Warbler (48).

SOURCES OF UNPUBLISHED INFORMATION

The sources of unpublished records cited within the species accounts are referenced by acronyms. In numerous cases this information comes from trip and expedition reports or summaries of research projects that are in manuscript form. We have cited them as such. In other cases the information was obtained through correspondence or some eclectic fashion and only the abbreviations of the reporter(s) or observer(s) are given. In several cases, further information was obtained from an observer(s) other than what was contained in a manuscript or unpublished report.

After the list of observer acronyms is a key to the acronyms of museums and private collections that specimen material is cited from. Also in the second list are acronyms for resource centres mentioned in the text.

Abbreviations of observers or reporters

AAS A. A. Salam
ACK A. J. Cnoop Koopmans
ADB A. D. de Boer
AeA A. G. A. el Attar
AEG A. E. Goodall
AGH A. G. Hurrell
AH A. Hollett
AJA A. J. Andriesse
AP A. Parker
AR A. Remeeus
AS A. Schramm
AvdB A. van den Berg (see also BH *et al.*)
AW A. Wassink
BB B. Bruun
BBe B. Behrouzi. Bird-ringing information supplied by the Department of the Environment, Tehran, Iran
BCD B. C. Dielissen
BH *et al.* B. J. M. Haase, A. B. van den Berg, C. W. Bosman, A. M. van der Niet, C. Smeenk. Notes on birds observed in Egypt during the Snellius I Expedition, 7 June–11 June 1984; B. J. M. Haase, A. B. van den Berg, C. W. Bosman, A. M. van der Niet, C. Smeenk. Notes on birds observed in Egypt during the Snellius II Expedition, 9–22 July 1985
BO B. Orchanian
BR B. Riedel. Beobachtungsbericht der Ägypten Studienreise vom 21.1–6.2.83 (unpublished report, 1983)
CAB *et al.* C. A. Bennett, R. G. Bijlsma, R. Stouthamer. Survey of waterbirds on Egyptian wetlands, autumn 1981 (unpublished report, 1982)
CB C. G. R. Bowden
CC C. Callsen. Ornithologische Beobachtungen auf einer Kulturellen Ägyptenreise vom 18.3–8.4.84 (unpublished report, 1984)
CHF C. H. Fry
CJG C. J. Gent
COM C. O. Maser. The vertebrate fauna around Gebel Adda (unpublished manuscript n.d.)

CT C. Thomas
CVH C. V. Haynes
DB D. Bradford
DES D. E. Sergeant
DJE D. J. Evans
DJF *et al.* D. J. Fisher, S. Gantlett, K. Harrison, R. B. H. Smith. Birds in Israel 17–31 March 1979 (unpublished report)
DJM D. J. Manns
DJO D. J. Osborn
DKT D. K. Thomas
DL D. Lavee
DM T. D. H. Merrie (see also SH)
DMe D. Meijer
DMi D. Mindell
DP D. Puleston
DSP D. S. Peters
EB E. Buchwald. Birds in Eilat, spring 1979 (unpublished report, 1979)
EE *et al.* E. Ebels, P. Gnodde, J. van der Laan. Report of an ornithological trip to Egypt, 5–30 August 1986 (unpublished report)
EH E. Hirschfeld
EK E. Khounganian
EKr E. Krabbe
EM E. Muller
EMN E. M. Nicholson
ES E. Schaumburg
FdR F. E. de Roder
FG F. Goethe
FRL F. R. Lambert
FV F. Vuilleumier. Ornithological observations during the fall 1979 Nile Cruise of the American Museum of Natural History, October 15 to October 30 (unpublished report, n.d.)
GCJ G. C. Johnson
GdB G. de Bièvre
GFB G. F. Barrowclough. Bird observations during the winter 1985 Nile Cruise of the American Museum of Natural History (unpublished report, 1985)
GIH G. I. Handrinos. Bird-ringing information supplied by the Hellenic Ornithological Society, Ringing Scheme, Zographos, Greece
GM G. Moll
GMA G. A. M. Atta
GS G. Sirotic. Bird-ringing information supplied by the Jugoslavenska Akademija Znanosti i Umjetnosti Zavod za Ornitologiju, Zagreb, Yugoslavia
GT G. A. Tyler
GWR G. W. Rayner
HB *et al.* H. Biebach, B. Biebach, D. Schmidl. Notes on birds observed and captured in the Western Desert, autumn 1980; H. Biebach, D. Schmidl, G. Heine, J. Resch, W. Friedrich. Notes on birds observed and captured in the Western Desert, autumn 1982; H. Biebach, H. Schüle, W. Friedrich, R. Mitreiter, K. Roth, G. Heine, T. Harth, E. Thielcke, J. Resch. Notes on birds observed and captured in the Western Desert, autumn 1983; H. Biebach, B. Biebach, W. Friedrich, P. Schülze, D. Schmidl, L. Jenni, S. Jenni. Notes on birds observed and captured in the Western Desert, autumn 1985
HBg H. Burgstaller

HBu H. Bult
HCK H. C. Kyllingstad
HD H. Deetjen
HGB H. G. Brownlow
HHg H. Hoogstraal
HL H. Lege
HM H. Mendelssohn
HR H. Riehmann
HS H. Sabry
HSh H. Shirihai; also H. Shirihai. Bird migration in Eilat. Observations summary autumn 1980 (unpublished report,1982)
HSw H. Swegen
HW H. Wohlmuth
HWu H. Wulffsberg
IH I. Helmy
IP I. Petersen
JB J. Bogaert
JBu J. Bull. Wildlife along the Nile—October 26 to November 7, 1978 (unpublished report, 1979)
JCC J. C. Cowley
JDG J. D. Gerhart
JG J. Gromadzka. Bird-ringing information supplied by the Polish Academy of Sciences, Institute of Zoology, Ornithological Station, Gdańsk, Poland
JH J. and A. Homfray
JJB J. J. Brinkman
JJH J. J. Hobbs
JO J. Ohlsson
JS J. Sultana. Bird-ringing information supplied by the Malta Ornithological Society, Valletta, Malta
JS *et al.* J. Sultana, J. Borg, V. Cilia, C. Gauci
JT J. Tennent
JV J. Visser
JvT J. van Tussenbroek
JW J. Wittenberg
JWW J. W. Wall
KLS K. E. L. Simmons
KO K. M. Olsen
LB L. E. Bouldin. The birds of Kasfareet. A contribution to Egyptian ornithological records (unpublished manuscript, 1953)
LBy L. Byrén
LGG L. G. Groen
LH L. Horváth
LNA L. N. Andersen
LPC L. P. Critchley
LRP L. R. Prince
MAl M. Alibone
MAM M. A. Mahmoud
MAS Magdy A. Saleh
MB M. A. Buise
MBe M. Beaman, S. C. Madge. Ornithological observations made during Birdquest visit to Egypt, 14–29 April 1984 (unpublished report)

MCJ *et al.* M.C. Jennings, D. Parr, P.C. Heathcote, R. Frost. Expedition to the Egyptian Red Sea Coast, 16 April to 18 May 1983: systematic list of the birds observed; M.C. Jennings, D. Parr, D.C. Heathcote, S.M. Baha el Din. Expedition to the Egyptian Red Sea Coast, 23 September–11 October 1984: systematic list of birds observed.

MCM M.C. McKitrick. Birds observed during the winter 1986 Nile Cruise of the American Museum of Natural History (unpublished report, 1986)

MK M. Kassas

MS M. Siering

MSm M. Smart

MSp M. Spagnesi. Bird-ringing information supplied by the Instituto Nazionale di Biologia della Selvaggina, Ozzano Emilia, Italy

MSt M. Stelzer

MU M. Ullmann

MuS Mustafa A. Saleh

MW M. Williams

MWa M. Wafaei

NdH N.A. de Haan, P.W. Boeren. Ornithological observations in Egypt, 22 September–4 October 1985. Stichting Natuurreizen (unpublished report)

NM N. Muddiman

OH O. Hochberg

OL O. Lewertoff

OT O. Thorup

PB P. Basterfield

PdB P. den Besten

PdG P. de Grissac

PDH P.A.D. Hollom

PE P.J. Ewins

PI P. Isenmann

PK P. Klaesson

PLA P.L. Ames

PLM P.L. Meininger

PW P. Waldeck

PWB P.W. Boeren (see NdH)

RD R. Dekker

RF R. Frumkin

RFe R. Fernea

RGB *et al.* R.G. Bijlsma, C.A. Bennett, B. van Poelgeest, T. Postma, F.E. de Roder, R. Stouthamer, J. Visser (see also CAB *et al.*)

RHG R.H. Greaves

RK R. Kinzelbach

RJM R.J. Mycock

RN R. Norman

RP R. Patapavičius. Bird-ringing information provided by the Lithuanian Ringing Centre, T. Ivanausko Zoologijos Muziejus, Kaunas, Lithuania

RR R. Rödel

RS R. Stouthamer (see also CAB *et al.* and RGB *et al.*)

RSc R. Schmidt. Bird-ringing information supplied by Vogelwarte Hiddensee, Greifswald, German Democratic Republic

RSS R.S.E. Swanquist

RSt R. Staav. Bird-ringing information supplied by the Naturhistoriska Riksmuseet, Ringmärkningscentralen, Stockholm, Sweden

RT R. Thoor

RV R. Völker

RvD *et al.* R. van Diggelen, H. Esselink, E. Oosterveld, B. van Os, B. van Poelgeest, C. Valkenburg, J. Visser

RW R. Weaving

RWa R. Wassenaar. Bird-ringing information supplied by Vogeltrekstation, Institute for Ecological Research, Heteren, The Netherlands

RWS R. W. Storer

SC S. Christensen

SFN S. F. Newton

SG S. Gysel. Reisebericht mit Faunalisten der ALA-Ägyptenrundfahrt, 18 Februar–10 März 1984 (unpublished report, 1984)

SH S. Howe, T. D. H. Merrie. Birds in Egypt, 1983–84 (unpublished report, 1984)

SK S. Keith. Birds observed during the winter 1980 Nile Cruise of the American Museum of Natural History (unpublished report, 1980)

SM S. C. Madge (see MBe)

SMB S. M. Baha el Din

SMG S. M. Goodman

SP S. Pihl. Birds of Eilat spring '77 (unpublished report, 1977)

SQ S. Quinn. Bird observations along the Egyptian Nile made during the American Museum of Natural History winter, 1982 Nile Discovery Cruise (unpublished report, 1982); S. Quinn. Bird observations along the Egyptian Nile made during the American Museum of Natural History winter, 1983 Nile Discovery Cruise (unpublished report, 1983)

SS S. Suaretz; also S. Suaretz. Survey of breeding birds on the Island of Tiran and the coast of southern Sinai (unpublished report, 1973)

SSa S. Salei

ST S. Tewfik

SV S. Vyas

SZ S. Zaher

TAW T. A. Walsh

TF T. Flaxman

TS T. Shariv

UA U. Andersen

UGS U. G. Sørensen

UP U. Paz; also U. Paz, Y. Paran. Summary of migration of water-birds in the north of Sinai 13.9.74–15.7.74 (unpublished report, n.d.) (see also YP)

VIM V. I. Massad

WaS W. Salama

WCM W. C. Mullié

WL W. Loode

WS W. Suter

WT W. Token. Bird-ringing information supplied by the Institut für Vogelforschung, Vogelwarte Helgoland, Wilhelmshaven-Rüstersiel, Federal Republic of Germany

YL Y. Leshem

YP Y. Paran; also Y. Paran, U. Paz. Autumn migration of water birds on the north coast of Sinai (unpublished report, n.d.); Y. Paran, U. Paz. Autumn migration of water birds on the north coast of Sinai (unpublished report, 1978) (see also UP)

YYT Y. Yom-Tov

Abbreviations for museums and private collections, and resource centres

AAUC Al Azhar University Collection, Cairo, Egypt
AMNH American Museum of Natural History, New York, USA
BEP Birds of Egypt Project (collective information of the authors)
BMNH British Museum (Natural History), Tring, UK
DMNH Delaware Museum of Natural History, Greenville, USA
EDB Euring Data Bank, Heteren, The Netherlands
FMNH Field Museum of Natural History, Chicago, USA; a portion of which is part of the Conover Collection (CC)
GZF Giza Zoological Museum Ringing Files, Giza, Egypt
GZM Giza Zoological Museum, Giza, Egypt
HGD private collection of Hassan Galal el Din, Cairo; part housed in the Alexandria Zoological Museum, Alexandria, Egypt
HH private collection of Haim Hovel, Haifa, Israel
IBIC Israel Bird Information Centre, Israel
IRIC Israel Raptor Information Centre, Israel
IRSNB Institut Royal des Sciences Naturelles de Belgique, Brussels, Belgium
KFMP King Farouk Collection, Manial Palace, El Roda, Cairo, Egypt
MCZ Museum of Comparative Zoology, Cambridge, Massachusetts, USA
NRA Nature Reserves Authority, Jerusalem, Israel
PMS Palearctic Migratory Bird Survey, records deposited in USNM, USA
RMS Naturhistoriska Riksmuseets, Stockholm, Sweden
RNHL Rijksmuseum van Natuurlijke Historie, Leiden, The Netherlands
TAUM Tel-Aviv University Museum, Tel-Aviv, Israel
UMMZ University of Michigan Museum of Zoology, Ann Arbor, Michigan, USA
UMZC University Museum of Zoology, Cambridge, UK
USNM National Museum of Natural History, formerly United States National Museum, Washington, DC, USA
YPM Yale Peabody Museum of Natural History, New Haven, USA
ZFMK Zoologisches Forschungsinstitut und Museum Alexander Koenig, Bonn, FRG
ZMA Instituut voor Taxonomische Zoölogie, Zoölogisch Museum, Universiteit van Amsterdam, The Netherlands
ZM-HUJ Zoological Museum, Hebrew University of Jerusalem, Jerusalem, Israel

CRITERIA USED TO ASSESS A SPECIES OCCURRENCE IN THE COUNTRY

With the ever-increasing skills and sophistication of bird watchers, reference materials available, and photographic documentation, the traditional basis of only using specimen records to admit a species to a national list is becoming archaic. It should be noted, however, that from a purely scientific point of view the only records that can truly be reassessed in several regards in the future are specimens.

Although there has been a surge of observational reports of species hitherto unknown from Egypt, several records have not been properly documented and are difficult to assess. To avoid any ambiguity in our admission of a species to the 'Egyptian list' we have set the following criteria:

1. Species are admitted to the Egyptian list and entered into the main text:

(*a*) when there is at least one specimen taken in the country. In most cases this material has been examined by one of the authors. It should be noted that for a few taxa there is evidence that the only known material from Egypt may have not actually been collected in the country and that the specimen labels were forged. Such cases of potentially bogus records are mentioned in the text.
(*b*) when there is photographic evidence for its occurrence in the country and the photographs have been examined by at least one of the authors.
(*c*) when there is a sight record by two or more observers for which a clear description was reviewed by the authors. However, not all cases fitting this criterion have been accepted.

2. Species are considered hypothetical to the Egyptian list and placed in brackets within the main text:
(*a*) when there is an observation by a single person for which a clear description was assessed by the authors and considered acceptable. In such cases we have tried to take into account the species occurrence in waters or countries adjacent to Egypt, the experience of the observer, and possible confusion with other species.

3. Species are rejected from the Egyptian list:

(*a*) when there is a published account of occurrence in Egypt, but for which specimen, photographic or observational details are lacking, vague, or indicating an identification error. These species are summarized in Appendix I. Sources are given for records not dealt with by Meinertzhagen (1930). Unpublished records of species reported from the country but rejected by the authors from the Egyptian list are generally not mentioned. Individually unacceptable records published after 1930 of species rare within the country are presented in the main text and generally mentioned as such therein.

ENGLISH VERNACULAR NAMES

The English names used in the text are generally those used or to be used in *The Birds of the Western Palearctic*, although there are a few deviations from this source. In a few cases alternative names are presented in parentheses.

ORDER AND NOMENCLATURE

We have generally followed the sequence presented by Voous (1973, 1977) for orders, families, genera, and species. The exceptions are for species not dealt with by Voous and for which more recent revisions have been proposed. In the latter case, for example, we have followed several of the changes proposed by Payne (1979) for the Ardeidae. At the subspecific level we have generally used the characters, sequence, and synonymies of Vaurie (1959, 1965), with a few modifications based on new information given in *The Birds of the Western Palearctic*. For the vast majority of species recent Egyptian specimen material is available, and subspecific designations are based on our own comparisons with geographically extensive reference collections. In a few cases, Egyptian material is lacking and the subspecies mentioned in the text as occurring in the country is inferred on geographical grounds. The scientific names and sequence of plants is after Täckholm (1974), mammals after Osborn and Helmy (1980), and reptiles and amphibians after Marx (1968).

STATUS ABBREVIATIONS

At the end of the first line of each species account is one or a series of abbreviations used to indicate the animals' status within the country. These are:

CB casual breeder
FB former breeder
IB introduced breeder
MB migrant breeder
RB resident breeder
AV accidental visitor or vagrant (up to and including five documented records for the country)
PV passage visitor
WV winter visitor
( ) abbreviation in parentheses is used to indicate that the status is variable or irregular, e.g. (PV) means 'irregular passage visitor'
? status uncertain, e.g. RB? means 'doubtful resident breeder'

ARABIC AND LOCAL NAMES

There is considerable regional variation in vernacular bird names used in the country. A single species may have different names in Upper and Lower Egypt, and among and between Bedouin tribes of the Eastern and Western Deserts. For a few species there are alternative names for different age and sex classes. One name may apply to several, often phylogenetically distant species. Not surprisingly, birds which are most 'conspicuous' by virtue of size, number, coloration, habits, habitat, or economic importance are those having the greatest range of specific names. The names cited herein have been principally compiled by Joseph J. Hobbs.

Pronunciation, spelling, and transliteration guide

1. The hyphen (-). This represents the Arabic letter 'hamza', except after the definite article 'al' or after the assimilation of 'al' with the 'sun letters' (n, l, ẕ, ṯ, ḏ, s, sh, z, r, dh, d, and th), e.g. 'an-naṯruun'. The hamza is omitted when it occurs in the initial position or following the definite article 'al'. The hamza or 'glottal stop' is pronounced by momentary closure of the glottis, 'stopping' the outgoing breath.

2. Consonant sequences sh, gh, kh, dh, and th. These represent the single Arabic letters shiin, ghayn, khaa-, dhaal, and thaa- (respectively).

3. Use of the oblique stroke (/). The oblique stroke distinguishes two independent letters from the consonant sequence noted above, e.g. ‘mus/har’.

4. Underlining of consonant sequences. This indicates that the single Arabic letter represented by the consonant sequence is doubled, e.g. ‘duk̲h̲ala’ would actually be pronounced ‘dukhkhala’.

5. Emphatic consonants s̲, d̲, t̲, z̲, and h̲. Underlining distinguishes s̲aad, d̲aad, t̲aa’, z̲aa-, and h̲aa- from their softer-spoken counterparts siin, daal, taa-, zaay, and haa- (s, d, t, z, and h, respectively).

6. The apostrophe (’). This is the consonant ‘ayn’, resembling a deep ‘a’ emanating from the throat cavity.

7. Alif maqs̲ura and alif maddah are written as ‘aa’.

8. Adjectival endings are ‘ii’ for masculine and ‘iyya’ for feminine.

Regional variation in pronunciation

The linguistic picture in Egypt is rather complex and it is difficult to offer generalizations. In particular, pronunciation of the literary Arabic letters ‘qaaf’ (q) and ‘jiim’ (j) varies considerably. Qaaf is probably not pronounced as ‘q’ anywhere in the country. In Cairo and other places in the Nile Delta and Valley it becomes a glottal stop, indistinguishable from the hamza, and in rural areas and among Bedouin it is often pronounced as a giim (a hard ‘g’). Thus the generic name for falcon, s̲aqr, may be heard as s̲a-r or as s̲agr, but probably not as s̲aqr. The jiim (j) in Cairo and other localities is pronounced as ‘g’. As a general rule when the letter qaaf is pronounced as ‘g’, the jiim will be retained as a soft ‘j’. The Arabic letter dhaal (dh) is often pronounced as a true ‘d’, but in literal pronunciation resembles the ‘th’ of ‘this’. Thaa- (th) is pronounced in literary Arabic as the ‘th’ in ‘thing’ but often becomes an ‘s’ in spoken Arabic.

Sequence within the heading ‘Local name(s)’

Although several ornithologists have written books on and lists of Egyptian birds with Arabic names, no attempt has been made to standardize Egyptian local common names. Using numerous published works, principally Anon. (1941), El Negumi *et al.* (1950), Al Hussaini (1954*a*), Etchécopar and Hüe (1964), and our own field experience as reference guides, we have given each species mentioned in the main accounts, excluding those in brackets, an ‘official’ Egyptian Arabic name. These names, which are always the first one under the heading, are presented in both the transliterated and original Arabic script forms.

After the official name, transliterated alternative names are presented, which are generally based on cultural or regional variations. Some of these alternative names are not derived from the Arabic. Their origin is identified using the following codes:

Ab Ababda-Bedouin tribe inhabiting the south–central Eastern Desert
Bi Bischarin-Bedouin tribe inhabiting the south-eastern Eastern Desert (Gebel Elba); names are not necessarily of Arabic origin, but rather Tu Bedawie
F Faiyum
Ma Ma'aza-Bedouin tribe inhabiting the central Eastern Desert
Mi El Minya
NC North Coast between Alexandria and Matruh
ND Nile Delta
RS Red Sea—names from local fishermen
SC St Katherine Monastery
Si Siwa—not necessarily of Arabic origin, but rather Berber
Su Suez
WN Wadi el Natrun

In many cases the alternative names are not culturally or geographically restricted in their usage and no identifying code is given. Often local people identify birds by no other than 'generic' names. When these names are generic for a broader group such as several species, a genus, or family, they are presented under the heading 'Generic local name(s)'.

GENERAL

We have attempted to summarize all of the available information on birds known to have occurred within the present Egyptian political boundaries between about 1850 and 1988. Excluding those forms placed in the main text within brackets, most species accounts begin with a summary paragraph outlining the breeding distribution within the country, periods of occurrence and passage routes for migrants, abundance, information on geographical variation, subspecies, synonymies, and breeding season.

In an effort to place Egypt in its proper ornithological context with regards to the Middle East and north-eastern Africa, particularly its position along migratory pathways and the occurrence of vagrants, we have freely used information from surrounding areas. Most notable in this respect is Elat, Israel, about 15 km from the Egyptian border and the site of intensive ornithological work since the late 1950s. The records from Elat augment information from the Sinai and the Gulf of Aqaba. To a lesser extent we have also cited information from the Sudanese Nile Valley and Red Sea coast when it emphasizes records from the Egyptian portions of these waters.

To simplify the notation used for the period of occurrence, the terms 'early, mid-, and late' have been used. These refer to the respective approximate 10-day periods of each month. Extreme ranges of dates are indicated within parentheses. Thus, 'in autumn from (mid-August) early September to mid- (late) October' means the main autumn passage is between early September and mid-October but birds have been observed as early as mid-August and up to late October. The exact dates of records obtained during systematic counts of migrants, e.g. water-birds at Zaranikh and raptors at Suez, were not always available to us. Thus, unless explicitly stated, the range of dates is inclusive of the actual study period and not necessarily when a given species was actually observed.

The abundance designations are generally based on the authors' subjective estimate of the maximum number present in the country on any given day. This includes the number of passage and winter visitors, and for residents the number of breeding pairs. Only in a few cases are estimates based on actual counts (e.g. water-birds and raptors). The following categories have been used:

rare 1–100
scarce 100–1000
fairly common 1000–10 000
common 10 000–100 000
abundant >100 000

When a species is locally restricted in its distribution we freely deviated from the above designations and use terms such as 'locally common'.

To help with organizational convenience we have often divided the information presented in a species account, particularly for species breeding within the country, into geographical areas. These include: **North Coast**—the area between Alexandria and the Libyan political border, generally within 30–40 km of the Mediterranean coast; **Western Desert**—the oases or depressions of Siwa, Qattara, Wadi el Natrun, Bahariya, Farafra, Dakhla, Kharga, Gebel Uweinat, Gilf Kebir, and the regions in between; **Nile Delta**—generally including the Suez Canal area south to at least Ismailiya; **Nile Valley**—including the Faiyum, Wadi el Rayan, and Lake Nasser; **Eastern Desert**—from the edge of the Nile Valley to the Red Sea (in a few cases the **Sudan Government Administration Area**, centred around Gebel Elba, is dealt with separately); and **Sinai**. Specific details on the geography of these areas are dealt with in the following chapter.

The phrases 'Nile basin' or 'Nile system' refer to the Nile Delta and Valley as defined above. The arable land surrounding Suez and cultivated areas between Burg el Arab and Alexandria are truly ecological extensions of the Nile Delta and are often considered as such herein. The regions 'Lower' and 'Upper' Egypt or Nile basin are variously demarcated in the literature. The reader should be aware that since the river flows north the upper portion is in the south and the lower portion in the north. We define 'Lower Egypt' as the section of the Nile from the Delta to El Minya and 'Upper Egypt' from El Minya to the Sudanese border. Some authors distinguish a portion of the river as 'Middle Egypt', generally the area between Beni Suef and El Minya; however, we only use this term when it is given in previously published information. The Red Sea proper is the body of water south of the southern tip of Sinai. The northern tributary Gulfs of Suez and Aqaba are extensions of this area and are considered herein a portion of the Red Sea. All references in the text to elevation are in respect to sea-level.

BREEDING-BIRD MAPS

Our original intention was to map the distribution of species breeding in Egypt using a grid system of half-degree latitude and longitude blocks. At the position of Egypt, such plots are approximately 47 × 55 km (Goodman *et al.* 1983). This type of system is now extensively used for breeding-bird atlas mapping in Africa (Ash and Pomeroy 1981) and Saudi Arabia (Jennings 1985). However, much of the data we have collected on Egyptian breeding birds allows a more detailed presentation than single symbols in half-degree blocks. To avoid loss of information we have chosen a system in which individual records are plotted on a map with a half-degree atlas grid. The precision of plotting individual records has been limited by the size of the symbols and the reduction of the map. Breeding records that are within a few kilometres of each other are shown as a single symbol, while those separated by at least 5 km appear as separate symbols.

Three distinct types of symbols have been used to plot individual breeding records: (●) = definite, (▲) = probable, and (✱) = possible. Generally the criteria used for these categories have been adapted directly from the codes devised by the European Ornithological Atlas Committee (Sharrock 1976), although there are a few deviations which are marked with an asterisk (*).

Definite breeding (●)

(1) nest building or excavating nest hole (*);
(2) distraction-display or injury-feigning;
(3) used nest or recent eggshells found;
(4) recently fledged young (nidicolous species) or downy young (nidifugous species);
(5) adults entering or leaving nest site in circumstances indicating occupied nest (including high nests or nest holes, the contents of which cannot be examined) or adult seen incubating;
(6) adult carrying food for young or faecal sac;
(7) nest containing eggs or young.

Probable breeding (▲)

(1) singing male of a resident species observed in suitable nesting habitat during breeding season (*);
(2) pair observed in suitable nesting habitat in breeding season;
(3) permanent territory presumed through registration of territorial behaviour (song, etc.) on at least two different days a week or more apart at the same place;
(4) courtship and display;
(5) visiting probable nest site;
(6) agitated behaviour or anxiety calls from adults.

Possible breeding (∗)

(1) species observed in breeding season in possible nesting habitat;
(2) singing male(s) of migrant breeding species present in nesting season (*).

The period from which breeding records have been used is not consistent between species. We have tried to present the currently known or presumed distribution of breeding species. Thus, if a bird was found breeding in the 1940s in an area not since visited by ornithologists, such as along the North Coast between Salum and Matruh, these records have been plotted. Approximately 95 per cent of the data presented on the maps are from the period between 1970 and 1987. In cases when older records have been plotted, details are generally mentioned in the text. For species known to have experienced a reduction in their Egyptian breeding range during the past few decades, e.g. Black-shouldered Kite, Black Kite and Pied Kingfisher, only post-1970 records have been plotted. For a selected group of species, e.g. Lappet-faced Vulture, it seemed useful to plot former nesting sites (indicated with [✪]). Post-1930 breeding records from the area of the Nile Valley now inundated by Lake Nasser are generally not plotted but only mentioned in the text. To visually emphasize the known nesting distribution of mapped species, grid plots have been shaded in which a given species has a definite or probable breeding status.

SOME NOTES ON OTHER GRAPHICS

For several species we have tabulated all recent dated records from the country. This information which illustrates the range of migration periods and influxes is presented as histograms divided into 10-day periods. (The last period in a month may have between 8 and 11 days.) Records of migrants noted along the north coast of Sinai in 1973 and 1978 (YP, UP) and birds captured at Bahig in the spring seasons of 1967 and between 1969 and 1973 (Horner 1977) have not been plotted in these histograms. The main reason for this is that exact dates were generally not available. It should be realized that the intensity of recent ornithological observations in Egypt is not consistent over the calendar year.

INDEXES

Four different indexes appear at the end of the book. The first is for scientific bird names. Entries are given for genera and species mentioned in the text. The second index is for common English bird names. We have split descriptive epithets so that, for example, Barred Warbler would be found under 'Warbler, Barred'. The third and fourth indexes are for the 'official' Egyptian Arabic names in both transliterated form and original script (respectively). In the case of the Arabic script index the official name is cross-referenced to the transliterated form, the scientific name, and page number of the beginning of the respective species account.

CITATION

The editors recommend that for references to this volume in scientific publications the following citation should be used: Goodman, S. M. and Meininger, P. L. (eds) (1988). *The birds of Egypt.* Oxford University Press, Oxford.

2. THE GEOGRAPHY OF EGYPT

Egypt occupies the north-east corner of Africa and the Sinai Peninsula, extending 1073 km south to north from about 22° to 32°N. and 1262 km west to east from about 25° to 35°E.; the total area is 1 019 600 km^2. Egypt's neighbours are Libya to the west, Israel to the north-east, and Sudan to the south. Since 1895, the Sudan has administered about 18 000 km^2 of the south-east corner of mainland Egypt, an area which includes Gebel Elba. The Mediterranean Sea provides the northern border, while the Gulf of Suez and Red Sea comprise the eastern border of mainland Egypt. Flowing from the Sudanese border north to its mouth on the Mediterranean, the Nile River separates the mainland's chief morphological regions, the Eastern and Western Deserts. The Sinai Peninsula, forming a narrow wedge between Africa and Asia, is defined by the Mediterranean on the north, the Isthmus and Gulf of Suez on the west, and the Gulf of Aqaba and Israeli border on the east.

Egypt's overall climate is that of a hot desert. The highest mean temperatures occur in July and August, except for a June maximum in the extreme south; the coldest period is December through February (Abdel Aziz Ali 1982; Ayyad and Ghabbour 1986). June through September are the driest months throughout Egypt. North of 28°N., 70 per cent of the annual precipitation falls between November and March. From north to south across Egypt, three general climatic zones may be distinguished: the Mediterranean coastal belt, with 70 to 200 mm annual rainfall, and temperatures ranging (at Alexandria) from a January mean minimum of 9.4°C to a July mean maximum of 29.7°C; the middle Egypt belt, with 29°N. as its southern boundary, receiving from less than 1 mm (Siwa) to 35 mm (Cairo) annual precipitation, and having only slightly higher temperatures than the Mediterranean zone; and the upper Egypt belt, where rainfall is scant and capricious, ranging from 3 mm (at Aswan) to none, with mean minimum and maximum temperatures (at Aswan) of 9.3°C in January and 41.8°C in July (Hurst 1952; Kassas 1955; Boulos 1985*a*). In Fig. 2.1 the mean daily maximum and minimum temperatures for January and July throughout Egypt are illustrated; Fig. 2.2 depicts rainfall patterns. North-westerly winds prevail through most of the year, but during the *khamsiin* ('fifty days') of March through May, hot Saharan winds breed sandstorms across the country.

Egypt's temperature and rainfall regimes are governed mainly by latitude and by the maritime effects of the Mediterranean and Red Seas. Altitudinal effects, however, are prominent in the highlands of the Sinai and Eastern and Western Deserts (Ayyad and Ghabbour 1986). Figure 2.3 depicts elevations throughout Egypt. Where orographic precipitation occurs, average annual rainfall may far exceed the averages for lowland locations at the same latitudes (Kassas 1953*a*, 1953*b*). However, averages are not reliable indicators of desert precipitation, which

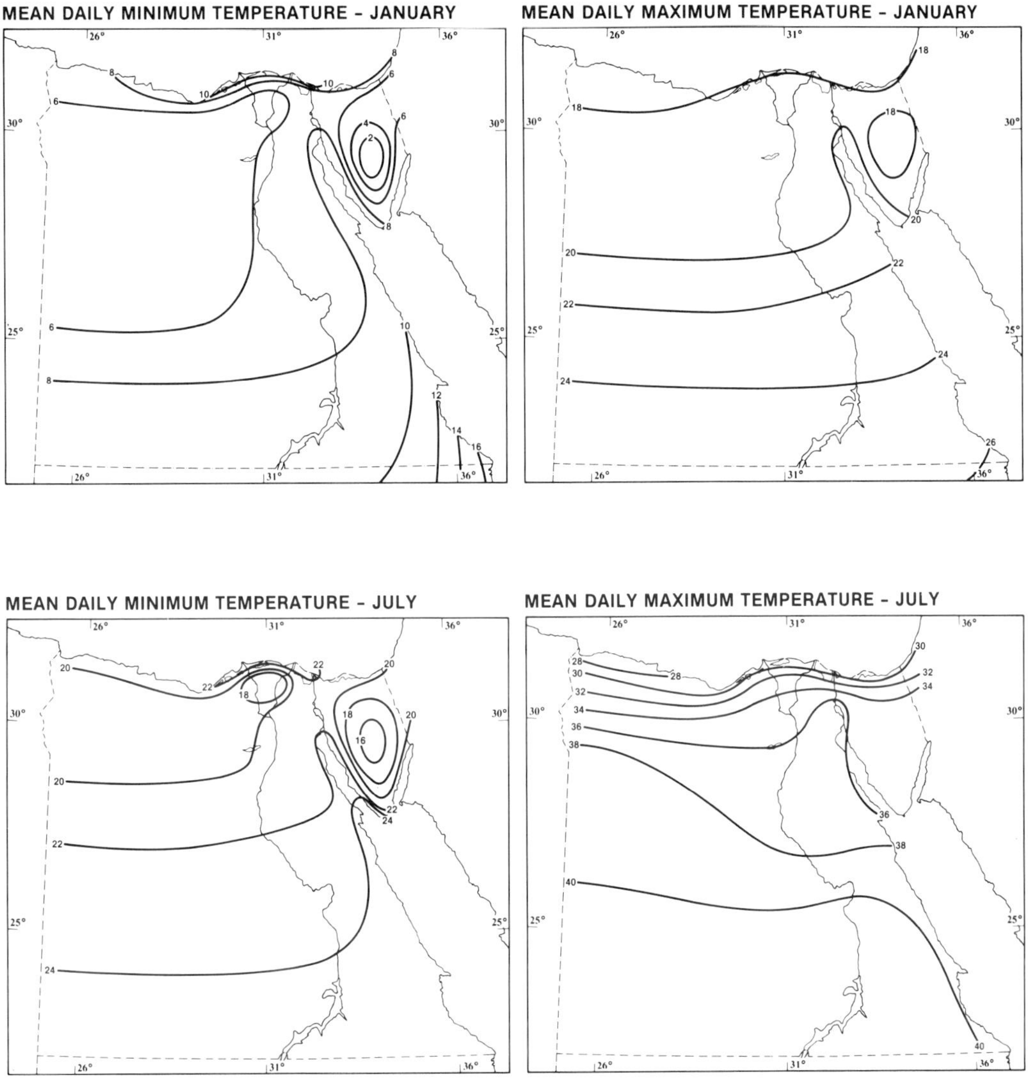

FIG. 2.1. Mean maximum and minimum daily temperatures in Egypt for two extreme months, January and July (after *Meteorological atlas of Egypt.* (1931). Government Press, Cairo).

is characterized by great temporal and spatial variability. These conditions create a broad range of habitats.

Egypt is a great desert in which the Nile is a thin, linear oasis. Deserts, the 'Red Land' of the ancients, occupy about 96 per cent of the country. Egypt has nevertheless always been synonymous with 35 000 km^2 of 'Black Land' along an extraordinary river. 'The Gift of the Nile', Herodotus (*c*.450 BC) called Egypt. Despite its vast

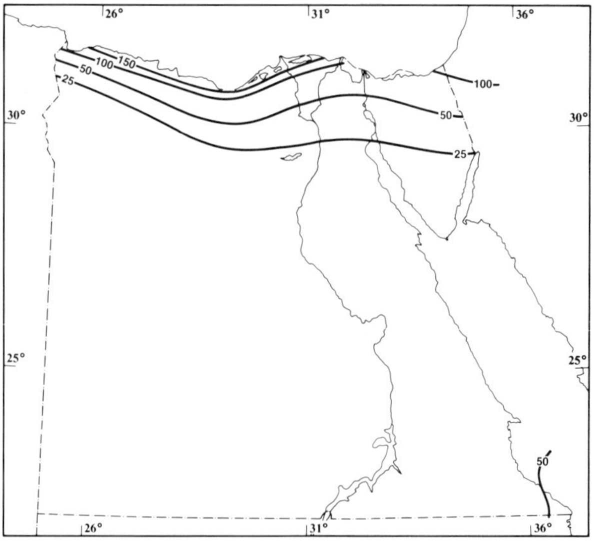

FIG. 2.2. Rainfall map of Egypt. Isohyets are in millimetres and represent average amount of precipitation per year. (After Osborn and Helmy 1980 and El Tom 1975.)

wilderness, Egypt is synonymous with civilization: its Arabic name, *Masr*, derives from the verb 'to found, build, settle, civilize, colonize' (Wehr 1976). The natural and cultural histories of Egypt are inseparable.

This chapter describes both the wild and domesticated environments of Egypt: riverine (Nile Valley and Delta), desert (North Coast, Western Desert, Eastern Desert, Gebel Elba, Sinai Peninsula), and marine (Red Sea, Gulf of Suez, and Gulf of Aqaba). Emphasis is on the environments in which birds occur, rather than on avian adaptations to those environments. Modern conditions are regarded as products of a long history of interaction between man and nature.

RIVERINE ENVIRONMENTS

The Nile Valley

The Nile River has two sources, one consolidating in highland Ethiopia at Lake Tana to feed the Blue Nile, the other in central Africa at Lake Victoria to feed the White Nile. Below their junction at Khartoum, Sudan, the Blue Nile accounts for about 70 per cent of the river's summer high-season volume (Arid Lands Information Center 1980). Until contained by a series of twentieth-century dams at Aswan, the surge of summer rain-water from the fertile Ethiopian highlands inundated much of Egypt's Nile flood-plain each August and September. The receding flood deposited about 1 mm of silt annually, building fertile sediments now averaging 8 m deep.

The cultivable land of Egypt is a 'river oasis' comprised of the Nile Valley (see Figs 2.4, 2.5), known traditionally as Upper Egypt, and Nile Delta (see Fig. 2.5), known as Lower Egypt (Butzer 1977). Along its 1800 km course through Egypt the Nile receives

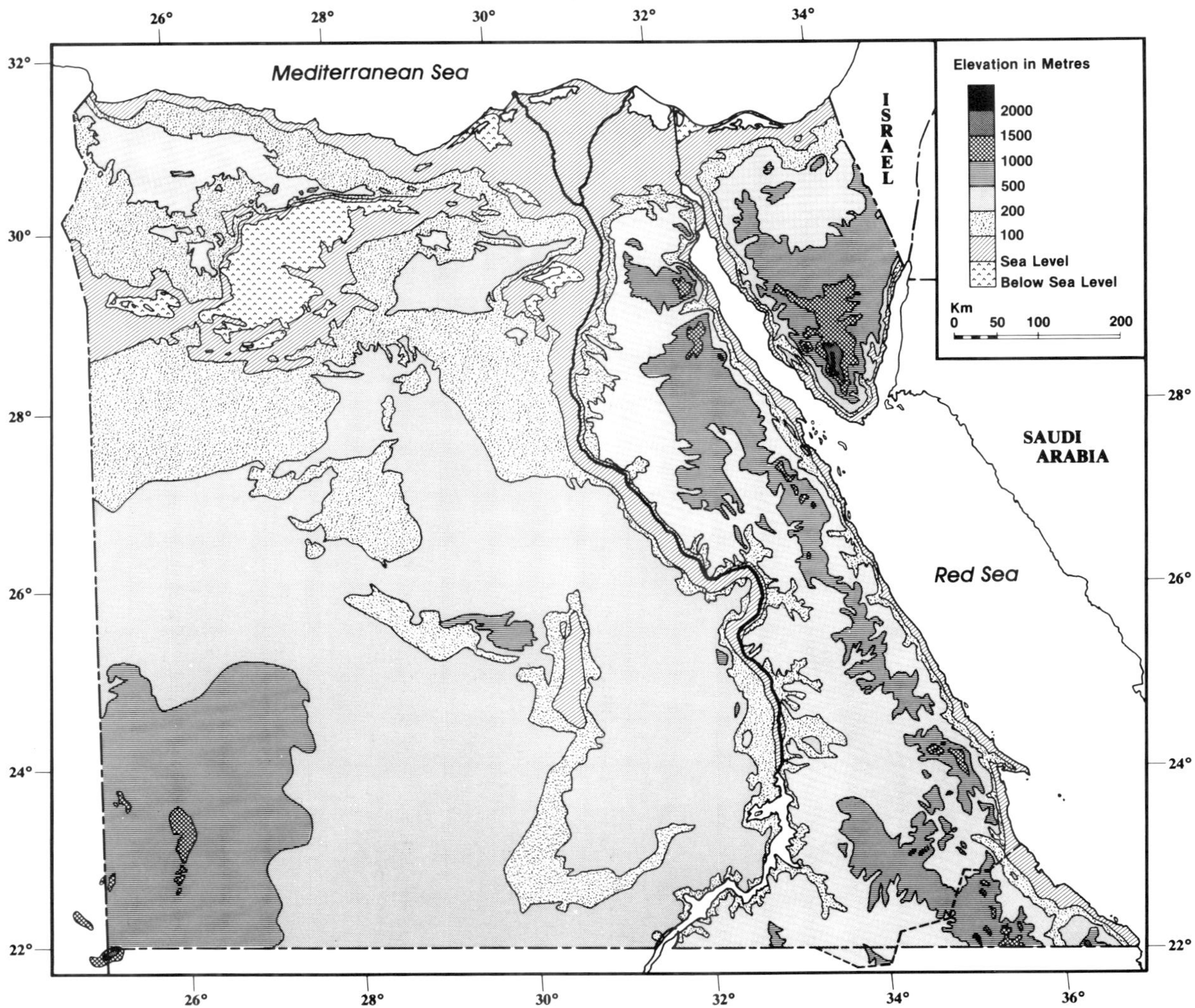

FIG. 2.3. Elevation map of Egypt (after *Africa North-east 1:500,000.* (1977). Bartholomew and Son, Edinburgh).

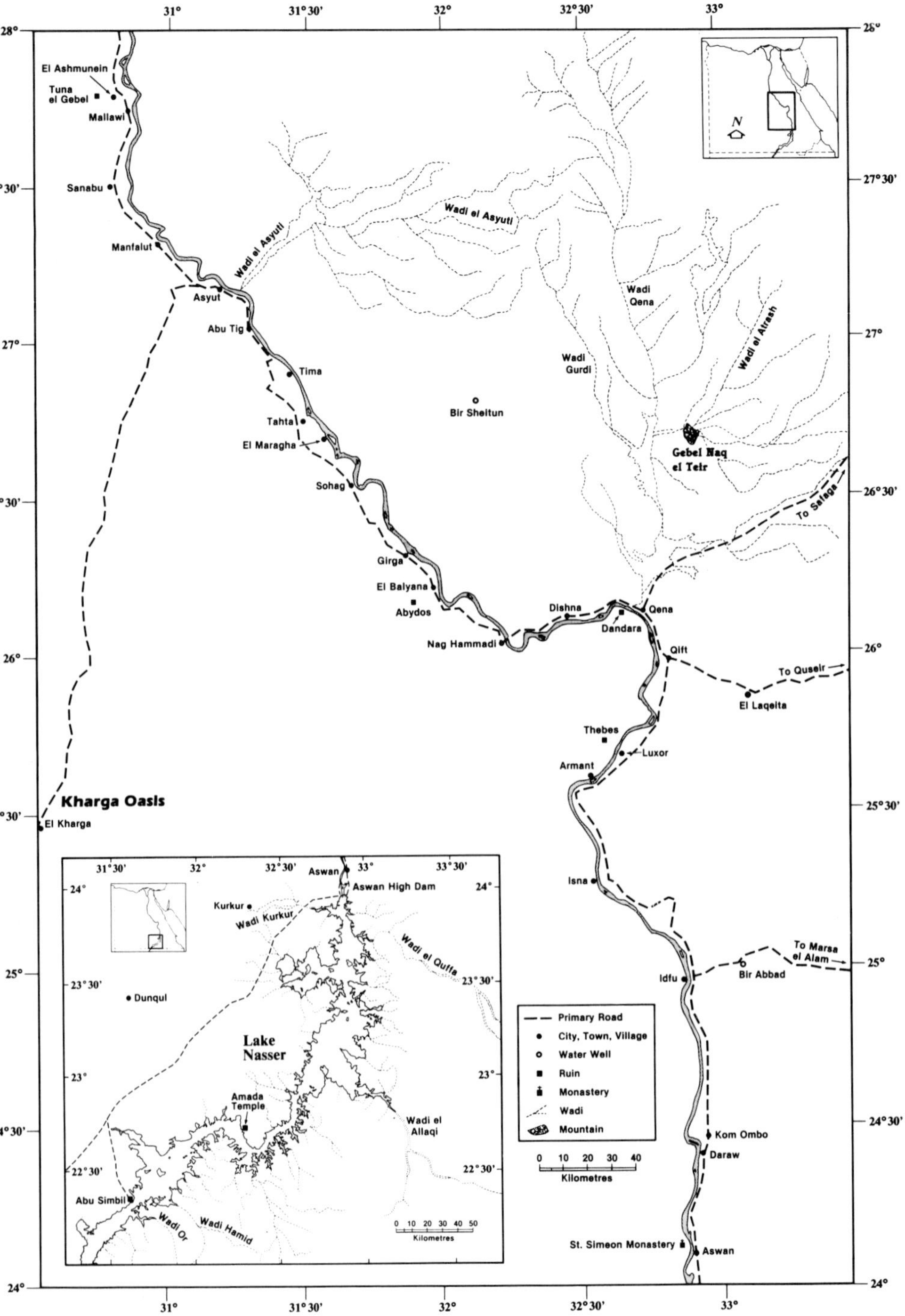

FIG. 2.4. Map of the Upper (southern) Nile Valley and surrounding areas; inset Lake Nasser.

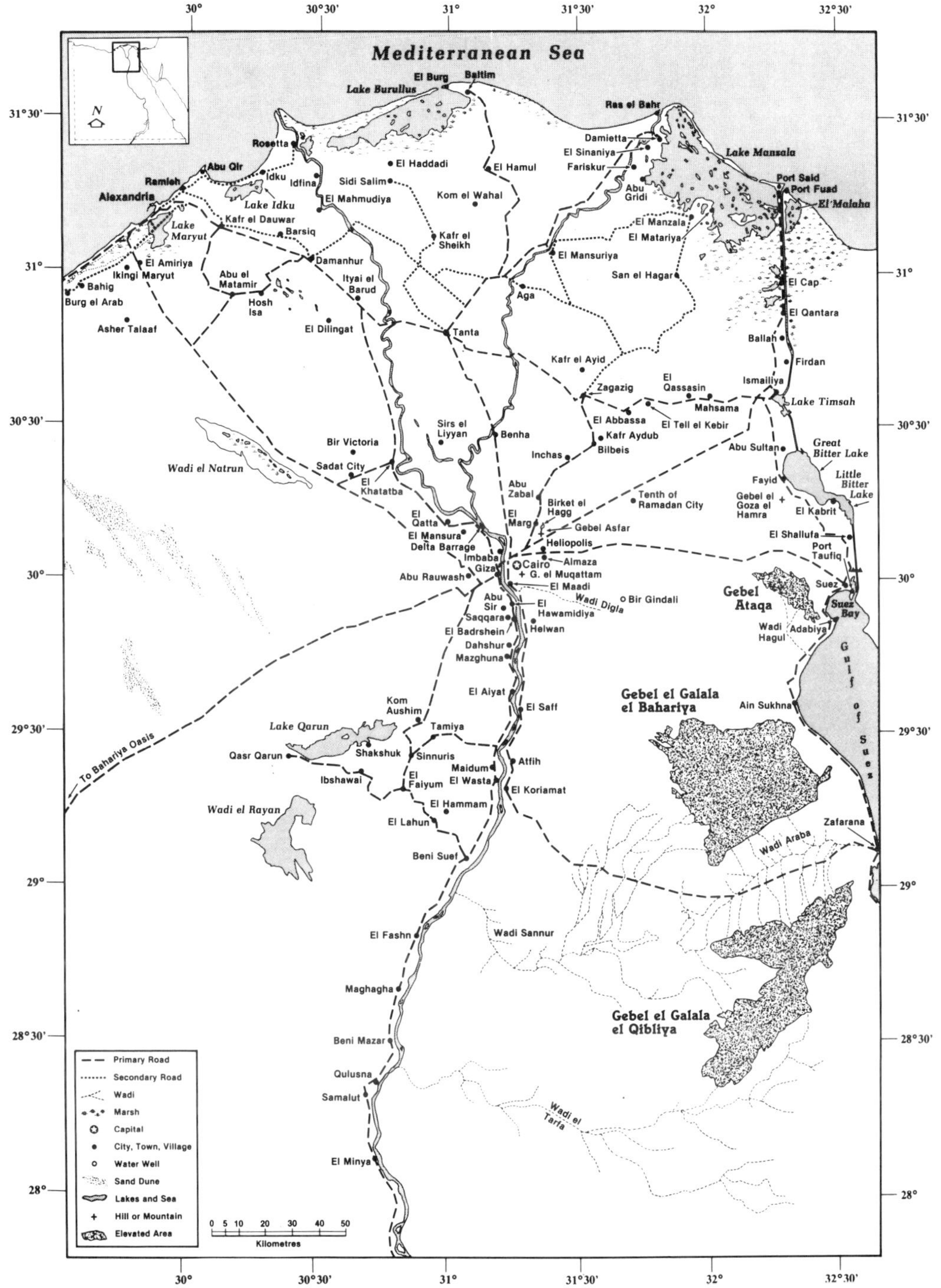

FIG. 2.5. Map of the Nile Delta, Lower (northern) Valley, and northern Eastern Desert.

FIG. 2.6. Nile Valley. Note abrupt transition between the vegetated flood-plain and the desert in background. Saqqara Step Pyramid is visible in the distance. (Photo P. L. Meininger, October 1980.)

no perennial tributaries. The river averages 0.75 km in width and tends to hug the east bank, creating a wider cultivable west bank (Fig. 2.6). The green ribbon of the Nile Valley has an average width of only 10 km. Escarpments of sandstone and limestone up to 300 m high define long stretches of the valley's desert borders (Abu al Izz 1971).

Adjoining the Nile's western flood-plain at 29°N. is a fertile 1700 km^2 depression, the Faiyum, watered from the Nile by the Bahr Yusef Canal. Excess water in the past drained to Lake Qarun, situated 45 m below sea-level at the northern end of the depression. Work began in 1974 to channel water exceeding Lake Qarun's capacity to the near-by desert depression of Wadi el Rayan. A new lake has formed there, and a natural succession is still in progress (Fig. 2.7; see p. 98).

The transformation of the Egyptian Nile from a natural to a cultural landscape has been in progress for over 7000 years. Neolithic people of the Faiyum (*c.*5000 BC) saw perhaps half the river's flood-plain vegetated with savanna and dry thickets. In such habitats people grazed domestic sheep, goats, pigs, and cattle, and hunted elephant (*Loxodonta africana*) and bubal hartebeest (*Alcelaphus buselaphus*). From the river they took Nile perch (*Lates niloticus*) and other fish, trapped birds, and hunted hippopotamus (*Hippopotamus amphibius*) and Nile crocodile (*Crocodilus niloticus*) (Kees 1961; Bender 1975; Butzer 1976; Gautier 1976; Brewer 1986). They grew barley and emmer using basin irrigation, sowing their crops once annual flood-waters had receded (usually late October). The land was left fallow after the May harvest. A

FIG. 2.7. Wadi el Rayan. Drainage between two of the lakes with dense reed-beds. For scale note person and motor vehicle above waterfalls. (Photo W. C. Mullié, April 1986.)

single crop (the winter crop) could be grown in this manner on about two-thirds of the Nile flood-plain (Butzer 1977).

With embellishments, the Neolithic technology of basin irrigation sustained the civilization of Pharaonic Egypt (3100–332 BC). The conversion to perennial irrigation began modestly around 1500 BC with the introduction of the *shaduf*, a device for lifting water to summer crops on levees. It brought increasing amounts of flood-plain wildlands into cultivation, especially in the middle portion of the Nile Valley (Butzer 1976). The pace of transformation quickened under Greek and Roman rule (332 BC to AD 324). Unexploited regions of Upper Egypt and the Faiyum were settled and planted with wheat, barley, broad beans, millet, sesame, lentils, clover, and flax. The Archimedean screw and water wheel domesticated more land along the Nile and its canals. The limit of cultivation set by the technology available was almost reached, with the planted area almost equal to that planted in 1880. Agricultural productivity was accompanied by population growth, with the number of Egyptians doubling in 300 years to almost 5 million in AD 50 (Butzer 1976; Waterbury 1979). New cities emerged in the Faiyum and near by in the Nile Valley.

The large-scale conversion of agricultural land from basin to perennial irrigation began after 1800. To retain water and raise and distribute it to summer crops, barrages were constructed on the Nile below Cairo and at sites on 30 000 km of new canals, and increasingly large dams were built on the Nile at Aswan in 1902, 1912, and 1933 (Waterbury 1979). In 1950 as much as 15 per cent of Egypt's cultivated land, mostly in Upper Egypt, remained under basin irrigation (Kashef 1981). The final conversion to perennial irrigation took place with the construction of the Aswan High Dam (beginning 1960). With this full-scale change the number of crops planted and the area under cultivation have increased. Egypt's leading agricultural products are now cereals (mainly wheat, maize, rice, and barley), cotton, sugar-cane, pulses, vegetables, fruits, and forage plants (Boulos 1985*a*). Cash crops such as cotton, rice, sugar-cane, and citrus fruits have been promoted at the expense of subsistence crops.

Agricultural expansion has brought radical changes in the environments and livelihoods of Egypt's Nile-dwellers. Most strikingly, Egypt's population has rocketed from 2.5 million in the early 1800s to 9.7 million in 1897, 18.8 million in 1946, 37 million in 1975, 46 million in 1984, and a projected 65 million in 2000 (Arid Lands Information Center 1980; *World Development Report* 1986; see Fig. 2.8). The official 1976 census lists the following populations for Egypt's principal cities: Cairo and Giza, 8 304 462; Alexandria, 2 317 705; Shubra el Kheima, 394 223; El Mahalla el Kubra, 292 114; Tanta, 283 240; Port Said, 262 760; El Mansura, 259 387; and Asyut, 213 751 (Central Agency for Public Mobilization and Statistics 1984). Notable are the primacy of Cairo, where 39 per cent of all Egyptian urbanites lived in 1980, and the concentration of secondary cities in the Delta rather than Nile Valley (*World Development Report* 1986). The great population sizes and densities (23 688 persons/km^2 in Cairo in 1976) are consequences not only of sheer growth but of the influx of people from rural areas. In 1907, urbanites accounted for 17 per cent of Egypt's population; by 1976 they made up 43 per cent (Central Agency for Public Mobilization and Statistics 1984). It is estimated that 1–2 per cent of Egypt's arable land is lost annually to the encroachment of human settlements (Hefny 1982). The historical

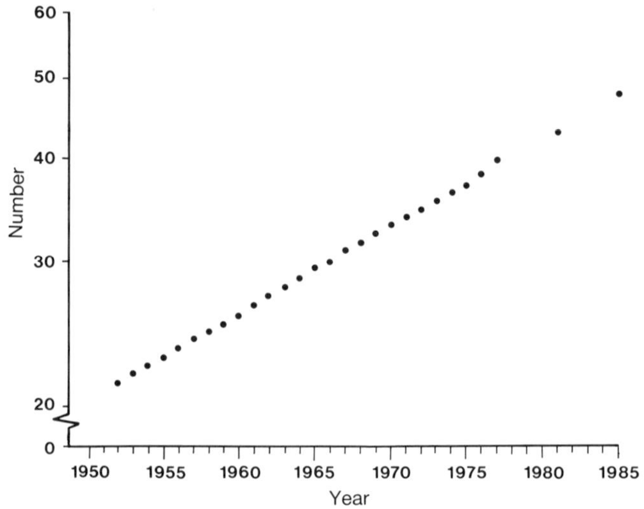

FIG. 2.8. Exponential growth of the human population in Egypt between 1952 and 1985. Number scale is in millions. (After Otto 1981, Botje 1985, and World Development Report 1986.)

reluctance of Egyptians to live outside the cultivated 4 per cent of their territory has persisted: 1976 population densities averaged 1034 persons/km^2 in the Nile Valley, Delta and Suez Canal governorates, versus 0.3, 0.2, and 0.5 persons/km^2 respectively in the Red Sea, Wadi Gedeed, and Matruh governorates (Central Agency for Public Mobilization and Statistics 1984).

The most enduring features of the Nile's ancient cultural landscape are those of its 4000 villages (Central Agency for Public Mobilization and Statistics 1984). The typical village is a densely packed conglomeration of mud-brick structures built in the midst of cultivation; as in ancient times, Egyptian peasants build only cemeteries on near-by desert margins. Waterworks, paved roads, railroads, and telephone and electric lines are becoming part of the traditional material culture and landscapes of rural Egypt. While man has in the past destroyed wildlife habitats along the Nile he has also created new ones, including the ancient temple complexes and pyramids which are among the world's oldest artificial bird roosting and nesting sites.

The modern fauna and flora of the Nile Valley attest to the thorough intervention of man. The characteristic mammals are those which farmers have accepted as neighbours and those best at avoiding contact with man: long-eared hedgehog (*Hemiechinus auritus*), giant musk shrew (*Crocidura flavescens*), Nile kusu (*Arvicanthis niloticus*), jackal (*Canis aureus*), red fox (*Vulpes vulpes*), Egyptian mongoose (*Herpestes ichneumon*), jungle cat (*Felis chaus*), and wild cat (*F. sylvestris*) (Osborn and Helmy 1980). The pesticides required for perennial irrigation agriculture have locally extirpated some birds of prey and other predators, allowing the introduced black rat (*Rattus rattus*) and Norway rat (*R. norvegicus*) to proliferate (Mullié and Meininger 1985). Characteristic Nile Valley birds are Cattle Egret, Black-shouldered Kite, Purple Gallinule, Senegal Thick-knee, Spur-winged Plover, Palm Dove, Pied Kingfisher, Little Green Bee-eater, Hoopoe, Sand Martin, Swallow, Fan-tailed Warbler, Graceful Warbler, Clamorous Reed Warbler, Nile Valley Sunbird, and Hooded Crow. Characteristic reptiles are the skinks *Chalcides ocellatus*, *Mabuya quinquetaeniata*, and *Scincus scincus*; the monitor lizard *Varanus niloticus*; the cobra *Naja nigricollis*; the now rare turtle *Trionyx triunguis*; and the Nile crocodile.

There is little non-domesticated indigenous vegetation in the Nile Valley. Trees grow mainly on the margins of fields, canals, and roads. Most are cultivated, including the native date palm (*Phoenix dactylifera*), native tamarisk (*Tamarix nilotica*), introduced eucalyptus (*Eucalyptus rostrata*), and introduced Australian pine (*Casuarina cunninghamiana*). Native acacia (*Acacia nilotica*) and sycamore (*Ficus sycomorus*) grow more spontaneously (El Hadidi and Boulos 1979). An exemplary consequence of human intervention is that some recently introduced weeds, such as *Aster squamatus*, have become widespread in the Nile Valley, while historically prolific native species such as *Ceruana pratensis* are nearly extinct (Boulos and El Hadidi 1984).

The most dramatic changes in the history of the Nile ecosystem have taken place since 1970, when the Aswan High Dam project was completed. The dam's effects have included a perennial increase in the amount of water applied to the land, a consequent increase in soil and lake salinity, the cessation of wide seasonal fluctuations in water levels of the Nile and its canals, and the reduction of silts passing below the dam. The ecological consequences have been especially strong for hydrophytic plants. With water perennially available, some species have spread rapidly and become pests: water hyacinth (*Eichhornia crassipes*), which was introduced as an ornamental and now chokes 80 per cent of Egypt's canal surfaces; water lettuce (*Pistia stratiotes*), previously extremely rare in Egypt; and water fern (*Salvinia auriculata*) (Montasir 1954; Täckholm 1974; Arid Lands Information Center 1980; El Hinnawi 1980). Correspondingly, once-common hydrophytes have become rare: *Ranunculus rionii*, *Utricularia inflexa*, *U. gibba*, *Ottelia alismoides*, *Najas minor*, and *N. graminea* (Täckholm 1974; Boulos 1985*b*). The lotus (*Nymphaea lotus*), formerly the prolific Pharaonic symbol of Upper Egypt, is now confined to small relict populations in Faiyum and Delta canals (El Hadidi 1971). Similarly, *Ceruana pratensis* was doomed by the containment of silts behind the Aswan High Dam. Used in ancient times for baskets and coffins and until the 1960s for brooms, this fibrous plant was almost extinct in Egypt by 1980 (Boulos 1985*b*).

Changes in bird life exemplify apparently positive as well as negative effects of the Aswan High Dam on Nile wildlife. The steady level of Nile water has created good conditions for the growth of *Phragmites* and *Typha* marshes where sand-bars had previously been repeatedly exposed and inundated. This growth has allowed reed-dwelling birds to expand their breeding ranges; however, it has apparently eliminated the sandy breeding habitat of the Egyptian Plover. The Aswan High Dam has resulted in a rising water-table in the Faiyum. *Phragmites* marshes have subsequently developed on Lake Qarun's margins, providing some new habitats for reed-dwelling species (see p. 97). Increased water supply has also raised the lake's salinity: once renowned as a habitat for freshwater flowering plants and rich microflora and fauna, Qarun is now as salty as the Mediterranean (Boraey 1980; Boulos 1985*b*). In near-by Wadi el Rayan, birds seem to have benefited from irrigation technology, as several species have colonized the area (see p. 98).

The Aswan High Dam has had very different effects upon its downstream and upstream environments. Above its 111 m wall a reservoir stretches 489 km south to the Sudan's Dal Cataract, drowning the ancient land of Nubia. Lake Nasser is narrow but has an extensive dendretic system of arms (locally called *khor*) which, when the reservoir's fluctuating volume is high, contains as much as 80 per cent of its surface area (Entz 1976).

Largely because of the effects of the dams built at Aswan before the High Dam, alien plants introduced by man represented more than two-thirds of Nubia's plant species by 1965 (El Hadidi 1976). The Aswan High Dam created a new set of artificial habitats and opportunities. Owing to fluctuating lake levels, it is difficult to predict the future of the succession now in progress. Nubian culture had been entirely uprooted and transplanted in advance of the Aswan High Dam flood. Recently, low water levels due to the African drought have threatened the livelihood of Nubians who returned to establish small farming communities on the lake's shores (RFe). In 1980, Lake Nasser's vegetation was distributed in three zones: in shallow water, with *Phragmites australis*, *Typha domingensis*, and *Potamogeton pectinatus* dominant; on sandy lake shores, with *Glinus lotoides*, *Senecio aegyptius*, *Oligomeris linifolia*, *Rumex dentatus*, *Amaranthus viridus*, *Crypsis schoenoides*, *Salsola baryosma*, and *Tamarix amplexicaulis* growing; and on some island interiors, with *Hyoscyamus muticus* and *Astragalus tribuloides* present (El Hadidi 1976; Entz 1976; Meininger and Mullié 1981*b*). The lake provides breeding habitats suitable for several species of Afrotropical birds; the Egyptian Goose was one of the first whose local populations increased substantially. The Nile crocodile had been almost extinct in Egypt but now proliferates in Lake Nasser.

The Nile Delta

The Nile Delta (see Fig. 2.5) fans out below Cairo, where the river divides into two distributaries, the Rosetta (Rashid) and Damietta (Dumyat), named for the towns situated at their respective mouths. Historical records indicate there were seven distributaries of the Delta in the first century AD and six in the twelfth century. The modern Delta forms a rough triangle with a height of 170 km from its 220 km long base on the Mediterranean to its apex near Cairo; the actual area is 22 000 km^2 (compared with the Nile Valley's 13 000 km^2) or 63 per cent of Egypt's inhabited terrain (Abu al Izz 1971). High ground in the Delta is limited to the Nile's natural levees and to 'turtlebacks', sandy ridges scattered between the river's distributaries (Holz 1969). Along the Delta coast is a chain of elongated lakes, from west to east: Maryut, Idku, Burullus, and Manzala. These are brackish and shallow bodies of water, generally less than 1 m deep, joined directly to the sea by narrow channels through sand-bars and limestone ridges (Abu al Izz 1971). The Delta enjoys a regular winter rainfall regime. Exposed first to the prevailing westerlies, the western Delta yearly receives up to 200 mm; Port Said, on the eastern Delta, averages 80 mm. Rainfall amounts also diminish southwards, with Cairo recording an average 34 mm annually.

Nile distributaries once snaked across the Delta to create a complex network of river arms, islands, seasonally inundated alluvial basins, and backswamps (Butzer 1959). Neolithic people exploited the Delta much as they did the Nile Valley, planting crops, herding animals, and fishing and fowling in marshes, lakes, and flooded lowlands (Hayes 1965). This pattern of exploitation persisted through early Pharaonic times, when Delta backswamps were called the 'bird tanks of pleasure' and the 'papyrus lands'. Papyrus and lotus were the most characteristic plants of these marshy, game-rich depressions.

Delta wildlands succumbed to man much later than their Nile Valley counterparts. Despite government incentives for colonization, the region was settled sparsely and slowly (Kees 1961). Only by about 1100 BC did the Delta's population equal and begin to surpass the Nile Valley's (Butzer 1976). Records dating from after that time make few references to the beloved 'papyrus lands', but many to Delta irrigation works (El Hadidi 1971).

Greek and Roman colonists effected an overall transfer of productivity and power from Upper Egypt to the Delta. They began to tame the northern Delta, turning marshes into vineyards and orchards (Kees 1961). Between 950 BC and AD 600, at least 35 new towns were established in the Delta (Butzer 1976). Pressure on wildlands intensified. The demise of papyrus (*Cyperus papyrus*), the Pharaonic emblem of Lower Egypt, was a major chapter in the Delta's transformation before the era of perennial irrigation. Conditions for the wild growth of this plant were especially good in Delta backswamps. The demand for papyrus products also resulted in its large-scale cultivation (Täckholm and Drar 1950). The plant was doomed once linen paper was substituted for papyrus in the tenth century. Untended stands of semi-domesticated papyrus were overtaken by aggressive competitors, the native perennial hydrophytes *Phragmites australis* and *Vossia cuspidata*, by reclamation for agriculture, and by the drying-out of wetlands when some Delta distributaries ceased to flow. After 1821, when a few stands were recorded near Damietta and Lake Manzala, *Cyperus papyrus* was considered extinct. In 1968 this plant was rediscovered in Wadi el Natrun, where a small thicket had survived as a relict of an ancient backswamp (Täckholm and Drar 1950; El Hadidi 1971; Boulos 1985*b*).

The northern lakes today comprise a vestige of the Delta's historic wildness. The vegetation which dominates them discourages human exploitation. Submerged aquatic plants, especially *Potamogeton pectinatus*, *P. crispus*, and *Ceratophyllum demersum* fill up to half the surface area of some lakes (notably Idku), and choke drains in adjacent cultivated fields (Meininger and Mullié 1981*b*). Lake margins and island shores sustain healthy marshes, including one on the south shore of Lake Burullus labelled the 'most virginal community of the Delta', comprised of *Suaeda fruticosa*, *Salicornia fruticosa*, *Inula erithroides*, *Phragmites australis*, and *Juncus acutus* (Fig. 2.9; George 1972; Samman 1974).

The Delta's more diverse environment and larger size yield to its human inhabitants a greater share of Egypt's riverine resources than the Nile Valley gives its people. This is partly responsible for the relatively greater economic prosperity of the Delta. The Delta lakes are among Egypt's most productive natural systems (Elkington 1975). Some 35 000–40 000 fishermen took about 25 per cent of Egypt's total fish catch for 1970–1 from a single Delta lake, Manzala, which has an area of about 1200 km^2 (Meininger and Mullié 1981*b*). Manzala yielded an estimated fish catch of 65 000 tons per annum in the early 1980s (Mullié and Meininger 1983). It has been suggested that fish production could be increased to about one ton per hectare per annum (Anon. 1977). This level of productivity is comparable with that of intensive aquaculture in lagoons and estuaries elsewhere in the world (Lasserre 1979; Barnes 1980).

FIG. 2.9. Nile Delta. Reed-beds along the edge of Lake Burullus, near Baltim. (Photo P. L. Meininger, May 1983.)

The ambitions, methods, benefits, and liabilities of perennial irrigation are stronger in the Delta than in Upper Egypt. The Delta's main summer crops of cotton, rice, and maize are Egypt's leading cash crops. Oranges, tangerines, and lemons, the country's most profitable citrus fruits, are cultivated mainly in the Delta and its reclaimed desert margins. From 1952 to 1979, about 930 000 feddans (or about 4000 km^2) of agricultural land were reclaimed in Egypt. Half this total was reclaimed from the Delta flood-plain, and most of the balance from the Delta's desert fringes (Euroconsult 1984). Due to subsequent salinity, waterlogging, and desertification, roughly half these reclaimed lands now yield 'above marginal' productivity. Comprising about 13 per cent of Egypt's total cultivated area, reclaimed lands contributed only 2 per cent of the nation's total agricultural output in 1980 (Anon. 1982).

Reclaimed desert and flood-plain lands represent new wildlife habitats. The area planted in citrus has increased substantially in recent years. These groves host prolific weed growth. Where desert lands have been reclaimed for orchards, as in the Beheira Governorate bordering the west Delta, *Casuarina cunninghamiana* and *C. glauca* trees have been planted as wind-breaks and sand-fences. These new habitats provide nest sites and food formerly not available and several species of bird have colonized these areas (see p. 99).

Perennial irrigation has had less favourable effects on wildlife in the intensively cultivated Delta flood-plain. By eliminating the natural pest controls provided by fluctuating water levels, the Aswan High Dam has required massive pesticide applications to crops (George 1972; Mullié and Meininger 1985). In 1953, about 1 948 000 kg of insecticides were used on Egyptian fields; in 1982/3, about 13 060 000 kg were applied (Central Agency for Public Mobilization and Statistics

1984). Pesticides in the food chain appear to be related to the decline of many bird species, particularly raptors (see p. 101).

Most current plans for agricultural development in the Delta call for large-scale reclamation of the northern lakes. Between 1800 and 1970, natural forces and human developments had already reduced the combined surface area of the Delta lakes from about 4545 to 2240 km^2 (Abu al Izz 1971). An agricultural five-year plan for 1982–7 targeted about 196 000 feddans (or 825 km^2) for reclamation from the lagoons, marshes, and salt-flats of Lakes Manzala, Burullus, and Idku (Euroconsult 1984). Brackish Lake Maryut and adjacent lands near Alexandria are also to be drained and cultivated.

Reclamation of the Delta lakes has few apparent benefits for wildlife. In the 30 per cent of Lake Manzala's former surface area already reclaimed, some wading birds feed from the shallow basins of 'fish farms' (Meininger and Mullié 1981*b*). However, local populations of several marsh-dwelling species have declined (Meininger *et al.* 1986).

Another region subjected historically to human modification is the Isthmus of Suez. Morphologically this area is an appendage of the Nile Delta. A canal carries Nile water to Ismailiya (1976 population 105 930), known as Egypt's 'garden city' (Central Agency for Public Mobilization and Statistics 1984). In ancient times El Malaha, now a hypersaline lake in the north-west corner of the Sinai, was connected to Lake Manzala and was drained by the Nile's Pelusium distributary. The Suez Canal (opened 1869) effectively severed El Malaha from the Delta; then a canal bypass (opened 1985) bisected and largely drained the lake itself (Meininger and Mullié 1981*b*).

On a continental scale, the Suez Canal physically separated Africa from Asia and via waterways linked both with Europe. The 150 km long canal system extends from Port Said on the Mediterranean to Suez on the Gulf of Suez. It is comprised of artificial excavations in sandy plains and (between Ismailiya and Suez) of the natural bodies of Lake Timsah and the Great and Little Bitter Lakes. The combined length of the Bitter Lakes is 37 km, and their maximum width is 12 km. Patches of *Dictyota dichotoma* occur in their shallows, and at depths to 5 m there is dense growth of *Halophila stipulacea*. Sparse beds of *Sargassum* occur at 5–10 m depths (Beets 1953). Lake Timsah, 7 km long and 3 km wide, supports marshes of *Typha* and *Phragmites*. Overall, the canal lake vegetation closely resembles that of Sinai's Lake Bardawil (Lipkin 1972).

As a link between oceans and an apparent barrier between continents, the Suez Canal (and before it, the Isthmic desert) has had unusual zoogeographical influences. Since the canal was completed there has been an open exchange of Red Sea and Mediterranean marine fauna (Fishelson 1980). The Isthmus has not inhibited movements of birds between Africa and Asia; even the flightless Ostrich colonized Asia. Most large mammals found east of the Nile are represented in Asia; distributions of small rodents also suggest no barrier across the Isthmus (Ayyad and Ghabbour 1986). The non-Sinaitic Egyptian reptile fauna has strong affiliations with south-west Asia

(Marx 1968). In sum, the Isthmus is more a gate than a fence. The canal's role in deterring future east–west exchanges is not clear.

The characteristic mammals of the Nile Delta and Isthmus of Suez are those of the Nile Valley, with the addition of Flower's shrew (*Crocidura floweri*), bandicoot rat (*Nesokia indica*), striped weasel (*Poecilictus libyca*), weasel (*Mustela nivalis*), and Egyptian mongoose (Osborn and Helmy 1980). Characteristic birds are Little Bittern, Purple Gallinule, Painted Snipe, Spur-winged Plover, Palm Dove, Senegal Coucal, Pied Kingfisher, Blue-cheeked Bee-eater, Hoopoe, Sand Martin, Fan-tailed Warbler, Graceful Warbler, and Clamorous Reed Warbler. Characteristic reptiles and amphibians are the gecko *Hemidactylus turcicus*, skink *Chalcides ocellatus*, chameleon *Chamaeleo chamaeleon*, snake *Natrix tessellata*, and frog *Bufo regularis*.

DESERT ENVIRONMENTS

Egypt's deserts were not always as hyperarid as they are today. Pleistocene era shifts between relatively wet and dry climates repeatedly opened and closed opportunities for wildlife and man away from the Nile (Sanford and Arkell 1933; Murray 1951). The Neolithic period 8000–3000 BC was relatively moist (Butzer 1977). During the Neolithic, as much as 100 to 300 mm yearly rainfall west of the Nile produced semi-arid 'Sahel shrub' habitats similar to those found in Kordofan (central Sudan) today (Haynes 1982). People probably kept cattle and sometimes grew crops on what is now completely barren range (Gautier 1984; Hassan 1986). They also hunted game including the giraffe (*Giraffa camelopardalis*) and knew the crested porcupine (*Hystrix cristata*) and striped ground squirrel (*Euxerus erythropus*), animals indicative of once mesic conditions in the Western Desert (Butzer 1976; van Noten 1978; Gautier 1984). Regular winter rains east of the Nile allowed hunters quarries of elephants and other savanna game (Murray 1951; Hayes 1965; Butzer 1976).

The Neolithic subpluvial period ended about 2400 BC and Egypt's climate has been hyperarid ever since (Butzer 1976). Savanna evolved from dry steppe to semi-desert, and finally hot desert. Animals with high water requirements became locally extinct. Some people retreated to permanent water in the Western Desert oases and along the Nile. Others, like the plants and animals which survive today, adapted themselves to the changing conditions by withstanding or avoiding drought in desert habitats.

The environments of Egypt's desert birds consist of both contiguous regions and disjunct habitats. On the one hand, there are some physical barriers against the exchange of many species between distinct biotic regions (the Eastern and Western Deserts, for example). On the other hand, many species occur in a particular type of habitat in different regions. Finally, there are often parallels between the habitats and adaptations of distinct species inhabiting different regions.

The following discussion examines separately the regions and habitats of desert birds. The regional section summarizes the climates, morphological units, habitats, biotic associations, and human impacts characteristic of each region. The section on

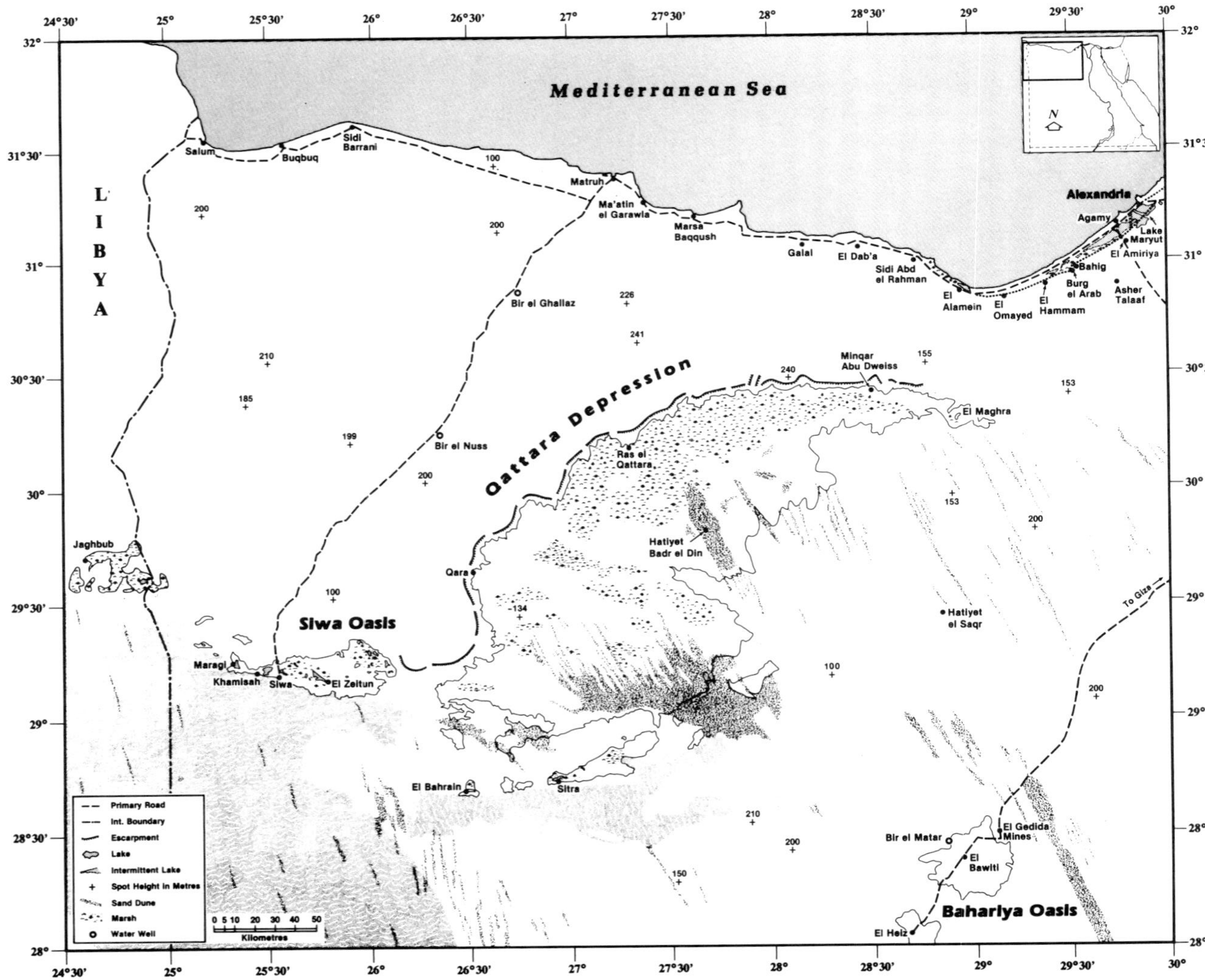

FIG. 2.10. North Coast and northern Western Desert.

habitats is intended to be a useful reference on Egyptian bird environments. A general outline of each habitat is followed by sketches of the plant communities which comprise that habitat in different regions.

Desert regions

The North Coast

The North Coast (western Mediterranean coastal desert, Mareotis, or western Marmarica), a natural region barely 50 km wide is quite distinct from the vast Western Desert to the south (see Fig. 2.10). Rainfalls along the 600 km coast from the Libyan border to Alexandria average 70 to 200 mm annually. This moisture nurtures Egypt's richest flora, with about 800 species comprising half of the total number of species recorded in Egypt (Tadros 1953; Kassas 1955). Natural vegetation is distributed in four zones: littoral oolitic sand-dunes, parallel ridges of oolitic limestone along the coast and inland, salt-marshes between these ridges, and inland plains. The leading crops of barley, olives, and dates are grown on the inland plains (Fig. 2.11) while figs are cultivated on the coastal dunes (Kassas 1955). Ptolemaic and Roman inhabitants of the Mareotis (332 BC–AD 324) planted the same crops but favoured viticulture. They knew Lake Maryut, now a shrunken saline depression, as a long freshwater body. After the seventh century the fecund North Coast environment deteriorated to an arid steppe (Moreau and Dolp 1970; Kassas 1972).

Characteristic mammals of the North Coast include the desert hedgehog (*Paraechinus deserti*), hare (*Lepus capensis*), Anderson's gerbil (*Gerbillus andersoni*),

FIG. 2.11. North Coast. Cultivated area near Burg el Arab. (Photo P. L. Meininger, March 1983.)

Simon's dipodil (*Dipodillus simoni*), Shaw's jird (*Meriones shawi*), mole rat (*Spalax ehrenbergi*), garden dormouse (*Eliomys quercinus*), four-toed jerboa (*Allactaga tetradactyla*), and greater Egyptian jerboa (*Jaculus orientalis*) (Osborn and Helmy 1980). Barbary Partridge, Houbara, Hoopoe Lark, Dupont's Lark, Temminck's Horned Lark, and Red-rumped Wheatear are the characteristic birds. The region's reptile fauna is typified by the lizards *Agama stellio*, *Eumeces schneideri*, and *Chamaeleo chamaeleon*; the snakes *Eryx jaculus*, *Macroprotodon cucullatus*, and *Malpolon monspessulanus*; and the tortoise *Testudo kleinmanni*.

The 1976 population of the North Coast (Matruh Governorate) was recorded as 112 547 (Central Agency for Public Mobilization and Statistics 1984). Roughly half live in small towns along the Alexandria–Salum road. The majority of North Coast people identify themselves as Bedouins, principally of the 'Awlad 'Ali and Jume'at tribes. They are no longer pastoral nomads. They maintain fixed residences while pursuing a variety of traditional and modern occupations, often simultaneously: herding sheep, goats, and camels; tending figs and small gardens; working for wages, trading, and making handicrafts. Like fishermen and farmers of the Delta coast, many of the men trap Quail and passerines during the autumn migration.

Tourism, heavy industry, traditional pastoralism, and land reclamation are altering the North Coast ecosystem. Several seaside tourist resorts are being developed between Alexandria and Matruh. If Egypt proceeds with plans to develop nuclear power, the first stations will be built on the North Coast between Alexandria and El Alamein (Ezzat 1981). The environment appears to be resilient to more traditional impacts: one experiment showed a remarkable increase in plant density and cover within a 50-hectare area fenced against grazing for three years (Ayyad 1980; Boulos 1985*b*). To modernize and expand agricultural production and attract Nile-dwellers to settle in the desert, a freshwater canal has been built between the west Delta and Bahig, 60 km west of Alexandria. This Nubariya Canal was to be extended another 120 km to El Dab'a; however, the overall scheme has proven defective (Hamdan 1980). Much of the irrigation water has percolated underground to reappear as about 2280 km^2 of stagnant saline lakes. Underground water levels have risen about 4 m yearly since the late 1960s (Boulos 1985*a*). Delta plants and animals (Pied Kingfishers, for example) have extended their ranges along the strip of new habitats created by the canal (ST).

The Western Desert

The Western or 'Libyan' Desert encompasses 681 000 km^2, or about two-thirds of Egypt's total area. East to west, it extends from the Nile Valley (roughly 31°E.) to the Libyan border (25°E.), and south to north from the Sudanese border (22°N.) to the North Coast zone (about 34°N.). Its largest morphological units are an Eocene limestone plateau and a Cretaceous (Nubian) sandstone plateau; the other principal natural regions are Gebel Uweinat, the Great Sand Sea, the Libyan Plateau, and the oases depressions.

The Western Desert lacks the spectacular relief of the Sinai Peninsula and Eastern Desert. The highest point is 1892 m on Gebel Uweinat (see Fig. 2.12), a 1500 km^2 sandstone and igneous massif spanning the extreme south-west corner of Egypt. This is an anomalous region in the sterile Western Desert. Orographic rain falls once every 7 to 10 years, feeding small springs at the base of the massif and supporting biota whose closest counterparts occur east of the Nile, in the Red Sea Mountains (Fig. 2.13; Abu al Izz 1971). Relicts of richer past environments also survive north of Gebel Uweinat at Gilf Kebir (see Fig. 2.12), a divided Nubian sandstone plateau rising to 1000 m. These Cretaceous sandstones continue as a broad band up to 300 km wide between the Libyan border and Nile Valley, forming the southern boundary of the Eocene limestone plateau.

The Western Desert's largest natural region is a 500 m high and as much as 500 km wide Eocene limestone plateau rising north of Gilf Kebir and stretching east to west from the Nile to Egypt's border with Libya (Abu al Izz 1971). Rains of 10 mm or less fall on the plateau at intervals of 10 years or more, producing patches of 'accidental' plant cover (Kassas 1955).

The Great Sand Sea overlies an area of the Eocene limestone plateau 800 km north to south from Siwa Oasis to Gilf Kebir, and 200 km west to east from the Libyan border to Farafra Oasis. Its waterless, almost lifeless sands are characteristically arranged as parallel *seif* ('sword') dunes 1 km or more wide running north-west to south-east with the prevailing winds, and separated by 'streets' up to 5 km wide (Abu al Izz 1971; Haynes 1982). A smaller sandy body, the Selima Sand Sheet, is a flat continuous expanse of 60 000 km^2 straddling the Egypt–Sudan border about half-way between the Nile and Gebel Uweinat (Haynes 1982).

North-west of the great Eocene plateau is a smaller one of Miocene limestone known as the Libyan Plateau, Marmarica, or El Daffa (about 30°N., 26°E.). It spans 300 km between Siwa Oasis and the eastern end of the Qattara Depression. The plateau tilts northwards and downwards, from elevations of 200 to 50 m, terminating along the inland frontier of the North Coast zone (Abu al Izz 1971). Scattered across the plateau are small sandy depressions where some trees and shrubs are able to grow.

Embedded in the Eocene limestone plateau are several great oases-depressions, the principal foci of life in the Western Desert. Their formation began during ancient periods of abundant rainfall, when water acted to widen fractures in the plateau floor. Subsequent winds excavated some of these fissures to the depths of water-bearing Nubian sandstone strata (Abu al Izz 1971). This process created the principal oases and depressions of the Western Desert: Wadi el Natrun (250 km^2); Siwa Oasis (1100 km^2); Qattara Depression (19 500 km^2); Bahariya Oasis (1800 km^2); Farafra Oasis (2800 km^2); Kharga Oasis (3000 km^2); and Dakhla Oasis (825 km^2).

The typical Western Desert depression is nearly or completely encircled by steep escarpments. Its floor is near or below sea-level (lowest points are: in Siwa −30 m; in Qattara Depression −134 m; in Bahariya +133 m; in Farafra +26 m; in Kharga

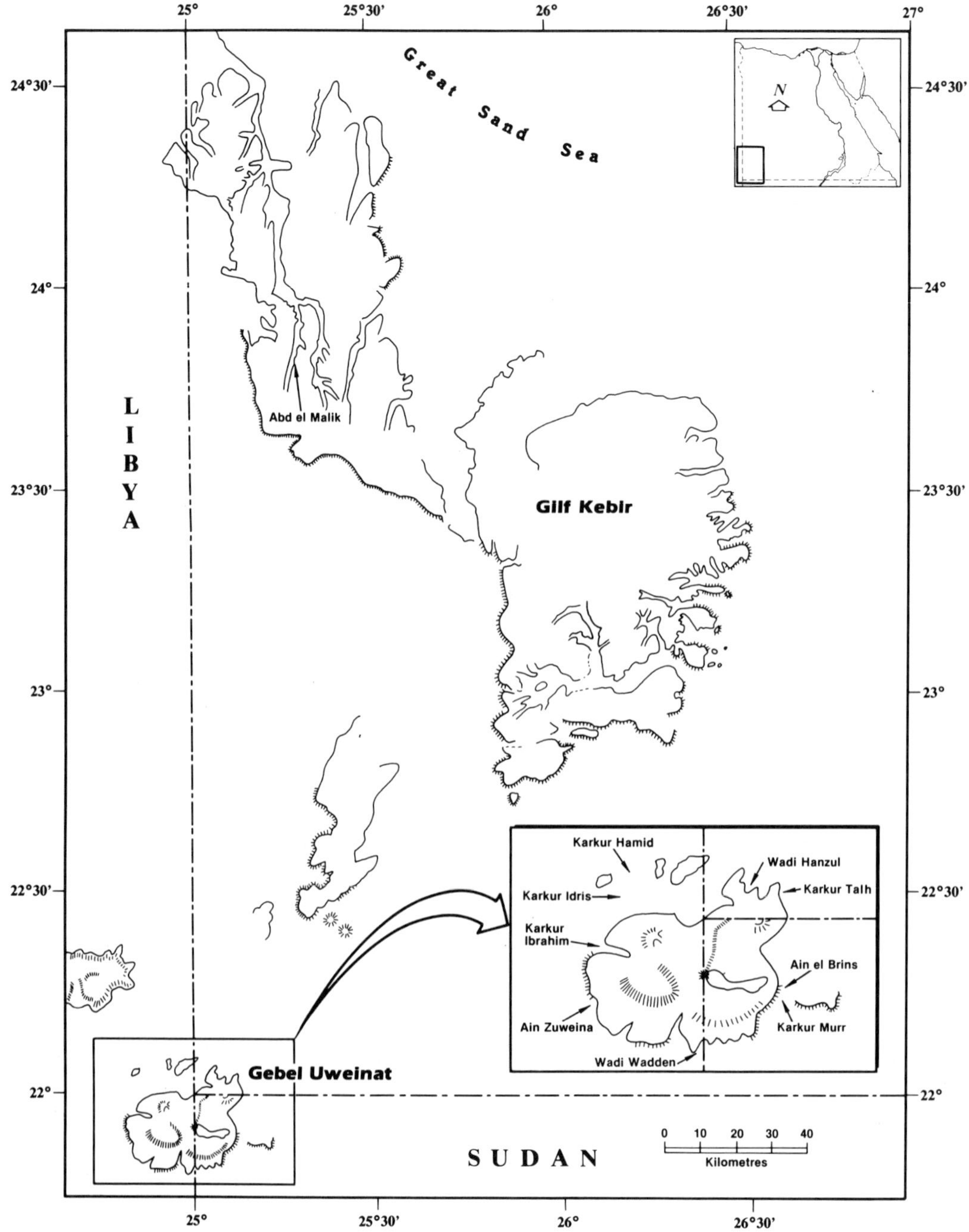

FIG. 2.12. Gebel Uweinat and Gilf Kebir region; inset Gebel Uweinat.

FIG. 2.13. Gebel Uweinat, south-east side. Characteristic plants are *Acacia ehrenbergiana* and *Fagonia* sp. (Photo X. Misonne, October 1968.)

−18 m; and in Dakhla +100 m) (Goodman *et al.* 1986). Deep alluvial deposits typically fill the depression floor.

In some locales, waters carried through the Nubian sandstone break to the surface as artesian springs. In poorly drained low-lying areas of depression floors, excess water accumulates in saline lakes surrounded by salt-marsh vegetation (Kassas 1955; Girgis *et al.* 1971; Zahran and Girgis 1972). On sandy plains between salt-marshes and the rocky plateau there is a distinct transitional flora (Omer-Cooper 1937; Migahid *et al.* 1960; Zahran 1972). Indigenous and introduced shrubs grow upon and between sand-dunes which have crept into the oases, threatening cultivated areas (Kassas 1955; Migahid *et al.* 1960).

The uninhabited Qattara Depression merits discussion, if only because it is by far the largest of the oases-depressions (see Fig. 2.10). About 25 per cent of its floor is occupied by dry lake-beds, sometimes filled by downpours. The rest of the depression is composed of sand, gravel, clay, and limestone outcrops (Abu al Izz 1971). Two small oases, Qara and El Maghra, are situated respectively at the extreme west (Qattara's deepest point, −134 m) and north-east of the depression. El Maghra has a small brackish lake bordered by a dense salt-marsh. On near by sandy plains there is little plant coverage and diversity, aside from isolated groves of *Acacia* trees called *hatiyet* (Girgis *et al.* 1971; Osborn and Helmy 1980).

Along the southern frontier of the Western Desert are several very small oases-depressions. To the east, the most prominent are Kurkur, Dunqul, and Dineiqil oases.

Their vegetation is characteristic of relatively well-drained reed oasis habitats (Reed 1964; Boulos 1966; Zahran 1966). To the west, the micro-oases are little more than the springs whose names they bear: Bir Tarfawi, Bir Kiseiba, Bir el Shab, and Bir Misaha. Their brackish waters sustain reed oasis plant types (Osborn and Helmy 1980).

Characteristic recent mammals of the Western Desert are lesser jerboa (*Jaculus jaculus*), Rüppell's sand fox (*Vulpes rueppelli*), fennec fox (*Fennecus zerda*), cheetah (*Acinonyx jubatus*, perhaps still present), scimitar-horned oryx (*Oryx dammah*, until 1975), addax (*Addax nasomaculatus*, until 1931), bubal hartebeest (until 1900), slender-horned gazelle (*Gazella leptoceros*), and dorcas gazelle (*G. dorcas*) (Osborn and Helmy 1980). Characteristic birds are Spotted Sandgrouse, Palm Dove, Hoopoe Lark, Rufous Bush Robin, and White-crowned Black Wheatear. Characteristic reptiles are the *Psammophis aegyptius* snake, *Varanus griseus* monitor lizard, and *Philochortus intermedius* lacertid lizard.

Fertile soils and fresh water have attracted people since Neolithic times to settle and cultivate the oases-depressions of Wadi el Natrun, Siwa, Bahariya, Farafra, Dakhla, and Kharga. Among the crops grown today are rice, maize, wheat, dates, olives, apricots, grapes, oranges, bananas, mangoes, sugar-cane, plums, tomatoes, egg-plant, cucumber, spinach, beans, okra, onions, Malta jute, and watermelons (Fig. 2.14; Fakhry 1974). Where high salinity or water shortages cause cultivation to be

FIG. 2.14. Western Desert oasis. Walled garden in Farafra Oasis. The central pathway is a thoroughfare for people and domestic animals. Cultivated plants in view include olive and date palm. (Photo S. M. Goodman, March 1984.)

abandoned, specialized wild plants succeed: *Hyoscyamus muticus*, *Sporobolus spicatus*, *Astragalus* spp., *Cyperus laevigatus*, and *Zygophyllum* spp. (Migahid *et al.* 1960).

The oases-depressions of Kharga, Dakhla, and Farafra comprise what Egyptians call the 'New Valley' (Wadi Gedeed). The 1976 census suggested that the Wadi Gedeed Governorate was underpopulated with only 85 175 inhabitants (Central Agency for Public Mobilization and Statistics 1984). Efforts are underway to draw settlers from the overcrowded Nile region by increasing the cultivable area of these oases (primarily by drilling new wells into the Nubian sandstone aquifer), and developing petroleum and mining industries. Existing infrastructure includes asphalt roads linking the Nile Valley at Cairo and Asyut with the oases of Bahariya, Farafra, Dakhla, and Kharga. Remote and culturally distinct Siwa remained outside the scope of Egyptian development schemes until the mid-1980s, when a 300 km road was paved between the oasis and Matruh, on the North Coast.

The Eastern Desert

The Eastern or 'Arabian' Desert is bounded by the Nile Valley to the west, the Nile Delta and Suez Isthmus to the north (about 30°50′N.), the Gulf of Suez and Red Sea to the east, and the Sudan border to the south. The total area is 223 000 km^2, equal to 21 per cent of Egypt's total area and 32 per cent of the area of the Western Desert (Abu al Izz 1971). The region is comprised of distinct morphological units, principally the Red Sea Mountains, coastal plain, Nubian Desert, and several sedimentary plateaux.

The Eastern Desert's backbone is a series of mountain chains trending parallel to the Red Sea coast from about 28°N. (roughly, the latitude of El Minya in the Nile Valley) southwards into the Sudan (see Figs 2.15, 2.16). Among the rugged igneous and metamorphic peaks is the highest in Egypt outside Sinai, Gebel Shayib el Banat (2187 m); others of note are Gebel Gharib (1751 m), Gebel Abu Dukhan (1662 m), Gebel Qattar (1963 m), Gebel Nuqrus (1505 m), Gebel Migif (1198 m), Gebel Zabara (1360 m), Gebel Abu Hamamid (1786 m), Gebel Umm Samiuki (1486 m), and Gebel Hamata (1977 m). Gebel Elba (1437 m), in the extreme south, comprises a distinct biotic region and is treated separately below.

Owing to its location between the Mediterranean winter rainfall belt and the tropical summer rainfall belt, where it benefits little from either, the Eastern Desert is a hyperarid region. Conditions on the coastal plain near Hurghada are indicative of the region's overall extreme aridity. Average annual rainfall is 3 mm. The vegetation of desert grasses, woody trees, scrub, and succulents is confined to drainage systems only (Kassas and Zahran 1965). The Eastern Desert's high mountains intercept cloud moisture, producing locally less xeric conditions (less pronounced on the western slopes, due to a rain-shadow effect; Ayyad and Ghabbour 1986). Plant communities dominated by hardy perennials flourish in the valleys between the high mountain blocks of the northern Eastern Desert.

North and west of the basement-complex ranges are dissected platforms of

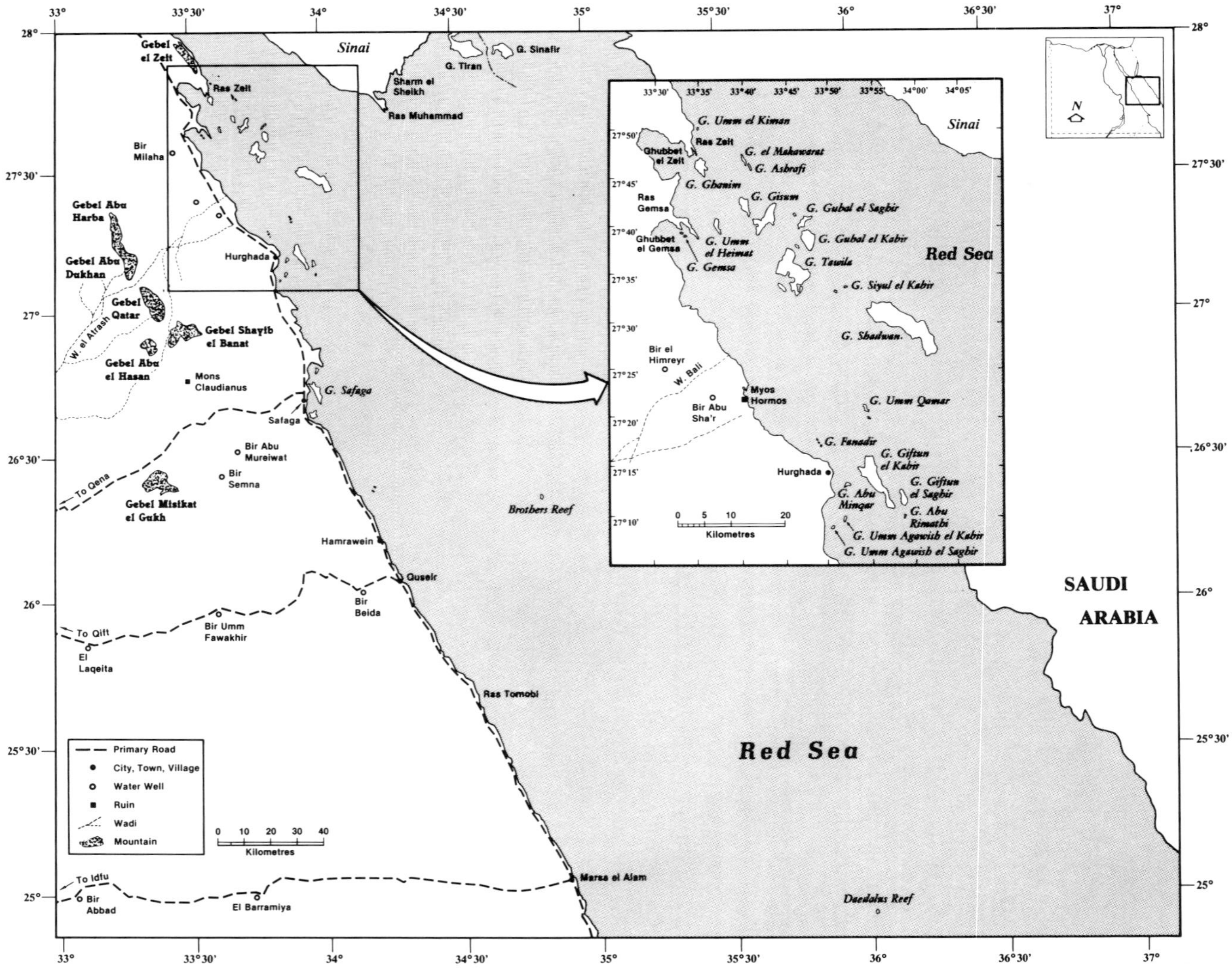

FIG. 2.15. Central Eastern Desert and northern Red Sea region; inset archipelago located at the mouth of the Gulf of Suez and off Hurghada.

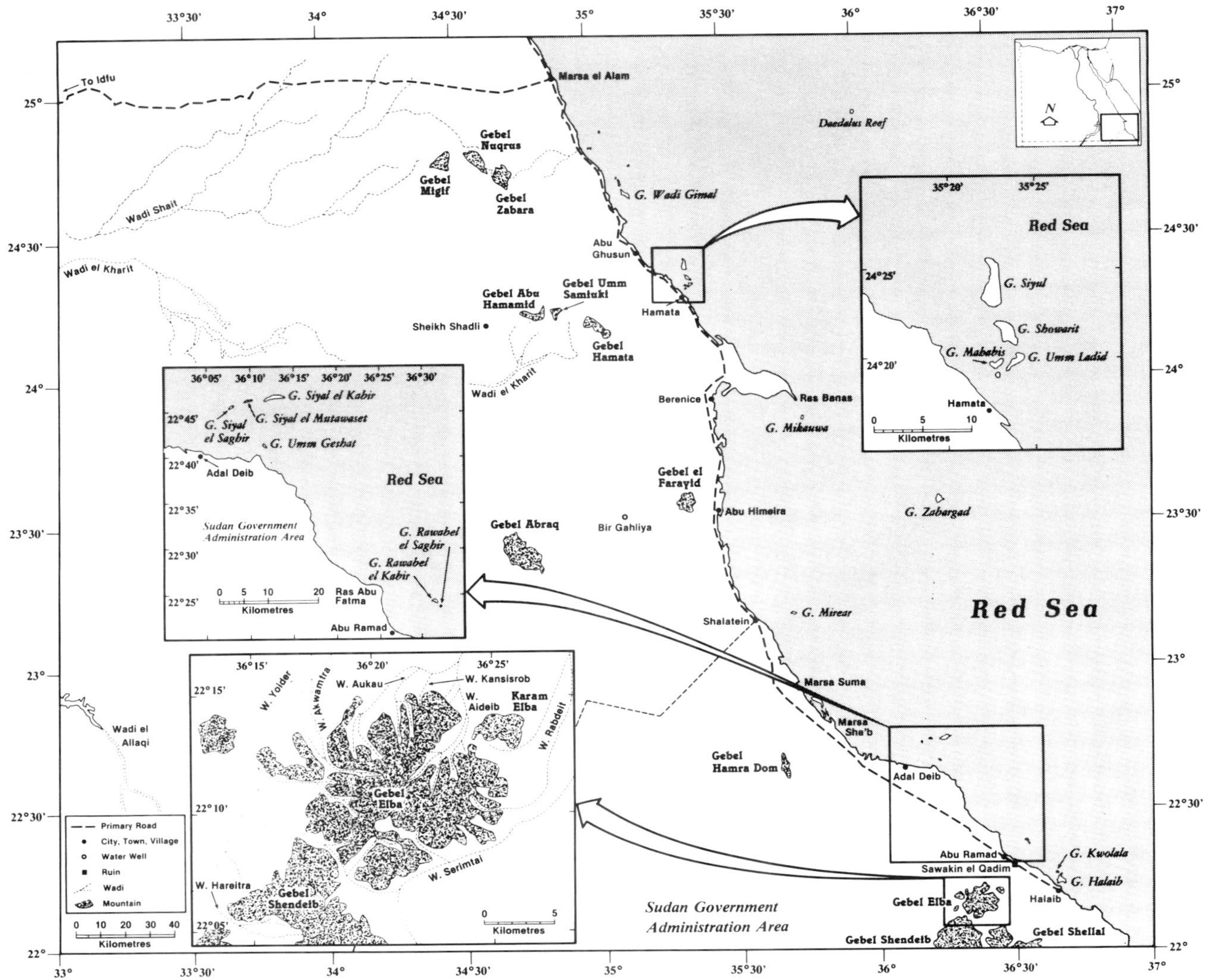

FIG. 2.16. Southern Eastern Desert and Red Sea region. Insets are of the Siyul Island group off Hamata (right upper), the Siyal and Rawabel groups off Adal Deib and Abu Ramad, respectively (left centre), and Gebel Elba (left bottom).

FIG. 2.17. Limestone country, northern Eastern Desert. Typical aspect of sedimentary highland habitat of Gebel el Galala el Qibliya. Note area is virtually devoid of vegetation. (Photo J. J. Hobbs, October 1986.)

sedimentary rock: the limestone plateaux of Gebel Ataqa (871 m), Gebel el Galala el Bahariya (to 1274 m), Gebel el Galala el Qibliya (to 1464 m; Fig. 2.17), and the Ma'aza Plateau (to 866 m, between 26°20′ and 28°N.); and the sandstone 'Ababda Plateau (to about 400 m between 24° and 26°N.). Wadi Araba, flowing east to the Gulf of Suez, drains much of the two Galalas. Wadi el Tarfa drains the southern flank of Gebel el Galala el Qibliya and the northern-most portions of the Ma'aza Plateau. Wadi Qena flows between the Red Sea Mountains and the Ma'aza Plateau over some 200 km of flat gravel plains from about 28° to 26°N. (Kassas and Girgis 1972). Wadi Qena is characteristic of the west slope of the Red Sea Mountains (Fig. 2.18), where drainages coalesce into a relatively small number of giant trunk channels; the eastern slope in contrast is drained by numerous short, independent channels (Ayyad and Ghabbour 1986). The principal wadis of the virtually rainless 'Ababda Plateau (el Kharit, Shait, and Zeidun) flow into the Nile.

The Nubian Desert, extending from the Sudan border north to about 24°N., consists of a sandstone fringe along the Nile Valley and a basement-complex portion stretching east to the Red Sea Mountains. The sandstone country is drained by numerous short wadis, while Wadi el Allaqi drains much of the basement-complex country. Allaqi's 44 000 km^2 watershed is the largest in Egypt but receives little moisture: *Acacia* scrub in the largest channels is the only significant perennial vegetation (Kassas and Girgis 1969).

FIG. 2.18. Central Eastern Desert in Wadi Qena. Characteristic plants are *Acacia ehrenbergiana* (medium-size tree in immediate background), *Aerva javanica*, *Stachys aegyptiaca*, *Haloxylon salicornicum*, and *Salsola baryosma*. (Photo J. J. Hobbs, November 1986.)

The characteristic recent mammals of the Eastern Desert are the hare, bushy-tailed dipodil (*Sekeetamys makrami*), Egyptian spiny mouse (*Acomys cahirinus*), Rüppell's sand fox, striped hyena (*Hyaena hyaena*), sand cat (*Felis margarita*), hyrax (*Procavia capensis*), dorcas gazelle, ibex (*Capra ibex*), and barbary sheep (*Ammotragus lervia*) (Osborn and Helmy 1980; Goodman and Helmy 1986). Characteristic birds are Sand Partridge, Crowned Sandgrouse, Spotted Sandgrouse, Desert Lark, Pale Crag Martin, Mourning Wheatear, White-crowned Black Wheatear, and Trumpeter Finch. Reptiles are characterized by the geckos *Tropicolotes steudneri* and *Ptyodactylus hasselquistii*; the agamids *Agama agama spinosa*, *A. sinaita*, and *Uromastyx aegyptius*; the lacertid *Acanthodactylus boskianus*; the *Varanus griseus* monitor; the viper *Cerastes cerastes*; and the rare black cobra *Walterinnesia aegyptiaca*.

The Eastern Desert has a rich environmental and cultural history. Outsiders have long been attracted to its strategic and geological resources. Pharaonic, Ptolemaic, and Roman quarry-men and gold miners built numerous settlements and burned great numbers of trees, especially *Acacia raddiana*, for smelting and smithing (Tregenza 1955, 1958). Egyptians today mine phosphate, gold, and uranium in the area. Before the Suez Canal, the region provided a link between the Mediterranean world and the East. Caravans followed what are now the El Koriamat–Zafarana, Qena–Safaga, Qift–Quseir, and Idfu–Marsa el Alam roads between the Nile Valley and Red Sea.

Routes across the Nubian Desert made Aswan a market for livestock and other desert produce of northern Sudan.

Pastoral nomads have exploited the Eastern Desert since the Neolithic period. Today, Ma'aza tribesmen range northwards from the Qena–Quseir road, rearing sheep, goats, and camels, harvesting plants for markets in the Nile Valley, hunting game, and collecting edible plants for their own use. Suffering the effects of a past in which they charcoaled trees and hunted game with abandon, these Bedouins now uphold a collective ethic to safeguard their resources. Ma'aza reports suggest that their southern neighbours, pastoralists of the 'Ababda tribe, are fewer in number but have even stronger measures for protecting tribal resources (Hobbs 1986).

Modern development projects have created new habitats and opportunities for Eastern Desert biota, including man. These are especially well developed along the desert road from Qena to Safaga. Water spills from several pumping stations, deliberate accesses and accidental fractures along a pipeline carrying Nile water to Safaga. *Acacia nilotica*, *Ricinus communis*, and other characteristic Nile Valley trees, and the introduced *Eucalyptus* have grown around these waters. Using the roadside waters with little fear of man and vehicles, populations of Crowned Sandgrouse, Desert Lark, and Trumpeter Finch have increased. Bedouins water herds and small gardens along the road. Pastoralists, herds, and wild animals are benefiting from an additional windfall this road and the Quseir–Qift road provide: spillage from the almost non-stop traffic of trucks carrying imported wheat from the Safaga harbour to Nile Valley mills. This resource will expire with the completion of the Qena–Safaga railway, when the grain will be carried in enclosed cars.

Gebel Elba

Gebel Elba, a single igneous mountain rising to 1437 m in Egypt's extreme south-east corner, comprises a distinct natural region of the country (see Fig. 2.16). Average annual rainfall in the area is 50 mm, but orographic precipitation drawn from moist north-easterly winds drops as much as 400 mm on Elba. Winter rains and less frequent summer showers contribute most of this, but dew falls regularly and mist and clouds often shroud and soak the mountain. Owing to an abundance of moisture and altitudinal effects, Gebel Elba has a higher biotic diversity than any other area of comparable size in Egypt: 458 plants (Fahmy 1936; Kassas and Zahran 1971); 23 mammals characterized by wild ass (*Equus asinus*), Ethiopian hedgehog (*Paraechinus aethiopicus*), common genet (*Genetta genetta*), hyrax, Rüppell's sand fox, and zoril (*Ictonyx striatus*) (Osborn and Helmy 1980); 22 reptiles and amphibians characterized by the frog *Bufo dodsoni*, geckos *Pristurus flavipunctatus* and *Tarentola annularis*, agamid *Uromastyx ocellatus*, and lacertids *Latastia longicaudata* and *Ophisops elbaensis*; and 40 breeding birds characterized by Ostrich, Lappet-faced Vulture, Pink-headed Dove, Namaqua Dove, Black-crowned Finch Lark, Blackstart, Arabian Warbler, Fulvous Babbler, Shining Sunbird, Rosy-patched Shrike, and Fan-

FIG. 2.19. Gebel Elba. 'Parkland' in Wadi Aukau to the north of summit. Characteristic plants are *Acacia raddiana*, *Euphorbia nubica*, and *E. cuneata*. Note relatively heavy ground vegetation and density of tree growth. (Photo P. L. Meininger, April 1985.)

tailed Raven (Goodman 1985). Many of these animals do not occur elsewhere in Egypt, and several are locally endemic.

Elba's biotic diversity is greatest at middle altitudes, and on northern and eastern slopes (Fig. 2.19). The coastal plain, rising from sea-level to 70 m, has a vegetation of scattered shrubs, notably *Euphorbia cuneata*. At elevations of 70–170 m, acacia trees abound (*Acacia tortilis*, *A. nubica*, *A. etbaica*, *A. mellifera*, and *A. laeta*): this is the 'only place in the Eastern Desert of Egypt where the desert looks like a forest' (Kassas and Zahran 1971). Associated plants are *Euphorbia nubica*, *Cocculus pendulus*, *Balanites aegyptiaca*, *Ochradenus baccatus*, and *Ephedra ciliata*. Between 170 and 700 m, *Acacia etbaica* is the dominant plant. The summit's mist oasis is a wet zone with non-xerophytic plants (Fig. 2.20): *Dracaena ombet*, *Euclea schimperi*, *Dodonaea viscosa*, *Rhus abyssinica*, *R. tripartita*, *Acacia etbaica*, *Ficus salicifolia*, *Pistacia khinjuk*, *Jasminum* spp., ferns, mosses, and liverworts (Hoogstraal *et al.* 1957; Kassas and Zahran 1971; Osborn and Helmy 1980).

Most human inhabitants of Elba and the adjacent coastal region are Beja-speaking Bischarin tribesmen. Many range freely between Egypt and the Sudan in their economic pursuits: breeding domestic animals, collecting edible and medicinal plants, producing charcoal, and hunting game. The Bischarin have a strong environmental ethic against misusing water, pasture, and game (Goodman 1985).

FIG. 2.20. Gebel Elba. Ombet 'forest' (*Dracaena ombet*) on upper reaches of ridge between Wadi Akwamtra and Wadi Aukau. Note relatively dense woodland, mostly *Acacia* and *Balanites*, about 600 m below in distant wadi. (Photo S. M. Goodman, June 1984.)

The Sinai Peninsula

The Sinai Peninsula is a triangle of 61 000 km^2 with its base along the Mediterranean at about 31°10′N. and its vertex, Ras Muhammad, jutting into the Red Sea at 27°44′N. (see Fig. 2.21). It is bordered on the west by the Suez Canal and Gulf of Suez, and on the east by the Gulf of Aqaba and Egypt's 200 km political boundary with Israel. The Mediterranean coast of Sinai combines features of the Nile Delta and North Coast, with shallow lagoons, sand-dunes and plains, and salt-marshes. The balance of the Peninsula is morphologically an extension of the Eastern Desert, with coastal plains along the Gulfs of Suez and Aqaba, a core of basement-complex mountains, table mountains of sedimentary rock, and aprons of gravel plains draining and flanking the highlands (Abu al Izz 1971).

The Mediterranean coastal region of north Sinai has less rainfall (97 mm annually at El Arish) and fewer plant species than its counterpart in non-Sinaitic Egypt, but has about the same temperature ranges (January mean minimum 8.5°C, July mean maximum 30.6°C at El Arish; Boulos 1985*a*). Almost half of Sinai's Mediterranean coastline is spanned by Lake Bardawil, a lagoon separated from the sea by a narrow sand-bar 85 km long. Although there are one natural and two artificial breaches in this sand-bar, there is little circulation or drainage of Bardawil's water. As a result the lake is hypersaline and its poor vegetation consists mainly of *Zostera noltii* (Meininger and Mullié 1981*b*). Sand accumulates to form long parallel ridges along segments of

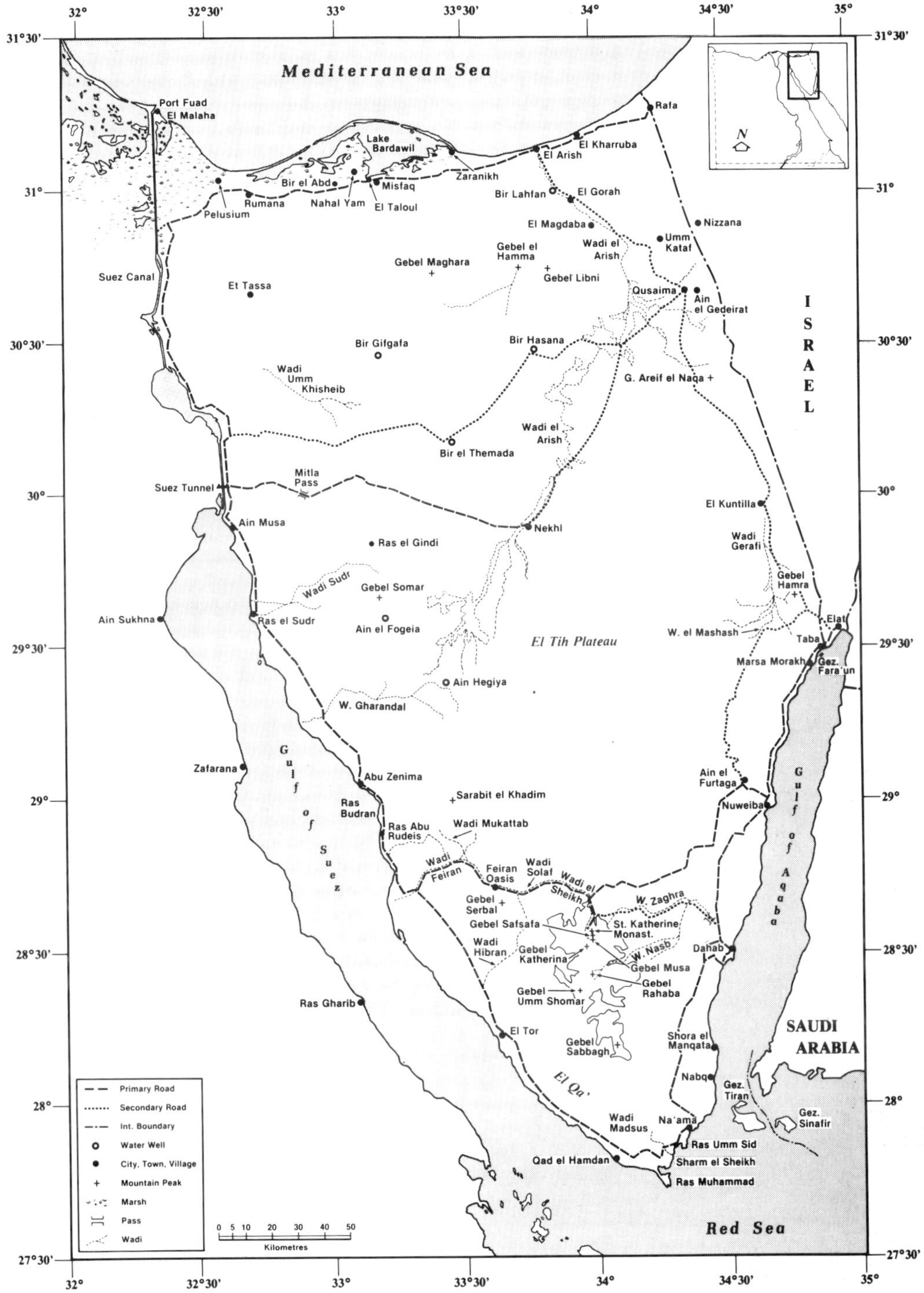

FIG. 2.21. Sinai Peninsula.

Sinai's north coast. Enough moisture collects between these ridges and in adjacent salt-marshes to support halophytic plants (Kassas 1955).

At about 31°N., the sandy plains of northern Sinai terminate at the base of a broken line of hills including Gebel Maghara (735 m), Gebel Yelleq (1090 m), and Gebel Halal (890 m). They are noteworthy for their relict populations of the conifer *Juniperus phoenicia*, which thrive there on about 100 mm annual rainfall (Boulos 1982; El Hadidi n.d.). South of this line is a low region of gravel plains incised by a vast network of north-flowing drainages, with Wadi el Arish the main trunk. These arid valleys drain two blocks of Cretaceous, Eocene, and Miocene limestones: the El Tih Plateau (800 m) and Gebel el Igma (1300 m). The orographic effects of 50–75 mm yearly rainfall support a Mediterranean semi-arid flora on these table mountains (Boulos 1982; Ayyad and Ghabbour 1986).

The interior of south Sinai is a fortress of Precambrian igneous and metamorphic mountains, with a northern rampart (about 29°N.) of Cretaceous (Nubian) sandstone. Habitats and biotic associations of Sinai's subalpine basement-complex slopes and drainages closely parallel those of the Eastern Desert (Zohary 1935; Danin 1979). However, the peaks of Gebel Katherina (2641 m, Egypt's highest), Umm Shomar (2586 m), el Thabt (2439 m), and Musa (2280 m) are higher than any in the Eastern Desert (Fig. 2.22). This alpine topography accounts for an exceptional variety of weather and plant and animal life. El Tor, on the southern Gulf of Suez, has an average

FIG. 2.22. High mountain region of southern Sinai. Sunrise from the summit of Gebel Musa (2280 m), a typical view of the 'roof of Egypt'. (Photo J. J. Hobbs, October 1982.)

rainfall of 13 mm per annum. The January mean minimum temperature is 8.8°C, and the July mean maximum 34.3°C. Only 50 km away at St Katherine Monastery (1600 m) the January mean minimum is 1°C; the August mean maximum 30°C; yearly rainfall is 64 mm. Annual precipitation in the mountains above the monastery may exceed 300 mm, contributed mostly by snow; winter temperatures drop to −10°C (Zohary 1944; Rabinowitz 1982; Boulos 1985*a*; Ayyad and Ghabbour 1986). The roof of Sinai has an extraordinary flora, including relict growths of formerly widespread Irano-Turanian species. About half of Sinai's 36 endemic species occur here (El Hadidi 1967; Danin 1979). This region and the lower Sinai massif (up to 1500 m) surrounding it also contribute about half of Sinai's inventory of just over 1000 plant species, which in turn represent 40 per cent of the total flora of Egypt (Hassib 1951; Danin 1973; Boulos 1982).

Characteristic mammals of south Sinai include the bushy-tailed dipodil, golden spiny mouse (*Acomys russatus*), hyrax, and ibex (Haim and Tchernov 1974; Wassif 1982). Mammals typical of north Sinai's sedimentary uplands and drainages are long-eared hedgehog, fat sand rat (*Psammomys terraesanctae*), crested porcupine (now locally extinct), red fox, fennec fox, and dorcas gazelle (Osborn and Helmy 1980; Wassif 1982). Reptiles characteristic of the Peninsula are the agamid lizards *Agama sinaita*, *A. stellio*, *Uromastyx aegyptius*, and *U. ornatus*; the lacertid lizard *Eremias brevirostris*; and the snakes *Leptotyphlops macrorhynchus*, *Coluber sinai*, *Eirenis coronelloides*, *Telescopus hoogstraali*, and *Atractapsis engaddensis* (Marx 1968). Typical birds of the southern Sinai mountains are Chukar, Sand Partridge, Lichtenstein's Sandgrouse, Crowned Sandgrouse, Hume's Tawny Owl, Black-capped Bulbul, Blackstart, Mourning Wheatear, White-crowned Black Wheatear, Fan-tailed Raven, Tristram's Grackle, and Sinai Rosefinch.

The Sinai is a mosaic of Bedouin tribal territories, with the largest areas claimed by the Terabin, Laheiwat, Haweitat, Aleiqat, and Muzeina tribes. Traditionally pastoral nomads, these Bedouins have adapted to changing seasonal and permanent opportunities, rearing livestock but also fishing and farming, and working for wages as mechanics, guards, and guides. The increasingly sedentary desert activities of Sinai Bedouins (who are themselves increasingly mobile, thanks to widespread adoption of pickup trucks) depend largely upon permanent springs in the basement-complex drainages of southern Sinai. In locales such as Wadi Feiran (Fig. 2.23), former nomads irrigate small crops of oranges, grapes, olives, apples, pears, lemons, apricots, almonds, dates, barley, wheat, onions, beans, lentils, cabbage, clover, and alfalfa (El Hadidi *et al.* 1970).

Sinai's economy and infrastructure are based largely on oil production and tourism. The only paved roads in the southern interior lead to St Katherine Monastery, where an airport is located. There are plans to diversify the economy in order to resettle one million Nile-dwellers in the Sinai (1976 population: 9717), half of them by the year 2000 (O'Brien 1982; Central Agency for Public Mobilization and Statistics 1984). The highest priorities are agricultural development projects, such as the reclamation

FIG. 2.23. Oasis in south Sinai mountains. Date palm grove in Wadi Feiran. Note the typical basement-complex mountains of this area in background. (Photo J. J. Hobbs, October 1982.)

of 1680 km^2 of cultivable land from the marshy plain of El Malaha in north-western Sinai (Meininger and Mullié 1981*b*).

Desert habitats

Basement-complex mountain habitat

General features

Local climatic conditions in the mountains are decided more by altitude than latitude. Owing to orographic rain and condensation averaging up to 100 mm annually, mountains higher than 1500 m have unusual and abundant plant species. Relict tree species, often represented by only one local specimen, survive on the slopes least accessible to man. Some mountains have *nakkaat* ('droplets') habitats, where water issues continuously to non-xerophytic biota sheltered by steep, shaded canyons (Kassas and Zahran 1965). These sources and the sometimes-brackish reed oases at the mountains' feet sustain the greatest concentrations of plant, animal, and human life in the Eastern Desert and Sinai.

Western Desert Dominant among Gebel Uweinat's 55 known plant species, which thrive because of orographic moisture, are *Acacia ehrenbergiana*, *A. raddiana*, and *Panicum turgidum* (Osborn and Krombein 1969).

Eastern Desert Relicts of *Rhus tripartita*, *Pistacia kninjuk*, *Olea indica*, *Acacia albida*, and *Maerua crassifolia* survive on the highest mountains (Tregenza 1958; Boulos and Hobbs 1986). These forms, as a whole, are more common on the upper reaches of Gebel Elba in the extreme south (Kassas and Zahran 1971). Typical *nakkaat* plants are *Adiantum capillus-veneris*, *Phragmites communis*, *Imperata cylindrica*, *Veronica beccabunga*, *Solanum nigrum*, *Ficus pseudosycomorus*, *Juncus arabicus*, *Cyperus laevigatus*, *Cynodon dactylon*, *Phoenix dactylifera*, mosses, and algae (Kassas and Zahran 1965).

Sinai The characteristic dwarf montane species of highest Sinai are *Hypericum sinaicum*, *Phlomis aurea*, *Primula boveana*, and *Veronica musa* (Boulos 1982). Lower slopes host the trees *Pistacia kninjuk*, *Ficus pseudosycomorus*, *Rhamnus disperma*, and *Cotoneaster orbicularis* (Zohary 1935; Danin 1979). *Crataegus sinaica* and *Scrophularia libanotica* grow as relicts (Danin 1983). The most common plant is *Artemisia herba-alba*, accompanied in different locales by *Gymnocarpos decander*, *Zilla spinosa*, *Fagonia mollis*, *Anabasis setifera*, *Halogeton alopecuroides*, *Atraphaxis spinosa*, *Stachys aegyptiaca*, and *Tanacetum santolinoides* (Danin 1983). *Nakkaat* plants at Wadi el Arbain include *Adiantum capillus-veneris*, *Juncus punctorius*, and *Equisetum ramosissimum*; and elsewhere *Scirpus holoschoenus*, *Veronica anagallis-aquatica*, and *Hypericum sinaicum* (Danin 1983; El Hadidi n.d.).

Sedimentary highland habitat

General features There are outstanding differences between the flora of Egypt's sedimentary plateaux and generally higher basement-complex mountains. In the former, water is available mainly as run-off from convectional rainfall, so that lower ground receives more moisture and sustains more plant growth (El Hadidi and Ayyad 1975). In the latter, orographic condensation produces greater precipitation and richer flora at higher elevations (Kassas and Zahran 1965).

North Coast On oolitic limestone ridges along and near the Mediterranean coast grow *Globularia arabica*, *Thymus capitatus*, *Helianthemum lippii*, *Asphodelus microcarpus*, and lichens (Kassas 1955); these benefit from regular winter rains exceeding 50 mm annually.

Western Desert The scant perennial vegetation of Gilf Kebir consists of relict populations of *Balanites aegyptiaca*, *Maerua crassifolia*, and *Acacia* spp. (Misonne 1970, 1974).

Eastern Desert On the slopes, cliffs, and dry waterfalls of Gebel el Galala el Bahariya, where an average 25 mm of rain falls yearly, *Capparis spinosa*, *C. cartilaginea*, and *Ficus pseudosycomorus* grow. Associated with these and occurring independently in small drainages are *Telephium sphaerospermum*, *Salsola tetrandra*, *Halogeton alopecuroides*, *Helianthemum kahiricum*, *Echinops galalensis*, *Ephedra alata*, *Erodium glaucophyllum*, *Hyparrhenia hirta*, *Stachys aegyptiaca*, *Reaumuria hirtella*, *Forsskalea tenacissima*, and *Fagonia mollis* (Kassas and Zahran 1962, 1965).

Sinai On the El Tih and El Igma plateaux a Mediterranean semi-arid vegetation type prevails; characteristic are *Ballota undulata*, *Capparis spinosa*, *Centaurea eryngioides*, *Kickxia aegyptiaca*, *Phagnalon rupestre*, *Teucrium polium*, and *Verbascum sinuatum* (Boulos 1982; Ayyad and Ghabbour 1986).

Basement-complex drainage habitat

General features Unlike the Western Desert, the Eastern Desert and Sinai are heavily dissected by drainage systems. Their erosion has been enhanced by the recurrence of heavy rains on arid soils lacking water-absorbing plant cover (Boulos 1985*a*). Below elevations of 1500 m, plant growth is largely confined to these torrent beds and their terraces, where vegetation is watered by floods and subsurface reservoirs.

Eastern Desert Plant communities are dominated by hardy perennial species which vary according to latitude. Community types are dominated along the entire Red Sea Mountain range by *Leptadenia pyrotechnica* and *Acacia raddiana*; in the north only by *Moringa peregrina* (at the feet of mountains higher than 1500 m), *Salvadora persica*, *Zilla spinosa*, *Zygophyllum coccineum*, and *Calligonum comosum*; in northern and central portions, by *Crotalaria aegyptiaca* and *Francoeuria crispa*; in the central portion only, by *Cassia italica*; in central and southern districts, by *Cassia italica*, *C. senna*, and *Salsola baryosma*; and in the south only, by *Indigofera argentea* (Kassas and Zahran 1965; Kassas and Girgis 1970).

Sinai Basement-complex wadis are confined to the south, and so their flora do not show latitudinal variations. Below 1500 m, the larger drainages are dominated by *Acacia raddiana*, *Calotropis procera*, and *Tamarix nilotica*. Common dominants of the smaller wadis are *Hammada elegans*, *Retama raetam*, *Aerva persica*, *Zilla spinosa*, *Artemisia judaica*, *Tephrosia apollinea*, *Fagonia mollis*, *Orchradenus baccatus*, *Hyoscyamus boveanus*, *Gomphocarphus sinaicus*, *Otostegia schimperi*, *Iphiona scabra*, *Capparis cartilaginea*, *Moricandia sinaica*, and *Launaea spinosa* (Danin 1983).

Sedimentary drainage habitat

General features Plant communities in wadis draining limestone and sandstone highlands are subject to many of the same conditions as basement-complex communities; the different soils, however, promote different species. Sedimentary drainages generally have less relief but larger watershed areas.

Eastern Desert Wadis of the 'Ababda Plateau support plant communties typical of Eastern Desert sedimentary drainages. Tributaries of larger wadis such as el Kharit are dominated by *Acacia ehrenbergiana* scrub with associated *A. raddiana*, *Capparis decidua*, *Zilla spinosa*, *Zygophyllum coccineum*, *Crotalaria aegyptiaca*, *Aerva persica*, *Colocynthis vulgaris*, *Cassia senna*, *Pulicaria undulata*, *P. crispa*, and *Fagonia bruguieri*. The main channels of the larger wadis also support *Acacia ehrenbergiana* accompanied by *Salsola baryosma*, *Zilla spinosa*, *Artemisia judaica*, *Pulicaria crispa*, *Aerva persica*, and *Leptadenia pyrotechnica* (Kassas and Girgis 1972).

Sinai Tributaries of Wadi el Arish, which comprise most of Sinai's sedimentary drainage habitat area, are dominated by *Anabasis articulata*, *Artemisia herba-alba*, *Astragalus spinosus*, *Atriplex hamilus*, *Cornacula monacantha*, *Gymnocarpos decandrum*, *Peganum harmala*, *Zilla spinosa*, *Salsola tetranda*, and *Hammada scoparia*; lower portions of the larger wadis have *Tamarix nilotica*, *T. aphylla*, *Acacia gerrardii*, and *A. raddiana* trees (Zohary 1935; Danin 1973; Boulos 1982; Ayyad and Ghabbour 1986).

Reed oasis habitat

General features 'Reed oasis' seems an appropriate designation for moist, moderately saline habitats transitional between those of the *nakkaat* and salt-marsh. Reed oases occur around springs in basement-complex and sedimentary highlands and wadis, on coastal plains near brackish water or wadi mouths, and in plateau depressions where the water-table is exposed in small, contained pools.

Western Desert Reed oasis vegetation thrives where water breaks the surface in the small depressions of Kurkur, Dunqul, and Dineiqil. Typical plants are *Imperata cylindrica*, *Salsola baryosma*, *Tamarix amplexicaulis*, *T. aphylla*, *Stipagrostis vulnerans*, *Alhagi mannifera*, *Juncus rigidus*, dom palm *Hyphaene thebaica*, and date palm *Phoenix dactylifera* (Reed 1964; Boulos 1966; Zahran 1966; Osborn and Helmy 1980). The brackish surface waters of Birs Tarfawi, Kiseiba, el Shab, and Misaha support a similar flora: dom and date palms, *Stipagrostis vulnerans*, *Sporobolus spicatus*, *Cynodon dactylon*, *Cornacula monacantha*, *Tamarix amplexicaulis*, *Alhagi mannifera*, *Imperata cylindrica*, *Phragmites australis*, and *Typha domingensis* (Beadnell 1931; Osborn and Helmy 1980).

Eastern Desert Reed oases on the Red Sea coastal plain, such as at Ain Sukhna, are dominated by *Phragmites australis* and *Typha domingensis*, the latter in less-saline locales. Common associates are *Samolus valerandi*, *Scirpus mucronata*, and *S. tuberosus* (Kassas and Zahran 1962; Ayyad and Ghabbour 1986). In basement-complex country, springs or near-surface waters create reed oases of *Typha domingensis*, *Phragmites australis*, *Imperata cylindrica*, *Juncus rigidus*, *Moringa peregrina*, *Ficus pseudosycomorus*, *Launea spinosa*, *Cynodon dactylon*, *Tamarix nilotica*, and date palms.

Sinai Mountain waters of south Sinai sustain date palms, *Moringa peregrina*, *Ficus pseudosycomorus*, *Holoschoenus vulgaris*, *Schoenus nigricans*, *Mentha longifolia*, and *Tamarix nilotica*. Near some permanent freshwater springs on the Gulf of Suez coastal plain (such as at Wadi Gharandal) grow *Typha domingensis*, *Phragmites australis*, *Juncus rigidus*, *Cyperus laevigatus*, and *Arundo donax* (Danin 1973; El Hadidi n.d.).

Salt-marsh habitat

General features Salt-marshes are widely distributed in the Egyptian deserts along coastal areas, where subsurface sea water permeates low-lying ground; and in the interior, wherever water sources are especially brackish, natural drainage is poor, or agriculture has been intensive. *Sebakha* is an Arabic term often used in scientific literature as a synonym for salt-marsh or salt-flats (Girgis *et al.* 1971; Boulos 1985*a*).

North Coast Salt-marshes occupy depressions between parallel limestone ridges along the Mediterranean; the dominant plants are *Halocnemum strobilaceum* and species of *Salicornia*, *Arthrocnemum*, and *Limoniastrum* (Kassas 1955).

Western Desert Marshes occupying poorly drained, low-lying areas of the oases typically contain *Phragmites australis*, *Scirpus litoralis*, *Typha elephantina*, *Juncus acutus*, *Suaeda monoica*, *Sporobolus spicatus*, *Cressa cretica*, *Panicum repens*, *Fuirena pubescens*, and *Samolus valerandi* (Kassas 1955; Zahran and Girgis 1972). The wet saline floor of the Qattara Depression at El Maghra is dominated by dense coverage of *Juncus rigidus*, with associated *Phragmites australis*, *Inula crithmoides*, *Tamarix nilotica*, *Arthocnemum glaucum*, and semi-wild date palms (Girgis *et al.* 1971).

Eastern Desert Along the Gulf of Suez coast are successive parallel zones of salt-marsh, each often comprised of only one of the following species: *Halocnemum strobilaceum*, *Nitraria retusa*, *Zygophyllum album*, *Limonium pruinosum*, and *Suaeda vermiculata*. Further south, *Halocnemum strobilaceum* is replaced by *Arthocnemum glaucum*, and *Limonium pruinosum* by *L. axillare* (Kassas and Zahran 1962, 1967; Ayyad and Ghabbour 1986).

Sinai Salt-marshes within 10 km of Sinai's Mediterranean coast are dominated by date palms, *Zygophyllum album*, *Halocnemum strobilaceum*, *Arthrocnemum macrostachyum*, *Suaeda aegyptiaca*, *S. vermiculata*, *Limoniastrum monopetalum*, and *Frankenia hirsuta* (Boulos 1982; Danin 1983). Many salt-marshes on the Gulf of Suez have pure associations of *Halopeplis perfoliata*, *Zygophyllum album*, *Arthrocnemum glaucum*, or *A. macrostachyum*. At Ain Musa, brackish springs issue from the coastal plain to water date palms, *Nitraria retusa*, *Tamarix nilotica*, and *T. passerinoides* (Danin 1983; El Hadidi n.d.).

Gravel plain habitat

General features The term 'gravel plain' here includes several land-form types often classified separately as *serir*, *hamada*, desert pavement, and gravel desert. *Serir* usually refers to a tightly packed veneer of large stones on compacted sand, the surface feature of most of the Western Desert and some northern portions of the Eastern Desert and Sinai. The rock cover is so complete that such plains are never vegetated, even after rain. The *hamada* is a layer of rock fragments dispersed less completely above a rock subsurface, as on the border of the coastal plain and basement-complex mountains of the Eastern Desert. Vegetation is able to grow in soil pockets after rain falls. The Eastern Desert coastal plain is largely another type, the gravel desert, a loose rocky surface on finer sediments and sand (Osborn and Helmy 1980; Ayyad and Ghabbour 1986).

Eastern Desert The stony sheets separating the tributaries of Wadi Qena comprise a gravel desert where the only plants are ephemerals such as *Fagonia indica* and *Morettia philaena* (Kassas and Girgis 1972). The gravel desert between Cairo and Suez is dominated by *Centaurea aegyptiaca*, with associated *Allium desertorum*, *Fagonia glutinosa*, and *Hammada elegans* (Kassas and Imam 1959).

Sinai Some places on the gravel plains of north Sinai are inhabited by a few opportunistic plants: shallow depressions and rivulets by *Anabasis articulata*; coarse sand deposits by *A. articulata* and *Zilla spinosa*; and shallow sand layers by *Hammada elegans* and *Panicum turgidum* (Ayyad and Ghabbour 1986).

Sand plain habitat

General features Sandy plains typically occur as long, narrow zones paralleling coasts; as 'streets' between parallel sand-dunes or stony ridges; as 'islands' surrounded by other habitats in the desert interior; and as beds of the broad, flat lower portions, and mouths of desert wadis.

North Coast Dominant plants on the sandy inland plains of the North Coast are *Thymelaea hirsuta*, *Anabasis articulata*, *Papaver dubium*, *Chrysanthemum coronarium*, and *Carthamus tenuis* (Kassas 1955).

Western Desert Sandy plains are home for most Western Desert plants. Sand flats of the Selima Sand Sheet and between some dunes of the Great Sand Sea support *Calligonum comosum* and *Stipagrostis* spp. (Osborn and Helmy 1980; Haynes 1982). On sandy plains between the salt-marshes and rocky margins of the oases-depressions, the dominant plants are *Alhagi maurorum*, *Nitraria retusa*, *Calotropis procera*, *Hyoscyamus muticus*, and *Tamarix nilotica* (Omer-Cooper 1937; Migahid *et al.* 1960; Zahran 1972). *Zygophyllum album*, *Nitraria retusa*, and *Tamarix nilotica* dominate sand formations on the floor of the Qattara Depression at El Maghra (Girgis *et al.* 1971). Within the Qattara Depression and on the El Daffa and north-central Eocene limestone plateaux are widely scattered, shallow basins of sand (*balata*). When rainfall accumulates in these, 'accidental' plant species germinate (Kassas 1955). Some support perennial shrubs such as *Capparis deserti*, *Zygophyllum coccineum*, and *Anastatica hierochuntica*; and incongruous groves of *Acacia* trees called *hatiyet* (Osborn and Helmy 1980).

Eastern Desert The wide, sandy beds of wadis leading to the Red Sea and Gulf of Suez support almost all the inland flora to be found on the coastal plain. Dominant are the desert grassland types *Hyparrhenia hirta*, *Lasiurus hirsutus*, *Panicum turgidum*, and *Pennisetum dichotomum*; the woody types *Iphiona mucronata*, *Cleome droserifolia*, *Zilla spinosa*, and *Launaea spinosa*; the succulent types *Zygophyllum coccineum*, *Hammada elegans*, and *Anabasis articulata*; and the desert scrubland types *Leptadenia pyrotechnica*, *Retama raetam*, *Tamarix aphylla*, and *Acacia raddiana* (Kassas and Zahran 1965).

Sinai Sandy plains near the Mediterranean coast are dominated by *Convolvulus lanatus*, *Artemisia monosperma*, *Cornulaca monacantha*, and *Stipagrostis plumosa*, shrubs possessing tap-roots adapted to volatile ground (Boulos 1982; Danin 1983).

Plains of deep sand along the northern half of the Gulf of Suez coast have a sparse cover of *Hammada elegans*, with associated *Zilla spinosa*, *Retama raetam*, *Hyoscyamus muticus*, *Zygophyllum coccineum*, and *Fagonia* spp. (El Hadidi n.d.). *Astragalus camelorum*, *Anabasis articulata*, and *Haloxylon persicum* inhabit nearby sands deposited on limestone hills and alluvial fans (Danin 1983).

Sand-dune habitat

General features Representing Egypt's poorest biotic habitat, sand-dunes comprise a very small proportion of the country's total desert area.

North Coast Psammophytic plant communities of the littoral oolitic dunes are dominated by *Euphorbia paralias*, *Pancratium maritimum*, *Ammophila arenaria*, and *Ononis vaginalis* (Kassas 1955).

Western Desert On the dunes lurking in the oases-depressions *Tamarix* spp., *Alhagi maurorum*, *Stipagrostis scoparia*, *Lagonychium farctum*, and *Calotropis procera* grow spontaneously (Kassas 1955; Migahid *et al.* 1960). People have cultivated various species of *Eucalyptus*, *Lagonychium*, *Saccharum*, *Tamarix*, and *Acacia* to slow the advance of dunes threatening settlements and farms (Migahid *et al.* 1960). Dunes of the Great Sand Sea are vegetated only by scattered individuals of *Stipagrostis scoparia*, *S. vulnerans*, and *Cornulaca monacantha* (Osborn and Helmy 1980).

Sinai Along the Mediterranean coast, the dominant and often only plant inhabiting mobile sand-dunes is the perennial grass *Stipagrostis scoparia* (Danin 1983).

MARINE ENVIRONMENTS

The Red Sea, Gulf of Aqaba and Gulf of Suez

Egypt's Red Sea region has traditionally been an underdeveloped frontier on both desert and sea. Modest fishing, mining, and trade industries have employed people from small coastal settlements: in the Sinai, Nuweiba, Dahab, Sharm el Sheikh, El Tor, Ras Abu Rudeis, and Ras el Sudr; in mainland Egypt, Ras Gharib, Hurghada, Safaga, and Quseir. South of Quseir even smaller communities such as Marsa el Alam, Abu Ghusun, Hamata, and Shalatein have served as security posts and as markets for pastoralists and fishermen.

A slow pace of development belies the great resource wealth of the Red Sea. Promoted by peace between Egypt and Israel, current exploitation of these resources is causing great economic and ecological changes. The reopening of the Suez Canal in 1976 restored the heavy maritime traffic that is a principal source of national income. Commercial fishing operations have returned to the Gulf of Suez, reportedly the richest grounds in the Red Sea region (Beltagy 1976). Since 1980, the Gulf of Suez oilfields have produced more for Egypt than ever before, earning the country much needed hard currency. The Sinai's return to Egyptian control and subsequent demilitarization of the Sinai and Red Sea coasts have attracted unprecedented numbers of tourists. Conflicting human interests are taking a toll of the region's natural resources. Pollution from oil-rigs and merchant ships is damaging the main tourist attraction, the coral reefs; ironically so are the tourists, by spear-fishing and shell collecting (Ormond 1981; Wells in prep.).

The coasts of the Egyptian Red Sea stretch 1100 km from the Sudan border to Suez. The Sinai forms the 320 km eastern shore of the Gulf of Suez and the 200 km western shore of the Gulf of Aqaba. Despite their symmetry, these two arms of the Red Sea differ greatly. Only 15–25 km wide, the Gulf of Aqaba is up to 1800 m deep. It is in effect the hinge of the Afro–Syrian rift system which created the great depths of the Red Sea (to 2500 m) and heights of East Africa (to 5879 m). Though slightly wider than the Gulf of Aqaba, the Gulf of Suez has an average depth of only 25 m. The Gulf of Suez has biotic affinities with the temperate Mediterranean, due to exchanges of water and fauna through the Suez Canal. Only 25 coral species are represented, and well-developed reefs are lacking. The Gulf of Aqaba, in contrast, shares with the Red Sea the rich tropical biota of Indo–Pacific waters. About 110 hermatypic and 120 soft coral species have been recorded from the Gulf of Aqaba (Fishelson 1980).

The high salinity (4 per cent) and relative warmth (surface average 21°C) of the Gulf of Aqaba and Red Sea favour coral development. Living coral polyps form large communities and their calcium carbonate skeletons accumulate to build reefs. Three reef types occur in the Red Sea region: shallow, sandy reefs where coral growth is limited; intermediate reefs adjacent to water deeper than 15 m; and deep-water reefs exposed to wave action of seas deeper than 100 m (as at Ras Muhammad). Intermediate reefs have the greatest number and diversity of coral and fish, while deep-water reefs are habitats of fewer but more unusual varieties, including large pelagic fish (Ormond 1981).

Each type of reef is separated from land by a shallow lagoon where rays, octopuses and sea-cucumbers are characteristic fauna. In some lagoons and in shallow water between reefs, sea-grasses including *Halophila stipulacea* and *Cymodocea ciliata* form beds. These scarce growths are the preferred pastures of the rare green turtle (*Chelonia mydas*) and dugong (*Dugong dugon*) (Barratt 1982; Frazier and Salas 1984). Low-tide transforms much of the lagoon into an exposed sand-flat where crabs retreat to coral debris and small fish to isolated pools. Low-tides south of Gebel el Zeit expose mud-flats sheltered by large bays. These provide important habitats for breeding, migrant, and winter visitor birds (Jennings *et al.* 1985). Tidal ranges vary considerably by locale in the Red Sea region: the variation is 2.1 m at Suez but only 0.3 m in the southern Gulf of Suez. There are also seasonal differences in the mean sea-level, as strong southerly winds from December to March raise northern Red Sea waters as much as 1 m (Admiralty of Great Britain 1946).

Except for tiny Gezira Fara'un (0.53 km^2) in the northern Gulf of Aqaba, there are no islands well within the Red Sea boundaries of the two gulfs. Gezira Tiran (30.4 km^2) occupies the mouth of the Gulf of Aqaba and forms the eastern wall of the navigable but treacherous Strait of Tiran (27°56′N.). Near the mouth of the Gulf of Suez is Egypt's largest archipelago (see Fig. 2.15), including the islands (Gezira) of Ghanim, Qeisum, Gubal el Kabir, Tawila, Shadwan (at 56 km^2 the largest and over 300 m the highest), Umm Qamar, Giftun el Kabir, Giftun el Saghir, Abu Rimathi, Umm Agawish el Kabir, Umm Agawish el Saghir, and Abu Minqar. Southern islands of the Egyptian Red Sea (see Fig. 2.16) include Gezira Safaga, off the port of Safaga; Gezira Wadi Gimal; a group off the coast near Hamata comprised of Gezira Siyul, Gezira Showarit, Gezira Mahabis, and Gezira Umm Ladid; the Siyul group off Adal Deib; the Rawabel group off Abu Ramad; and the lone islands of Gezira Mikauwa and Gezira Halaib.

The most common type of Red Sea island has an igneous core ringed by fossil coral reefs which were exposed by uplifting of the core; even the remote deep-sea island of Gezira Zabargad (famous as St John's Reef for its peridot) has this structure. Notably, most of these islands have high bluffs over their north-eastern shores but low, sandy slopes which merge with the tidal zones of the south-western shores. Before oil polluted the archipelago at the mouth of the Gulf of Suez, the north-east coasts provided ideal feeding and breeding habitats for such birds as Kentish Plover and Striated Heron; the relatively clean south-western shores suit many feeding and resting birds (Jennings *et al.* 1985). The Umm Agawish islands represent another island type, the islet or atoll comprised exclusively of fossil coral. Brothers Reef (Gezira el Akhawein) is a rare deep-sea example of this variety. Gezira Ghanim represents yet another island type, a coastal headland cut off from the mainland by erosion (Abu al Izz 1971; Goodman and Storer 1987).

FIG. 2.24. Red Sea island. Sandy beach with halophytic vegetation on Gezira Rawabel el Kabir. Characteristic plants include *Nitraria retusa* and *Tamarix nilotica*. (Photo S. M. Goodman, February 1985.)

FIG. 2.25. Red Sea mangrove swamp. Dense growth of *Avicennia marina* in bay of Gezira Abu Minqar, off Hurghada. (Photo S. M. Goodman, May 1984.)

Plant populations and species diversity on island and mainland bay shores are low, often representing only one or two of the following halophytic species (Fig. 2.24): *Suaeda monoica*, *S. vermiculata*, *Limonium axillare*, *Halocnemum strobilaceum*, *Arthrocnemum glaucum*, *Nitraria retusa*, and *Tamarix nilotica* (Kassas and Zahran 1962, 1965; Danin 1983; Goodman and Storer 1987). In some protected bays and lagoons of island and mainland shores grow thickets of two mangrove species: *Rhizophora mucronata* (south of 23°N.; Zahran 1965) and *Avicennia marina* (as far north as the southern Gulf of Aqaba; Fig. 2.25). There are notable populations of *Avicennia marina* along the coast south of Safaga, particularly between Abu Ghusun and Halaib; on Gezira Wadi Gimal, Gezira Safaga, Gezira Qeisum, Gezira Umm el Heimet, and Gezira Abu Minqar; at Ras Muhammad; and as far north as Shora el Manqata. The latter locality at 28°13′N. is the northern-most occurrence of mangrove on the globe. Mangrove thickets provide breeding and/or feeding habitats for such Egyptian birds as Western Reef Heron, Striated Heron, Spoonbill, and Osprey.

Terrestrial reptiles characteristic of the Red Sea islands are a colubrid snake, *Psammophis aegyptius*; a gecko, *Hemidactylus turcicus*; and a lacertid lizard, *Mesalina guttulata* (Werner 1973; Goodman *et al.* in press). An estimated 500 hawksbill turtles (*Eretomochelys imbricata*) and 100 green turtles nest on island and mainland sands of the Egyptian Red Sea (Frazier and Salas 1984). There is also speculation that loggerheads (*Caretta caretta*) and leatherbacks (*Dermochelys coriacea*) lay eggs on Sinai beaches (Marx 1968; Edwards *et al.* 1981; Ormond 1981; Ross and Barwani 1981; Groombridge 1982). Populations of all Red Sea turtles have apparently declined due to oil pollution, seismic explosions for oil prospecting, dynamite fishing, and human predation on the eggs for food and the adults for the tourist trade (Frazier and Salas 1984).

Characteristic birds of the Red Sea islands and coasts are the Brown Booby, Western Reef Heron, Spoonbill, Osprey, White-eyed Gull, Caspian Tern, and White-cheeked Tern.

3. NATURE CONSERVATION AND BIRD PROTECTION IN EGYPT

Few areas in the world have been manipulated by man for as long as the Egyptian Nile basin. The earliest evidence for artificial irrigation is the mace-head of the Scorpion King (*c*.3100 BC) showing one of the last Predynastic kings cutting an irrigation ditch (Butzer 1976). Large mammals, such as elephant, rhinoceros, and giraffe, inhabited the Nile Valley north of Aswan until *c*.2600 BC but through a process of climatic changes and human hunting pressure they disappeared from the area (Butzer 1976). A decrease in rainfall, especially after *c*.2040 BC, caused groups of pastoral nomads to settle in the Nile Valley (Butzer 1976) and thus increased the human ecological impact on the Nile flood-plain. The cutting of acacia forests in ancient times in places like Wadi Qena or Wadi el Qubbaniya may have resulted in ecological changes similar to those in the Sahel in the past two decades. During the Greek, Ptolemaic, Roman, Mameluke, and Ottoman periods people certainly exploited the environment but with the arrival of modern Europeans to Egypt the degradation of the environment was accelerated (cf. Ghabbour 1975).

In 1798 the French were the first Europeans to interrupt the centuries of classical rulers. The French were expelled from the country in 1801. Four years later Pasha Mohammed Ali invited French scientists back to the country to continue their work and essentially remodel Egypt as a 'modern' European country. A portion of the results were published between 1808 and 1825 in the 24 volumes of the *Description de l'Égypte*, an important contribution to the knowledge of Egypt's (natural) resources. After this period the human impact on the environment became more prominent. Dams were built across the Nile and irrigation systems installed to help with the production of cash crops. As early as 1843 a barrage was constructed across the Nile, north of the present Delta Barrage, to irrigate fields for cotton production (Clawson *et al.* 1971; Kashef 1981). After Egypt became a British protectorate in 1882, bird hunting was intensified and subsequently commercialized. The increased impact on the environment, primarily through Western influence, led to the first attempts for nature conservation and bird protection in Egypt. In this chapter the history of nature conservation in Egypt, particularly in relation to bird protection, is reviewed.

HISTORICAL OVERVIEW

The first modern attempts to protect birds in Egypt were at the beginning of the twentieth century. After successful action for better legislation concerning the trapping of small birds with nets and limesticks (Innes 1902, 1909), the Ministry of Interior forbade in a 1905 decree the hunting of small birds on state-owned and some private lands. During this same period, the Egyptian Cattle Egret population declined substantially, at least in part due to European plume hunters (Bonhote 1918). The 1905 decree was followed by Law 9 for 1912, in which 39 species of birds useful to agriculture were protected (Flower and Nicoll 1918). This last legislation was not enacted until Lord Kitchener personally took up the matter (Berney 1919; Flower in Bonhote 1919). Except for the recovery of the Cattle Egret (El Bakri 1927–8; Kadry 1942), the 1912 law was found to be generally ineffective. This was principally due to the lack of enforcement. Bonhote (1919) reported to the Ministry of Public Works that bird liming continued in the Nile Delta despite this legislation.

Law 13 for 1922 and a ministerial decree of 1928 augmented the list of protected bird species and prohibited the use of bird lime and cultivation of the bird lime tree *Cordia myxa* (Meinertzhagen 1930). Quail trapping was reputedly regulated as well after successful intervention of the Director of the Société International de Chasse (Greaves 1936), but it is doubtful if any serious enforcement took place (cf. Moreau 1951). Even though a number of birds were officially protected, there was still no law concerning the hunting of raptors, water-fowl, and most waders. Al Hussaini (1954*b*) suggested a further extension of Law 13 for 1922 and 1928 by including more protected taxa. On the other hand, there was a well-documented call for a modern law, i.e. protecting most species (Meinertzhagen 1930). The next governmental action was in 1967 when Decree 28 of the Ministry of Agriculture stated some 'sorts of birds useful for cultivation' were protected from hunting. Most of Al Hussaini's suggestions (i.e. better protection of water-fowl and game birds) were not followed, although the list of protected taxa was almost doubled. This decree, the current bird protection law in Egypt:

(1) protects a number of bird and mammal taxa;
(2) prohibits hunting or exporting Quail in spring;
(3) gives regulations for the use of Quail nets and 'nests' (presumably *munsaab* style trap, see p. 88 for description).

Decree 66 for 1983 of the Ministry of Agriculture added all birds of prey and the Houbara to the list of protected species. The taxa of birds currently protected by law are summarized in Table 3.1.

In the early 1900s some areas of the country were protected, although this was mainly to safeguard hunting places. One reserve was Wadi Rishrash, approximately 80 km south-east of Cairo in the Eastern Desert. Prince Kamel el Din Hussein found Nubian ibex frequented the wadi and in the 1920s he acquired the rights to an area of approximately 100 km^2 including and surrounding the wadi. This block of land was guarded by Albanian keepers and used as the prince's private hunting reserve. After his death the wadi was taken over by the government as a game sanctuary (Jarvis 1942; Ghabbour 1975) but it was destroyed by 1952 (Ayyad and Ghabbour 1986).

TABLE 3.1. Bird taxa protected by Egyptian Law 28 for 1967 or by Decree 66 for 1983 (*)

Ardeola spp.	*Centropus senegalensis*	*Cercomela melanura*
Egretta spp.	Tytonidae (*)	*Saxicola* spp.
Ardea spp.	Strigidae (*)	*Oenanthe* spp.
Ciconia spp.	*Caprimulgus* spp.	*Monticola* spp.
Plegadis falcinellus	*Apus* spp.	*Turdus* spp.
Phoenicopterus ruber	*Merops* spp.	*Prinia gracilis*
Accipitridae (*)	*Coracias* spp.	*Scotocerca inquieta*
Pandionidae (*)	*Upupa epops*	*Locustella* spp.
Falconidae (*)	*Jynx torquilla*	*Acrocephalus* spp.
Gruidae	Alaudidae	*Hippolais* spp.
Chlamydotis undulata (*)	*Riparia riparia*	*Sylvia* spp.
Burhinus spp.	*Hirundo* spp.	*Phylloscopus* spp.
Pluvianus aegyptius	*Delichon urbica*	*Regulus* spp.
Cursorius cursor	*Anthus* spp.	*Muscicapa striata*
Glareola spp.	*Motacilla* spp.	*Ficedula* spp.
Hoplopterus spinosus	*Pycnonotus* spp.	*Turdoides* spp.
Vanellus vanellus	*Cercotrichas galactotes*	*Oriolus oriolus*
Chettusia spp.	*Erithacus rubecula*	*Lanius* spp.
Columba spp.	*Luscinia* spp.	*Sturnus* spp.
Clamator glandarius	*Phoenicurus* spp.	

Another small reserve existed in Ain el Gedeirat, eastern Sinai. Here Jarvis (1942), the Governor of Sinai between 1922 and 1936, restored an ancient Byzantine scheme of irrigation which watered about 1 km^2 of wadi. This area attracted both Sand Partridge and Chukar. It was guarded and kept as a 'partridge preserve' during the 1920s and 1930s. Jarvis also made some efforts, with limited success, to preserve Nubian ibex on certain Sinai mountains by employing Arab keepers to keep watch over them.

A duck shooting reserve was established as early as 1923 at Tel el Kebir (Hindmarsh 1949), an area presently administered by the Egyptian Shooting Club and referred to as El Abbassa A(SZ). This estate was rented by the Tel el Kebir Shooting Club from the Ministry of Waqfs. In 1943 the ministry sold the land to King Farouk (Hindmarsh 1949). This is the location of the so-called 'Royal Blind' (Williams 1981). Since the establishment of the Egyptian Shooting Club in 1939–40 several lakes have been rented and ecologically manipulated to maintain conditions for water-fowl. Some of the existing reserves are Lakes Ikyad, El Abbassa, El Amia, Ba'alwa, and Gezira, all in the Sharqiya and Ismailiya Governorates. Lake Dahshur, originally rented by the Egyptian Shooting Club, was taken over by the government in the 1970s and is presently being used as a military hunting reserve (SZ).

INTERNATIONAL CONVENTIONS

The post-colonial period conservation movement in the Middle East commenced in the 1950s. From 1954, when Unesco held a conference in Beirut on nature protection, through the early 1970s, international efforts to stir the interests of local people in conservation largely ended as academic exercises (Hobbs 1980). The Egyptian botanist Montasir (1954) was definitely ahead of his time when he proposed at the Beirut conference that 'Ecologists should be consulted before beginning any cultivation, irrigation or drainage projects'. The 1961 Aruska conference on 'Conservation of Nature and Natural Resources in Modern African States' became a turning point (Curry-Lindahl 1974). Eventually this movement led to the 'African Convention on Conservation of Nature and Natural Resources' (1968) of which Egypt became a party. In recent years Egypt has signed five international conventions dealing with nature conservation. These conventions are mentioned below and where appropriate their consequences are briefly discussed.

African Convention on Conservation of Nature and Natural Resources (Algiers 1968) The basic premise of this convention is the conservation of certain animals at various levels of protection as well as habitat. This document provides a uniform base for international conservation action (Eidsvik 1980). However, as pointed out by Curry-Lindahl (1974), without being fully implemented the African Convention cannot create miracles. All African countries were invited to sign it. To date 41 countries have adopted it, including Egypt. However, within the country no action has been taken to uphold the goals of the convention. The African Convention is essentially the complementary equivalent to the Bern Convention (1985) for Europe (Broggi 1985).

Convention on Wetlands of International Importance especially as Waterfowl Habitat (Ramsar 1971) As mentioned by Etherington (1983), starting in the early 1960s attempts were made to define criteria by which important wetlands could be identified. The Project MAR organized by the International Union for the Conservation of Nature and Natural Resources (IUCN), International Council for Bird Preservation (ICBP), and International Waterfowl Research Bureau (IWRB), stressed the need for international conventions. Simultaneously, the International Biological Programme (IBP) together with the Societas Internationalis Limnologiae (SIL) and the IUCN published 'Project Aqua; a source book of inland waters proposed for conservation', a list of lakes and rivers of high limnological interest (Luther and Rzóska 1971). The lakes of Idku, Qarun, and Nasser, and the Siwa Oasis were proposed for conservation, but no Egyptian governmental action was undertaken.

The Ramsar Convention, the direct result of these projects, requires countries to designate (at least) one wetland of international importance at the time of their respective signing (Eidsvik 1980), and Egypt's choice was Lake Bardawil. More wetlands meet the requirements, but none has yet been proposed for conservation within the framework of this convention.

Convention Concerning the Protection of the World Cultural and Natural Heritage (Paris 1972) This convention includes only areas of 'outstanding universal significance' (Eidsvik 1980). As such, a number of ancient Egyptian cultural monuments or sites have been designated to the World Heritage List. This convention leaves room for the inclusion of natural areas. To date only the general area surrounding the St Katherine Monastery has been proposed (Rabinowitz 1980).

Convention on International Trade in Endangered Species of Wild Fauna and Flora (CITES) (Washington 1973) This convention has led to the establishment of a world-wide network of national management authorities to uniformly enforce animal trade controls. It is the most widely accepted conservation treaty today (UNEP 1985). The Egyptian Wildlife Service is the responsible authority for the implementation of CITES within the country. At the Cairo airport there are special officers to control the trade of endangered and threatened species.

Convention on the Conservation of Migratory Species of Wild Animals (Bonn 1979) As stated in Article II of this convention: 'The parties... [will be] paying special attention to migratory species the conservation status of which is unfavourable, and taking individually or in co-operation appropriate and necessary steps to conserve such species and their habitat'. In Appendix 1, endangered species are listed, of which six occur in Egypt. These are White Pelican, Dalmatian Pelican, Houbara, Slender-billed Curlew, White-eyed Gull, and Tristram's Serin. It should also be mentioned that the Houbara and White-eyed Gull breed in Egypt. The latter species is a Red Sea endemic and may have its population stronghold in Egypt (Jennings *et al.* 1985). Houbaras are now generally rare in the country due to over-hunting. Except for the protection of Houbara by a ministerial decree, no appropriate measures have been taken in Egypt to safeguard these species and their habitat. In Appendix 2, species are listed which have an 'unfavourable conservation status' and which require international agreements for their conservation and management. Birds falling into this class and occurring in Egypt include Anatinae, Quail, and waders. All of these forms are heavily hunted in Egypt, even those for which protective legislation exists.

NATIONAL LEGISLATION

In 1971 the Egyptian Academy of Scientific Research and Technology was established and in turn created in 1976 a National Committee for Conservation of Nature (Kassas 1981). This committee was considered the principal 'pressure group' to promote local research in conservation, lobby for government approval, and seek financial and technical assistance from national and international agencies (Hobbs 1980). During this same period several non-governmental societies were formed such as the Tree Lovers (1973), the Egyptian Society for Protection of Nature (1975), and the Egyptian Society for Environmental Sciences (1976) (Kassas 1981). The establishment in 1982 of the Ornithological Society of Egypt (OSE) was a major leap forward in promoting public interest in the conservation of birds. An important impetus on the governmental level came from the joint US Fish and Wildlife Service—Egyptian Ministry of Agriculture 'Endangered Species Conservation Project', which was initiated in 1977.

The provisional Egyptian office for this project as authorized by Decree 349 for 1979 of the Ministry of Agriculture is the Egyptian Wildlife Service (EWS).

On 5 March 1980 the late President Anwar Sadat established a policy for the protection of natural resources in Egypt, which was based upon the World Conservation Strategy (IUCN 1980). This was a major success for the Egyptian and international conservation movement. The agenda for this policy was to be worked out by the Egyptian Academy of Scientific Research and Technology and implemented by the Egyptian Wildlife Service in co-operation with local authorities. It was proposed to establish nature reserves throughout the country with varying degrees of protection. Further, at this time Professor Mohammed Kassas, Cairo University, was President of IUCN and the progress and problems in Egyptian conservation were exposed to a larger international audience. The enactment of Law 102 for 1983 concerning 'natural protectorates' (= nature reserves) heralded a new conservation era in Egypt. This law offers a legal framework for the establishment and management of natural areas and the protection of resources within the reserves. As of December 1986 the following areas have been designated by Prime Ministerial Decrees:

No. 1067 for 1983: Ras Muhammad and Gezira Tiran;
No. 1429 for 1985: Areas of Zaranikh and of Lake Bardawil proper and the coastal sand-dunes, known as al Ahrash (scrub), between El Arish and the international border at Rafa;
No. 450 for 1986: Islands in the [southern] Red Sea; the coastal mangrove swamps; and the areas surrounding Gebel Abraq, El Deib, and Gebel Elba.
No. 928 for 1986: Gezira Saluga and neighbouring islands.

Decree 40 for 1982 of the Matruh Governorate and 701 for 1982 of the Ministry of Agriculture established a nature reserve at El Omayed, between El Alamein and El Hammam. This fenced area is officially designated as a Unesco Man and Biosphere (MAB) reserve and as such has a legal status as a protected area. El Omayed was originally the experimental plot of the University of Alexandria Systems Analysis of Mediterranean Desert Ecosystems of Northern Egypt project (SAMDENE) for studying the effects of controlled grazing on a local ecosystem (Ayyad and Ghabbour 1977).

The protection of areas in the Sinai was a direct result of the Camp David Peace Accord. During the occupation of the Peninsula between 1967 and 1979 the Israeli Nature Reserves Authority (NRA) was responsible for local conservation matters (HM; UP). Although reluctant to interfere too much with local traditions, they successfully established a nature reserve on Gezira Tiran, minimized the trapping of migrant Quail, and substantially reduced the robbery of coral reefs and hunting of large mammals (UP). During the autumns of 1968–71 Quail netting along the north Sinai coast was restricted and in 1972 and 1973 (until the October War) it was completely forbidden (Zuckerbrot *et al.* 1980). Along Lake Bardawil the robbing of bird eggs by fishermen was controlled with limited success (UP).

In 1979, just before the signing of the Camp David Accord, the Society for the Protection of Nature in Israel (SPNI) who operated two field study centres in Sinai at Nahal Yam and St Katherine, composed a 'Proposed Nature Conservation Plan in Sinai' (Sagi *et al.* 1979). In this document very specific recommendations were made for the protection of south Sinai. A series of reserves was suggested and each defined in accordance with the local inherent characteristics and the role they could play in fulfilling the needs of the public. The 'wilderness reserves' selected were the Sinai high mountain area around St Katherine Monastery, Wadi Feiran Oasis, Gebel Umm Shomar, and its western canyons; the sandstone hills of eastern Sinai; the mining region near Abu Rudeis; and the Elat Gulf Beach. Ras Muhammad was proposed as a 'strict nature reserve'. Relevant information concerning the conservation of Sinai was passed to the Egyptian authorities by SPNI and NRA upon the return of the Sinai to Egypt. Action taken by the Egyptian government to date includes the establishment of nature reserves at Ras Muhammad and on Gezira Tiran.

The protection of the general Gebel Elba region, as urged earlier by Ghabbour (1975), was in part a result of the recommendations made by an IUCN expedition to the area in 1985 (Goodman 1985). Another reserve called Asyut University Protected Area, which covers about 200 acres in the Wadi el Asyuti, has no legal status. Following the successful operation of a wildlife office in the North Sinai Governorate, a presidential directive in early 1983 established wildlife offices in all governorates (Bruun 1982, 1983). Decree 1068 for 1983 of the Egyptian Council of Ministers established the Environmental Affairs Agency (EAA) as the national authority responsible for implementing the provisions of Law 102 for 1983. Decree 101 for 1985 levied a 10 per cent tax on all tickets for international aeroplane travel issued in the country and paid for in Egyptian currency. These funds are to be used for national pollution prevention and nature conservation (Bruun 1986) but no permanent formula has yet been devised to divide these monies.

BEDOUIN CONSERVATION SYSTEMS

In the discussion about conservation in Egypt, the practices of some Bedouin tribes should not be ignored. To the contrary, the knowledge about their 'conservation systems' should form the basis of any strategy to be implemented in their territory. Some groups of pastoral nomads, contrary to a widely held view, do not inevitably destroy the natural environment, particularly in vast portions of the Egyptian Eastern Desert (Goodman 1985; Hobbs 1986). The following is a quotation from Hobbs (1986), who extensively studied the conservation practices of Khushmaan clansmen of the Ma'aza tribe in the northern Eastern Desert. 'Without outsiders' guidance, the Khushmaan have developed their own ethics and practices of conservation ... The most important law is that green limbs of woody trees, particularly the sayaal *Acacia raddiana* may not be cut ... Only when no other food is available should a man take acacia leaves for his herd ... The nomads apply similar principles but different methods in managing game . . . A man should hunt infrequently, and then kill no more than one or two animals ... Khushmaan culture provides means of enforcing the nomads' environmental code.' Within the clan, verbal castigation is the punishment for resource abuse by which the violator's reputation is attacked. However, the nomads are often unable to reprimand non-Bedouin violators, and although the Red Sea Governorate has passed some decrees which prohibit desert hunting, tree cutting, and charcoaling, its bureaucracy prevented officials from reacting adequately to Khushmaan requests.

Some other tribes in the Eastern Desert have similar tribal laws. A strong environmental ethic exists among the Bischarin tribe, inhabiting the general Gebel Elba area. Tribal laws pose severe punishments for the misuse of freshwater resources, herbage, and the cutting of green wood for fuel and charcoal. Some Bischarin hunt local vertebrates. However, they generally use classical methods of trapping and hunting (i.e. throwing stones), and firearms are rarely available (Goodman 1985), a situation comparable with that for the Khushmaan (Hobbs 1986).

4. BIRD HUNTING IN EGYPT

Bird hunting and trapping have been a part of the Nile Valley cultural tradition since time immemorial and fowling scenes are depicted in many ancient Egyptian tombs (e.g. Hobbs 1980; Houlihan and Goodman 1986). However, some marked changes have occurred in the preferred species and in the means of obtaining them. Bonhote (1919) proposed that bird liming had been carried on in Egypt, possibly since Roman times. Greaves (1936), however, suspected that the use of lime-sticks did not occur in Egypt before 1890, when the tree the substance is obtained from was introduced from Syria.

Quail netting in Egypt for local consumption had been a well-known phenomenon since the Old Kingdom (*c.*2650–2150 BC) (Darby *et al.* 1977; Houlihan and Goodman 1986). After 1882, the year the British occupied Egypt, trapping was intensified and birds were exported to European markets (Anon. 1898). Moreau (1927–8) mentioned 'that never has the capture of Quail been systematized to the same extent as at present'. The netting of small birds also increased during this period and spread as far south as Cairo (Greaves 1936).

British hunters, guided by Shelley's (1872) book, habitually invaded Egypt in winter to have 'a little sandgrouse shooting' (Whymper 1909). Birds were also relentlessly pursued for millinery purposes. Innes (1902) noted that in the Nile Delta and the Faiyum, Little Terns, Little Egrets, and Great White Herons, and to a lesser extent pelicans, flamingos, and grebes were killed in large quantities. Kingfishers and apparently Rollers and Turtle Doves were also shot for the same reason (Whymper 1909; Bonhote 1919). The decline of Egyptian Cattle Egrets to near extinction is well documented (Bonhote 1918), although Whymper (1909) doubted if persecution by plume hunters was indeed the cause. Whatever the reason, the decrease of Cattle Egrets prompted the Egyptian government to undertake legislative action (see p. 68).

Duck hunting was a very popular sport among Egypt's European community. In *Justine*, book one of the *Alexandria Quartet*, Durrell (1957) exalts a hunt which took place in Lake Maryut, just before World War II: 'Their wings purr, as they feather the sky, their necks reach. Higher again in mid-heaven there travel the clear formations of Mallard, grouped like aircraft against the light, ploughing a soft slow flight. The guns squash the air and harry them as they pass, moving with a slow curling bias towards the open sea . . .'. Meinertzhagen (1930) warned against the commercialization of duck hunting in Egypt. He proposed that a reserve should be established on one of the Delta lakes and netting should be restricted. However, only hunting reserves were established. It seems reasonable to presume that bird trapping was intensified to meet the demand of the local European community. In Meinertzhagen's (1930) words:

'There can be no excuse for slaughtering small birds to serve as appetisers for bar-loafers, lounge-lizards and froth-blowers.'

THE HUNTING OF WATER-BIRDS

Water-bird hunting occurs throughout the Nile Delta and Valley and the Faiyum. Two main groups of hunter can be distinguished: market hunters and sport hunters. A detailed description of the methods used to secure water-birds in Lake Manzala by market hunters is given in Meininger and Mullié (1981*b*). The devices used in the other coastal lakes are generally the same. One method, also used during dynastic times, is a large clap-net. This technique remained through the ages one of the most successful methods to trap ducks and Coots alive and is widely used on Lake Burullus (CAB *et al.*; RS) and Lake Manzala (Mullié and Meininger 1983). Throwing sticks, a widely used weapon of ancient Egyptian fowlers (Schäfer 1919; Clark 1948), are no longer used. Despite the greater profit in selling live birds in the market many more are shot, particularly waders in the eastern portion of Lake Manzala (Mullié and Meininger 1981*a*, 1983). Recently it has been found that groups of children near Port Said catch terns using fishing lines with hooks, baited with small fish, and sell the birds locally (Mullié 1988).

The study of Egyptian bird markets made by Mullié and Meininger (1983) was based on 18 visits to the Port Said market and 8 to the Damietta market. Additional information has since been gathered. Including the visits reported upon by Mullié and Meininger, the Port Said market was visited 70 times throughout the fowling season

TABLE 4.1. Numbers of water-birds found for sale in the Port Said market by season

Year	1978–9	1979–80	1980–1	1981–2	1982–3	1983–4	1984–5	1986–7	1978–87
(Number of visits)	(3)	(10)	(5)	(6)	(18)	(17)	(5)	(6)	(70)
Species									
Little Grebe	0	0	0	0	3	4	1	0	8
Great Crested Grebe	10	5	1	0	26	18	6	10	76
Black-necked Grebe	0	0	0	1	4	26	2	4	37
Cormorant	0	0	0	0	2	0	0	0	2
White Pelican	0	0	1	0	0	0	0	0	1
Bittern	0	0	0	0	1	2	0	1	4
Little Bittern	2	1	0	0	24	37	3	8	75
Night Heron	0	1	0	4	0	0	0	1	6
Squacco Heron	0	0	0	0	0	0	1	0	1
Little Egret	0	0	2	6	3	2	0	6	19
Great White Heron	0	3	0	0	0	0	2	0	5
Grey Heron	1	4	9	6	19	4	5	32	80
Purple Heron	0	1	0	3	0	0	1	0	5
Greater Flamingo	0	5	0	0	11	0	8	0	24
Ruddy Shelduck	0	0	0	0	1	0	0	0	1
Shelduck	11	26	0	0	222	39	9	16	323

TABLE 4.1. Continued

Year	1978–9	1979–80	1980–1	1981–2	1982–3	1983–4	1984–5	1986–7	1978–87
(Number of visits)	(3)	(10)	(5)	(6)	(18)	(17)	(5)	(6)	(70)
Wigeon	0	4	0	0	16	99	1	11	131
Gadwall	0	0	0	0	0	13	0	5	18
Teal	0	187	22	10	21	36	21	69	366
Mallard	0	1	0	0	8	9	8	4	30
Pintail	1	4	0	2	15	3	18	1	44
Garganey	0	0	0	0	0	0	1	2	3
Shoveler	8	88	0	0	38	21	42	39	236
Pochard	0	7	0	0	247	223	132	439	1048
Ferruginous Duck	0	0	0	0	5	1	39	6	51
Tufted Duck	0	0	0	0	38	0	62	46	146
Water Rail	4	12	0	1	13	33	9	8	80
Spotted Crake	1	0	0	0	0	3	0	0	4
Little Crake	0	0	1	0	0	7	1	0	9
Baillon's Crake	0	0	0	0	0	0	1	0	1
Moorhen	145	250	358	260	1959	1104	261	627	4964
Purple Gallinule	3	13	17	9	40	52	34	28	196
Coot	186	967	4	90	11 509	3377	395	833	17 361
Painted Snipe	1	0	0	0	0	0	0	0	1
Avocet	5	46	52	2	9	48	6	32	200
Ringed Plover	0	3	5	8	3	20	0	30	69
Kittlitz's Plover	1	0	0	0	0	0	0	0	1
Kentish Plover	0	342	36	56	43	303	0	7	787
Grey Plover	0	6	0	2	1	12	5	1	27
Spur-winged Plover	1	1	0	0	0	1	0	0	3
Lapwing	1	1	0	0	0	0	0	0	2
Little Stint	0	988	138	268	632	355	270	75	2726
Curlew Sandpiper	0	0	0	4	0	0	0	0	4
Dunlin	0	808	378	487	1212	675	420	71	4051
Ruff	5	3	0	0	3	302	1	0	314
Snipe	0	2	2	1	0	1	0	0	6
Black-tailed Godwit	0	0	0	0	3	2	0	0	5
Whimbrel	0	0	0	0	1	0	0	0	1
Redshank	0	24	6	34	47	269	25	29	434
Marsh Sandpiper	1	1	0	0	0	0	0	1	3
Greenshank	0	2	1	1	0	1	0	0	5
Wood Sandpiper	1	0	0	0	0	0	0	0	1
Great Black-headed Gull	0	0	0	0	2	0	0	0	2
Black-headed Gull	19	9	0	0	62	239	125	45	499
Slender-billed Gull	4	0	2	11	1	11	1	0	30
Lesser Black-backed Gull	0	2	0	0	0	0	0	0	2
Yellow-legged Gull	0	20	0	0	3	2	2	0	27
Gull-billed Tern	0	0	0	0	0	0	0	1	1
Little Tern	0	0	0	1	0	0	0	20	21
Whiskered Tern	10	0	0	4	7	0	1	1	23
Black Tern	0	0	0	0	0	0	0	1	1
Total number	421	3837	1035	1271	16 254	7351	1919	2510	34 601

Year	Sept.		Oct.		Nov.		Dec.		Jan.		Feb.		Mar.		Apr.		Total
	I	II	I	II	I	II	I	II	I	II	I	II	I	II	I	II	
1978–9										2	1						3
1979–80								5	4	1							10
1980–1		2	2	1													5
1981–2	1	2	1	1	1												6
1982–3								8	3	4			2		1		18
1983–4								2	5	4			6				17
1984–5					1	1				2						1	5
1986–7		2	2				2										6
No. visits	1	6	5	2	2	1	2	15	12	13	1	–	8	–	1	1	70
No. seasons	1	3	3	2	2	1	1	3	3	5	1	–	2	–	1	1	8

TABLE 4.2. Timetable of visits to Port Said market based on half-month totals

between January 1979 and December 1986 and the Damietta market 14 times (RvD *et al.*; SMG; PLM; WCM; RS; JV). Information on the numbers and species of water-birds actually found for sale in the Port Said market each season is presented in Table 4.1, while the timetable of the visits is given in Table 4.2. For each half-month period and for each species the average daily counts of birds for sale were calculated. These numbers were used to calculate the total numbers of birds for sale each half-month period. These figures, as well as our estimates in which some apparent artefacts have been eliminated, are presented in Table 4.3. We consider the figures given in the last column of Table 4.3 to be good estimates of the average annual numbers of water-birds sold in the Port Said market in the period between January 1979 and December 1986. Graphic illustrations are provided for ten common taxa (Fig. 4.1) as well as for the average daily total numbers of water-birds found for sale (Fig. 4.2).

Information on the Damietta market (Fig. 4.3) is insufficient to present a similar analysis. Therefore, we have only presented the basic data and calculated average monthly totals for each species (Table 4.4).

It should be kept in mind that there are considerable differences in both the species composition and numbers taken between successive years. This may be partly influenced by severe weather in Eurasia and by differences in the hunters' activities. For example, the numbers of Coot taken in December 1982 and January 1983 were exceptionally high, almost certainly reflecting a greater than average number wintering in the area. This phenomenon was also noticed in the wintering areas in Israel in January 1983 when the highest numbers of Coot were counted for the period 1965–85 (UP; NRA). Taking into account the birds offered for sale in some smaller markets around the periphery of the lake (estimated at 4500–16 200) and the birds consumed by the hunters and their relatives (estimated at 9000–18 000) (Mullié and Meininger 1983), the combined average annual bag in Lake Manzala between 1979 and 1986 is estimated to be 122 500–149 000.

The situations on Lakes Burullus, Idku, and Maryut are less well known. Based on a survey of Lake Burullus in October and November 1981 (RS), it was estimated that annual catches of water-birds, mainly Coot, may range from 28 000 to 57 600 birds. Total annual catches of water-birds on Lake Maryut were estimated to be 22 500–45 000 and on Lake Idku 36 000–48 000 birds. The estimates were based on interviews with hunters and middlemen and not on market sales.

There are two principal public shooting clubs in Egypt, the Alexandria Shooting Club in Alexandria and the Egyptian Shooting Club in Cairo. The latter club, the largest of the two, carefully maintains a number of hunting reserves in Sharqiya and Ismailiya Governorates. Duck shooting at these reserves is allowed only on 16 days per year, falling approximately between early December and mid-March. Hunting is from blinds. In recent years the total numbers of duck shot at the lakes of the Egyptian Shooting Club ranged from 20 000 to 30 000 per season. No records of the exact species composition have been kept but they appear to be predominantly Teal. Incomplete bag records for the period 1971–85 are presented in Figure 4.4.

For the Alexandria Shooting Club no bag records are available but the number of hunting members is between 200 and 300, a portion of which only shoot clay pigeons (SSa). Except for Lakes Idku and Mallahet Maryut where this group have their principal fowling areas (EK), most of the shooting grounds in the western Nile Delta have been destroyed since 1980 by land reclamation projects (VIM). For military officers there is a separate shooting club at the Dahshur lake, for which no information is available. A considerable number of Egyptian sport hunters, perhaps up to several thousand, have no affiliation with shooting clubs. They hunt in various areas such as the southern fringes of the Delta lakes or the Faiyum. Their annual bag is estimated to be 30 000–45 000 birds. The total number of duck secured in the Nile Delta excluding the northern lakes is tentatively estimated to be between 50 000 and 75 000. There is no information available on the numbers of other species taken, such as rails and Coots, but presumably it is also considerable.

The annual numbers of water-birds bagged in the Nile Delta are given in Table 4.5. A conservative estimate of the total number of water-birds taken annually in this area is between 261 000 and 375 000. To this figure the annual total taken in the Nile Valley and Faiyum should be added. However, for these two regions virtually no quantitative information is available. Based on field observations of hunters and hunting parties in these areas, consisting of Egyptians as well as of foreigners, and of their bag, it is plausible that the average annual total of water-birds taken in the whole of Egypt reaches at least half a million.

Meinertzhagen (1930) presented 'official figures for the number and species of duck caught yearly on the different lakes of Egypt'. Most likely his figures concern market hunting only and it is not mentioned whether these data are from a single year or the average for a number of years. As stated earlier, the annual totals vary greatly from year to year and therefore it is impossible to compare in detail Meinertzhagen's (1930) data with ours. However, several changes are clear. There has been a decline in the proportion of diving ducks (Pochard and Tufted Duck) taken annually from almost

TABLE 4.3. Calculated average numbers of water-birds for sale in the Port Said market per half-month period and per season (based on 70 visits to the market throughout the season in the period between January 1979 and December 1986)

	Total number calculated per half-month period																Total number calculated	Estimated average number per season
	Sep. I	Sep. II	Oct. I	Oct. II	Nov. I	Nov. II	Dec. I	Dec. II	Jan. I	Jan. II	Feb. I	Feb. II	Mar. I	Mar. II	Apr. I	Apr. II	Sept.–April	
Number of visits	1	6	5	2	2	1	2	15	12	13	1	0	8	0	1	1	70	
Species																		
Little Grebe	0	0	0	0	0	0	0	3	0	4	0	2	4	2	0	0	14	15–20
Great Crested Grebe	0	0	0	8	0	0	75	6	23	46	60	30	0	0	0	0	247	250–260
Black-necked Grebe	0	3	0	0	0	15	30	5	29	4	0	0	0	0	0	0	85	85
Cormorant	0	0	0	0	0	0	0	0	0	0	0	2	4	2	0	0	8	8
White Pelican	0	3	0	0	0	0	0	0	0	0	0	0	0	0	0	0	3	3–5
Bittern	0	0	0	0	0	0	8	1	0	0	0	2	4	2	0	0	16	20
Little Bittern	0	0	18	0	0	0	15	26	19	23	0	8	17	8	0	0	134	150–200
Night Heron	0	10	3	0	0	0	0	1	0	0	0	0	0	0	0	0	14	15
Squacco Heron	0	0	0	0	0	0	0	0	0	1	0	0	0	0	0	0	1	1
Little Egret	0	0	6	16	45	0	30	0	0	6	0	0	0	0	0	0	103	125–150
Great White Heron	0	0	0	0	0	30	0	0	0	4	0	0	0	0	0	0	34	15–25
Grey Heron	0	43	54	64	38	0	23	11	6	15	15	8	2	1	0	0	278	300–325
Purple Heron	0	5	0	8	0	0	0	0	0	1	0	0	0	0	0	15	29	25
Greater Flamingo	0	0	0	0	53	15	0	6	11	1	0	0	0	0	0	0	86	90–100
Ruddy Shelduck	0	0	0	0	0	0	0	1	0	0	0	0	0	0	0	0	1	1
Shelduck	0	0	0	0	0	0	120	145	96	113	0	2	4	2	0	0	482	500–600
Wigeon	0	0	0	0	0	0	83	16	123	9	0	0	0	0	0	0	230	230
Gadwall	0	0	0	0	0	0	38	0	16	0	0	0	0	0	0	0	54	54
Teal	0	70	0	24	15	225	518	206	41	23	0	4	8	4	0	0	1137	1200
Mallard	0	0	3	0	0	0	23	7	6	17	0	0	0	0	0	0	56	56
Pintail	0	0	3	0	30	75	0	16	4	20	0	0	0	0	0	0	147	200
Garganey	0	0	6	0	0	0	0	0	0	0	0	0	0	0	0	15	21	20–25
Shoveler	0	30	6	0	23	0	188	94	21	108	0	1	2	1	0	0	473	600–650
Pochard	0	0	0	0	0	285	3293	262	289	139	0	0	0	0	0	0	4268	2500–3500
Ferruginous Duck	0	0	15	0	0	0	8	3	1	50	0	0	0	0	0	0	77	80–110
Tufted Duck	0	0	0	0	0	405	345	35	0	49	0	0	0	0	0	0	834	850
Water Rail	0	0	6	0	8	90	45	19	24	30	0	4	8	4	0	0	236	250–260

Spotted Crake	0	0	0	0	0	0	0	0	0	1	0	3	6	3	0	0	12	15
Little Crake	0	0	0	8	0	0	0	0	0	1	0	7	13	7	0	0	35	40–45
Baillon's Crake	0	0	0	0	0	0	0	0	0	1	0	0	0	0	0	0	1	1
Moorhen	375	383	879	2800	375	420	2993	1425	776	1516	420	622	823	472	120	30	14 429	14 000–15 000
Purple Gallinule	30	23	63	112	105	135	60	32	49	48	45	30	15	8	0	0	754	750–800
Coot	0	48	708	232	1238	2850	4486	7512	5597	5167	2175	1333	491	246	0	0	32 082	30 000–32 000
Painted Snipe	0	0	0	0	0	0	0	0	0	1	0	0	0	0	0	0	1	1
Avocet	0	105	84	64	0	0	60	16	73	46	60	30	0	0	0	0	538	700–750
Ringed Plover	0	90	12	24	0	0	0	14	6	0	0	8	15	8	0	0	176	200–225
Kittlitz's Plover	0	0	0	0	0	0	0	0	0	0	15	8	0	0	0	0	23	5–10
Kentish Plover	0	48	132	160	120	0	0	267	383	79	0	64	128	64	0	0	1442	1750–1850
Grey Plover	0	3	0	0	15	0	0	13	6	7	0	1	2	1	0	0	48	75
Spur-winged Plover	0	0	0	0	0	0	0	1	0	0	15	8	2	1	0	0	27	25–30
Lapwing	0	0	0	0	0	0	0	1	0	0	15	8	0	0	0	0	24	25–30
Little Stint	225	385	228	848	975	300	0	1186	718	490	0	132	264	132	0	0	5883	7000
Curlew Sandpiper	45	3	0	0	0	0	0	0	0	0	0	0	0	0	0	0	48	50
Dunlin	150	730	1197	1296	413	1200	135	988	821	1538	0	189	379	189	0	0	9225	10 000
Ruff	0	0	0	0	0	0	0	3	1	11	0	282	564	282	0	0	1144	1250–1350
Snipe	0	8	0	0	0	0	0	2	0	1	0	0	0	0	0	0	11	10–15
Black-tailed Godwit	0	0	0	0	0	0	0	0	0	6	0	0	0	0	0	0	6	6
Whimbrel	0	0	0	0	0	0	0	0	1	0	0	0	0	0	0	0	1	1
Redshank	150	58	45	24	0	300	135	74	105	116	0	92	184	92	0	0	1373	1400–1500
Marsh Sandpiper	0	3	0	0	0	0	0	1	0	1	0	0	0	0	0	0	5	5–10
Greenshank	0	3	3	0	0	0	0	1	0	1	0	1	2	1	0	0	12	10–15
Wood Sandpiper	0	0	0	0	0	0	0	0	0	1	0	0	0	0	0	0	1	1
Great Black-headed Gull	0	0	0	0	0	0	0	2	0	0	0	0	0	0	0	0	2	2
Black-headed Gull	0	0	0	0	0	1605	338	93	159	97	240	156	71	36	0	0	2794	1500–2000
Slender-billed Gull	0	30	0	8	8	0	0	1	11	2	60	30	0	0	0	0	150	150–175
Lesser Black-backed Gull	0	0	0	0	0	0	0	2	0	0	0	0	0	0	0	0	2	2
Yellow-legged Gull	0	0	0	0	8	0	0	22	1	5	0	0	0	0	0	0	36	50–60
Gull-billed Tern	0	3	0	0	0	0	0	0	0	0	0	0	0	0	0	0	3	3
Little Tern	0	50	0	0	8	0	0	0	0	0	0	0	0	0	0	0	58	60–70
Whiskered Tern	0	3	0	32	0	0	0	7	0	14	0	0	0	0	0	0	56	80–100
Black Tern	0	3	0	0	0	0	0	0	0	0	0	0	0	0	0	0	3	3
Total number	975	2135	3471	5728	3473	7950	13 043	12 528	9415	9815	3120	3065	3009	1565	120	60	79 471	76 813–82 170

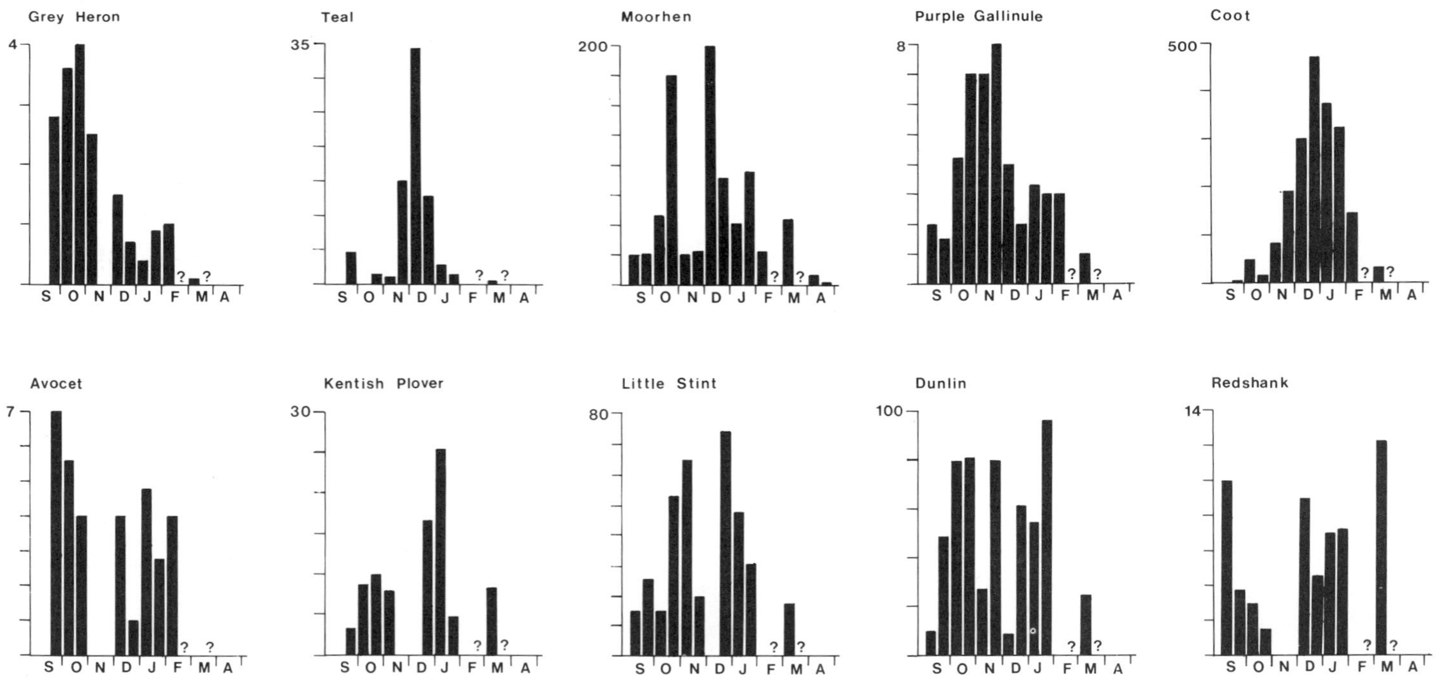

FIG. 4.1. The average daily number of ten commonly captured species found for sale in the Port Said market per half-month period between January 1979 and December 1986.

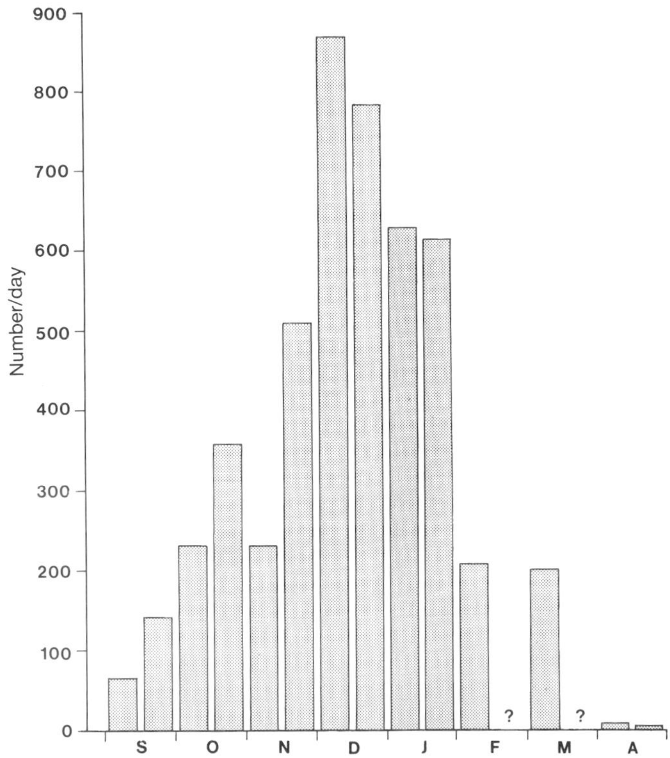

FIG. 4.2. The average daily number of water-birds found for sale in the Port Said market per half-month period between January 1979 and December 1986.

FIG. 4.3. Live water-birds offered for sale in the Damietta market. Small sign in Arabic on back of table lists prices for individual species. (Photo W. C. Mullié, winter 1979–80.)

TABLE 4.4. Numbers of water-birds for sale in the Damietta market per visit, and calculated number offered per month and per season*

Date	Oct.						Nov.†	Dec.				Jan.								Feb.		Total
	6th 1986	9th 1981	10th 1981	16th 1981	aver. /day	total calc.	total calc.	24th 1979	26th 1979	aver. /day	total calc.	4th 1980	19th 1980	26th 1980	29th 1979	31st 1979	17th 1985	aver. /day	total calc.	8th 1979	total calc.	Sept.–Feb.
Number of visits						4					2								6		1	14
Species																						
Great Crested Grebe					0.0	0	8		1	0.5	16			1		8		1.5	47	4	112	182
White Pelican					0.0	0	31		4	2.0	62							0.0	0		0	93
Little Bittern		1			0.3	8	4			0.0	0				1	2		0.5	16	1	28	55
Cattle Egret				6	1.5	47	23			0.0	0							0.0	0		0	70
Purple Heron					0.0	0	0			0.0	0			1				0.2	5		0	5
Shelduck					0.0	0	0			0.0	0			20				3.3	103		0	103
Wigeon					0.0	0	0			0.0	0					2		0.3	10		0	10
Teal	1				0.3	8	988	60	67	63.5	1969	55	2	39	95	41	154	64.3	1994	6	168	5127
Mallard					0.0	0	0			0.0	0						23	3.8	119		0	119
Pintail	18			10	7.0	217	333		29	14.5	450	14	28	68	6	31	264	68.5	2124	202	5656	8779
Garganey				8	2.0	62	31			0.0	0							0.0	0		0	93
Shoveler				1	0.3	8	229	10	19	14.5	450		5	44		21	23	15.5	481	7	196	1362
Pochard					0.0	0	310	40		20.0	620					45	194	39.8	1235		0	2165
Ferruginous Duck			3	10	3.3	101	50			0.0	0							0.0	0	1	28	179
Tufted Duck					0.0	0	0			0.0	0				3	44	62	18.2	563	1	28	591
Water Rail					0.0	0	0			0.0	0				20			3.3	103	5	140	243
Moorhen	7	152	8		41.8	1294	1073	40	15	27.5	853	6	2		49	45	16	19.7	610	13	364	4194
Purple Gallinule			1	7	2.0	62	31			0.0	0						2	0.3	10	1	28	131
Coot		13	10	26	12.3	380	391	26		13.0	403			3	40	104	19	27.7	858	159	4452	6484
Ringed Plover					0.0	0	0			0.0	0				1			0.2	5		0	5
Little Stint					0.0	0	233	30		15.0	465				10			1.7	52		0	749
Dunlin					0.0	0	233	30		15.0	465							0.0	0		0	698
Ruff					0.0	0	0			0.0	0					5		0.8	26	1	28	54
Redshank					0.0	0	8	1		0.5	16	1						0.2	5	2	56	84
Black-headed Gull					0.0	0	8	1		0.5	16					1		0.2	5	12	336	364
Slender-billed Gull					0.0	0	0			0.0	0					3		0.5	16		0	16
Larus spp.					0.0	0	0			0.0	0							0.0	0	2	56	56
Whiskered Tern					0.0	0	0			0.0	0						1	0.2	5		0	5
Total number	26	166	22	68	71	2186	3984	238	135	187	5782	76	37	176	225	352	758	271	8391	417	11 676	32017

* The market was visited once on 25 September 1980 and no birds were for sale
† November data have been obtained by interpolation between October and December

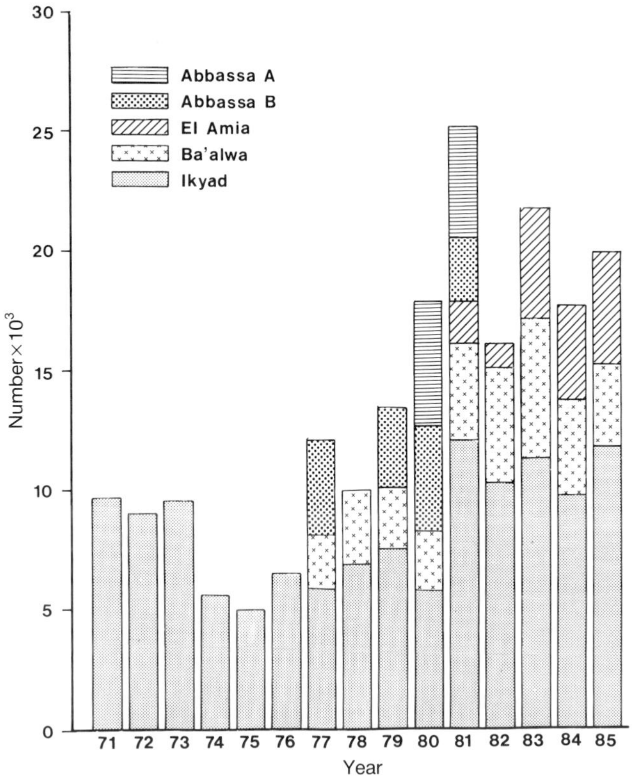

FIG. 4.4. Numbers of duck shot annually between 1971 and 1985 on the lakes rented by the Egyptian Shooting Club.

40 per cent to less than 10 per cent (cf. Mullié and Meininger 1983). Some other species like Mallard are now generally scarce in the catches. In Israel, where hunting was banned in 1968 in a number of areas (UP) it was found that the number of Mallard wintering increased substantially after this date (Paz 1986). A shift to other near-by wintering areas may therefore account for some of the apparent differences between 1930 and recent years.

Our present estimate for the number of ducks shot and caught annually on the Delta lakes is less than Meinertzhagen's. Without a study of long-term trends in the population dynamics of Eurasian breeding birds and the relationship between climatic variables and migratory movements it is hard to account for this difference. However, human impact on the local ecosystem is certainly one of the important factors.

FALCONRY AND HOUBARA HUNTING

Although the practice of falconry is a centuries-old tradition among groups of desert Arabs, the trapping of large falcons is becoming increasingly popular among non-

TABLE 4.5. Estimated average numbers of water-birds taken annually in the Nile Delta between 1979 and 1986

Area	ducks	Coot	Moorhen	waders	other	total
Lake Maryut						22 500–45 000
Lake Idku						36 000–48 000
Lake Burullus						29 000–57 500
Lake Manzala	28 000–34 000	41 000–50 000	20 500–25 000	27 000–32 000	6000–8000	122 500–149 000
Hunting reserves	20 000–30 000					20 000–30 000
Elsewhere Delta	30 000–45 000					30 000–45 000
Total Nile Delta						260 000–374 500

Arabs because of the high prices paid for a single bird by falconers. It is of interest to note that historically no native Egyptian ethnic group practised the art of falconry, i.e. the complete process of capturing, training, and hunting. There are probably only a few dozen professional falcon trappers in Egypt. They mainly catch birds along the North Coast between Bahig and Matruh, along the Mediterranean coast of Sinai, in Western Desert oases, in the general Faiyum and Wadi el Rayan area, and in a few other places. Two Delta villages in particular, Abu Hammad in the east, and Abu Rauwash in the south-west, are well known for their falconers and falcon trappers. The latter village is the main centre in Egypt for the trade of wild animals. Would-be falcon trappers augment the professionals many times over and literally hundreds of people try their luck during the autumn migration. As a result of their lack of experience and knowledge this group of trappers is not as successful as the professionals but nevertheless the total number of falcons they take should not be neglected.

Several methods are used to trap falcons. Along the Mediterranean coast of Sinai many Bedouin families involved in Quail netting also try to capture falcons. Each of them has two or three Bal-chatri traps with tethered doves as bait (Petersen and Sørensen 1980, 1981*a*). Another technique is to tether a blinded raptor, grasping a dove or even a tight tuft of feathers to a Bal-chatri trap. The idea here is that when a falcon passes over it will be enticed to 'steal' the food from the stool-pigeon and become entangled in the trap. Petersen and Sørensen (1980) saw Pallid Harrier, Hobby, Kestrel, and Lanner being used in this way as decoys. A third method used by professional falconers in Sinai and along the North Coast employs a large falcon, e.g. a Barbary Falcon (Petersen and Sørensen 1981*a*; RR). The decoy falcon which has had its eyelids sewed together and partially perforated along the top margin, is placed on top of a Bal-chatri trap with some prey. When a wild falcon is noted, the tethered and almost blind bird is thrown into the air and it glides 200–300 m before settling on the ground. If the technique is successful the wild falcon will swoop at the decoy in an attempt to steal the food and become entangled in the trap.

It is difficult to estimate the number of falcons being trapped in Egypt per year. Five to seven falcons each autumn along the entire north coast of Sinai was the number given by Bedouins to Petersen and Sørensen (1981*a*). In the autumn of 1981, five or six falconers were active along the North Coast west of Alexandria. Two of these trappers claimed to have captured approximately 35 large falcons between them in 15 years (RS). Mullié and Meininger (1985) estimated that for the whole of Egypt 30–40 large falcons are trapped on the average per year but in exceptional years the number may be over 100 (Mullié 1985). Most birds are sold to middlemen and then find their way (illicitly) to Saudi Arabia and the Gulf States. Before 1952 several Bedouin living in the northern Negev of Israel made a living from catching and training falcons, most of which were sold to King Farouk of Egypt (Mendelssohn 1972).

In the past in Egypt large falcons were often used to hunt Houbara. Houbaras are rare in desert areas with less than 100 mm of annual rainfall (Mendelssohn 1975;

HM) and until recently the main stronghold in Egypt was the Mediterranean coastal desert between Alexandria and Salum, an area that receives over 100 mm of precipitation per year (Fig. 2.2). Rainfall conditions in the western Mediterranean coastal desert have apparently not changed significantly during the last 2000 years (Kassas 1972). There has been though a severe degradation of range vegetation due to overgrazing by domestic herds, uncontrolled cutting, and farming (Ayyad and Ghabbour 1977). This presumably has had a negative impact on Houbara populations as well. In this area there are still annual hunting parties coming from the Gulf States and Saudi Arabia. Bedouin living near El Omayed, who often hunt Cream-coloured Coursers, reported that Kuwaiti falconers took up to four or five Houbara per day on November trips to the area (WCM); however, this figure seems exaggerated. In northern Sinai, despite marginal conditions for Houbara, they increased in numbers during the 1970s when hunting was restricted and regulations were enforced (HM). In the mid-1980s, however, there were organized Houbara shooting trips of Egyptians to northern Sinai (SZ). Thus, there is evidence that the decline of the Egyptian Houbara populations is a direct result of hunting with guns and falcons.

THE CAPTURE AND MARKETING OF QUAILS

The autumn Quail season in Egypt falls between late August and mid-October, with the main period in the first three weeks of September. Quail are generally trapped by two different methods. In the coastal area west of Baltim, tent-like structures called *munsaab* are used, while to the east trammel-nets are employed. At several places along the Delta coast both methods are used. Additional techniques include lime-sticks, guns, and drop-nets. The last procedure employs recycled fishing nets, called *romaka ghazl*, which are operated by two men and dropped over a place where a Quail is hiding.

The *munsaab* trap is extensively used west of Alexandria (probably to the Libyan border), particularly in the littoral sand dunes near El Gharbaniyat between El Hammam and Bahig. The trap is usually composed of a bunch of *Ammophila arenaria*, a tall sea-shore grass, which is plucked and arranged in a tent-like structure with the entrance facing south and a piece of fishing net closing off the seaward opening. In other areas different trees and shrubs are utilized, and near Lake Burullus sun-baked mud tunnels have also been used (Moreau 1927–8). Literally thousands of *munsaab* are operated in autumn. It was proposed by Ayyad and Ghabbour (1977) that the reason vegetation in the dune area near El Gharbaniyat is not heavily grazed by domestic animals, as it is further inland, is to save the local habitat for Quail trapping.

According to Moreau (1927–8) it was only since 1906 that Quail taken along the North Coast were exploited for export. The numbers of birds estimated to have been collected along the 55 km stretch between Alexandria and El Alamein were 10 000–20 000 in 1980 (WCM) and 18 000–25 000 in 1981 (RS). In the late 1960s and early

1970s about 30 000–40 000 Quail were taken on average in this area per annum (ST) and in the mid-1970s about 30 000–35 000 (EK). In 1980 the number of Quails taken additionally between El Alamein and Matruh was estimated at 6500–13 000 (WCM). According to native hunters living between Matruh and Salum all Quail shot or trapped in the area west of Matruh were for local consumption.

Most birds trapped between El Dab'a and Alexandria are sold to middlemen for marketing in Alexandria (see Fig. 4.5), although a portion is sold at roadside stands. In September and October 1980 there were generally about nine such stands in this area (BEP). Smaller markets exist in Matruh and in some villages such as El Dab'a. In Alexandria there are some exclusive restaurants which open only during the netting season to serve Quails.

Presently the number of Quails taken along the North Coast differs sharply from before World War II. Between 1906 and 1913 the numbers of Quails exported from Egypt ranged from one to two million (Moreau 1927–8) with a good portion of these birds coming from the general Alexandria area. As late as 1926 about 535 000 Quails were still exported, 250 000 came from the environs of Alexandria (Moreau 1927–8). Since the 1920s and 1930s (Moreau 1927–8, 1951) the numbers of Quails trapped along the North Coast have declined dramatically.

FIG. 4.5. Merchant in Alexandria market with basket of plucked Quail. (Photo W. C. Mullié, September 1980.)

The method which nowadays accounts for most of the Quail captured in Egypt is the trammel-net. The first mention was by the Greek historian Diodorus (first century BC), who described it being practised by the residents of the village of Rhinocolura (= El Arish). He also noticed that 'they caught a sufficient number to provide themselves with food'. This type of net consists of two panels, a small mesh (3 × 3 cm) which is stretched in front of a larger mesh (17 × 17 cm), and is generally about 4 m high and 16 m long (Jarvis 1932; Mendelssohn *et al.* 1969). As prescribed by Egyptian Law 28 for 1967 the length of each net should not exceed 30 m, and an opening of 20 m must be kept between successive nets. Further, the distance from the net to the sea-shore should be greater than 500 m. In September 1980 an estimated 61 km of the possible 130 km of shore-line between Baltim and Port Said was spanned by nets. Banks ranged in length from 400 m to 6 km (Petersen and Sørensen 1981*a*; WCM), while the nearest distance to the sea-shore was sometimes only some tens of metres. Petersen and Sørensen (1981*a*) found that between El Arish and Zaranikh there was 34 km of almost continuous nets. Further, they recorded nets on the barrier islands west of Zaranikh 'as far as the eye could reach' and estimated the bank to span at least 10 km of shore-line (UGS). The entire sand-bar that separates Lake Bardawil from the Mediterranean is likely to be netted (GMA). A conservative estimate of the total length of the Sinai coast spanned by Quail nets in the mid-1980s is 70 km. Significant numbers of other bird species are also trapped in these nets.

In Table 4.6 information is presented on the numbers of Quails netted in a monitored stretch of coast between Zaranikh and El Arish. The large differences between the years are particularly striking. Assuming that presently an estimated 70 km of Sinai coast is netted, the annual catch of Quails may range from 50 000 to 170 000. These figures are in the same range as other recent estimates for the Sinai: 1967—150 000 (Mendelssohn *et al.* 1969); 1968—200 000 (Woldhek 1979); 1981—55 000–75 000 (RS). Compared to the number of Quails exported during the period 1888–1934 (Anon. 1912; Moreau 1927—8, 1951) the present Sinai catch is presumably of the same general order as between 1925 and 1934 but significantly lower than between 1906 and 1913. Quails netted in Sinai are sold to middlemen who transport them to the retail markets in Port Said and to a lesser extent in El Arish. Here they are sold to mongers who deal directly with the consumer. Some hunters sell their catches

TABLE 4.6. Numbers of Quail netted along a monitored stretch of coast between Zaranikh and El Arish

Year	Period	No. of days	Length of nets (m)	No. of Quail	Quail/m net	Quail/km/day
1972	4/9–19/10	46	2000	4863	2.43	52.9
1973	13/8–6/10	55	2500	1823	0.73	13.3
1984	8/9–8/10	31	9220	8117	0.88	28.4

Figures for 1972 and 1973 based on Zuckerbrot *et al.* (1980) and for 1984 on GMA directly to customers or to market merchants, often fishmongers, in El Arish (Mendelssohn *et al.* 1969). Some of the Quails captured west of Port Said are sold directly by the trappers to market sellers in coastal settlements. Since 1979, when the western part of the Sinai was returned to Egypt, precise records on the numbers of Quails caught are kept by the netters and the middlemen, as payment only takes place at the end of the netting season. However, the data have not been gathered on a regular basis by the government. In 1985 a Quail farm was established in El Arish, in which Japanese Quail (*C. c. japonica*) are bred for consumption.

The 1980 catch for the area from Baltim to Ras el Bahr was estimated to be 21 000–48 000 based on field observations and 15 000–36 000 based on market sales (WCM). An additional 6000–12 000 were taken in the region between Ras el Bahr and Port Said. In 1981 the total catch from Port Said to Gamasa was estimated to be between 6500 and 16 200 (RS), which is considerably lower than our 1980 estimates. In the early 1980s the area between Baltim and Alexandria was of little importance for Quail trapping, which may in part have been due to restricted access to the beach imposed by the Egyptian military authorities. However, in the past this stretch of shore had considerable importance, particularly the area between Rashid and Alexandria (Moreau 1927–8).

Since spring migratory movements of Quails through Egypt are more diffuse than in autumn other methods are used to secure them. The number taken in spring appears to be significant. Moreau (1951) mentioned that between 1930 and 1934 the numbers of Quails exported from Egypt varied between 285 000 and 580 000, 'more than half of the birds being captured as they were travelling north to breed'. One method used to trap Quails in spring is a fine-mesh fishing net which is held nearly horizontally over the ground by four men, each gripping one of the corners, and drawn during a dark night over a growing crop where birds are suspected to be resting (Hornell 1937). The Quails rise in alarm and become entangled in the mesh. This technique is likely to be the same as is depicted in the tomb of Mereruka (*c.*2325 BC) at Saqqara, and from the New Kingdom at Thebes (Darby *et al.* 1977; Houlihan and Goodman 1986). A comparable modern method was observed in lentil fields near Qift. A large net, of which one side was weighed down by the attachment of stones and clay, was pulled forward by two men grasping ropes attached to each end of the net, often considerably damaging the crop (Schäfer 1919). Other techniques used were small holes in the ground in which some grain was put. Over it a 'Teller aus Nilschlamm' was loosely set up against a small stick (Schäfer 1919). This type of trap, which is widely used in modern Egypt, is unknown from ancient Egypt. A comparable method with flat stones is employed by Bedouin in the Sinai to secure Sand Partridge (HM). A favourite sport in the first half of this century among the Europeans in Egypt (Maratos 1938) and presently among the members of shooting clubs (SZ) is Quail shooting.

Except for Moreau's (1951) estimates, we do not have any more recent figures on the numbers of Quails being taken in spring. In the 1980s thousands were being sold in Cairo; the origin of some was reputed to have been the Faiyum (SMG).

THE CAPTURE AND MARKETING OF PASSERINES AND NEAR-PASSERINES

Various birds are taken for food on a large scale during autumn migration. Some species, like House Sparrow, are taken all year round, and winter visitors such as White Wagtail are obtained with some regularity from roosts.

Along the North Coast different methods are used to secure birds. The main techniques are large nets placed over trees, air rifles, and shotguns. There is some local use of lime-sticks and other devices, such as steel traps for shrikes. Based on field observations and interviews with hunters in September and October 1980 and 1981 (BEP; RS) a rough estimate can be made of the total number of small migratory birds being bagged (Table 4.7). During the PMS between 1966 and 1973 the annual numbers of birds taken in this area were estimated to be 260 000–270 000 (ST), which is in accordance with our minimum figures. Since the 1960s this region has been extensively planted with olive and fig trees, as part of a programme to settle about 100 000 Bedouins (Zottola 1969). As of the mid-1970s there were over 300 000 olive and 500 000 fig trees in the area (Ayyad and Ghabbour 1977) but only several hundred of these trees are netted. As pointed out by Moreau and Dolp (1970), this increase in local vegetation may provide improved ecological conditions for migratory birds. The most commonly captured small birds are (in order of abundance): Willow Warbler, Lesser Whitethroat, Red-backed Shrike, Redstart, Whitethroat, Thrush Nightingale, and Spotted Flycatcher (Hoogstraal *et al.* 1963, 1964; PMS; also see Fig. 4.6).

The hunting of small birds in the Nile basin and along the Mediterranean coast of Sinai is less common than along the North Coast. A considerable number are trapped as non-target species in Quail nets, mainly in the coastal area between Baltim and Rafa. Lime-sticks and other techniques are also widely used in the region. No exact information is available on the extent of small bird trapping. Based on visits to various portions of the coast in September and October we estimate that at least several tens of thousands of small migratory birds are secured annually. Meinertzhagen (1930) mentioned the coastal zone of the Delta as the area with the most intensive bird trapping. He estimated the total bag of small birds in this region to be well over one

TABLE 4.7. Estimated numbers of small Palearctic migrants taken in the autumns of 1980 and 1981 between Alexandria and Matruh

Method	Birds/day	Season (days)	Total number of birds
Nets (n=200)	10–20	45	90 000–180 000
Air rifle, shotgun (n=300)	20–40	30	180 000–360 000
Total/season			270 000–540 000

FIG. 4.6. Live passerines offered for sale in Rosetta market. (Photo W. C. Mullié, September 1980.)

million. Even if our present estimates are conservative, it is still clear that the importance of the Nile Delta coast for trapping small Eurasian migrants has decreased.

In the Ataba market, Cairo, up to 50 plucked White Wagtails have been found on a single day (SMG) and it is presumed that these birds were taken at roosts. The numerous recoveries of White Wagtails banded near Cairo in the early 1940s and captured by 'netters' in the Delta (GZF) seems to confirm that there has been a regular wagtail market for several decades. House Sparrows (in winter also Spanish Sparrows) are taken in large numbers. These birds are sold in a variety of ways including fried snacks in street markets and bazaars throughout the country to pre-packaged frozen items in exclusive supermarkets of Cairo. Golden Orioles are persecuted in the oases of the Western Desert. In Bahariya they were considered a pest to ripening mulberry (*Morus* sp.) and date crops, and are consequently hunted (Goodman *et al.* 1986; Woldhek 1979).

THE HUNTING OF OTHER BIRD SPECIES

Several other types of bird are taken in Egypt and a few examples will be given here. White Storks are hunted with some regularity, especially when roosting. Baron (1966) saw tens being shot from a flock of exhausted birds south of Suez on 12 May 1964. At 'Gamble', 40 km south of El Shallal in the southern Nile Valley, an area presently inundated by Lake Nasser, as many as 8–10 White Storks were caught per person by hand on dark nights (Scharff 1928). At least 119 rings of White Storks have been collected between Idfu and the Sudanese border (see also species account for ringing recoveries), a further indication of local hunting pressure. The southern Egyptian Nile Valley may act as a bottleneck for migrating storks.

Special mention should be made of Turtle Dove shooting. According to information provided by the Egyptian Shooting Club (SZ), the autumn Turtle Dove season is from 1 September to 15 October and the spring season from 15 April to 10 May. In autumn a hunter's daily bag may reach 55–60 birds and in spring 110–115. The total number taken each year is unknown but may easily exceed several tens of thousands. The statement of Woldhek (1979) that spring Turtle Dove shooting in Delta citrus plantations has decreased due to a reduction in the number of birds passing through the area cannot be substantiated. The area of citrus plantations within the Delta has increased considerably (Fig. 5.3) and Turtle Doves moving through this area on spring migration may be more diffused.

In many parts of the Nile Delta and Valley, including the Faiyum, birds are netted or shot and sold at local markets. We have no precise information on the total number of birds involved per year but it presumably involves tens of thousands and perhaps even more than 100 000. In Damietta alone a daily average of 200 Palm Doves were offered on eight days in the winters between 1978 and 1981 (WCM, PLM) and up to 1500 birds on 17 January 1985 (SMG).

The Faiyum is frequented in the winter by large numbers of hunters, particularly Egyptians and Italians. According to information provided in May 1986 by a manager of one of the main hotels at Shakshuk along Lake Qarun (WCM), each winter since 1984, 20 groups, each composed of 8 to 14 hunters, stayed at that hotel alone. Before 1984 these groups were often composed of 20 to 30 people. Meininger and Mullié (1981*b*) presented information on the seasonal bags of these hunters, who literally shoot every bird within range, and a conservative estimate is from 30 000 to 50 000 per year; the bulk of which are Palm Doves. Large numbers of bee-eaters, Hoopoes, waders, and ducks are also taken. Many birds collected by organized shooting trips to the Faiyum are sold to local middlemen but small birds are usually discarded. Some species, particularly doves and ducks, are often shipped by aeroplane to the hunter's native country.

Foreign hunting groups, often Italians, travel to other parts of Egypt, such as Lake Nasser (IH). The Egyptian Ministry of Tourism widely promotes sport hunting and accordingly it is easy for visiting foreigners to bring guns into the country from abroad. Some of the large airlines collaborate with Egyptian tourist offices abroad to advertise and co-ordinate shooting trips to Egypt (Williams 1981). Competitions are sometimes held to publicize this type of tourism, and prizes are given by the organizers for the largest numbers of ducks shot at such events (Ghabbour 1971).

Wanton slaughter of Ostriches, particularly by Egyptian army officers, is a threat to this species in the general Gebel Elba area (DJO). Birds of prey are seldom shot compared to other groups (e.g. Innes 1902), although there is some demand for taxidermy (Mullié and Meininger 1985). Cecil (1904) mentioned that Nile basin dwellers found Black Kites useful and protected them. Owls, on the other hand, are hunted relentlessly for superstitious reasons. Another species which is not currently hunted with regularity is the Cattle Egret. Darby *et al.* (1977) proposed that the reason for this is it is considered bad luck to kill one. Probably this belief is better protection than any official law!

EGG COLLECTING

Although there is no precise information available to assess the extent of egg collecting in Egypt for food, it is considered a permanent threat to vulnerable sea-bird colonies as it is elsewhere in Africa (e.g. Cooper *et al.* 1984). Baha el Din (1984*e*) found that during August and September fishermen from the mainland collected large numbers of eggs and chicks on Gezira Wadi Gimal for food. On a visit to two islands off Hurghada in July 1951 it was noted that boatmen took all the Lesser Crested Tern's eggs, the state of incubation being of no concern, but ignored the smaller eggs of the White-cheeked Tern (Anon. 1981). In 1977, when the beaches of the Gulf of Aqaba were reopened to fishing, virtually all Osprey nests on Gezira Tiran were robbed of eggs. After this incident the Israelis guarded the nests in subsequent breeding seasons (Zimmerman 1984). Colonies in El Malaha of Slender-billed Gull and Greater Flamingo have often been plundered when they could be reached (Meininger and Mullié 1981*b*). Eggs of Yellow-legged Gull near El Alamein (Meininger and Baha el Din 1986) and Little Tern and Kentish Plover along the shore of Lake Bardawil (UP) regularly fall prey to fishermen. According to Jarvis (1942) Bedouins in the eastern Sinai collected eggs of Chukar and Sand Partridge, 'the age and condition of the eggs being quite immaterial'. Ostrich eggs are removed from nests with some regularity for food, medicinal purposes, and storage containers by Bedouins living in the southern Eastern Desert. Taylor (1867) found wild bird eggs for sale in the Damietta market and Sonnini (1800) saw fresh Ostrich eggs on sale in Egypt. We found no recent evidence of any commercial trade in wild bird eggs for personal consumption or sale.

5. RESPONSES OF BIRDS TO CHANGING ENVIRONMENTS

Since the publication of Meinertzhagen (1930) some major ecological changes have occurred in Egypt which have had an enormous impact on the distribution and populations of many bird species. These include the building of the Aswan High Dam and the subsequent inundation of the southern Nile Valley by Lake Nasser; changes in Nile basin irrigation; the exponential growth of the human population (see Fig. 2.8); reclamation of new lands for cultivation; and intensified agriculture practices. This chapter briefly reviews how these vicissitudes have affected birds.

NILE RIVER CONTROL

The building of the Aswan High Dam allowed the level of water further downstream to be regulated. Most importantly, the seasonal inundation of the Nile basin between August and November no longer takes place (Fig. 5.1). Before the construction of the dam, only a few species of wild plants occurred along the edge of the river (Hurst 1952) and the total biomass was generally low. During the former yearly inundation, the quick-moving waters cleared channels and banks of reed-beds, and only stunted tamarisks or other deep-rooted shrubs remained (Whymper 1909). A more or less comparable situation can still be found in Lake Nasser where various plants along the shore, including *Phragmites australis* and *Tamarix niloticus*, are often swept away during floods (Latif 1984). The regulation of the Egyptian Nile below the dam allowed *Phragmites* and *Typha* marshes to develop along the river edge. Reed-bed-dwelling species such as the Purple Gallinule and Clamorous Reed Warbler have recently established themselves along the whole length of the Nile below the dam.

Since the completion of the Aswan High Dam, the silt load delivered by the river has been reduced to only about 1.5 to 4 per cent of the previous mean annual amount (Hefny 1982). This has resulted in greater Nile water transparency. Compared with the situation before the Aswan High Dam (Elster and Vollenweider 1961) Secchi values (a measurement for water transparency) between Aswan and Cairo increased 30 to 40 per cent (Saad 1980). This in turn has resulted in the rapid spread of aquatic macrovegetation, and several plant species have become pests (see p. 32). The better water transparency may have improved feeding conditions for several piscivorous birds such as Pied Kingfisher and Little Egret in Upper Egypt. The creation of Lake Nasser offered new ecological niches for birds. The local breeding population of the Egyptian Goose, which in the early 1950s was thought to be decreasing due to human

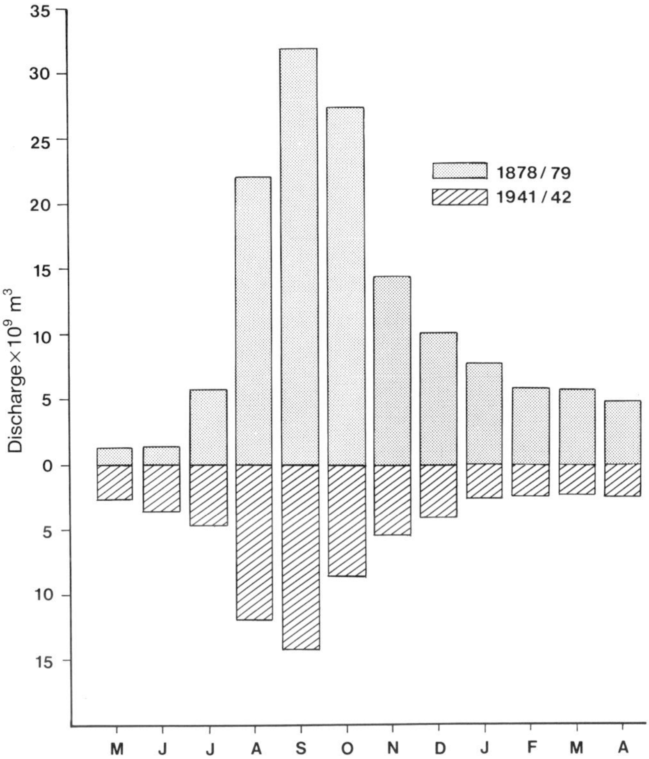

FIG. 5.1. Variation in the discharge of Nile waters downstream from Aswan before the construction of the Aswan High Dam. 1941–2 was a dry year and 1878–9 was a wet year. (After Hurst *et al.* as quoted by Clawson *et al.* 1971.)

interference (Al Hussaini 1954*b*), seems to have increased substantially since the formation of the lake. On the other hand, the Fulvous Babbler disappeared from the Nile Valley after its breeding grounds were permanently flooded by the lake.

As a side effect of improved irrigation, new marshlands were created by lifted groundwater tables. One of the best examples is former agricultural lands due south of the eastern end of Lake Qarun. This area, which has no direct connection with the lake, was inundated with freshwater and *Phragmites* marsh quickly developed. In recent years Purple Gallinules have established themselves as breeding birds in the reed-beds and Painted Snipes found a habitat in *optima forma* along the periphery of the marsh.

The permanent inflow of slightly saline drainage water from the agricultural areas of the Faiyum into the lake has had another extreme effect. Lake Qarun has undergone a slow and progressive increase in chlorosity from 5.6 g/l in 1906 to 19.97 g/l in 1978 because evaporation is the only outflow (Meshal 1977; Bishai and Kirollus 1980). The salinity of Lake Qarun has reached the level of the Mediterranean Sea (Boraey 1980) and obligate freshwater fish have completely disappeared. In recent years marine fishes such as sole (*Solea vulgaris*) have been introduced into the lake (El Zarka 1963). The coastal vegetation and reed-beds along the edge of the lake have disappeared. The presence there of large numbers of Black-necked Grebes and Slender-billed Gulls, in Egypt generally confined to saline habitats, is a good indication of these changed conditions. Freshwater-birds such as Long-tailed Cormorant and Purple Gallinule, and species nesting in coastal vegetation, such as the locally endemic form of the Sardinian Warbler, have disappeared from the complete region of Lake Qarun proper. The Marbled Duck, although not strictly a freshwater species, also disappeared from the area after its breeding habitat was lost.

Another by-product of changes in the water system of the Faiyum has been the creation of new aquatic habitat in Wadi el Rayan. Excess drainage water from the Faiyum, which formerly was directed into Lake Qarun but now exceeds the storage capacity of the lake, is being channelled into Wadi el Rayan. The filling of the area started in 1968 and has been operating regularly since early 1974 (Bishai and Kirollus 1980). In about 1980 the area of the Wadi el Rayan basin was about 15 750 acres (64 km^2). At present there are two lakes connected at their closest ends by a canal; along it there are waterfalls dropping about 10 m (see Fig. 2.7). The final area of the Wadi el Rayan lake is estimated to be 190 km^2 at a contour of −13 m (Zahran 1970–1). Filling is still continuing and the succession of lake-shore vegetation has given rise to extensive reed-bed habitat and several marsh-dwelling species like Little Egret, Purple Gallinule, and Little Bittern have invaded the area. Except for some commercial fishing activities and minor reclamation efforts, Wadi el Rayan remains essentially unspoilt and with little effort could be managed as an important nature preserve.

CHANGING AGRICULTURAL PRACTICES

Basin irrigation was gradually converted to a perennial system, mainly after the construction of the 'old' dam at Aswan in 1902 (as described on p. 29). With the completion of the Aswan High Dam in 1964, the seasonal inundations further downstream no longer occurred, and the conversion of the irrigation system was accelerated. Agricultural practices based on perennial irrigation depend heavily upon pesticides (George 1972), and most insecticides used in Egypt are applied to cotton fields (Ghabbour 1974; Attiah 1977; Isa 1980). The 1985 distribution of cotton-growing areas in Egypt is illustrated in Fig. 5.2. The large-scale application of pesticides is likely to have been one of the main factors which resulted in a substantial decline of Black Kite and Black-shouldered Kite populations inhabiting the cotton-growing regions of the Nile Delta and parts of the Nile Valley, mainly between Sohag and Beni Suef (Mullié and Meininger 1985). In the Nile Delta there has also been a strong decline in the numbers of Cattle Egret and Hooded Crow (Ghabbour 1976; Mullié and Meininger 1985). Senegal Thick-knee, Pied Kingfisher, Kestrel, and perhaps Barn Owl have almost disappeared from the same general area. Pied Kingfishers have also declined in numbers in the northern Nile Valley. In May 1986, several farmers were interviewed in the Beheira Governorate and without exception they blamed the excessive use of pesticides as the main factor for these apparent changes (WCM). In a 1986 spring survey, Palm Dove, Fan-tailed Warbler, Graceful Warbler, House Sparrow, and Goldfinch were found breeding in 75–100 per cent of 26 randomly chosen plots in the agricultural area of the central Nile Delta. Senegal Coucal, Swallow, Common Bulbul, and Olivaceous Warbler were found in 50–75 per cent of these plots and 17 other species were found in less than 50 per cent (WCM).

After the construction of the Aswan High Dam, there was a shift away from traditional crops, such as cotton, wheat, and rice to fruits, vegetables, and livestock, mainly because of higher profits (Badder *et al.* 1985). The total area planted in the Nile basin with orange, mandarin, and salt lemon, the three main fruit crops in Egypt, increased from 32 000 feddans (approximately 135 km^2) in 1950 to 233 000 feddans (approximately 980 km^2) in 1985 (Fig. 5.3). These figures do not include the Sinai, where several thousand feddans of orange groves have recently been planted between El Arish and Rafa (Yohe 1982). Currently the orange-price ceiling is a problem for farmers, and if higher prices were allowed, more new lands would be planted to citrus. Farmers in the Delta must get a permit before planting new orange trees (Yohe 1982). The 1985 distribution of citrus-growing areas in Egypt is illustrated in Fig. 5.2. The main regions are situated along the periphery of the Nile Delta, often in newly reclaimed desert land. These citrus groves, especially if over 15 years old, often provide good habitat for seed-eating birds such as Goldfinch, Greenfinch, and Serin. The last two species have recently colonized the area and are exclusively found in these orchards. The Blackbird has also recently colonized the Delta. This species is found breeding mainly in mango orchards over 20 years old, but it will probably take over other suitable habitats in the near future, as it has done in Israel (OH; HM). Starting in the early twentieth century the Common Bulbul, another frugivorous bird species, also substantially increased its Egyptian breeding range.

A second factor which has favoured the spread of several birds is the planting of casuarina trees. This tree was introduced into Egypt some time before the middle nineteenth century. Until about 1920 it was mainly found near Cairo, Alexandria, and the Suez Canal area. Large-scale planting of casuarina commenced around 1920, and they can now be found all over the country, particularly in the Delta and the middle portion of the Nile Valley. It is more sparsely distributed south of Asyut. The best-adapted and most vigorous species are *Casuarina cunninghamiana* and *C. glauca* (Täckholm and Drar 1969). These trees are used extensively as wind-breaks, especially bordering sown lands and orchards, in the periphery of the Nile basin and for road-side plantings. They provide breeding habitat for a number of bird species and their edible seeds usually ripen between the middle of March and the end of the summer (Täckholm and Drar 1969). In the Western Desert there seems to be a

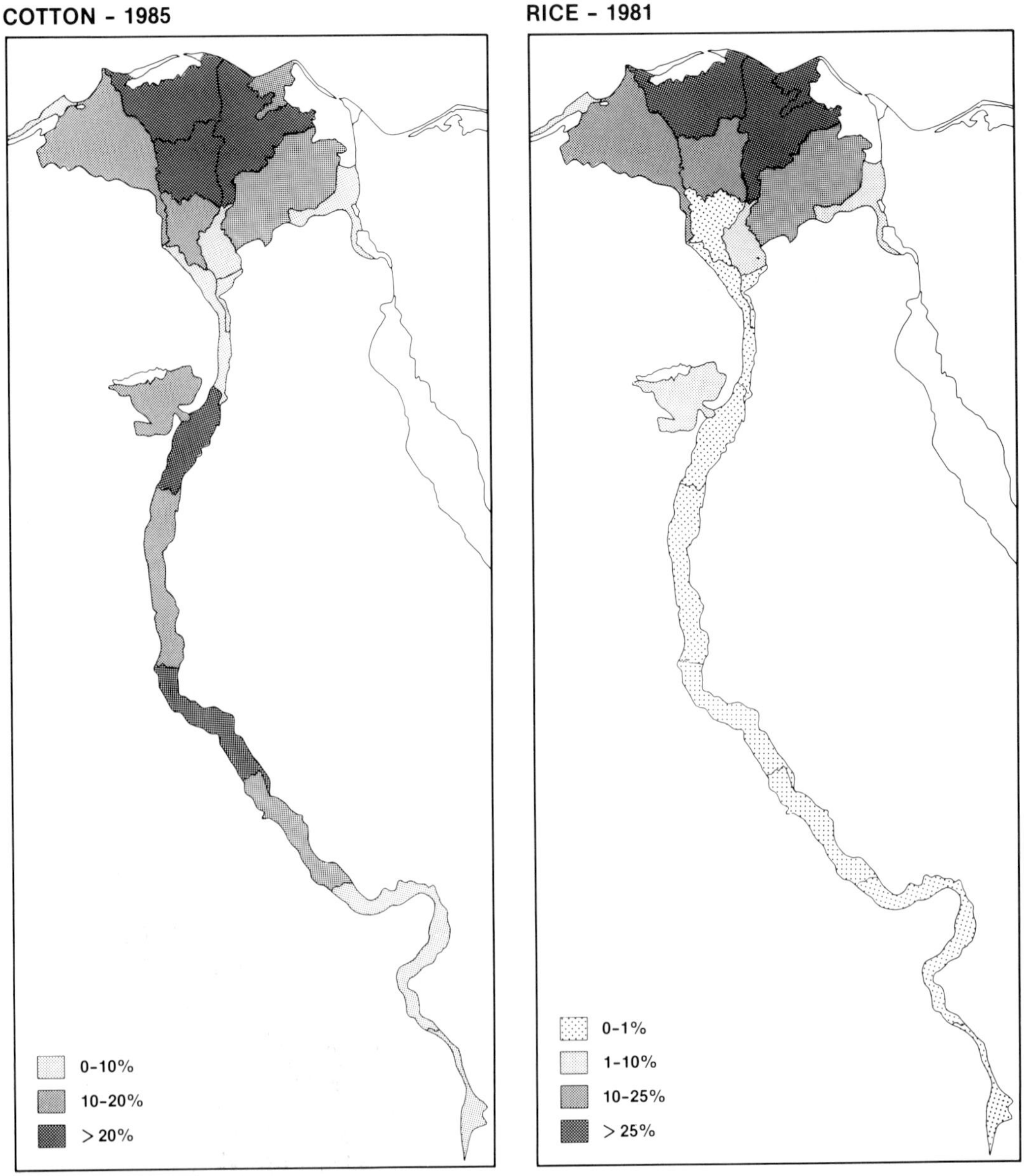

FIG. 5.2 (above and facing). Distributional maps of the total areas grown with cotton, rice, mango, and citrus (orange, mandarin, and salt lemon). The percentages are the total cropped area per district. (After Moheyi el Din 1986 and figures supplied by the Ministry of Agriculture (AAS).)

relationship between the planting of casuarina and the subsequent colonization by the Goldfinch (Goodman *et al.* 1986).

Some changes caused by the obstruction of annual Nile floods and the subsequent

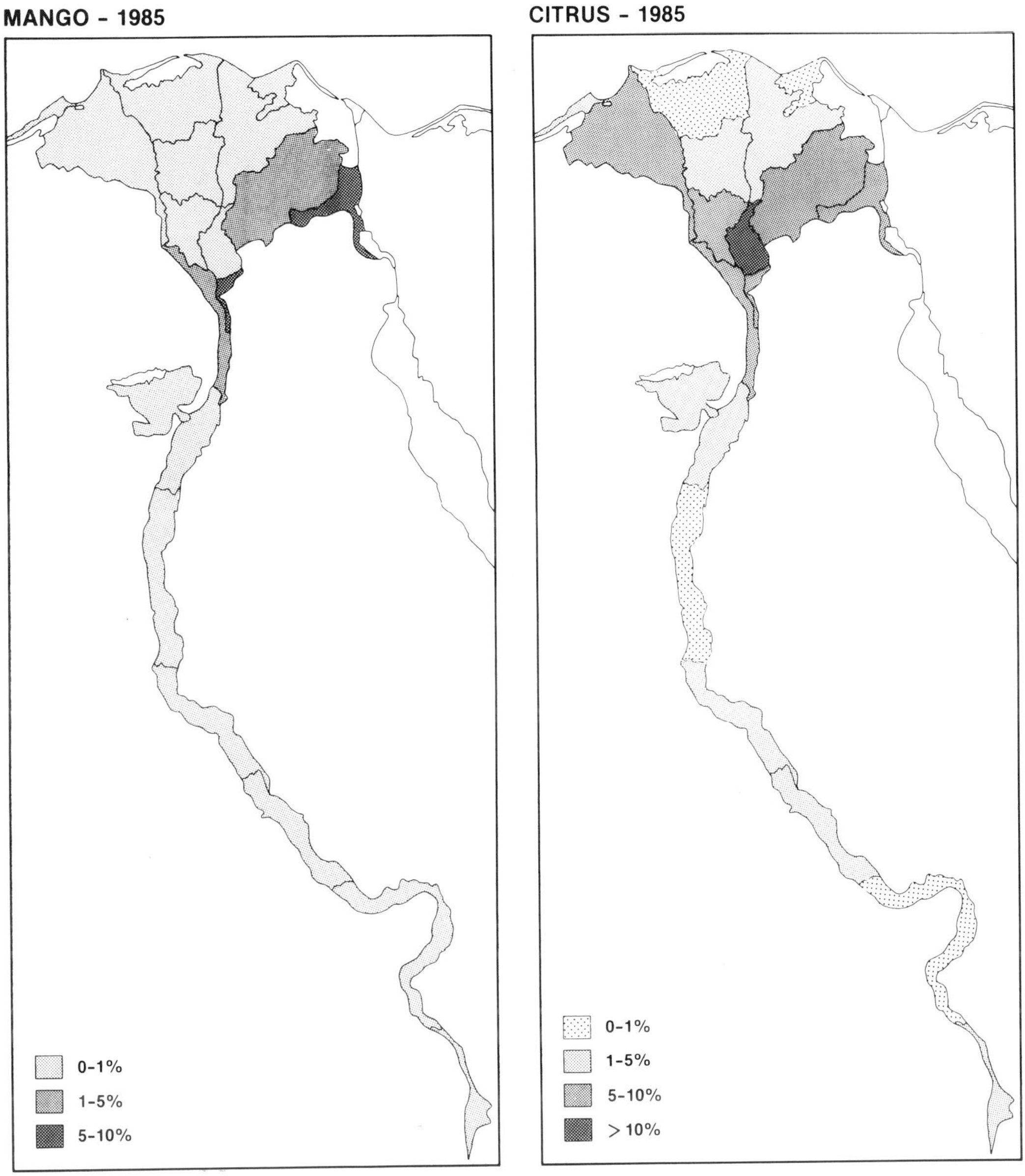

large-scale conversion from basin to perennial irrigation, created good opportunities for population expansion of murid rodents (*Arvicanthis niloticus*, *Rattus rattus*, *R. norvegicus*, and *Mus musculus*). The problem this caused was further aggravated by the indiscriminate use of pesticides, which caused local extirpation of birds of prey and other predators (Mullié and Meininger 1985). In 1983 rodents destroyed substantial portions of crops in several governorates, including 13.4 per cent of the

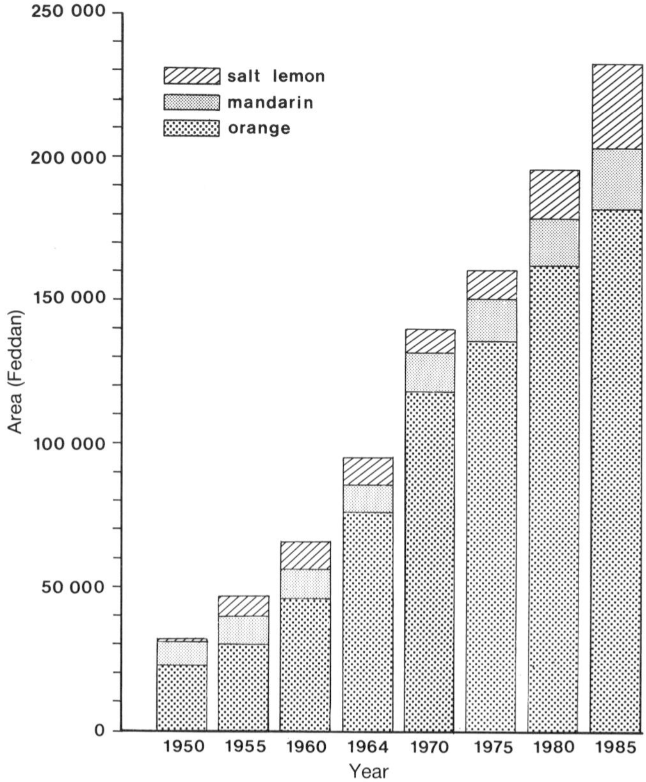

FIG. 5.3. Changes between 1950 and 1985 in the total area planted with citrus trees (orange, mandarin, and salt lemon) in Egypt, excluding the Sinai. (Based on figures supplied by the Ministry of Agriculture (AAS).)

expected wheat crop, 28.4 per cent of the soya beans, and 10.1 per cent of the watermelons (Amr and Burgstaller 1984). In September 1982 the Egyptian–German Field Rat Control Project was inaugurated. One of the project's most vital aims was to help Egyptian farmers with preventative rat control measures in their fields. Another point of the project was to transfer the responsibility for rat control from the government to the local farmers' co-operatives (Kurylas 1985). Before 1983 there were intensive campaigns initiated by the government to kill these animals using zinc phosphide, a highly toxic acute rodenticide. In 1986 such campaigns were still practised in Upper Egypt and locally in other portions of the Nile basin. At several localities this rodenticide caused considerable secondary poisoning of non-target organisms, i.e. birds of prey (Mullié and Meininger 1985), goats and donkeys (HBg), and small carnivores. We have information from several groups of trappers/hunters that carnivore populations along the Egyptian Nile basin crashed after the application of this rodenticide (BEP). The Egyptian–German project has stimulated interest in the use of the so-called second generation anticoagulant rodenticides and the role of natural predators in regulating vermin. As a first attempt, nest boxes for Barn Owls have been placed in the Faiyum (HBg). An analysis of pellets collected between 1983 and 1985 from Barn Owl roosts in Karnak, revealed that the prey of this species was composed (by weight) primarily of rats (Goodman 1986). The Egyptian government has taken an interest in the protection of owls and other birds of prey. Decree 66 for 1983 of the Ministry of Agriculture prohibits the hunting and killing of all birds of prey (see p. 68). Furthermore, the government has expressed its concern about the destruction of (Barn) owls by superstitious farmers (*Egyptian Mail*, 3 August 1985).

House Sparrows and in winter Spanish Sparrows are reputedly important agricultural pests (Ghabbour 1972, 1976). The intensified use of arable lands and the increase of grain spillage near storage places has allowed House Sparrows to increase in population size. This species has also colonized new areas like the Western Desert oases (Goodman *et al.* 1986). These factors combined with an apparent reduction in bird-eating falconiforms have aggravated sparrow impact on the consumption of agricultural products. Spilt grains are also an important food item for Palm Doves. It is conceivable that their recent increase in number (e.g. Al Hussaini 1954*b*), is related to improved feeding and nesting conditions. Both House Sparrow and Palm Dove were found to be important prey of Barn Owls in Upper Egypt, together representing almost 50 per cent (by weight) of the total prey taken (Goodman 1986).

IMPACT OF WETLAND RECLAMATION ON BIRDS

If the proposed reclamation schemes of wetlands are realized (see p. 36) they will have a major impact on birds using these habitats. This is particularly true for Lakes Manzala and Burullus, which are of international significance for wintering Palearctic water-birds (Meininger and Mullié 1981*a*, 1981*b*). For example, up to 7400 Whiskered Terns have been counted in winter on Lake Manzala and up to 17 500 on Lake Burullus, and these lakes are the main wintering area for the populations breeding in eastern Europe and western Asia (Meininger and Mullié 1981*a*). The Nile Delta is the most important wintering area in the eastern Mediterranean–Black Sea region for Shoveler, the second most important for the Ferruginous Duck, and the third most important for Coot (Rüger *et al.* 1986). The Delta, particularly Lake Manzala, is of great importance as a staging and wintering area for waders. Thousands are collected each winter in this region (Table 4.5). The results of a survey of the local breeding birds showed that significant populations of Little Bittern, Purple Gallinule, Painted Snipe, Collared Pratincole, Little Tern, and Clamorous Reed Warbler occur in the area and would be seriously threatened by the loss of habitat due to reclamation (Meininger *et al.* 1986). More research is urgently needed to quantify local breeding bird numbers, and the effects of these changes on them and wintering and staging Palearctic migrants.

When Egypt signed the Bonn and Ramsar Conventions (see p. 70), it accepted the responsibility for safeguarding important bird habitats. The northern wetlands are presently one of the most threatened areas in the country and they should be protected under the vigilance of these conventions and of Law 102.

SUMMARY OF CHANGES IN THE EGYPTIAN BIRD LIST

There are a considerable number of changes in the species we present for the Egyptian bird list compared with Meinertzhagen (1930). Some of these differences are simply the result of taxonomic shuffling, while others are a reassessment of the pre-1930 literature and specimens, and the majority are based on more recent information. The following species which were rejected from the Egyptian list by Meinertzhagen have been added in the present work: Pink-backed Pelican, Yellow-billed Stork, Gabar Goshawk, and House Crow. Six species accepted by Meinertzhagen have been rejected or considered hypothetical to the national list herein: Allen's Gallinule, Glaucous Gull, Palm Swift, Small Skylark, Black Wheatear, and Jackdaw. Although Meinertzhagen's information on the occurrence of Verreaux's Eagle proved to be unacceptable, there have been subsequent records of this species in Egypt (see p. 194 for details). Including these alterations, Meinertzhagen's tally came to 370 species. Subsequently, the occurrence of an additional 51 species has been satisfactorily proven, bringing the Egyptian list to 421 species. Of these 51 'new species', 20 are accidental visitors, 17 of more or less regular occurrence, and 14 have been found breeding in the country (Table 5.1).

Accidental visitors	More or less of regular occurrence	Breeding in the country
Red-necked Grebe	Wilson's Storm Petrel	Striated Heron
Shy Albatross	Red Kite	Goliath Heron
Sooty Shearwater	Eleonora's Falcon	Bateleur
Leach's Storm Petrel	Lesser Sand Plover	Verreaux's Eagle
Darter	Caspian Plover	Collared Dove
Whooper Swan	Bar-tailed Godwit	Namaqua Dove
Greylag Goose	Terek Sandpiper	Nubian Nightjar
Barnacle Goose	Red-necked Phalarope	Syrian Woodpecker
Velvet Scoter	Grey Phalarope	Arabian Warbler
African Fish Eagle	Arctic Skua	Great Tit
Saunder's Little Tern	Long-tailed Skua	Orange-tufted Sunbird
White-throated Robin	Kittiwake	Sudan Golden Sparrow
Firecrest	Little Swift	Streaked Weaver
Desert Sparrow	Barred Warbler	African Silverbill
Tree Sparrow	Yellow-browed Warbler	
Red-fronted Serin	Penduline Tit	
Cinnamon-breasted Rock Bunting	Scarlet Rosefinch	
Rustic Bunting		
Little Bunting		
Yellow-breasted Bunting		

TABLE 5.1. Species admitted to the Egyptian bird list since 1930

Species	Discovery probably due to better exploration	Probably colonized Egypt since 1930
Striated Heron	+	
Squacco Heron	+	
Western Reef Heron	+	
Goliath Heron	+	
Spoonbill	+	
Lappet-faced Vulture	+	
Bateleur	+	
Verreaux's Eagle	+	
Bonelli's Eagle	+	
Hobby		+
Sooty Gull	+	
Slender-billed Gull	+	
Yellow-legged Gull		+
Lesser Crested Tern	+	
White-cheeked Tern	+	
Bridled Tern	+	
Little Tern	+	
Collared Dove		+
Namaqua Dove	+	
Nubian Nightjar	+	
White-breasted Kingfisher		+
Bee-eater		+
Syrian Woodpecker		+
Black-crowned Finch Lark	+	
Blackbird		+
Reed Warbler	+	
Arabian Warbler	+	
Great Tit		+
Orange-tufted Sunbird		+
Sudan Golden Sparrow	+	
Streaked Weaver		+
African Silverbill	+	
Serin		+
Greenfinch		+

TABLE 5.2. Species found breeding for the first time in Egypt since 1930

Meinertzhagen (1930) listed 116 extant breeding species for Egypt, including casual nesters, and seven additional species believed to have been locally extirpated: Ostrich, Long-tailed Cormorant, Sacred Ibis, Greater Flamingo, White-tailed Eagle, Griffon Vulture, and Egyptian Plover. Between 1930 and 1987, 153 species were found nesting (definite or probable breeding status, see pp. 19–20 for criteria) in the country, including 34 'new' breeding species. The discovery of 22 'new' breeding species is probably due to better exploration of the country by ornithologists, while 12 species have apparently colonized Egypt since 1930 (Table 5.2). There is no post-1930 documentation in Egypt for the breeding of the Great Crested Grebe, Grey Heron, Avocet, or Short-toed Lark. However, since then both the Ostrich and Greater

Flamingo have been found nesting. Several other species may breed in the country on occasion but there is insufficient evidence to include them as breeding birds. These are: Mallard, Little Crake, Greater Sand Plover, Swift Tern, Kingfisher, Blue Rock Thrush, and Black-eared Wheatear. These species have not been included in Table 5.3, in which the status and distribution of all Egyptian species with definite and probable breeding status are summarized.

TABLE 5.3. Distribution by region of breeding birds between 1930 and 1987*

		Regions							
Species	Breeding status†	North Coast	Western Desert	Nile Delta	Nile Valley	Eastern Desert	SGAA‡	Red Sea	Sinai
Ostrich	RB	A	A			D	D		A
Little Grebe	RB		C	C	C				
Great Crested Grebe	FB			B					
Red-billed Tropicbird	RB							C	
Brown Booby	RB							C	
Long-tailed Cormorant	FB				A				
Little Bittern	RB		C	C	C				
Night Heron	RB			B	C				
Striated Heron	RB							D	
Squacco Heron	RB				D				
Cattle Egret	RB		?	C	C				
Western Reef Heron	RB							D	
Little Egret	RB			B	C				
Grey Heron	CB				A				
Goliath Heron	RB							D	
Sacred Ibis	FB			A					
Spoonbill	RB							D	
Greater Flamingo	RB			A					D
Egyptian Goose	RB			A	C				
Marbled Duck	RB		C		A				
Black-shouldered Kite	RB		B	C	C				
Black Kite	RB			C	C				
White-tailed Eagle	FB?			A					
Lammergeyer	RB					D	D		C
Egyptian Vulture	RB				C	C	D		C
Griffon Vulture	FB					A			
Lappet-faced Vulture	RB		B			D	D		D
Short-toed Eagle	CB		B			B			B
Bateleur	RB						D		
Long-legged Buzzard	CB					B			B
Golden Eagle	RB								C
Verreaux's Eagle	RB					D	D		?
Bonelli's Eagle	RB					D	D		C
Osprey	RB							C	
Lesser Kestrel	CB			B					
Kestrel	RB	C	C	C	C	C			C
Hobby	MB								D
Sooty Falcon	MB	?	D	C	D	D		D	D

TABLE 5.3. Continued

Species	Breeding status†	Regions							
		North Coast	Western Desert	Nile Delta	Nile Valley	Eastern Desert	SGAA‡	Red Sea	Sinai
Lanner	RB	?	D	C	C	D	D	D	C
Barbary Falcon	RB				C	D	?	D	D
Chukar	RB								C
Barbary Partridge	RB	C							
Sand Partridge	RB					C	C		C
Quail	(RB)			C	C				
Water Rail	RB		C	C	?				
Baillon's Crake	RB			C					
Moorhen	RB		C	C	C				
Purple Gallinule	RB		D	C	D				
Coot	RB		?	B	C				
Houbara	RB	C	C			C			C
Painted Snipe	RB		C	C	D				
Black-winged Stilt	MB		B	C					
Avocet	CB		B		B				?
Stone Curlew	RB	D	C	C	C	D	?		C
Senegal Thick-knee	RB		C	C	C				
Egyptian Plover	FB				A				
Cream-coloured Courser	RB	C	C	C	C	C	C	?	C
Collared Pratincole	MB	C	?	C	C				
Little Ringed Plover	MB		C	?					
Kittlitz's Plover	RB		C	C	C				
Kentish Plover	RB	D	C	C	C			C	C
Spur-winged Plover	RB		C	C	C				B
Sooty Gull	RB							D	
White-eyed Gull	RB							C	
Slender-billed Gull	RB								C
Yellow-legged Gull	RB	D							
Caspian Tern	RB							C	
Lesser Crested Tern	MB							C	
White-cheeked Tern	MB							C	
Bridled Tern	MB							C	
Little Tern	MB	C		C	C				C
Lichtenstein's Sandgrouse	RB					D	C		C
Crowned Sandgrouse	RB		?			D	D		C
Spotted Sandgrouse	RB		C	C	C	C			C
Chestnut-bellied Sandgrouse	FB				B				
Rock Pigeon	RB	C	C	C	C	C	C	C	C
Pink-headed Dove	RB						C		
Collared Dove	RB		D	D	D	D			D
Turtle Dove	MB		C	C	C	D	?		D
Palm Dove	RB		C	C	C	C	C		C
Namaqua Dove	RB						D		
Ring-necked Parakeet	IB			C	C				
Great Spotted Cuckoo	MB			C	C				
Senegal Coucal	RB			C	D				
Barn Owl	RB	?	C	C	C				C

TABLE 5.3. Continued

Species	Breeding status†	Regions: North Coast	Western Desert	Nile Delta	Nile Valley	Eastern Desert	SGAA‡	Red Sea	Sinai
Eagle Owl	RB	D	C	C	C	C	D		C
Little Owl	RB	C	C	C	C	?	C		C
Hume's Tawny Owl	RB					?			C
Nubian Nightjar	MB?						D		
Egyptian Nightjar	RB		C	C	C				
Pallid Swift	RB MB	D	C	C	C				?
White-breasted Kingfisher	RB?			D					?
Pied Kingfisher	RB			C	C				
Little Green Bee-eater	RB			C	C				?
Blue-cheeked Bee-eater	MB		D	C	C				D
Bee-eater	MB								D
Hoopoe	RB	D	D	C	C	?			D
Syrian Woodpecker	RB								D
Black-crowned Finch Lark	RB					?	D		
Bar-tailed Desert Lark	RB	D	C		C	C	C		C
Desert Lark	RB		C	D	C	C	C	D	C
Hoopoe Lark	RB	C	C		D	C	C		C
Dupont's Lark	RB	C							
Short-toed Lark	CB		B						
Lesser Short-toed Lark	RB	D	D	C					C
Crested Lark	RB	C	C	C	C				C
Thekla Lark	RB?	C							
Temminck's Horned Lark	RB	D	C			C			C
Sand Martin	MB		D	C	C				
Pale Crag Martin	RB		C		C	C	C		C
Swallow	RB	D	D	C	C				
Yellow Wagtail	RB	?	C	C	C				D
African Pied Wagtail	RB				C				
Black-capped Bulbul	RB								C
Common Bulbul	RB		D	C	C		C		
Rufous Bush Robin	MB	C	C	C	C	D	?		D
Blackstart	RB		D		?		C		C
Desert Wheatear	RB	C	D	D	?	?			C
Red-rumped Wheatear	RB	C							C
Mourning Wheatear	RB	?			C	C	C		C
Hooded Wheatear	RB				D	C			C
White-crowned Black Wheatear	RB		C		C	C	C		C
Blackbird	RB			D					D
Fan-tailed Warbler	RB		C	C	C				?
Graceful Warbler	RB		C	C	C				D
Scrub Warbler	RB					C			C
Reed Warbler	MB			D					
Clamorous Reed Warbler	RB		C	C	C				?
Olivaceous Warbler	MB	?	C	C	C	?		?	D
Spectacled Warbler	RB?					?			C
Sardinian Warbler	RB				B				D

TABLE 5.3. Continued

		Regions							
Species	Breeding status†	North Coast	Western Desert	Nile Delta	Nile Valley	Eastern Desert	SGAA‡	Red Sea	Sinai
Arabian Warbler	RB						D		
Brown Babbler	RB								C
Fulvous Babbler	RB				B		C		
Great Tit	RB								D
Nile Valley Sunbird	RB			?	C				
Shining Sunbird	RB						C		
Orange-tufted Sunbird	RB								D
Rosy-patched Shrike	RB						C		
Great Grey Shrike	RB	C	C	C	C	D	C		C
House Crow	IB			D		D			?
Hooded Crow	RB			C	C				D
Brown-necked Raven	RB		C	D	C	C	C	D	C
Raven	RB	C							
Fan-tailed Raven	RB						C		C
Tristram's Grackle	RB								C
House Sparrow	RB	C	D	C	C	D			C
Sudan Golden Sparrow	RB?						D		
Streaked Weaver	IB			D	D				
Avadavat	IB			C	C				
African Silverbill	RB?						C		
Serin	RB?			?					
Greenfinch	RB		D	D					D
Goldfinch	RB		D	C	D				D
Trumpeter Finch	RB		C		C	C	C		C
Sinai Rosefinch	RB								C
House Bunting	RB		?			D	D		C

* Status symbols: A, extinct before 1930; B, extinct after 1930; C, breeding (definite or probable, see p. 19 for definitions); D, breeding not mentioned by Meinertzhagen (1930); ? = status uncertain

† See p. 18 for definitions of regions

‡ SGAA = Sudan Government Administration Area

The total number of species known to breed in each of the half-degree grid squares is summarized in Fig. 5.4. Not surprisingly, squares with the highest number of breeding species (over 40) are those with the greatest habitat diversity. Moreover, many of these areas have been heavily visited by persons interested in natural history because of their touristic attractions (e.g. Cairo, Giza, Luxor, and Aswan), 'ornithological fame' (e.g. Wadi el Natrun, Faiyum, and Suez), or strategic position (e.g. Suez). The squares with the greatest number of breeding species are those including Cairo and Giza with 61 species each, and Aswan and Luxor with 48 species each. In many cases more exploration will certainly reveal species not currently known to occur locally; although in some habitats the presumed low species diversity actually reflects the true status of the region. Squares without any marks have not been visited and are

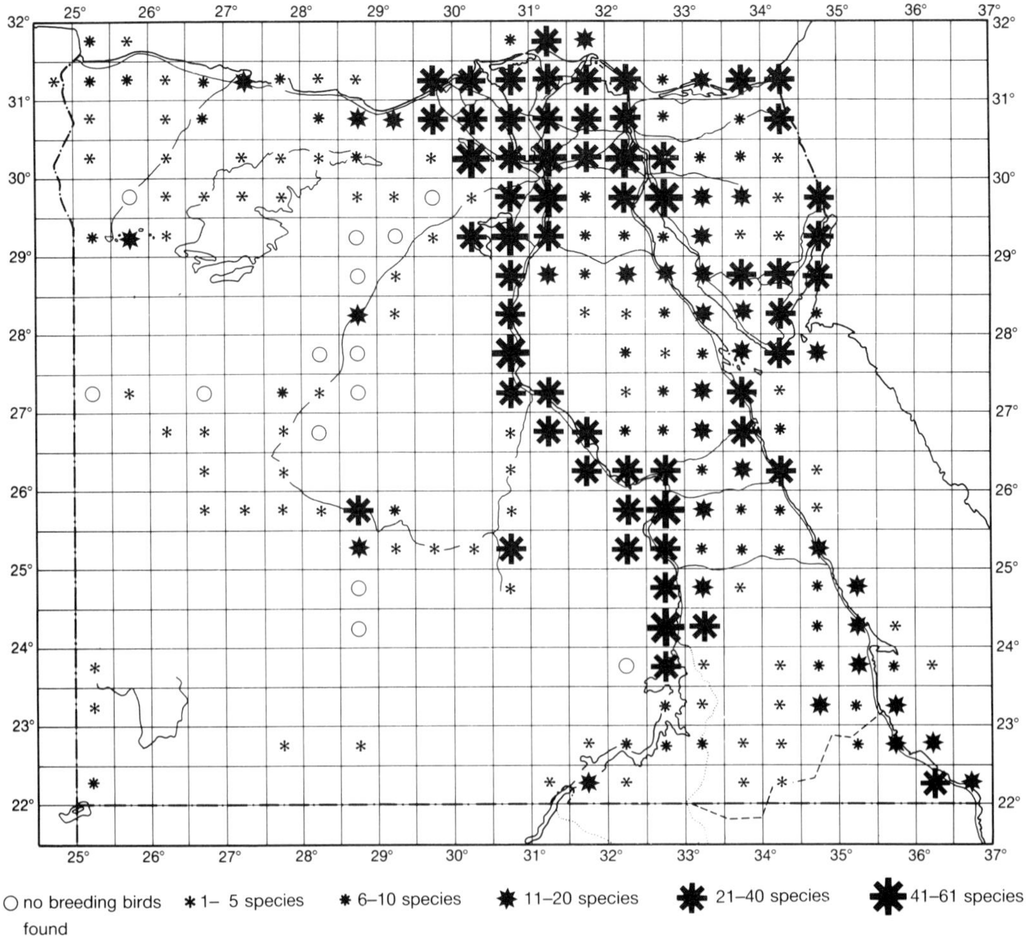

FIG. 5.4. Numbers of species found breeding (definite and probable status) between 1930 and 1987 per half-degree grid squares.

ornithologically unknown. However, many of these locales, particularly in the Western Desert, are unlikely to hold many breeding birds if any at all. Exploration of little studied or unknown areas around Lake Nasser, in the Eastern Desert, and in north or central Sinai will undoubtedly reveal much interesting new information.

Part 2

6. THE BIRDS OF EGYPT

ORDER STRUTHIONIFORMES

Family Struthionidae

Ostrich *Struthio camelus* RB

Local name Na'aam نعام.

Local breeding resident in south-eastern Egypt. The subspecies occurring in the country is *S. c. camelus*. On the evidence of remains of Ostrich eggshell fragments and historical accounts, this species formerly inhabited the North Coast, Western and Eastern Deserts, and northern Sinai. Lowe (in Moreau 1934), presumably using pore size and pattern, referred eggshell fragments collected in the Western Desert to *S. c. massaicus*. However, Sauer (1972) has shown that pore pattern characters are often not reliable on the subspecific level.

North Coast Several eighteenth-century explorers observed Ostrich in the hills south-west of Alexandria (Goodman *et al.* 1984). Non-mineralized eggshell fragments have also been found in the 'sandhills east of Matruh' and 72 km south-east of Matruh (GZM). A diminutive egg of indeterminate age was found in the Qattara Depression in 1979. The size of this egg falls within the range of the Middle Eastern form *S. c. syriacus* (Walters 1982); however, the possibility that it is a runt egg of *S. c. camelus* cannot be eliminated.

Western Desert There are numerous observations of the Ostrich from the 1700s to 1900s (Goodman *et al.* 1984), and eggshell fragments of indeterminate age are common in some areas. The most recent record is one captured in 1935 at 'Abou-al-Oql' (= Ezbet Zayat), midway between Kharga and Dakhla (Al Hussaini 1959).

Eastern Desert Numerous eighteenth- and nineteenth-century explorers found this species at various localities throughout the Eastern Desert, particularly the south-eastern corner. Between the turn of the twentieth century and the early 1960s there appears to be no authentic record from this region. In 1962 it was noted between Gebel Elba and the Red Sea, and several were killed during the winter of 1961–2 near Gebel Hamata (Goodman *et al.* 1984). In May 1962, one was shot and photographed by a group of geologists about 80 km east of Kom Ombo (HCK). Since the 1960s this species has been noted somewhat regularly, but in varying numbers, throughout the southern half of the Eastern Desert, particularly in the Sudan Government Administration Area. Eggs were collected near Gebel Elba on 14 February 1967 in

Wadi Yoider and a chick on 20 February 1967 in a tributary of Wadi Hareitra (FMNH). There is also a breeding record from 1982 in the desert about 60 km north-east of Aswan (Goodman *et al.* 1984).

North Sinai Ostrich eggshell fragments of indeterminate age have been found across the sandy plain of northern Sinai and east of the Suez Canal. The last known record is two near Gebel Maghara on 22 April 1816 (Burckhardt 1822). (This locality was incorrectly reported by Goodman *et al.* 1984 as between Cairo and Suez.)

[ORDER GAVIIFORMES

Family Gaviidae

Red-throated Diver *Gavia stellata* AV?

Von Heuglin (1873) stated that the Red-throated Diver occurs in January and February on the lagoons of Lower Egypt. This record was rejected by Meinertzhagen (1930). On 28 and 29 December 1952, five Red-throated Divers, together with six Great Crested Grebes, were observed near Deversoir and on 12 January 1953, one was noted at El Kabrit (AP). A specimen was collected at Ein Gedi, Israel, on 18 December 1982 (Goodman and Storer 1987). A diver, thought to be this species or *G. arctica*, was observed on 1 March 1979 at Elat, Israel (EB).]

ORDER PODICIPEDIFORMES

Family Podicipedidae

Generic local names for grebe and diver Ghaṯaas, ghawwaaṣ.

Little Grebe *Tachybaptus ruficollis* RB WV

Local name Zahuut زهوت.

Fairly common breeding resident in marshes of the Western Desert oases and the Nile Delta. In the non-breeding season birds are seen at sea and on inland waters, often in areas far from known breeding localities. Until recently it was assumed that all Egyptian breeding birds were referable to *T. r. capensis*. However, two specimens collected on Lake Manzala are referable to *T. r. ruficollis*. These include a female obtained on 8 January 1971 and a male (in breeding condition) on 23 January 1985 (Goodman and Watson 1983; UMMZ). The subspecific identity of the birds breeding on the Nile Delta lakes needs further investigation. Birds collected in breeding condition in Western Desert oases are referable to *T. r. capensis* (UMMZ; USNM). Northern migrants, including birds referable to *T. r. ruficollis*, winter in Egypt. This species commences breeding in early winter. A chick was found for sale in the Port Said market on 21 December 1982 (UMMZ).

North Coast Only one record of the Little Grebe is known from this area. Four birds were noted on 9 March 1985 in pools east of El Alamein (DM).

Western Desert The breeding of Little Grebes has been proven or suspected in the oases of Wadi el Natrun, Bahariya, Dakhla, and Kharga (Goodman *et al.* 1986). In winter local breeding populations are augmented by migrants. For example, on 26 January 1981 at least 200 Little Grebes were noted on a newly created lake at Dakhla (AH).

Nile Delta Little Grebes breed in marshes and reed-beds of the Delta lakes, with an estimated breeding population of at least several hundred pairs (Meininger *et al.* 1986). This species also nests at several scattered localities in the southern Nile Delta and the Suez Canal area. In winter it is often observed on the Suez Canal.

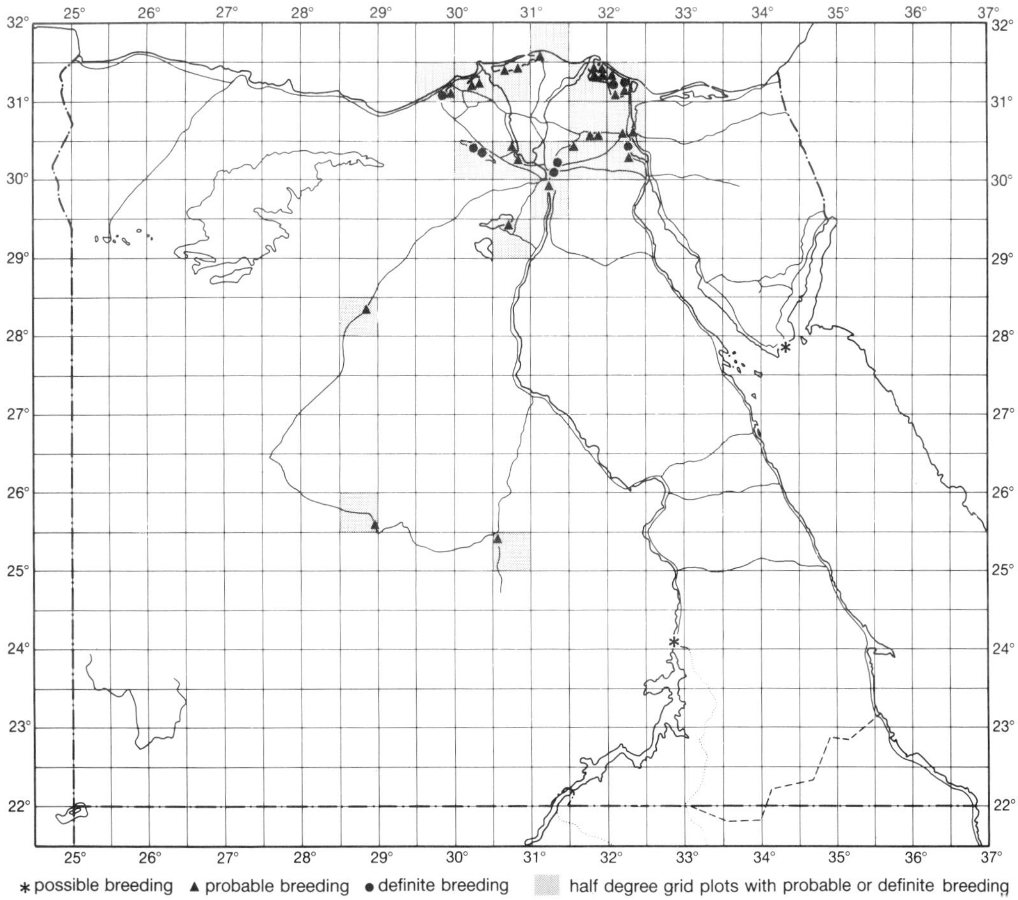

Map B1. Breeding distribution of Little Grebe *Tachybaptus ruficollis* between 1970–87.

Nile Valley Between October and April small numbers are irregularly observed along the complete length of the Nile. It formerly bred in the Faiyum (Gurney 1876), and probably still does so in the marshes south of Lake Qarun. It is likely that Little Grebes breed locally in the Nile Valley proper but this needs confirmation. The latest recent spring record is one at Aswan on 22 April 1985 (DKT).

Sinai Small numbers are seen in winter on Lake Bardawil. A few are regularly observed between December and April in freshwater pools near Sharm el Sheikh.

Great Crested Grebe *Podiceps cristatus* FB WV

Local names Ghaṯaas mutawwaj غطاس متوج, shaa'ir (NC).

Former breeding resident of the Nile Delta lakes. Presently it is a fairly common winter visitor between (early) mid-September and early April. *P. c. cristatus* is the form occurring in Egypt.

Meinertzhagen (1930) considered the Great Crested Grebe a rare resident in Egypt, occurring in small numbers on Lake Qarun and the Nile Delta lakes. The only documented breeding record appears to be of chicks being fed by two adults in June on Lake Manzala (Meinertzhagen 1930). Pairs were also noted on 4 March 1935 in Lake Manzala (Jourdain and Lynes 1936). Small numbers were seen at this locality in July and August 1979 (EM), which may have been early migrants or summering birds. There are no recent nesting records of Great Crested Grebes in Egypt and it is presumed to be a locally extirpated breeding bird (Meininger *et al.* 1986).

An early migrant was noted at Zaranikh on 3 September 1982 (Baha el Din and Salama 1984). The Great Crested Grebe is a fairly common winter visitor, especially on saline lakes. About 3400 were counted on Lake Qarun in January 1980 (Meininger and Mullié 1981*a*). Small numbers are found between mid-October and early April along the Mediterranean coast, on the Nile Delta lakes, in Wadi el Rayan, the Bitter Lakes, Lake Bardawil, and the Gulfs of Suez and Aqaba. The species is rare on the Nile above Cairo, and the southern-most report is four near Asyut on 18 March 1985 (Grieve 1985). The only known record from the Western Desert is one at Dakhla on 21 February 1986 (AH).

Red-necked Grebe *Podiceps grisegena* AV

Local name Ghaṯaas aẖmar ar-raqaba غطاس أحمر الرقبة.

Accidental visitor. No Egyptian specimens have been examined but on geographical grounds the subspecies occurring in the country is presumably *P. g. grisegena*.

Von Heuglin (1873) considered the Red-necked Grebe a probable ('Wahrscheinlich') winter visitor to the Nile Delta. There are four relatively recent sight records from the country: two with five Great Crested Grebes at El Kabrit on 30 January 1953 (AP), two in winter plumage along the Nile near Beni Suef on 13

November 1977 (Short and Horne 1981), one at Sharm el Sheikh on 18 September 1980 (DJM, DES), and four south of Isna on 2 January 1986 (AS).

Black-necked Grebe *Podiceps nigricollis* WV

Local name Ghaṯaas aswad ar-raqaba غطاس أسود الرقبة.

Common winter visitor from mid-September to mid-April, especially on saline lakes. The form occurring in the country is *P. n. nigricollis*.

Large winter concentrations of Black-necked Grebes have been recorded in Wadi el Natrun—200 on 20 February 1984 (SG); on Lake Qarun—10 800 between 10 and 12 January 1979 and 10 300 between 1 and 2 January 1980; Lake Bardawil—280 on 7 January 1980; El Malaha—1330 on 9 January 1980; and Bitter Lakes—450 on 18 January 1980 (Meininger and Mullié 1981*a*). Small numbers are regularly seen on other Nile Delta lakes, in Wadi el Rayan, the Gulfs of Suez and Aqaba, and the Nile Valley south to Aswan. The only record from the Western Desert is a bird taken in Kharga on 14 January 1972 (Goodman *et al.* 1986).

ORDER PROCELLARIIFORMES

Family Diomedeidae

Shy Albatross (White-capped Albatross) *Diomedea cauta* AV

Local name Albaṯras al-khajuul البطرس الخجول.

Accidental visitor to Egyptian Red Sea waters. The subspecies of the single Egyptian specimen has not been determined.

A Shy Albatross was noted in the area from Elat, Israel, to Gezira Fara'un between 20 February and 7 March 1981 (Paz 1987). During this period it was also observed within Egyptian waters off Taba (GWR). On 7 March 1981, one Shy Albatross, presumably the same bird, was found dead at Elat (TAUM; Goodman and Storer 1987).

Family Procellariidae

Cory's Shearwater *Calonectris diomedea* PV WV

Local name Jalam al-maa- al-kabiir جلم الماء الكبير.

Scarce passage migrant and winter visitor in the Mediterranean and Red Sea generally from mid-August to December, although it has been recorded throughout the year. *C. d. diomedea* is the form occurring in the country.

Mediterranean coast Small numbers of Cory's Shearwaters have been recorded throughout the year. In autumn they are frequently observed passing off the coast of Lake Bardawil, with a daily maximum of 38 on 23 September 1980 (Petersen and Sørensen 1981*a*).

Red Sea Records from Egyptian waters include: one in the Gulf of Suez at 29°00′N., 32°51′E. and two at Suez on 15 August 1980 (Chapman 1982); two off Taba on 2 September 1980 (JB); one group of five and another of three in the Suez Bay on 24 September 1982 (JV); one south of Ain Sukhna on 31 October 1982, one found dead 35 km south of Safaga on 5 November 1982, one near Gezira Siyul (off Hamata) on 27 November 1982, and one at Ain Sukhna on 3 December 1982 (Baha el Din and Saleh 1983); six off Gezira Fara'un on 24 April 1985 (JvT); and five near Suez on 9 June 1984, three at sea off Quseir (26°03′N., 35°00′E.) on 10 June 1984, two north of Brothers Reef on 10 July 1985, and up to 10 per day in Suez Bay between 11 and 19 July 1985 (BH *et al.*). At Elat, Israel, it is noted throughout the year but appears to be more regular between July and November, when up to 20 have been observed together (Paz 1987; Shirihai 1987). There are a few spring records of single birds from this locality.

Sooty Shearwater *Puffinus griseus* PV

Local name Jalam al-maa- al-as_h_am جلم الماء الأسحم.

Rare passage migrant along the Mediterranean coast in autumn and in the Red Sea in spring. Monotypic.

Mediterranean coast There are a few records of the Sooty Shearwater from the north Sinai coast. These include one on 2 September 1974 and two on 14 September 1974 and on 12 November 1975 (Paran 1979), and single birds on 30 August and 12 September 1981 (Petersen and Sørensen 1981*b*), and 16 September 1982 (Baha el Din and Salama 1984).

Red Sea There are a few sight records from the northern Red Sea coast: one on 2 April 1973 off Taba, one on 5 April 1974 at Ras Muhammad (SC), and one on 9 April 1979 at Gezira Fara'un (RSS). A female was taken on 5 May 1980 near Sharm el Moiya (Goodman and Storer 1987). An apparent Sooty Shearwater was observed near Hurghada on 28 April 1982 (Frazier *et al.* 1984). At Elat, Israel, this species is observed regularly from mid-March to early August, with a maximum count of 17 on 31 May 1982 (HSh; Shirihai 1977).

Manx Shearwater *Puffinus puffinus* PV WV

Local names Jalam al-maa- جلم الماء, umm gharnaaya.

Scarce but regular passage migrant and winter visitor between mid-August and early May. All Egyptian material examined is referable to *P. p. yelkouan*.

Small numbers of Manx Shearwaters are observed along the Mediterranean coast, particularly in autumn. The species is a regular migrant off Lake Bardawil, with up to 249 passing between 15 August and 5 October 1978 (YP, UP) and daily maxima of 13 on 30 August 1981 and on 26 September 1980 (Petersen and Sørensen 1981*a*, 1981*b*). A male was taken on 20 August 1973 at Zaranikh (ZM-HUJ). Other specimens include a bird found dead on 7 February 1979 near Misfaq (EB) and one collected on 9 December 1917 at Rafa (BMNH). The only known record from the Egyptian Red Sea is two observed on 8 May 1981 near Suez after a heavy sand-storm (SMB, SV). This species has not been reported from Elat, Israel (Krabbe 1980; HSh).

Family Hydrobatidae

Wilson's Storm Petrel *Oceanites oceanicus* PV

Local name T̲aa-ir an-naw- طائر النوء.

Rare visitor to the Red Sea. The subspecies occurring in the country has not been determined.

While travelling north in the Red Sea in June 1944, Maclaren (1946) noted a few Wilson's Storm Petrels within 800 km of Suez. Sight records of this species in Egyptian waters in 1985, by observers experienced with the bird, include: three on 9 July at 24°15′N., 36°08′E., and 14 on 11 July and one on 16 July in Suez Bay (BH *et al.*). A single Wilson's Storm Petrel was seen near Elat, Israel, on 1 June 1983 (Paz 1987; Shirihai 1987). Five petrels observed in Sudanese Red Sea waters, just south of the Egyptian border, on 28 June 1945, were thought to have been this species (Phillips 1947). Unidentified petrels have been noted off Ras Muhammad on 4 April 1974 (SC) and near the Brothers Reef on 10 July 1985 (BH *et al.*).

Leach's Storm Petrel *Oceanodroma leucorhoa* AV

Local name T̲aa-ir an-naw- abyad̲ al-'ajuz طائر النوء أبيض العجز.

Accidental visitor to Egyptian waters. All Egyptian specimens examined are referable to *O. l. leucorhoa*.

The only positive records of the Leach's Storm Petrel in Egypt are birds collected at Suez on 15 June 1955 (RMS) and near Sharm el Sheikh on 8 August 1978 (Goodman and Storer 1987; TAUM). Etchécopar and Hüe (1967) mentioned it as sometimes occurring in the Red Sea between October and March but presented no details.

ORDER PELECANIFORMES

Family Phaethontidae

Red-billed Tropicbird *Phaethon aethereus* RB

Local names Ra-iis al-bahr رئيس البحر, abuu guhar (RS), kurraan (RS).

Rare breeding resident on islands in the Red Sea. Occasionally observed in the Gulfs of Suez and Aqaba. The form occurring in Egypt is *P. a. indicus.*

In the spring of 1893 an egg of the Red-billed Tropicbird was collected on Gezira Shadwan (Kaiser, in le Roi 1923) and on 25 May 1926 a nestling was taken on Gezira

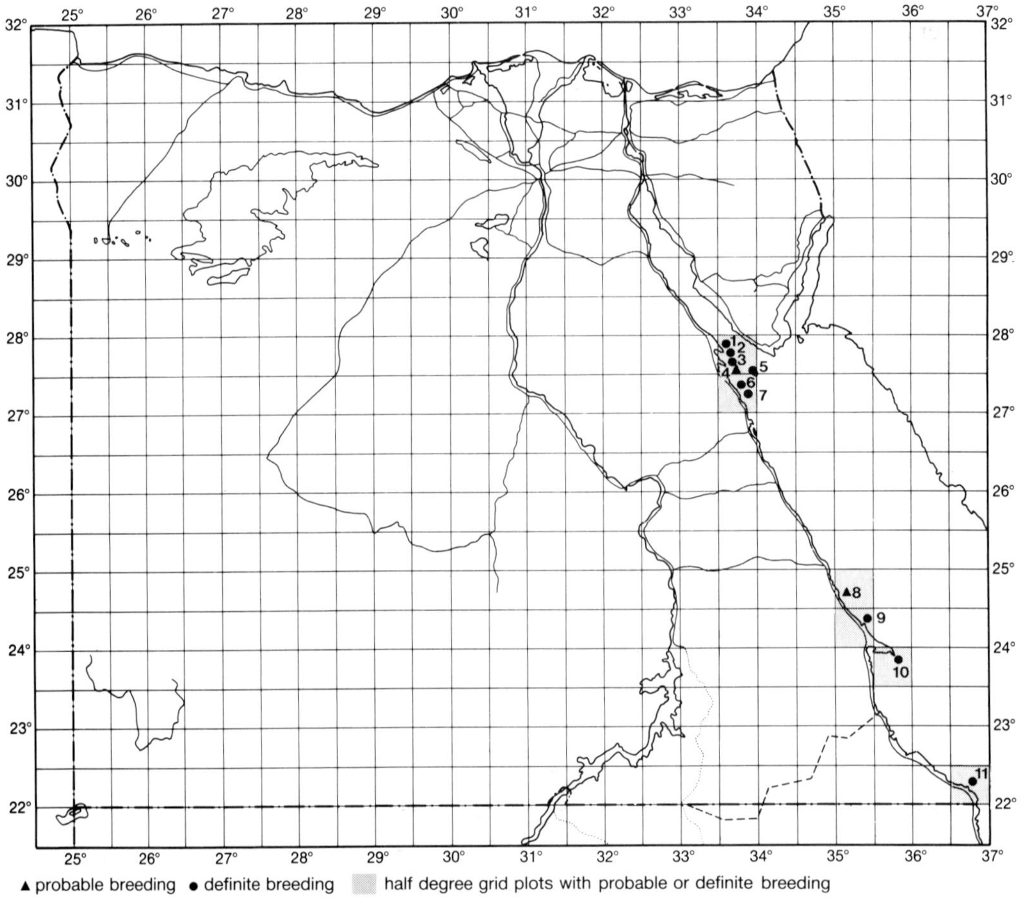

Map B2. Breeding distribution of Red-billed Tropicbird *Phaethon aethereus* between 1920–87. *Breeding localities* (see text for details): (1) Gezira el Makawarat. (2) Gezira Ashrafi. (3) Gezira Qeisum. (4) Gezira Gubal el Kabir. (5) Gezira Shadwan. (6) Gezira Abu Minqar. (7) Gezira Umm Qamar. (8) Gezira Wadi Gimal. (9) Gezira Showarit. (10) Gezira Mikauwa. (11) Gezira Halaib.

Ashrafi (Borman 1929). Baha el Din and Saleh (1983) found evidence of breeding on Gezira Abu Minqar and Gezira Showarit, and learned from local fishermen that nests and nestlings had been noted on Gezira Umm Qamar, Gezira Qeisum, and Gezira el Makawarat. Skulls of this species were picked up in rock crevices, perhaps nesting sites, on Gezira Qeisum, Gezira Gubal el Kabir, and Gezira Wadi Gimal (Jennings *et al.* 1985; Goodman and Storer 1987). In 1984 and 1985 local fishermen from the southern Egyptian Red Sea reported that this species breeds on Gezira Mikauwa and on Gezira Halaib (SMG, PLM).

The species has been noted in the Gulf of Suez in May 1945 (Maclaren 1946), in April (Zedlitz 1912; Alexander 1929), and on 10 January 1913 (Chapman 1921). Records from southern Sinai include single birds off Ras Muhammad on 27 March 1976 and 2 km south-west of Gezira Tiran on 22 April 1977 (Murphy and Redman 1983). Along the Egyptian Red Sea coast, between Hurghada and Halaib, this species is generally rare and observed only during the summer months. At Elat, Israel, this species has been recorded in January, April, June, and at least twice in May (Krabbe 1980; Ben Dov and Golan 1983; Murphy and Redman 1983; Paz 1987).

Family Sulidae

Brown Booby *Sula leucogaster* RB

Local names Aṯiish أطيش, saamaadawaii (Ab), sumad (RS), samuuṯ (RS).

Scarce breeding resident on islands in the Red Sea. Observed with some regularity in the Gulf of Aqaba and rarely in the Gulf of Suez. The form occurring in Egypt is *S. l. plotus*.

The Brown Booby has been found breeding on the following islands at the mouth of the Gulf of Suez: Gezira Ashrafi—young were collected in August 1925 and an egg on 13 June 1926 (Borman 1929), and seven breeding pairs were found in spring 1983 (Jennings *et al.* 1985); Gezira Qeisum—46 breeding pairs were noted in spring 1983 and 18 chicks in autumn 1984 (Jennings *et al.* 1985); Gezira Umm el Heimat—three pairs were recorded in spring 1983 and one chick in autumn 1984 (Jennings *et al.* 1985); and Gezira Umm Qamar—two nests were discovered on 11 July 1943, one with two eggs and the other with one chick (Abdel Megid 1945), and naked chicks were found on 23 August 1938 (Al Hussaini 1939). The breeding population in this region has been estimated at between 56 and 70 pairs (Jennings *et al.* 1985). The only island it is known to nest on in the southern Egyptian Red Sea is Gezira Zabargad, where downy young were discovered in May sometime during the 1980s (DP).

This species is noted in small numbers throughout the year along the Red Sea coast. A few birds are regularly observed throughout the year near Elat, Israel. In the Gulf of Suez it is rare north of the line between Ras Zeit and El Tor. The species often forages out to sea and not within viewing distance from the shore.

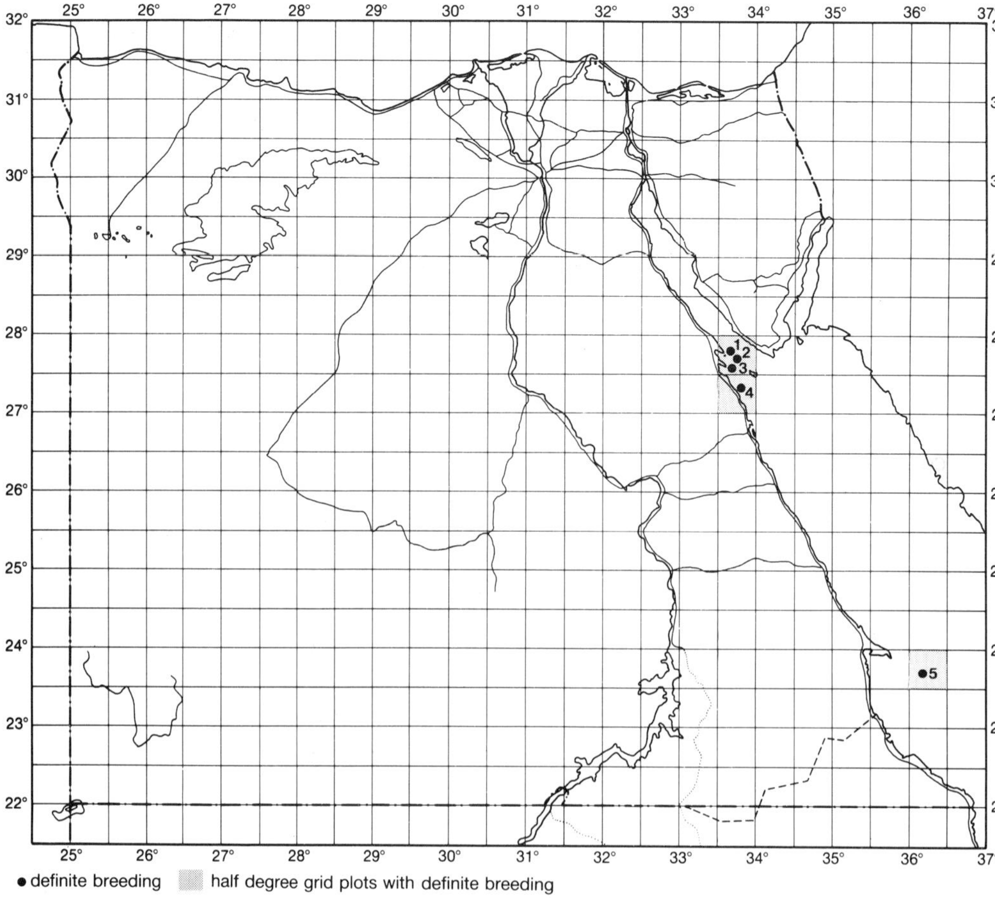

Map B3. Breeding distribution of Brown Booby *Sula leucogaster* between 1940–87.
Breeding localities (see text for details): (1) Gezira Ashrafi. (2) Gezira Qeisum. (3) Gezira Umm el Heimat. (4) Gezira Umm Qamar. (5) Gezira Zabargad.

Gannet *Sula bassana* PV WV

Local names Atiish abyad أطيش أبيض, salaanit.

Rare passage migrant and winter visitor along the Mediterranean coast. It may occur more frequently than presently known. The single Egyptian specimen examined is referable to *S. b. bassana*.

Records of Gannets from Port Said include an adult taken on 20 September 1905 (Koenig 1932; ZFMK) and 19 adults and one immature observed in early February 1961, 8 km off the coast (Jones 1961). On 17 February 1961, one was noted off the Nile Delta at 32°10′N., 30°20′E. (Bourne and Radford 1962). Near Alexandria an immature bird was observed on 6 March 1959 (Kumerloeve 1962), another on 31

March 1983 (Chapman 1984), and an adult on 21 August 1964 (Bourne 1967). Records from the north Sinai coast, off Lake Bardawil, include single birds on 22 March 1976 (SC), 16 October 1979 (Petersen and Sørensen 1980), and 30 April 1985 (Fry *et al.* 1985).

Family Phalacrocoracidae

Generic local names for cormorant Ghuraab al-bahr, 'ajaaj, 'aqaaq, ghaaq, ghuraab al-maaya.

Cormorant *Phalacrocorax carbo* PV WV

Local names Ghuraab al-bahr غراب البحر, abuu ghataas (NC).

Fairly common passage and winter visitor from early October to early (late) April. It is rare in summer. All Egyptian material that we have examined is referable to *P. c. sinensis*. It is plausible that the subspecies *P. c. lucidus*, which breeds in sub-Saharan Africa, may occasionally wander down the Nile into Egypt.

Large groups of Cormorants are regularly observed in Egypt. On 10 January 1980, 2200 were noted at El Malaha (Meininger and Mullié 1981*a*), and on 3 February 1983, 2000–3000 at Lake Manzala (BR). Smaller numbers are regularly noted on the other Delta lakes, Lake Qarun, along the Mediterranean coast, the Nile Valley, Lake Nasser, the Suez Canal, and in the Red Sea and northern gulfs. The only Western Desert records are from Dakhla and include eight on 21 February 1986 and seven between 23 January and 8 March 1987 (AH). A migrating flock of 100 Cormorants was noted off Ain Musa on 19 March 1986 (AS). The southern-most Red Sea coastal record is one on 12 February 1984, 60 km north of Berenice (SMG).

Apparently, the species was more common along the Nile in the nineteenth century. Between November 1853 and January 1854, Taylor (1859) found it abundant and roosting on cliffs along the river. Adams (1864) mentioned this species as 'plentiful' north of Luxor and flocks roosting in date palm trees; however, his statement that it bred at Manfalut cannot be substantiated. There are many other early references or inferences of colonies in Egypt (e.g. Whymper 1909; Chapman 1930) but there is no modern documentation of this species breeding in the country. Abdel Rahman (1955) reported that one was noted on a rock off Marsa Matruh on 15 July 1955 and that others were seen on small islands further out to sea. The possibility that these records concerned Shags cannot be excluded. Remains of Cormorants have been recovered from a late Pleistocene site (15 000–10 000 BC) near Kom Ombo (Churcher 1972).

Ringing A bird ringed as a nestling in Yugoslavia in 1981 was recovered at Matruh in 1984 (BEP, GS).

Shag *Phalacrocorax aristotelis* AV

Local names Ghuraab al-bahr al-akhdar غراب البحر الأخضر, qaaq al-maaya.

Accidental visitor, although possibly a rare winter visitor to the Mediterranean coast west of Alexandria. The form occurring in the country is *P. a. desmarestii*.

Meinertzhagen (1930) saw a few Shags at Salum and Sidi Omar in February and March and collected two dead birds. The only other record from the country is one oiled bird on 20 November 1982 on Gezira Abu Minqar, off Hurghada (Baha el Din and Saleh 1983).

Long-tailed Cormorant *Phalacrocorax africanus* FB

Local names Ghuraab al-bahr al-afriiqii غراب البحر الأفريقي, fasiisha.

A former breeding resident in the Faiyum. All Egyptian material examined is referable to *P. a. africanus*.

In the early literature on the birds of Egypt the Long-tailed Cormorant was confused, at least in several cases, with the Pygmy Cormorant (*P. pygmaeus*). The latter species has not been documented in Egypt. Long-tailed Cormorants formerly bred on the western shore of Lake Qarun. The species also occurred in the Nile Delta and Valley. The last known record from Egypt appears to be birds for sale in the Damietta market on 5 March 1903 (Loat 1906); although the records of Whymper (1909) of two small colonies of 'Cormorants' in April 1908 on islands in Lake Manzala may have concerned Long-tailed Cormorants.

Von Heuglin (1873) found a small cormorant, that he identified as *P. pygmaeus*, to be common in the lagoons of Lower Egypt. Adams (1864) frequently noticed a cormorant along the Nile which was 'possibly *P. pygmaeus*'. The statement of Etchécopar and Hüe (1967) that the Long-tailed Cormorant breeds in Upper Egypt is erroneous. A specimen examined by Koenig (1932) in the Medical School, Cairo, and identified as a Pygmy Cormorant is clearly a Long-tailed Cormorant. Kaiser's (1890) record of *P. pygmaeus* near El Tor on 29 November 1890 cannot easily be dismissed as *P. africanus*, since the bird was out of the range of the Long-tailed Cormorant and closer to known wintering areas of the Pygmy Cormorant.

Family Anhingidae

Darter *Anhinga melanogaster* AV

Local name Zuqq زق.

Accidental visitor to the Upper Nile Valley. No modern Egyptian specimens are known but on geographical grounds *A. m. rufa* should be the form wandering to the country.

In March and October 1950 Darters were collected in Upper Egypt, near Ineiba (Zain el Din 1951). The specimens cannot be traced. More recently single birds have been observed near Idfu on 26 September 1985 (NdH, PWB), on 27 December 1985 (AS), and on 15 April 1985 (JvT). Bones of this species have been identified in archaeological remains from the Faiyum dated to probably the Predynastic period (4500–3000 BC) (Ballmann 1980).

Family Pelecanidae

Generic local names for pelican Baja', baqr al-baẖr.

White Pelican *Pelecanus onocrotalus* PV WV

Local names Baja' abyaḏ بجع أبيض, abuu ẖilb, abuu jiraab, ẖawṣal, jamal al-baẖr, qawq, su'aad (ND).

Fairly common passage visitor in autumn from (late August) early September to late November and in spring from early March to mid-May (Fig. 6.1). It is rare in winter. Monotypic.

According to Meinertzhagen (1930) the White Pelican was common on the larger bodies of water from autumn through spring, particularly in the Faiyum, Suez Canal area, and along the Nile south to Wadi Halfa. Nicoll (1908) saw large numbers on

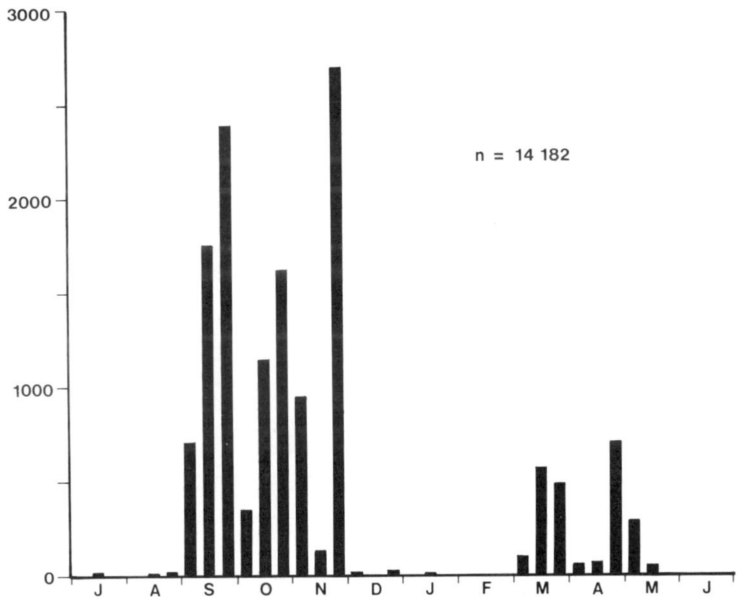

FIG. 6.1. Histogram of the numbers of White Pelicans recorded during 10-day periods throughout the year.

Lake Manzala in January 1908. Since the publication of Meinertzhagen (1930) the wintering status and distribution of this species in Egypt has drastically changed. In recent years only small numbers have been observed between December and February in the southern Nile Valley (including Lake Nasser). There are no recent winter records from the northern Nile Valley, the Faiyum, or Nile Delta lakes, except for a few captive birds in the markets of Port Said and Damietta, which were presumably obtained locally during migration, and up to nine birds at El Kabrit between 1 and 30 January 1953 (AP).

It is still a regular autumn migrant along the north coast of Sinai, the coastal plain of the Eastern Desert south to Hurghada, and the southern Nile Valley. Resting birds are frequently seen on Lake Bardawil and in small numbers on the banks of the Upper Nile and Lake Nasser. Autumn records of large numbers include: 725 on 22 September

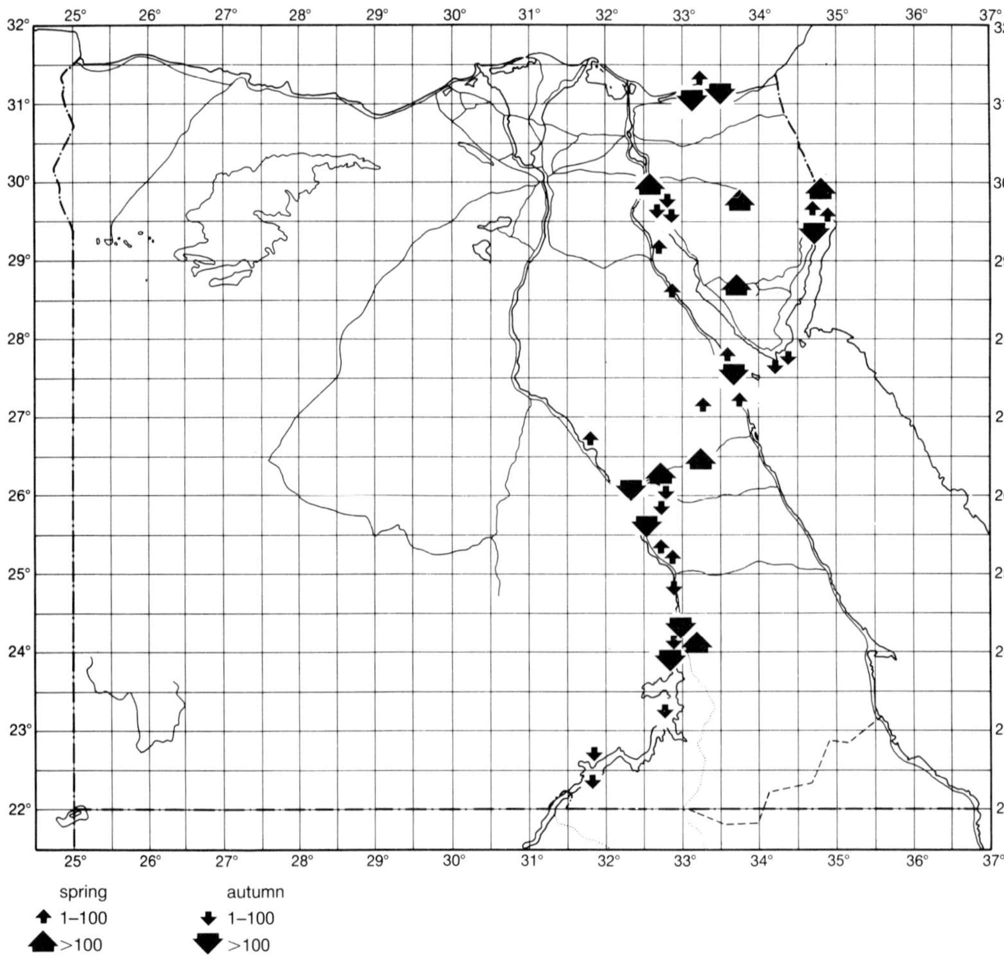

FIG. 6.2. Sight records divided by season of White Pelican flocks.

1982 over Lake Bardawil (Baha el Din and Salama 1984); 1000 on 21 October 1957 at Nag Hammadi (Horváth 1959); 750 on 4 November 1977 in three flocks between Idfu and Kom Ombo (Short and Horne 1981); 1000 on 22 November 1983 flying south near Luxor (Bezzel 1987); and 1200 on 25 November 1982 resting in the desert near Ras Gemsa and later flying in a south-western direction (Baha el Din and Saleh 1983). The earliest autumn record appears to be one at Na'ama on 20 August 1976 (TAUM).

In spring it is regularly seen in the southern Nile Valley, northern Red Sea, and Gulf of Suez. Flocks are generally smaller than in autumn (Fig. 6.2). On 22 April 1982 a group of 211 was observed at Suez (Wimpfheimer *et al.* 1983). A flock of about 420 was noted on 24 April 1984 passing over the Safaga–Qena road in an easterly direction, 81 km west of Safaga (SMG). Large flocks in Sinai include about 500 on 18 March 1981 east of the Mitla Pass (SMB, BB) and 3500 on 26 March 1981 over Moon Valley (Paz 1987). The only summer record is a group of 18 at Suez on 16 July 1985 (BH *et al.*).

Ringing A bird ringed in Iran was found dead at Ras Muhammad (Paz 1987). Another pelican, either this species or the Dalmatian Pelican, ringed in Romania was recovered in the central Nile Delta in March 1935 (GZF).

Dalmatian Pelican *Pelecanus crispus* WV

Local name Baja' balqaanii بجع بلقاني.

Rare winter visitor between mid- (late) October and mid-March. Monotypic.

Meinertzhagen (1930) stated that the Dalmatian Pelican occurred in large flocks in the Faiyum from autumn to spring, and often on the Nile and in the Suez Canal area. Nicoll (1908) saw birds on Lake Manzala in January 1908. In recent years this species has become markedly rare in Egypt, which is presumably related to the sharp decline in Eurasian breeding populations (Crivelli and Vizi 1981). Recent records include: Sohag—three between 31 October and 1 November 1978 and an unspecified number on 10 November 1977 (Short and Horne 1981); El Malaha—seven on 10 January 1980 (Meininger and Mullié 1981*a*); Suez—25 on 3 December 1982 (PB) and 19 on 10 March 1980 (SMB, BB); Ain Musa—two on 11 October 1984 (RvD *et al.*); Luxor—one on 11 November 1984 (SH, DM); and near Idfu—one on 2 January 1986 (AS). A single bird was noted during the winter of 1982–3 at Elat, Israel (Paz 1987).

Pink-backed Pelican *Pelecanus rufescens* PV

Local name Baja' ramaadii بجع رمادي.

Rare passage and summer visitor to the Upper Nile Valley and Red Sea. Monotypic.

There are a few records of the Pink-backed Pelican from the Egyptian Red Sea and northern tributary waters. These birds presumably wander from breeding areas further south along the Saudi Arabian (Jennings *et al.* 1982) and African (Cramp and Simmons 1977) Red Sea coasts. A single bird was captured on the Bitter Lakes (Nicoll 1919) and a flock of 23 was noted near Ain Sukhna on 12 April 1965 (HCK). One was observed at Elat, Israel, between 31 March and 1 April 1974 (Krabbe 1980).

Flower (1933) mentioned this species as a summer visitor, singly or in parties of up to 22, from May to September in Upper Egypt, and exceptionally as far north as Cairo. He noted it in seven years between 1902 and 1920. A single bird was observed on 8 or 9 August 1905 on the Nile between Beni Suef and Giza (Nicoll 1909). One was collected out of a flock of five near Luxor on 22 October 1957 (Horváth 1959). In spring and summer this species has been regularly recorded in Lake Nasser at Wadi Halfa, just across the border in Sudan. Up to 24 individuals were seen at this locality on 25 August 1980 (JJB). One was collected at Abu Handal, in the area now inundated by Lake Nasser, on 8 November 1947 (Al Hussaini 1954*a*). Recent records from Egyptian Lake Nasser include two at Abu Simbil on 24 August 1980 (JJB); six on 26 April 1984 and 15 on 27 April 1984 (MBe, SM); and one, 50 km north of Wadi Halfa on 22 April 1983 (SC, PLM, IP). Nowadays, the Pink-backed Pelican seems to be a regular visitor to Lake Nasser.

ORDER CICONIIFORMES

Family Ardeidae

Generic local names for heron and egret   Balasaan, balashuun, waaq.

Bittern *Botaurus stellaris* WV

Local names   Waaq واق, 'ajjaaj, bughbugh (ND), buubuu.

Rare winter visitor between (early September) mid-October and early May, mainly to the northern portion of the country. Occasional in summer. All Egyptian specimens examined are referable to *B. s. stellaris.*

Meinertzhagen (1930) mentioned the Bittern as common in Egypt during the winter, particularly in the Faiyum and Wadi el Natrun. However, his record of 'several hundreds' resting in reeds near El Qantara on 19 April seems unlikely.

Presently the Bittern is a rare winter visitor to areas with reed-beds in the Nile Delta and Valley (south to Kom Ombo), and generally only solitary birds are observed. There are several recent observations from Suez, Lake Qarun, and Wadi el Rayan. The southern-most Nile Valley record is one at Kom Ombo on 16 February 1982 (SQ). This species may be more locally common than known, as six birds were found in the Port Said market on 2 January 1984 (SMG) and three observed in Wadi el Rayan on 8 December 1985 (SMB, DM). The only known summer record is one at Lake Maryut on 21 June 1909 (GZM).

Little Bittern *Ixobrychus minutus* RB PV WV

Local names Waaq saghiir واق صغير, jamal (ND), maliiha, wibba.

Fairly common breeding resident in Wadi el Natrun, the Nile Delta and northern Valley, the Faiyum, and the Suez Canal area. Fairly common passage migrant and probable winter visitor throughout portions of the country between mid-August and early May. All Egyptian specimens examined are referable to *I. m. minutus*.

The Little Bittern is a common breeding resident of the Nile Delta lakes. This region holds a breeding population of many hundreds, and perhaps well over a thousand pairs (Meininger *et al.* 1986). Small numbers nest locally in reed-beds elsewhere in the Delta, Wadi el Natrun, Suez Canal area, the Faiyum, and Nile Valley (at least south to El Badrshein). It probably also breeds along the newly created lakes in Wadi el Rayan.

In autumn there is a considerable influx of migrants. Along the north coast of Sinai,

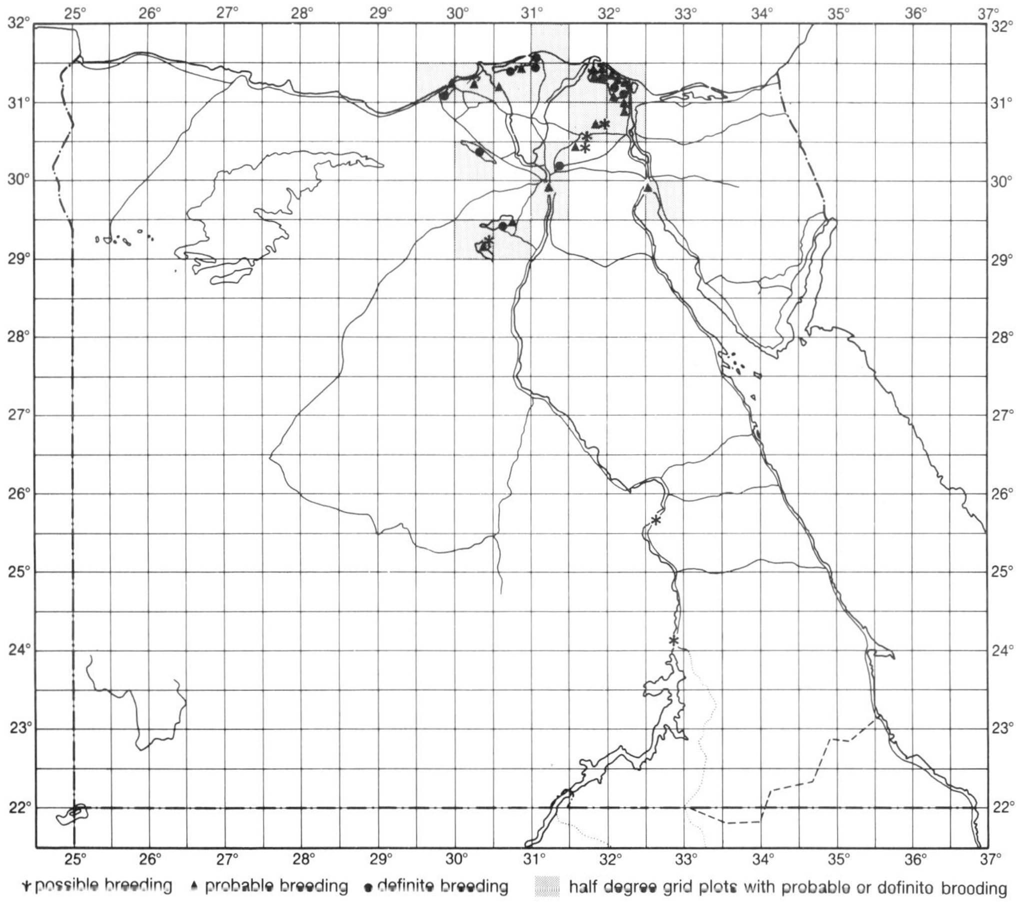

Map B4. Breeding distribution of Little Bittern *Ixobrychus minutus* between 1970–87.

near Zaranikh, a heavy diurnal westward movement has been noted in some years: 1861 on 11 September 1976 and 4172 from 15 August to 10 October 1978, with 2693 on 17 September (Paran and Schlüter 1981); 2426 from 16 August to 24 September 1981, with 793 on 29 August (Petersen and Sørensen 1981*b*); but only 15 from 2 to 24 September 1982 (Baha el Din and Salama 1984). Most of these birds probably go inland, as suggested by an observation of a flock of 40–45 moving south at Ballah on 9 September 1944 (Bodenham 1945*b*), or continue west to the Nile Delta lakes. Migratory movements have also been noted along the North Coast—on 23 September 1980, 15 flying west off El Alamein (WCM) and on 3 October 1980, three were noted 35 km east of Matruh (HB *et al.*).

This species is present throughout the year on the Nile Delta lakes, and the local population is augmented in winter by migrants. Little Bitterns have been reported between late September and late April in small numbers along the Nile Valley south to Aswan, and the species may breed locally. One was noted at Luxor on 22 May 1985 (JS *et al.*). There are a few records, mainly from autumn and spring, from the Western Desert oases of Bahariya, Siwa, and Farafra (Goodman *et al.* 1986; HB *et al.*), and from the Red Sea coast. Some migrants cross the Eastern Desert. Tregenza (1955) found one dead in Wadi Qena in spring and mentioned a record of ten in Wadi Gallaba on an unspecified date. Saleh (1984) reported four birds at Bir Abraq on 3 May 1983. The record of one on 18 February 1983, 101 km east of Qena on the Qena–Safaga road (JJH), may be indicative of early spring movements.

Night Heron *Nycticorax nycticorax* RB? PV WV

Local names Ghuraab al-layl غراب الليل, gharag (Ma), kutub kaali (Ma), waaq ash-shajar.

Probably a rare breeding resident in the Upper Nile Valley. Fairly common passage and winter visitor from mid-August to late May. *N. n. nycticorax* is the form occurring in the country.

The present breeding status of the Night Heron in Egypt is not clear. Rudolf of Austria (1889) noted that on 23 March 1881 'incredible numbers of heavy-looking Night-Herons nested in the tall pines that adorned an artificial mound near the palace' in Shubra, Cairo. Raw (1921) noted a wild breeding colony on the grounds of the Giza Zoological Gardens in 1916 and 1917, and suspected the existence of another colony near Kafr Aydub, Zagazig. Wild nestlings were collected in the Giza Zoological Gardens on 2 November 1915, 1 October 1916, 17 October 1916, 13 July 1917, and 6 October 1922 (GZM). Koenig (1928) mentioned it as one of the most abundant birds in Egypt. On 12 April 1899 he observed a dense concentration of Night Herons, including full-grown young at 'Béllianah' (= El Balyana ?), but made no mention of nests. At the turn of the twentieth century a colony of Night Herons occupied trees in a Luxor hotel garden (Whymper 1909). Meinertzhagen (1930) mentioned breeding colonies in the Faiyum, near Giza, the Delta Barrage, Beni Suef, Luxor, Aswan, and 'other localities', without giving any details. In June 1944 the species was fairly plentiful at Giza and along the Nile near Cairo (CJG). White (1949) observed up to 100 free-living birds, including immatures, in the Giza Zoological Gardens throughout the spring and summer of 1945 but made no mention of a nesting colony.

Presently the Night Heron may breed in small numbers on islands and along the shores of the Nile near Aswan, where adults are observed throughout the year. It may also nest elsewhere along the Nile between Qena and Aswan. Juveniles have been observed south of Nag Hammadi on 20 March 1985, near Luxor on 21 March 1985, and south of Idku on 24 March 1985 (Grieve 1985). These records are of birds fledged in late winter and/or early spring and may be indicative of local breeding or possibly post-fledging dispersal from sub-Saharan breeding grounds. There is no recent evidence that the species breeds in Lower Egypt.

The earliest autumn record of migrants is a group of 35 on 14 August 1955 at Fayid (RN). In September large numbers are observed moving west across northern Sinai. Records from Zaranikh include: 2489 between 15 August and 5 October 1978 (YP, UP); 2546 between 7 and 30 September 1980, with a maximum of 977 on 24 September (Petersen and Sørensen 1981*a*); 2742 between 16 August and 24 September 1981, with a maximum of 377 on 8 September (Petersen and Sørensen 1981*b*); and 1899 between 2 and 24 September 1982 with 531 on 7 September and a maximum of 574 on 21 September (Baha el Din and Salama 1984). There are several September and October observations from the North Coast and the Western Desert, presumably of trans-Saharan migrants. There are also several autumn records of small numbers along the Red Sea coast. Night Herons can be observed, sometimes in considerable numbers, in autumn and winter in the Nile Valley between Cairo and Aswan, in the Nile Delta, and along the Suez Canal. Nicoll (1912*a*) noted 360 on the grounds of the Giza Zoological Gardens on 11 December 1902. Up to 120 were observed near Ismailiya between December 1944 and February 1945 (LB).

Spring migration commences in mid-March and continues at least until mid-May. Tuck (1963) presented a record of a Night Heron observed on 27 May 1962 at sea off Port Said. During spring migration single birds or small flocks are noted throughout the country, including the Western Desert, North Coast, and the Red Sea.

Ringing Single birds ringed in Romania and Czechoslovakia were trapped near Burg el Arab in autumn 1962 (Hoogstraal *et al.* 1964), four from Hungary (Keve and Pátkai 1959; GZF), and two from Yugoslavia (Stromar 1971; Watson 1971) were also recovered in Egypt. One bird ringed as a chick in Israel on 4 July 1979 was recovered in the Port Said market on 26 December 1979 (BEP).

Striated Heron (Green-backed Heron) *Ardeola striata* RB PV

Local names Waaq akhdar واق أخضر, abuu jaqbaan (Ab).

Local breeding resident in mangrove swamps along the coast and on islands of the Red

Sea (*A. s. brevipes*) and uncommon visitor to the Upper Nile Valley (presumably *A. s. atricapilla*).

Nile Valley In recent years the Striated Heron has been recorded with increasing frequency in the Upper Nile Valley, particularly between October and April. These birds presumably originate from Nile Valley populations south of Egypt, although local breeding cannot be ruled out. The following summarizes the majority of known records: Qena—one on 22 May 1984 (JS *et al.*); Luxor—two on 4 October 1984 (Miles 1984), and one on 15 January 1979 (SMG); Isna—one on 2 January 1986 (AS); between Isna and Idfu—four on 9 October 1983 (Hume 1983); south of Kom Ombo—one on 9 October 1983 (Hume 1983) and three on 7 October 1985 (Hume

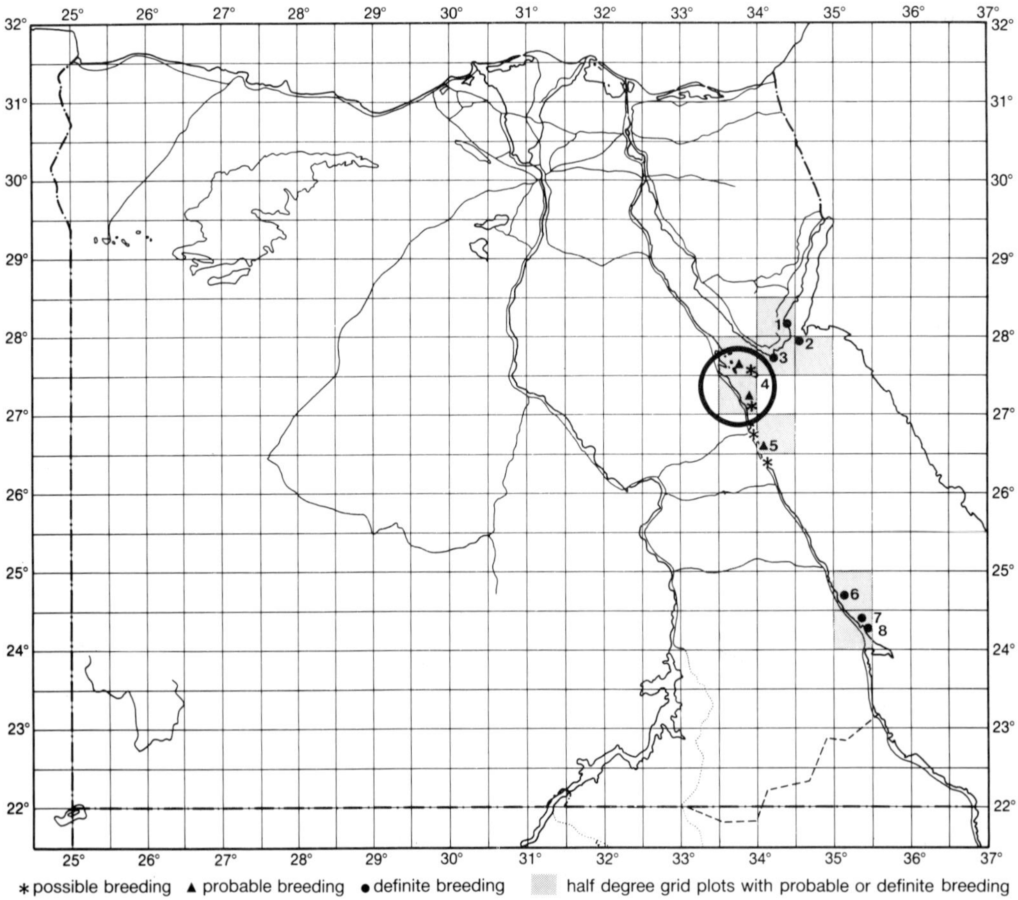

Map B5. Breeding distribution of Striated Heron *Ardeola striata* between 1970–87.
Breeding localities (see text for details): (1) Shora el Manqata. (2) Gezira Tiran. (3) Ras Muhammad. (4) Islands mouth Gulf of Suez. (5) Mangroves 23 km south of Safaga. (6) Gezira Wadi Gimal. (7) Gezira Showarit. (8) Gezira Umm Ladid.

1985); and Aswan—single birds on 11 October 1983 (Hume 1983), 3 November 1982 (Davies 1982), 9 December 1985 (HD), 25 February 1986 (MCM), 25 March 1986 (JW), and 24 April 1983 (SC, PLM, IP), three on 13 August 1986 (EE *et al.*), four on 8 October 1985, and two on 9 October 1985 (Hume 1985) and 9 December 1985 (HD).

Red Sea This species has been found breeding in the Red Sea and Gulf of Aqaba at the following localities: Shora el Manqata (north of Nabq)—one egg collected on 24 August 1976 (Goodman and Storer 1987), at least 20 active nests in 1973 (SS), and a nest with three young on 10 September 1980 (JB); Gezira Tiran—one or two pairs and one old nest with two unhatched eggs between 14 and 30 March 1979 (RF); Ras Muhammad—two used nests and three newly fledged young on 25 August 1982 (SMB); and Gezira Wadi Gimal—two nests on 26 June 1984, Gezira Showarit—one nest on 18 June 1984, and Gezira Umm Ladid—one nest on 18 June 1984 (Goodman and Storer 1987). An estimated 40–60 pairs breed along the Egyptian Red Sea coast. After the nesting season Striated Herons disperse from the immediate vicinity of nesting localities. The species is frequently recorded throughout the year in small numbers along the Red Sea coast but rarely at the northern limits of the Gulfs of Suez and Aqaba.

Squacco Heron *Ardeola ralloides* RB PV WV

Local names Waaq abyad واق أبيض, sabiisa, waaq abyad sughayyar.

Rare and local breeding resident in the Upper Nile Valley. Fairly common on passage and scarce winter visitor from mid-August to mid-May. Monotypic.

There is no documentation of the Squacco Heron breeding in the Nile Delta. Ticehurst (1912) saw 50 birds on 15 May 1909 in a 'tamarisk swamp' near Lake Maryut and collected a female that was in breeding condition. He mentioned that according to native people this species bred locally; however, they may have confused them for Cattle Egrets. Meinertzhagen (1930) mentioned that Squacco Herons were observed in the Nile Delta in every month of the year but no nest had been found. A bird taken on 25 May 1936 at Lake Idku was 'breeding' (BMNH). Etchécopar and Hüe (1967) erroneously indicated that this species nests throughout the Nile Delta and Valley.

Presently, the only area in Egypt where Squacco Herons are known to breed is in Upper Egypt, near Aswan. Between 1978 and 1986 up to ten pairs nested in a colony of Cattle Egrets and Little Egrets on Kitchener's Island. A colony of 40 nests was discovered near the Old Dam, Aswan, on 25 June 1981; by April 1983 this colony appeared to have been abandoned (Meininger *et al.* 1986). Small numbers of Squacco Herons can be observed throughout the Nile Valley and Delta lakes during the summer months and it presumably breeds at other localities.

In autumn considerable numbers are observed at Zaranikh migrating in a westerly direction: 2128 between 13 August and 29 September 1973 and 4458 between 15 August and 5 October 1978 (YP, UP); 1287 between 7 and 30 September 1980 (Petersen and Sørensen 1981*a*); 5487 between 16 August and 24 September 1981 (Petersen and Sørensen 1981*b*); and 2711 between 2 and 24 September 1982 (Baha el Din and Salama 1984). Starting in mid-August large numbers are present in the Nile Delta and Valley and there is a substantial increase in September and October. On 20 September 1980, 14 were observed flying west near El Alamein (WCM). The only autumn records from the Western Desert are single birds at Bahariya between 21 August and 16 September 1983 (HB *et al.*). A flock of 16 was noted resting in the desert 70 km south-west of Aswan on 10 October 1986 (EH, PK). It is rare in autumn along the Red Sea coast.

Most Squacco Herons winter further south in Africa and between November and February only small numbers are present in the Nile Delta and Valley, and the Faiyum. Notable spring movements begin in late February, peak in March and April, when many hundreds are present in the Nile Valley, and continue into May. In spring small numbers have been recorded in most Western Desert oases, along the North Coast, the Red Sea, and in Sinai.

Cattle Egret *Egretta ibis* RB PV WV

Local names Abuu qurdaan أبو قردان, gharnuug (Ma).

Common breeding resident in the Nile Valley and locally in the Nile Delta and Suez Canal area. It is also a passage and winter visitor in unknown numbers throughout the country. *E. i. ibis* is the form occurring in Egypt. Eggs are laid throughout the year, though generally between February and July.

Western Desert In recent years Cattle Egrets have been observed in all major oases. Hundreds of birds are often noted in Kharga and Dakhla. On 25 April 1981, 250 flew from a night roost or breeding colony in Kharga, where according to local inhabitants this species breeds in trees (Goodman *et al.* 1986). Breeding colonies in Bahariya and Dakhla cannot be excluded but to date there appears to be no documentation of breeding in the Western Desert (*contra* Cramp and Simmons 1977).

Nile Delta and Valley Around the turn of the twentieth century Cattle Egrets were thought to be almost extinct in Egypt due to wanton hunting but after legal protection was enacted and enforced, they recovered in number (Raw 1921; Koenig 1928). Raw (1921) and Meinertzhagen (1930) mentioned several large colonies in the southern Nile Delta, without giving exact localities. On 17 July 1917 a colony composed of 2672 nests was found near Zagazig (Ticehurst 1931). Up until 1978 there was a large colony near the Alexandria Zoo (Kattinger 1970; BO), and until the 1970s another near Ityai el Barud, 25 km south-east of Damanhur (Meininger *et al.* 1986). At present there are several Cattle Egret breeding colonies in the southern Nile Delta, near Suez,

TABLE 6.1. Locality and number of nests of recent Cattle Egret colonies

Locality	1977	1978	1979	1980	1981	1982	1983	1984	1985	1986	1987
1. Alexandria Zoo*	+	+	–	–	–	–	–	–	–	–	
2. Suez				+	30	100	+	+	+		
3. Delta Barrage				>200	>200						
4. El Qalag				+	>1000						
5. Giza Zoo	2500			2636	1733	1850	1400	1110	1242	1439	1604
6. Shakshuk										20	
7. El Faiyum	600				>1000					1445	
8. Beni Suef–El Wasta				>400							
9. East of Maghagha							'100s'				
10. Samalut					750–1000	>500	+				
11. Abu Tig								>300			
12. Tima							100				
13. Nag Hammadi						>300	>50				
14. Qus								>40			
15. South of Luxor						50–100					
16. Kitchener's Island		385	300	180	200–300	200–300	800	+	+	+	
17. Gezira Saluga									>100		
18. Aswan, old dam					50		–				

* 'Thousands' bred until 1978 (BO)
\+ = breeding number unknown; – = colony abandoned; blank space = no information available

in the Faiyum, and throughout the Nile Valley (Table 6.1). In the past decade or so this species has declined in numbers in the greater part of the Nile Delta, which appears to be the result of local extensive use of pesticides (Mullié and Meininger 1985). However, the number of colonies in the Nile Valley and the Faiyum seems to have increased. The Cattle Egret is observed throughout the year in the Nile basin but is distinctly more common in the Nile Valley than in the Delta. There are several records from Lake Nasser, e.g. 20 at Abu Simbil on 27 April 1984 (MBe, SM). Roosting sites, often containing large numbers of birds, are common, e.g. 8000–9000 on 2 October 1985 south of Sohag (Hume 1985).

Eastern Desert and Red Sea Small numbers are regularly recorded throughout the year along the entire Red Sea coast, generally near human settlements. There are several records of migrating birds from the central Red Sea (Bulman and Beale 1944; Elliott and Monk 1952; Tuck 1961). They are also noted during the winter months in remote and barren areas of the Eastern Desert, often in association with herds of domestic and wild animals (SMG, JJH).

Sinai In autumn the Cattle Egret is a regular migrant along the north coast of Sinai. Near Zaranikh records include: 324 between 15 August and 5 October 1978 (YP, UP); 295 between 7 and 30 September 1980 (Petersen and Sørensen 1981*a*); and 577 between 16 August and 24 September 1981 (Petersen and Sørensen 1981*b*). These migrants, at least in part, presumably augment Egyptian populations during the non-breeding season.

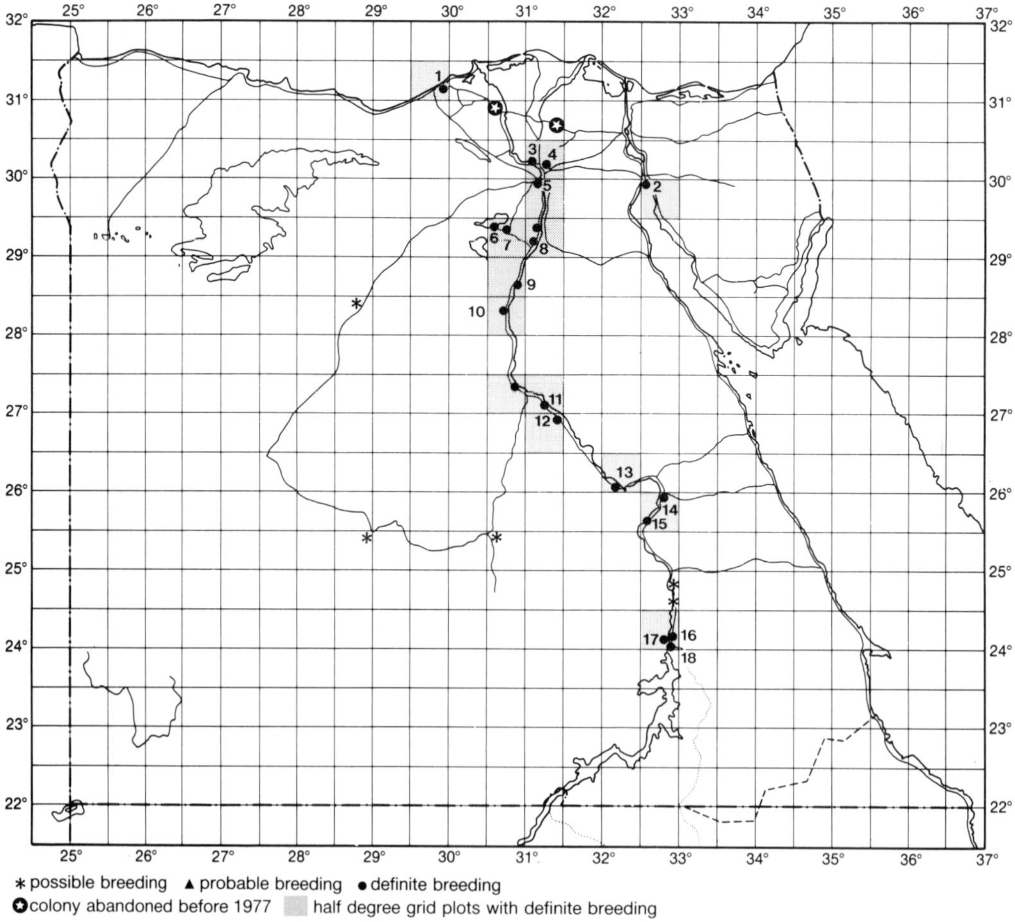

Map B6. Breeding distribution of Cattle Egret *Egretta ibis* between 1977–87. Numbers refer to colonies mentioned in Table 6–1.

Western Reef Heron *Egretta gularis* RB

Local names Balashuun as-sakhr بلشون الصخر, abuu saah, abuu sarraah, balashuun bahrii, gharnuug (Ma).

Scarce and local breeding resident, often in mangrove swamps, along the Red Sea coast, and on islands. The form occurring in Egypt is *E. g. schistacea*.

Western Reef Herons breed in numerous small colonies along the Egyptian Red Sea coast and on islands. The largest known colony is at Shora el Manqata, north of Nabq, where 40–60 pairs breed annually. Egg laying commences in the south in mid-March and in the north in late April and continues to late July. The total Egyptian breeding population is estimated to be 130–200 pairs.

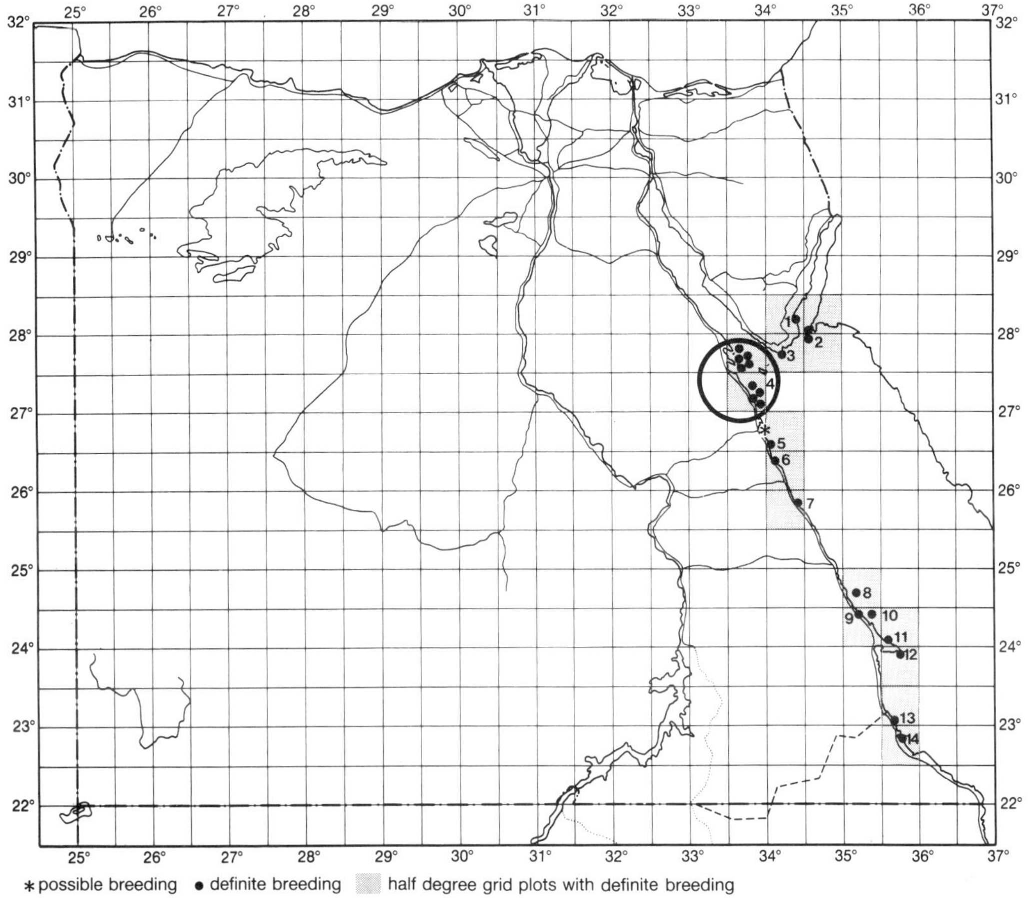

Map B7. Breeding distribution of Western Reef Heron *Egretta gularis* between 1965–87. *Breeding localities* with most recent estimate of breeding pairs: (1) Shora el Manqata (1978: 60). (2) Gezira Tiran (1973–79: 2–3). (3) Ras Muhammad (1985: at least 1). (4) Islands mouth Gulf of Suez (1983: 25–30). (5) Mangroves 23 km south of Safaga (1983: 15). (6) Mangroves 42 km south of Safaga (1983: 3). (7) Sharm el Bahari (1982: at least 2). (8) Gezira Wadi Gimal (1984: 3). (9) Mangroves 13 km north of Hamata (1984: 2). (10) Gezira Showarit (1982: 1). (11) Ras Banas north (1965: present). (12) Ras Banas east (1965: present). (13) Mangroves 15 km south of Shalatein (1985: 5). (14) Marsa Sha'b (1985: at least 1).

This species is observed throughout the year along the entire Egyptian Red Sea coast, with a few birds migrating in the non-breeding season as far north as Suez and Elat. Based on field records no difference was found in the ratio of white : intermediate : dark morphs between northern and southern populations inhabiting the Egyptian coast (Table 6.2). Along the Sudanese coast dark-phase birds outnumber white birds, by a ratio of 37:15:48 ($n = 27$), respectively (Madden 1927).

TABLE 6.2. Distribution of Western Reef Heron colour phases along the Egyptian Red Sea coast and the Gulfs of Aqaba and Suez

Locality	White	Intermediate	Dark
Gulf of Aqaba	182(86.6%)	3(1.4%)	25(12.0%)
Gulf of Suez	21(91.3%)		2(8.7%)
Hurghada to Quseir	465(83.1%)	7(1.3%)	87(15.6%)
Quseir to Halaib	242(82.3%)	13(4.4%)	39(13.3%)

Little Egret *Egretta garzetta* RB PV WV

Local names Balashuun abyad بلشون أبيض, abuu baliiqa, bayaadii, shin'aal (ND).

Scarce and local breeding resident. Fairly common passage migrant and winter visitor throughout the country from mid-August to mid-May. The form occurring in Egypt is *E. g. garzetta*.

Meinertzhagen (1930) considered the Little Egret an abundant breeding resident throughout the Nile Delta, Upper Egypt, and the Faiyum. However, he only specifically mentioned a wild colony on the grounds of the Giza Zoological Gardens. An egg labelled 'Giza' and taken on 17 April 1922 is presumably from this colony (BMNH). In the GZM there is a nestling taken in July 1913 at Asirat. White (1949) counted small numbers of wild birds on the Giza Zoo grounds during the summer of 1945 but did not mention breeding. In May 1950 a few Little Egrets were also reported to breed in the Cattle Egret colony near the Delta Barrage (JCC). There is no recent evidence of Little Egrets breeding in the Nile Delta (Meininger *et al.* 1986), and the distributions mapped by Etchécopar and Hüe (1967) and Cramp and Simmons (1977) are incorrect.

At present the only known Egyptian colonies are in Wadi el Rayan—13 active nests in mid-July 1984 (SMB); Gezira Saluga, near Aswan—at least 100 pairs in September 1985 (AeA, MAM); and Kitchener's Island, Aswan—27 nests in 1978, 120 in 1983, and 200 in 1984. A colony near the Old Dam, Aswan, that contained 600 nests in 1981, was abandoned by 1983. In April 1983 about 75 Little Egrets were observed on a rocky island just south of Abu Simbil, Lake Nasser, and may have been a breeding colony (Meininger *et al.* 1986).

Small numbers are observed throughout the summer months in the Nile Delta lakes, the Faiyum, the Suez Canal area, and along the Nile Valley. In autumn substantial migratory movements have been noted along the north coast of Sinai. Records from Zaranikh include: 1548 between 13 August and 29 September 1973 and 3519 between 15 August and 5 October 1978 (YP, UP); 2367 between 7 and 30 September 1980 (Petersen and Sørensen 1981*a*); 4239 between 16 August and 24 September 1981 (Petersen and Sørensen 1981*b*); and 2743 between 2 and 24 September 1982 (Baha el Din and Salama 1984). From late August to mid-May large numbers are present in the Nile Delta and Valley, including the Faiyum and Lake Nasser. Thousands pass through these areas on autumn and spring migration and many remain for the winter. Small numbers are regularly recorded during migration in the Western Desert and along the Red Sea coast.

Ringing At least two birds ringed in the northern Black Sea have been recovered in autumn in Egypt (Kistchinski 1978). One ringed on 31 July 1964 at Tel Aviv, Israel, was recovered in Egyptian Nubia sometime during 1965–6 (Watson 1971).

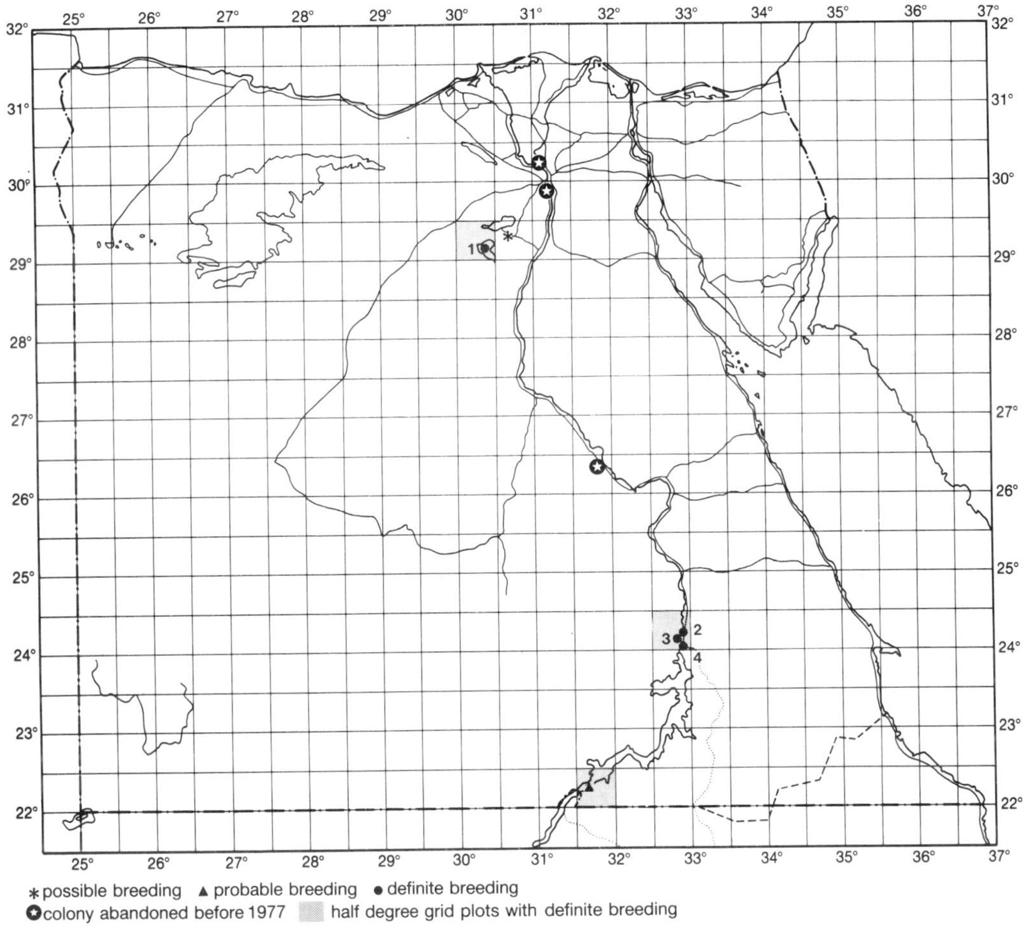

Map B8. Breeding distribution of Little Egret *Egretta garzetta* between 1977–87.
Breeding colonies (see text for details): (1) Wadi el Rayan. (2) Aswan, Kitchener's Island. (3) Aswan, Gezira Saluga. (4) Aswan, Old Dam.

Great White Heron *Ardea alba* PV WV

Local names Balashuun abyad̲ kabiir بلشون أبيض كبير, ghayt̲ii, ibn al maaya, riit̲ii (NC), waaq abyad̲.

Scarce passage migrant and winter visitor from early September to late March (mid-May), mainly in the northern portion of the country. *A. a. alba* is the form found in Egypt.

The Great White Heron is a rare migrant during September and October along the north coast of Sinai, with a maximum of 32 birds between 11 and 20 October 1979 (Petersen and Sørensen 1980). The earliest autumn record is six at El Kabrit on 5 September 1952 (AP). Some birds apparently cross the northern Red Sea (Tuck 1961). Between October and March birds concentrate in the Nile Delta lakes, where 110 were counted between December 1979 and January 1980 (Meininger and Mullié 1981*b*). In the winter a few occur in the Nile Valley and along the Red Sea coast. Small numbers have been noted in autumn and spring along the Suez Canal, the Nile Valley south to Aswan, the Faiyum, and along the Red Sea coast. There are no reports from the Western Desert (Goodman *et al.* 1986). The latest spring records are single birds on 24 April 1964 at Matruh (HCK) and on 16 May 1985 in the Faiyum (JS *et al.*).

Ringing Three birds ringed in the northern Black Sea have been recovered in winter in Egypt (Watson 1971; Kistchinski 1978; GZF).

Grey Heron *Ardea cinerea* CB PV WV

Local names Balashuun ramaadii بلشون رمادي, naghaaj (ND).

Casual breeding resident. Fairly common migrant and winter visitor throughout the country from mid-August to early May, and scarce non-breeding summer visitor. The subspecies occurring in Egypt is *A. c. cinerea.*

Between 1909 and 1911 and in 1918 a wild pair of Grey Herons bred on the grounds of the Giza Zoological Gardens (Nicoll 1912*a*; Raw 1921). This is the only documented nesting record of this species in Egypt. There is no proof of breeding on Red Sea islands or along the coast (cf. Nicoll 1919; Meinertzhagen 1930; El Negumi *et al.* 1950). Frazier *et al.* (1984) found a nest containing four eggs on 27 May 1984 in a mangrove swamp at Marsa Suma, which they identified as this species. However, it seems likely that the nest belonged to a Goliath Heron (Goodman and Storer 1987). On 21 May 1984, 23 km south of Safaga, an adult Grey Heron flew out of a single dense clump of mangrove where both Western Reef Herons and Spoonbills were breeding; however, no Grey Heron nest was located (SMG).

Between mid-August and early May this species is common throughout the country, including Western Desert oases and along the Red Sea coast. Large numbers migrate along the north Sinai coast. Records from Zaranikh include: 1563 between 13 August and 29 September 1973 and 6149 between 15 August and 5 October 1978 (YP, UP); 3190 between 7 and 30 September 1980 (Petersen and Sørensen 1981*a*); 4273 between 16 August and 24 September 1981 (Petersen and Sørensen 1981*b*); and 2808 between 2 and 24 September 1982 (Baha el Din and Salama 1984). Along the North Coast some autumn passage has been noted: 78 on 20 and 120 on 23 September 1980 flying west at El Alamein (WCM). In October 1979 up to 5000 were recorded along the Nile between Cairo and Aswan (FV).

Grey Herons winter in the Nile Delta and Valley, where they are present in substantial numbers until late March. Thousands concentrate around portions of Lake Manzala that have been diked and turned into temporary fish farms. In such areas they are heavily persecuted, at least in part because of the large number of fish they reputedly consume. Between May and early August small numbers, mainly immatures, are regularly recorded in the Nile Delta and Valley and along the Red Sea coast.

Ringing There are a number of recoveries in Egypt of birds ringed in and between the Astrakhan and Darwin Reserves, USSR (Watson 1971; Kistchinski 1978; BEP; GZF).

Purple Heron *Ardea purpurea* PV WV

Local names Balashuun urjuwaanii بلشون أرجواني, hajaf, jaha̲af (Su), jah̲fa, maalik al-h̲aziin, naghaaj (ND).

Fairly common passage visitor throughout the country in autumn from (mid-) late August to late October (late November) and in spring from late February to mid-May (early June). It is a rare winter and summer visitor to the Nile Valley. *A. p. purpurea* is the form occurring in Egypt.

The Purple Heron is a regular autumn migrant, mainly to northern Sinai, and the Nile Delta and Valley. The earliest autumn record is of one on 16 August 1981 at Zaranikh (IP, UGS). In autumn considerable numbers migrate along the north coast of Sinai. Counts at Zaranikh include: 1031 between 13 August and 29 September 1973 and 5349 between 15 August and 5 October 1978 (YP, UP); 2401 between 7 and 30 September 1980 (Petersen and Sørensen 1981*a*); 3437 between 16 August and 24 September 1981, with a maximum of 985 on 11 September (Petersen and Sørensen 1981*b*); and 1168 between 2 and 24 September 1982 (Baha el Din and Salama 1984). Small numbers have been recorded migrating westward along the Mediterranean coast west of Alexandria. The only autumn record from the Western Desert is one bird between 2 and 4 October 1962 in Dakhla (Scharlau 1963). In September and October parties of up to ten are regularly observed in the Nile Delta and Valley, and the Faiyum, and generally single birds occur along the Red Sea coast.

Most migrants pass to areas south of Egypt, although a few are observed throughout the winter in the Nile Valley, mainly along the upper portion. In spring Purple Herons migrate through Egypt in small flocks from late February to mid-May, peaking in April. At this time of the year, the species is observed throughout the country, including small numbers in the Western Desert and along the Red Sea coast. Late spring records include single birds on 31 May 1986 at Lake Qarun (DM, WCM) and on 2 June 1959 at Imbaba (FMNH).

Ringing At least four birds have been recovered in Egypt in September and March which were ringed in the northern Black Sea (Watson 1971; Kistchinski 1978) and western Turkey (EDB).

Goliath Heron *Ardea goliath* RB PV

Local names Balashuun jabbaar بلشون جبار, abuu al-'ayzaar, mirra, subayṯir.

Rare and local breeding resident along the southern Egyptian Red Sea coast, and rare visitor to the northern portions of the Red Sea and Upper Nile Valley. Monotypic.

Nile Valley There are several recent observations of Goliath Herons in the Upper Nile Valley, which are presumably birds dispersing from breeding areas south of Egypt. These records include: Hawatka, near Asyut—one on 15 April 1978 (RJM, LRP); Luxor—one on 29 September 1981 (Bijlsma and de Roder 1982); Idfu—one on 23 March 1986 (UA); Aswan—one on 22 July 1981 (CB, GT), two on 1 October 1981 and one on 3 October 1981 (Bijlsma and de Roder 1982), one on 28 February 1987 (PdB), one on 14 March 1985 (Kinzelbach 1986), one on 21 March 1986 (UA), and three on 27 March 1982 (Meilstrup and Sørensen 1983).

Red Sea Until recently there were only a few records of this species from Egypt, all from the Red Sea coast (Von Heuglin 1873; Jourdain and Lynes 1936; Bulman 1944). Kaiser (in le Roi 1923) observed one Goliath Heron on 10 April 1892 at Ras Muhammad and in March 1895 discovered a nest with two young on Gezira Umm el Heimat. Meinertzhagen (1930) rejected these records. However, considering the recent evidence of breeding along the southern Egyptian Red Sea coast, Kaiser's observations are acceptable.

More recently, nesting has been documented in mangrove swamps along the southern Egyptian Red Sea coast. Records include: 15 km south of Shalatein—a nest with one nestling (about two weeks old) on 6 April 1985 (SMG, PLM); Marsa Suma—a nest with two young on 27 May 1982 (Goodman and Storer 1987); and Marsa Sha'b—a nest with three eggs on 19 June 1984 (Goodman and Storer 1987). It is also thought to breed along the Sudanese Red Sea coast (Madden 1927).

In recent years this species has been observed with increasing frequency along the Red Sea coast and Gulf of Suez. Generally single birds are recorded; although at Suez up to five Goliath Herons were observed between 14 September and 27 October 1981 (Bijlsma and de Roder 1982) and eight on 5 April 1982 (Wimpfheimer *et al.* 1983). There are several recent sightings from the southern Sinai coast. Intensive field work at Elat, Israel, has not produced a single verifiable record (HSh; *contra* Krabbe 1980).

Family Ciconiidae

Generic local names for stork Laqlaq, 'anaz.

Yellow-billed Stork (Wood Stork) *Mycteria ibis* (PV)

Local name Laqlaq asfar al-minqaar لقلق أصفر المنقار.

Occasional summer visitor to Egypt. The type specimen was taken in the country. Monotypic.

Magaud d'Aubusson (1892) mentioned that the Yellow-billed Stork could occasionally be met with in Upper Egypt during the inundation. Flower (1933) noted that before 1913 this species was observed with some frequency in the Upper Nile Valley between May and September and small flocks were occasionally found in the 'Qena and Girga [= Sohag] Provinces' in July and August. Twelve were noted on 21 October 1957 along the Nile opposite El Balyana and four the next day near Luxor (Horváth 1959). On 29 November 1982, one was found dead on Gezira Qeisum (Baha el Din 1984*a*). A specimen housed in the Institute of Zoology, Asyut University, Aswan Office, was found dead near the Aswan High Dam (*c.*1980), and one was observed near Aswan on 14 March 1985 (Kinzelbach 1986). Flocks of 21 birds were noted flying north near Abu Simbil on 26 and 27 April 1984 (MBe, SM). A single bird was noted near Abu Simbil on 19 August 1986 (EE *et al.*).

Black Stork *Ciconia nigra* PV (WV)

Local names Laqlaq aswad لقلق أسود, abuu butayna (Ma), 'anaz (Ma), 'anaz aswad, balazaan, wizz 'iraaqii.

Scarce migrant in autumn from (mid-) late September to mid-November and in spring from late February to late May (Fig. 6.3). Rare winter and summer visitor. Monotypic.

In autumn small flocks of Black Storks are regularly observed in the Upper Nile Valley, Eastern Desert, and along the east coast of Sinai. Apparently most birds cross from southern Sinai to the African continent via the Gulf of Suez. During systematic counts of migratory birds at Suez between 4 September and 5 November 1981, only nine Black Storks were noted (Bijlsma 1983). Three were observed from 18 October to 30 November 1952 at El Kabrit (AP). The largest flocks of autumn migrants recorded in recent years were 53 on 25 September and 99 on 3 October 1984 at Ghubbet el Gemsa (MCJ *et al.*), 43 on 9 October 1985 near Aswan (Hume 1985), and 46 on 26 October 1986, 17 km south of Hurghada (SMG, JJH). In the autumn this species is rare in the Nile Delta and northern Valley. The only relatively recent winter records are one on 20 December 1952 at Cairo (HCK) and one on 20 January 1979 near Aswan (Meininger *et al.* 1979).

The earliest spring records are single birds on 22 February 1953 at El Kabrit (AP) and on 29 February 1984 between Qift and Quseir (SG). Spring migration continues at least until mid-May and most records are from the southern Nile Valley, the Eastern Desert, Suez, and Sinai. The only Western Desert record is one at El Maghra on 18 May 1975 (Goodman *et al.* 1986). Between 19 March and 14 May 1982, 265 were

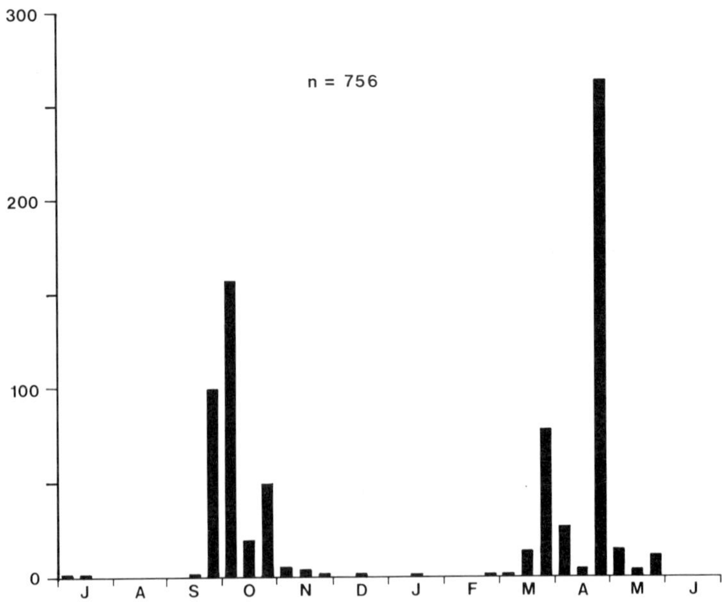

FIG. 6.3. Histogram of the numbers of Black Storks recorded during 10-day periods throughout the year.

seen at Suez, 225 of which passed between 16 April and 5 May (Wimpfheimer *et al.* 1983). Other records of large spring flocks include: Suez—43 flying north on 28 March 1983 (PLM) and at least 60 flying north on 29 April 1916 (Boyd 1917); Ghubbet el Zeit—69 on 29 April 1983 (MCJ *et al.*); and Hurghada—104 on 24 March 1982 and 117 the following day (Wimpfheimer *et al.* 1983). Eleven were seen at San el Hagar in the Nile Delta on 25 May 1984 (RJM, LRP). Single birds noted in the Faiyum on 10 July 1981 (CB, GT) and one north of the Little Bitter Lake on 20 July 1985 (BH *et al.*) were presumably summering locally.

Ringing Two birds ringed in the Soviet Baltic region were recovered near Lake Burullus and in the central Nile Valley (Kistchinski 1979).

White Stork *Ciconia ciconia* PV (WV)

Local names Laqlaq abyad̲ لقلق أبيض, al-h̲aajj laqlaq, 'anaz, baaja, balaaraj, badshaa, jiija (ND), naadsha, raahu (Ma).

Common autumn migrant from late July to late October (mid-November) and in spring from mid-February to late May (mid-June). There are a few winter and summer records. The nominate form is the one occurring in the country.

Egypt is situated along a major migratory pathway between the White Stork's Eurasian breeding and African wintering grounds (Schüz 1959; Reed and Lovejoy 1969). In autumn large flocks of White Storks move across central and southern Sinai, the southern Gulf of Suez, and the Eastern Desert to the general Qena region, where they follow the Nile Valley south (Fig. 6.4). The bulk pass between mid-August and mid-September. Records of large flocks include: 10 000 at Ras Gemsa on 18 August 1979 (SMB), an estimated 100 000 crossing the southern Gulf of Suez on 31 August 1957 (Tuck 1964), 25 000–30 000 crossing the Gulf of Suez north of Hurghada on 5 September 1951 (Stamation 1951), and 18 600, 24 km north of Hurghada on 12 September 1980 (Meininger *et al.* 1981). Only eight birds were recorded during systematic bird counts at Suez between 4 September and 5 November 1981 (Bijlsma 1983). Small numbers of White Storks have also been recorded in autumn in the Western Desert, and the Nile Delta and northern Valley. Relatively recent midwinter records include one bird on 6 December 1970 in 'New Nubia' (USNM), nine on 12 December 1980 at the Delta Barrage (SV), two on 13 December 1980 near Dahshur

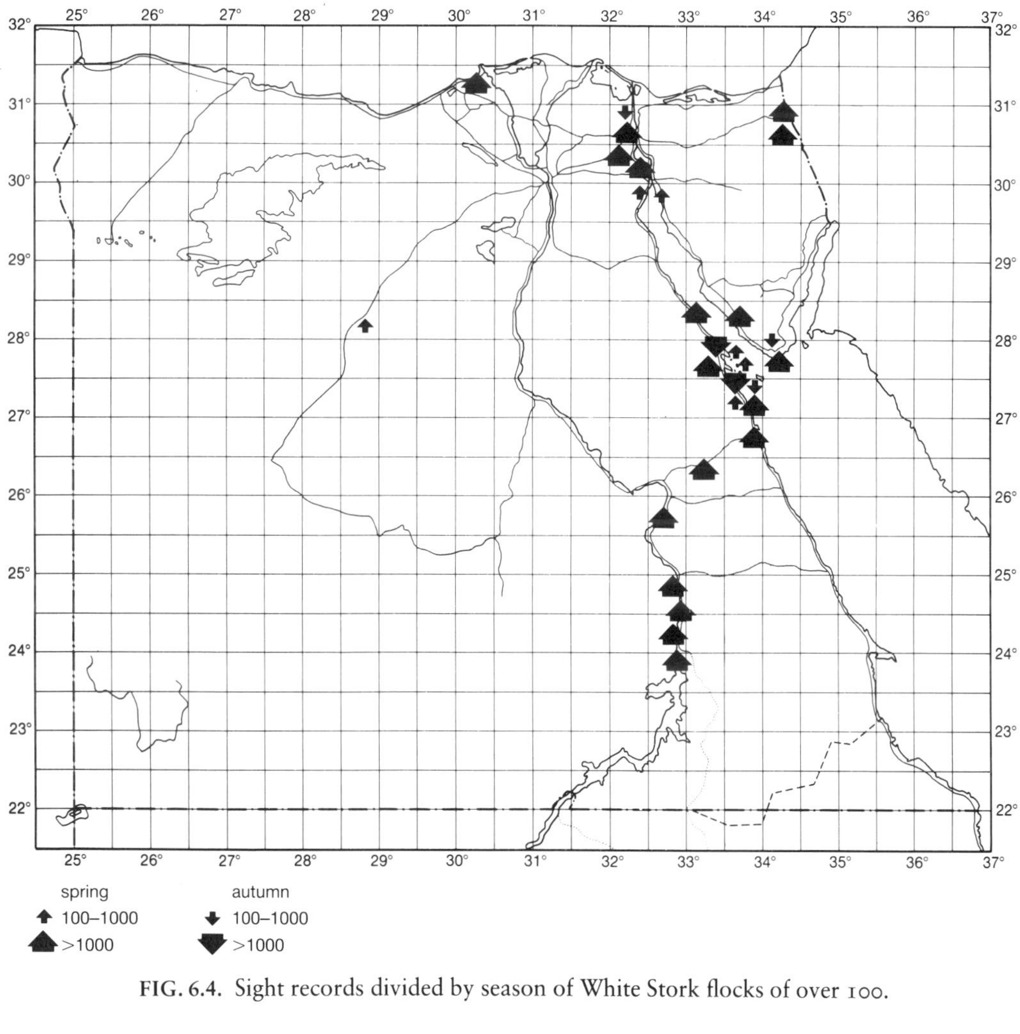

FIG. 6.4. Sight records divided by season of White Stork flocks of over 100.

(RW), seven in a group of Grey Herons on 13 January 1954 at Qena (HCK), and two on 31 January 1965 at Tushka (Reed and Lovejoy 1969).

The earliest known record of spring migratory movements is 600 birds on 14 February 1984 at Ras Budran (SH, DM). Spring migration peaks in March and April, and continues into May. The main spring pathway comes up through the southern Nile Valley, and then across the Eastern Desert, Gulf of Suez, and Sinai. The spring route through the Eastern Desert and Sinai appears to be slightly further north than in autumn (Koch *et al.* 1966). In spring substantial numbers are observed in the Western Desert and northern Nile Valley. Large concentrations, sometimes thousands, are also noted along the Suez Canal, mainly in April and May. Some recent records of

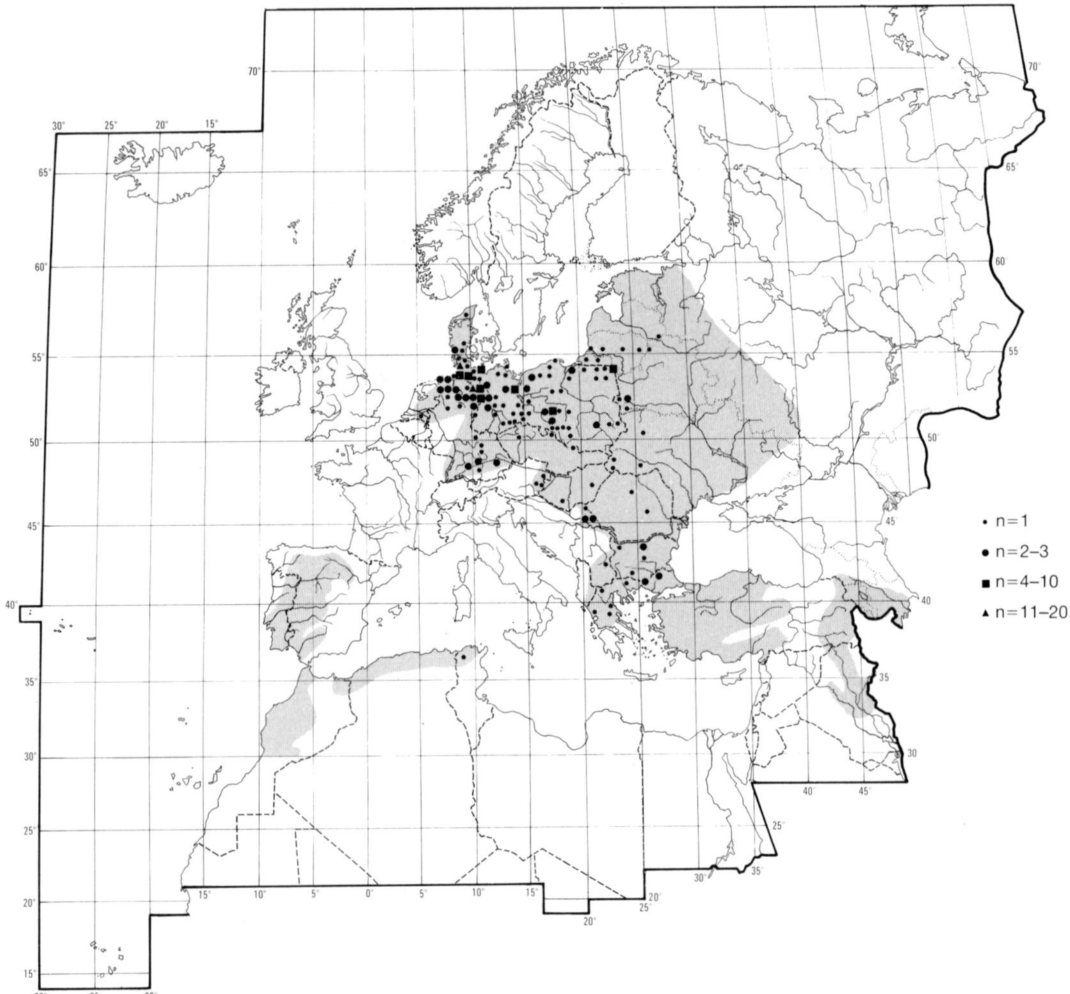

FIG. 6.5. Origin of White Storks (n=227) recovered in Egypt. All birds ringed as nestlings. Shaded area is breeding range based on Cramp and Simmons (1977).

exceptionally large spring flocks are 12 000 near Hurghada on 25 March 1980 (JV), 7300 at Aswan on 26 March 1982 (Meilstrup and Sørensen 1983), 5300 moving north-east over Wadi Feiran on 1 April 1983 (PLM), 8000 at Suez on 2 April 1982 (Wimpfheimer *et al.* 1983), 5000–10 000 crossing the southern Gulf of Suez on 3 April 1962 (Tuck 1964), 5000–10 000 on the beach near Ras Dib on 17 April 1956 (Phillips and Rosenberg 1959), and 3840 at Lake Bardawil on 30 April 1985 (Fry *et al.* 1985). There is no evidence that significant numbers of White Storks summer in the Nile Delta as has been proposed by Kania (1985) based on ringing recoveries. Presumably the time-lag between the true recovery date of the ringed bird and reporting date produced this temporal artefact. A few birds have been recorded throughout the summer in the Nile Delta and Valley.

Ringing There are well over 300 recoveries in Egypt of birds ringed in Austria, Bulgaria, Denmark, The Federal Republic of Germany, German Democratic Republic, Greece, Hungary, Latvia, Lithuania, Poland, Romania, The Netherlands, Tunisia, western USSR, and Yugoslavia (Fig. 6.5). Bedouins in Sinai and residents in Upper Egypt collect rings from exhausted or dead White Storks for use as amulets and for their magical properties (Mackintosh 1941; Keve and Pátkai 1959; Johansen 1962; Preuss 1963; Stromar 1970; Watson 1971; Muzinic 1984–5; EDB; GS; GZF; JG; RP; RSc).

Family Threskiornithidae

Generic local name for ibis Abuu minjal.

Glossy Ibis *Plegadis falcinellus* PV (WV)

Local names Abuu minjal aswad ابو منجل أسود, ma'aza al-maaya.

Scarce passage visitor in autumn from mid-August to mid-November and fairly common in spring from early February to early (late) May (Fig. 6.6). Irregular winter visitor. Monotypic.

There is no evidence that Glossy Ibis breed in the country (*pace* Etchécopar and Hüe 1967). The only known recent summer record is one at Aswan on 21 July 1981 (CB, GT).

In autumn this species has been noted in small numbers migrating along the north Sinai coast. Records from Zaranikh include: 121 between 15 August and 5 October 1978 (YP, UP), and 81 between 16 August and 24 September 1981, with a maximum of 24 on 25 August (Petersen and Sørensen 1981*b*). The majority of other autumn records are from the Nile Valley, where flocks have been recorded between mid-August and mid-November, with a peak in October. The species is rarely observed in the Nile Delta, along the Suez Canal, and south Sinai. There are no known autumn records from the Western Desert and Red Sea coast south of Hurghada.

In December and January small flocks are occasionally observed in the Upper Nile

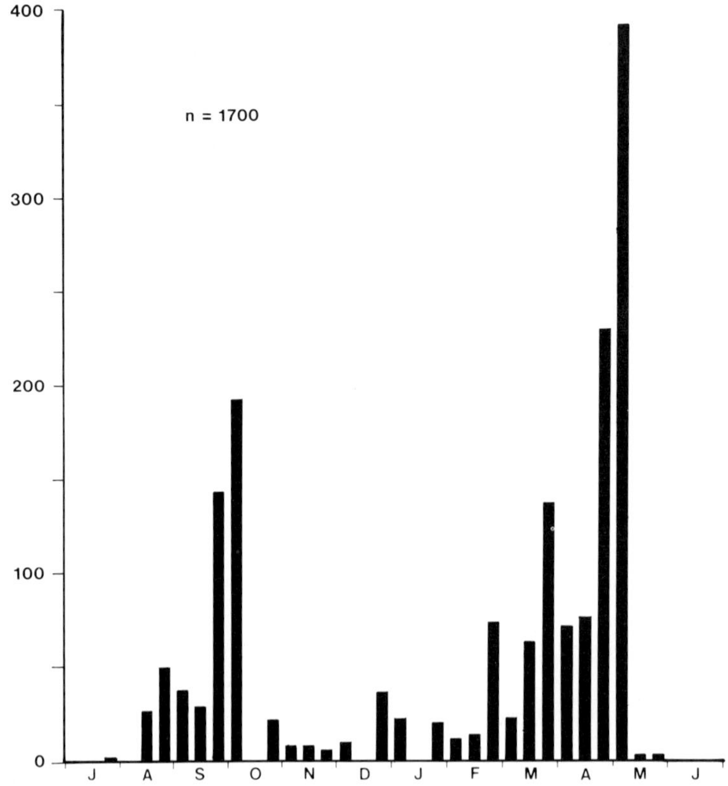

FIG. 6.6. Histogram of the numbers of Glossy Ibis recorded during 10-day periods throughout the year.

Valley. In the north it is rarely seen in winter; one was present between early January and 26 February 1955 at Fayid (RN), another on 24 January 1940 at Dahshur (GZM), and six on 4 January 1985 at Bilbeis (DM). Spring migratory movements commence in February and continue to late May, with a distinct peak in late April and early May. The species is more common in spring than autumn, with flocks of up to 150 recorded in the Nile Delta and Valley, the Suez Canal area, and the northern portions of the Red Sea. It is also observed in small numbers in the Western Desert (Goodman *et al.* 1986).

Ringing There are at least six recoveries in Egypt of birds ringed in Hungary (Keve and Pátkai 1959; Watson 1971; GZF).

[Bald Ibis (Hermit Ibis) *Geronticus eremita* AV?

On 8 May 1921 R. E. Moreau observed eight Bald Ibis near Giza. This record was not accepted by Meinertzhagen (1930) and he considered this species hypothetical in the country. The only more recent record is one adult on 31 March 1962 in Wadi Feiran (Kyllingstad 1986). There are two recent records from Elat, Israel, both from the spring (Krabbe 1980) and one at Suakin along the Red Sea coast of Sudan (Nikolaus 1984). This species is common in pictorial art from Dynastic Egypt (Kumerloeve 1983; Houlihan and Goodman 1986).]

Sacred Ibis *Threskiornis aethiopicus* FB

Local names Abuu minjal al-muqaddas ابو منجل المقدّس, abuu h̲annash, abuu h̲innaa-, naaja.

Former breeding resident. All modern Egyptian specimens examined are of *T. a. aethiopicus*.

During the Dynastic periods of ancient Egypt the Sacred Ibis was apparently abundant. Mummies of this species were placed in several Late Dynastic and Greco–Roman necropoli, and a conservative estimate of the number entombed in two catacombs at Saqqara has been given at 'well over 1½ million' (Smith 1974).

This species was apparently common until the early nineteenth century. Savigny (1805) noted solitary birds or parties of eight to ten near Damietta. Their last stronghold seems to have been Lake Manzala (Meinertzhagen 1930). In December 1864, one was shot at El Qatta (Von Heuglin 1873). A specimen in the BMNH was collected in November 1877 at Damietta (Gurney 1876, post-publication note). Schrader (1892) mentioned that this species was very rare in the Nile Delta. A clutch of four eggs obtained in May 1896 at Damietta (BMNH) is of enigmatic origin. Kaiser (in le Roi 1923) recorded this bird near El Tor, Sinai, on 28 September 1886 and 5 November 1891. There are no known subsequent records from Egypt. With the recent change in the ecology of the Upper Nile Valley above the Aswan High Dam, it is plausible that there may be some post-breeding dispersal to Lake Nasser from the Sudanese nesting grounds.

Spoonbill *Platalea leucorodia* RB PV WV

Local names Abuu mil'aqa ابو ملعقة, abuu mil'aga (Ma), dawwaas, madwaas.

Rare and local breeding resident in Red Sea coastal mangrove swamps and on islands (*P. l. archeri*). Scarce passage and winter visitor from (late August) early September to late May (*P. l. leucorodia*) (Fig. 6.7). Rare summer visitor to the Nile Delta and Valley.

Western Desert There are only a few records of Spoonbills from this area. On 27 May 1910, one was obtained in Wadi el Natrun (Meinertzhagen 1930). Records from Dakhla include single birds between 2 and 4 October 1962 (Scharlau 1963) and in early October 1982 (Goodman *et al.* 1986), and eight on 21 February 1986 (AH).

North Sinai and Nile Delta and Valley In autumn this species is noted migrating in small numbers along the north coast of Sinai (Petersen and Sørensen 1981*a*, 1981*b*). The earliest autumn record is one on 31 August 1981 at Zaranikh (IP, UGS). In the Nile Delta and Valley (including the Faiyum), Spoonbills are regular migrants in

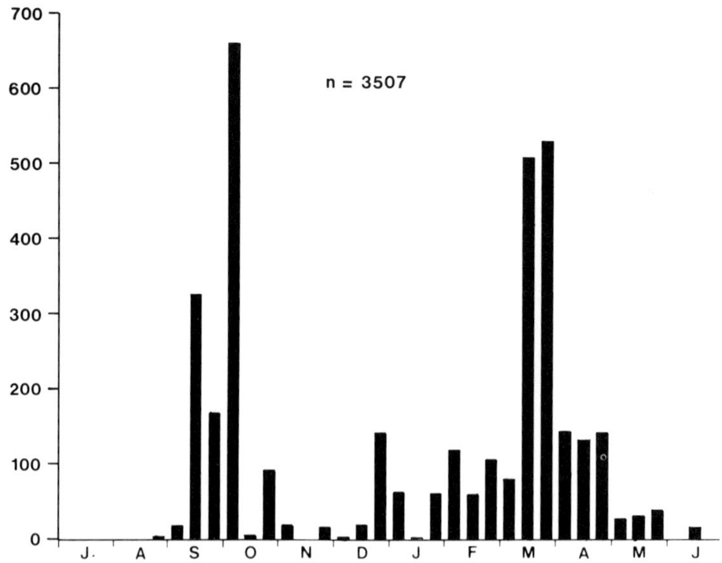

FIG. 6.7. Histogram of the numbers of Spoonbills recorded during 10-day periods throughout the year (excluding the Red Sea coast).

flocks of up to 70, starting in early September and peaking in late September and October. On 14 September 1944 between 250 and 300 birds were noted on Lake Manzala (Bodenham 1945*b*). Most migrate further south but throughout the winter months small flocks are infrequently observed in the Upper Nile Valley. There appear to be no midwinter records from the Nile Delta lakes (Meininger and Mullié 1981*b*).

Spring movements begin as early as January and continue at least until mid-May. In late winter and early spring individuals or flocks of up to 250 are seen in the Nile Valley (Short and Horne 1981). On 30 May 1980, 30 were seen flying north at El Maadi (SV). Flower (1933) found it numerous in the southern Nile Valley in July 1909 and in June 1920. There are no recent summer records from the Nile proper, although 17 were noted at Lake Qarun on 15 June 1985 (DM) and up to 17 in Wadi el Rayan in July 1984 (SMB).

Red Sea Several small colonies are known in mangrove swamps along the Red Sea coast and on islands (see map for details). The total Egyptian breeding population is estimated at 30 to 50 pairs. On Gezira Tiran breeding commences in March and fresh eggs have been found up to 9 May (Su-Aretz 1979). Small flocks of Spoonbills can be observed throughout the year along the entire Red Sea coast. The resident population is augmented by Eurasian migrants between autumn and spring.

Ringing In Egypt there are six recoveries of birds ringed in Hungary, three from the general Hungary/Yugoslavia area, four from western Turkey, and singles from Austria and Romania (Mackintosh 1941; Keve and Pátkai 1959; Watson 1971; Kistchinski 1978; EDB; GZF).

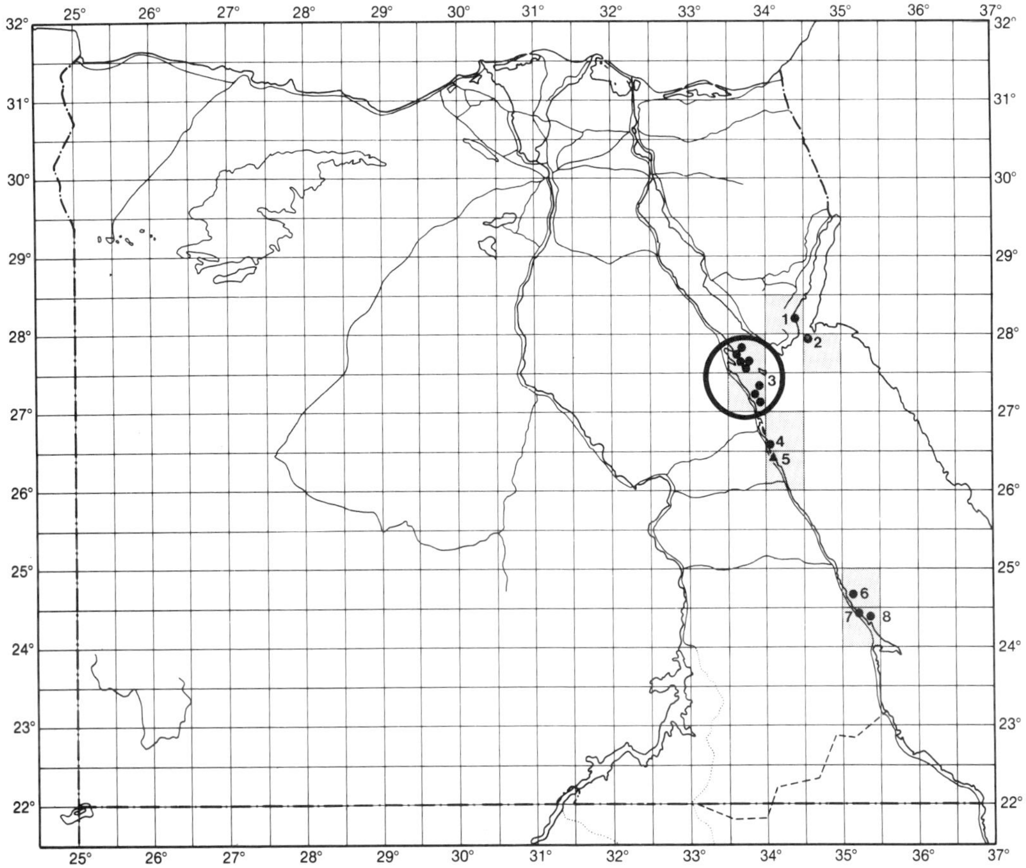

▲ probable breeding ● definite breeding ▒ half degree grid plots with probable or definite breeding

Map B9. Breeding distribution of Spoonbill *Platalea leucorodia* between 1965–87. *Breeding localities* with most recent estimate of breeding pairs: (1) Shora el Manqata (1973–79: 1–2). (2) Gezira Tiran (1973–79: 5–6). (3) Islands mouth Gulf of Suez (1983–84: 11–20). (4) Mangroves 23 km south of Safaga (1984: 2). (5) Mangroves 42 km south of Safaga (1983–86: 0–2). (6) Gezira Wadi Gimal (1984: 4). (7) Mangroves 13 km north of Hamata (1965: at least 2). (8) Gezira Showarit (1984: 4).

ORDER PHOENICOPTERIFORMES

Family Phoenicopteridae

Greater Flamingo *Phoenicopterus ruber* RB PV WV

Local names Bashaarush بشاروش, nuh̲aam.

Scarce breeding resident in north-western Sinai. Fairly common migrant and winter visitor, mainly to northern Sinai and Lake Manzala, between early September and late April. The form occurring in the country is *P. r. roseus*.

Until the beginning of the twentieth century the Greater Flamingo was a breeding resident on several Nile Delta lakes, but presently it is unknown to nest in this area. Nicoll (1908) interviewed local Delta fishermen and was unable to ascertain for certain whether Greater Flamingos nested on Lake Manzala. On 14 May 1909 Ticehurst (1912) observed a flock of about 100 flying over Lake Maryut and local people accurately described its nest and egg and assured him that they bred on the lake. (The actual breeding site was most likely in a depression west of the lake, where hypersaline habitat still occurs.) According to Meinertzhagen (1930) this species 'formerly bred on both Lake Menzaleh and Lake Bardawil, whence eggs have been examined, taken in May 1894', but it had ceased to breed in Egypt by 1929.

Information from north Sinai during the past century is spotty and it is not clear if this species regularly nested in this area. Two eggs were collected at Lake Bardawil on 10 May 1898 (Koenig 1932). After the statement of Meinertzhagen mentioned above the next record is from 1970 when a colony of 500–600 pairs was (re-) discovered at El Malaha (Mendelssohn 1975), where it has irregularly bred since. Aerial surveys of this area found 2000 birds on 20 April 1985 (Fry *et al.* 1985; DM) and 750–1000 adults with 350–400 fledglings on 7 July 1986 (DM, WCM). Until about 1978 up to several tens nested or attempted to do so at Zaranikh and Bir el Abd (Fry *et al.* 1985; UP).

Russell (1905) and Koenig (1932) reported flocks of thousands wintering on Lake Manzala. This species is still a regular winter visitor to Lake Manzala and the salt-lagoon west of Lake Maryut. Large flocks have been noted at El Malaha, including at least 8000 in December 1972 and 11 000 in January 1973 (UP), and 6300 in January 1980 (Meininger and Mullié 1981*a*). At Lake Bardawil 800–1500 were noted in mid-June 1967, 3000 in December 1972, and 8000 in January 1973 (UP). In Wadi el Natrun several thousand were seen between 21 January and 24 February 1903 (Loat 1905) and a 'huge flock' in March 1910 (Nicoll 1912*b*) but no large numbers have subsequently been recorded. Small numbers are also occasionally observed between mid-August and late April in Siwa, Sitra, El Bahrain, the Bitter Lakes, the Nile Valley, Lake Qarun, and all along the Red Sea coast. An exceptional record is of two Greater Flamingos in a flock of White Pelicans flying south near Aswan in the late autumn of 1974 (Prendergast 1975–6).

Ringing There are eight recoveries from Egypt of birds ringed at Lake Rezaiyeh, Iran, and three from Lake Tengiz, USSR (Meininger and Mullié 1981*b*; BBe). One bird ringed in Egypt was recovered at Lake Rezaiyeh (GZF).

ORDER ANSERIFORMES

Family Anatidae

Generic local names for swan Timm, wizz 'iraaqi (NC).

Mute Swan *Cygnus olor* (WV)

Local name Timm abyad تم أبيض.

Rare and irregular winter visitor from November to mid-May, mainly to the northern Nile Delta. Monotypic.

An immature Mute Swan taken alive in March 1856 on Lake Manzala was described as a species new to science, *Cygnus pelzelni* (Stejneger 1882). In January 1899, three Mute Swans were captured in the Nile Delta and subsequently presented to the Giza Zoological Gardens (Flower 1933). An adult was collected in early February 1928 near Alexandria (Meinertzhagen 1930) and another on 7 December 1931 at Lake Maryut (GZM). On 14 May 1943 a flock of 14 was noted passing over Abu Qir Bay (Bodenham 1945*a*).

Recent records include an incapacitated adult in heavily worn plumage in November 1982 at Hurghada (UMMZ); a single bird on 1 December 1981 at Sharm el Sheikh (UP); one shot from a flock of three on 12 December 1984 at El Abbassa (BO); an immature on 27 February 1985 near Beni Hassan, two on 1 March 1985 south of Asyut, and four on the same day south of Abu Tig (AS); one bird near Beni Hassan on 17 March 1985 (Hume 1985); and one decomposed bird in April 1985 at Lake Bardawil (Fry *et al.* 1985). A single bird was noted on Lake Qarun on 26 February 1987 (PdG). In the winter of 1985–6 about ten Mute Swans were shot in the Nile Delta by members of the Egyptian Shooting Club (SZ).

Three were noted on 9 January 1979 at Elat, Israel (Krabbe 1980). Since 1979 the number wintering in Israel has gradually increased, and in the winter of 1984–5 several hundred, mostly immatures, were noted (Paz 1987).

Whooper Swan *Cygnus cygnus* AV

Local name Timm saafir تم صافر.

Von Heuglin (1873) reported the Whooper Swan as a winter visitor to the lagoons of Lower Egypt. Schrader (1892) mentioned it as rare in November and December in Lake Manzala. The only known subsequent records of this monotypic species are one collected in the autumn of 1948 on Lake Idku (El Negumi *et al.* 1950) and another seen on 17 February 1976 along the Nile about 14 km north of Abu Tig (Short and Horne 1981). Three were observed at Elat, Israel, on 17 March 1985 (Paz 1987).

Generic local names for goose Awazz, wizza, wizz.

Several types of *Anser* geese are common domestic birds in Egypt, and some phenotypically resemble *A. albifrons*, *erythropus*, *fabalis*, or *anser*. Thus, caution needs to be used when assessing records of accidental species, particularly when observed near villages.

Bean Goose *Anser fabalis* AV

Local names Awazz al-fuul أوز الفول, wizz 'iraaqii.

Accidental visitor. The single known complete Egyptian specimen has not been examined and it is not clear which subspecies occurs in the country.

There are two wings in the GZM of a Bean Goose collected in December 1922 at 'Hod el Raml' (Behera Land Coy) (Meinertzhagen 1930). El Negumi *et al.* (1950) mentioned another specimen taken in the Nile Delta on 15 June 1942. The only recent record appears to be two observed and photographed in October 1982 near Sharm el Sheikh (Baha el Din 1984*a*).

White-fronted Goose *Anser albifrons* (WV)

Local name Awazz abyad̲ al-jabha أوز أبيض الجبهة.

Rare and irregular winter visitor between late November and late March. *A. a. albifrons* is the form occurring in the country.

Taylor (1859) and Adams (1864) mentioned the White-fronted Goose as a common winter visitor, mainly to northern Egypt, and rarely south of Idfu. One was taken at Qena on 16 January 1875 (FMNH). Koenig (1932) travelled extensively up and down the Nile Valley in the 1890s and the only observation he noted of this species was a large flock near El Minya on 9 February. According to Meinertzhagen (1930) it was the most regular wintering 'true' goose in Egypt; sometimes recorded in flocks, although never abundant and rarely common. Two were shot at El Tell el Kebir on 27 February 1944 (Hindmarsh 1949).

There are only a few recent records from Egypt. Small numbers have been noted from December to March in the Nile Delta and Valley, and it seems unlikely that substantial numbers still regularly winter in Egypt (Meininger and Mullié 1981*a*). Eight were noted near Asyut on 14 February 1983 (SQ). An exceptionally southern record is three at Abu Simbil on 8 March 1982 (OL). The only recent report from Sinai is one on 4 January 1983 at Sharm el Sheikh (PW). On 24 December 1982 a flock of 12 was noted at Port Sudan, Sudan (Nikolaus 1984). At Elat, Israel, a single bird was present from late November to mid-December 1978 (Krabbe 1980).

Lesser White-fronted Goose *Anser erythropus* AV

Local name Awazz abyad̲ al-jabha s̲aghiir أوز أبيض الجبهة صغير.

Accidental visitor. No recent records. Monotypic.

Lesser White-fronted Geese were collected at Damietta in January 1875 (Gurney 1876) and on 27 February 1886 (USNM). Meinertzhagen (1930) stated that this species was probably not as rare as supposed, since most 'sportsmen' would not differentiate it from the White-fronted Goose; he further noted that six were offered for sale in the Cairo market in February 1928. One bird was captured near Giza on 30 January 1907 (Nicoll 1909). We are unaware of any recent records of this species in Egypt. A single bird was seen at Elat, Israel, on 19 May 1983 (Paz 1987).

Greylag Goose *Anser anser* AV

Local name Awazz ramaadii أوز رمادي.

Accidental visitor. We have not examined any modern Egyptian material of this species, and it is not clear which subspecies occurs locally.

A Greylag Goose was collected out of a flock of six in January 1942 at Lake Maryut (El Negumi *et al.* 1950). The specimen cannot be traced. This species is known from Dynastic Egypt in the form of bone remains and art representations (Houlihan and Goodman 1986). The only known record from Elat, Israel, is five birds in early spring 1966 (Krabbe 1980).

Barnacle Goose *Branta leucopsis* AV

Local name Awazz abyad̲ al-raa-s أوز أبيض الرأس.

A female Barnacle Goose was collected on 16 April 1931 near Helwan (Jourdain and Lynes 1936). This specimen, which is the only known record of this monotypic species from Egypt, is housed in the Entomological Society of Egypt collection, Cairo.

Brent Goose *Branta bernicla* AV

Local name Awazz aswad أوز أسود.

Accidental visitor. No Egyptian specimens have been properly examined, but based on geographical grounds *B. b. bernicla* is the form likely to occur.

Von Heuglin (1873) noted many Brent Geese in February and early March in the bird markets of Alexandria and Damietta. Other records include: two individuals near Aswan on 25 February 1917, a flock near Sanabu, south of Mallawi, in February 1918, and a small flock near Cairo in March 1922 (Meinertzhagen 1930). On 30 December 1977, two domestic Brent Geese were noted in a village between El Balyana and Nag Hammadi (AS). The origin of these birds is unknown. We are unaware of any recent records of wild birds from Egypt.

Red-breasted Goose *Branta ruficollis* AV

Local name Awazz ah̲mar as̲-s̲adr أوز أحمر الصدر.

The monotypic Red-breasted Goose has not been recorded in the country since the nineteenth century. The 'modern' records include an immature bird shot near Alexandria on 2 December 1874 (Meinertzhagen 1930) and another near Damietta in the winter of 1882 (Schrader 1892). There is pictorial evidence that this species occurred in ancient Egypt (Houlihan and Goodman 1986).

Egyptian Goose *Alopochen aegyptiacus* RB

Local names Awazz masrii أوز مصري, shayyiqa.

Fairly common breeding resident along Lake Nasser and winter visitor to the Nile Valley almost as far north as Cairo. Monotypic.

Meinertzhagen (1930) mentioned the Egyptian Goose as a former abundant resident throughout the Nile Valley and scarce in Lower Egypt. Documentation of its breeding in the Nile Delta includes a clutch of six eggs collected at Damietta on 14 April 1894 (Koenig 1932) and a single egg collected in 'Lower Egypt' in 1862 (BMNH). Raw (1921) noted the breeding of 'semi-domesticated' birds in an old Black Kite nest near Abu Zabal and further added that flocks of over a hundred were observed locally during the winter. This species formerly bred on cliffs along the Nile at Gebel Abu Foda (Gurney 1876).

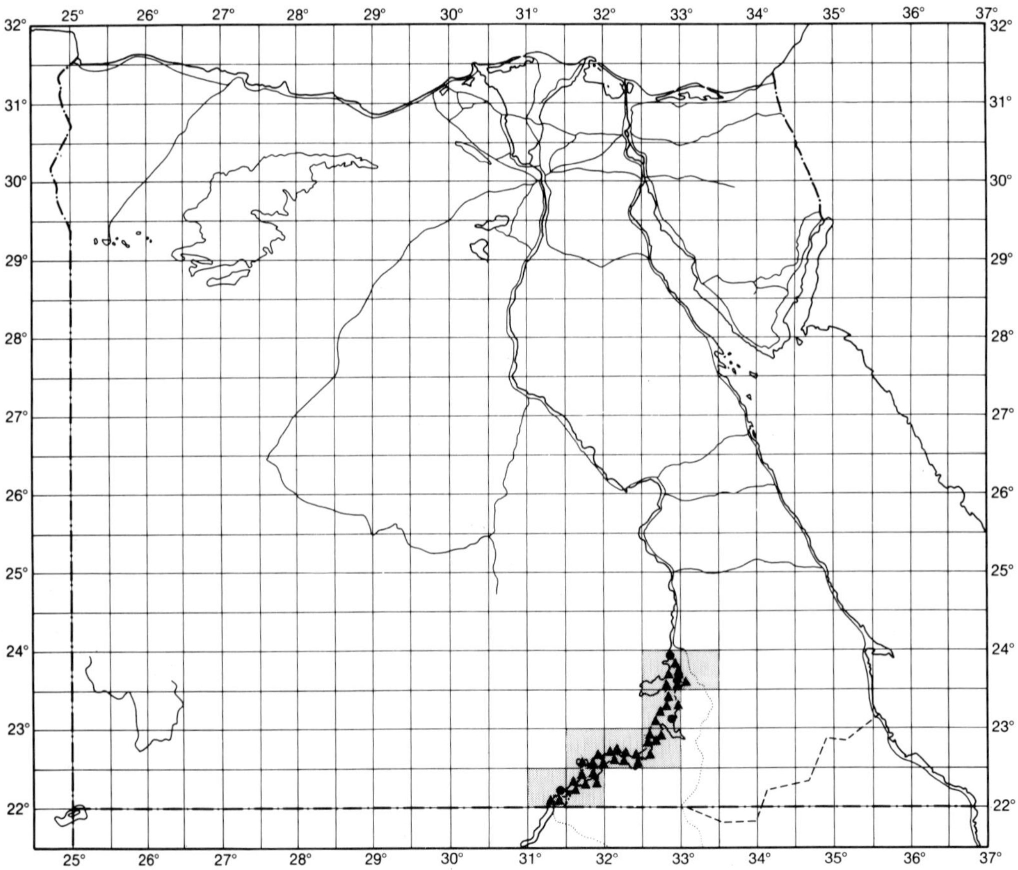

▲ probable breeding ● definite breeding ░ half degree grid plots with probable or definite breeding

Map B10. Breeding distribution of Egyptian Goose *Alopochen aegyptiacus* 1970–87.

Nowadays Egyptian Geese appear to have abandoned former breeding areas in the Nile Valley north of Aswan. After the completion of the Aswan High Dam and the subsequent filling of Lake Nasser, a considerable amount of suitable breeding habitat was created. In 1974 the Lake Nasser population was estimated at 5–10 birds per km of shore-line (Entz 1980). There are probably thousands of nesting birds in Lake Nasser, a portion of which migrate north in the non-breeding season into the Nile Valley.

Flocks of Egyptian Geese numbering up to 2000 are found on the banks of the Nile, mainly between late September and late March. The species is common between Aswan and Sohag, scarce between Sohag and Asyut, and rare between Asyut and Cairo. As many as 11 000 were observed in the Nile Valley south of Sohag in February 1976 (Short and Horne 1981). An exceptional record is two in Wadi el Rayan on 19 January 1985 (SMB).

Relatively recent records from the Nile Delta and Suez Canal include 28 on 22 September 1952 and two on 29 November 1952 at El Kabrit (AP) and a few in a mixed flock of Pintails and Shovelers in early December 1974 in Lake Maryut (Prendergast 1975–6). The only records of the Egyptian Goose outside the Nile basin are one at Hurghada on 30 January 1981 (SMB) and a possible record of six near Mut, Dakhla Oasis, on 25 January 1982 (Goodman *et al.* 1986).

Ruddy Shelduck *Tadorna ferruginea* PV WV

Local names Abuu farwa ابو فروة, ba<u>hh</u> (Mi), wizz 'iraaqii (NC).

Rare passage migrant and winter visitor from late October to early May. There is no evidence of this species breeding in Egypt. Monotypic.

Jourdain and Lynes (1936) mentioned two eggs of the Ruddy Shelduck taken at Gebel el Muqattam in '5.62' (= May 1862) by S. S. Allen and housed in the Southampton University Collection. Allen (1863, 1864), however, did not mention these eggs in his papers on Egyptian birds and they are not acceptable as evidence that this species bred in the country (cf. Etchécopar and Hüe 1967).

Meinertzhagen (1930) mentioned it as a fairly common winter visitor from November to May in the Faiyum, Wadi el Natrun, and Upper Egypt. Koenig (1932) saw large flocks near Asyut in 1899. Nowadays, it is a rare winter and passage visitor to the Nile Delta and Valley (south to Aswan), and the Faiyum. The largest group observed in recent years is six between Beni Hassan and El Amarna on 15 February 1976 (Short and Horne 1981). Three were seen at Suez on 28 March 1982 (MU) and at Ras Muhammad on 22 April 1979 (HW). A similar decrease in the number of wintering birds has also been noted in the Sudan (Nikolaus and Hamed 1984). At Elat, Israel, it is a rare winter visitor from late October to April (Krabbe 1980).

Shelduck *Tadorna tadorna* WV

Local names Shahramaan شهرمان, bahh abyad (Mi).

Fairly common winter visitor from (early September) late November to late April. Monotypic.

The Shelduck is a winter visitor to the northern part of the country, mainly on salt and brackish lakes. The number of birds recorded in Egypt is variable from year to year and presumably related to weather fluctuations in Eurasia. On Lakes Bardawil, Burullus, Maryut, Idku, and Qarun the numbers are usually small. Large concentrations have been irregularly recorded on Lake Manzala—1000 on 16 January 1979 (Meininger *et al.* 1979); El Malaha—2650 in January 1973 (Paz 1987) and 825 on 9 January 1980 (Meininger and Mullié 1981*b*); and Wadi el Natrun—3000–5000 in February 1983 (Goodman *et al.* 1986). The only other record from the Western Desert is two on 23 January 1987 at Dakhla (AH). The species is occasionally observed in the Nile Valley, in the Suez Canal area, and the Gulfs of Suez and Aqaba. The southernmost record in Egypt is two between Kom Ombo and Aswan on 22 March 1986 (UA). On 11 December 1980, 44 were seen at Suez (SMB); on 30 January 1983, one between Safaga and Hurghada (BR); and on 11 February 1984, two south of Hurghada (SMG). The latest known spring record is three at Zaranikh on 29 April 1985 (Fry *et al.* 1985).

[**Spur-winged Goose** *Plectropterus gambensis* AV?

Presumably accidental visitor. No modern Egyptian material has been examined but on geographical grounds *P. g. gambensis* should be the form occurring in the country.

Meinertzhagen (1930) ascribed several records of the Spur-winged Goose in Egypt to feral domestic birds. Today, particularly in Upper Egypt, domesticated Spur-winged Geese are not uncommon in poultry flocks. Ripley (1963) observed solitary birds on four occasions at Abu Simbil in March 1962. It is not clear if these birds were wild or feral domesticates. It is interesting to note that a bone recovered from a late Pleistocene site (15 000–10 000 BC) near Kom Ombo may be referable to this species (Churcher 1972).]

Generic local name for duck Batt.

Wigeon *Anas penelope* PV WV

Local names Siwwaay صوّاي, marawaanii (Mi), zaaw, zaayy.

Common passage migrant and winter visitor to most inland waters from late September to late April. Monotypic.

In autumn Wigeons usually begin to arrive in late September, numbers increase in

October and November, and peak in midwinter. Some birds continue on migration to wintering areas south of Egypt. Flocks of hundreds to several thousand have been recorded on the Nile Delta lakes, Lake Qarun, Lake Bardawil, El Malaha, and in the Nile Valley (south to Aswan). On Lake Burullus, 35 600 Wigeons were counted between 14 and 16 January 1980 (Meininger and Mullié 1981*b*). This species has also been observed, generally during winter and in small numbers, in several Western Desert oases (Goodman *et al.* 1986) and along the Red Sea coast of Sinai. Two were noted near Ras Banas on 14 November 1982 (Baha el Din and Saleh 1983). In spring there is a notable increase in numbers, particularly along the Red Sea. Up to 200 were present near Elat, Israel, between 30 March and 10 April 1978 (AR, AW).

Ringing Three birds ringed on 2 January 1940 near Cairo were recovered that same month at various places in the Nile Delta (Mackintosh 1941).

Gadwall *Anas strepera* PV WV

Local names Samaarii سماري, marawaanii asmar (Mi).

Scarce passage migrant and winter visitor from (late August) mid-October to mid-April. The form occurring in Egypt is *A. s. strepera.*

Gadwalls are regular winter visitors to most inland waters but generally in small numbers. The earliest known autumn record is 15 near Baltim on 26 August 1986 (EE *et al.*). The largest concentration noted in recent years is 235 at Lake Burullus on 1 February 1979 (Meininger *et al.* 1979). About 500 were observed in February 1977 along the Nile between Cairo and Aswan (Short and Horne 1981). Smaller numbers have been recorded in several Western Desert oases and on Lake Qarun. Fourteen were noted at Taba on 9 March 1987 (PLM). At Elat, Israel, this species is a rare winter visitor and spring migrant (Krabbe 1980).

Teal *Anas crecca* PV WV

Local names Sharshiir شرشير, abuu faashika (Mi), karkaj, sharshiir shatawii, shashiir (ND).

Common passage migrant and winter visitor to most inland waters from (mid-August) early September to mid- (late) April. *A. c. crecca* is the form occurring in the country.

In some autumns a considerable number of Teal migrate along the north coast of Sinai, e.g. 3251 at Zaranikh between 16 August and 24 September 1981, with a maximum of 335 on 15 September (Petersen and Sørensen 1981*b*). Concentrations are found in winter on the Nile Delta lakes, Lake Qarun, and along the Nile. Recent records of large numbers include: 1150 on 13 October 1980 and 4700 in December 1979 on Lake Manzala (Meininger and Mullié 1981*b*; PLM); thousands on 8 December 1982 on the Nile near Idfu (PW); 1000 between 10 and 12 January 1979 on

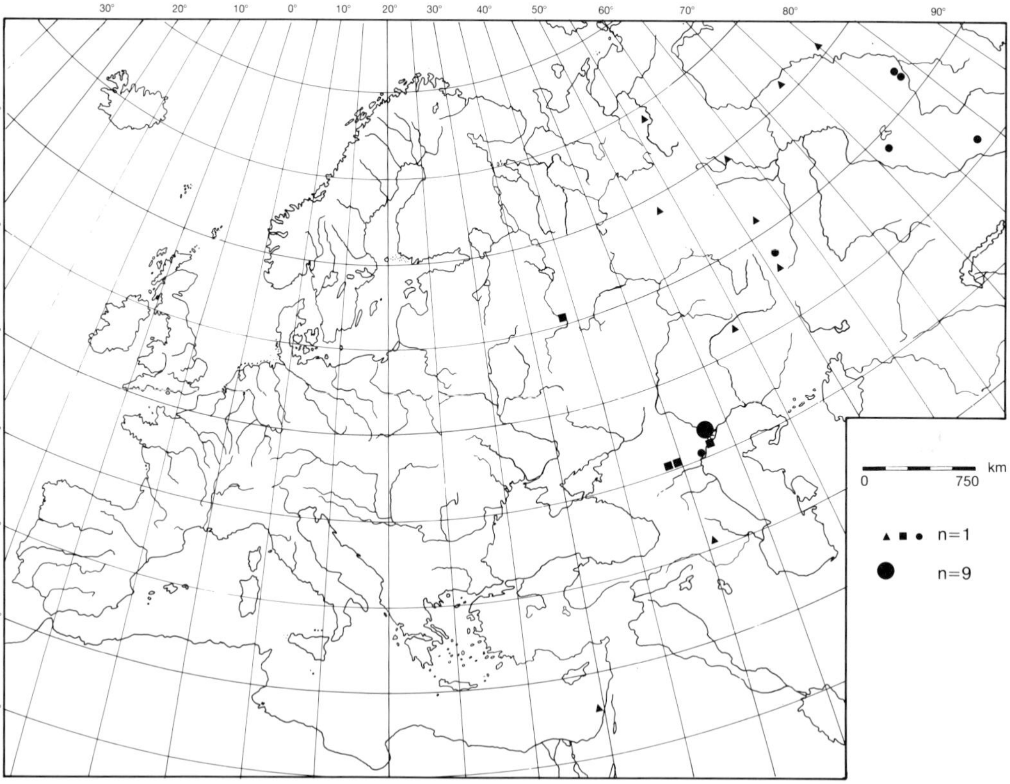

FIG. 6.8. Ringing or recovery sites of Teals (▲, n=10), Mallards (■, n=4), and Shovelers (●, n=15) occurring in Egypt.

Lake Qarun (Meininger *et al.* 1979); and 2100 on 15 March 1983 between Tima and El Balyana (Welch 1983). The species has been recorded in all major Western Desert oases (Goodman *et al.* 1986). Teal also occur in small numbers along the Mediterranean and Red Sea coasts, mainly on autumn and spring migration.

Ringing Eight birds ringed in Cairo have been recovered in the Caucasus and east of the Urals (Fig. 6.8; Mackintosh 1941; GZF). One ringed at Ma'agan Michael, Israel, on 13 February 1968 was shot at Damietta on 17 February 1970 (Watson 1971). Another ringed in the USSR was recovered in Beheira Governorate (GZF).

Mallard *Anas platyrhynchos* CB? PV WV

Local names Khuḏaarii خضاري, burkaawaa, sulṯaanii (female).

Possible occasional breeding resident. Fairly common passage and winter visitor from (late August) mid-September to late March (mid-May). *A. p. platyrhynchos* is the form occurring in the country.

There is no positive evidence of wild Mallards breeding in Egypt. According to Meinertzhagen (1930), a few may nest on Lake Manzala, where they have been noted throughout the year. A female shot in Dakhla on 25 March 1928 had an enlarged ovary and appeared about to lay eggs (Meinertzhagen 1930). Domestic Mallards are common in poultry flocks.

From mid-September to late March Mallards are regularly observed on all Nile Delta lakes, in the Nile Valley (south to Aswan), and on Lake Qarun. In recent years, the largest concentration recorded is 407 at Lake Burullus on 31 January 1979 and 163 at El Malaha on 9 January 1980 (Meininger and Mullié 1981*b*). The numbers seen on the Nile vary from year to year. Mallards are irregularly observed along the Red Sea coast, mainly in the northern portion of the Gulfs of Suez and Aqaba. The species has also been noted in several Western Desert oases (Goodman *et al.* 1986). An exceptionally late record is several at Lake Maryut between 15 and 18 May 1909 (Ticehurst 1912).

Ringing There are three recoveries in Egypt of birds ringed in the Caucasus and Moscow regions, USSR, and two recoveries in the Caucasus of birds ringed in Egypt (Fig. 6.8; GZF).

Pintail *Anas acuta* PV WV

Local names Balbuul بلبول, abuu sabala (ND), abuu sibiiba (Mi).

Fairly common passage and winter visitor from (late August) mid-September to late April (mid-May). *A. a. acuta* is the form occurring in the country.

In autumn large numbers of Pintail migrate through Egypt along a broad front to sub-Saharan wintering grounds. Some remain in the country for the winter. Autumn passage, which continues into December, is noticeably heavy along the north coast of Sinai and in the Nile Delta and Valley. On 13 October 1986 about 350 were seen resting in the sea off Hurghada (EH, PK). In winter Pintails are sometimes recorded in relatively large numbers on inland lakes and the Nile. For example, 700 at Dahshur Lake on 6 January 1976 and 400 at El Abbassa on 12 January 1976 (MSm); 250 at Lake Idku on 27 January 1979 (Meininger *et al.* 1979); and 200 in Dakhla on 26 January 1982 (AH). On the Nile it is one of the most common dabbling ducks but numbers vary from year to year (Short and Horne 1981).

Spring passage is noted from late February to mid-April. In some years substantial numbers of spring migrants concentrate at the northern limits of the Gulfs of Suez and Aqaba.

Ringing There are 13 records of birds ringed at the Astrakhan Reserve, Caspian Sea, which were recovered in Egypt (Fig. 6.9). Thirty-three birds ringed in Egypt have been recovered in the USSR east of the Urals and Caucasus, and single birds in Albania, Finland, Iraq, The Netherlands, and Turkey (Wuczeticz and Tugarinov 1937; Mackintosh 1941; Watson 1971; GZF; RWa). Six birds ringed in Egypt were

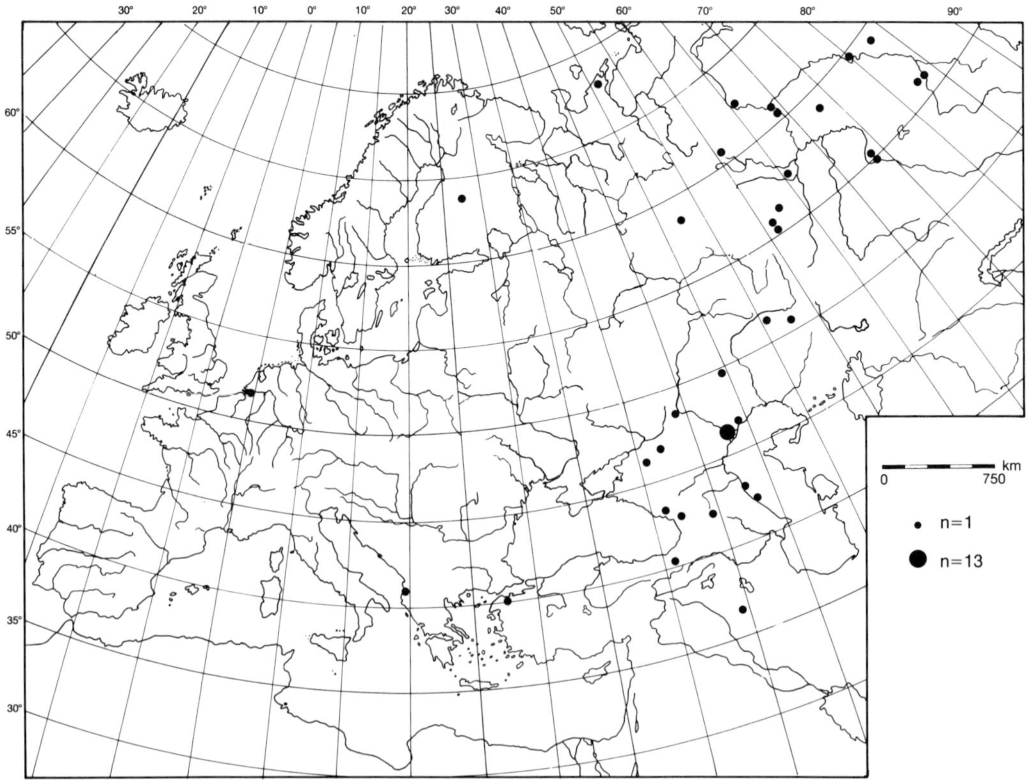

FIG. 6.9. Ringing or recovery sites of Pintails (n=48) occurring in Egypt.

recovered within the country—two in the same season and four in subsequent seasons (one after two years).

Garganey *Anas querquedula* PV

Local names Sharshiir s̱ayfii شرشير صيفي, abuu ariida (Mi), araash, karkaj.

Common passage migrant in autumn from late July to late October (late November) and in spring from late February to mid- (late) May. Monotypic.

In Egypt Garganey appear in autumn as early as late July (Meinertzhagen 1930), and from early August to mid-October are regularly seen on most inland waters. There is a marked passage along the north coast of Sinai. During systematic counts of migrating birds at Zaranikh the following totals were recorded: 96 062 between 13 August and 29 September 1973 and 203 437 between 15 August and 5 October 1978 (YP, UP); 2410 between 11 and 20 October 1979 (Petersen and Sørensen 1980); 85 056 between 7 and 30 September 1980 (Petersen and Sørensen 1981*a*); 221 616 between 16 August and 24 September 1981, with a maximum of 22 150 on 9 September (Petersen and Sørensen 1981*b*); 94 440 between 2 and 24 September 1982, with a maximum of 19 900 on 3 September (Baha el Din and Salama 1984); and 112 883 between 22 August and 29 September 1986, with a maximum of 9275 on 6 September (El Badrawy 1986).

It is not clear which route(s) Garganeys take after leaving northern Sinai to their tropical African wintering grounds. There are no records of large flocks resting on Nile Delta lakes (though autumn records from this area are somewhat limited), or from the northern portion of the Nile Valley. Small numbers are regularly noted in the Suez Canal area, and along the Gulfs of Suez and Aqaba. Between 13 and 14 September 1964 small flocks of migrants were seen in the Gulf of Suez and northern Red Sea (Bailey 1966) and on 3 October 1984 about 600 were counted at Ras Shukheir (MCJ *et al.*). There are only a few scattered autumn records from the Faiyum and Western Desert. Some concentration seems to occur in the Upper Nile Valley. Near Aswan 1700 Garganey were counted flying south in 12 hours of observation between 29 September and 3 October 1981 (FdR). On 10 October 1983 about 450 were noted between Isna and Aswan (Hume 1983). However, the numbers noted on autumn passage in northern Sinai are not subsequently found anywhere else in the country and it seems likely that Garganeys cross over Egypt along a broad front and generally do not stop until they reach tropical Africa (Moreau 1967).

The latest autumn records are 20 near Safaga on 2 November 1982 (Baha el Din and Saleh 1983) and 131 on Lake Nasser between 20 and 22 November 1981 (CAB *et al.*). Kaiser (1892) reported the species throughout the winter months on the west coast of Sinai near El Tor but there are no recent winter records from Egyptian waters. At Elat, Israel, most autumn migrants have departed by mid-November, although a few apparently remain for the winter (Krabbe 1980).

The first spring migrants arrive in late February and passage continues throughout the country until late April. In spring some concentration of migrants occurs but it does not parallel autumn numbers. Up to 300 were observed near Baltim on 29 March 1981 (WS), 400 at Suez on 30 March 1982 (Wimpfheimer *et al.* 1983), and 150 on Lake Qarun on 17 April 1964 (HCK). A total of 717 was noted along the Nile between Cairo and Aswan from 14 to 24 March 1985 (Grieve 1985). The latest spring record is one on 31 May 1938 at Abu Rauwash (HGD). Large numbers funnel through Elat, Israel, and during the heaviest migration, early March to mid-April, up to 1400 have been observed per day (Krabbe 1980). Presumably these birds move north through the Red Sea into the Gulf of Aqaba.

Ringing Two birds ringed in the Province of Kalinin, USSR, on 21 August were recovered in Egypt (Rjabow 1959 in Impekoven 1964).

Shoveler *Anas clypeata* PV WV

Local names Kiish كيش, abuu maalik, araash, h̠iqq, luugat.

Common passage migrant and winter visitor throughout the country from late August to mid- (late) May. Rare in summer. Monotypic.

In autumn Shovelers arrive as early as late August and the numbers gradually increase until midwinter. Small flocks have been recorded migrating along the north coast of Sinai in September and October (Petersen and Sørensen 1981*a*, 1981*b*; Baha el Din and Salama 1984). The species is a regular winter visitor to all inland waters, with counts of up to 8800 on Lake Manzala, 63 500 on Lake Burullus, and 11 700 on Lake Qarun (Meininger and Mullié 1981*a*). Lake Burullus is the most important wintering area for Shoveler in the eastern Mediterranean and Black Sea region (Rüger *et al.* 1986). Shovelers are also common during the winter on the Nile south to Aswan and near Suez. Along the Red Sea coast this species is mainly noted during migration, although there are a few winter records.

The number of Shovelers seems to increase in March, signalling an influx of migrants. In spring it has been recorded in small numbers in most Western Desert oases (Goodman *et al.* 1986). An exceptional record is one at Hurghada on 12 June 1981 (SMB). Up to 700 per day have been observed in spring at Elat, Israel (HBu).

Ringing There are 14 recoveries from Egypt of birds ringed north of the Caspian Sea and east of the Urals, including Siberia (Fig. 6.8; Mackintosh 1941; Treous 1957; Anon. 1961; Watson 1971; BEP) and vice versa (GZF). Seven birds ringed in Egypt were recovered within the country, two in subsequent seasons (Mackintosh 1941).

Marbled Duck *Marmaronetta angustirostris* RB (WV)

Local names Sharshiir mukhat̲t̲at̲ شرشير مخطط, abuu sawaana, h̲adhaf, minya, silsil.

Rare breeding resident on inland waters. Often observed between late August and mid-March in areas where it is not known to breed. Monotypic.

Western Desert According to Meinertzhagen (1930) the Marbled Duck bred in Wadi el Natrun and Dakhla; however, his basis for considering it a breeding bird in Dakhla is ambiguous. The only Marbled Duck eggs from Egypt examined by Meinertzhagen were a clutch of four taken in May in Wadi el Natrun. Four specimens were collected at this locality between 16 May and 5 June 1910 (Goodman *et al.* 1986), and Nicoll (1912*a*) mentioned that young were observed locally. The only recent reports from the Western Desert are a pair on 27 April 1986 in Wadi el Natrun (DM, WCM), three birds on 19 March 1984 in Farafra, and a few on 7 September 1984 in Dakhla (Goodman *et al.* 1986).

Nile Delta Meinertzhagen (1930) stated that in winter Marbled Duck is often obtained in the Nile Delta and the Suez Canal area. At El Marg, near Abu Zabal, single birds were collected on 24 August 1917 (FMNH CC) and on 24 September 1917 (Raw 1921). A female was taken on 19 April 1920 at El Qantara (BMNH). The only post-Meinertzhagen (1930) records are three near Cairo on 24 August 1984 (DM), single birds at El Kabrit on 29 October 1952 (AP), in the Port Said market on 16

January 1979 (Meininger *et al.* 1979), and at El Tell el Kebir on 11 February 1941 (HGD).

Nile Valley Gurney (1876) found Marbled Ducks to be quite common on Lake Qarun in June and collected several males. He suggested that although no nests were found, females were breeding in the surrounding reed-beds. The only post-1930 record from the Faiyum is two on 22 February 1965 (HCK). Meinertzhagen (1930) mentioned the absence of this species from the Nile Valley; although a specimen was taken near Dahshur on 23 December 1923 (BMNH). Short and Horne (1981) noted that it was observed every winter between 1976 and 1979 on the Nile in 'Middle Egypt' as far south as Nag Hammadi. Several were seen at Isna on 31 October 1982 (Davies 1982) and two near Aswan on 24 December 1983 (PE).

Red-crested Pochard *Netta rufina* WV

Local names Wanas ونس, abuu farwa (NC), abuu t̲arbuush (Mi).

Scarce winter visitor to the Nile Delta and Valley from (mid-September) late October to mid-April. Monotypic.

The Red-crested Pochard is a regular winter visitor, but in small numbers, to the Nile Delta and Valley (south to at least Isna), and the Faiyum. An exceptionally early record is one at Zaranikh on 10 September 1982 (Baha el Din and Salama 1984). The largest winter concentrations recorded in recent years are 80 south of Isna on 2 January 1986 (AS), 32 on Lake Maryut on 12 January 1980 (Meininger and Mullié 1981*b*), 70 on Lake Burullus on 1 February 1979 (Meininger *et al.* 1979), and 20 between Isna and Idfu on 22 February 1981 (Short and Horne 1981). The only record from the Western Desert is a male collected in Wadi el Natrun on 20 January 1911 (Nicoll 1912*b*). No records are known from the Egyptian Red Sea coast. There are three spring records from Elat, Israel, between 1966 and 1979, including a maximum of 20 birds on any one day (Krabbe 1980).

Pochard *Aythya ferina* PV WV

Local names H̲umraay حمراي, abuu haggaar (Mi), batt̲ ah̲mar, h̲amruun, h̲umaari, h̲umr, h̲umraan (NC).

Common passage and winter visitor from late September to mid-April. Rare summer visitor. Monotypic.

The Pochard is a common winter visitor to the lakes of the Nile Delta, Faiyum, and Wadi el Rayan. In recent years the largest numbers counted in the Nile Delta have been at Lake Burullus—8300 on 1 February 1979 (Meininger *et al.* 1979) and 8200 between 15 and 16 January 1980 (Meininger and Mullié 1981*b*). This species is sometimes common on the Nile south to Aswan but numbers vary from year to year (Short and Horne 1981). It has been recorded in several Western Desert oases

(Goodman *et al.* 1986). Pochards are irregularly seen in small numbers along the Red Sea coast and northern gulfs, mainly during autumn and spring migration. An exceptional record is three on 14 June 1954 at the Sacred Lake, Karnak (HCK).

Ringing A bird ringed as a pullus on 15 June 1972 at Lake Kalmykia, USSR, was recovered on 10 January 1980 at the Port Said market (BEP).

Ferruginous Duck *Aythya nyroca* PV WV

Local names Zurqaay a<u>h</u>mar زرقاي أحمر, furfur, sharshiir shukraf (ND), shatruuri, zariiqal.

Fairly common passage and winter visitor from (mid-August) late September to mid-April (mid-May). Monotypic.

Ferruginous Ducks are common in winter on the Nile Delta lakes and Nile Valley south to Aswan. The species is rarely noted on salt-water lakes such as Lake Qarun. An exceptionally early autumn record is a male at the Delta Barrage on 14 August 1945 (White 1949). Small numbers have been observed migrating along the north coast of Sinai in September and October, with the earliest record being two on 10 September 1982 (Baha el Din and Salama 1984). The numbers present in the Nile Delta and Valley vary from year to year (Meininger and Mullié 1981b; Short and Horne 1981). The largest winter concentration recorded in recent years is 6580 on Lake Burullus on 1 February 1979 (Meininger *et al.* 1979). According to Rüger *et al.* (1986) Lake Burullus is one of the main wintering areas for Ferruginous Duck in the eastern Mediterranean and Black Sea region. In the Westen Desert it has been noted in Wadi el Natrun and Dakhla (Goodman *et al.* 1986). One was taken at Sharm el Sheikh on 7 February 1982 (TAUM). The latest spring record is a single bird on 16 May 1985 in Wadi el Natrun (JS *et al.*). At Elat, Israel, this species is an uncommon migrant and winter visitor from early November to early April (mid-May) (Krabbe 1980).

Tufted Duck *Aythya fuligula* PV WV

Local names Zurqaay زرقاي, abuu haggar asmar (Mi), farfuur (WN), summiyya, zurq, zurqaa (WN).

Common passage and winter visitor from (early September) late October to late April (early May). Monotypic.

The Tufted Duck is a common passage migrant and winter visitor to the Nile Delta lakes and Valley (south to Aswan), and Lake Qarun. Small numbers have been observed migrating along the north coast of Sinai in September and October, with the earliest autumn record being 5 September 1982 (Baha el Din and Salama 1984). The largest winter concentrations recorded in recent years include: Lake Burullus—13 400 between 10 and 15 November 1981 (CAB *et al.*); and Lake Qarun—5180 between 10 and 12 January 1979 (Meininger *et al.* 1979) and 5650 between 1 and 2 January 1980

(Meininger and Mullié 1981*b*). On the Nile the numbers vary from year to year. In February 1976 the Tufted Duck was the most common species of the approximately 70 000 diving ducks observed along the river (Short and Horne 1981). In the Western Desert it has been noted in Wadi el Natrun and in Dakhla (Goodman *et al.* 1986). Tufted Ducks are rare migrants and winter visitors to Red Sea waters, although up to 130 have been noted in one day at Elat, Israel (Krabbe 1980).

Ringing A bird ringed at Lake Mukhartukha, western Siberia, on 16 July 1937 was recovered near Imbaba on 30 January 1938 (GZF).

Velvet Scoter *Melanitta fusca* AV

Local name Ba<u>tt</u> ba<u>h</u>rii akh<u>d</u>ar بط بحري أخضر.

Accidental visitor. No Egyptian material has been examined, but on geographical grounds *M. f. fusca* is the form presumed to occur in the country.

Von Heuglin (1873) mentioned that the Velvet Scoter was accidental in winter to the lagoons of Lower Egypt. Magaud d'Aubusson (1891) mentioned that single birds were occasionally observed in Lower Egypt but provided no details. Cholmley (1897) saw two birds in mid-January 1896 along the southern Red Sea coast, 22 km north of Halaib. More recent well-documented sight records include two males at Fayid on 12 November 1952 (AP) and a female in Suez Bay from 24 February to 12 March 1982 (Baha el Din 1984*b*). This species is unknown from Elat, Israel (HSh).

Smew *Mergus albellus* (WV)

Local name Balqasha bay<u>d</u>aa- بلقشة بيضاء.

Rare and irregular winter visitor to the Nile Delta and the northern Gulf of Aqaba. Monotypic.

Koenig (1932) shot a male Smew on the Nile in the spring of 1899. Meinertzhagen (1930) considered this species a rare winter visitor and mentioned seven records from the Nile Delta between December and January. This species was occasionally noted in the western Nile Delta between 1947 and 1979 (VIM) and one was observed near Alexandria on 21 December 1953 (HCK). At Elat, Israel, it is a rare and irregular winter visitor between November and December (Safriel 1968; Krabbe 1980). A portion of a corocoid recovered from a late Pleistocene site (15 000–10 000 BC) near Kom Ombo was tentatively assigned to this species (Churcher 1972).

Red-breasted Merganser *Mergus serrator* (WV)

Local name Balqasha <u>h</u>amraa- a<u>s</u>-<u>s</u>adr بلقشة حمراء الصدر.

Rare and irregular winter visitor to the country. Monotypic.

There are only a few reports of Red-breasted Mergansers in Egypt and none since 1953. Records from Suez include one shot from a flock of 12 on 13 January 1921 (Meinertzhagen 1930) and one immature obtained on 15 June 1926 (GZM). A specimen in the RNHL obtained by Dr Clot is labelled Egypt. Five were seen on 11 February 1953 near El Kabrit (AP). Wyatt (1870) noted it at Ain Musa on an unspecified date. Both Meinertzhagen (1930) and Hardy (1946) mentioned that this species had been recorded in the northern Gulf of Aqaba. At Elat, Israel, it is a scarce and irregular winter visitor, with a maximum of three birds recorded at any one time (Krabbe 1980).

White-headed Duck *Oxyura leucocephala* (WV)

Local names Batt abyad al-wajh بط أبيض الوجه, abuu karaka (Mi), abuu mirwaha, abuu nawaama (F).

Formerly a rare winter visitor. Monotypic.

Meinertzhagen (1930) mentioned that a few White-headed Ducks were obtained every winter between early November and March in the northern Nile Delta, with as many as 40 in 1925. No recent records are known of this species in Egypt.

ORDER ACCIPITRIFORMES

Family Accipitridae

Honey Buzzard *Pernis apivorus* PV

Local names Huwwaam an-nahl حوّام النحل, ihdayii (Ma).

Scarce passage migrant in the autumn from (late August) early September to late October and common in spring from (mid-March) mid-April to late May (early June). Rare winter visitor. Monotypic.

Western Desert Records of Honey Buzzards include Wadi el Natrun—one seen on 7 May 1981 and one collected on 2 June 1910 (Nicoll 1912*b*; Goodman *et al.* 1986); Dakhla—two noted on 2 October 1962 (Scharlau 1963); and Bahariya—one recorded on 23 September 1981 (HB *et al.*).

Nile Delta and Valley During autumn migration the Honey Buzzard is seen in small numbers, mainly in the Upper Nile Valley. In spring it is distinctly more common, particularly between mid-April and late May, in the Nile Delta and Valley, the Faiyum, and Suez Canal area. The largest reported concentration is 219 flying west over Lake Manzala, near Port Said, on 6 May 1983 (PLM, UGS).

Eastern Desert and Sinai In autumn relatively small numbers of Honey Buzzards migrate along the eastern shore of Sinai, across the southern Gulf of Suez and then south through the Red Sea Mountains to wintering grounds in tropical and southern Africa. Small numbers are recorded at Suez: 79 between 13 September and 27 October 1981 (Bijlsma 1983) and 16 between 19 September and 30 October 1984 (RvD *et al.*). A few pass along the north coast of Sinai and the maximum number recorded during recent autumn counts is 34 between 7 and 30 September 1980 (Petersen and Sørensen 1981*a*). Winter records are exceptional, and the only well-documented observation is one on 20 January 1981 at Hurghada (SMB).

In spring substantial numbers are observed and they seem to follow the same pathway as autumn migrants. Passage is generally from mid-April to early June, peaking in mid-May. The majority of birds presumably cross the southern Gulf of Suez. The species appears to be uncommon at Suez or perhaps passage is temporarily concentrated: on 5 May 1978 about 850 were recorded (Meininger and Dielissen 1979), and between 16 March and 16 May 1982, 630 were counted, of which 500 passed on 11 May (Wimpfheimer *et al.* 1983). Hundreds of Honey Buzzards were observed crossing the Gulf of Suez near Hurghada on 28 April 1982 (Ferguson 1984) and about 3400 over Ghubbet el Zeit on 6 May 1983 (MCJ *et al.*). Between 16 April and 17 May 1977, 226 000 were recorded at Elat, Israel (Christensen *et al.* 1981). The maximum daily count at this locality is about 220 000 on 6 May 1983 (Paz 1987).

Generic local names for kites H̲id-a, h̲idaaya.

Black-shouldered Kite (Black-winged Kite) *Elanus caeruleus* RB Plate 1

Local names Kuuhiya كوهية, abuu sharraaqa, s̲aqr abyad̲, s̲aqr ash-shaaduuf, sharq, siriqa, zurraq.

Fairly common breeding resident in cultivated fields of the Nile Valley and Faiyum, and less common in the Nile Delta. The Egyptian population is referable to *E. c. caeruleus*. This species is an opportunistic breeder and eggs are laid almost throughout the year (Meinertzhagen 1930; Horner *et al.* 1982).

Western Desert The Black-shouldered Kite formerly bred in Wadi el Natrun. Meinertzhagen (1930) mentioned that nestlings were found on 22 January. There are no recent records from this area and it is presumed to be locally extirpated.

Nile Delta and Valley Numerous nineteenth-century authors mentioned this species as 'fairly common' to 'very abundant' in the Nile Delta and Valley (Taylor 1867; Shelley 1872; Gurney 1876). According to Meinertzhagen (1930) it was a local breeding resident in the Nile Delta and Valley and the Faiyum, 'nowhere very common, but fairly well distributed'. Until about 1985 this species was a rare and local breeding resident in the Nile Delta and many of its former haunts had been abandoned; in subsequent years it has greatly increased in numbers. It is apparently absent from the coastal area. There are only a few records from the Suez Canal region: single birds were noted along the western shore of Great Bitter Lake on 7 August and 2 October 1944 (CJG).

Formerly it was uncommon in the Nile Valley between Cairo and Sohag, but in recent years it has increased in numbers. Presently this species is relatively common in the Faiyum, between Sohag and Qena, and Kom Ombo and Aswan; and common between Qena and Kom Ombo. Horner *et al.* (1982) found the density of Black-shouldered Kites between Qena and Kom Ombo to be 1 bird/0.8 or 0.9 km, the highest reported in the literature anywhere in this species' range.

The number and distribution of birds in the Nile Valley seems to be constant throughout the year and there is no indication of migration. Some regional reduction in numbers appears to be related to the local heavy use of pesticides, particularly in cotton fields (Mullié and Meininger 1985). However, there is only circumstantial evidence since no residue analysis has been carried out. The total Egyptian population presently numbers over 1000 breeding pairs.

Sinai Kaiser (1892, and in le Roi 1923) noted it in eastern Sinai, but there is not a

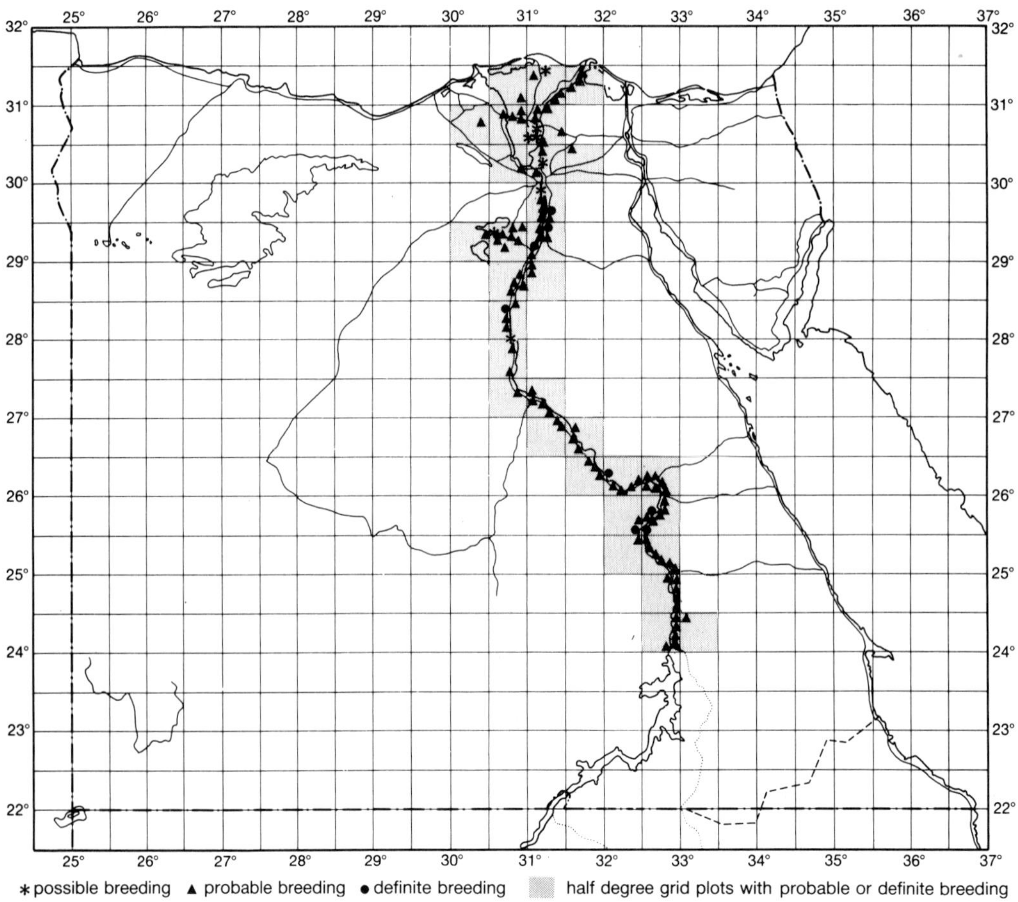

Map B11. Breeding distribution of Black-shouldered Kite *Elanus caeruleus* between 1970–87.

more recent report from the Peninsula. Four sight records are known from the region near Elat, Israel (Christensen *et al.* 1981; Paz 1987).

Black Kite *Milvus migrans* RB PV WV

Local names H̲id-a suudaa- حدأة سوداء, abuu khut̲t̲aaf, abuu s̲illat.

Scarce and local breeding resident in the Nile Delta and Valley. Fairly common passage migrant in autumn from mid-August to late October (early November) and more common in spring from (late February) March to May (mid-June), mainly along the Red Sea coast and in Sinai. Adults of the Egyptian breeding form, *M. m. aegyptius*, have yellow bills and are distinguishable in the field from adults of Eurasian migrants, *M. m. migrans*, which have black bills. The Egyptian breeding population nests between December and June.

Western Desert The Black Kite has been noted in Wadi el Natrun, Qattara Depression, Bahariya, and Dakhla. All specimens examined are of *M. m. migrans* and are presumed to be migrants (Goodman *et al.* 1986).

Nile Delta and Valley Shelley (1872) stated that this species was abundant throughout this area, particularly near refuse dumps in Cairo and Alexandria. Several sets of eggs were collected at the turn of the twentieth century at Damietta (Koenig 1907). Between 1914 and 1924 this species colonized portions of Lower Egypt it had not previously been known to breed in but 'had not reached Alexandria, Port Said or Suez' (Flower 1933). Bartlett (1942) saw a few near Alexandria in June 1917. Between 1940 and 1960 it was fairly common in the Suez Canal area. For example, several pairs bred near Ballah in 1944 (Bodenham 1945*b*) and a bird was noted carrying nest material at Ismailiya on 6 February 1945 (CJG). Nowadays it is decidedly rare in the northern portion of the Delta and there are no recent breeding records from the Faiyum or the Suez Canal area. It is still fairly common near Cairo, particularly near El Abbasia and Heliopolis. In the Nile Valley it is uncommon south to Luxor but distinctly more common from this point southwards. Hundreds of Black Kites occupy a night roost throughout the year on Kitchener's Island, Aswan. The species has been found breeding at Abu Simbil and presumably does so elsewhere along the shores of Lake Nasser.

The statement of Jourdain (1935) that this species is 'very common in most [Nile Delta] towns; present in thousands' is no longer true. During the past few decades the Egyptian breeding population of Black Kites has decreased dramatically. This decline is almost certainly a result of the wide use of insecticides and herbicides to control crop damage and rodenticides to curb massive outbreaks of rodents (Mullié and Meininger 1985).

Eurasian migrants move through the Nile corridor on both autumn and spring passage and some overwinter, particularly in Upper Egypt. Short and Horne (1981) reported large flocks in winter of black-billed birds near Aswan.

Eastern Desert and Sinai The Black Kite is a common migrant through the Eastern Desert and eastern Sinai. Some migrants fly directly across the northern Red Sea between Africa and Arabia. In autumn most Black Kites presumably cross between the Sinai and the African continent via the southern Gulf of Suez; however, there are no reports of concentrated movements across this area. During systematic counts at Suez, only 106 were observed between 13 September and 3 November 1981 (Bijlsma 1983) and 71 between 5 September and 26 November 1984 (RvD *et al.*). This species is seldom recorded on autumn passage along the north coast of Sinai.

In spring most birds follow a similar pathway through Egypt as in autumn, although some continue northwards in the Red Sea Mountains to Suez. On 22 March 1982, 684 were counted passing over Safaga (Sørensen 1983). Spring migration at Hurghada has been recorded from 27 February to 12 June, with the bulk passing from late March to early April (Marchant 1941). On 29 March 1980, 560 birds were

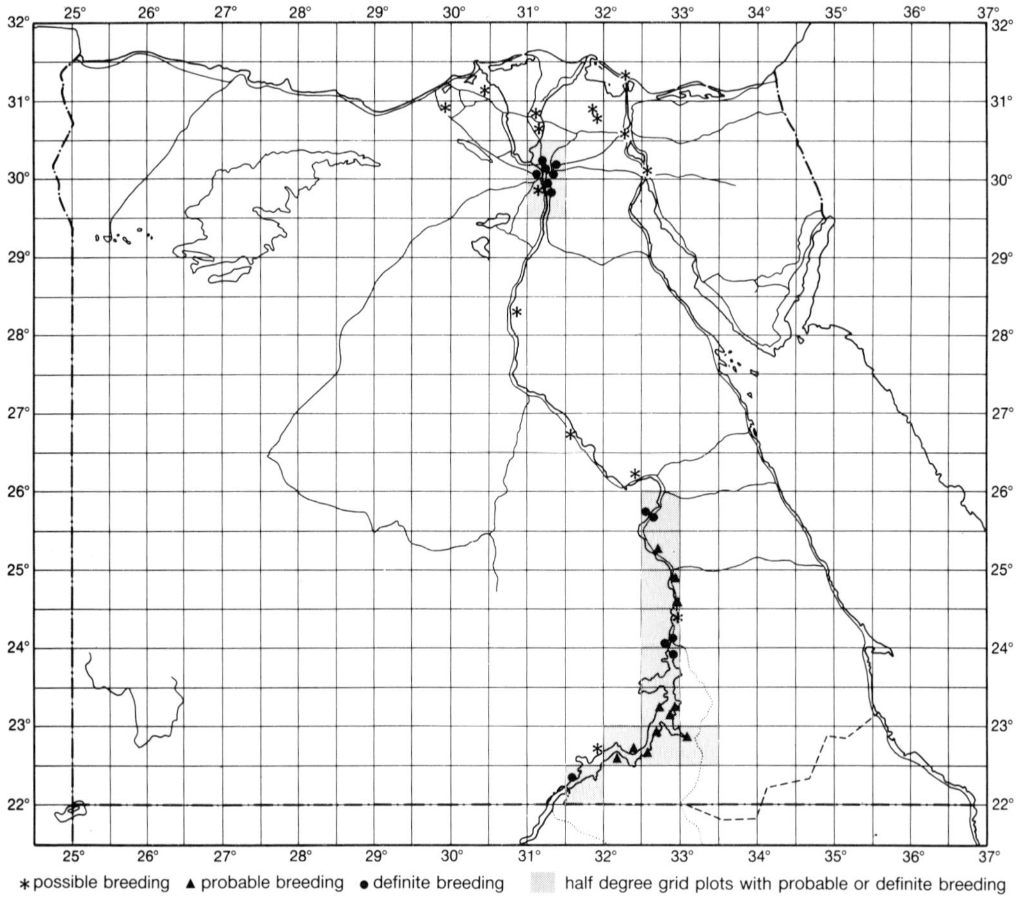

Map B12. Breeding distribution of Black Kite *Milvus migrans* between 1970–87.

counted 15 km south of Hurghada (JV). During systematic raptor counts at Suez, a total of 3861 was noted between 1 March and 16 May 1982 (Wimpfheimer *et al.* 1983). Spring migrants that cross the Gulf of Suez appear to proceed up the eastern side of Sinai towards Elat, Israel. On 8 April 1977, 1250 Black Kites were counted near Nuweiba. At Elat, 36 680 were recorded between 26 February and 5 April 1980 (Christensen *et al.* 1981).

Red Kite *Milvus milvus* (PV)

Local name H̲id-a h̲umraa- حدأة حمراء.

Rare and irregular passage visitor. The single Egyptian specimen examined is referable to *M. m. milvus*.

Nicoll (1908) reported that two 'undoubted examples' of Red Kites were observed near Tanta in January 1908 and Moreau (1928) noted one at El Shallufa, along the Suez Canal in May. El Negumi *et al.* (1950) mentioned that this species was collected near Abu Rauwash in January and December 1939 and January 1943. One of these birds is presumably the specimen reported by Goodman and Watson (1983). During systematic counts of migrating raptors at Suez, nine Red Kites were noted between 27 September and 7 October 1981, four of which were on 27 September (Bijlsma 1983); five were seen between 19 March and 17 April 1982, two of which were on 15 April (Wimpfheimer *et al.* 1983); and single juveniles were observed on 12 and 19 September 1984 (RvD *et al.*). There are also observations of single birds at Suez on 7 May 1981 (SMB, SV) and on 26 March 1984 (SG), and at Sharm el Sheikh on 10 September 1980 (JB).

African Fish Eagle *Haliaeetus vocifer* AV

Local name 'Uqaab as-samak al-afriiqii عقاب السمك الأفريقي.

Accidental visitor. Monotypic.

An adult African Fish Eagle was collected at Abu Handel, south of Aswan, on 1 November 1947 (El Negumi *et al.* 1950; GZM). Three were also observed in Sinai in November 1967 (Cramp and Simmons 1980). The Sinai records were rejected by Hovel (1987). Bones of this species have been identified from archaeological remains recovered from a fifth-dynasty site (*c.*2400 BC) at Elephantine, south of Aswan (Boessneck 1985).

White-tailed Eagle *Haliaeetus albicilla* FB? (WV)

Local names 'Uqaab abyad̲ adh-dhanab عقاب أبيض الذنب, shumiit̲a, 'uqaab al-bah̲r.

Apparently a former breeding resident. Presently irregular winter visitor. Monotypic.

According to Von Heuglin (1869*b*) a diminutive form of the White-tailed Eagle formerly bred in Egypt. In a letter from Murray Finch-Hatton to Howard Saunders, it was stated that 'several nests of this species were found upon the ground in the reedy marshes of Lake Menzaleh, and two adults were shot ... The nests were large structures, to which it was necessary to wade knee-deep in mud and water.' The young fledged in January (Meinertzhagen 1930).

Meinertzhagen (1930) considered it a somewhat scarce winter visitor and former breeding resident in the country. An adult male was taken at Wadi el Natrun on 28 February 1911, and others have been seen on occasion in December and February along the Suez Canal and on Lake Manzala. Single birds were present at Abu Zabal in January 1918 (Raw 1921) and at Sidi Abd el Rahman in February 1939 (Al Hussaini 1938–9). The only more recent record is one bird on 28 May 1954 flying north along the Suez Canal near Firdan (RN).

Generic local names for vulture Nasr, nisr, rakhama.

Lammergeyer (Bearded Vulture) *Gypaetus barbatus* RB (WV)

Local names Nasr abuu dhiqan نسر ابو ذقن, abuu li<u>h</u>ya, bawj, kaasir al-'i<u>z</u>aam, silaawi (Ma), shumiiṯa.

Rare breeding resident and migrant in southern Sinai and the Red Sea Mountains. A specimen from the Gebel Elba area appears to be referable to the East African form *G. b. meridionalis* (Goodman 1984*a*) and those from the central Red Sea Mountains and Sinai are of *G. b. aureus* (Vaurie 1965) or *G. b. barbatus* (Hiraldo *et al.* 1984).

Nile Valley and Eastern Desert Antinori (1862) stated that Lammergeyers bred at Gebel el Muqattam near Cairo, where they could be seen throughout the year. Adams (1864) saw a single bird at the Giza Pyramids on 14 November 1862. An egg of this species was reputed to have been collected 'at a cliff' near Damietta on 2 February 1860 (Jourdain 1934). Since Damietta is a most unlikely place for Lammergeyers to breed and there are no local cliffs, it is assumed that the collection locality is incorrect. Early in the twentieth century Lammergeyers were occasionally observed in Wadi Hof near Helwan, and along the Cairo–Suez Road (Meinertzhagen 1930). There are no recent records from the Nile Delta or Valley.

Von Heuglin (1869*b*) saw solitary birds at several places along the Red Sea coast, e.g. near Suez. Between 1980 and 1986 Lammergeyers were occasionally noted in the northern Red Sea Mountains near Gebel Ataqa, Gebel el Galala el Bahariya, and Gebel el Galala el Qibliya. Tregenza (1951, 1955, 1958) noted the species somewhat regularly during his travels of 1947 and 1949 in the central Red Sea Mountains at Gebel Abu Harba (nest), Gebel Abu Dukhan, Gebel Qattar, Gebel Ruweished, and Gebel Shayib el Banat (perhaps breeding). An expedition to the last locality in May 1981 did not find this species (JJH) but local Bedouins mentioned that they are still occasionally seen in the area. More recently it has been observed in January 1983 at

Gebel Abu el Hasan and on 2 May 1983 in Wadi Abu Khasiba (JJH). It is observed with some regularity in the southern Red Sea Mountains and is presumed to breed at least in the vicinity of Gebel Abraq and Gebel Elba. According to local Bedouin informants this species also nests near Gebel Hamra Dom, where an adult was noted several times in February 1984 (SMG).

Lammergeyers are mainly residents in the area and records of migrating birds are exceptional. Near Suez, one immature was noted on 8 March 1950 (Simmons 1951), and one, possibly immature, on 10 October 1981 (Bijlsma 1983).

Sinai In the nineteenth and early twentieth century Lammergeyers were not uncommon in the mountains of southern Sinai (Von Heuglin 1856; Zedlitz 1912). According to Meinertzhagen (1930) it had been recorded all over the southern portion of the Peninsula. He observed a pair copulating on Gebel Musa in March. Nowadays a few pairs still breed in the mountains of southern Sinai near Gebel Sabbagh, Gebel

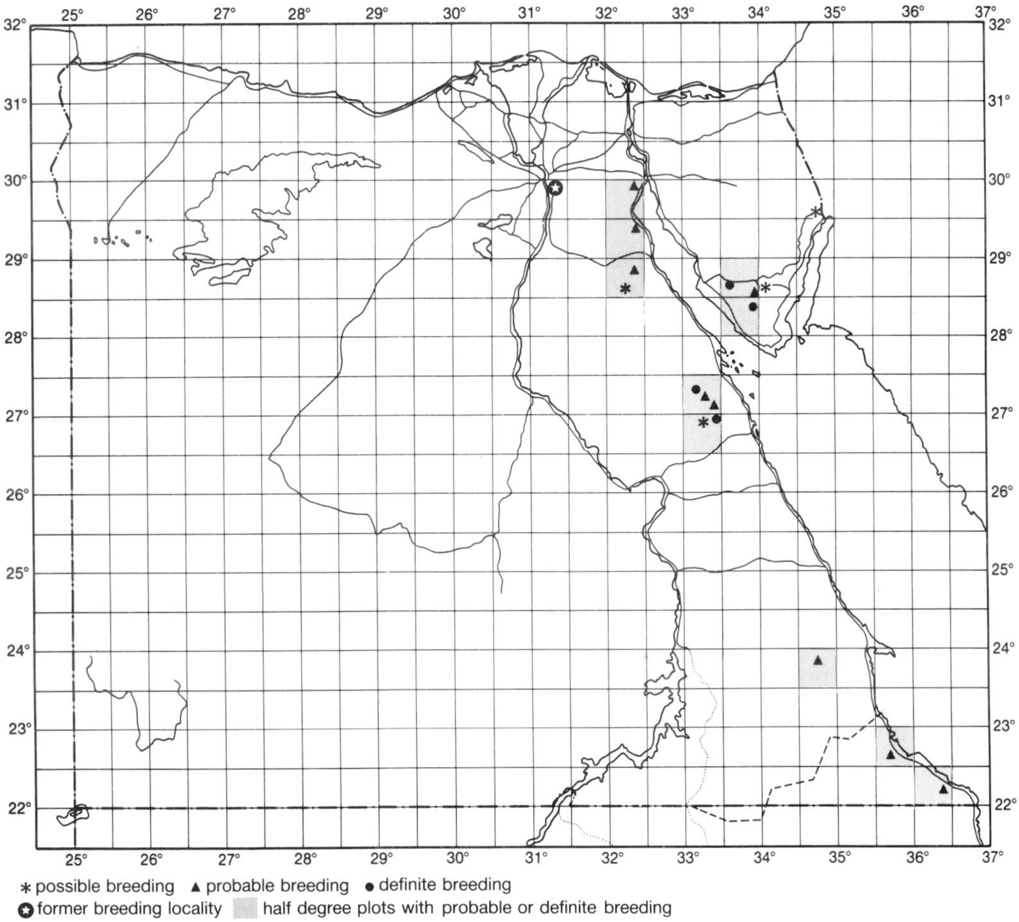

Map B13. Breeding distribution of Lammergeyer *Gypaetus barbatus* between 1945–87.

Serbal, Gebel Musa, and Gebel Umm Shomar (Mendelssohn and Leshem 1983). A pair has recently been noted with some frequency in the mountains of north-eastern Sinai but there is no proof of breeding (YL; UP). An immature was observed on 16 April 1983 flying north near Ras el Sudr (SFN); this bird may well have been a migrant.

Egyptian Vulture *Neophron percnopterus* RB PV (WV)

Local names Rakhama mas̲riyya رخمة مصرية, abyad̲ 'abaar rakhaam (Ma), anuuq, ar'i (Bi), iykiit (Bi), rakhaam (Ma), umm rakham allah.

Rare breeding resident in parts of the country, except in the extreme south-eastern corner where it is locally common. Passage visitor from late August to early June. A few migrants presumably winter in the country. *N. p. percnopterus* is the form occurring in Egypt.

The breeding status of the Egyptian Vulture in most areas of the country is not clear. This species breeds in March and April (Meinertzhagen 1930), coinciding with the period that spring migrants are passing through the country. Meinertzhagen's (1930) claim that it is solely a resident is not true.

Western Desert The statements of Meinertzhagen (1930) that this species is a breeding resident of Wadi el Natrun, Dakhla, and Kharga lack documentation. Small numbers have been observed on a few occasions in spring and autumn in several Western Desert oases, including Gebel Uweinat, but no evidence of breeding is known (Goodman *et al.* 1986).

Nile Delta and Valley Hasselquist (1766) noted 'All the places round Cairo are filled with the dead bodies of asses and camels; and thousands of these birds fly about . . .'. Gurney (1876) mentioned the Egyptian Vulture as common in Cairo, and Taylor (1867) and Shelley (1872) also mentioned it as very common in the Nile Valley. Koenig (1936) collected eggs around the turn of the twentieth century from numerous nests in Upper Egypt. Meinertzhagen (1930) noticed a decrease in numbers, especially in the more cultivated areas, and its local disappearance from the Cairo region. At present this species probably breeds in small numbers in the Nile Valley south of Luxor, particularly in the Lake Nasser area, where it is observed throughout the year. There are a few post-1930 reports from the Cairo area of possible breeding: an immature was noted on 18 June 1944 (CJG), another at Gebel el Muqattam on 10 August 1945 (White 1949), and a recently fledged bird near Heliopolis on 18 June 1979 (SMB). There is a single breeding record from the Faiyum, which is questionable. R. Sparrow (in Raw 1921) noted, 'I was told of five fresh eggs, from different nests, being taken at Hash Medibah, Fayum, on 27.iii.1909'. A nest was found on Gebel el Goza el Hamra, near Fayid, on 6 March 1952 (Mullié and Meininger 1985).

In autumn and spring small numbers of migrants are observed throughout the Nile Valley. There are a few reports from the Nile Delta, and the only summer record is one on 16 July 1942 at Abu Qir (Bodenham 1945*a*). In the northern Suez Canal area it is a regular migrant, mainly in spring from early March to early June. Late records of migrating birds include 25 at Fayid on 3 June 1954 and eight at Wadi el Abyad on 5 June 1955 (RN).

Eastern Desert This species is observed throughout the year in the mountainous areas of the Eastern Desert, and small numbers breed at various localities. In May, June, and July 1942 it was 'common' at Safaga (Bulman 1944). Breeding has been documented at Gebel Elba (El Negumi *et al.* 1950; Goodman 1984*a*), at Gebel Abraq (FMNH), at Gebel Abu Hareigal (SMB), in Wadi el Allaqi (IH; SMB), and near Marsa el Alam (MCJ *et al.*).

Sinai Zedlitz (1912) mentioned the Egyptian Vulture as relatively common in southern Sinai, and that he shot a female with an incubation patch on 18 April in Wadi Gharb (= Wadi Gharba?). The inscription 'shot on nest' is on the label of a specimen

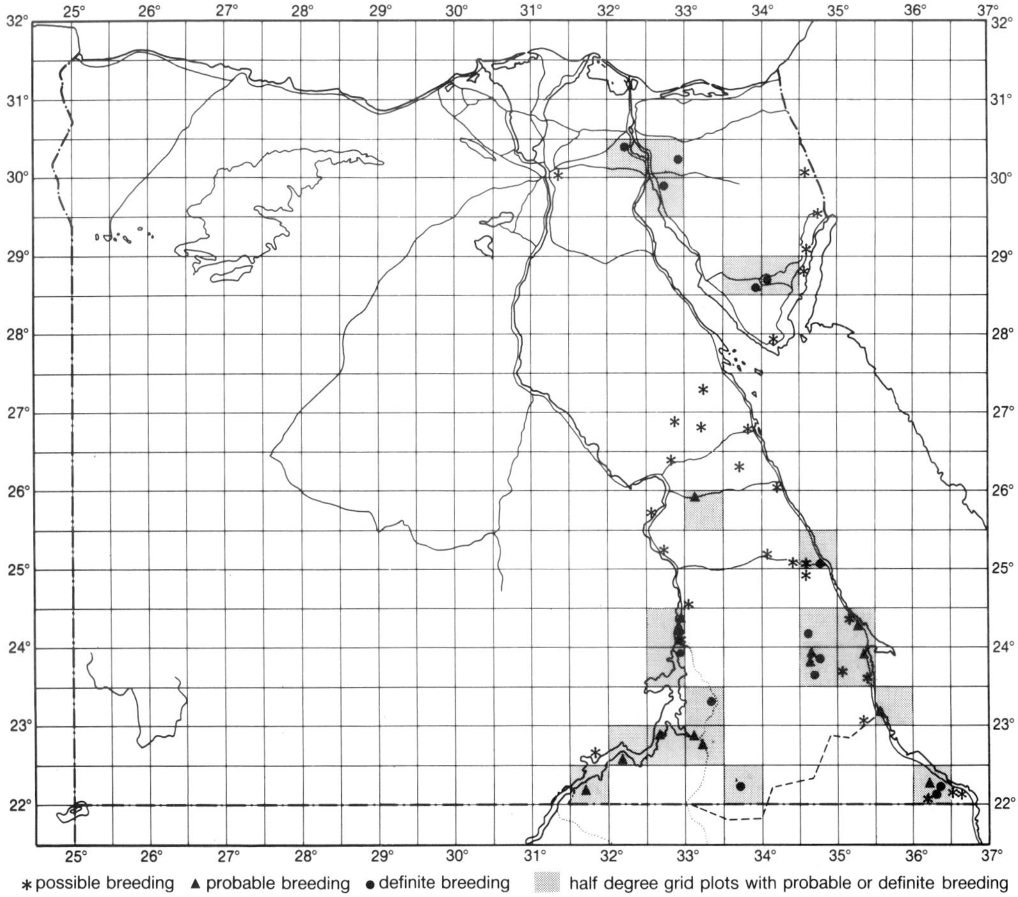

Map B14. Breeding distribution of Egyptian Vulture *Neophron percnopterus* between 1945–87.

taken on 6 May 1939 in Wadi Feiran (GZM). Recent breeding records include a nest found in approximately 1975 in Wadi Umm Hashiba (YL), a pair with fledglings on 16 August 1982 at Ain Musa (SMB), and a pair noted with a fledgling in August 1983 near St Katherine (SMB). It is also thought to breed between Qusaima and Elat, Israel (IRIC).

Considerable numbers of Egyptian Vultures migrate through Egypt in the autumn and spring, mainly via the Sinai and Eastern Desert. Autumn migration commences in late August, peaks in September, and continues into November. At Suez a total of 437 was recorded between 4 September and 31 October 1981 (Bijlsma 1983) and 1002 between 5 September and 30 October 1984 (RvD *et al.*). Small numbers regularly cross the Gulf of Suez (Marchant 1941). Spring movements begin as early as January. Between mid-February and early June small groups are regularly observed migrating north through the Red Sea Mountains. Some birds cross the southern Gulf of Suez towards Sinai but most apparently continue towards Suez or further north along the Suez Canal. A total of 1189 was counted at Suez between 24 February and 16 May 1982, with the bulk passing between mid-March and mid-April (Wimpfheimer *et al.* 1983). Near Elat, Israel, 802 were recorded between 21 February and 17 May 1977 (Christensen *et al.* 1981).

Griffon Vulture *Gyps fulvus* FB PV (WV)

Local name Nasr asmar نسر أسمر.

Former breeding resident. Scarce passage visitor in autumn from mid-September to mid-November and in spring from (late January) mid-February to mid-May, mainly in the eastern portion of the country. It is a rare winter visitor. *G. f. fulvus* is the form occurring in Egypt.

Western Desert Meinertzhagen (1930) noted that the Griffon Vulture had been observed in Wadi el Natrun, Dakhla, and Kharga but did not present any documentation for these records. There are no recent reports of this species in the Western Desert (Goodman *et al.* 1986).

Nile Valley, Eastern Desert, and Sinai In the nineteenth century this species was described as 'plentifully distributed, assembling in great numbers at carcasses' (Adams 1864) and 'very abundant' (Taylor 1867). Gurney (1876) mentioned that in 1875 a few inhabited the east bank opposite El Minya but it is not clear if this was indeed a breeding or roosting site. The only definite record of breeding in the country is given by Kaiser (1890), who stated that 'until four years ago' there was a breeding colony in Wadi Digla near Cairo, where young birds were collected from several nests. Meinertzhagen (1930) considered it a fairly common resident in Egypt (including Sinai). He stated that breeding sites were not all known but reported a colony in the hills south-west of Suez and in Sinai about 32 km south of Qusaima. El Negumi *et al.* (1950) also reiterated the claim that it resides in the country. An egg housed in the

KFMP Collection was taken on 28 March 1946 at an unspecified locality, presumably within the country. However, there appears to be no positive documentation of this species breeding anywhere in Egypt after 1900. Tregenza (1951, 1958), who explored the northern and central Red Sea Mountains in 1947 and 1949, stated that none of his Bedouin informants knew anything about a colony south-west of Suez. There is also no positive evidence for its breeding in Sinai and presumably Meinertzhagen's records were of migrant flocks. This species, however, is known to nest in the Negev Desert of Israel (Mendelssohn and Leshem 1983; Frumkin 1986), not far across the Sinai border. The only recent summer records are one at Sharm el Sheikh on 2 July 1974 (TAUM) and one at Aswan on 20 July 1981 (CB).

At present the Griffon Vulture is a regular passage migrant in Egypt, mainly in Sinai and the Eastern Desert. It is more common in autumn than spring. During a systematic study of migrating raptors at Suez between 23 September and 5 November 1981, 1284 Griffon Vultures (97 per cent juveniles and immatures) were counted, with numbers peaking in late October and passage continuing at least into early November (Bijlsma 1983). A total of 605 was recorded at Suez between 16 September and 18 November 1984 (RvD *et al.*). This species is fairly common between October and November at Elat, Israel (Krabbe 1980).

An early spring migrant was seen at Suez on 24 January 1984 (Baha el Din 1984*b*). During a count of raptors passing over Suez in the spring of 1982, 141 were noted between 24 February and 10 May (Wimpfheimer *et al.* 1983). Small flocks of migrants have been reported in autumn and spring in the southern Nile Valley. This species apparently does not cross the Gulf of Suez. At Elat, Israel, it is scarce on spring passage (Christensen *et al.* 1981).

The number of Griffon Vultures passing through Egypt seems to have drastically declined in the past few decades. Goodwin (1949) estimated 19.5 per cent of the 20 000 raptors (or about 3900 individuals) he noted at Suez between 4 and 12 October 1947 were this species. Since the majority of Griffon Vultures move through Suez at the end of October, it is clear that there is a significant difference in the number observed by Goodwin compared with recent counts. This decline is also reflected in this species' winter status (Mullié and Meininger 1985). Until about 1930 it was a regular winter visitor in considerable numbers. The only recent winter records are 12 near Na'ama on 25 December 1980 (DMe) and over 20 feeding on dead camels near Gebel Abu Hareigal, between Aswan and Wadi el Allaqi, on 29 January 1986 (SMB).

Rüppell's Vulture *Gyps rueppellii* (PV)

Local name Nasr abqa' نسر أبقع.

Irregular visitor to Egypt from sub-Saharan breeding areas. We have not examined any specimens of Rüppell's Vultures taken in Egypt but based on geographical grounds the form occurring in the country should be *G. r. rueppellii.*

There are only a few records of the Rüppell's Vulture in the country. Von Heuglin (1861) reported several immatures near Cairo in 1857. Pairs were housed in the Giza Zoological Gardens that were obtained in the desert west of Giza on an unknown date (Sclater 1895) and near Giza on 29 August 1901 (Flower 1933). Other records include single birds on an unspecified date at Aswan and in February 1928 near Quseir (Meinertzhagen 1930), on 27 March 1983 at Gebel Elba (SMG), on 15 April 1965 at Ain Sukhna (HCK), and on 24 April 1933 at Safaga (Greaves in Marchant 1941). A mixed group of one Rüppell's Vulture and several Griffon and Lappet-faced Vultures was noted at a carcass just south of Suez on 29 March 1962 (HCK). There are a few unconfirmed observations from southern Sinai (Ilani 1974, 1980).

Lappet-faced Vulture *Torgos tracheliotus* RB

Local names Nasr idhuun نسر أذون, abuu wudaan, ba'aaniib (Bi), shawb.

Rare and local breeding resident in the Eastern Desert, and probably a former breeding resident of the northern Western Desert and Sinai. A considerable amount of variation has been noted in soft part and plumage coloration of the Lappet-faced Vulture population inhabiting Gebel Elba, which appears to be related to age, sex, and individual differences. This variation appears to be continuous and includes the characters used to diagnose the recently described subspecies *T. t. negevensis*, of the Negev Desert, Israel (Bruun *et al.* 1981). Thus, we do not consider this subspecies distinctive and subsequently do not recognize any geographical forms of the Lappet-faced Vulture.

Western Desert There is evidence that Lappet-faced Vultures bred or breed in the Qattara Depression. In 1984, three old nests, of approximately 1.5 m in diameter, were found in a clump of large acacia trees at Hatiyet el Saqr, Qattara Depression (SMB). *Torgos tracheliotus* is the only large vulture in Egypt known to breed in trees and it seems likely that these nests belonged to this species. In the 1950s 'large dark vultures' were numerous near Alexandria, with flocks of up to 20 seen daily. Beginning in the early to mid-1960s they drastically decreased in number and subsequently disappeared from the area (Mullié and Meininger 1985). These birds may well have been Lappet-faced Vultures.

Nile Valley and Eastern Desert Meinertzhagen (1930) mentioned this species as probably the most common vulture to be seen in Egypt, although it had not been recorded north of Cairo and no nest had been found in the country. He also stated that sometimes hundreds were seen near Helwan and large numbers near Luxor and Quseir. One was collected in the Faiyum on 3 September 1927 (GZM). Sparrow (in Raw 1921) mentioned a nest near Asyut with one egg and placed in a small *sont* tree (*Acacia* sp.) as possibly the Black Vulture. In the 1940s local Bedouins of the central and northern Red Sea Mountains gave convincing descriptions of two pairs of vultures nesting in the northern Eastern Desert. One nest was in an acacia at the head of Wadi el Tarfa, Gebel el Galala el Qibliya, and the other in a *yasar* tree (*Moringa peregrina*) in Wadi el Sheikh (Tregenza 1951). In two different years a nest was found in an acacia at the divide between the Red Sea watershed of Wadi Araba and the Nile Valley watershed of Wadi Sannur. The first year the nest contained one egg and the next year one nestling (Tregenza 1958). It seems likely that all of these records concern Lappet-faced Vultures. On 29 March 1962 a single bird was observed near Suez feeding on a carcass (HCK).

In recent years this species has been regularly recorded in small numbers in the southern portion of the Eastern Desert. Most records are from the Idfu–Marsa el Alam road, near Aswan, Gebel Abraq, and throughout the Sudan Government Administration Area. Breeding has been documented in several wadis around Gebel Elba. On 23 June 1984, four active nests were found in *Acacia raddiana* trees within a 2 km area in Wadi el Kharit, just west of Sheikh Shadli (SMG). This species

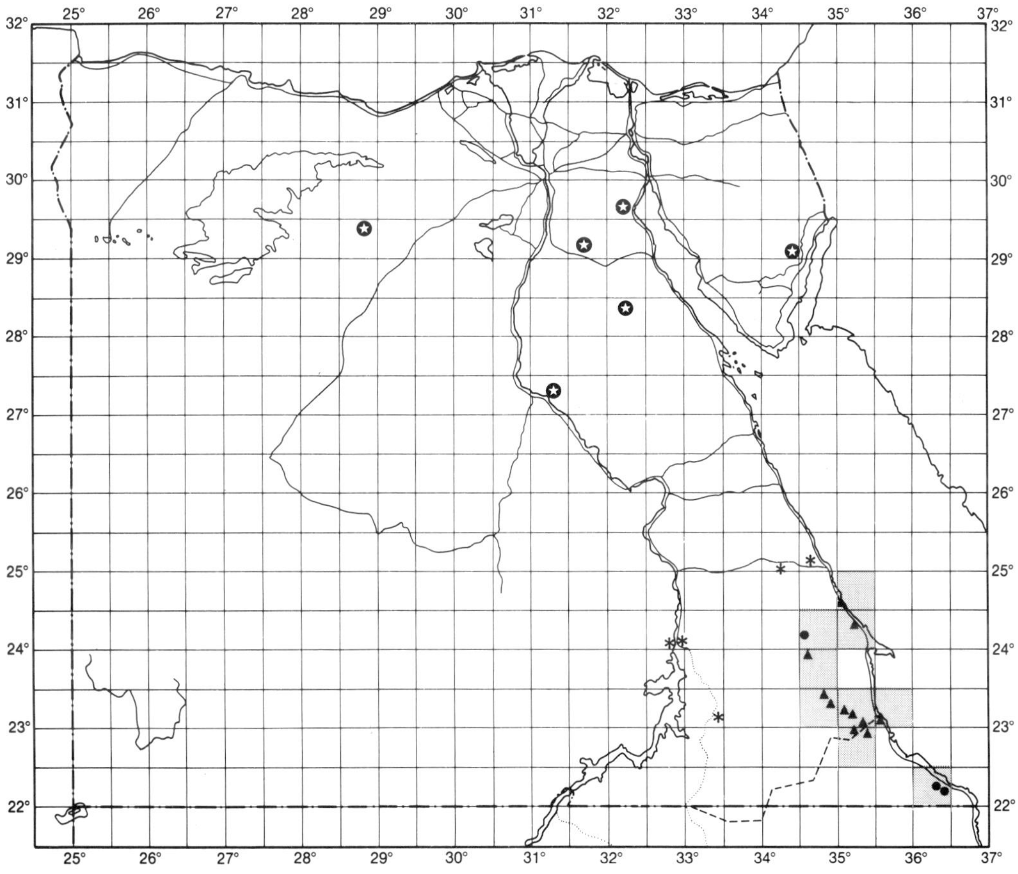

Map B15. Breeding distribution of Lappet-faced Vulture *Torgos tracheliotus* between 1970–87.

presumably nests in other areas of the southern Eastern Desert but has retreated from the northern portion. The large numbers recorded in the early twentieth century indicate that a considerable breeding population once inhabited the Egyptian Eastern Desert, and possibly this population was augmented during the non-breeding season by birds moving north from the Sudan. In Sudan, vulture populations dropped significantly which could 'almost certainly be related to the indiscriminate use of strychnine to poison hyenas' (Wilson 1982).

Sinai In 1971 an old and unoccupied nest was found in Wadi Watir (UP). A few pairs still breed in Wadi Arava, north of Elat, Israel (Mendelssohn 1975; Cramp and Simmons 1980; Frumkin 1986). We are not aware of any recent Sinai records.

Black (Cinereous) Vulture *Aegypius monachus* (PV) (WV)

Local name Nasr aswad نسر أسود.

Rare passage migrant and winter visitor to Egypt. Monotypic.

There are several sight records from the nineteenth and early twentieth century of single or small numbers of Black Vultures in the Nile Valley (Taylor 1867; Shelley 1872; Gurney 1876; Sherman 1915). Meinertzhagen (1930) stated that this species infrequently occurred throughout the year and that it possibly bred in the country. One specimen was collected at El Mansuriya on 4 March 1911 (GZM), and single birds observed east of Suez, and between the Nile Valley and Kharga in March 1928 (Meinertzhagen 1930). It is plausible that some of the above sight records are actually of misidentified Lappet-faced Vultures.

There are only a few recent observations of this species in Egypt: Suez—single juveniles on 17, 25, and 27 October 1981 (Bijlsma 1983) and on 7 November 1984 (RvD *et al.*), and Gulf of Suez—two on 18 April 1960 and one on 11 April 1965 (HCK). This species is a rare migrant at Elat, Israel (Krabbe 1980).

Short-toed Eagle *Circaetus gallicus* CB PV (WV)

Local names 'Uqaab ṣaraara عقاب صرارة, 'uqaab abyaḏ.

Casual breeding resident. Fairly common migrant in autumn from (mid-August) early September to early (late) November and in spring from late February to mid-May. Occasional summer and winter visitor. *C. g. gallicus* is the form occurring in Egypt.

There are only a few breeding records of Short-toed Eagles in Egypt. A nest containing two young was reputedly found in Wadi el Natrun in 1911 (Nicoll 1912*b*), and a nest with a single egg was found at Gebel el Muqattam on 5 April 1923 (Meinertzhagen 1930; BMNH). Short-toed Eagles were regularly observed in the Red Sea Mountains in June and July 1947 and 1949 (Tregenza 1951, 1955) but no evidence of breeding was found. According to Meinertzhagen (1930) it 'doubtless' breeds in Sinai but the only confirmation of this is a nest found in the early 1970s west of Elat, Israel (UP).

Autumn passage generally begins in early September, peaks between late September and mid-October, and continues into November. It is most abundant in Sinai and the Red Sea Mountains. Most birds avoid crossing the Gulf of Suez and pass through Suez, where during systematic raptor counts 2476 birds were recorded between 6 and 12 October 1980 (Meininger *et al.* 1981), 9447 between 6 September and 5 November 1981 (Bijlsma 1983), and 12 136 between 5 September and 25 November 1984 (RvD *et al.*). Small numbers are observed in autumn in the Nile Valley and Faiyum. The only recent winter record is one in Wadi Buhluk, between Qena and Quseir on 17 January 1986 (SMB).

In spring, as in autumn, small numbers of birds migrate through the Nile Valley but the bulk of passage is through the Red Sea Mountains and Sinai. This species generally avoids crossing the Gulf of Suez and considerable numbers funnel through Suez, e.g. 3063 were recorded between 24 February and 10 May 1982 (Wimpfheimer *et al.* 1983). Along the North Coast and in the Western Desert it has been recorded in small numbers on spring passage. At Elat, Israel, it is relatively scarce in spring (Christensen *et al.* 1981).

Bateleur *Terathopius ecaudatus* RB

Local name 'Uqaab h̲akiim عقاب حكيم.

Rare and local breeding resident at Gebel Elba. Monotypic.

In June 1984 a Bateleur was observed tending a nest in a tributary of Wadi Aideib, Gebel Elba. In the winter and spring of 1985 this species was still present at this locality but the nest had been partially destroyed. A maximum of two birds has been observed in the area at any one time, and a local Bischarin informant mentioned that only a single pair nests in the immediate Gebel Elba area (SMG). The only other record from Egypt appears to be a first-year male observed at the Abu Simbil airport on 23 December 1985 (DM). This species was observed at Elat, Israel, on 15 April 1982 (Ekins and Hendriks 1983).

Generic local names for harrier Murza, s̲aqr.

Marsh Harrier *Circus aeruginosus* PV WV

Local names Murza durii'a مرزة دريعة, darraa'a, h̲idim.

Scarce passage and winter visitor throughout the country between late August and late May (early June). All Egyptian material examined is referable to *C. a. aeruginosus*.

Adult Marsh Harriers have been observed in proper breeding habitat on several occasions in May near the Nile Delta lakes (Ticehurst 1912; Meininger *et al.* 1986) and at various times of the year in Wadi el Natrun (Nicoll 1912*b*; Maclaren 1944). Although this species has never been found breeding in the country, it is plausible that it might do so (sporadically?).

Between late August and November small numbers of Marsh Harriers migrate along a broad front through the Nile Valley, Western and Eastern Deserts, and Sinai. At Suez, 35 were counted between 13 September and 2 October 1981 (Bijlsma 1983) and 25 between 10 September and 24 October 1984 (RvD *et al.*). In winter small numbers have been observed throughout the Nile Delta and Valley, in Wadi el Natrun, and Dakhla (Goodman *et al.* 1986). At least several hundred winter in the vicinity of the Nile Delta lakes (Mullié and Meininger 1985).

Spring migration begins in late February, peaks in mid-March to mid-April, and continues at least until early June. Spring passage, like autumn, also proceeds along a broad front throughout the country but there is a distinct concentration in the Nile Valley and along the Red Sea coastal plain.

Hen Harrier *Circus cyaneus* PV WV

Local names Murza ad-dajaaj مرزة الدجاج, abuu huṣn, ṣaqr al-fayraan, 'uqaab ad-dajaaj.

Rare passage and winter visitor between late September and mid-April. *C. c. cyaneus* is the form occurring in the country.

Between late September and mid-April Hen Harriers are observed in small numbers throughout the Nile Delta and Valley, with some concentration near the Delta lakes. There are a few records of this species in the Western Desert between late September and 17 April (Goodman *et al.* 1986; HB *et al.*). In both autumn and spring small numbers migrate through Sinai, the Red Sea Mountains, and along the coastal plain.

Pallid Harrier *Circus macrourus* PV WV

Local names Murza baahita مرزة باهتة, murza baghthaa-.

Scarce passage and winter visitor between (early) mid-September and early May. Monotypic.

Autumn passage of Pallid Harriers is along a broad front, including the Western and Eastern Deserts, and begins in early September and continues into November. The majority of migrants reaching Egypt move to areas further south in Africa but some remain for the winter in the Nile Delta and Valley. The number wintering in Egypt seems to have decreased since the turn of the twentieth century (Meinertzhagen 1930).

Spring migration is also along a broad front and proceeds from mid-March to early May, peaking between late March and mid-April. Some concentration of migrants occurs along the Red Sea coastal plain and in eastern Sinai. On both passages this species freely crosses the Red and Mediterranean Seas.

Montagu's Harrier *Circus pygargus* PV (WV)

Local names Murza muuntaaguu مرزة مونتاجو, abuu sharada.

Rare passage visitor in autumn from late August to late October (late November) and in spring from mid-March to late April (mid-May) (Fig. 6.10). Occasional in winter. Monotypic.

Small numbers of Montagu's Harriers are noted on both autumn and spring migration in the Nile Delta and Valley, the Eastern Desert, and Sinai. Some concentration of migrants occurs along the Red Sea coastal plain and in Sinai. Between 3 and 23 September 1982, 27 were observed at Zaranikh (Baha el Din and Salama 1984). A late migrant was noted at Suez on 23 November 1984 (RvD *et al.*). Records from the Western Desert include birds in Wadi el Natrun on 2 April 1981 (Goodman *et al.* 1986), in Bahariya on 29, 30, and 31 August and 3 September 1983 (HB *et al.*), and at East Uweinat on 8 April 1984 (Baha el Din in press). The only known winter record is an adult male shot at Inchas on 28 January (Meinertzhagen 1930).

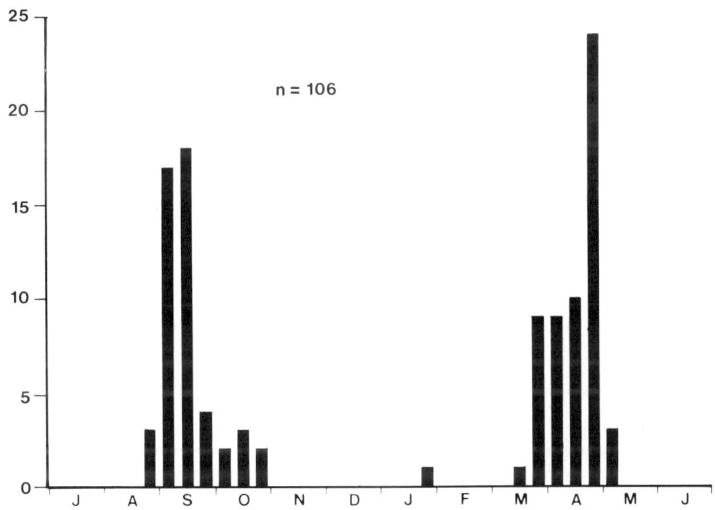

FIG. 6.10. Histogram of the numbers of Montagu's Harriers recorded during 10-day periods throughout the year.

Gabar Goshawk *Micronisus gabar* AV

Local name Baaz jabbaar باز جبّار.

Accidental visitor. Monotypic.

Shelley (1872) mentioned a specimen of the Gabar Goshawk reputed to have been collected in Upper Egypt. In the RNHL there is an old mounted specimen taken in

'Égypte' by Dr Clot. Schlegel (1862) reported that there was a specimen in the Berlin Museum taken at Suez. In January and February 1904, two pairs of Gabar Goshawks were observed flying along the banks of the Nile at El Shallal, south of Aswan (Russell 1905). In the early twentieth century Flower (1933) sometimes saw this species in August and September in the Upper Nile Valley as far north as Aswan. He was under the impression that they were breeding birds from the Sudan that followed the river northwards during the inundation. There are no recent records from the country. Mummified material of this species has been identified from Dynastic Egypt (Lortet and Gaillard 1903).

Generic local names for *Accipiter* spp. Baashiq, baaz.

Goshawk *Accipiter gentilis* WV

Local name Baaz باز.

Rare winter visitor between mid-September and early May to the northern portion of the country. *A. g. gentilis* is the form occurring in Egypt.

The majority of Goshawk records are from the Suez Canal area. Single birds have been noted at Suez on 12 September and 8 October 1984 (RvD *et al.*), 18 September 1981 (Bijlsma 1983), and 22 April 1983 (JDG). During the winter of 1952–3 a maximum of two were found at Fayid and one at El Kabrit (AP). Birds were also noted south of Port Said on 6 March 1935 and at Kafr Hemaid on 13 March 1935 (Jourdain and Lynes 1936). The southern-most Nile Valley record appears to be one collected on 24 March 1868 at Beni Suef (Shelley 1872). Goshawks have also been observed at Hurghada on 21 April 1983 and in the Red Sea Mountains along the Marsa el Alam–Idfu road on 27 April 1983 (MCJ *et al.*). The latest spring record is one at Fayid on 5 May 1955 (RN). There are only a few records of migrants at Elat, Israel (Krabbe 1980; Christensen *et al.* 1981).

Sparrowhawk *Accipiter nisus* PV WV

Local names Baashiq باشق, 'allaam, ṯuuṯ.

Fairly common passage visitor throughout the country and scarce winter visitor from late September to mid-May. The majority of birds occurring in Egypt are of *A. n. nisus*, although a few specimens appear referable to *A. n. nisosimilis* (Meinertzhagen 1930).

Sparrowhawks have been regularly recorded in the Nile Delta and Valley, the Suez Canal area, and in Sinai. There are a few records from Wadi el Natrun and Kharga, almost all in autumn (Goodman *et al.* 1986). Up to three birds were regularly seen at Bahariya between 28 September and 29 October 1983 (HB *et al.*). The main migratory pathway in Egypt appears to be through Sinai and the Red Sea Mountains. It also crosses the Gulf of Suez and Red Sea. Single migrants were observed at the Daedalus Reef, central Red Sea, on 29 February and 6 March 1936 (Bulman 1944). Some concentration occurs at Suez, e.g. 23 were counted between 23 September and 4 November 1981 (Bijlsma 1983), 51 between mid-September and late October 1984 (RvD *et al.*), and 116 between 14 April and 11 May 1982 (Wimpfheimer *et al.* 1983). At least 75 migrants were noted in the vicinity of Gebel Elba between 7 and 24 April 1985 (SMG, PLM). Birds are usually observed singly or rarely in small flocks. Winter records are generally confined to the Nile Delta and Valley. Near Elat, Israel, small numbers of migrants have been recorded in spring, with a maximum of 155 between 3 March and 9 May 1977 (Christensen *et al.* 1981).

Ringing A bird ringed at St Katherine on 20 April 1972 was recaptured at the same locality on 30 June 1977 (Hovel 1987).

Levant Sparrowhawk *Accipiter brevipes* PV

Local name Baydaq بيدق.

Fairly common passage visitor, almost exclusively in the Eastern Desert and Sinai, in autumn from mid-September to early (late) November and in spring from (mid-March) mid-April to mid-May. Monotypic.

In autumn large numbers of Levant Sparrowhawks migrate through Kfar Kasem, Israel, where 16 479 were counted from 30 August to 15 October 1981 (Dovrat 1982). Presumably, these birds follow a predominantly southerly course through Sinai. However, recent observations do not support this, for only small numbers of Levant Sparrowhawks have been recorded between late September and early November in southern Sinai and near Hurghada. During systematic raptor counts at Suez, 41 Levant Sparrowhawks were observed between 27 September and 24 October 1981 (Bijlsma 1983) and 29 between 23 September and 22 October 1984 (RvD *et al.*). The earliest autumn record is one in Wadi Feiran on 11 September 1920 (AMNH) and the latest is one at Sharm el Sheikh on 26 November 1978 (TAUM).

In spring considerable numbers pass through Egypt, generally in large flocks. The principal route appears to be through the Eastern Desert, across the Gulf of Suez, along the eastern coast of Sinai, and north to Elat, Israel. On 11 May 1961, thousands of 'Sparrowhawks' were seen crossing the Gulf of Suez near Gezira Ashrafi (De Openshaw 1961); on 22 April 1979, 1440 were observed at Ras Muhammad (HW); and on 25 April 1983 about 500 were noted over Ghubbet el Zeit (MCJ *et al.*). Almost 6000 Levant Sparrowhawks were counted at Elat, Israel, between 30 March and 10 May 1977, with peaks of 2558 on 20 April and 1182 on 26 April (Christensen *et al.* 1981). Some flocks follow the Red Sea coast north to Suez; between 12 April and 12 May 1982, 141 were counted, of which 120 passed in one flock on 16 April (Wimpfheimer *et al.* 1983). A flock of 45 was noted on 21 April 1985 at El Arish (Fry *et al.* 1985).

There are a few records of this species in the Nile Valley during migration but no documented winter records. The earliest spring records are a female at Giza on 19 March 1928 (Meinertzhagen 1930) and one at Luxor on 22 March 1984 (Durdin 1984). The only report from the Western Desert is one collected in Wadi el Natrun on 30 April 1910 (Nicoll 1912*a*; Goodman *et al.* 1986). One was taken at Bahig on 5 May 1971 (USNM).

Generic local names for buzzards Hamiimaq, ihdayii (Ma), saqr jarraah.

Buzzard *Buteo buteo* PV WV

Local name Saqr huwwaam صقر حوّام.

Fairly common passage visitor in autumn from early September to mid-November and common in spring from mid-February to mid- (late) May, mainly in the Eastern Desert and Sinai. Uncommon winter and summer visitor. All recent Egyptian material is referable to *B. b. vulpinus*, except for one specimen taken on 28 September 1962 in Wadi Qiseib, which approaches the plumage coloration of *B. b. buteo* (FMNH).

We are not aware of any evidence of Buzzards breeding in Egypt (cf. Nicoll 1919; Sparrow in Raw 1921). Presumably the breeding records mentioned by these authors were of Long-legged Buzzards.

Relatively small numbers of Buzzards have been recorded on autumn migration in Sinai and in the Eastern Desert. Many of these birds cross the southern Gulf of Suez and others funnel through Suez. During systematic counts at Suez, 640 Buzzards were recorded between 13 September and 4 November 1981, with a maximum of 66 on 9 October (Bijlsma 1983), and 857 between 5 September and 25 November 1984 (RvD *et al.*). Small flocks also migrate in autumn along the North Coast, through the Western Desert, and the Nile Valley. This species is rarely recorded along the Nile in winter.

In spring, the Buzzard is a distinctly more common migrant than in autumn. From late February to mid-May large numbers migrate through the Eastern Desert, many of which cross the southern Gulf of Suez, and the rest continue north to Suez. A total of 80 887 was counted at Suez between 24 February and 16 May 1982, with the bulk passing in April (Wimpfheimer *et al.* 1983). At Elat, Israel, about 316 000 were recorded between 20 February and 16 May 1977, with a peak between late March and early April (Christensen *et al.* 1981). Other reports of concentrated spring migration include: 14 300 at Nuweiba from 8 to 10 April 1977 (RSS); 6000 at Suez on 28 March 1983 (PLM, HR); 3000 in one flock over St Katherine on 31 March 1983 (PLM); 3370 in the mountains west of Safaga between 21 and 23 March 1982 (Sørensen 1983); and about 2100 at Gebel Elba between 6 and 7 April 1985 (SMG, PLM). In spring smaller flocks of migrating Buzzards have been regularly recorded in the Nile Delta and Valley, the Faiyum, along the North Coast, and in the Western Desert. At Lake Bardawil, 1650 were counted between 20 April and 3 May 1985 (Fry *et al.*

1985). Spring passage is generally completed by late May. Relatively recent summer records include one at Hurghada on 12 June 1981 (SMB) and at Abu Rauwash on 22 July 1969 (USNM).

Long-legged Buzzard *Buteo rufinus* CB PV WV

Local names S̲aqr jarraah̲ صقر جرّاح, h̲uwaam, siqaawa.

Casual breeding resident. Scarce passage and winter visitor between mid-September and mid-May. The form breeding in Egypt is *B. r. cirtensis* and the majority of migrants *B. r. rufinus*.

There are a few breeding records of the Long-legged Buzzard in Egypt. From April to July 1910, a pair was observed daily at Giza, and in June they were accompanied by two fledglings (Nicoll 1912*a*). Two recently hatched nestlings were found 40 km east of Cairo on 4 April 1923 (Meinertzhagen 1930). In May 1939 Galal el Din and Fouad (1939–40) observed 'Buzzards ... circling over their young in nests on the rock cliffs' at Ain Sukhna and 'Buzzards nesting in the cliffs' at Ras el Naqb. It is not clear if Galal el Din and Fouad actually found nests or just presumed they were present based on the adults' behaviour, and thus, confusion with migrants cannot be ruled out. Tregenza (1951) suspected that a buzzard, presumably Long-legged, was resident in the central Red Sea Mountains but presented no details. The only recent breeding record is a nest found on 27 June 1978 in Wadi Tinia, near St Katherine (YL). Until 1979 there was probably a breeding pair in the Wadi el Giddi and Gebel Umm Khisheib area, and it was presumed to nest in the mountains between Qusaima and Elat, Israel (IRIC). One was captured in Wadi Umm Khisheib on 13 June 1975 (TAUM). A 'displaying pair' was noted between Beni Suef and El Minya on 17 March 1984 (Durdin 1984).

Considerable numbers of Long-legged Buzzards migrate through central and southern Sinai and the Eastern Desert, some crossing the southern Gulf of Suez but most passing through Suez. Between 18 September and 5 November 1981, 1816 were observed at Suez, with numbers sharply tapering in the last 10 days of October, and a few passing into November (Bijlsma 1983). From mid-September to mid-May small numbers of Long-legged Buzzards have been observed in all portions of the country. In winter the species is most frequently observed on the edge of the Nile Delta and Valley and specimens taken during this period are referable to both *B. r. rufinus* and *B. r. cirtensis*.

Spring migration is most notable in the Eastern Desert and Sinai, and fewer birds pass than in autumn. At Suez, 53 were counted between 24 February and 16 April 1982 (Wimpfheimer *et al.* 1983). At Elat, Israel, small numbers of migrants are noted from mid-February to late April (Christensen *et al.* 1981).

Generic local names for eagle 'Uqaab, ih̲dayii (Ma).

Lesser Spotted Eagle *Aquila pomarina* PV (WV)

Local name 'Uqaab sa'faa- s̲ughraa عقاب سعفاء صغرى.

Fairly common passage visitor in autumn from early September to late November and in spring from (late January) late February to mid-May. Rare winter visitor. *A. p. pomarina* is the form occurring in the country.

On autumn migration a considerable number of Lesser Spotted Eagles pass through Sinai and the Eastern Desert. They generally avoid crossing the Gulf of Suez. At Suez, nearly 22 000 were counted between 10 September and 31 October 1981, the bulk passing between 24 September and 20 October, with daily maxima of 2708 on 29 September and 3193 on 9 October. An estimated 40 000–45 000 Lesser Spotted Eagles were believed to have actually passed Suez in the autumn of 1981 (Bijlsma 1983). Between 5 September and 19 November 1984, 31 559 birds were counted at Suez (RvD *et al.*). Small numbers have been reported from the Upper Nile Valley in September and October, and a specimen was collected in the Faiyum on 11 November 1909 (Meinertzhagen 1930). On 15 October 1976, 150 were noted at Nuweiba (LNA). The species is a rare autumn migrant at Elat, Israel (Krabbe 1980). The only Egyptian winter records appear to be single birds at Qulusna on 6 January 1875 (FMNH), and at Suez between 9 and 11 December 1980 and near El Tor on 9 December 1983 (SMB). Early spring migration commences in late January: over 300 birds were counted at Suez on 24 January 1984 (Baha el Din 1984*a*).

Spring passage at Suez is rather pronounced. Over 10 000 birds were counted on five days between 26 February and 10 March 1981 (Baha el Din and Bruun 1981). A total of 7755 was observed from 24 February to 11 May 1982, with the bulk passing between late February and early March and a smaller peak in late March (Wimpfheimer *et al.* 1983). Between late February and early May small numbers of Lesser Spotted Eagles have been reported from the Nile Delta and Valley, the northern Suez Canal area, and the mountains of south-central Sinai. The only definite record from the Western Desert is one on 24 April 1981 at Dakhla (Goodman *et al.* 1986). At Elat, Israel, it is considered a scarce migrant from early March to early May, with most birds passing from late March to early April (Christensen *et al.* 1981).

Spotted Eagle *Aquila clanga* PV WV

Local name 'Uqaab sa'faa- kubraa عقاب سعفاء كبرى.

Rare passage and winter visitor between mid-September and early May. Monotypic.

In the nineteenth century the Spotted Eagle was one of the most common wintering *Aquila* in Egypt (Adams 1864; Taylor 1867; Shelley 1872). In the first third of the twentieth century it was still common in winter (Meinertzhagen 1930), being the most abundant eagle in the Nile Delta (Meinertzhagen 1954). It has since dramatically decreased in numbers, particularly as a winter visitor (Mullié and Meininger 1985). There are only a few recent records from the Nile Valley during the migration periods and none in winter from the Nile Delta. In four winters between 1976 and 1979 only six birds were observed along the Nile Valley (Short and Horne 1981).

This species migrates through Sinai and the Red Sea Mountains. Records at Suez include 86 between 16 September and 1 November 1981, with a maximum of 12 on 9 October (Bijlsma 1983), and 42 between 17 September and 16 November 1984 (RvD *et al.*).

On spring migration, a total of 74 was observed at Suez between 27 February and 10 March 1981 (Baha el Din and Bruun 1981) and 20 between 26 February and 5 May 1982 (Wimpfheimer *et al.* 1983). It is a rare spring migrant at Elat, Israel (Christensen *et al.* 1981).

Steppe Eagle *Aquila (rapax) nipalensis* PV (WV)

Local name 'Uqaab as-suhuul عقاب السهول.

Fairly common autumn migrant from early September to late November and in spring from (late January) early February to mid-May. It is rare in winter and summer. The form *A. n. orientalis* is the one occurring in the country.

In autumn large numbers of Steppe Eagles migrate through Sinai and the Red Sea Mountains. They generally appear to avoid crossing the Gulf of Suez. During systematic counts of raptors at Suez from 4 September to 5 November 1981, 64 880 were observed, and the total passing through the area may have been over 100 000 (Bijlsma 1983). Peak migration was between mid-October and early November, and the highest daily total was 8240 on 25 October. In autumn most Steppe Eagles pass through Egypt via the Eastern Desert and only small numbers have been observed in the Nile Valley. The only autumn record west of the Nile Valley is an immature between Matruh and El Alamein on 20 September 1980 (WCM). There is a heavy autumn migration of Steppe Eagles at Elat, Israel, with 24 246 counted on 55 days between 24 September and 29 November 1980 (Shirihai 1982).

The only recent Egyptian winter records include single birds at Sharm el Sheikh on 25 December 1980 and from 23 to 31 December 1981 (BCD), and on 2 December 1983 (RvD *et al.*).

Spring migration may commence as early as late January, as one was noted between Safaga and Hurghada on 30 January 1983 (BR). At Suez, 15 778 were observed between 25 February and 16 May 1982, with the bulk passing between late March and mid-April (Wimpfheimer *et al.* 1983). About 610 Steppe Eagles were counted moving north over the mountains west of Safaga between 21 and 23 March 1982 (Sørensen 1983). From late February to early May small numbers of Steppe Eagles are recorded in the Nile Delta and Valley, and the Faiyum. There are a few spring records from the Western Desert, the earliest being two at Dakhla on 2 February 1983 (Goodman *et al.* 1986). The only known summer records are one near Ain Sukhna on 16 June 1980 and an immature 50 km west of Safaga on 22 June 1981 (SMB).

At Elat, Israel, this species is a common migrant between mid-February and late April, with the bulk passing between late February and early March (Christensen *et al.* 1981). It is striking that the majority of Steppe Eagles pass through Elat earlier than Suez.

Tawny Eagle *Aquila rapax* AV

Local name 'Uqaab lum'a عقاب لموعة.

Apparently an accidental visitor. The single Egyptian specimen examined is referable to *A. r. belisarius.*

The only verifiable record of the Tawny Eagle in Egypt is a specimen taken at Giza on 20 January 1924 (Meinertzhagen 1930; GZM). Published sight records of Tawny Eagles (e.g. Short and Horne 1981) are more likely misidentified Steppe Eagles.

Imperial Eagle *Aquila heliaca* PV (WV)

Local names 'Uqaab malakii عقاب ملكي, khaaṯiya.

Rare migrant in autumn from mid-September to mid-November and in spring from mid-February to mid-May. It is an irregular winter visitor. *A. h. heliaca* is the form occurring in Egypt.

Small numbers of Imperial Eagles are found in Egypt during autumn, particularly in Sinai and the Red Sea Mountains. At Suez, 556 migrants were recorded between 20 September and 5 November 1981. During this period, peak migration was from late October to early November, with 97 birds on 27 October (Bijlsma 1983). A total of 178 was counted between 16 September and 20 November 1984 at Suez (RvD *et al.*). There are several old autumn and a few midwinter sight records from the Nile Valley, many of which may be of other eagle species. The only known post-Meinertzhagen (1930) winter records include single birds at Abu Rauwash on 4 December 1939 (HGD), at Aswan on 24 December 1983 (PE), and one, perhaps an early migrant, between Safaga and Hurghada on 26 January 1983 (BR).

Thirty-six Imperial Eagles were observed on spring passage at Suez between 27 March and 12 May 1982 (Wimpfheimer *et al.* 1983). At Elat, Israel, the species was found to be a regular spring migrant in small numbers from mid-February to mid-April, with adults passing before immatures (Christensen *et al.* 1981).

Golden Eagle *Aquila chrysaetos* RB (WV)

Local names 'Uqaab dhahabii عقاب ذهبي, 'ajuz, lammaa'a, 'uqaab al-kabiir.

Rare breeding resident in north-eastern Sinai. It is occasionally observed elsewhere in the country. All Egyptian material examined is referable to *A. c. homeyeri*.

Nile Valley and Eastern Desert Adult Golden Eagles have been collected near Helwan in December and January (years not given) (Meinertzhagen 1930). It is plausible that these birds were breeding locally. In the summers of 1947 and 1949 several 'large eagles' were seen in the central Red Sea Mountains (Tregenza 1951). There are only a few recent records from this area. Solitary immatures were seen at El Balyana on 9 February and Sohag on 31 October 1978 (Short and Horne 1981).

Passage through Suez appears to be minimal. During a systematic raptor count at Suez between 4 September and 5 November 1981, two sub-adults were recorded on 18 September, two on 16 October, and one on 25 October (Bijlsma 1983).

Sinai Meinertzhagen (1930) reported that a Golden Eagle was taken from a nest at Ain el Gedeirat on 4 March 1927, and that this species was observed at Gebel

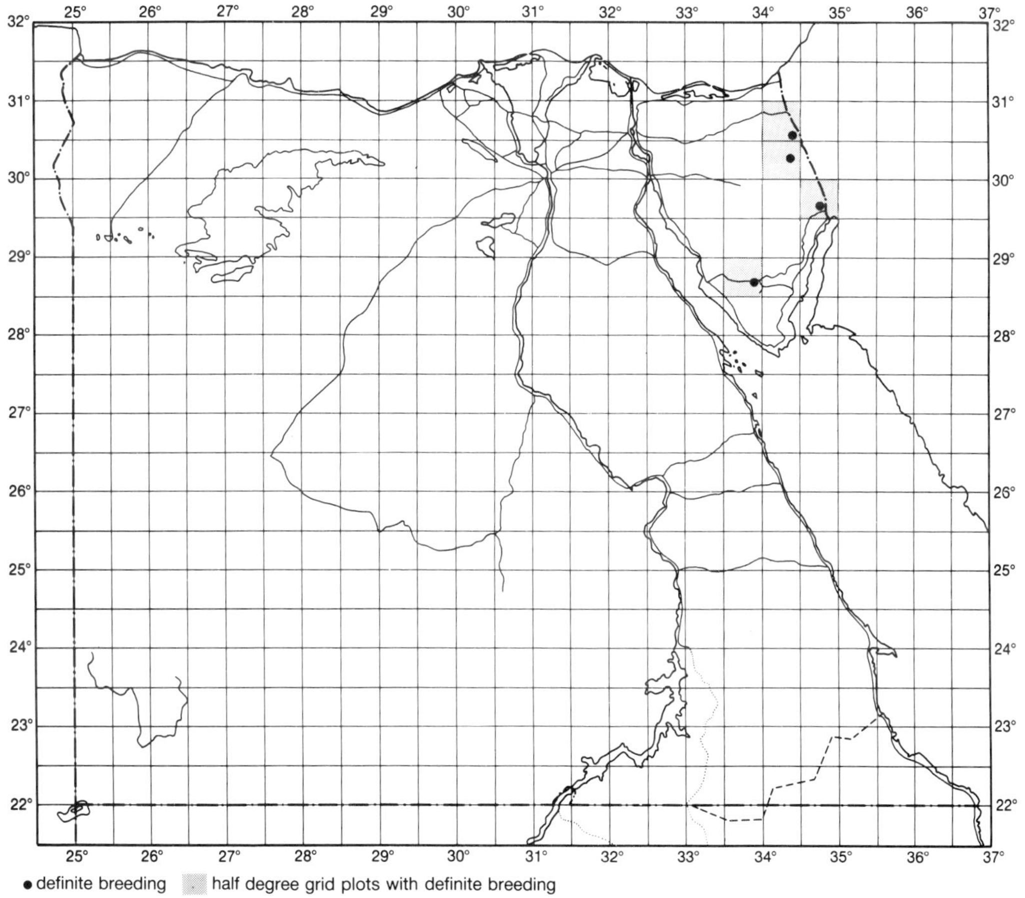

Map B16. Breeding distribution of Golden Eagle *Aquila chrysaetos* between 1970–87.

Musa and Gebel Katherina in March. In 1967, three old nests and one active nest attended by an adult were found between Wadi Feiran and St Katherine (UP). A nest near Ain el Gedeirat was active from at least 1968 to 1981 and another near Gebel Areif el Naqa had two chicks in 1981 (IRIC; UP). At Gebel Hamra, along the Egyptian/Israeli border near Elat, an active nest with three chicks was discovered in 1982 (IRIC).

Verreaux's Eagle *Aquila verreauxii* RB

Local names 'Uqaab suudaa- عقاب سوداء, bit (Bi).

Rare breeding resident in south-eastern Egypt and presumably Sinai. Rarely observed outside known breeding areas. Monotypic.

Eastern Desert The range of the Verreaux's Eagle is closely linked to the distribution of hyrax, a favourite food item. In June 1984 and at various times in the winter/spring 1985 at least one pair of adults was observed at both Gebel Abraq and Gebel Elba (SMG, PLM). It is almost certain that pairs breed at these two localities. An adult reputed to have been captured along the Suez Canal on 1 November 1916 (Meinertzhagen 1930) was actually from the Zagazig Zoo (Flower 1933).

Sinai A single bird was observed in Wadi Feiran on 23 September 1963 (Undeland 1964). Ferguson (1970) found a carcass at El Tor on 20 October 1967 that he believed belonged to this species; unfortunately the specimen was lost and the record is not verifiable. Between March and May 1975 and 1986 a few adults and immatures were regularly observed in Wadi Taba and surroundings (HSh; RSS; SC; UP). Presumably one or two pairs nest in this area. Two adults with a fledgling were seen in Wadi el Arbain in 1978 (HSh). Recently single adult birds have been noted on 21 July 1983 near Gebel Rahaba, on 14 September 1983 in Wadi el Sheikh, on 30 September 1983 in Wadi el Arbain, and on 10 February 1984 at Gebel Safsafa (SMB). A pair of adults was seen near Gebel Katherina on 16 October 1983 (Hovel 1987).

Booted Eagle *Hieraaetus pennatus* PV (WV)

Local names 'Uqaab masiira sughraa عقاب مسيرة صغرى, hidaaya abyad, masiira saghiira.

Scarce passage visitor in autumn from (early) mid-September to mid-November (early December) and in spring from mid-March to early June. Rare in winter. Monotypic.

Booted Eagles are regularly observed in Egypt on autumn passage, mainly in Sinai and the Red Sea Mountains. Small numbers occur elsewhere in the country. During a raptor count at Suez, 761 Booted Eagles were counted between 13 September and 5 November 1981, with the majority passing in the second half of September (Bijlsma 1983). During another survey at Suez, 1104 birds were noted between 5 September and 4 November 1984 (RvD *et al.*). Autumn migration may continue into December, for one was noted at Suez on 9 December 1980 and two, three days later (SMB). Small numbers are observed in the Nile Valley in autumn and there are a few October records from Bahariya (HB *et al.*) and Dakhla and Kharga (Goodman *et al.* 1986). The only post-Meinertzhagen (1930) winter records appear to be one at Abu Rauwash on 25 January 1939 (HGD) and up to three along the Nile in four winters between 1976 and 1979 (Short and Horne 1981).

Spring migration takes place between early March and early June and is concentrated in the eastern portion of the country, including Sinai. At Suez, 457 were noted between 15 March and 13 May 1982, with a peak in mid-April (Wimpfheimer *et al.* 1983). In spring small numbers have been observed in the Nile Delta and Valley, and the Western Desert. The latest spring records are of solitary birds moving north at Fayid on 22 May and 5 June 1955 (RN). At Elat, Israel, it is considered a regular spring migrant in small numbers from early March to at least mid-May (Christensen *et al.* 1981).

Bonelli's Eagle *Hieraaetus fasciatus* RB PV WV

Local name 'Uqaab masiira kubraa عقاب مسيرة كبرى.

Rare and local breeding resident. It is also a rare passage and winter visitor between early September and late April. All Egyptian material examined is referable to *H. f. fasciatus*. It is plausible that birds occurring in southern Egypt, particularly near Gebel Elba, are referable to the sub-Saharan breeding form *H. (f.) spilogaster*.

On 2 April 1985 Rashida Bedouin camping near Gebel Abraq had in captivity a Bonelli's Eagle nestling that they had removed from a local nest (SMB, SMG, PLM). A pair was found attending a nest on 9 April 1985 in Wadi Aukau, Gebel Elba (SMG, PLM). In 1979 an active nest was located in Wadi Taba (HSh). These records are the only documentation of this species breeding in Egypt. In spring 1978 a pair was noted at Ain el Furtaga performing courtship displays and this species presumably bred locally (IRIC).

Between early September and mid-April single Bonelli's Eagles are occasionally observed in the Nile Valley, the Eastern Desert, and Sinai. Small numbers pass through Sinai and the Red Sea Mountains during autumn and spring migration. During systematic raptor counts at Suez, six were identified between 18 September and 7 October 1981 (Bijlsma 1983) and seven between 20 September and 21 October 1984 (RvD *et al.*). A single bird was observed on 19 October 1980 crossing the Gulf of Suez and reaching land north of Hurghada (Meininger *et al.* 1981). In spring, solitary birds have been seen migrating over Suez on 26 February 1981 and 9 March 1981 (SMB); and on 28 February, 3 March, and 4 April 1982 (Wimpfheimer *et al.* 1983).

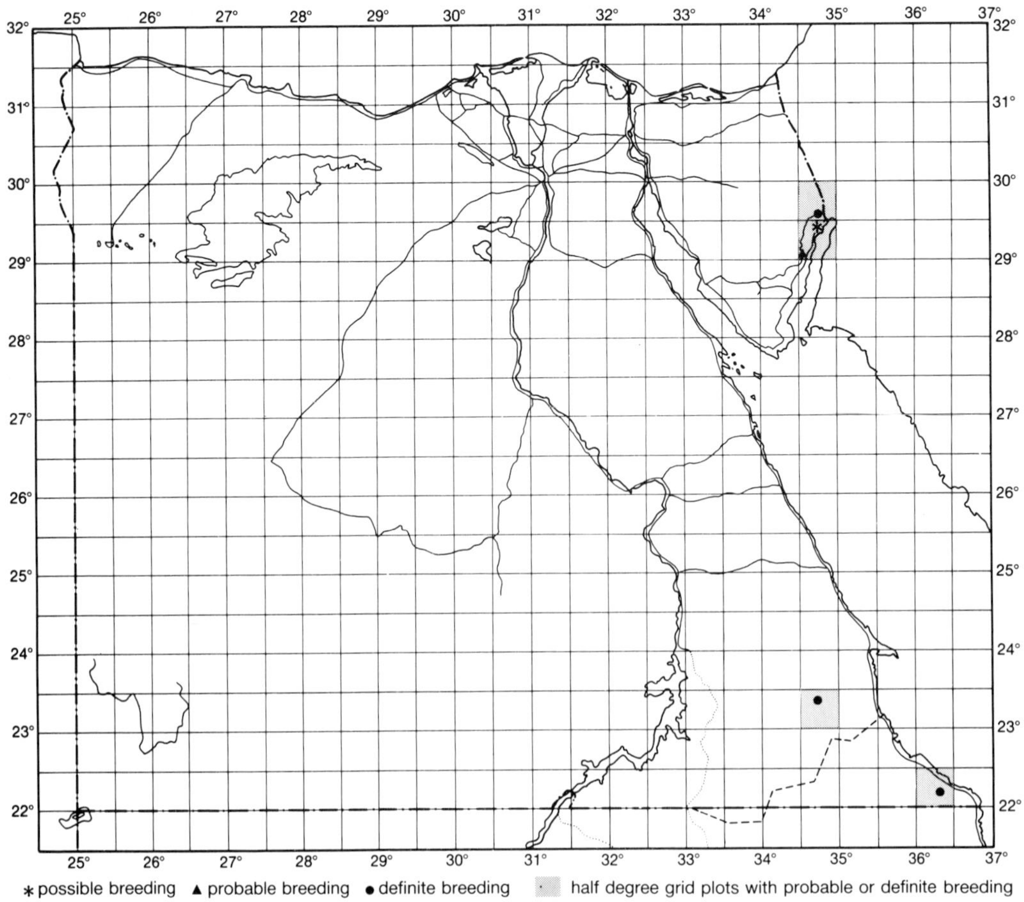

Map B17. Breeding distribution of Bonelli's Eagle *Hieraaetus fasciatus* between 1970–87.

Family Pandionidae

Osprey *Pandion haliaetus* RB PV WV

Local names 'Uqaab nasaariyya عقاب نسارية, abuu gadaaf (RS), abuu mukdaaf, al-mansuur, gadaafii (RS), mansuurii, naasuurii.

Breeding resident along the Red Sea coast and on islands. Scarce passage migrant and winter visitor between early September and early May. Migrants and winter visitors are referable to *P. h. haliaetus*. However, Red Sea breeding birds tend to be distinctly smaller and lighter in plumage coloration than typical *P. h. haliaetus*. There is considerable variation in the laying dates of Red Sea Osprey and non-addled eggs have been noted between late January and April. A few cases of polyandry have been documented (Suaretz 1983). This species was reported to have nested in northern Sinai near El Malaha (Paz 1987) but details are unknown.

Ospreys breed at numerous localities along the Red Sea coast and on islands. Nests are placed on rocky outcrops, on makeshift fishermen's huts, on sandy beaches, and in mangroves. On some islands they are found in loose colonies and in relatively high densities. The total Egyptian breeding population is estimated at 150–200 pairs. Based on inference of incomplete counts made in the area early in this century, the Egyptian Red Sea Osprey population seems to be stable. This population appears to be non-migratory, although some birds may wander locally along the Red Sea coast.

Although Cowley (1953) mentioned a possible nest site in February 1950, 48 km south of Suez, the northern-most known nest in the Gulf of Suez was found 30 km north of Ras Gharib (Marchant 1941). In 1973, 21–30 nests were active on Gezira Tiran (SS; Safriel *et al.* 1985), and in other years 18–20 pairs bred on this island (SS;

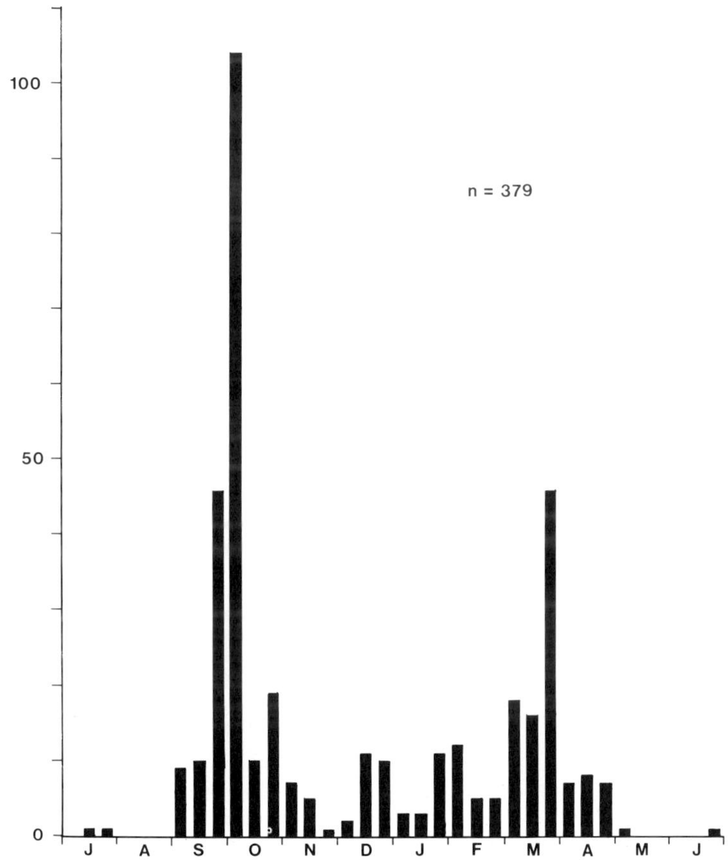

FIG. 6.11. Histogram of the numbers of Ospreys recorded during 10-day periods throughout the year (excluding the Red Sea).

YL). Along the coast of Sinai an estimated 10–15 pairs breed. In April and May 1983 at least 38 breeding pairs were located on the islands in the mouth of the Gulf of Suez and near Hurghada (Jennings *et al.* 1985). In June 1984, 11 nests were found on Gezira Wadi Gimal that had been used during the previous breeding season (SMG). Ten active nests were counted on the four islands of the Siyul group, off Hamata, in 1982 and 1983 (Frazier *et al.* 1984; SMG), and nine on the three islands of the Siyal group, off Adal Deib, in 1985 (SMG, PLM). Another 55–90 pairs are estimated to nest along the coast and on other islands south of Hurghada. The species is also known to occur on reefs in the central Egyptian Red Sea; several pairs breed annually on Gezira Zabargad (JJH; DP).

Birds belonging to Eurasian breeding populations are regular passage and winter visitors to the North Coast, Nile Delta and Valley (including Lake Nasser), the Suez Canal area, the Faiyum, and northern Sinai. Notable influxes of migrants have been

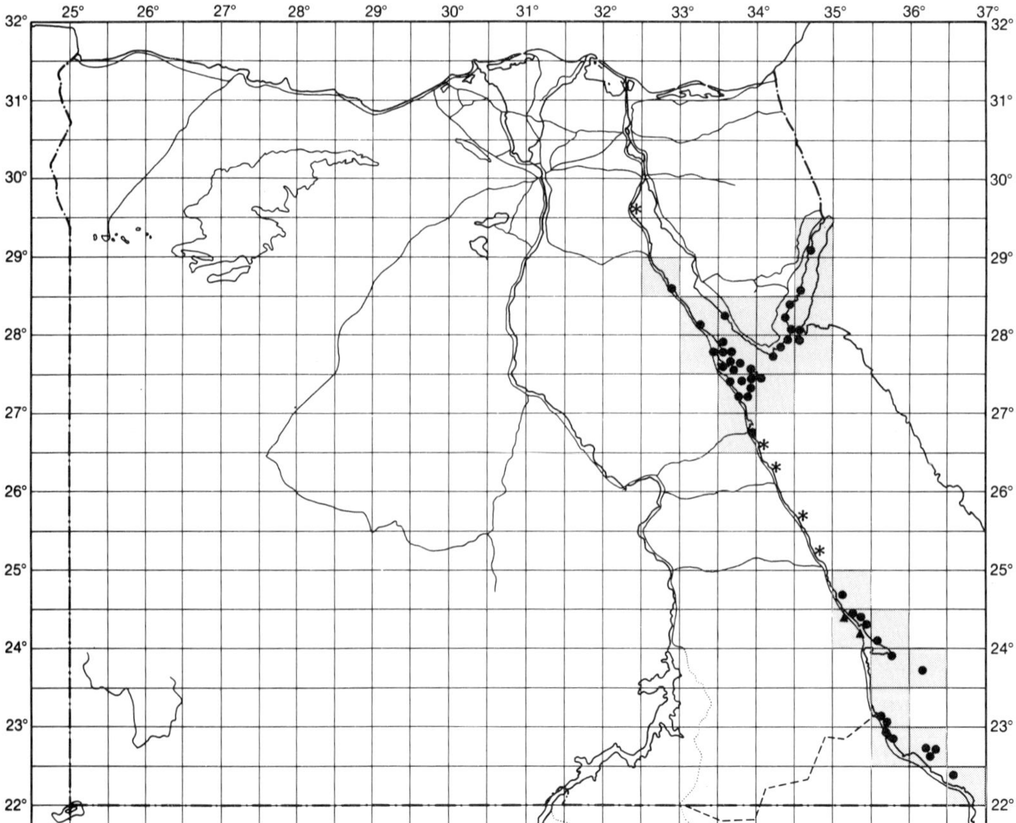

Map B18. Breeding distribution of Osprey *Pandion haliaetus* between 1950–87.

recorded in October and March (Fig. 6.11). At Elat, Israel, Osprey are regular spring migrants (in small numbers) between late February and early May, peaking in the first half of April. The maximum number recorded at this locality on spring passage is 122 between 20 February and 10 May 1977 (Christensen *et al.* 1981). In winter the species is most frequently observed in the Upper Nile Valley and Lake Nasser but there are numerous records from further north, including the Delta lakes and Lake Bardawil. Shelley (1872) mentioned that Osprey were plentiful in the country during the winter months. Nowadays, the species is not 'plentiful' at any time of the year in the Nile system and there has clearly been a sharp decline in the number wintering along the Nile.

There are a few records from the Nile Valley in June and July, presumably of summering immatures. According to Meinertzhagen (1930) Osprey could be observed on Lake Qarun throughout the year but this is no longer the case. The only record from the Western Desert is two or three birds observed south-east of El Maghra on 13 April 1977 (Goodman and Ames 1983).

Ringing There are several reports of birds ringed as nestlings in Sweden being recovered in Egypt in September and October (Österlöf 1977; EDB). One ringed in Finland on 4 July 1954 as a nestling was recovered at El Umbarakab on an unknown date (Watson 1971). Another Finnish bird was recovered near Matruh (Österlöf 1977).

ORDER FALCONIFORMES

Family Falconidae

Generic local names for falcon Ṣaqr, qish (Ma), sagr (Ma).

Lesser Kestrel *Falco naumanni* CB PV (WV)

Local names 'Awsuq ṣaghiir عوسق صغير, abuu saraqa, 'awiisuq, burgayz, ṣaqr al-jaraad.

Casual migrant breeder in the northern portion of the country. Scarce autumn migrant from early September to mid-October (mid-November) and fairly common in spring from (early February) early March to late April (mid-May). Occasional winter visitor. Monotypic.

There are a few breeding records of Lesser Kestrels in Egypt. Von Heuglin (1869*b*) found a pair breeding in a wall at Alexandria. In June 1952, three pairs bred at Deversoir, near the Great Bitter Lake; two nests were placed in an old mud wall cavity and the other in a date palm. Pairs with recently fledged young have been observed at Cairo on 10 September 1981 and at Suez between 15 and 19 September 1981 (Mullié and Meininger 1985).

In autumn the species normally migrates at high altitudes and apparently non-stop to the sub-Saharan winter quarters (Moreau 1961). In Egypt only small numbers are observed between early September and late October in the Nile Delta and Valley, the Eastern Desert, and Sinai. During systematic raptor surveys at Suez, 26 Lesser Kestrels were counted between 5 September and 31 October 1981 (Bijlsma 1983) and 10 between 17 September and 9 October 1984 (RvD *et al.*). Only one September and two October records are known from the North Coast near Matruh (Hoogstraal *et al.* 1963; HB *et al.*; PLM), and there are no autumn reports from the Western Desert. In the Nile Valley the species is occasionally observed during the winter months (e.g. Short and Horne 1981).

In spring the first Lesser Kestrels usually arrive in early March; although they have been noted at Idfu on 9 February 1977 (Short and Horne 1981). Lesser Kestrels are fairly common throughout the country, often in small flocks, between mid-March and mid-April. During the third week of April, Brewster (1966) observed about 500 at one locality in northern Sinai, and on 23 March 1941 it was numerous at Matruh (Bulman 1942). During a systematic raptor count in the spring of 1982 at Suez, only three Lesser Kestrels were noted (Wimpfheimer *et al.* 1983) but on 8 April 1966 at the same locality many were observed feeding gregariously and resting on the ground (JT). The latest spring record is one at El Maghra on 15 May 1975 (Goodman and Ames 1983). At Elat, Israel, it is considered a fairly common migrant from late February to early May but never in large flocks (Christensen *et al.* 1981).

Ringing A bird ringed at El Abbasia on 20 April 1967 was shot in Lebanon on 23 March 1970 (Watson 1971).

Kestrel *Falco tinnunculus* RB PV WV

Local names 'Awsuq عوسق, 'aasuuq, abuu murkays (NC), abuu saraqa (ND), saggaawii, s̱aqr al-jaraad, s̱aqr baladii, saqsaaqa.

The form *F. t. rupicolaeformis* is a fairly common breeding resident in the Nile Delta and Valley and in several Western Desert oases. It is unknown if the subspecies breeding in Sinai is *rupicolaeformis* or *tinnunculus*; the latter is known to nest across the border in Israel. Between (early) mid-September and mid-May (early June) the resident population is augmented by migrants belonging to *F. t. tinnunculus*.

North Coast The only positive breeding record of the Kestrel from this area is a nest found west of Matruh on 24 May 1981 (JJH). A specimen taken on 10 October 1962 at Bahig is referable to *F. t. rupicolaeformis* (YPM). Kestrels are not uncommon in autumn, winter, and spring between Salum and Burg el Arab (Rowntree 1943). The subspecies thought to breed in Cyrenaica, Libya, is *F. t. tinnunculus* (Vaurie 1965).

Western Desert Kestrels have been found breeding or suspected of doing so in Wadi el Natrun, Bahariya, Farafra, Dakhla, and Kharga (Goodman *et al.* 1986). They may also breed in Siwa (Al Hussaini 1938–9).

Nile Delta and Valley The Kestrel is a common breeding resident throughout the Nile Delta and Valley. In recent years it has declined in numbers in the Nile Delta and along the Suez Canal, and to a lesser extent in the Nile Valley. This decline is most likely related to the extensive use of pesticides (Mullié and Meininger 1985). A few pairs presumably breed along Lake Nasser. Nests with young have been noted in early January near Aswan (AS).

Eastern Desert Definite breeding records from this area are all from localities in close proximity to the Nile Valley. Kestrels have been found breeding in wadis near Helwan (Meinertzhagen 1930) and in March 1983 near El Amarna (SV). Between mid-September and mid-May Kestrels have been reported from numerous localities in the Eastern Desert. Breeding along the Red Sea coast, particularly in villages and towns, cannot be excluded.

Some migrants cross the Red Sea. At the Daedalus Reef, 100 km off the coast from

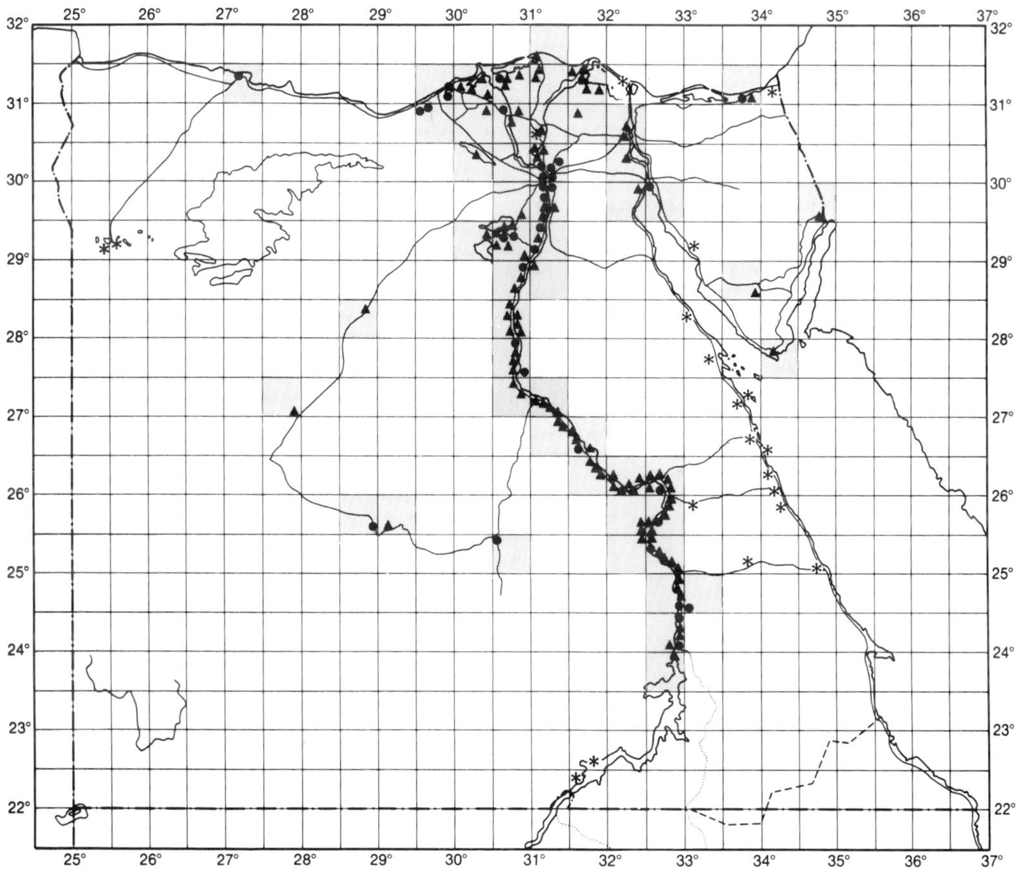

Map B19. Breeding distribution of Kestrel *Falco tinnunculus* between 1970–87.

Marsa el Alam, Kestrels have been noted on 11 January 1946 (Elliott and Monk 1952), 21 March 1936 (Bulman 1944), and 21 April 1913 (Ticehurst 1924). Small numbers of migrating Kestrels have been noted at Suez during the autumn and spring. An exceptionally late record is a bird referable to *F. t. tinnunculus* taken at Gebel Qattar on 3 June 1984 (UMMZ).

Sinai In recent years evidence has been found of this species breeding at El Arish (WaS). Kestrels probably also nest at St Katherine, Sharm el Sheikh, and Wadi Taba. Zedlitz (1912) observed pairs displaying in Wadi Hibran and 'Wadi Gharb' (= Wadi Gharba?).

Red-footed Falcon *Falco vespertinus* PV (WV)

Local names Luziiq لزيق, kuubij.

Scarce passage visitor in autumn between (early) late August and early (late) November and rare in spring between mid-March and early June (Fig. 6.12). There is a single winter record and a few from the summer months. All Egyptian material examined is referable to *F. v. vespertinus*. An egg in ZFMK identified as this species and labelled Egypt cannot be accepted as evidence for its breeding in the country.

The migration of Red-footed Falcons from the Eurasian breeding grounds to the sub-Saharan wintering areas takes place along a broad front and at high altitudes.

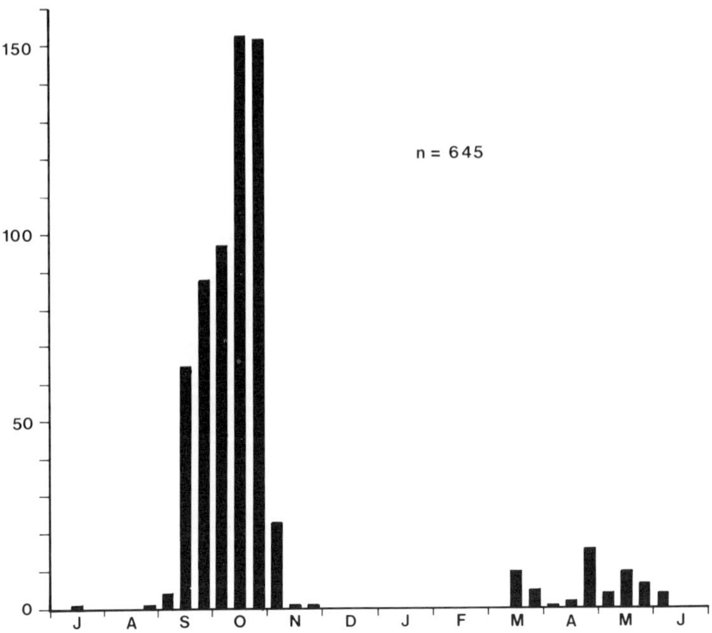

FIG. 6.12. Histogram of the numbers of Red-footed Falcons recorded during 10-day periods throughout the year.

This species is notorious for its elusiveness during migration, both in Europe and Africa (Brown *et al.* 1982). It is known to cross the northern Red Sea (Tuck 1966).

The earliest autumn migrants in Egypt have been observed on 7 August in Sinai (Meinertzhagen 1930) and in late August 1963 near Hurghada (Kiepenheuer and Linsenmair 1965). Small numbers of resting birds, often in flocks, have been observed throughout the country, including the Western Desert, between mid-September and early November. The largest reported concentration in the Nile Valley in recent years is 15–20 near Beni Suef on 28 September 1985 (Hume 1985). During systematic autumn counts of raptors at Suez, 462 Red-footed Falcons were recorded between 17 September and 3 November 1981, generally at high altitudes and in flocks composed of up to 61 birds (Bijlsma 1983). The latest autumn record is an immature at Suez on 24 November 1984 (RvD *et al.*). One was noted in Sinai on 21 December (Meinertzhagen 1930).

In spring the species is distinctly less common than in autumn. The earliest spring record is 10 at Port Said on 15 March 1981 (SMB, BB). Between mid-March and early June small numbers of Red-footed Falcons have been observed throughout the country. Migration peaks in mid-May. During systematic counts of raptors at Suez between 23 February and 16 May 1982, three Red-footed Falcons were noted between 17 April and 7 May (Wimpfheimer *et al.* 1983). An exceptionally late spring record is four at Fayid on 5 June 1955 (RN). A pair was seen daily from May to July 1910 near Giza (Nicoll 1912*b*). A single bird observed some time between 6 and 22 July 1937 at Hurghada may well have been a Sooty Falcon (Al Hussaini 1938*a*).

Merlin *Falco columbarius* WV

Local names Yu-yu- يؤيؤ, abuu riyaah̲, jaraadiya.

Rare winter visitor, mainly to the northern portions of the country, between (late September) mid-October and mid- (late) April. The majority of Merlins visiting Egypt are referable to *F. c. aesalon*, and a few to *F. c. insignis*.

The earliest autumn records of Merlin in Egypt are one near Rafa on 21 September (Meinertzhagen 1930) and two between Matruh and Sidi Barrani on 1 October 1940 (Bulman 1942). Between mid-October and mid-April Merlins have been observed in small numbers along the North Coast, in Wadi el Natrun, Siwa, Farafra, the Nile Delta and Valley (to Aswan), the Suez Canal area, the Faiyum, along the Red Sea coast, and in Sinai. The latest spring record is two on Gezira Giftun el Kabir on 28 April 1982 (Ferguson 1984). Meinertzhagen's (1930) statement that 'birds move north in large parties during the first fortnight in April, as many as forty having been seen together . . .' was based on his observation of about 40 resting in trees in the Delta Barrage on 11 and 18 April. This record seems doubtful and there is no recent evidence to support the occurrence of such large flocks in Egypt.

Hobby *Falco subbuteo* MB PV (WV)

Local name Kuunj كونج.

Apparently a rare migrant breeder in north-eastern Sinai. Scarce migrant throughout the country in autumn from early September to early (late) November and in spring from (early February) early April to mid-May. Rare summer visitor. *F. s. subbuteo* is the form occurring in the country.

Recent evidence of the Hobby breeding near El Arish includes two recently fledged young near an old Hooded Crow nest on 21 August 1982 and three adults observed in this area between 25 and 27 June 1984 (SMB). There are a few other summer records from the country. Single birds were noted at Giza on 23 June 1937 (HGD) and at Bahariya on 18 June 1974 (Goodman *et al.* 1986).

Passage between Eurasian breeding grounds and the sub-Saharan winter quarters takes place along a broad front and birds do not avoid crossing large bodies of water. The species is regularly noted on autumn passage throughout most of Egypt between early September and early November, with the bulk passing between mid-September and mid-October. There is considerable variation in the number that are recorded from year to year. During systematic counts at Suez, 31 were recorded between 14 September and 30 October 1981 (Bijlsma 1983), and only five between 12 September and 19 October 1984 (RvD *et al.*). The latest autumn records are 23 November at Rafa (Meinertzhagen 1930) and 26 November 1911 at Wadi el Natrun (Nicoll 1912*b*).

An adult noted at Luxor on 8 February 1977 and an immature at Karnak on 1 March 1979 (Short and Horne 1981) were presumably early spring migrants. Between early April and early May small numbers have been observed throughout the country. A late spring record is one at El Maadi on 15 May 1927 (GZM).

Ringing A bird ringed on 30 July 1939 in Latvia was recovered on 1 October 1939 at Lake Idku (Mackintosh 1941).

Eleonora's Falcon *Falco eleonorae* PV

Local names S̱aqr as̱ham صقر أسحم, abuu ẖuwwaam (NC).

Rare passage visitor in autumn from (early September) early October to mid-November and in spring from mid-April to mid-May. Monotypic.

Walter (1979) hypothesized that the migration pathway of Eleonora's Falcon from the breeding grounds in Morocco and the Mediterranean to the wintering quarters in Somalia and Madagascar was via the Mediterranean and Red Sea, rather than overland. Until recently there was hardly any information on this species in Egypt. It is now clear that Eleonora's Falcons are regular migrants along the coastal waters of Egypt, in both autumn and spring, lending support to Walter's proposed migration route. All of the dated records we are aware of are summarized below by season.

Autumn These records include: Suez—one on 26 October, five on 27 October, three on 31 October, and 18 on 2 November 1981, all at high altitudes and flying due south (Bijlsma 1983); three on 21 October 1983 (SMB); and three between 1 and 3 November, one on 7 November, and one on 19 November 1984 (RvD *et al.*); Zaranikh—one on 7 September and one on 21 September 1982 (Baha el Din and Salama 1984); and one on 1 October 1985 (SMB); El Arish—a dispersed flock of eight on 8 October 1985 (SMB); and Elat, Israel—at least four birds between 3 and 8 November 1986 (EH).

Spring These records include: Matruh—one in a flock of Red-footed Falcons on 24 April 1964 (HCK); Suez—single birds on 15 April, 17 April, and 18 April and three on 9 May 1982 (Wimpfheimer *et al.* 1983); and one on 18 April 1984 (MBe, SM); Gulf of Suez—one was taken at sea on 12 April 1948 (BMNH; Meinertzhagen 1954); Wadi Taba—one on 18 April 1978 (RSS); El Taloul—one on 28 April 1985 (Fry *et al.* 1985); 43 km south of Safaga—two adults on 21 April 1984 (SMG); Ras el Naqb—three on 3 May 1978 (RT, LBy); and Elat, Israel—six between 20 April and 11 May 1977 (Christensen *et al.* 1981).

Ringing A bird ringed as a nestling at Akrotiri, Cyprus, on 15 September 1967 was shot in September 1969 at Matruh (Watson 1971).

Sooty Falcon *Falco concolor* MB (WV)

Local names S̱aqr al-ghuruub صقر الغروب, tiita (Ab).

Scarce but locally common migrant breeder in desert areas throughout the country and on islands in the Red Sea. It is generally present in Egypt from mid-April to late November; although there are a few midwinter records. The total Egyptian breeding population is tentatively estimated to be 150–300 pairs. On Red Sea islands the majority of birds breed in loose colonies and at desert sites tend to be of a few or solitary pairs. Monotypic.

North Coast Raw (1921) mentioned that F. W. Borman obtained some information about Sooty Falcons breeding near Salum; however, no details were given and this statement remains enigmatic.

Western Desert A Sooty Falcon nest was found between 1940 and 1945 at El Maghra (Galal el Din 1959). At Bahariya, three adult females were collected on 14 and 17 June 1974 and there were at least two breeding pairs in the mid-1970s (Goodman *et al.* 1986). Pairs were observed in Bahariya between 21 August and 21 September 1983 and north-east of the oasis between 1 August and 2 October 1982 (HB *et al.*). In late February or early March 1933, four pairs were found occupying rocky outcrops between 110 and 190 km south and west of Farafra, and one was collected at 26°35′N., 26°53′E. (Moreau 1934; Wingate 1934). A pair was observed south of Farafra between 13 September and 1 October 1982 (HB *et al.*). On 21 March 1984 an adult was observed at Abu Minqar, where a falconer was reputed to have taken a nestling the previous autumn (Goodman *et al.* 1986). One bird was collected on 2 October 1932, 110 km south of Dakhla at 24°30′N., 28°58′E. (Moreau 1934) and another was observed 60 km west of Kharga on 7 September 1984 (Goodman *et al.* 1986). The species was noted at Gebel Uweinat in autumn 1954 (Booth 1961).

Nile Delta and Valley A pair was observed displaying and copulating near a nest site in Wadi el Rayan between 7 and 14 July 1984 (SMB). There are many archaic and recent records of Sooty Falcons in the general environs of Cairo between mid-May and late November. A pair bred near El Maadi in 1928 (Meinertzhagen 1930), and a number of recent midsummer records are indicative that it is still a local migrant breeding bird. A single bird was noted in Cairo every day between 14 October and 26 November 1971 (Haensel 1975). One was taken at Abu Zabal on 6 August 1918 (BMNH). Between 19 July and 22 August 1944, one, probably of a breeding pair, was observed daily near Kasfarit, along the Suez Canal (LB). Other records from this area include one at Ballah on 6 August 1944 (Bodenham 1945*b*), and four on 5 June, one on 28 July, and one on 14 August 1955 at Fayid (RN). The only known record from the Nile Delta proper is an adult collected at Damietta in October 1895 (ZFMK). There are many records from the Faiyum and the Nile Valley south to Aswan, where this species presumably breeds at various localities. Most records are from the period between mid-April and late November. The only winter records are single birds at El Amarna on 26 February 1979 (Short and Horne 1981) and near Luxor on 28 January 1986 (CT), and two near Aswan on 27 February 1982 (AS). These winter observations lack documentation and it is not clear if Sooty Falcons occasionally winter in the Nile Valley.

Eastern Desert Several birds were present at Gebel Ataqa on 8 October 1947 (Goodwin 1949), and the species was also noted at this locality in October (Tregenza 1951). A single bird was observed at Wadi Garn el Kabsh, Gebel el Galala el Qibliya, on 1 May 1983 (JJH). A nest was discovered on 15 August 1983 in Wadi Muweih (JJH). In the summers and early autumns of 1947 and 1949 Tregenza (1951) recorded this species at 19 localities in the central Red Sea Mountains, including several breeding sites. Two nests were discovered in a wadi near Marsa el Alam in 1980 (IH), and one was observed at Gebel Abraq on 23 June 1984 (SMG).

Red Sea coast Three or four pairs breed on Gezira Tiran (RF). On the islands at the southern end of the Gulf of Suez and off Hurghada at least 114 adults were seen in the autumn of 1984, including at least 44 presumed pairs and 10 nests (Jennings *et al.* 1985). On Red Sea islands birds generally commence breeding in the first week of August and most young fledge by mid-November. The species was noted as scarce at Safaga in November 1942 (Bulman 1944). On 22 May 1982 about 10 pairs were present on Gezira Wadi Gimal and on 31 May 1982 at least five pairs on Gezira Showarit (Frazier *et al.* 1984). Between 12 and 13 November 1982 about 70 birds were observed on Gezira Wadi Gimal and Gezira Showarit, 60 per cent of which were recent fledglings and some still in family groups near nest sites (Baha el Din 1984*c*). Long (1961) observed one on 2 May 1956 in the Red Sea south of the Egyptian border at 18°59′N., 38°45′E., about 160 km from land, which would indicate that some migrants do not always follow the coastline. One observed on 14 February 1984 at Shalatein (SMG) was either wintering locally or an early spring migrant.

Sinai Sooty Falcons have been found nesting at several places on the Peninsula. The majority of summer records are from localities in the south (Frumkin and Pinshow 1983). Along the north coast near Zaranikh, single birds were observed on 19 and 22 August and on 4 September 1981 (IP, UGS). A single bird was observed at Ain Musa on 4 October 1984 (RvD *et al.*). The total Sinai population, excluding Gezira Tiran, is estimated to be at least 20 breeding pairs.

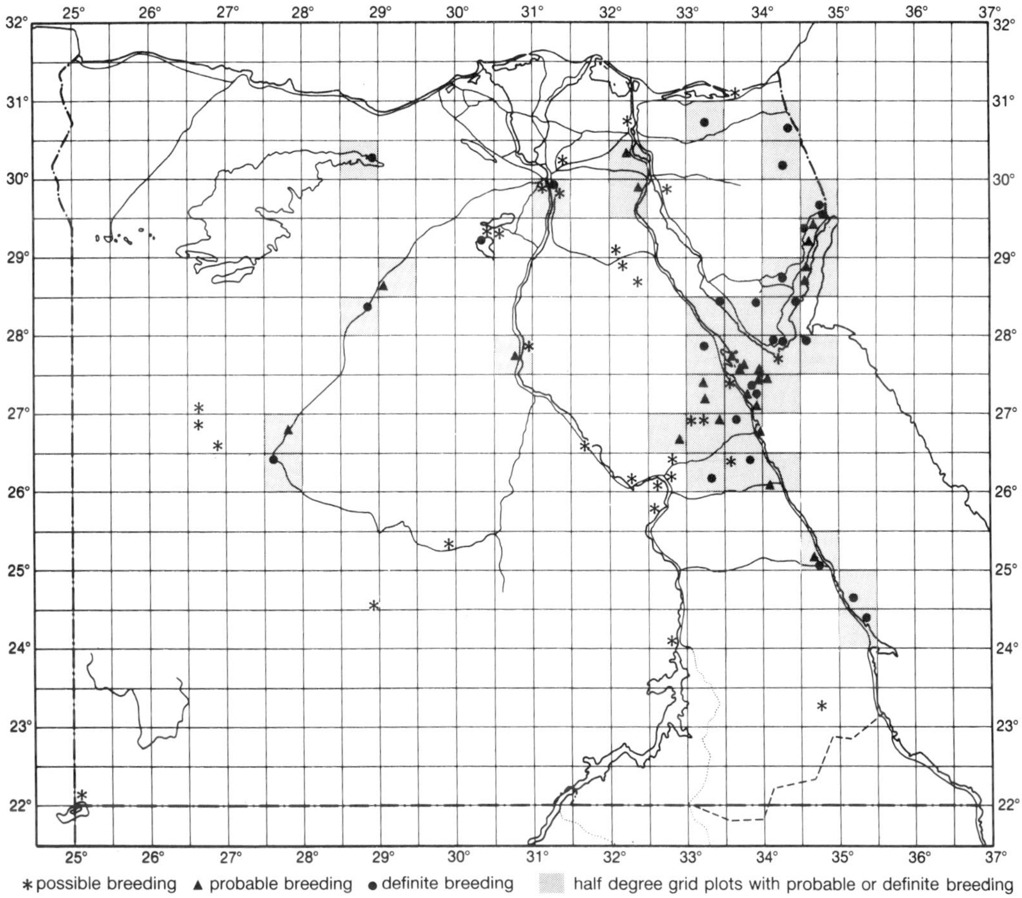

Map B20. Breeding distribution of Sooty Falcon *Falco concolor* between 1930–87.

Lanner *Falco biarmicus* RB WV

Local names S̲aqr h̲urr صقر حرّ, baaw (Bi), s̲aqr al-ghazaal, s̲aqr shaahiin.

Scarce breeding resident in desert areas and along the edge of cultivation throughout the country. The resident population is augmented by migrants during the winter. The Egyptian breeding population is referable to *F. b. tanypterus*. A few specimens, presumably migrants, have been identified as *F. b. feldeggii*.

North Coast Lanners are occasionally noted throughout the year between Salum and Bahig; although no nest has been found in the region. One was taken on 12 June 1963, 64 km south of Bahig (FMNH).

Western Desert Pairs have been observed in most Western Desert oases, desolate areas of the Great Sand Sea, and in isolated vegetated spots, such as Abu Minqar, Bir Tarfawi, and East Uweinat (Baha el Din in press; Goodman *et al.* 1986). A nesting pair was found on 23 February 1988 near Dakhla (AH).

Nile Delta and Valley Lanners have occasionally been observed in the Nile Delta and the Suez Canal area. The only relatively recent breeding record from this area is a pair in March and April 1952 using an old *Corvus* nest at Gebel el Goza el Hamra

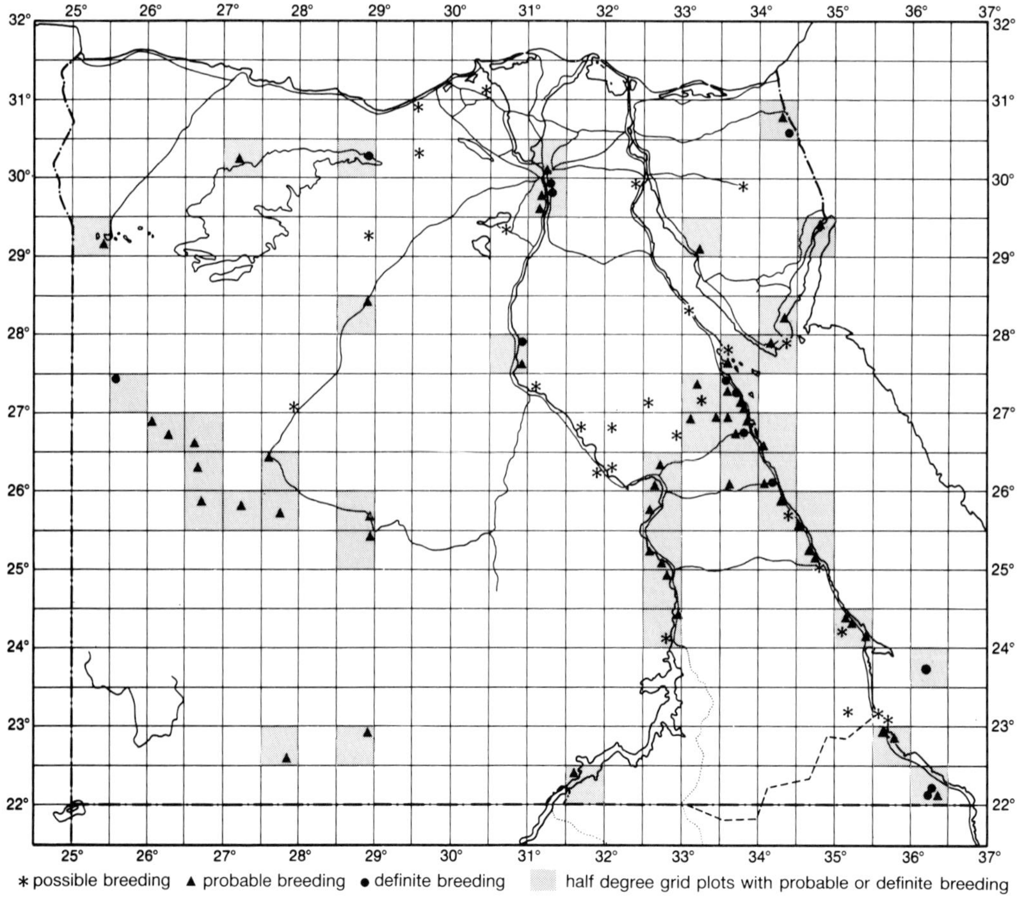

Map B21. Breeding distribution of Lanner *Falco biarmicus* between 1930–87.

(AP). Three eggs were taken near Damietta on 9 April 1898 (BMNH). The species is regularly observed throughout the year in the Nile Valley between Cairo and Abu Simbil, where it breeds on cliffs skirting the flood-plain, and in the Faiyum. Lanners formerly bred on the Giza Pyramids, where four young were taken on 30 April 1893 and four eggs on 18 March 1894 (Sparrow in Raw 1921) and 18 March 1919 (BMNH). In June 1952 a pair was suspected of nesting on ruins at Saqqara (HHg). An active nest was found at Beni Hassan on 22 May 1981 (SMG).

Eastern Desert and Red Sea Small numbers of Lanners are resident in the central Red Sea Mountains (Tregenza 1951, 1955). Higher densities seem to be present along the Red Sea coastal plain between Ras Gharib and the Sudanese political border. A portion of the birds occurring along the Red Sea coast, may actually nest further inland, an area poorly known, and only visit the coastal plain to hunt. This species generally nests on cliffs. A nest containing two young was found on an abandoned oil derrick at Abu Sha'r on 8 May 1939 (Marchant 1941). In May 1982 and 1983 a pair of Lanners successfully raised broods of one and two (respectively) on Gezira Zabargad (DP). Migrating Lanners have occasionally been observed at Suez in autumn and spring. The species is also known to cross the Gulf of Suez (Marchant 1941).

Sinai This species breeds at a few localities in eastern and southern Sinai. Single migrants were observed near Zaranikh on 26 September 1980 (Petersen and Sørensen 1981*a*) and on 12 September 1982 (Baha el Din and Salama 1984). A few birds, presumably migrants, are observed in spring at Elat, Israel (Christensen *et al.* 1981).

Saker *Falco cherrug* PV WV

Local names S̲aqr al-ghazaal صقر الغزال, burgayz (NC), s̲aqr h̲urr, sharq.

Rare passage and winter visitor throughout the country from late September to late April. All Egyptian material examined is referable to *F. c. cherrug*.

Many of the Saker sight records from Egypt, particularly older ones, are open to question because of possible confusion with Lanners. Specimens have been taken between late November and late February (Meinertzhagen 1930). The species has been observed a few times in the Western Desert (Goodman *et al.* 1986) and along the North Coast. It is apparently more regular in the Nile Delta and Valley, along the Red Sea coast, and in Sinai. Most records from the Red Sea coast are from October, November, and March, suggesting migration. One bird was observed flying aboard a ship on 18 February 1964 in the Red Sea at 20°22′N., 38°47′E., just south of Egyptian waters (Tuck 1964). The earliest autumn record is two at Giza on 20 September 1957 (Horváth 1959) and the latest spring record is an immature between Quseir and Safaga on 29 April 1983 (SC).

Ringing In March 1982 a dried-out Saker carcass was found at Sharm el Sheikh with a ring of King Abdullah Khaled of Saudia (Hovel 1987).

Peregrine *Falco peregrinus* PV WV

Local names Shaahiin شاهين, ṣaqr at-ṭayr.

Rare passage and winter visitor throughout the country from mid-September to late April. Birds referable to *F. p. peregrinus* and *F. p. calidus* have been taken in Egypt but the majority of specimens are intermediate in size and plumage coloration between these two forms.

Peregrines are uncommon passage visitors to Egypt and apparently migrate along a broad front. Small numbers can be observed in winter along the entire Mediterranean coast, in the Nile Delta and Valley (occasionally), and along the Red Sea coast. A single bird was observed in downtown Cairo from October 1979 to March 1980 and from December 1980 to March 1981 (SV).

Peregrines are usually observed singly or in small groups while migrating. In autumn small numbers have been observed migrating along the north coast of Sinai (Petersen and Sørensen 1981*a*). Nine were observed at Zaranikh between 13 and 24 September 1982 (Baha el Din and Salama 1984) and there are also records from along the Suez Canal (Bodenham 1945*b*). Three were noted between Matruh and Siwa on 28 September 1963 (Kiepenheuer and Linsenmair 1965) and three near Matruh on 22 September 1980 (WCM). About 10 migrants were seen at Hurghada between 4 and 6 April 1964 (Kiepenheuer and Linsenmair 1965) and three near Lake Bardawil between 21 April and 29 April 1985 (Fry *et al.* 1985).

Barbary Falcon *Falco pelegrinoides* RB PV? Plate 1

Local name Shaahiin maghrabii شاهين مغربي.

Scarce local breeding resident in the Nile Valley, Red Sea Mountains and islands, and eastern Sinai. It is infrequently observed in other portions of the country. All Egyptian material examined is referable to *F. p. pelegrinoides*. The only exception is one taken at Luxor on 9 March 1852 which resembles specimens of *F. p. babylonicus* from India (AMNH).

North Coast The only definite record of the Barbary Falcon is one on 19 April 1954, 24 km east of Sidi Barrani (FMNH).

Western Desert Well-documented reports of Barbary Falcon include single birds at Wadi el Natrun on 3 April 1981 (Goodman *et al.* 1986) and 140 km south-west of Cairo on 4 September 1983 (HB *et al.*).

Nile Delta and Valley This species is not known to breed in the Nile Delta. Single birds have been found wintering at Alexandria from 30 January to 12 April 1965 and from 22 September 1965 to 18 January 1966, and at Cairo in the winter of 1965–6 (JT; JWW). At the Dahshur Pyramids nests with eggs have been found on 10 April 1903 (Koenig 1907) and on 28 March 1909 (Sparrow in Raw 1921), and a nest with young on 12 April 1918 (Raw 1921). Ripley (1963) found an apparent nest site at

Gebel Adda, Egyptian Nubia. Small numbers have been regularly observed throughout the year in the entire Nile Valley and the species is known to breed near Luxor and Aswan.

Eastern Desert and Red Sea islands This species is regularly observed at Gebel Ataqa and at several localities between Hurghada and the Sudanese political border. Two birds, which by their behaviour were presumed to be nesting, were found on Gezira Giftun el Saghir in spring 1983; the following autumn, a total of eight birds was observed on the islands off Hurghada. These may have been in part birds coming from the mainland to hunt migrating songbirds (Jennings *et al.* 1985). A presumed migrant was observed at Suez on 3 November 1981 (Bijlsma 1983).

Sinai Several pairs are resident along the Sinai coast of the Gulf of Aqaba. A nest with two chicks was found on Gezira Tiran on 27 March 1979 (RF).

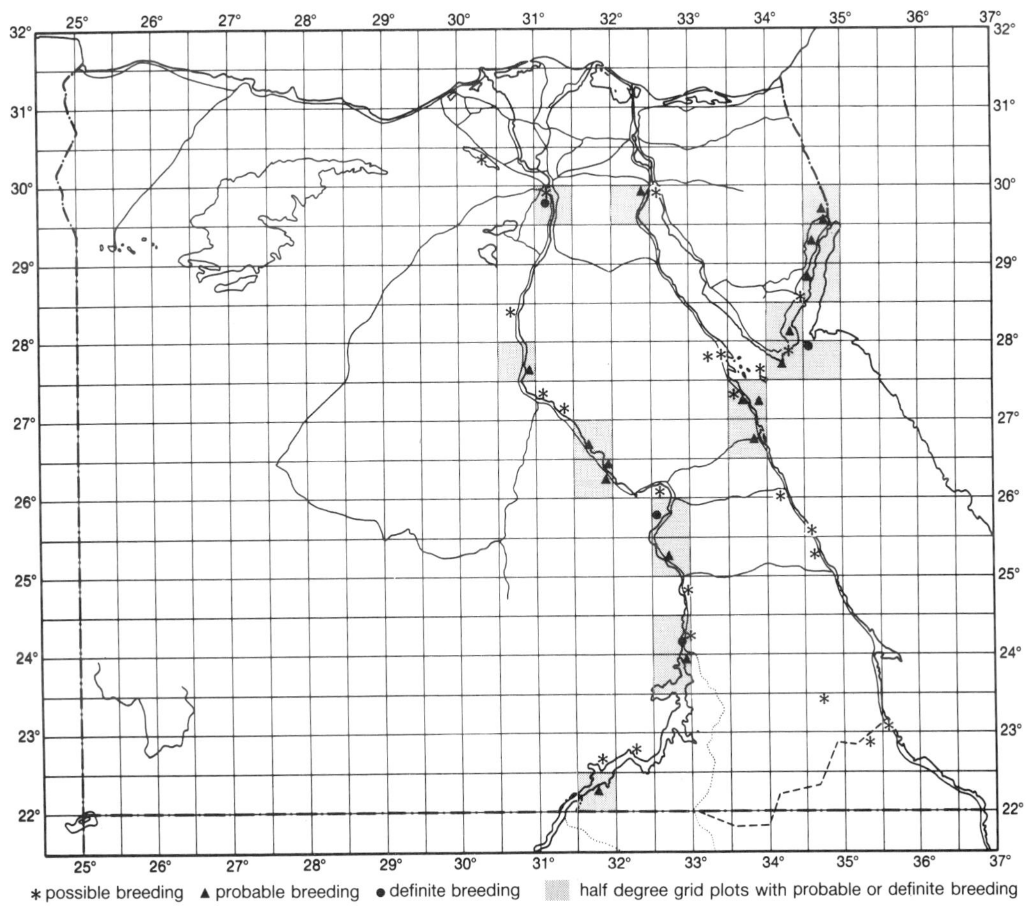

Map B22. Breeding distribution of Barbary Falcon *Falco pelegrinoides* between 1930–87.

ORDER GALLIFORMES

Family Phasianidae

Chukar *Alectoris chukar* RB

Local name Shinaar شنار.

Fairly common breeding resident in portions of Sinai. The local form is *A. c. sinaica.*

In Sinai the Chukar is a local breeding resident in rocky wadis of the southern mountains and in orchards and sparsely vegetated desert in the north-east. On 2 May 1985, 6 adults and 10 chicks were noted 5 km west of Rafa (SMB). One specimen was

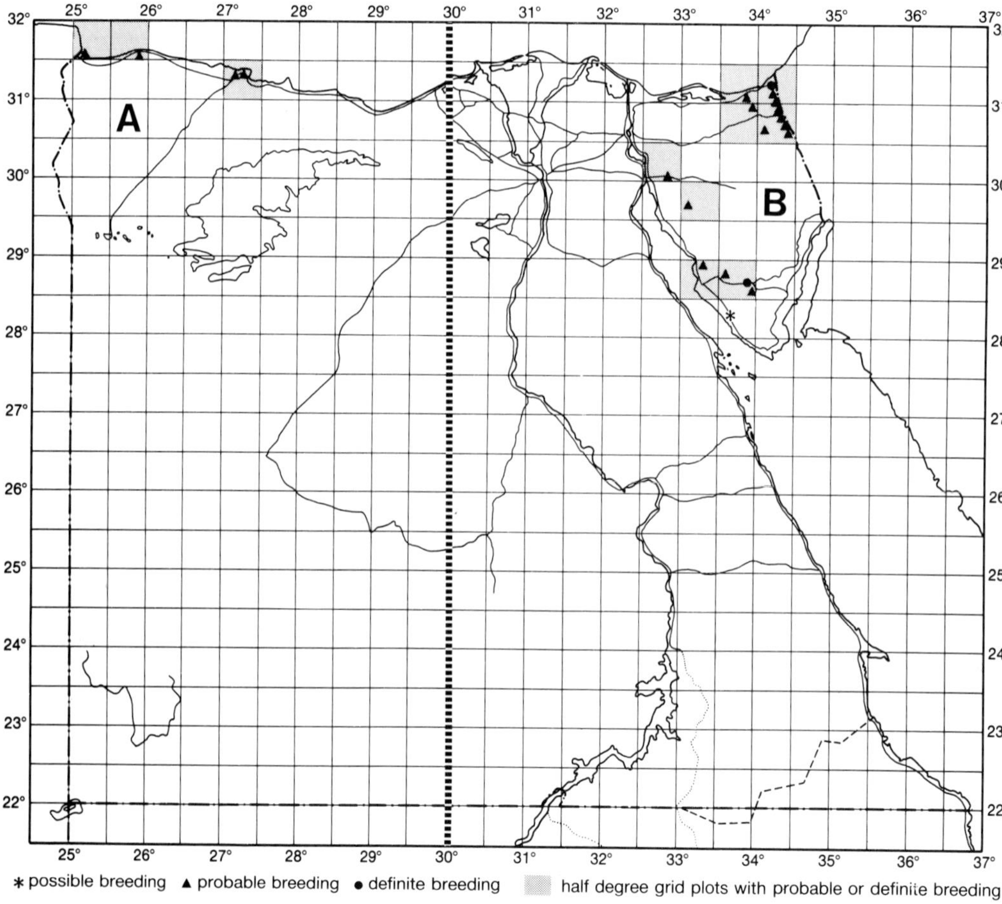

Map B23. Breeding distribution of Barbary Partridge *Alectoris barbara* (A); and Chukar *A. chukar* (B) between 1930–87.

collected at El Tor on 1 May 1911 (RMS). Ticehurst's (1912) record of this species in Wadi Hof, near Helwan, is not acceptable.

Barbary Partridge *Alectoris barbara* RB

Local names Shinaar maghrabii شنار مغربي, ya'quub.

Rare breeding resident in extreme north-western Egypt. The form found in the country is *A. b. barbata*.

According to Meinertzhagen (1930) Barbary Partridges occurred in the Mediterranean coastal desert near Salum, Sidi Barrani, and Matruh. He noted that it was formerly common but by the late 1920s locally extinct except in a few wadis. An adult and nestling were collected on 30 January 1928 at Matruh (BMNH). The most recent record from the Matruh area is one observed on 24 April 1964 (HCK). In recent years no ornithological surveys have been conducted west of Matruh, and thus its current status in this area is unclear. Bedouin living at Matruh and El Alamein, however, are familiar with the bird in the Salum area and across the Libyan border in Cyrenaica (SMG, PLM).

Sand Partridge *Ammoperdix heyi* RB Plate 5

Local names H̱ajal as̱-s̱akhr حجل الصخر, ẖajal (Ma, Ab), quhayba, quhbii, tuwwa (Bi).

Common breeding resident in the foothills, mountains, and wadis of the Eastern Desert, and central and southern Sinai. Birds from the northern Eastern Desert are referable to *A. h. nicolli* and those from the southern Eastern Desert to *A. h. cholmleyi*. Specimens from the central zone are intermediate between these two subspecies. The Sinai population is referable to *A. h. heyi*.

The Sand Partridge is usually seen in pairs or small groups of generally not more than 15 birds. Pairs with young chicks have been recorded between late March and late May.

Eastern Desert This species is locally common in rocky areas from Wadi Hof in the north to Gebel Elba in the south, at altitudes of up to 1500 m. In midwinter it is occasionally observed along the Red Sea coastal plain, where it presumably does not breed. The northern-most record is a single bird observed at Gebel Asfar, near Abu Zabal (Raw 1921). In recent years it seems to have become more common along several roads in the Red Sea Mountains, particularly near passes where grain is spilt by lorries (e.g. near Wadi el Hammamat).

Sinai This species is particularly common in the rocky wadis and mountainous areas in the southern third of the Peninsula. Wimpfheimer *et al.* (1983) reported that 10 were observed near El Gorah on 6 May 1982; this locality is a considerable distance from any known Sand Partridge breeding area and the record is not acceptable.

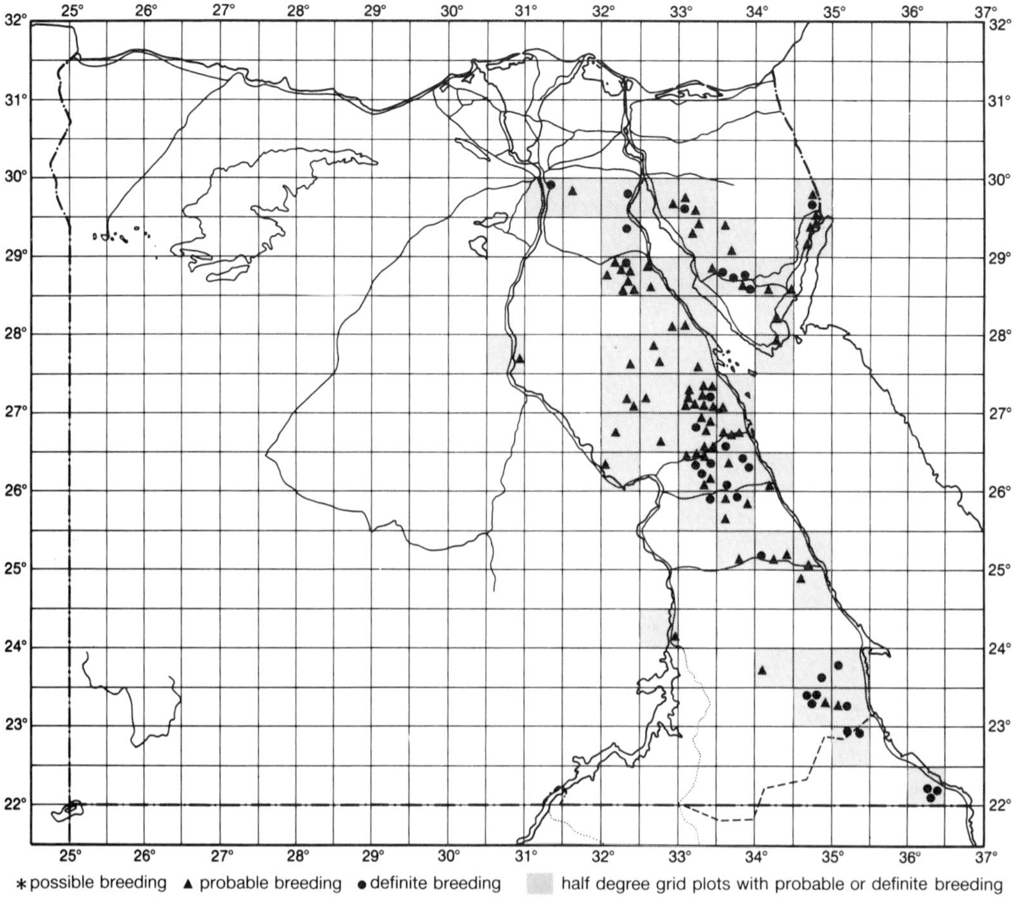

Map B24. Breeding distribution of Sand Partridge *Ammoperdix heyi* between 1930–87.

Quail *Coturnix coturnix* (RB) PV WV

Local names Summaan سمّان, firra (Ma), salwaa.

Scarce and local breeding resident in the Nile Delta and Valley. Common autumn migrant from (mid-) late August to early October (mid-November) and in spring from (early February) early March to mid- (late) April. Uncommon in winter and there are a few summer records from areas where it is not known to breed. All Egyptian material examined is referable to *C. c. coturnix*. Breeding is from mid-January to late May.

The present breeding status of the Quail in Egypt is unclear. There are several nesting records from the late nineteenth century and early twentieth century near

Luxor (Raw 1921; Koenig 1926; BMNH) and in the northern Nile Valley and southern Delta (Raw 1921; Meinertzhagen 1930; BMNH). The only recent documentation of this species breeding in Egypt is three nests found near Bahig in May 1967 (Hubbard and Seymour 1968). It is unknown if birds heard calling in various portions of the Nile Delta and Valley during spring are migrants or local breeding residents and if the Egyptian nesting population is migratory.

Autumn migration begins in mid-August, peaks in September, and continues into November. This species is observed in large numbers along the entire Mediterranean coast, where it is regularly captured for human consumption (see p. 88). Single birds or small flocks are also noted in Western and Eastern Deserts, the Red Sea Mountains and along the coastal plain, and Sinai. There has been a substantial decline since the early twentieth century in the number of Quails netted each autumn. In 1967 about 150 000 were taken along the north Sinai and Gaza coastline (Mendelssohn *et al.* 1969) and in both 1980 and 1981 about the same number were captured along the complete Egyptian Mediterranean coast (Mullié 1985). This is in sharp contrast to the estimated 1.2–1.8 million birds that were annually exported from the country between 1906 and 1913 (Moreau 1927–8). In recent years substantial differences have been found in the number of Quail reaching the north Sinai coast; these vicissitudes are thought to be the result of varying weather conditions on the European breeding and sub-Saharan wintering grounds (Zuckerbrot *et al.* 1980).

There are scattered December and January records from the Nile Delta and Valley, the Faiyum, the Suez Canal area, and along the Red Sea coastal plain. It is unknown if these records are of resident breeding birds, particularly in the Nile basin, or migrants.

Spring migration is from early February to late April, peaking between mid-March and mid-April. Along the Red Sea coast this species is distinctly more common in spring than in autumn. On spring passage it is also noted in the Faiyum, Western and Eastern Deserts, the Suez Canal area, and in Sinai. In six spring seasons (between 1967 and 1973) of systematic study along the Mediterranean coast about 1.0 per cent (522) of the birds captured were Quails (Horner 1977), compared with three autumn seasons (between 1968 and 1972) in the same area when Quails made up 25.4 per cent (8426) of the local catch (GZF). A few summer records are known from non-breeding areas, e.g. in June 1944 at Safaga (Bulman 1944).

Ringing There are almost 83 records of Quails ringed in Egypt and recovered elsewhere in the western Palearctic (Fig. 6.13). These include (number of recoveries for each country in parentheses): Austria (1), Bulgaria (23), France (1), Greece (4), Israel (1), Italy (1), Lebanon (11), Rumania (1), Syria (12), Turkey (7), and various portions of USSR (19). Birds ringed in Israel, Italy, and the USSR have been recovered in Egypt (Watson 1971; Glutz von Blotzheim *et al.* 1973; Paz 1987; GZF; MSp). The only extralimital African record is one ringed at Bahig on 19 September 1971 and recaptured at Outtam-Bac, Bossangoa, Central African Republic, on 26 December 1973 (GZF). There is some evidence that within Egypt some birds show migration route tenacity: six Quails ringed in autumn were recovered in the same area in

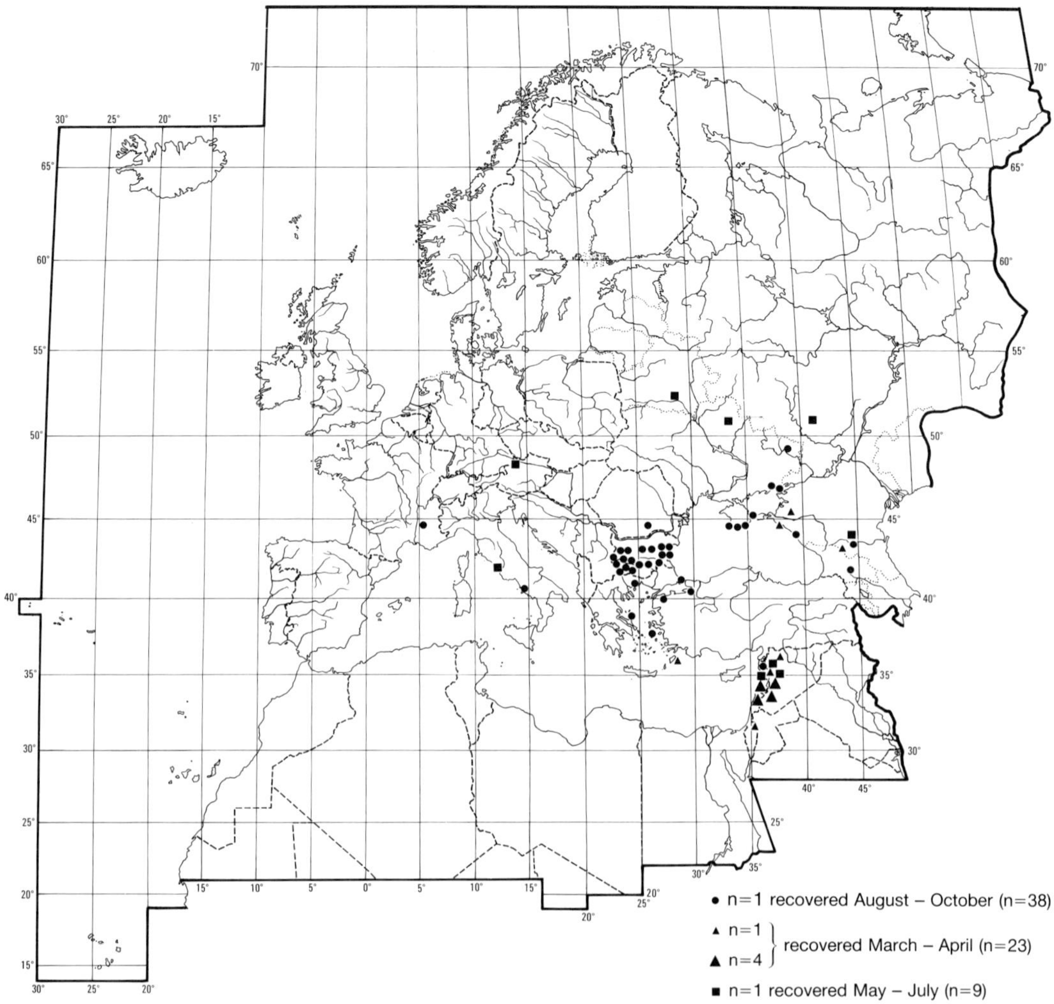

FIG. 6.13. Recovery sites of Quails (n=70) ringed in Egypt.

subsequent autumns (up to three years later), three ringed in autumn were recaptured the following spring, and one ringed in spring was recaptured the following autumn (Watson 1971; GZF). In the eastern Mediterranean this species apparently has a circular migration route. Recoveries from the Levant are almost exclusively from March and April, whereas those from south-eastern Europe are confined to August and September. In the first few decades of the twentieth century Quails were exported from Egypt to France and Italy, ringed, and released (for hunting purposes). Of these Quails 86 were recovered, including single birds in the Federal Republic of Germany, Poland, Sweden, Turkey, and Yugoslavia (Duse 1935; Toschi 1956).

ORDER GRUIFORMES

Family Rallidae

Generic local names for rail Mur'a, fitkaawii (ND), mur'aay.

Water Rail *Rallus aquaticus* RB WV

Local names Mur'at al-maa- مرعة الماء, farkha al-maaya (WN), kalb al-bahr, 'usfuur kalb al-maaya, zanaaziin, zanzuun (ND).

Fairly common breeding resident in the north of the country. It is a common winter visitor from late September to mid-April. Both residents and migrants are referable to *R. a. aquaticus*. Eggs are laid between late March and late May.

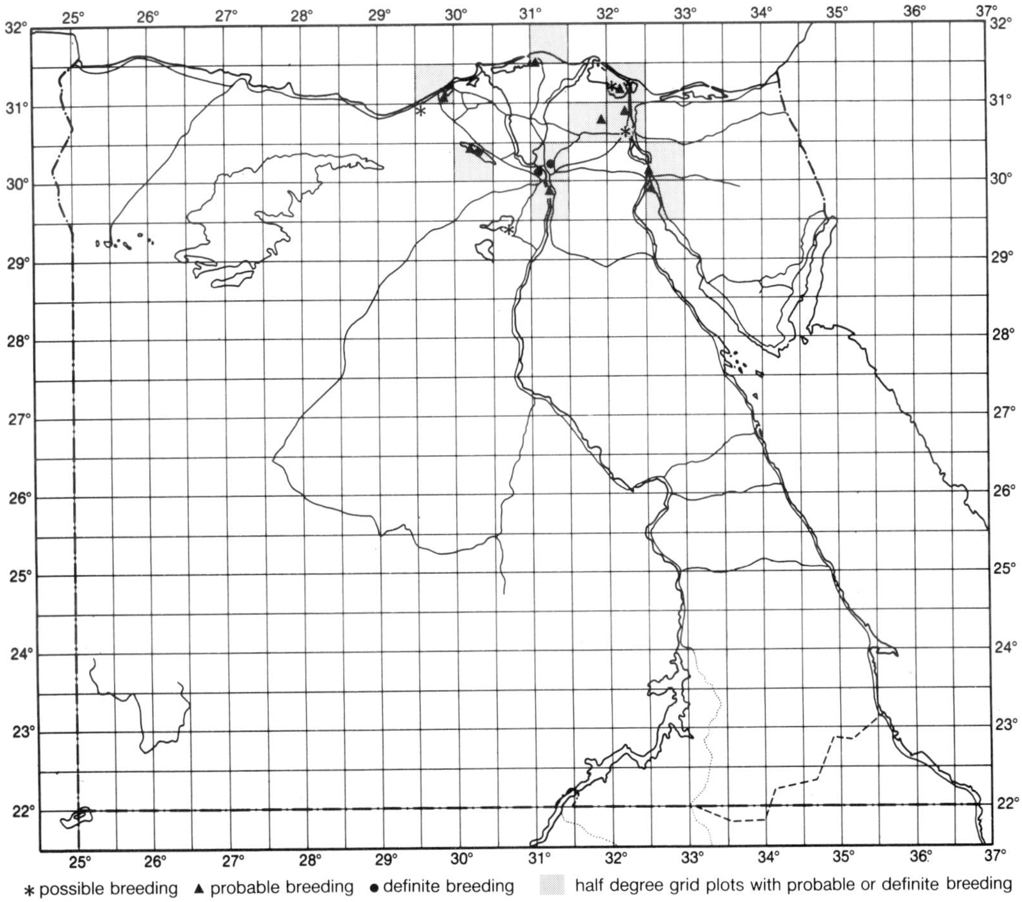

Map B25. Breeding distribution of Water Rail *Rallus aquaticus* between 1930–87.

Meinertzhagen (1930) stated that the Water Rail was abundant in Wadi el Natrun and at suitable localities in the northern Nile Delta. Recent information indicates that this species breeds in the marshes of Wadi el Natrun, the Nile Delta, and the Suez Canal area. There is considerable temporal overlap between the departure of winter visitors and the commencement of local breeding. Thus, this species is potentially a more widespread breeding resident than is currently known. It may nest along the Nile between Cairo and Aswan, e.g. one was heard calling near Helwan on 14 May 1983 (MCJ *et al.*).

Autumn passage has been reported from late September to early November (Meinertzhagen 1930) but the influx of northern migrants may continue into December and January. In the winter it is common in the aforementioned breeding areas and the complete Nile Valley. The largest winter concentrations are found near the Delta lakes. An estimated annual average of 500 birds are offered for sale in the Port Said and Damietta markets alone (Mullié and Meininger 1983; tables 4.3 and 4.4).

Presumed winter visitors are noted until mid-April in areas where there is no evidence of breeding, e.g. one at El Maghra, Qattara Depression, on 15 April 1977 (Goodman and Ames 1983). Water Rails have been observed on a few occasions, mainly during the periods of autumn and spring passage, along the North Coast, in several Western Desert oases, and Sinai.

Spotted Crake *Porzana porzana* PV WV

Local name Mur'a munaqatta مرعة منقطة.

Fairly common passage migrant and scarce winter visitor from early September to late March (late April). Monotypic.

Due to the skulking habits of the Spotted Crake it is difficult to estimate its abundance in the country. Between September and October and in March this species appears to be fairly common and widespread in numerous portions of Egypt. The earliest autumn record is one at Bahig on 10 September 1962 (FMNH). The only Western Desert records appear to be single birds in Wadi el Natrun on 1 October 1965 (HCK) and on 29 October 1910 (Nicoll 1912*b*). There are also records from the migration periods in Sinai and the southern Nile Valley. They are distinctly scarcer during the winter months. In recent years there have been several winter records of small numbers in the Nile Delta, the Suez Canal area, and the Faiyum. A maximum of three per day have been found in the Port Said bird market (Mullié and Meininger 1983; SMG). The latest spring records are single birds at Bahig on 25 and 27 April 1971 (USNM).

Little Crake *Porzana parva* CB? PV WV

Local name Mur'a saghiira مرعة صغيرة.

Possibly a casual breeding resident in the Nile Delta. Scarce passage and winter visitor from (early) mid-September to late April. Monotypic.

According to Meinertzhagen (1930) the Little Crake is not uncommon in Egypt during the winter and several were collected from October to May. Between 1916 and 1918, Raw (1921) collected several specimens and noted possible breeding near Abu Zabal. A clutch of eggs taken at this locality in April 1916 is either this species (Meinertzhagen 1930) or Baillon's Crake (Raw 1921). Breeding in the Nile Delta lakes seems probable but excluding the ambiguous Abu Zabal record there is no evidence (Meininger *et al.* 1986).

In recent years the species has been noted with some frequency, mainly in autumn and spring, along the North Coast, in the Nile Delta lakes and Valley (south to Aswan), the Suez Canal area, and southern Sinai. The earliest autumn record is one on 4 September 1983 in Bahariya (HB *et al.*), which is also the only known report from the Western Desert. There are winter records from the Nile Valley above Aswan before the formation of Lake Nasser, e.g. one on 17 February 1964 at Nag Misaw (FMNH). Generally Little Crake is seen singly or in pairs; an exception is up to six at Lake Qarun between 15 and 17 April 1984 (MBe, SM). The latest recent spring record is 26 April 1983 at Aswan (UGS, HR).

Baillon's Crake *Porzana pusilla* RB PV WV

Local name Mur'a baayluun مرعة بيلون.

Rare breeding resident in the Nile Delta and Suez Canal area. Rare passage and winter visitor between early October and mid-April. All of the Egyptian material examined, including the two specimens (GZM) Meinertzhagen (1930) identified as *P. p. pusilla*, are referable to *P. p. intermedia*.

Meinertzhagen (1930) collected an adult Baillon's Crake with downy young on 1 May 1920 at El Qantara (BMNH) and mentioned a clutch of six eggs taken in the Delta (no details given). Raw (1921) obtained a bird on 19 April 1917 at Abu Zabal in breeding condition. There are no recent nesting records from the Nile Delta but the species is presumed to still nest there (Meininger *et al.* 1986). This species is a rare passage and winter visitor to Egypt. The only known Egyptian records after 1930 are single birds on 4 October 1942 at Abu Qir (Bodenham 1945*a*), 21 January 1983 in the Faiyum (BR), 23 January 1985 in the Port Said market (UMMZ), 25 March 1969 at Abu Rauwash (USNM), 4 April 1983 at Suez (SC, IP), 22 March 1979 (DJF *et al.*), 10 April 1979 at Sharm el Sheikh (EB), and 17 April 1984 along Lake Qarun (MBe, SM). Three were noted at Sharm el Sheikh on 27 March 1979 (DJF *et al.*).

Corncrake *Crex crex* PV

Local names Mur'a al-ghalla مرعة الغلّة, sifrad.

Scarce passage visitor presumably throughout the country in autumn from (early) late August to mid- (late) October and in spring from (mid-February) early March to late April (late May). Monotypic.

According to Meinertzhagen (1930) Corncrakes were regularly noted in autumn on the north coast of Sinai 'about one out of every fifteen parties of Quails arriving at dawn having a single Landrail [Corncrake] among them'. He further stated that it was common in Egypt during the first two weeks of September. In recent years this species has been rarely noted in Egypt and based on Meinertzhagen's statement it has decreased in numbers since 1930.

During systematic counts of migrating birds at Zaranikh, six Corncrakes were observed between 7 and 30 September 1980 (Petersen and Sørensen 1981*a*), eight between 16 August and 24 September 1981 (Petersen and Sørensen 1981*b*), and five between 2 and 24 September 1982 (Baha el Din and Salama 1984). Since 1954 approximately 30 specimens have been taken on migration along the North Coast (FMNH, USNM, YPM). Recent records from Bahariya include single birds on 16 and 22 September 1983 and 9 on 21 September 1985 (HB *et al.*). This species is known to cross the Red Sea on passage; one hit the Daedalus Reef lighthouse on 27 August 1913 (Ticehurst 1924). There are few autumn records from other portions of the country. The earliest autumn record is one at Bahariya on 4 August 1983 (HB *et al.*) and the latest records are single birds at Bahig on 24 October 1966 (USNM) and at Ain Musa on 30 October 1984 (RvD *et al.*).

Two observed near Beni Hassan on 11 February 1978 (Short and Horne 1981) were presumably early spring migrants. In March and April this species is rarely observed and most records are from the Nile Delta and Valley. The only known spring record from the Western Desert is one Corncrake in Siwa on 9 May 1985 (Goodman *et al.* 1986). The latest spring records are single birds in Wadi Feiran on 24 May 1938 (GZM) and near Cairo on 9 June (Meinertzhagen 1930).

Ringing A bird ringed in Sweden on 13 June 1971 was trapped at El Arish on 3 October 1971 (Anon. 1972).

Moorhen *Gallinula chloropus* RB PV WV

Local names Farkha al-maa- فرخة الماء, ba<u>tt</u> maa-ii (Su), dajaaja al-maaya, firaakh (ND), zunzuun (ND).

Common breeding resident. Abundant passage and winter visitor from mid-September to mid- (late) April. All Egyptian material examined is referable to *G. c. chloropus*.

The Moorhen is a common breeding resident in marshes and along canals in the Nile Delta and Valley (south to Aswan), the Suez Canal area, and the Faiyum. The Nile Delta breeding population is estimated at several thousand pairs (Meininger *et al.* 1986). In the Western Desert the Moorhen is known to breed in Kharga, is suspected of doing so in Dakhla, and information is inconclusive from Siwa, Bahariya, and Farafra (Goodman *et al.* 1986).

Although Koenig (1928) travelled widely in Egypt between 1897 and 1899, particularly in the Upper Nile Valley, he seldom observed this species and stated that there was no proof of its breeding in Egypt. However, according to Meinertzhagen (1930) it was a common resident in numerous areas of the country, including the Nile basin.

Between mid-September and late April the Egyptian breeding population of Moorhens is augmented by Eurasian migrants. This species is regularly observed during migration along the North Coast, in the Western and Eastern Deserts, and Sinai. In winter up to 200 have been recorded in Dakhla (Goodman *et al.* 1986). An estimated 20 500–25 000 Moorhens are shot or trapped annually on Lake Manzala (Mullié and Meininger 1983; table 4.5). During migration this species is noted at

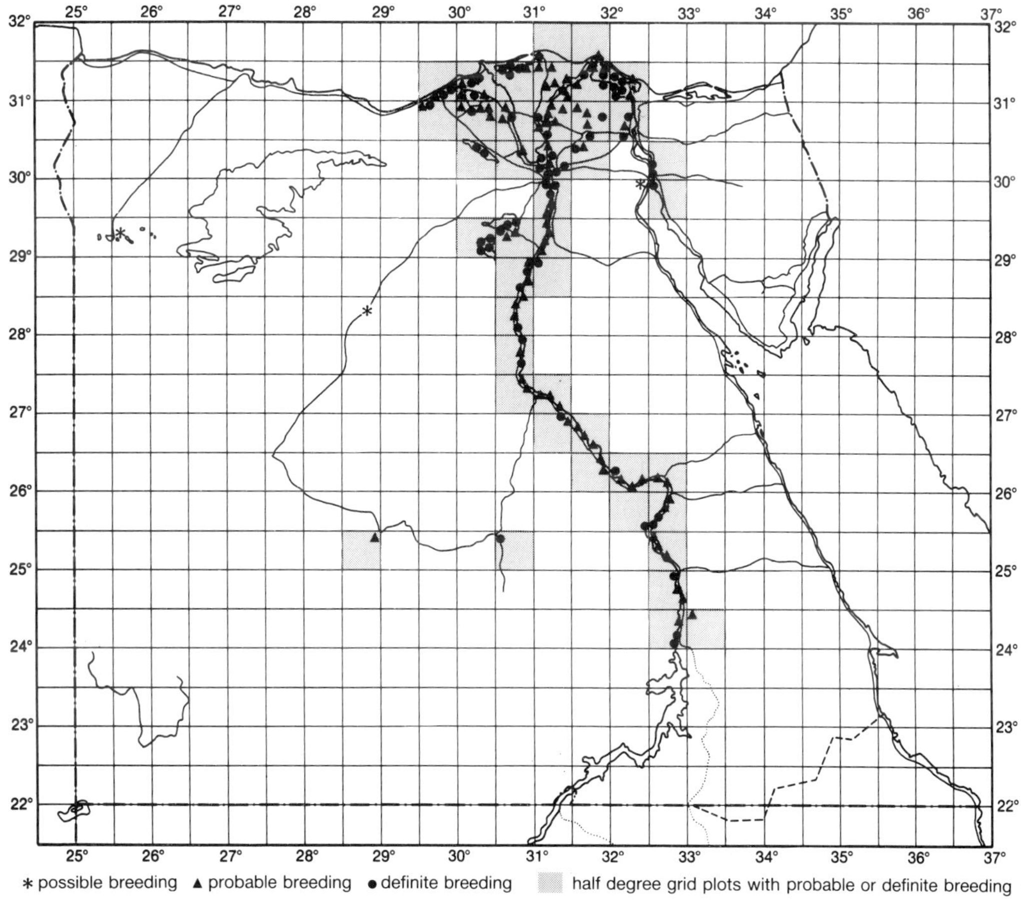

Map B26. Breeding distribution of Moorhen *Gallinula chloropus* between 1970–87.

unusual places, e.g. at Bir Beida, west of Quseir, on 2 February 1983 (SMG) and 30 April 1983 (PLM).

Purple Gallinule *Porphyrio porphyrio* RB Plate 2

Local names Farkha sulṯaaniyya فرخة سلطانية, dajaaja sulṯaaniyya, digima, diik (ND), diik al-ghaab (Su), diik buurḏii (NC), diik sulṯaanii, sulṯaanii.

Fairly common breeding resident in the Nile Delta and Valley, Suez Canal area, Faiyum, Wadi el Rayan, and probably Wadi el Natrun. All Egyptian material examined is of the green-backed form, *P. p. madagascariensis*.

Western Desert In late January 1985 a local resident hunter of Wadi el Natrun mentioned that he had observed a Purple Gallinule 'several months ago' in the oasis (SMG). The only other record from the region is from Dakhla where the remains of one bird were found in mid-March 1987 (AH).

Nile Delta The Purple Gallinule is a characteristic breeding bird of the Nile Delta lakes. Lake Manzala and Lake Burullus appear to be the most important breeding sites for this species in the western Palearctic and may hold several thousand breeding pairs (Meininger *et al.* 1986). On Lake Manzala alone 800–900 birds are shot or caught annually and offered for sale in local markets (Mullié and Meininger 1983; tables 4.3 and 4.4). Lakes Maryut and Idku have small breeding populations. Meinertzhagen's (1930) statement that 'their main [breeding] station is Maryut', appears to be no longer the case. The species also nests in the marshes of the southern and central Nile Delta and at the sewage settling ponds of Tenth of Ramadan City.

This species is a resident in the Suez Canal area. Raw (1921) noted a nest in the marshes near El Qantara; the same locality where a dead bird was found on 17 April 1984 (Siering 1986). In recent years this species has bred at several places near Ismailiya. Recent records from Suez include one on 19 April 1984 (SMB), up to five calling between 22 March and 4 April 1985 (UGS), and three in May 1985 (UGS).

Nile Valley According to Meinertzhagen (1930) the Purple Gallinule was a former breeding resident in the Faiyum. In the 1970s and early 1980s the Faiyum was regularly visited by bird-watchers but no records of this species are known from this period. The first recent record was of one on 15 November 1983 along the shore of Lake Qarun (Bezzel 1987). Since the early 1980s there has been a substantial increase in the size, number, and density of reed-beds along southern Lake Qarun, where this species is now regularly noted. The earliest known record for Wadi el Rayan is 1982 (MuS) and it is now common in this area. The colonization of Wadi el Rayan by the Purple Gallinule also appears to be correlated with the spread of reed-beds.

Neither Meinertzhagen (1930) or Cramp and Simmons (1980) mentioned the occurrence of this species in the Nile Valley south of Cairo. There were no modern records from this region until the mid-1970s; although Cecil (1904) noted that it 'has scarcely ever been seen on the river'. In the past 20 years this bird has been recorded throughout the year along the complete length of the Egyptian Nile Valley and there are confirmed breeding records as far south as Aswan. At least several hundred pairs breed along the Nile and their numbers seem to be increasing. A total of 300 was counted along the river between Cairo and Aswan from 14 to 24 March 1985 (Grieve 1985). There is a single recent record along the White Nile, central Sudan, near Kosti (Nikolaus 1987).

Since the completion of the Aswan High Dam there has been a notable increase in the distribution and extensiveness of reed-beds in the Nile Valley, which is presumably directly related to the Purple Gallinule's recent colonization of the area. A nearly complete right carpometacarpus recovered in the Upper Nile Valley and radio-carbon dated at 25 000 to 30 000 BC was identified as this species (Tchernov 1968). Thus, it would appear that it has actually recolonized the Nile Valley in recent years.

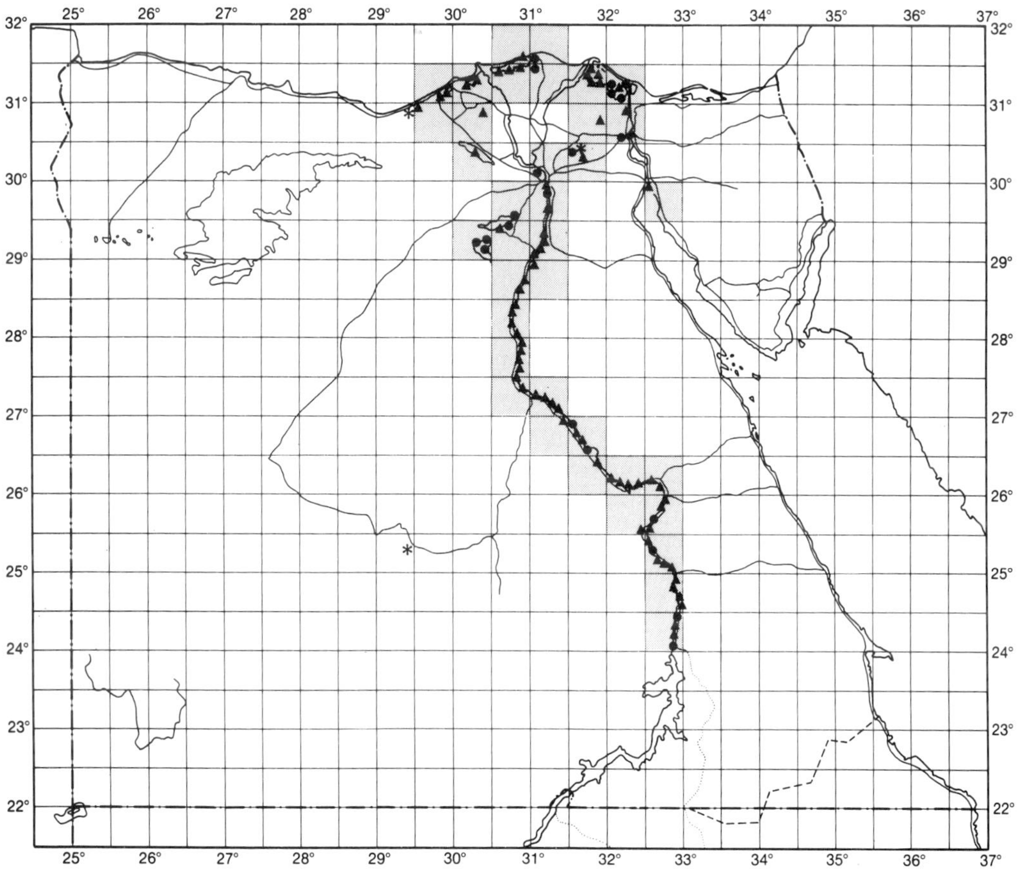

∗ possible breeding ▲ probable breeding ● definite breeding ▒ half degree grid plots with probable or definite breeding

Map B27. Breeding distribution of Purple Gallinule *Porphyrio porphyrio* between 1970–87.

Coot *Fulica atra* RB WV

Local names Ghurr غرّ, ghurra (ND), ghararıyya (Ma).

Rare breeding resident. Abundant winter visitor from mid-September to early April (mid-May). Rare summer visitor. All Egyptian material examined is of *F. a. atra*.

Based on observations of adults with chicks, Meinertzhagen (1930) concluded that the Coot was a common breeding resident on the larger lakes of the Nile Delta. He also thought that they bred at Lake Qarun. Coots have been observed carrying nest material in Dakhla (Meinertzhagen 1930). During a survey of the Nile Delta lakes in May 1983 not a single Coot was observed and it presumably no longer breeds in this region (Meininger *et al.* 1986). Recent observations supporting possible breeding in the Nile Valley include a pair at El Badrshein in April, May, and June 1980 (SV), one between Beni Suef and El Minya on 25 April 1983 (UGS, HR), a pair between Beni Suef and Beni Mazar on 30 April 1978 (Meininger and Dielissen 1979), and one south of Giza on 25 May 1982 (Wimpfheimer *et al.* 1983). In Dakhla, 10 Coots were noted in suitable nesting habitat on 24 April 1981 (Goodman *et al.* 1986). Several Coots were noted at Fayid on 26 June 1949 (JCC).

This species is an abundant winter visitor throughout the country, particularly in areas with open water such as the Nile Delta and Valley, and lakes in the Western Desert. The first autumn migrants arrive in mid-September, numbers increase throughout November and December, and the largest concentrations are present in midwinter. By mid-March most Coots have left Egypt and the latest spring records of large flocks are in early April. In the winters of 1978–9 and 1979–80 about 154 000 Coots were counted on Lake Burullus, 51 300 on Lake Manzala, and 18 000 on Lake Qarun (Meininger and Mullié 1981*b*). The Coot is the most numerous bird offered for sale in the Port Said market with up to 11 500 birds in 18 days during the winter of 1982–3 (SMG; table 4.1). In most winters Coot is common along the Nile south to Aswan in flocks of up to 1000, while in other winters few are observed in this region. It is regularly seen in small numbers along the entire Red Sea coast and northern gulfs.

Ringing A bird ringed near Odessa, USSR, on 12 June 1981 was recovered in the Damietta market on 10 October 1981 (BEP).

Family Gruidae

Generic local names for crane Ghurnuuq, kurk, kurkii (WN), rahuu.

Crane *Grus grus* PV (WV)

Local names Kurkii كركي, wizz 'iraag (Ma).

Fairly common passage visitor in autumn from late September to early December and in spring from early March to late April (mid-May). Occasional in winter. All Egyptian material examined is referable to *G. g. grus*.

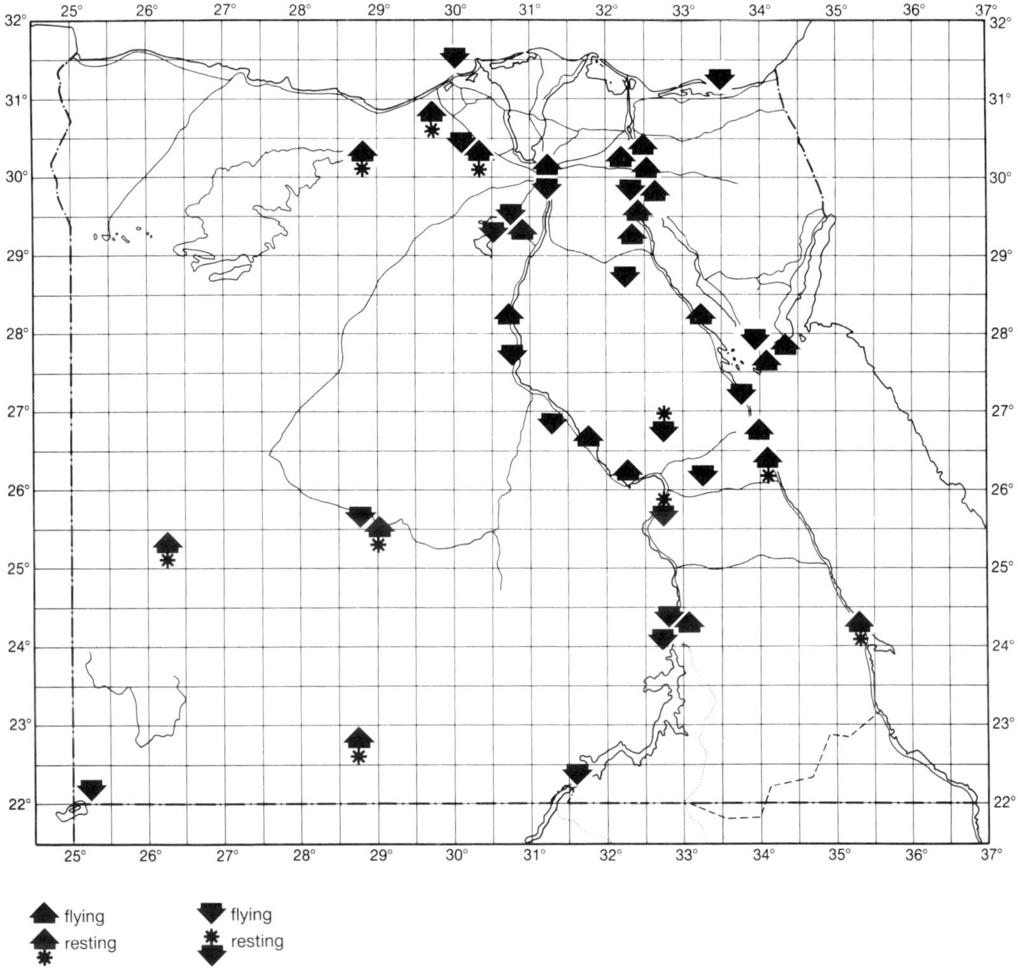

FIG. 6.14. Sight records divided by season of Crane flocks.

Cranes migrate along a broad front and appear to be more common in the east of the country in autumn and the west in spring (Fig. 6.14). Autumn migration usually commences in late September, peaks between mid-October and mid-November, and small numbers are noted into December. Cranes enter Egypt by crossing the Mediterranean (Alexander 1929; Flint and Stewart 1983) and via the Levant–Sinai pathway. Autumn records from the Western Desert include 150 birds over Kharga on 12 October 1978 (Goodman *et al.* 1986), 88 at Wadi el Natrun on 17 October 1986 (SMG), and flocks moving over Gebel Uweinat in December 1968 (Misonne 1974). Small flocks of migrants are regularly seen or heard at night in the Cairo area and throughout the Nile Valley (including Lake Nasser).

The only known autumn record from northern Sinai is seven at Zaranikh some time between 15 August and 5 October 1978 (YP; UP); however, it should be noted that counts in this area have usually ceased before Crane migration reaches Egypt. During a systematic count at Suez from 4 September to 5 November 1981, 4694 Cranes were observed between 9 October and 4 November, the highest daily numbers being 644 on 19 October, 1450 on 27 October, and 1554 on 3 November (RGB). Flocks are known to fly over the Gulf of Suez: on 19 October 1980, 60 were seen crossing near Hurghada (Meininger *et al.* 1981); on 22 October 1983, 130 flying west into Wadi Araba (JJH); and on 2 December 1984, 137 over Sharm el Sheikh (RvD *et al.*). The record of flocks flying over Almaza on 25 and 26 August 1945 (White 1949) seems exceptionally early for Cranes, and these dates match the period for peak migration of Demoiselle Cranes over Cyprus (Flint and Stewart 1983).

Meinertzhagen (1930) stated that the only place in Egypt where Cranes winter is Lake Maryut. Flower (1933) noted that some are found in Upper Egypt in the winter and that he saw the species near Aswan in December. Twelve were observed in Siwa between 7 and 13 February 1939 (Al Hussaini 1938–9). The only recent winter records are single birds on 25 January 1985 at Birket el Hagg (DM) and on 18 February 1985 north of Asyut (GFB).

In spring most records of large groups are from the Western Desert and generally in March. Moreau (1927) mentioned that 'immense numbers pass by way of Dakhla in spring, flying due north'. At this oasis about 300 were noted on 5 and 6 March 1986 (AH) and 290 on 17 March 1986 (AS). Flocks of resting Cranes have been reported in March at Bir Tarfawi about 300 km south of Dakhla (Moreau 1928), in the desert about 300 km west of Dakhla (Moreau 1967), east of Gebel Uweinat (Bagnold 1939), south of Burg el Arab (PLM), and in Wadi el Natrun (BO). The largest number of Cranes recently counted in the Nile Valley during spring migration is 1741 between 14 and 24 March 1985 (Grieve 1985). During a systematic count of migrants at Suez between 23 February and 16 May 1982, 668 Cranes were recorded in six days between 2 and 16 March 1982, with a maximum of 474 on 9 March (Wimpfheimer *et al.* 1983). A flock of 27 was noted resting along the southern Red Sea coastal plain, near Abu Ghusun, on 20 March 1983 (Goodman and Atta 1987). On 19 March 1988 a total of 700 (in flocks of up to 60) was seen migrating north between Suez and Zafarana (JS *et al.*). Late spring records include 50 over the Bitter Lakes on 27 April 1967 (JT), three at Birket el Hagg on 4 May 1985 (DM), and one at El Maghra on 12 May 1975 (Goodman and Ames 1983).

Demoiselle Crane *Anthropoides virgo* PV (WV)

Local names Rahuu رهو, kurkii habashii (WN).

Rare passage and winter visitor to Egypt. Little recent information from the country. Monotypic.

Flower (1933) saw a large flock of Demoiselle Cranes in September 1894 flying over the Suez Canal. One was found dead at El Tell el Kebir on 24 October 1899 (Flower 1933). Older records of spring concentrations include about 1000 birds over Aswan on 28 February 1901 and 'perhaps several thousand' along the Nile near Abu Girgeh on 24 March 1874 (Flower 1933). Shelley (1872) saw a large flock near Beni Suef on 2 April. Koenig (1928) reported this species as 'vereinzelt, so doch nicht selten' along the Nile between Asyut and Aswan in 1899.

It can be inferred from these records that this species was probably a relatively common passage migrant in Egypt until the beginning of the twentieth century. There are only a few subsequent reports. Several birds were caught near Alexandria in late September 1912 (Meinertzhagen 1930). One was observed at Fayid on 11 October 1950 (LPC) and two in a flock of migrating Cranes at Suez on 3 November 1981 (RGB). In Bahariya two Demoiselle Cranes were seen on 24 March 1925 and in Dakhla the species was noted on an unspecified spring date (Moreau 1927, 1967). In March 1928 a small detached group was noted at Kharga in a large flock of Cranes (Meinertzhagen 1930) and on 18 May 1944 a single Demoiselle Crane was seen at Ismailiya (CJG). One was shot by a hunter on 5 May 1985, 70 km east of Cairo along the Cairo–Ismailiya road and brought into the GZM (SMG). The only known relatively recent winter records are one bird on 22 December 1953 south of Alexandria (HCK) and one in January 1978 at Sharm el Sheikh (Hovel 1987). At Elat, Israel, this species is a rare migrant in March and April and was recorded five times between 1966 and 1979, with a maximum flock size of 14 birds (Krabbe 1980).

The paucity of recent Demoiselle Crane records from Egypt is striking in the light of the large concentrations that pass over Cyprus (Flint and Stewart 1983) and the considerable numbers wintering in the Sudan (Urban *et al.* 1986).

[**Crowned Crane** *Balearica pavonina* AV?

Von Besserer (1908) observed 15 Crowned Cranes on an island in the Nile near El Kab on 10 February and three days later another flock of 15 further south (year not stated). Unconfirmed records of this species near Gebel Elba include five observed in 1964 (IH) and three or four shot in 1971 or 1972 (MWa).]

Family Otididae

Little Bustard *Tetrax tetrax* (WV)

Local names H̲ubaaraa s̱aghiira حبارى صغيرة, ra''aad, s̱uww.

Formerly rare and irregular winter visitor to the Nile Delta and Sinai. No recent records. Monotypic.

Von Heuglin (1873) considered the Little Bustard an irregular winter visitor to Egypt. He noted this species several times near Pelusium and found several specimens in the Alexandria market. It was also found on 12 February 1896 in the Cairo market

(Taylor 1896). Beadnell (1927) shot one near Gebel Maghara some time between 1921 and 1924. A bird collected at El Qatta on 8 December 1922 (GZM) appears to be the last verifiable record from the country. The disappearance of this species as a winter visitor in Egypt is presumably related to the drastic decline of eastern European and USSR breeding populations (Cramp and Simmons 1980).

Houbara *Chlamydotis undulata* RB (WV)

Local names H̱ubaaraa حبارى, ẖubjar, ẖubraj.

Rare breeding resident in the Mediterranean coastal desert, Sinai, and presumably in the Eastern Desert. Two subspecies occur in the country. *C. u. undulata* is the form breeding west of the Nile and generally thought to be sedentary. The populations of Sinai and the Eastern Desert are referable to *C. u. macqueenii*; nothing is known about the dispersal and migration of the breeding populations occurring in these two areas. Migratory Houbaras from western Asia and the Middle East (*C. u. macqueenii*) augment Egyptian populations during the winter months.

North Coast and Western Desert Houbaras were formerly common breeding residents in the Mediterranean coastal desert between Salum and Alexandria and near Wadi el Natrun (Meinertzhagen 1930). According to Magaud d'Aubusson (1892) this species was frequently found in the Alexandria market. Between 1965 and 1967 it was observed 20 to 30 km south of Bahig (Goodman *et al.* 1986). Around 1975, one was shot 6 km south of Alexandria, an area where this species appears to be now locally extirpated (BO). In February 1920 many were observed between Salum and Siwa (Meinertzhagen 1930). From mid-September to early December 1940, Bulman (1942) regularly observed this species 15 to 50 km from the coast between Sidi Barrani and Matruh but never between Sidi Barrani and Salum. In October 1940, the number of Houbaras living in the Matruh area was estimated to be about 50 (Bulman 1942). Specimens were collected from a 'flock of five on open plateau desert' on 23 October 1953, 32 km south-east of Salum; 'in barley field on desert' on 20 April 1954, 4 km south of Sidi Barrani; and (two) 'on desert at night' on 23 April 1954, 16 km east of Sidi Barrani (FMNH). In 1980 there were still 'many' Houbaras up to 80 km south-west of Matruh on the road to Siwa (Goodman *et al.* 1986) and in 1985–6 it was reported to be not uncommon near Salum (MAM). Currently the area between Matruh and Salum appears to be the only remaining stronghold of this species in Egypt, although its actual present status west of Matruh is not clear.

This species' current distribution in portions of the Western Desert is ambiguous. It formerly nested in Wadi el Natrun, where an egg was collected on 10 April 1914 (GZM). The most recent record from this oasis is one taken on 5 April 1937 (HGD). A specimen referable to *C. u. macqueenii* was collected on 4 December 1910 west of Cairo (Koenig 1928; ZFMK). The species was also reputed to have formerly nested in the region between Bahariya and Siwa. In 1977, one was shot 25 km north of Bahariya but there are no other recent confirmed reports from this oasis or Farafra (Goodman *et al.* 1986). According to Meinertzhagen (1930) the Dakhla population was migratory, being absent during the winter months and returning in spring to nest on the edge of the oasis. However, there is no evidence that it actually bred locally nor recent records from this oasis. Jarvis (in Moreau 1927) noted that Houbaras were observed from 'time to time in winter in the open sandy country at the north end of Kharga'. These birds were presumably winter visitors; however, Meinertzhagen (1930) thought they might breed locally on the desert edge. The only recent information from Kharga is an observation in 1978 reported by a local hunter (Goodman *et al.* 1986). Fresh tracks of one Houbara were photographed on 31 March 1982 in the desert west of Aswan, near the St Simeon Monastery (JW).

Eastern Desert Koenig (1928) mentioned that fresh eggs were collected on 7 March 1917 in Wadi Digla (ZFMK). Specimens referable to *C. u. macqueenii* have

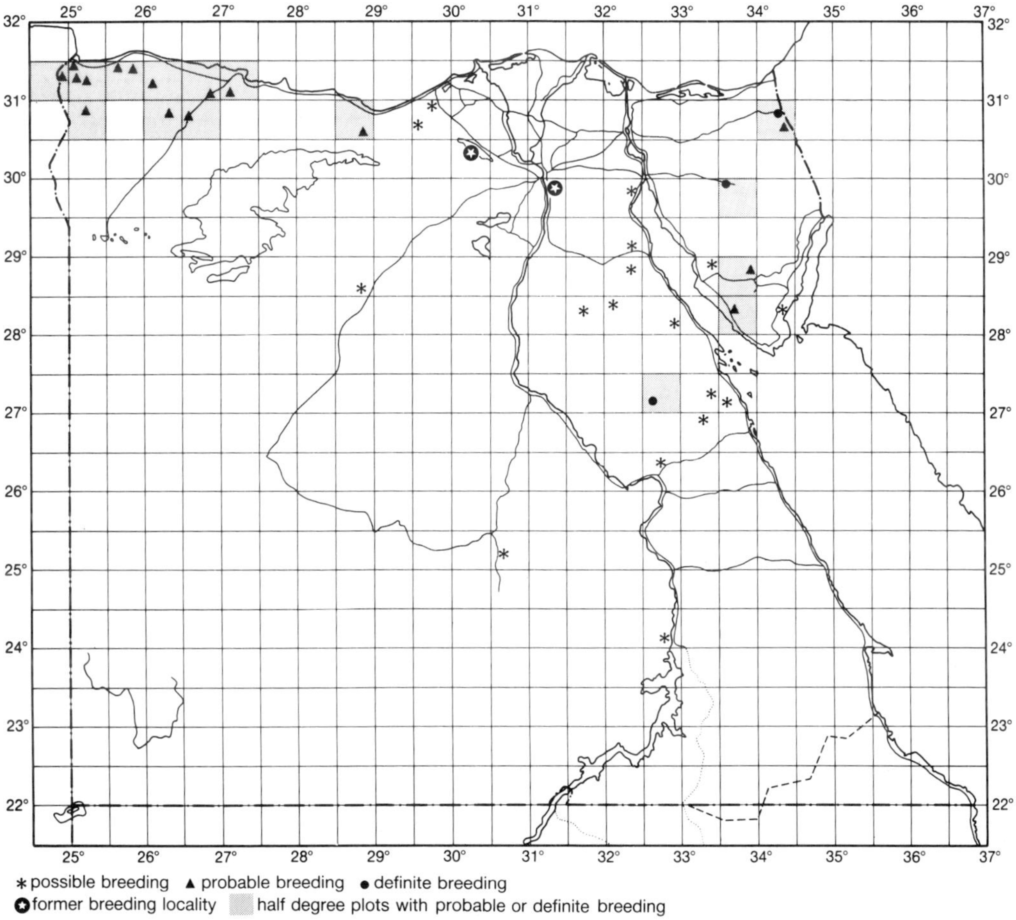

Map B28. Breeding distribution of Houbara *Chlamydotis undulata* between 1940–87.

been taken east of Cairo in June and December (Meinertzhagen 1930). Two birds were observed on 17 October 1981 in the desert just outside old Cairo (PW). Other recent records from the northern portion of the Eastern Desert include one shot by a Cairene hunter in Wadi Hagul in the late winter or early spring of 1981 (IH), and fresh tracks in Wadi el Tarfa on 30 April 1983 (JJH) and on two occasions in late November 1986 (SMG, JJH). In July or August 1985 a single bird was noted between Suez and Ain Sukhna (IH). Tregenza (1955) mentioned that tracks were found in the winter of 1948 in Wadi el Atrash. A small population still apparently inhabits portions of Wadi Qena, where on 18 October 1986 a hunter from Alexandria shot one adult male and on 30 November 1986 another adult male was observed (SMG, JJH).

Sinai Meinertzhagen (1930) mentioned this species as a rare resident and winter visitor to the sandy plains of northern Sinai, although he presented no evidence of local breeding. On 10 May 1939 Galal el Din and Fouad (1939–40) observed a bird with young between the Mitla Pass and Nekhl. In recent years a few pairs have been reported from the area between Qusaima and the Israeli border near Nizzana. A female with two chicks was noted in this region on 3 April 1973 and one bird was killed by an automobile near Ain el Gedeirat on 21 March 1976 (SC). Recent records from southern Sinai include: tracks in Wadi Kid on 27 December 1978 (EB), three birds along the coastal plain north of El Tor on 16 August 1982 (Baha el Din 1984*a*), one bird and numerous tracks at Ilwa el Agramiya on 5 November 1983 (SMB), and one in Wadi Mukattab on 26 January 1985 (SMG).

ORDER CHARADRIIFORMES

Family Rostratulidae

Painted Snipe *Rostratula benghalensis* RB Plate 2

Local names Bakaashiin muzawwaq بكاشين مزوّق, khawlii, shunqub, shuusha hindii.

Fairly common breeding resident in portions of the Nile Delta and Lower Valley, Wadi el Natrun, the Faiyum, Suez Canal area, and perhaps locally in the Upper Valley. *R. b. benghalensis* is the form occurring in Egypt.

Western Desert The Painted Snipe is a relatively common breeding resident in Wadi el Natrun. There are several archaic and recent breeding records from this oasis (Goodman *et al.* 1986). The species is unknown from other oases in this region.

Nile Delta This species is a fairly common breeding bird of marshes and wet agricultural areas. However, because of this species' skulking habits it is difficult to estimate the number breeding locally. It is often found on the edges of the Delta marshes (Meininger *et al.* 1986) but also in other areas; for instance: Asher Talaaf—chicks observed in 1982 (BO); El Mahmudiya—nest found on 11 May 1978 (RJM,

LRP); San el Hagar—one male and two females on 25 May 1978 (RJM); and El Khatatba—one on 14 June 1985 (DM). In the Suez Canal area a pair was seen with young on 19 September 1942 in the Great Bitter Lake (Maclaren 1944), a pair was noted on 1 June 1955 near Abu Sultan (RN), and a single bird on 27 September 1952 near El Kabrit (AP). In recent years the species has been regularly seen near Suez and Ismailiya.

Nile Valley The Painted Snipe is fairly common in marshes along the southern shore of Lake Qarun and south of Cairo near Saqqara and Giza. In the nineteenth century it was collected in winter near Asyut (Adams 1864) and Qena (Taylor 1859). The only recent records from the Upper Nile Valley are one near Kom Ombo on 30 September 1980 (JJB) and two south of Luxor on 21 January 1986 (CT).

Sinai The only records we are aware of are single birds observed on 4 and 8 September 1986 at Zaranikh (El Badrawy 1986).

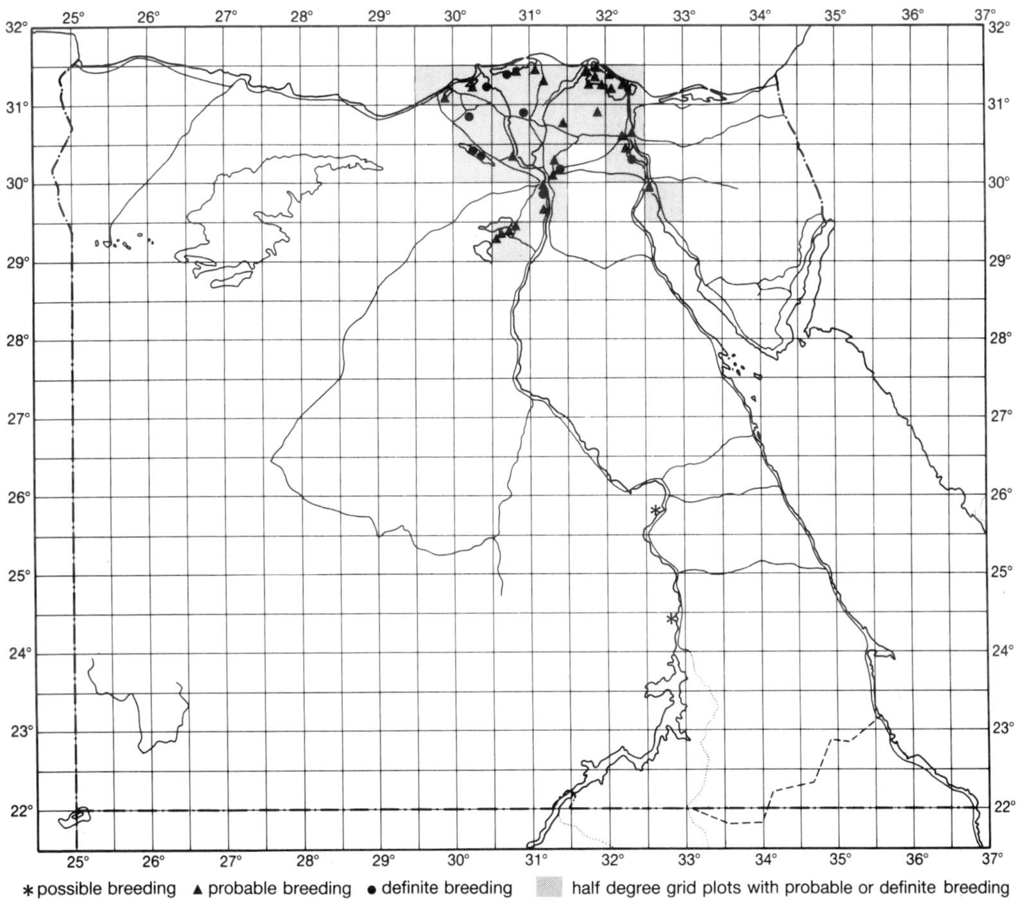

Map B29. Breeding distribution of Painted Snipe *Rostratula benghalensis* between 1970–87.

Family Haematopodidae

Oystercatcher *Haematopus ostralegus* PV WV

Local names Aakil al-ma<u>h</u>aar آكل المحار, ba'ja al-ba<u>h</u>r.

Scarce passage and winter visitor along the coasts of the Mediterranean and Red Sea, rarely found inland, from mid-August to mid- (late) May. Rare summer visitor. Virtually all of the Egyptian material examined is of *H. o. ostralegus*. The only exception is a specimen referable to *H. o. longipes* collected along the southern Red Sea coast at Halaib (GZM).

The Oystercatcher is a local winter visitor, usually in small flocks, to the Egyptian sea coasts. Concentrations of more than 50 birds are rare and known from the Mediterranean coast near Lake Manzala, and the Suez and Hurghada areas. The number wintering in the country probably does not exceed 1000 birds. There are a few summer records, mainly from the Red Sea coast (e.g. Searight 1952*c*) and rarely from north Sinai (Shariv 1970). One was noted in a flooded field near Alexandria on 30 May 1986 (JH). Inland reports include: Wadi el Natrun—one shot on 15 August 1910, a flock noted in November 1911, two on 7 May 1981, and one on 6 January 1982 (Goodman *et al.* 1986); south of El Fashn—one observed on 12 March 1983 (Welch 1983); and Lake Qarun—three on 25 October 1985 (DM).

Family Recurvirostridae

Black-winged Stilt *Himantopus himantopus* MB PV WV

Local names Abuu maghaazil ابو مغازل, abuu qa<u>s</u>aba, abuu saaq, bijji, suqdaa<u>h</u>, <u>t</u>uwwal.

Rare migrant breeder, fairly common passage migrant throughout the country, and rare winter visitor from early August to early (late) May (Fig. 6.15). The nominate subspecies is the one occurring in Egypt.

Meinertzhagen (1930) stated that the Black-winged Stilt bred in Wadi el Natrun, and possibly the Faiyum and several places in the northern Nile Delta. Six pairs were observed and three nests located in Wadi el Natrun on 25 May 1918 (Raw 1921). The only recent documentation of this species breeding in Egypt is a colony found on 18 July 1986 of approximately 20 pairs, 30 immatures, and two one-day-old chicks near the Tenth of Ramadan City (DM). Up to five birds were noted in July and August 1979 at Lake Manzala (EM) but no evidence of nesting was found.

This species is a fairly common passage visitor throughout the country, including the Nile Delta and Valley, Western Desert, Red Sea coast, and Sinai. Autumn passage is from early August to mid-November. The earliest autumn record is 34 birds on 7 August 1955 at Fayid (RN). A total of 206 migrants was recorded at Zaranikh

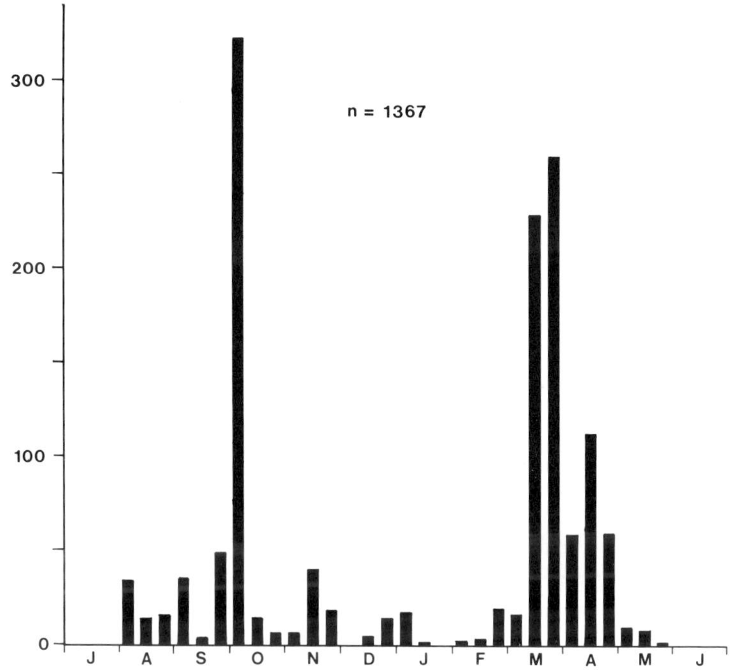

FIG. 6.15. Histogram of the numbers of Black-winged Stilts recorded during 10-day periods throughout the year.

between 15 August and 5 October 1978 (YP, UP). In winter this species is rare, and infrequently observed in the Nile Delta lakes, Lake Bardawil, Wadi el Natrun, Nile Valley, and Faiyum. Spring passage is from late February to late May, peaking in mid-March to late April. Fifty-two were observed at Lake Qarun on 17 April 1984 (MBe, SM). On 24 March 1985, two birds were noted in a desert pool in Wadi Araba, 25 km from the Red Sea coast (SMB, SMG, PLM). The number of autumn migrants appears to be less than in spring.

Avocet *Recurvirostra avosetta* CB PV WV

Local names Nakkaat نكّات, h̲alaabii (Su), h̲aliibii (ND), kaharamaan (WN).

Rare and probably casual breeding resident. Common winter visitor from mid-August to mid-May. Rare summer visitor. Monotypic.

According to Meinertzhagen (1930) Avocets formerly bred in 'some numbers', but by the late 1920s the only known colonies were a few birds in Wadi el Natrun and the Faiyum. There is no definite proof of breeding along the Delta lakes (Meininger *et al.* 1986). The only recent evidence is a female shot in spring 1979 at El Malaha with an egg in its oviduct (Meininger and Mullié 1981*b*). In 1968 up to 11 Avocets were seen in the late spring and summer near Misfaq but no evidence of nesting was found

(Shariv 1970). According to Y. Paran (in Cramp and Simmons 1983) this species has recently bred in southern Sinai, but details are lacking and this record is not acceptable. A pair was noted at a salt lake near Ras Gharib on 29 April 1983; one bird was sitting on the ground in suitable nesting habitat but the area could not be inspected (MCJ *et al.*). Between 15 and 20 were observed on 13 July 1950 at the Gebel Maryam lagoon (GCJ) and two on 19 July 1952 at the same locality (AP).

The Avocet is a regular visitor to Wadi el Natrun, the Nile Delta lakes, Lake Qarun, the Suez Canal area, and northern Sinai. During autumn migration considerable numbers pass along the north coast of Sinai, e.g. 4426 from 15 August to 5 October 1973 and 5585 from 13 August to 29 September 1978 (YP, UP), and 6828 from 16 August to 24 September 1981 (Petersen and Sørensen 1981*b*). Recently large concentrations, although varying in number from year to year, have been recorded in the northern Nile Delta and northern Sinai. At El Malaha about 15 000 Avocets were noted in early January 1972 (Paran 1980) and at Lake Manzala about 2700 on 13 October 1980 (PLM), 1800 between 25 and 29 December 1979 (Meininger and Mullié 1981*b*), and 800 on 7 March 1984 (SMG). There are several records from the Western Desert indicating trans-Saharan migration: Siwa—two on 3 May 1935 (Moreau 1940–1); and Dakhla—five in mid-October 1982 and three on 27 October 1980 (Goodman *et al.* 1986). In the Nile Valley and along the Red Sea coast this species is irregularly recorded in small numbers, almost exclusively in autumn (generally October) and spring (generally mid-March to mid-April).

Family Dromadidae

Crab Plover *Dromas ardeola* WV PV

Local names H̱unkuur حنكور, abuu baliiẖa (Su).

Scarce but locally common visitor to the southern Red Sea coast, mainly between mid-October and late May, and occasionally to the Nile Delta. Monotypic.

Red Sea There is no evidence of Crab Plovers breeding in Egyptian waters. El Negumi *et al.* (1950) reported that in October 1946, one was shot at Shalatein from a large flock. Madden (1929) noted at least eight in the last portion of May 1928 at Halaib. In recent years there has been an increase in the number of ornithologists visiting the area, and it has become clear that this species is a regular visitor to the coastal coral reefs and islands of the southern Egyptian Red Sea, mainly between Shalatein and Halaib. The largest concentrations noted in Egypt are 180 on 30 March 1983 and 110 on 6 April 1985 near Shalatein (SMG, PLM). In June 1984 the species was not observed in the vicinity of Shalatein (SMG). Records from further north include one bird on 22 February 1979 on Gezira Tiran (Hovel 1987), one on 25 May 1976 at Ras el Millan (UP, YYT), three on 13 May 1982 on Gezira Abu Minqar, three on 31 May 1982 north-west of Gezira Mahabis (Frazier *et al.* 1984), and one on 17 July 1984 on Gezira Abu Minqar (SH, DM).

Nile Delta The Crab Plover is occasionally observed in this area. Meinertzhagen (1930) mentioned records of two in August 1916 at Lake Burullus, a large flock in August 1918 near Rosetta, and one on 5 November 1920 that struck the Damietta lighthouse. More recent observations include: one at Lake Manzala on 10 October 1943 (Meiklejohn 1944), three near Ismailiya on 16 April 1958 (Fischer 1963), and five at Lake Maryut on 13 May 1978 (RJM, LRP).

Family Burhinidae

Generic local names for *Burhinus* spp. Karawaan, karawaan jabalii.

Stone Curlew *Burhinus oedicnemus* RB PV WV

Local name Karawaan s̱ah̲raawii كروان صحراوي.

Fairly common breeding resident in several areas of the country. It also occurs on passage and as a winter visitor from (early) mid-September to mid-April. Two subspecies occur in the country, *B. o. saharae* is the local breeding form and *B. o. oedicnemus* a winter visitor.

The Stone Curlew is a fairly common, but local, breeding resident of semi-desert, desert/cultivation ecotone, and occasionally open desert. Its breeding status in several areas is not clear because of possible confusion of migrants and winter visitors with breeding birds.

North Coast A nest with two young was discovered on 15 June 1942 at Sidi Barrani (Tomlinson 1943). In recent years it has been found breeding along the coastal strip south of El Dab'a, El Alamein, and Bahig. Between October 1940 and March 1941, Bulman (1942) found this species to be common, usually in flocks, along the North Coast between Matruh and Salum; however, he did not observe it in the Salum area during the summer months. A single bird was noted aboard a ship off Alexandria on 17 April 1962 (Tuck 1963).

Western Desert Two eggs were collected in Wadi el Natrun on 21 May 1910 (Nicoll 1912*b*), as was a newly hatched chick on 15 April 1923 (Meinertzhagen 1930). Moreau (1927) found a pair nesting in Kharga in April 1925. The most recent Western Desert records are from Wadi el Natrun: several on 10 January 1964 (HCK) and one on 3 November 1972 near the rest-house along the Cairo–Alexandria desert road (Haensel 1980). A total of six birds, presumably migrants, was noted at East Uweinat between 9 and 11 April 1984 (SMB).

Nile Delta and Valley The Stone Curlew is a fairly common breeding resident along the ecotone between desert and cultivation in the Nile Delta and Valley, the Suez Canal area, and the Faiyum. Koenig (1928) reported several nests on islands in the Nile. However, he made no mention of the Senegal Thick-knee, the locally common breeding *Burhinus*, and it seems likely that he confused the two species. Two pairs of Stone Curlews and several nest scrapes were found near Beni Hassan on 17 March

1985 (Grieve 1985). Breeding records from the Suez Canal area include: nests on an undisclosed date near Ismailiya (Kennedy 1918) and on 5 June 1955 at Fayid (RN). In winter small flocks are observed, e.g. over 50 in the Faiyum on 29 November 1917 (Flower 1933). It is not clear if these flocks are extralimital migrants or congregations of local Egyptian breeding birds.

Eastern Desert This species probably breeds locally along the Red Sea coastal plain. At Hurghada two were observed on 9 April 1982 (De Bièvre 1983), two on 12 May 1983 (MCJ *et al.*), and one heard on 10 May 1984 (SMG, RWS). Three were seen at Safaga on 9 May 1983 (MCJ *et al.*). On 16 April 1985 a single bird was calling in Quseir (SMG, PLM). At Marsa el Alam one was vocalizing the night of 28 April 1983 (PLM) and one was taken on 3 May 1964 (YPM). There are a number of winter records in the vicinity of Gebel Elba (Cholmley 1897; Goodman 1984*a*). Spring records from this region include one in Wadi Yoider on 8 March 1984 (PDH, EMN)

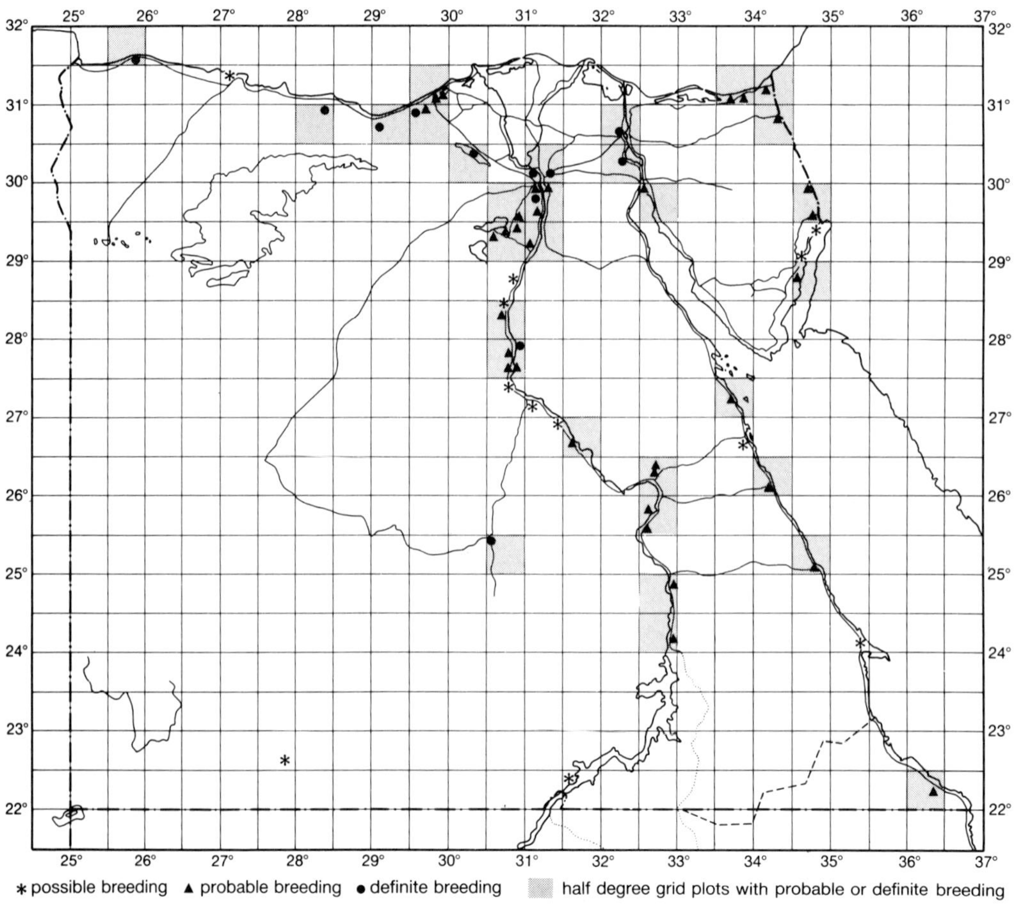

Map B30. Breeding distribution of Stone Curlew *Burhinus oedicnemus* between 1930–87.

and one in Wadi Aideib on 14 April 1985 (SMG, PLM). This species has been recorded crossing the northern Red Sea (e.g. Tuck 1964).

Sinai Local resident in semi-desert areas near Qusaima and El Kuntilla (Galal el Din and Fouad 1939–40) and near El Arish and Rafa (SMB). Birds observed in southern Sinai between October and late March are presumably winter visitors and migrants rather than local breeding birds. Along the north coast, small numbers were seen migrating or resting throughout September 1982 (Baha el Din and Salama 1984). On 27 April 1985, 28 were noted flying east near El Arish (DM).

Senegal Thick-knee *Burhinus senegalensis* RB Plate 1

Local name Karawaan sanaghaalii كروان سنغالي.

Fairly common breeding resident in Wadi el Natrun, the southern Nile Delta, and

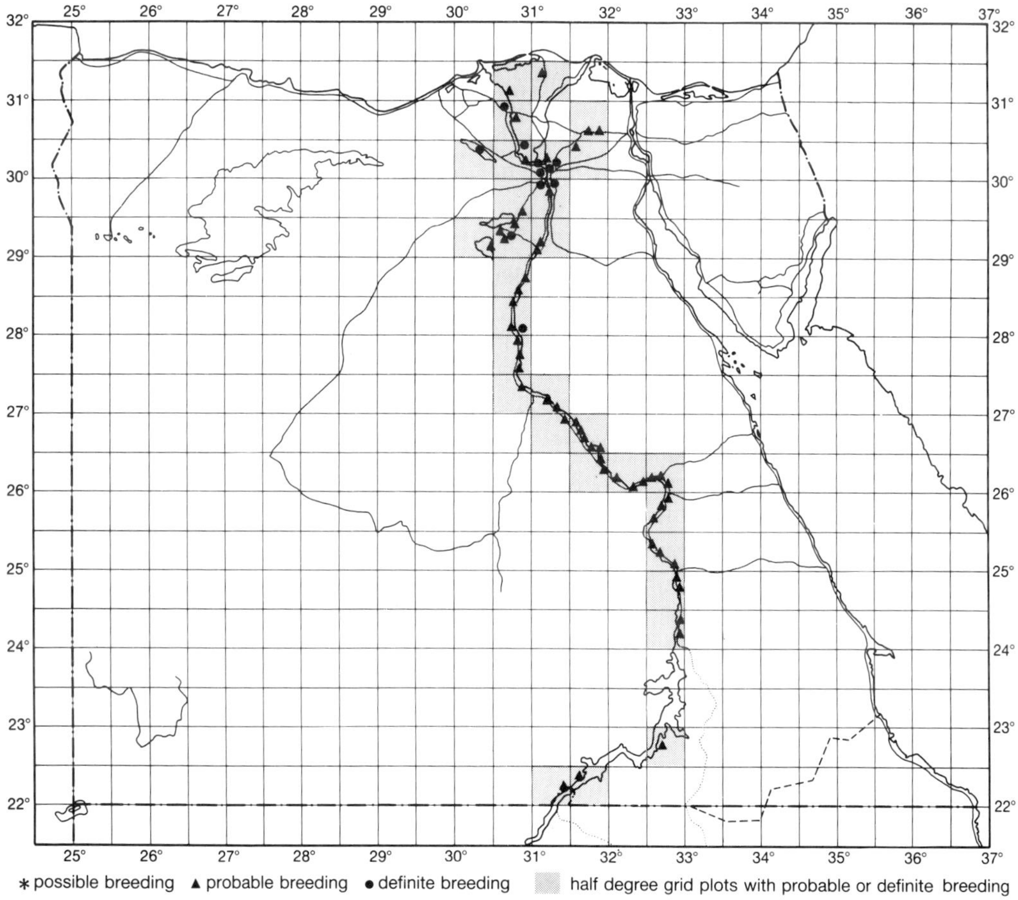

Map B31. Breeding distribution of Senegal Thick-knee *Burhinus senegalensis* between 1930–87.

complete Nile Valley. The form occurring in the country is *B. s. inornatus*. Records of this species from the Red Sea and Mediterranean coasts probably concern misidentified Stone Curlews.

Western Desert In Wadi el Natrun eggs of the Senegal Thick-knee were found in May 1910 (Nicoll 1912*b*) and one was heard calling on 21 April 1984 (SG). This species is locally sympatric with the Stone Curlew and both appear to breed at the same time of the year.

Nile Delta and Valley Senegal Thick-knees are widely distributed breeding residents of cultivated areas, river banks, on buildings in the southern Nile Delta and complete Valley, and the Faiyum. The northern-most record in the Nile Delta appears to be at El Hamul (SMB). It is a common breeding bird on flat roof-tops in Cairo and local environs. There are several records from the Suez Canal area near Fayid, e.g. two on 22 and 24 May 1955 (AEG) but possible confusion with the Stone Curlew cannot be excluded. The species is common along the Nile Valley between Cairo and Aswan, where small flocks of up to 20 are observed in winter. Senegal Thick-knees have been recorded during the breeding season at several places along Lake Nasser (PLM) and a group of 60 was noted at Abu Simbil on 2 November 1982 (Davies 1982).

Family Glareolidae

Egyptian Plover *Pluvianus aegyptius* FB AV?

Local names Qaṯqaaṯ maṣrii قطقاط مصري, qaṯqaaṯ, ṯayr at-timsaah̲, tuurim.

Former breeding resident in the Nile Valley. There are no documented records for the country since 1937. Monotypic.

The Egyptian Plover was formerly a regular breeding resident on the banks and islands of the Nile Valley. Taylor (1859) mentioned it as numerous everywhere in the Nile Valley south of Cairo. In 1897, Koenig (1926) found this species common along the Nile between Cairo and Aswan and collected at least 12 clutches of eggs. In the first few decades of the twentieth century it apparently declined in numbers in the Nile Valley. Russell (1905), who sailed on the Nile in January and February 1905, only saw a pair at El Wasta and another at Sohag. Nicoll (1919) mentioned that the only reliable record he was aware of since 1907 was a pair on 15 May 1916 between El Aiyat and El Badrshein. Hachisuka (1924) noted four on 16 February 1924 near Qena. According to Meinertzhagen (1930) it was scarce with 'four or five records during the last 20 years'. However, Jourdain and Lynes (1936) stated that 'the species has apparently re-established in Lower Egypt and occurs regularly on the Nile within 40 km of Cairo, where several have been shot. There is little doubt that it breeds at the present time on the sand-banks between Helwan and Qena.' Unfortunately, they presented little information to support these statements. El Negumi *et al.* (1950) considered it rare in Upper Egypt and still breeding locally but they also presented no information to support their claim. The most recent documented Egyptian Nile Valley record is one collected at Helwan on 7 February 1937 (HGD). Two were seen on 23 November 1981 at Wadi Halfa, just across the border in Sudan (HL). The only record from Sinai is one on 15 January 1979 at Qad el Hamdan (Ilani 1979–80), which cannot be accepted without further details.

Cream-coloured Courser *Cursorius cursor* RB PV WV Plate 3

Local names Al-jaliil الجليل, garawaan (Ma), karawaan jabalii.

Fairly common breeding resident in desert areas. Local populations are augmented between autumn and spring by migrants. Late spring migrants may in some cases be mistaken for presumed breeding birds. *C. c. cursor* is the subspecies occurring in Egypt.

North Coast The Cream-coloured Courser is common along the coastal strip between Salum and Alexandria. Breeding has been documented at several places (e.g. Bulman 1942; Maclaren 1944). In winter it is usually found in small flocks.

Western Desert Confirmed breeding is only known from Wadi el Natrun (Raw 1921). Meinertzhagen (1930) listed it as a breeding resident in Kharga and Dakhla but presented no supporting evidence. There are several recent records, mainly in spring and autumn, from the Qattara Depression, Bahariya, Farafra, and Kharga, but no documentation of breeding is known (Baha el Din in press; Goodman *et al.* 1986).

Nile Delta and Valley This species breeds locally along the desert edge of the Nile Delta, the Suez Canal area, and the Faiyum. It is rare in the Nile Valley above the Faiyum. The southern-most records are three at El Amarna on 14 March 1983 (Welch 1983) and 11 flying north on 18 March 1985 (Grieve 1985). At Abu Zabal, Raw (1921) found a substantial increase in local densities in August and September, with flocks of up to 45 being noted. It is assumed that a portion of this influx was extralimital migrants.

Eastern Desert and Red Sea Islands Cream-coloured Coursers are frequently recorded along the Red Sea coastal plain, mainly from September to late April. Some of these birds are certainly wintering migrants. Tregenza (1951) observed a party of six in late July 1947 in Wadi Himreyr, near Bir Milaha, and Marchant (1941) noted probable breeding in 1939 near Hurghada. Birds showing distraction display were observed near Bir Abu Sha'r on 25 April 1983 and south of Safaga on 9 May 1983 (MCJ *et al.*). A male and a female collected in Wadi Naam, 30 km east of Gebel Abraq, on 27 February 1985 were approaching breeding condition (UMMZ). This species has been observed on several islands off Hurghada and in the mouth of the Gulf of Suez.

Sinai This species is a locally common breeding resident in northern Sinai. In the south, between Nabq and Sharm el Sheikh, it is regularly observed until late April and probably breeds locally. One pair was noted on 8 March 1987 near Nuweiba in suitable breeding habitat (PLM). A chick was collected at El Qa' on an unknown date (GZM). This species has been observed in August at Abu Rudeis (Hovel 1987). On 7 September 1982, three were seen migrating at Zaranikh (Baha el Din and Salama 1984).

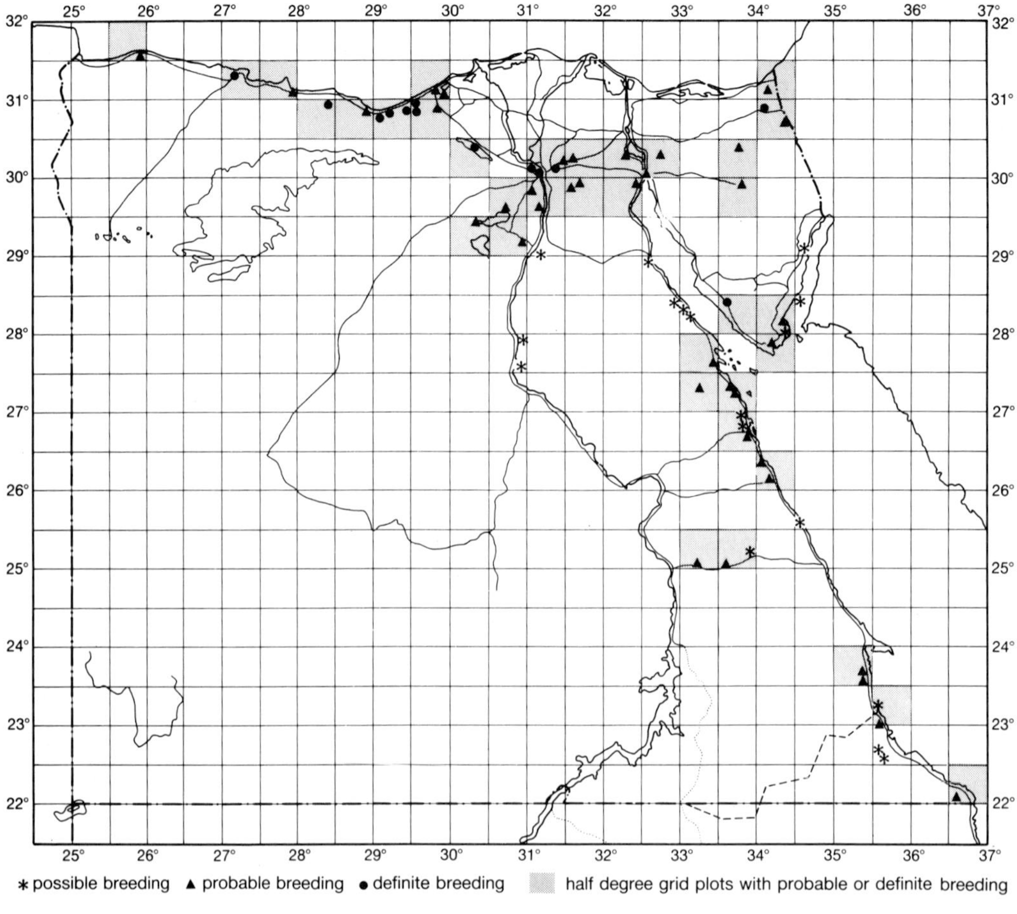

Map B32. Breeding distribution of Cream-coloured Courser *Cursorius cursor* between 1930–87.

Collared Pratincole *Glareola pratincola* MB PV (WV)

Local names Abuu al-yusr ابو اليسر, khuulii al-iwazz.

Fairly common migrant breeder in the Nile Delta, the Faiyum, and possibly in Wadi el Natrun and near El Alamein. Common passage visitor throughout the country in autumn from mid-August to late October (late November) and in spring from (mid-February) late March to mid-May. Rare in winter. All Egyptian material examined is referable to *G. p. pratincola*. The total Egyptian breeding population is estimated to be over 1000 pairs.

North Coast Two pairs of Collared Pratincoles were observed on 17 May 1986 at a coastal lagoon near El Alamein performing courtship displays (GMA, MAM).

Western Desert According to Meinertzhagen (1930) Collared Pratincoles have been obtained in Wadi el Natrun throughout May and June and it may nest locally. However, there is no recent information indicating breeding in this area. The only autumn records from the Western Desert are single birds in Bahariya on 16 August and 20 September 1985 and five on 7 September 1983 (HB *et al.*), and three on 8 September 1984, 140 km north-east of Bahariya (SH, DM). In spring the species is common between mid-April and mid-May (Goodman *et al.* 1986).

Nile Delta and Valley In the Nile Delta this species commonly breeds in salt-marshes, particularly around Lake Burullus and Lake Manzala (Meininger *et al.*

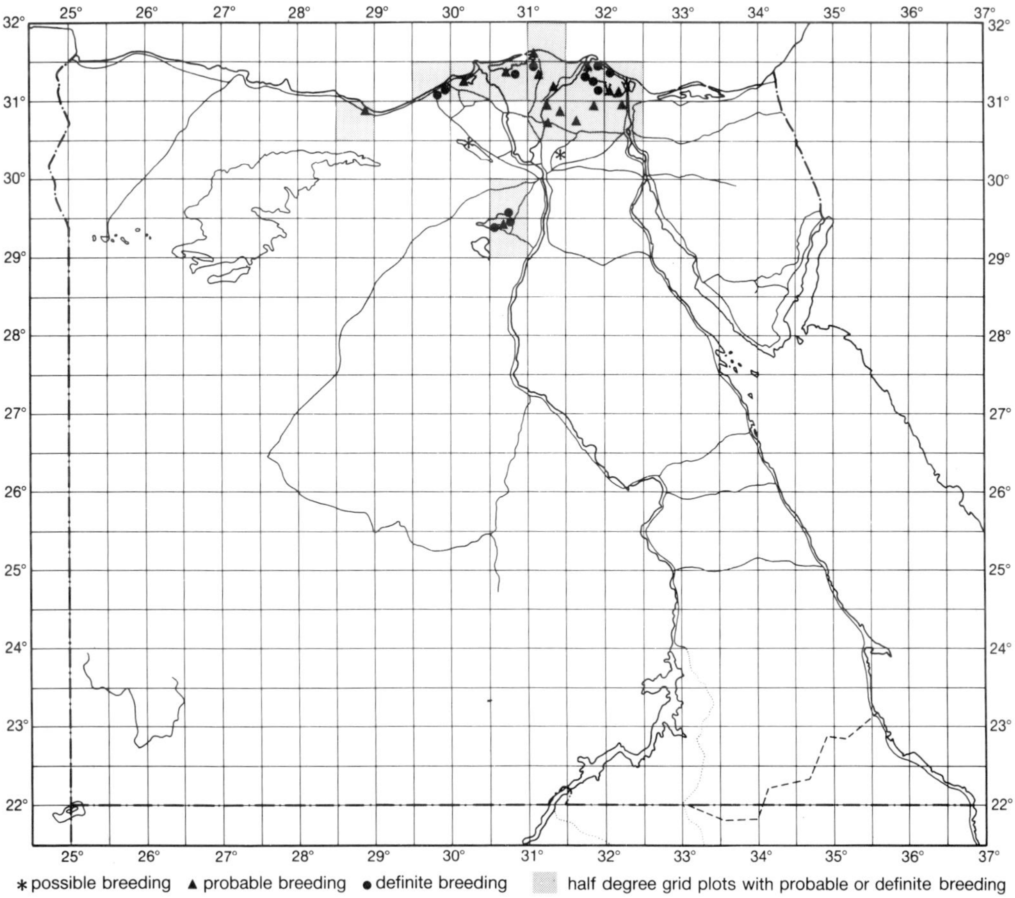

Map B33. Breeding distribution of Collared Pratincole *Glareola pratincola* between 1970–87.

1986). It was reported to nest locally in the salt-marshes of the eastern Delta (Flower 1933) but there is little recent information from this area. In 1986 about 20 pairs bred along the eastern side of Lake Qarun (DM, WCM).

The latest autumn observation is 40 at El Kabrit on 20 November 1984 (RvD *et al.*). The only midwinter record is one on 24 December 1980 between Idfu and Isna (RW). Spring passage is more pronounced than autumn, with the bulk of birds moving through between mid-April and mid-May. The earliest spring record is a large flock near Giza on 14 February 1906 (Flower 1933).

Sinai Autumn migration is from late August to late October. Records from Zaranikh include: 167 between 13 August and 29 September 1973 and 490 between 15 August and 5 October 1978 (YP, UP), and 84 between 4 and 21 September 1982 (Baha el Din and Salama 1984).

Black-winged Pratincole *Glareola nordmanni* PV

Local name Abuu al-yusr aswad al-janaah̲ .ابو اليسر أسود الجناح

Rare passage visitor in autumn from late August to mid-October and in spring from late March to late April (early May), mainly to the eastern portion of the country. Monotypic.

Meinertzhagen (1930) mentioned the Black-winged Pratincole as a migrant on both passages, temporally coinciding with the Collared Pratincole, but presented no details. The following records summarize information from Egypt by season.

Autumn At Cairo one was noted on 24 August 1984 (SH, DM); at Zaranikh 14 between 10 and 24 September 1980, with nine on 16 September (Petersen and Sørensen 1981*a*); at Sharm el Sheikh one on 18 September 1980 (DJM, DES), two on 24 September 1980 (GWR), and one on 11 October 1978 (Ilani 1979–80); at Giza immatures on 19 September 1906 and 25 September 1907 (GZM); at Suez on 19 September 1981 (RGB *et al.*); at Abu Zabal an immature on 4 October 1916 (Raw 1921); and at Abu Rauwash on 10 October 1985 (SMB, DM).

Spring Records include single birds on 30 March 1982 (MU) and 4 April 1983 (SC) at Suez; unspecified number on 5 April 1950 at Fayid (HGB); one on 6 April 1953 and six on 3 May 1953 at El Kabrit (AP); and six on 20 April 1952 at Deversoir (AP). There are a few mid-May sight records from Elat, Israel (Paz 1987).

Family Charadriidae

Generic local names for plover Qat̲qaat̲, abuu ar-ru-uus, abuu h̲agar (ND), abuu qat̲qaat̲ (NC), abuu ra-s (ND), damshaq, gumayya (WN), saqsaaq, wudwud (NC), zaqzaaq.

Little Ringed Plover *Charadrius dubius* (MB) PV WV

Local name Qaṯqaaṯ muṯawwaq ṣaghiir قطقاط مطوق صغير.

Rare migrant breeder in the north of the country. Scarce passage and rare winter visitor throughout the country from late July to mid-May. All Egyptian material examined is referable to *C. d. curonicus*.

Meinertzhagen (1930) remarked that two Little Ringed Plover nests were discovered in Wadi el Natrun in April. More recent evidence from this locality includes breeding pairs on 7 July 1984 (Goodman *et al.* 1986) and on 9 August 1985 (with fledglings) (SMB). A pair with fledglings was observed on 1 August 1955 at the Gebel Maryam lagoon (RN). Although Meinertzhagen (1930) mentioned this species as a not uncommon resident in Egypt, the records mentioned here are the only known breeding records from the country.

From late July to late October and from (late February) mid-March to late April small numbers of migrants are observed in Western Desert oases, along the North Coast, in the Nile Delta and Valley, along the Red Sea coast, and in Sinai. In winter this species is rare and records from December and January include small numbers in Dakhla, the Nile Delta lakes, Nile Valley, and southern Sinai. The statements of Adams (1864), Nicoll (1908), and Meinertzhagen (1930) that it is a common winter visitor are unsubstantiated.

Ringing A bird ringed in Greece in August was recovered near Matruh three years later in March (Michelson 1985).

Ringed Plover *Charadrius hiaticula* PV WV

Local names Qaṯqaaṯ muṯawwaq قطقاط مطوق, ṯa-ṯii- (Su).

Fairly common passage and winter visitor throughout portions of the country. Autumn passage is from mid-August to late November and spring passage from late February to early (mid-) June. The total number wintering in Egypt is estimated to be 2000–3000. Irregular summer visitor. All Egyptian material examined is referable to *C. h. tundrae*.

Western Desert There are a few records of small numbers of Ringed Plovers in autumn, winter, and spring from various Western Desert oases (Goodman *et al.* 1986).

Nile Delta This species is a regular passage and winter visitor to the mud-flats of the Nile Delta and the Suez Canal area (Table 6.3). Considerable numbers are present in the Delta lakes, and the largest midwinter concentrations are on Lake Idku (up to

TABLE 6.3. Maximum numbers of Ringed Plovers noted in Suez Bay between 1978 and 1986

Jan.	Feb.	March	April	May	June	July	Aug.	Sept.	Oct.	Nov.	Dec.
140	208	200	100	50	?	?	200	1900	1530	900	360

280), Lake Burullus (up to 200), and Lake Manzala (up to 720) (Meininger and Mullié 1981*a*).

Nile Valley This species is a regular migrant and winter visitor in small numbers to the entire Nile Valley, including Lake Nasser, from mid-September to late April. The bulk of observations are from the migration periods. Concentrations are sometimes noted, e.g. 225 along Lake Qarun on 20 October 1980 (SMB).

Red Sea coast Regular migrant and winter visitor along the entire Red Sea coast, generally in small numbers. A considerable number funnel through Suez Bay on passage, with a maximum of 1900 in September 1982 (CAB *et al.*).

Sinai In autumn the Ringed Plover is a common migrant along the north coast. Observations of migrants at Zaranikh include: 1909 between 15 August and 5 October 1978 (YP, UP); and 1460 between 16 August and 24 September 1981 (Petersen and Sørensen 1981*b*).

Kittlitz's (Sand) Plover *Charadrius pecuarius* RB

Local name Qaṯqaaṯ bunnii قطقاط بنّي.

Scarce breeding resident in Wadi el Natrun, the Nile Delta, the Suez Canal area, the Faiyum, and probably along Lake Nasser. Nicoll (1921) described the Faiyum population as a distinct subspecies, *C. p. allenbyi.* However, there appear to be no clear patterns of geographic variation in this species (Vaurie 1965). It breeds from late March to late June in saline, brackish, and fresh water habitats.

Western Desert There are numerous records of the Kittlitz's Plover throughout the year in Wadi el Natrun (Goodman *et al.* 1986), with up to 60 on 24 August 1986 (EE *et al.*). It is unknown from any other Western Desert oasis.

Nile Delta This species is an uncommon local resident along the Nile Delta coastal lakes. Breeding has been documented or strongly suspected at Lakes Manzala, Burullus, and Maryut (Raw 1921; Meininger *et al.* 1986). An adult with fledgling were noted at San el Hagar on 25 May 1978 (RJM, LRP). On 25 April 1986 a few pairs were noted at Ikyad (SMB, WCM). Small numbers are observed throughout the year along the Nile Delta lakes and in the northern Suez Canal area. Meinertzhagen (1930) noted that this species was absent from the Suez Canal area. However, relatively recent breeding records from the Gebel Maryam lagoon include: 'young' on 15 April 1943 (Molteno 1944) and a pair with young on 24 April 1955 (RN). At Deversoir up to 11 were observed between 18 May and 22 July 1952 (AP). The only known record from Suez is three on 19 October 1985 (DM). In the southern Nile Delta and in the environs of Cairo this species is often recorded in small numbers between August and March (e.g. Raw 1921; Horváth 1959) but there is no evidence of local breeding.

Nile Valley A few pairs have recently been found breeding along Lake Qarun, where numbers of up to 10 are regularly observed. In 1899 Koenig (1928) found this species fairly common on the Nile banks but did not locate any nests. Meinertzhagen (1930) mentioned it as a resident in the Nile Valley, south to at least Luxor; however, we are unaware of any documented breeding between Cairo and Aswan. The only post-Meinertzhagen (1930) record from this area is two on 1 November 1982 at Kom Ombo (Davies 1982). This species may well be a resident in the Lake Nasser area. Eight were observed on 23 January 1980 in northern Lake Nasser (Meininger and Mullié 1981*b*) and a few breeding pairs were noted on 21 April 1983 at Wadi Halfa, just across the border in Sudan (PLM).

Red Sea There are no documented records of this species along the Egyptian Red Sea coast. Two birds were observed at Elat, Israel, between 31 October 1986 and January 1987 (Shirihai and van den Berg 1987).

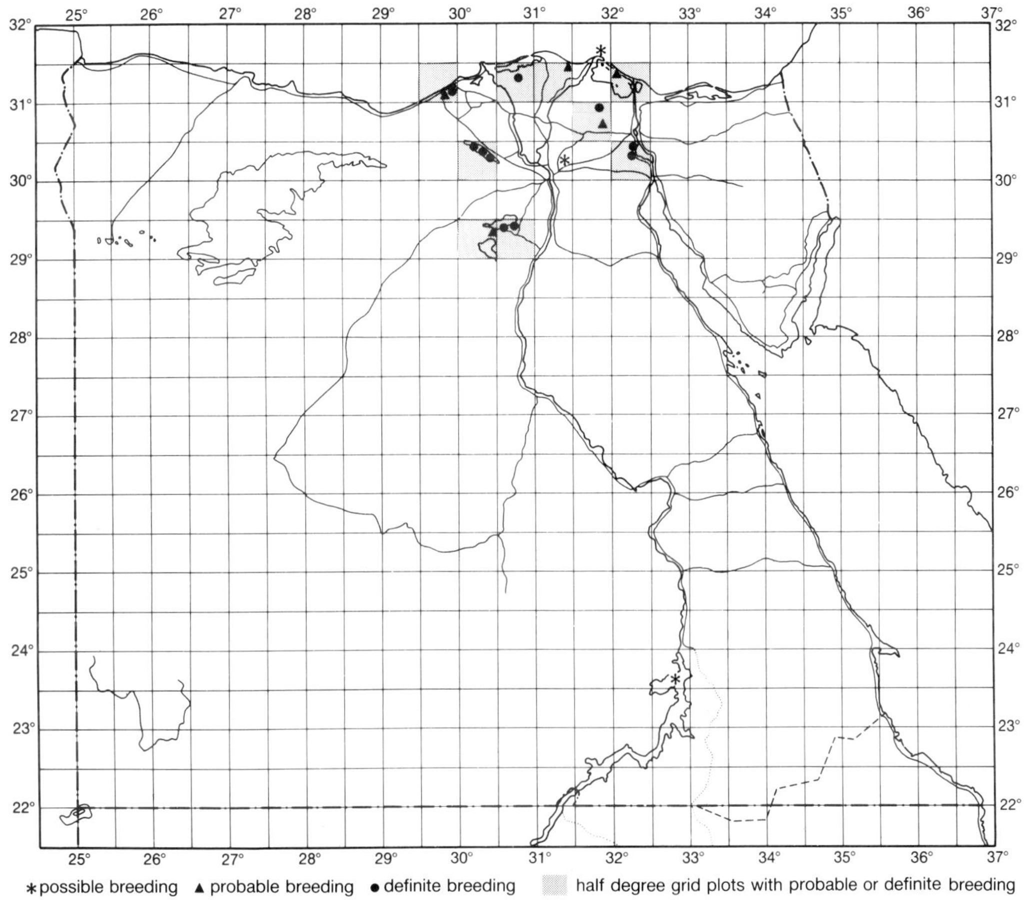

Map B34. Breeding distribution of Kittlitz's Plover *Charadrius pecuarius* between 1970–87.

Kentish Plover *Charadrius alexandrinus* RB PV WV

Local names Qatqaat abuu ar-ru-uus قطقاط ابو الرؤوس, abuu hamaara.

Fairly common breeding resident in various portions of the country. Common passage and winter visitor, mainly in coastal habitats from mid-August to late May. Breeding is from mid-March to mid-June. All Egyptian material examined is referable to *C. a. alexandrinus*.

The highest concentrations in Egypt of breeding Kentish Plovers are found along the Nile Delta lakes, with probably over 1000 pairs (Meininger *et al.* 1986), and along Lake Bardawil, with about 1900 pairs (Dunnet *et al.* 1986). The species is also known to nest at inland portions of the Nile Delta. Small numbers breed locally along the Mediterranean coast between Alexandria and Salum, in Wadi el Natrun, Siwa, the Faiyum, Wadi el Rayan, the Suez Canal area, and along the Red Sea coast and on islands.

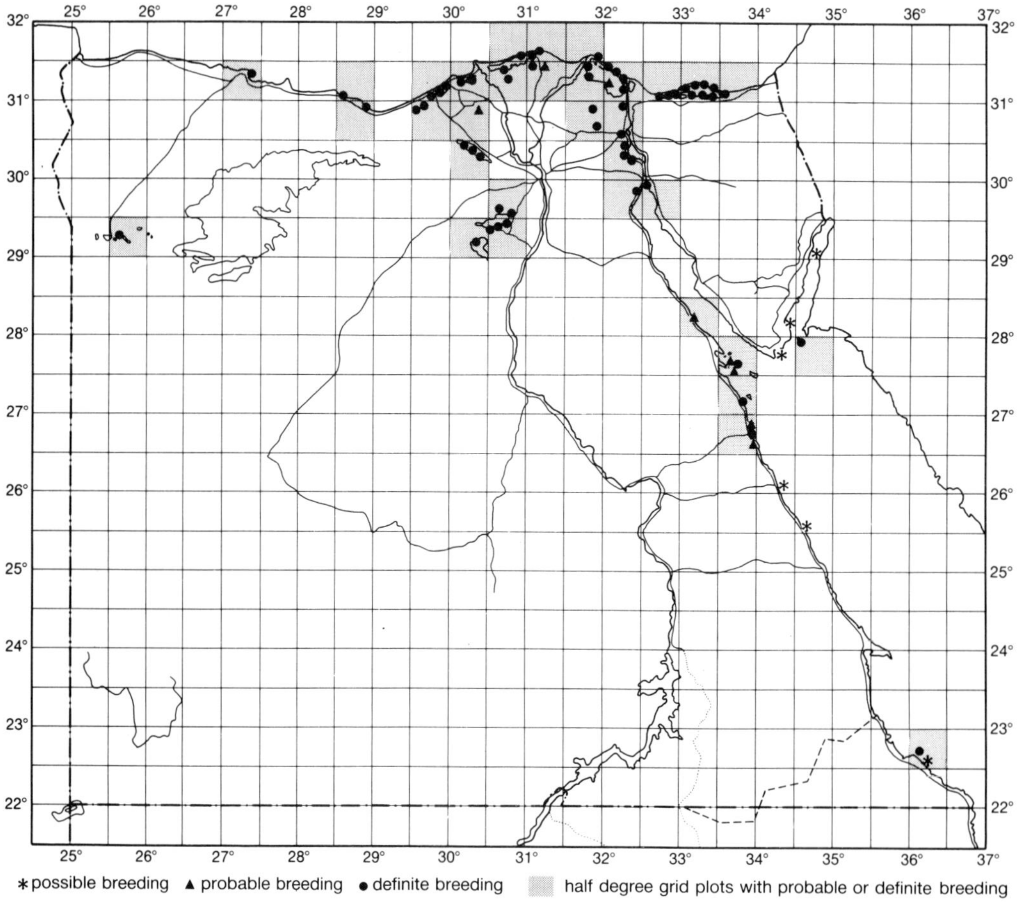

Map B35. Breeding distribution of Kentish Plover *Charadrius alexandrinus* between 1970–87.

In autumn, from mid-August into October, there is a significant influx of migrants, particularly along the north coast of Sinai. Records of migrants passing Zaranikh include: 5687 between 7 and 30 September 1980 (Petersen and Sørensen 1981*a*), 5504 between 16 August and 24 September 1981 (Petersen and Sørensen 1981*b*), and 3844 between 2 and 24 September 1982 (Baha el Din and Salama 1984). In September up to 1250 have been noted in Suez Bay on a single day (Table 6.4). The largest winter concentrations are noted in the Nile Delta lakes, where approximately 15 000–20 000 birds winter (Meininger and Mullié 1981*a*). On Lake Manzala alone an estimated 1750–1850 are shot annually, although in some years substantially more are taken (Mullié and Meininger 1983; table 4.3). Up to several hundred have been recorded in the Faiyum during the migration periods and winter. In the Nile Valley, from Cairo to Lake Nasser, the species is recorded in small numbers between mid-August and mid-April. Spring passage peaks in March and is most notable along the Red Sea coast. There are a few March and April records from the southern Western Desert oases, which are presumably of trans-Saharan migrants (Goodman *et al.* 1986).

TABLE 6.4. Maximum numbers of Kentish Plovers noted in Suez Bay between 1978 and 1986

Jan.	Feb.	March	April	May	June	July	Aug.	Sept.	Oct.	Nov.	Dec.
60	50	150	20	10	10	40	70	1250	635	750	40

Lesser Sand Plover *Charadrius mongolus* PV

Local name Qaṯqaaṯ ar-raml as-ṣaghiir قطقاط الرمل الصغير.

Rare passage visitor in autumn from mid-July to early September and in spring from (late February) mid-March to mid-June, mainly along the Red Sea coast. No Egyptian specimens are known but on geographical grounds the subspecies occurring in the country could be either *C. m. pamirensis* (Cramp and Simmons 1983) and/or *C. m. atrifrons* (Vaurie 1965).

Only in the past few years has it become clear that the Lesser Sand Plover is a rare, but regular, migrant in Egypt. Along the southern Red Sea coast it is probably more common than currently realized. The following summarizes all known records from the country.

Autumn Three were observed on 19 July 1984 at Zaranikh (SH, DM), 6–10 on 20 July 1984 at El Taloul (SH, DM), one on 25 August 1985 at Nuweiba (DM), two on 27 August 1954 at Gebel Maryam lagoon (RN), and two on 7 September 1985 at Lake Qarun (DM).

Spring Two were noted on 22 February 1987 at Nabq (PdG) and 24 February 1985 at Hamata (SH, DM); one on 6 March 1984 in Wadi Lahami (PDH, EMN); one on 23 March 1978 at Nabq (KO); three on 5 April 1982 at Lake Idku (De Bièvre 1983); 20 on 6 April 1985 at Shalatein (SMG, PLM); three on 25 April 1985 on the

Gezira Siyal group off Adal Deib (SMG, PLM); and one on 23 June 1980 at Suez (SMB). Records from Hurghada include one on 13 March 1982 (ES, HWu), one on 14 March 1982 (RV), and five on 17 June, 12 on 20 June, and two on 21 June 1980 (SMB).

Greater Sand Plover *Charadrius leschenaultii* PV WV CB?

Local names Qaṯqaaṯ ar-raml al-kabiir قطقاط الرمل الكبير, damshaq zahraawii.

Fairly common passage and winter visitor, mainly along the coastal waters, from late July to May (mid-June). Possibly an occasional breeding resident along the Mediterranean and Red Sea coasts. Monotypic.

Although there is no positive documentation, the Greater Sand Plover may occasionally breed in Egypt. On 12 July 1942 Bodenham (1945*a*) noted 10 Greater Sand Plovers, 'including four young probably about a fortnight old' at Abu Qir. He also strongly suspected that it bred in this same area during the summer of 1943. El Negumi *et al.* (1950) mentioned that recently fledged birds were collected in July at Matruh and along the Red Sea coast. Marchant (1941) noted that it 'definitely . . . did not breed at Hurghada in 1939', although displaying birds were noted a few times in the spring. In recent years, small numbers have been observed during the breeding season in the Nile Delta, the Suez Canal area, Wadi el Natrun, along the Red Sea coast, and in northern Sinai.

This species is a passage and winter visitor to the Mediterranean coast, mainly east of Alexandria, the Suez Canal area, and the entire Red Sea coast. Autumn migratory movements begin in late July, or in some years earlier, and local numbers increase into January. At Zaranikh 71 birds were noted between 2 and 24 September 1982 (Baha el Din and Salama 1984). Relatively large influxes of migrants are known, e.g. at Suez 220 on 11 December 1980 (SMB), at Hurghada 79 in August 1938 (Al Hussaini 1939) and 100 on 30 January 1981 (SMB), at Quseir 50 on 15 August 1986 (EE *et al.*), at Nabq 84 on 18 September 1980 (DJM, DES), and at Deversoir 50 on 27 July 1952 (AP). The number wintering in the country is estimated to exceed 1000. Spring passage is from late February to May. An exceptional record is a flock of about 200 on 17 June 1984 on Gezira Mahabis (SMG).

Inland reports are scarce. There are several records between late September and late March from Wadi el Natrun (Goodman *et al.* 1986; DM) and up to 13 have been noted between mid-July and late February on Lake Qarun. The only known observations from the Nile Valley proper are one at Beni Hassan on 13 November 1977 (Short and Horne 1981) and two near Kom Ombo on 25 March 1985 (Grieve 1985).

Caspian Plover *Charadrius asiaticus* PV

Local name Qaṯqaaṯ qazwiinii قطقاط قزويني.

Rare passage visitor in autumn from early September to late October and in spring from mid-March to late April (mid-June), mainly along the Red Sea coast. Monotypic.

Western Desert and North Coast There are only single records of the Caspian Plover in each of these two areas. Two were photographed at Dakhla on 24 April 1981 (Goodman *et al.* 1986) and one was collected at Bahig on 14 October 1962 (Goodman and Watson 1983).

Nile Delta and Valley and Suez Canal Most records in these regions are from autumn. These include: El Kabrit—two on 7 September, one on 12 September, and one on 22 October 1952 (AP); Great Bitter Lake—one on 7 September 1952 (AEG); and Ras el Bahr—three on 8 October 1943 (Meiklejohn 1944). The only Nile Valley report appears to be a single bird at Heliopolis on 22 April 1980 (Baha el Din 1984*b*).

Red Sea All records from this area are from the spring. These include: Safaga—three on 13 March 1985 and one the next day (UGS); Hurghada—one in summer plumage on 15 March 1985 (RV); Ghubbet el Gemsa—one on 21 March 1983 (NM); Nabq—one on 3 April 1979 (HBu); Taba—one on 3 April 1979 (HBu); Shalatein—two on 6 April 1985 (SMG, PLM); Sharm el Sheikh—three to five on 7 April (JO), four on 10 April (EB), and one on 12 April 1979 (RSS); east coast of Sinai—one on 9 April 1981 (Heinze 1981); and Hamata—one on 17 June 1984 (SMG). At Elat, Israel, the Caspian Plover is a fairly common spring migrant from (late February) late March to late April (mid-May) and a rare winter visitor (Krabbe 1980).

Dotterel *Charadrius morinellus* WV

Local names Qaṯqaaṯ aghbar قطقاط أغبر, duwwa (NC).

Scarce winter visitor to the north of the country from (late August) early November to late March (late May). Monotypic.

The earliest autumn records of the Dotterel in Egypt are from Zaranikh and include one taken on 28 August 1973 (TAUM), three observed on 13 September 1980 (Petersen and Sørensen 1981*a*) and one on 13 September 1982 (Baha el Din and Salama 1984). A single bird was taken at Marsa Matruh on 13 September 1958 (FMNH). This species is rarely recorded south of Cairo or along the Red Sea coast. Kaiser (1892) observed solitary birds near El Tor on 23 November 1890 and 21 December 1891; Bulman (1944) recorded it at Safaga in November 1942; and Horváth (1959) saw one near Helwan on 17 November 1957.

According to Meinertzhagen (1930) the Dotterel is a winter visitor to the semi-desert of north-eastern Sinai and 'on the poor land bordering the desert' west of the Suez Canal, fairly common in some winters and entirely absent in others. There are only a few recent north Sinai records but winter observations from this region in the past decade or so are somewhat limited. Nowadays, it appears to be a scarce winter visitor to the edge of the Nile Delta, the semi-desert west of Alexandria, and northern Sinai. A specimen was collected in the Faiyum on 3 January 1918 (GZM). More recent records include: one on 14 November 1983 near Wadi el Natrun (Bezzel 1987*a*), seven on 8 February 1986 near Baltim (SMB) and one on 12 December 1980 at Sharm el Sheikh (GWR).

There are a few spring records of substantial concentrations, e.g. about 70 on 25 March 1983 near Burg el Arab (PLM). The latest spring record is two near Ismailiya on 25 May 1950 (JCC).

Golden Plover *Pluvialis apricaria* PV WV

Local name Qaṯqaaṯ dhahabii قطقاط ذهبي.

Scarce winter visitor to the Nile Delta and northern Sinai from (late August) late October to late February (early May). Rare in the Western Desert and Sinai. We follow Cramp and Simmons (1983) in considering this species monotypic.

The Golden Plover is often observed in small flocks in the cultivated portions of the Nile Delta and Suez Canal area but the numbers vary from year to year. Substantial concentrations are occasionally noted, e.g. 300–400 on 20 January 1875 near Damietta (Gurney 1876) and 120 on 14 January 1980 near Tanta (PLM, WCM). The only known record from the Faiyum is five on 2 October 1986 (EH, PK). There are numerous records of small numbers in northern Sinai. Western Desert records include several seen and one collected in Wadi el Natrun on 26 November 1911 (Nicoll 1912*a*) and a single specimen collected in Farafra (Goodman *et al.* 1986). Rowntree (1943) observed a flock of 30 on 10 January 1942 at El Dab'a. The earliest autumn record is one at Zaranikh on 21 August 1981 (Petersen and Sørensen 1981*b*) and the latest spring record is one at the St Katherine airport on 6 May 1983 (SMB).

Grey Plover *Pluvialis squatarola* PV WV

Local name Qaṯqaaṯ ramaadii قطقاط رمادي.

Fairly common passage and scarce winter visitor mainly to the northern portion of the country and along the Red Sea from mid-August to mid- (late) May. Uncommon summer visitor. Monotypic.

The Grey Plover is a passage migrant and winter visitor to the Mediterranean coast, the Nile Delta lakes, the Suez Canal area, northern Sinai, and the Red Sea coast. Autumn passage begins in mid-August and peaks in September and October. A total of 260 Grey Plovers was counted at Zaranikh between 15 August and 5 October 1978 (YP, UP) and 164 between 2 and 24 September 1982 (Baha el Din and Salama 1984). Substantial influxes are sometimes noted, e.g. up to 760 at Suez in mid-October (RvD *et al.*). The total population wintering in Egypt is estimated to be 500–1000 birds, mainly in the Delta lakes and Suez Bay, although small numbers are also noted elsewhere, such as Lake Qarun (Meininger and Mullié 1981*a*).

The bulk of spring passage is from mid-March to early May but small numbers are seen until late May. Spring migration is particularly notable near Suez: 400 were counted on 1 March 1982 (MCJ), 400–500 on 26 March 1982 (MU), and 400 on 28 March 1983 (PLM). Western Desert records include two on 26 April 1981 in Farafra and one on 8 May 1985 in Siwa (Goodman *et al.* 1986). The only records from the Nile Valley south of Cairo appear to be one on 9 October 1983 at Idfu (Hume 1983), two on 24 December 1983 at Aswan (PE), and one on 12 March 1983 at El Fashn (Welch 1983). A few immature birds are known to summer in the Nile Delta and along the Red Sea. An exceptional record is 72 on 28 June 1986 in Suez Bay (DM, WCM).

Spur-winged Plover *Hoplopterus spinosus* RB PV WV

Local names Zaqzaaq زقزاق, abuu maghaa̱sin, abuu ẕufur, galiila (NC), ṯa-ṯuu (ND).

Fairly common breeding resident in cultivated and marshlands of Wadi el Natrun, the Nile Delta and Valley, Suez Canal area, the Faiyum, and Wadi el Rayan. It occasionally breeds in northern Sinai. During the winter some local breeding populations are not completely sedentary and are augmented by migrants. The Egyptian population breeds between early April and mid-July and comprises many thousands of pairs. Monotypic.

According to Meinertzhagen (1930) Spur-winged Plovers were abundant residents in cultivated areas of the Nile Delta, Upper Egypt, in the Suez Canal area, the Faiyum, and Wadi el Natrun. Flower (1933) gave a more detailed description of the Delta range and mentioned that this species was absent in the north-western and north-eastern portion, between Ismailiya and Port Said, and near Suez. Presently it breeds in most portions of the Delta, including cultivation and salt-marshes around the lakes (Meininger *et al.* 1986) but densities are variable. The Suez Canal area has been colonized in recent years and tens of pairs breed at Suez. This species is also found in varying densities at Wadi el Natrun, the Faiyum, Wadi el Rayan, and in the Nile Valley south to Aswan. It was known to occur above Aswan before the formation of Lake Nasser (YPM) and still breeds in the area just south of the Aswan High Dam. At Abu Simbil two were seen on 17 April 1984 (MBe, SM) and 10 on 12 August 1986 (EE *et al.*). Several pairs were found nesting in April 1983 near Wadi Halfa, Sudan (PLM). This species is apparently an irregular breeding bird in north Sinai. On 29 July 1968, eight individuals, including chicks, were noted west of Bir el Abd (TS).

Outside the breeding season small flocks are observed in or near breeding areas. In some regions it appears that during the non-breeding season local birds sometimes become gregarious and migrate. During systematic counts near Suez in the autumn of 1984 the number of local Spur-winged Plovers decreased from 370 in mid-September to about 100 in November (RvD *et al.*). A flock of 150 was noted along the southern shore of Lake Qarun on 27 April 1984 (SMG). The species is observed throughout the year in Wadi el Natrun. Other records from the Western Desert include three on 20

October 1980 at Dakhla and one on 23 April 1981 at Kharga (Goodman *et al.* 1986). A single bird was taken along the North Coast at Marsa Baqqush, between Matruh and El Dab'a, on 22 September 1941 (UMZC). Small numbers of migrating birds are regularly observed along the north coast of Sinai in September and October (Petersen and Sørensen 1981*a*, 1981*b*; Baha el Din and Salama 1984; HL). Cheke (1967) reported that on 26 September 1966 a flock of 12 Spur-winged Plovers were observed crossing the Mediterranean in a south-westerly direction off Alexandria. There are a few records from southern Sinai and several reports from along the Red Sea coast, mainly during autumn and spring.

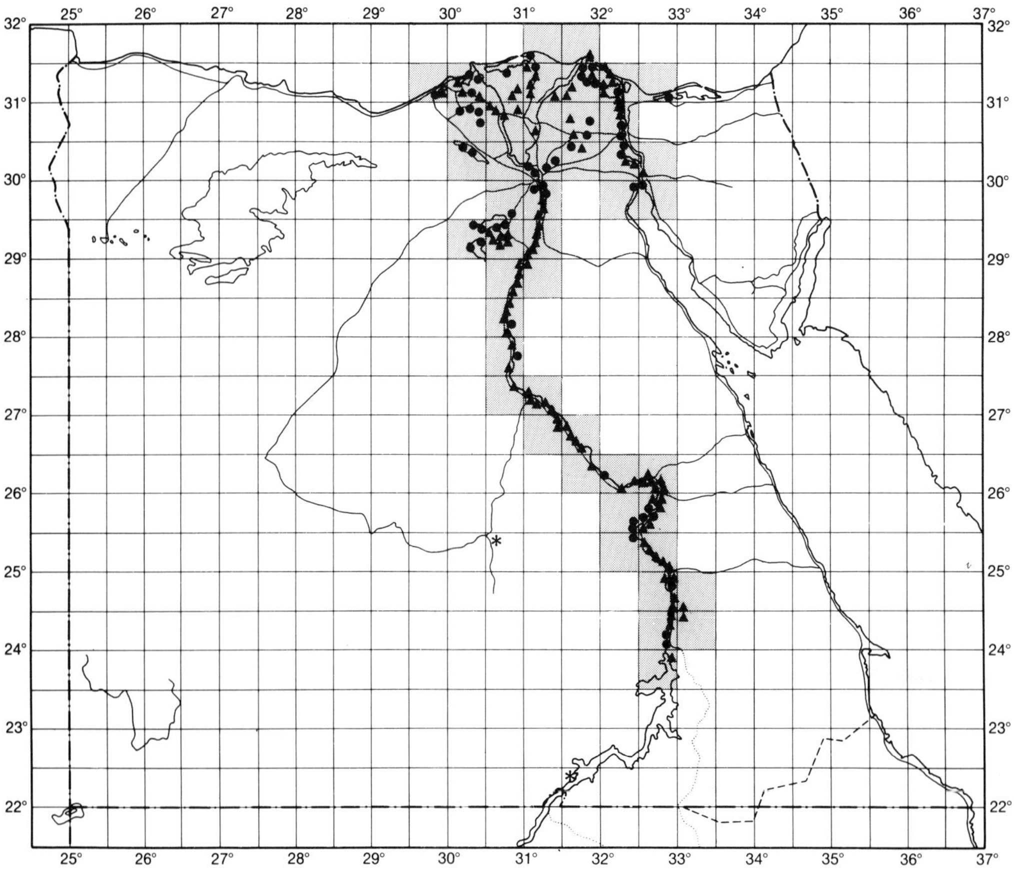

* possible breeding ▲ probable breeding ● definite breeding ░ half degree grid plots with probable or definite breeding

Map B36. Breeding distribution of Spur-winged Plover *Hoplopterus spinosus* between 1967–87.

Sociable Plover *Chettusia gregaria* PV (WV)

Local names Zaqzaaq ijtimaa'ii زقزاق اجتماعي, abuu gatgaat (NC), t̲aqt̲iiqa.

Rare passage visitor to the Nile Delta and Valley, and Red Sea coast in autumn from early September to mid-November and in spring from late February to mid-April. Occasional in winter. Monotypic.

Meinertzhagen (1930) noted that the Sociable Plover was a 'not uncommon' winter visitor to Egypt and occasional in Sinai. He examined specimens taken at Giza on 26 March, at El Qantara on 12 October, in the Faiyum from October to March, and Upper Egypt in March. Since this species is hardly mentioned by other ornithologists in their discussion of Egyptian birds, it seems doubtful that it was indeed 'not uncommon' in the country. El Negumi *et al.* (1950) considered it fairly rare in winter. There are relatively few recent records from Egypt, and almost all during the migration periods. It is striking that there are no post-1950 records from the Sudan (Nikolaus and Hamed 1984), although Cave and Macdonald (1955) considered it a non-breeding visitor to the country. The following is a listing of all known Egyptian records after 1930.

Autumn and winter El Kabrit—two on 7 September 1952 and three on 28 September 1952 (AP); Faiyum—two on 6 October 1980 (MSt), one on 15 November 1977 (Short and Horne 1981), and one with six Lapwings on 31 December 1963 (HCK); Idfu—two on 4 October 1981 (FdR); Suez—one on 26 October 1984 (RvD *et al.*); Cairo–Alexandria desert road near Wadi el Natrun—one on 3 November 1972 (Haensel 1980); Abu Ghusun—12 on 12 November 1982; Gezira Wadi Gimal—seven on 12 November 1982; Hamata—24 on 13 November 1982; Gezira Mahabis—two on 13 November 1982 (Baha el Din and Saleh 1983); and Cairo—one on 21 December 1984 (DM).

Spring Hurghada—one on 2 March 1984 (SG); and Ras Gharib—one on 15 March 1982 (ES, HWu).

White-tailed Plover *Chettusia leucura* PV (WV)

Local names Zaqzaaq abyad̲ adh-dhanab زقزاق أبيض الذنب, t̲aqt̲iiqa, shumiit̲a.

Rare passage visitor to the eastern portion of the country in autumn from early September to early November and in spring from late February to late March (mid-April) (Fig. 6.16). Occasional in winter and summer. Monotypic.

Meinertzhagen (1930) considered the White-tailed Plover an uncommon winter visitor to Egypt and noted that he examined specimens taken between November and March. In recent years there have been several observations of small numbers in the Nile Valley and the Faiyum, mainly during autumn migration. The largest numbers recorded are 10 at Idfu on 29 September 1981 (FdR), five in the Faiyum on 6 October 1980 (MSt) and at least 12 on 20 October 1950 (Searight 1952*a*), and eight south of Qena on 6 October 1983 (Hume 1983). Small numbers have been recorded in autumn along the north coast of Sinai. Records from Zaranikh include one on 17 September 1980 (Petersen and Sørensen 1981*a*), three on 6 September, one on 13 September, and one on 16 September 1982 (Baha el Din and Salama 1984).

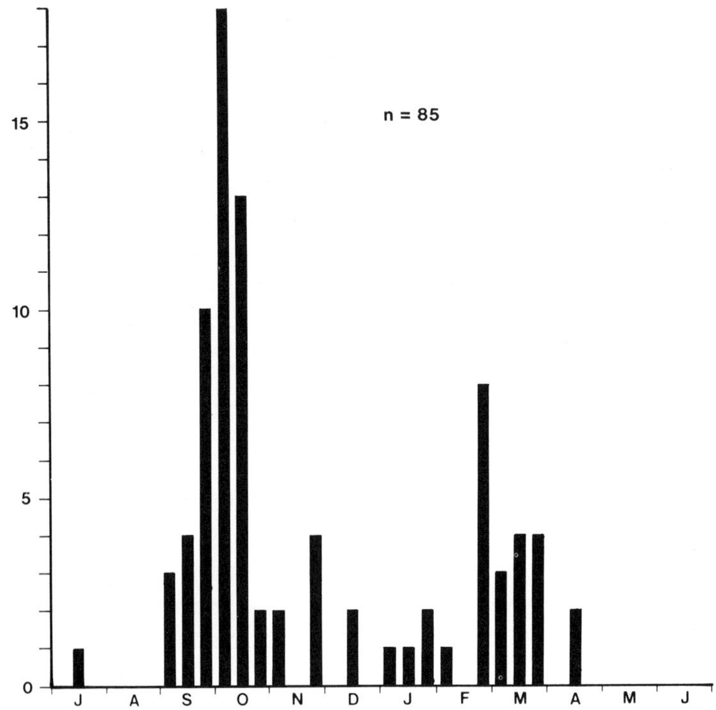

FIG. 6.16. Histogram of the numbers of White-tailed Plovers recorded during 10-day periods throughout the year.

Recent winter records include single birds collected in 'New Nubia' on 12 December 1970 and 1 January 1971 (USNM), one at Abu Rauwash on 17 December 1960 (FMNH), two south of Luxor on 17 December 1987 (ADB, AJA), three near Luxor on 21 and 22 January 1986 (CT), and one near Baltim on 8 February 1986 (SMB).

There are a few records from the Eastern Desert and along the Red Sea coast. Bulman (1944) recorded the species at Safaga in October 1942. At Sharm el Sheikh single birds were observed on 23 November 1984 (JDG) and 23 and 27 March 1979 (DJF *et al.*), and two were taken on 23 March 1981 (TAUM; HH). Single birds were noted near Gebel Elba on 15 April 1985 (SMG, PLM) and at Hurghada on 20 April 1984 (MBe, SM). One observed in July 1974, along the shore of Lake Nasser near Abu Simbil, was presumably summering locally (HCK). At Elat, Israel, this species is an uncommon migrant from late October to mid-November and from February to mid-April (mid-May) (Krabbe 1980).

Lapwing *Vanellus vanellus* PV WV

Local names Zaqzaaq shaamii زقزاق شامى, abuu t̲iit̲, al-farkha umm furuun, t̲iibat̲.

Fairly common passage and winter visitor, mainly to the northern portion of the country, from (mid-September) mid-October to mid-April (early May). Monotypic.

The earliest autumn records of Lapwings in Egypt are two at Zaranikh on 12 September 1982 (Baha el Din and Salama 1984) and one at Bahariya on 22 September 1983 (HB *et al.*). The bulk of migrants begin to arrive in late October and the species is common during the winter in several Western Desert oases, the Nile Delta and Valley, the Suez Canal area, and Sinai. The numbers recorded in the country vary from year to year. Flocks of up to several hundred birds are regularly noted in the Nile Delta and the Faiyum. Smaller numbers are found in the Nile Valley and it is scarce in Upper Egypt. On 21 January 1980, 16 Lapwings were observed at Aswan flying in a southerly direction (PLM, WCM), suggestive of passage into Sudan. This species is rarely observed along the gulf coasts of Sinai. Most migrants leave the country by mid-April. The latest spring record is five flying north over Cairo on 7 May 1978 (Meininger and Dielissen 1979).

Family Scolopacidae

Generic local names for *Calidris* spp. Durrayja, fuṯayra, karawaan al-maaya.

Knot *Calidris canutus* (PV)

Local name Durrayja ash-shamaal دريجة الشمال.

Rare and irregular passage visitor in autumn between mid-September and (early) mid-October and in spring from early March to mid-May. Occasional in winter and summer. All Egyptian material examined is referable to *C. c. canutus*.

Nicoll (1919) mentioned records of Knots from Suez, the Faiyum, and Bilbeis. The following summarizes all other known information from Egypt.

Autumn and winter At Zaranikh single birds were seen on 18 and 24 September 1981 (IP, UGS), on 20 September 1982 (Baha el Din and Salama 1984), and two on 6 October 1985 (DM). The only known Egyptian winter records are 20 at Suez on 3 December 1982 (PB), two at El Kabrit on 10 February 1953 (AP), and two at Quseir in February (Meinertzhagen 1930).

Spring and summer One was collected out of a small group at Suez on 5 March 1903 (Koenig 1928; Flower 1933). More recent records from Suez include two on 30 March and one on 23 April 1982 (Wimpfheimer *et al.* 1983). A flock of 20 was noted in late March 1928 at El Arish (Meinertzhagen 1930), 10 on 15 April 1984 in the Faiyum (MBe, SM), and 8 on 14 May 1909 at Lake Maryut (Ticehurst 1912). Two were seen in May 1985 near Elat, Israel (Paz 1987). Summer records include a bird in breeding plumage on 8 June 1952 at Deversoir (AP) and one on 19 July 1984 at Zaranikh (DM).

Sanderling *Calidris alba* PV WV

Local name Midrawaan مدروان.

Fairly common passage and winter visitor, generally along the Mediterranean and Red Sea coasts, from (early) mid-August to mid-May. It is rare on inland waters. Monotypic.

In autumn Sanderlings begin to arrive in mid-August. Considerable numbers have been recorded migrating along the north coast of Sinai: 4042 between 13 August and 29 September 1973 and 5223 between 15 August and 5 October 1978 (YP, UP); 744 between 7 and 30 September 1980 (Petersen and Sørensen 1981*a*); 5776 between 16 August and 24 September 1981 (Petersen and Sørensen 1981*b*); and 2585 between 2 and 24 September 1982 (Baha el Din and Salama 1984). It is not clear where these birds go once they pass Sinai. Only small numbers are observed on passage and in the winter along the Suez Canal and Red Sea coast.

The only recent inland records are: Wadi el Natrun—one on 29 November 1985 (DM) and three on 19 March 1985 (JW); Bahariya—two on 22 August 1983 (HB *et al.*); and Lake Qarun—three on 6 August 1966 (Kattinger 1970), two on 27 September 1984 (Miles 1984), three on 13 January 1982 (YL), and three on 17 May 1985 (JS *et al.*). According to Meinertzhagen (1930) this species has been observed in Upper Egypt in autumn and winter but there is no recent information to support this. Sanderlings are abundant in winter along the Mediterranean coast between Matruh and Alexandria (EK) and between Damietta and El Malaha (PLM, WCM). Several thousand Sanderlings probably winter along the Mediterranean coast. Spring passage is not clearly defined and birds seem to filter out of the country. The latest spring record is two on 18 May 1952 at Deversoir (AP). The only summer report is one at El Arish on 19 July 1984 (DM) that may have been a summering bird or perhaps early autumn migrant.

Little Stint *Calidris minuta* PV WV

Local name Fuṭayra فطيرة.

Abundant passage and winter visitor throughout the country from (mid-July) mid-August to mid-May (early June). Uncommon summer visitor. Monotypic.

In autumn the first Little Stints arrive in mid-July. The numbers increase substantially between August and October. Counts of migrating Little Stints at Zaranikh include: 9935 between 13 August and 29 September 1973 and 15 503 between 15 August and 5 October 1978 (YP, UP); and 10 266 between 16 August and 24 September 1981 (Petersen and Sørensen 1981*b*).

Between September and April this species is one of the most common waders in Egypt, with over 100 000 estimated to winter in the country. Between 7000 and 8000 Little Stints are estimated to be shot annually on Lake Manzala alone, although in

TABLE 6.5. Maximum numbers of Little Stints noted in Suez Bay between 1978 and 1986

Jan.	Feb.	March	April	May	June	July	Aug.	Sept.	Oct.	Nov.	Dec.
4100	5000	600	1200	1000	?	?	500	4100	5000	5000	?

some years considerably more are taken (Mullié and Meininger 1983; tables 4.3 and 4.4). The largest concentrations noted in recent years are 6200 on Lake Manzala in January 1980 (Meininger and Mullié 1981*a*), 2000 on Lake Qarun on 24 October 1980 (PLM), and regularly up to 5000 in Suez Bay in the autumn and winter (Table 6.5). Smaller numbers, peaking during autumn and spring migration, are frequently observed in Western Desert oases, the Nile Valley, along the Red Sea coast, and even in the mountains of south-central Sinai. The species is fairly common until mid-May and most have left the country by early June. An exceptional record is a few near Hurghada in early July 1951 (Searight 1952*c*).

Temminck's Stint *Calidris temminckii* PV WV

Local name Fuṯayra timmink فطيرة تمنك.

Scarce passage and rare winter visitor throughout the country from early August to mid- (late) May. Monotypic.

Temminck's Stints are regularly recorded, singly or in small flocks, in suitable habitat throughout the country. Flocks of more than 10 are exceptional and reports of over 100 birds are more likely to be Little Stints. The largest flocks recorded in recent years are 15 in Dakhla on 24 April 1981 (Goodman *et al.* 1986) and 13 at Suez on 12 September 1981 (FdR).

This species usually avoids large saline mud-flats, such as Suez Bay and the Delta lakes, and is generally found on the muddy edges of small bodies of slightly brackish or fresh water. The earliest autumn record is a single bird on 1 August 1952 at Deversoir (AP). Autumn migration peaks between mid-September and mid-October and spring migration is mainly in late April. The latest spring records are 19 May 1967 at Bahig (USNM) and 'late May' (Meinertzhagen 1930). Records from the Red Sea coast are almost exclusively confined to autumn and spring.

Curlew Sandpiper *Calidris ferruginea* PV (WV)

Local name Durrayja karawaaniyya دريجة كروانيّة.

Scarce migrant in autumn from mid-July to mid-November and fairly common in spring from (late February) mid-March to late May (mid-June). Rare in winter. Monotypic.

Between mid-July and mid-November small flocks of Curlew Sandpipers pass through Egypt. Numbers appear to peak between mid-September and early October.

In autumn this species is regularly observed along the Mediterranean coast, in the Nile Delta, Lake Qarun, and near Suez. In Suez Bay a maximum of 74 were noted in September and 154 in October 1981 (CAB *et al.*). The species is rarely noted in the Western Desert, the Nile Valley, and along the Red Sea coast. The maximum number counted during any autumn migratory bird survey at Zaranikh was 638, between 16 August and 24 September 1981 (Petersen and Sørensen 1981*b*).

In mid-winter the Curlew Sandpiper is rarely observed in Egypt. A few were noted in January 1908 on Lake Manzala (Nicoll 1908), one on 4 January 1980 at Suez (PLM, WCM), and four on 1 February 1983 near the Great Bitter Lake (BR).

The species is generally more abundant on spring migration than autumn. Spring passage commences in late February and continues into late May. A few birds are present to mid-June (e.g. Marchant 1941). In late April to mid-May flocks of up to about 200 have been seen on the Nile Delta lakes and in Suez Bay. Smaller numbers are regularly noted in Western Desert oases, the Nile Valley, and along the Red Sea coast.

Ringing A bird ringed in Cape Province, South Africa, on 29 November 1975 was recovered in Egypt before mid-April 1985 (BEP; GZF). It is not clear when and where the bird was actually found. A bird ringed in Bulgaria on 22 August 1982 was recovered in the Port Said region in early spring 1987 (BEP).

Dunlin *Calidris alpina* PV WV

Local name Durrayja دريجة.

Abundant passage migrant and winter visitor throughout the country from mid-July to mid-May. The majority of Egyptian specimens are referable to *C. a. alpina*. However, based on size, some material appears to be of *C. a. schinzii* (FMNH CC).

The first autumn Dunlins, some still in breeding plumage, arrive in mid-July. Autumn migration generally peaks in September, when large numbers are observed in various portions of the country. For example, at Zaranikh 8070 migrants were counted between 13 August and 29 September 1973 (YP, UP) and 8134 between 16 August and 24 September 1981 (Petersen and Sørensen 1981*b*). Between mid-August and mid-April flocks of several thousand are present in the Nile Delta lakes and in Suez Bay (Table 6.6). The total number of Dunlins wintering in Egypt is estimated to be over 100 000. Approximately 10 000 are shot annually on Lake Manzala alone, although in some years considerably more are taken (Mullié and Meininger 1983; tables 4.3 and 4.4).

This species is regularly seen in small numbers, particularly in autumn and spring, in

TABLE 6.6. Maximum numbers of Dunlins observed in Suez Bay between 1978 and 1986

Jan.	Feb.	March	April	May	June	July	Aug.	Sept.	Oct.	Nov.	Dec.
(15)	2000	3000	2000	30	0	0	30	300	3440	1100	1350

the Faiyum (up to 500), the Nile Valley, and along the Red Sea coast. It is rare in the Western Desert. Migrants leave the country by mid-May.

Ringing Two birds ringed in Israel, seven in Poland, and one in Norway were recovered in the Lake Manzala area. One bird ringed at Suez on 5 November 1984 was recovered in the Port Said market in early spring 1987 (Watson 1971; BEP).

Broad-billed Sandpiper *Limicola falcinellus* PV (WV)

Local names T̲iyt̲awaa 'ariid̲ al-minqaar طيطوى عريض المنقار, t̲iyt̲awaa minjalii.

Rare passage visitor in autumn from early August to mid-October and in spring from late March to late May (Fig. 6.17). Occasional in winter. All Egyptian material examined ($n = 19$) is referable to *L. f. falcinellus* and, excluding one, female.

Meinertzhagen (1930) stated that the Broad-billed Sandpiper was 'regular and not uncommon ... on the Mediterranean coast, where it has been obtained from mid-

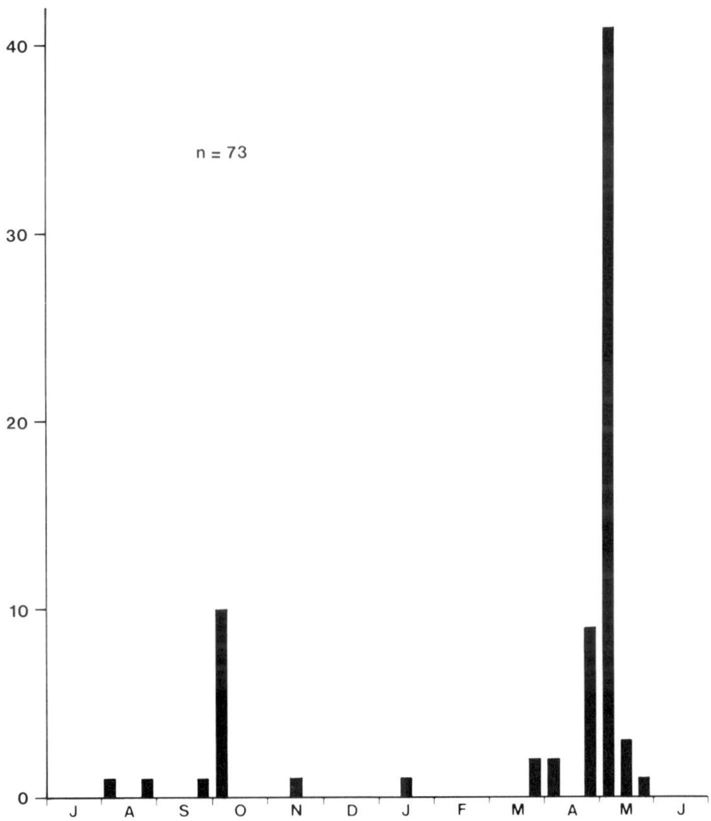

FIG. 6.17. Histogram of the numbers of Broad-billed Sandpipers recorded during 10-day periods throughout the year (1930–87).

October to early April'. A specimen was collected along the Suez Canal on 22 May 1922 (Meinertzhagen 1930). Four specimens were collected at Rosetta on 16 and 17 October 1917 (GZM), two at El Arish on 19 October 1917, and one at Port Said on 8 April 1928 (BMNH). The earliest autumn record appears to be 5 August in north Sinai (Meinertzhagen 1954).

Despite intensive wader counts in spring and autumn at Suez and in winter in the Nile Delta, relatively few Broad-billed Sandpipers have been recorded in Egypt since 1930. The majority of these records are from Suez and along the Red Sea coast, and fall within the period of spring migration. The largest concentrations noted in recent years are 14 on 2 May 1982 at Suez (Wimpfheimer *et al.* 1983) and 11 on 6 May 1983 at Ghubbet el Zeit (MCJ *et al.*). The single winter record is a bird collected at Lake Timsah on 14 January 1932 (HGD).

Meinertzhagen (1930) suggested that this species winters along the Mediterranean coast; however, there are no recent records confirming this. At Elat, Israel, the numbers in spring vary, and the species is uncommon in some years to fairly common in others from (early) late April to mid-May; 44 records were tabulated (maximum 18 birds/day) between 1966 and 1979 (Krabbe 1980).

Buff-breasted Sandpiper *Tryngites subruficollis* AV

Local name T̠iyt̠awaa as̠far as̠-s̠adr طيطوى أصفر الصدر.

The only record of the monotypic Buff-breasted Sandpiper in Egypt is a female collected at Quseir on 21 February 1928 (Meinertzhagen 1930).

Ruff *Philomachus pugnax* PV WV

Local names H̠ajwaala حجوالة, bayyaad, shaqiiy, t̠awt̠aw, t̠iyt̠awaa.

Common passage visitor throughout the country and fairly common winter visitor, mainly in the Nile Delta and Valley, from mid-July to mid- (late) May. Rare in summer. Monotypic.

Autumn migrational movements of Ruffs in Egypt are mainly noted between mid-July and early November, and peak in September. Generally small flocks can be observed throughout the country, including Western Desert oases and along the Red Sea coast. At Zaranikh this species is a regular migrant, e.g. 702 were counted passing between 15 August and 5 October 1978 (YP, UP) and 418 between 16 August and 24 September 1981 (Petersen and Sørensen 1981*b*).

In December and January small flocks of Ruffs are found on the marsh and cultivated lands of Wadi el Natrun, the Nile Delta, the Suez Canal area, the Faiyum, and locally in the Nile Valley. In recent years flocks composed of hundreds have been noted during the winter in the Nile Delta. Spring passage is from early February to mid-May, peaking between mid-March and mid-April. The Ruff is more abundant in spring than in autumn and is observed throughout the country. In late March thousands of migrants were noted during a few days in Wadi el Natrun (Nicoll 1912*a*) and on 1 March 1946 hundreds were seen in the Faiyum (PDH). Three were observed at El Qalag on 27 May 1945 (White 1949). A flock of 150 noted at Lake Qarun on 27 June 1977 (Meininger and Dielissen 1979) was either summering locally or migrating early in the autumn. At Elat, Israel, the Ruff is more common in spring than in autumn and up to 1850 have been noted in one day (Krabbe 1980).

Ringing A bird ringed in Italy on 10 April 1968 was recovered at Lake Manzala in February 1969 (Watson 1971) and another ringed in Finland on 19 August 1970 was recovered in Egypt on 18 April 1971 (EDB). A third bird ringed in Sweden in August was recovered in Egypt in October (Michelson 1985).

Generic local names for snipe Bakashiin, shunqub.

Jack Snipe *Lymnocryptes minimus* PV WV

Local names Bakaashiin saghiir بكاشين صغير, kaniisa, kinnis.

Scarce passage and winter visitor between (early September) early October and mid-April. All Egyptian specimens examined ($n = 26$), save two, are females. Monotypic.

Small numbers of Jack Snipes are regularly recorded in the marshy habitats of Western Desert oases, the Nile Delta, the Suez Canal area, the Faiyum, and north Sinai. The species is rarely noted in the Nile Valley. However, due to its skulking behaviour it is difficult to obtain an impression of the actual numbers passing through and wintering in the country.

The earliest autumn record is four at Zaranikh on 4 September 1982 (Baha el Din and Salama 1984). In January 1908, Nicoll (1908) found it to be common at Lake Manzala. Recent information shows that numbers peak in late February and March, suggestive of early spring passage. Other recent records include: Suez—up to five noted at one time between 12 and 17 March 1982 (Wimpfheimer *et al.* 1983) and at least 12 on 22 March 1985 (UGS); Ismailiya—one on 3 February 1974 (ZM-HUJ); Wadi el Natrun—six on 20 February 1984; and Lake Qarun—five on 22 February 1984 (SG) and 12 on 26 February 1987 (PdG). The only known post-1930 record from the Upper Nile Valley is one at Qena on 13 January 1954 (HCK). An exceptional record is one at St Katherine on 6 October 1971 (DL). This species is considered a scarce migrant at Elat, Israel, between late March and early May (Krabbe 1980).

Snipe *Gallinago gallinago* PV WV

Local names Bakaashiin بكاشين, bigay afriiqii (ND), shuush.

Common passage and winter visitor from (late July) mid-August to mid-April (early May). All Egyptian material examined is referable to the nominate subspecies.

Snipes are locally common between mid-August and mid-April in marshes of Western Desert oases, the Nile Delta and Valley, the Suez Canal area, and the Faiyum. Autumn migration peaks between mid-September and mid-October. During this period small numbers are observed along the north Sinai and Red Sea coasts. In winter concentrations of up to several hundred have been found in the Nile Delta and in the Suez Canal area. Several hundred migrants were noted in Dakhla in mid-March (Meinertzhagen 1930). The latest spring records are single birds on 8 May 1955 at Fayid (RN) and on 13 May 1909 at Lake Maryut (Ticehurst 1912). A bird observed on 23 July 1966 at Lake Maryut (JWW) was presumably an early autumn migrant.

Great Snipe *Gallinago media* PV (WV)

Local name Bakaashiin kabiir بكاشين كبير.

Rare passage visitor in autumn from late August to late October and in spring from early March to mid-April (late May). Occasional in winter. Monotypic.

Meinertzhagen (1930) considered the Great Snipe a rare passage visitor and unknown in the country during winter. The only records he cited were from the Nile Delta and included one in October and several between April and late May. There are only a few recent records from Egypt but due to this species' reclusive habits and possible confusion with the Snipe, it may be more common than currently known.

Nicoll (1912*a*) considered it a common passage visitor to Wadi el Natrun. Recent records from this locality include three on 17 January 1979 (Meininger *et al.* 1979) and five on 8 March 1985 (Goodman *et al.* 1986). Single birds were noted on 25 September 1984 at Suez (RvD *et al.*), 20 October 1979 at Luxor (FV), 8 January 1979 at Lake Timsah (Meininger *et al.* 1979), and on 12 March 1982 at Suez (Wimpfheimer *et al.* 1983). Two were observed at Zaranikh on 26 August and one on 16 September 1981 (IP, UGS) and two at Suez on 25 October 1981 (CAB *et al.*). The Great Snipe is a scarce migrant at Elat, Israel, in autumn from mid-September to mid-November and in spring from mid-March to late April (Krabbe 1980).

Woodcock *Scolopax rusticola* WV

Local names Diik al-ghaaba ديك الغابة, dajaaja al-ar<u>d</u>, <u>h</u>imaar al-<u>h</u>ajal.

Rare winter visitor to the north of the country from early November to early March. Monotypic.

Woodcocks have been recorded in orchards, palm groves, and gardens of the Nile Delta and Valley (south to Giza), and northern Sinai, particularly near El Arish. There are a few exceptional records from other areas of the country. One albinistic specimen was collected at Matruh on 13 November 1969 (USNM). Remains of Woodcock have been found on 14 October 1983 in Wadi el Arbain (SMB), on 13 November 1982 on Gezira Umm el Showra (Baha el Din and Saleh 1983), and on 25 December 1980 at

Sharm el Sheikh (BCD). In the spring of 1910, Woodcock feathers were noted in the nest of a Scrub Warbler in Wadi Hof, near Helwan (Sparrow in Raw 1921). This species is a rare winter visitor to Elat, Israel, between December and early March (Krabbe 1980).

Black-tailed Godwit *Limosa limosa* PV WV

Local names Biqwayqa suudaa- adh-dhanab بقويقة سوداء الذنب, al-agaat, biqa, biqwayqa sulṯaaniyya.

Fairly common passage visitor and rare in winter from late July to mid-May. Rare summer visitor. The nominate subspecies is the one occurring in Egypt.

Between late July and early November small numbers of Black-tailed Godwits are observed in marshy areas throughout the country. Flocks of more than 50 are exceptional. Along the north coast of Sinai, at Zaranikh, it is a regular autumn migrant with counts of 382 passing between 15 August and 5 October 1978 (YP, UP); 342 between 7 and 30 September 1980 (Petersen and Sørensen 1981*a*); and 1254 between 16 August and 24 September 1981, with a maximum of 154 on 2 September (Petersen and Sørensen 1981*b*). This species is uncommon along the Red Sea in both spring and autumn. It is rarely observed in December and January in Wadi el Natrun, the Faiyum, and the Nile Delta.

Spring movements are mainly in February and March, and involve considerably more birds than in autumn. Flocks of up to 200 are not unusual in the Delta lakes in February, and smaller numbers are regularly observed in the Western Desert oases, the Faiyum, Nile Valley, and near Suez. The latest spring records are from mid-May. The record of a bird at Hurghada on 11 June 1939 in full breeding plumage is exceptional (Marchant 1941).

Bar-tailed Godwit *Limosa lapponica* PV WV

Local name Biqwayqa mukhaṯṯaṯ adh-dhanab بقويقة مخطط الذنب.

Rare passage and winter visitor along the Mediterranean and Red Sea coasts from late August to mid-May (Fig. 6.18). Occasionally noted inland. Although no Egyptian material has been examined, on geographical grounds the subspecies occurring in the country is presumably *L. l. lapponica.*

Taylor (1896) recorded several Bar-tailed Godwits in the Cairo market on 5 February 1896. Ticehurst (1912) noted 'a small flock ... on some flooded fields near Alexandria on April 28th [1909] I am pretty sure belonged to this species'; however, this record is not acceptable. Meinertzhagen (1930) did not admit the Bar-tailed Godwit to the Egyptian bird list. The first documented record of this species in Egypt was apparently one collected on 26 February 1941 at the Bitter Lakes (El Negumi *et al.* 1950).

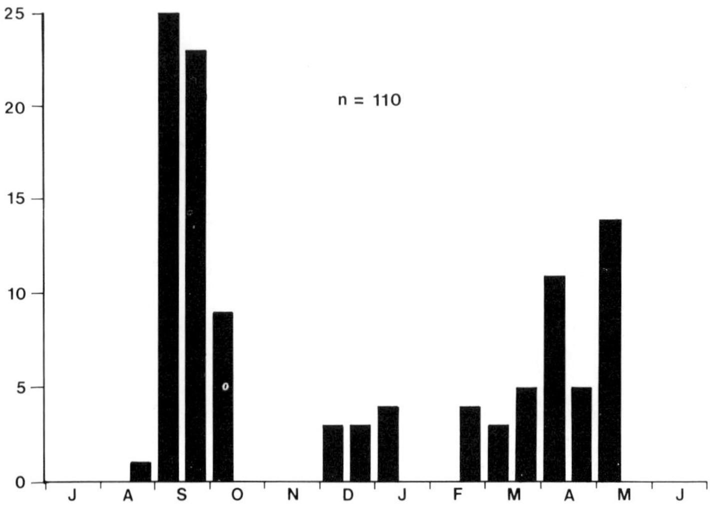

FIG. 6.18. Histogram of the numbers of Bar-tailed Godwits recorded per half-month period throughout the year.

There are several recent sight records from Egypt, and it appears that this species occurs regularly in small numbers. Observations from Zaranikh include: one on 14 September 1974 and three between 15 August and 5 October 1978 (YP, UP); 16 between 7 and 30 September 1980, with 10 on 8 September (Petersen and Sørensen 1981*a*); 10 between 1 and 24 September 1981 (IP, UGS); 10 between 14 and 23 September 1982 (Baha el Din and Salama 1984); and three on 6 October 1985 (DM). Between late August and mid-October flocks of up to four birds have been irregularly recorded along the Suez Canal and Suez Bay. Five were noted in Ghubbet el Gemsa on 24 September 1984 (MCJ *et al.*). The only known inland records are single birds at Lake Qarun on 27 September 1984 (Miles 1984) and 18–19 April 1986 (SMB, DM, WCM), and at Kom Ombo on 8 October 1984 (Miles 1984).

There are five winter records: one observed on 23 December 1978 at Ras Muhammad (EB); several on 24 December 1924 at the Bitter Lakes (Lynes and Bolam, in Chapman 1930); three on 11 December 1980 at Suez (SMB); at least four in early January 1928 on Lake Manzala (Chapman 1930); and one collected on 26 February 1941 at Great Bitter Lake (El Negumi *et al.* 1950). Between early March and mid-May small numbers have been recorded in Suez Bay and numerous places along the Red Sea coast. Maxima include five on 7 April 1980 at Suez (SMB) and seven on 3 May 1953 at El Kabrit (AP). At Elat, Israel, the species is rare between early March and late April (Krabbe 1980).

Generic local names for *Numenius* spp. Karawaan al-maa-, abuu hiirman (NC), karawaan (ND).

Whimbrel *Numenius phaeopus* PV (WV)

Local names Karawaan al-maa- aṣ-ṣaghiir كروان الماء الصغير, al-fayyuub, karawaan ghaytii ṣughayyar.

Scarce passage visitor in autumn and spring, mainly in coastal areas, and rare winter visitor from (early August) early September to mid- (late) May (Fig. 6.19). Rare summer visitor. All Egyptian material examined is referable to *N. p. phaeopus*.

The Whimbrel is a regular autumn migrant along the north coast of Sinai and the Red Sea between early August and mid-November. The numbers that pass through Zaranikh seem to vary from year to year: 164 between 13 August and 29 September 1973 and 152 between 15 August and 5 October 1978 (YP, UP); five between 7 and 30 September 1980 (Petersen and Sørensen 1981*a*); 72 between 16 August and 20 September 1981 (IP, UGS); and eight between 2 and 24 September 1982 (Baha el Din and Salama 1984). The Whimbrel is rare in the Nile Delta and Valley. Single birds have been observed on 30 October 1978 near El Amarna and on 15 November 1977 in the Faiyum (Short and Horne 1981), and one on 21 November 1978 at Kom Ombo (HW). There are no known records from the Western Desert.

A few winter along the Red Sea coast. One was collected on Lake Qarun on 25 January 1961 (FMNH CC). Between late February and mid-May small numbers are regularly observed along the Red Sea coast, in Suez Bay, and along the Suez Canal. The

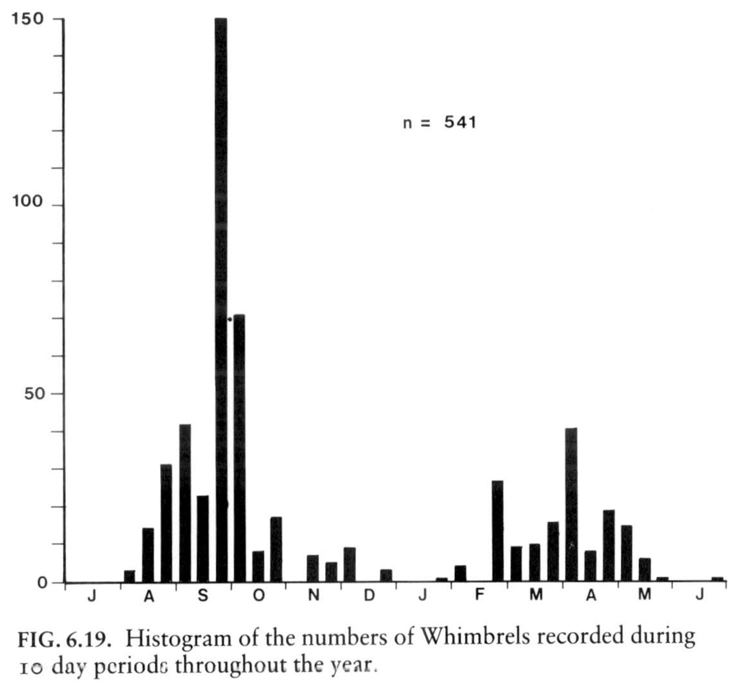

FIG. 6.19. Histogram of the numbers of Whimbrels recorded during 10 day periods throughout the year.

largest spring flocks noted in recent years are 27 at Safaga on 29 February 1984 (SG), 11 at Nabq on 1 April 1980 (KO), and 11 at Suez on 7 April 1980 (SMB). The only inland spring records are single birds on 19 March at Asyut and on 20 March 1984 at El Balyana (Durdin 1984), and on 8 April 1981 at Lake Qarun (WS). A bird observed on 22 June 1980 at Hurghada (SMB) represents an exceptional summer record from the country.

Slender-billed Curlew *Numenius tenuirostris* (PV) (WV)

Local name Karawaan al-maa- rafii' al-minqaar كروان الماء رفيع المنقار.

Rare and irregular passage and winter visitor. Monotypic.

The type specimen of the Slender-billed Curlew is from Egypt. In the RNHL there is an old specimen labelled 'Égypte'. Von Heuglin (1856) claimed to have seen a large flock in the desert near Alexandria, and Meinertzhagen (1930) saw what he believed to be a flock of this species in the Faiyum in January 1920. On 4 October 1917 this species was collected along the Mediterranean, just across the Sinai border, in present-day Israel (Meinertzhagen 1930). Other Egyptian specimens include single birds taken in September 1903 at Abu Qir (Koenig 1928; ZFMK); in 1933 at Giza and El Tell el Kebir (El Negumi *et al.* 1950); on 6 December 1930 at Port Taufiq (HGD); and on 18 September 1955 at Ras Gemsa (HGD). There are three relatively recent observations—single birds on 20 September 1982 at Zaranikh (Baha el Din 1984*b*) and 25 December 1980 at Ras Umm Sid (DMe), and seven on 30 December 1965 in the Faiyum (HCK). The records of Kattinger (1970) of this species in August 1966 at Lake Qarun are not acceptable. The only record from Elat, Israel, is one on 12 April 1977 (SP; Krabbe 1980); however, this observation has recently been brought into question (Paz 1987).

Curlew *Numenius arquata* PV WV

Local names Karawaan al-maa- كروان الماء, al-layl wal-nahaar, karawaan al-ghayt̲, karawaan ghayt̲ii.

Scarce passage and winter visitor to coastal areas and inland saline lakes from early August to late May. A few summer along the Red Sea coast. Both *N. a. orientalis* and *N. a. arquata* have been taken in Egypt.

In autumn the Curlew is a regular migrant along the north coast of Sinai and the Red Sea. Counts at Zaranikh include 114 between 13 August and 29 September 1973 and 35 between 15 August and 5 October 1978 (YP, UP); 20 between 7 and 30 September 1980 (Petersen and Sørensen 1981*a*); 50 between 16 August and 24 September 1981 (Petersen and Sørensen 1981*b*); and 52 between 2 and 24 September 1982 (Baha el Din and Salama 1984).

Small numbers are regularly observed between August and April along the

PLATES

Plate 1. Birds of the Nile Valley. (1) Barbary Falcon *Falco pelegrinoides*. (2) Nile Valley Sunbird *Anthreptes platurus* (a, male; b, female). (3) Little Green Bee-eater *Merops orientalis cleopatra*. (4) Black-shouldered Kite *Elanus caeruleus* (a, adult; b, immature). (5) Graceful Warbler *Prinia gracilis*. (6) Great Grey Shrike *Lanius excubitor*. (7) Senegal Thick-knee *Burhinus senegalensis*.

Plate 2. Birds of the Nile Delta. (**1**) Pied Kingfisher *Ceryle rudis* (a, male; b, female). (**2**) Streaked Weaver *Ploceus manyar* (a, female; b, male). (**3**) Avadavat *Amandava amandava* (a, immature; b, female; c, male). (**4**) Purple Gallinule *Porphyrio porphyrio madagascariensis* (a, immature; b, adult). (**5**) Senegal Coucal *Centropus senegalensis*. (**6**) Clamorous Reed Warbler *Acrocephalus stentoreus*. (**7**) Painted Snipe *Rostratula benghalensis* (a, male; b, female).

Plate 3. Birds of the Western Desert. (1) Spotted Sandgrouse *Pterocles senegallus* (a, male; b, female). (2) Cream-coloured Courser *Cursorius cursor*. (3) Hoopoe Lark *Alaemon alaudipes*. (4) White-crowned Black Wheatear *Oenanthe l. leucopyga* (a, adult; b, immature). (5) Bar-tailed Desert Lark *Ammomanes cincturus*. (6) Desert Lark *Ammomanes deserti*.

Plate 4. Birds of Gebel Elba. (1) Rosy-patched Shrike *Rhodophoneus cruentus* (a, male; b, female), (2) Shining Sunbird *Nectarinia habessinica* (a, female; b, male). (3) Arabian Warbler *Sylvia leucomelaena blanfordi* (a, female; b, male). (4) Fulvous Babbler *Turdoides fulvus acaciae*. (5) Nubian Nightjar *Caprimulgus n. nubicus*. (6) Lichtenstein's Sandgrouse *Pterocles lichtensteinii* (a, female; b, male). (7) Black-crowned Finch Lark *Eremopterix nigriceps* (a, female; b, male).

Plate 5. Birds of Sinai. (**1**) Tristram's Grackle *Onychognathus tristramii* (a, female; b, male). (**2**) Hume's Tawny Owl *Strix butleri*. (**3**) Black-capped Bulbul *Pycnonotus xanthopygos*. (**4**) Orange-tufted Sunbird *Nectarinia osea* (a, female; b, male). (**5**) Sand Partridge *Ammoperdix h. heyi* (a, male; b, female). (**6**) Scrub Warbler *Scotocerca inquieta*. (**7**) House Bunting *Emberiza s. striolata* (a, female; b, male). (**8**) Sinai Rosefinch *Carpodacus synoicus* (a, female; b, male).

Plate 6. Birds of the Red Sea. (**1**) Sooty Gull *Larus hemprichii* (a, adult; b, immature). (**2**) White-eyed Gull *Larus leucophthalmus* (a, adult; b, immature). (**3**) Swift Tern *Sterna bergii* (a, adult; b, immature). (**4**) Lesser Crested Tern *Sterna bengalensis* (a, adult; b, immature). (**5**) Bridled Tern *Sterna anaethetus* (a, adult; b, immature). (**6**) White-cheeked Tern *Sterna repressa* (a, adult; b, immature).

Mediterranean coast, in the Nile Delta (mainly on coastal lakes), the Suez Canal area, and Lake Qarun. A few are occasionally noted in the Nile Valley south to Aswan. Flocks of up to 60 are regularly reported from along the Red Sea coast. A total of 283 was noted in the northern Red Sea area between 23 September and 10 October 1984 (SMB). In January 1941, 70 to 100 were present in Ghubbet el Gemsa (Marchant 1941). This species has been recorded in Wadi el Natrun, where 62 were noted on 17 January 1979 (Meininger *et al.* 1979) and over 50 on 18 March 1986 (JW). There are also a few records from Siwa (Moreau 1928, 1940–1). The total number wintering in Egypt is estimated at 500–1000 (cf. Meininger and Mullié 1981*a*). Six were noted on 31 May 1986 at Lake Qarun (WCM). Small numbers of non-breeding birds summer along the Red Sea coast (e.g. Marchant 1941). An exceptionally large flock of 70 was noted at Hurghada on 6 July 1951 (Searight 1952*c*).

Generic local names for *Tringa* spp. T̲iyt̲awaa, faariikii (ND), nasnaasa, t̲awt̲aw, wiiba, zamaar.

Spotted Redshank *Tringa erythropus* PV WV

Local name T̲iyt̲awaa ah̲mar as-saaq arqat̲ طيطوى أحمر الساق أرقط.

Fairly common passage visitor and scarce in winter from (late July) early August to mid-May. Monotypic.

On 23 July 1966, 23 Spotted Redshanks, presumably early autumn migrants, were observed at Lake Maryut (JWW). Between early August and early November small flocks are regularly observed throughout the country, including Western Desert oases, the Nile Valley, and the Red Sea coast. This species is an uncommon migrant along the north coast of Sinai.

Records of large winter concentrations include 150 at El Abbassa on 5 January 1976 (MSm) and 139 on Lake Qarun between 10 and 12 January 1979 (Meininger *et al.* 1979). Small numbers also winter in Wadi el Natrun, Dakhla Oasis, and along the Red Sea coast. An estimated 400 to 800 winter in the country, mainly in the Nile Delta (Meininger and Mullié 1981*a*).

During spring passage, between late February and mid-May, this species is more common than in autumn and observed throughout the country. The largest known spring concentration is 500 on 19 April 1987 along the eastern shore of Lake Manzala (WCM).

Redshank *Tringa totanus* PV WV

Local name T̲iyt̲awaa ah̲mar as-saaq طيطوى أحمر الساق.

Fairly common passage and winter visitor from late July to late May. Occasional in summer. Based on size the majority of Egyptian material examined is referable to *T. t. totanus*, and only a few specimens to *T. t. eurhinus*.

The Redshank is a fairly common autumn migrant throughout the country. Counts of migrating birds at Zaranikh include: 850 between 13 August and 29 September 1973 and 561 between 15 August and 5 October 1978 (YP, UP); 227 between 7 and 30 September 1980 (Petersen and Sørensen 1981*a*); 476 between 16 August and 24 September 1981 (Petersen and Sørensen 1981*b*); and 280 between 2 and 24 September 1982 (Baha el Din and Salama 1984). Considerable concentrations occur in the autumn. At Lake Manzala 1700 were noted on 13 October 1980 (PLM), and in Suez Bay 860 on 8 September and 1250 on 25 October 1984 (RvD *et al.*). Small flocks are also observed between August and November throughout the country, including the Western Desert and Nile Valley.

An estimated 8000–10 000 Redshanks winter in the country, with midwinter counts of up to 114 in Wadi el Natrun, 500 on Lake Idku, 520 on Lake Burullus, 3400 on Lake Manzala, 160 on Lake Qarun, and 400 in Suez Bay (Meininger and Mullié 1981*a*). Small numbers winter along the Mediterranean and Red Sea coasts. The species is a rare winter visitor to the Western Desert, except Wadi el Natrun, and in the Nile Valley. Spring passage is from March to late May. Throughout the summer small numbers are present in the Nile Delta, Suez Canal area, and along the Red Sea coast. An exceptionally large number of 80 was noted in Suez Bay on 28 June 1986 (DM, WCM).

Ringing One bird ringed in the northern Black Sea in June was recovered at Port Said in winter (Michelson 1985) and another ringed in Hungary some time before 1931 was recovered west of Salum (Hale 1973).

Marsh Sandpiper *Tringa stagnatilis* PV WV

Local name T̲iyt̲awaa al-mustanqa' طيطوى المستنقع.

Scarce passage and winter visitor from early August to early May. Monotypic.

On autumn migration, from early August to November, single Marsh Sandpipers or small flocks are observed throughout the country, including the Western Desert oases, the Nile Valley, and along the Red Sea coast. Concentrations of more than 10 birds are exceptional. This species is a regular autumn migrant along the north coast of Sinai. Counts of migrating birds at Zaranikh include 14 between 13 August and 29 September 1973 and 42 between 15 August and 5 October 1978 (YP, UP); 44 between 7 and 30 September 1980, with 28 on 11 September (Petersen and Sørensen 1981*a*); 209 between 16 August and 24 September 1981 (Petersen and Sørensen 1981*b*); and 143 between 2 and 24 September 1982, with 72 on 10 September (Baha el Din and Salama 1984). During systematic counts of waders in Suez Bay from 5 September to 26 November 1984, the largest number of Marsh Sandpipers recorded was about 45 between early September and early October (RvD *et al.*).

An estimated 200 to 400 birds winter in the marshy areas of Western Desert oases, the Nile Delta and Valley (south to Aswan), the Suez Canal area, and the Faiyum

(Meininger and Mullié 1981*a*). Spring passage is throughout the country from mid-March to early May. In recent years the largest reported spring flock is 10 and Meinertzhagen's (1930) statement that 'in some springs immense numbers pass through in April' lacks confirmation.

Greenshank *Tringa nebularia* PV WV

Local names Ṯiyṯawaa akhḏar as-saaq طيطوى أخضر الساق, abuu manaazaa (Su).

Fairly common passage visitor and scarce winter visitor from (late July) early August to mid- (late) May. Rare in summer. Monotypic.

From early August to November small flocks of Greenshanks are observed throughout the country, including Western Desert oases, the Nile Delta and Valley, and the Red Sea coast. This species is regularly noted migrating along the north coast of Sinai with counts of 69 between 15 August and 5 October 1978 (YP, UP); 78 between 7 and 30 September 1980 (Petersen and Sørensen 1981*a*); and 100 between 16 August and 24 September 1981 (Petersen and Sørensen 1981*b*). A total of 150 was recorded from 29 September to 10 October 1983 along the Nile between Cairo and Aswan (Hume 1983) and 60 were seen on 19 September 1981 at Hurghada (FdR).

An estimated 300 to 500 Greenshanks winter in the country, mainly in north Sinai, the Nile Delta lakes, and Lake Qarun (Meininger and Mullié 1981*a*). Flocks of more than 20 birds are exceptional in all seasons. Spring migration is most notable from late February to mid- (late) May, and takes place along a broad front across the complete country. Observations of either summering birds or early autumn migrants include: one on 25 June 1939 at Hurghada (Marchant 1941); four on 6 July 1951 at Hurghada (Searight 1952*c*); 13–14 on 13 July 1950 at the Gebel Maryam lagoon (GCJ); and one on 19 July 1981 at Aswan (CB).

Ringing A bird ringed at Port Alfred, South Africa, on 15 March 1985 was recovered in Bahariya Oasis on 14 July 1987 (Tree 1987).

Green Sandpiper *Tringa ochropus* PV WV

Local names Ṯiyṯawaa akhḏar طيطوى أخضر, farkha al-ghayṯ.

Fairly common passage migrant and scarce winter visitor from early August to late May. Rare in summer. Monotypic.

Between early August and November small flocks (generally up to 40) of Green Sandpipers are regularly observed throughout the country. An exceptionally large flock of 80 was noted in Bahariya on 10 September 1983 (HB *et al.*). This species is a common autumn migrant along the north coast of Sinai. Counts of migrating birds at Zaranikh include: 549 between 13 August and 29 September 1973 and 987 between 15 August and 5 October 1978 (YP, UP); 138 between 7 and 30 September 1980 (Petersen and Sørensen 1981*a*); 675 from 16 August to 24 September 1981 (Petersen and Sørensen 1981*b*); and 498 between 2 and 24 September 1982 (Baha el Din and Salama 1984). This species has been noted several times at the St Katherine airport, with a maximum of 15 on 23 August 1983 (SMB).

An estimated 200–500 Green Sandpipers winter in the country; however, this number is conservative as this species is found in a variety of habitats and never in large concentrations. Winter records from the Red Sea coast are rare. Spring migration is from early March to mid- (late) May and proceeds along a broad front across the country, including the Western Desert and the Red Sea coast. There are several late June and July records, which are either locally summering birds or early autumn migrants.

Wood Sandpiper *Tringa glareola* PV WV

Local name T̲iyt̲awaa al-ghiyaad̲ طيطوى الغياض.

Fairly common passage and scarce winter visitor from late July to mid-May (early June). Rare in summer. Monotypic.

The bulk of Wood Sandpiper autumn migration in Egypt is between late July and late October and proceeds along a broad front, including the Western Desert, Red Sea coast, and the mountains of south Sinai. It is a regular autumn migrant along the north Sinai coast and counts of migrants at Zaranikh include: 80 between 13 August and 29 September 1973 and 112 between 15 August and 5 October 1978 (YP, UP); 52 between 7 and 30 September 1980 (Petersen and Sørensen 1981*a*); and 255 between 16 August and 24 September 1981 (Petersen and Sørensen 1981*b*). On migration this species is generally noted in small numbers but a flock of up to 50 was recorded on 17 September 1980 at Lake Idku (WCM).

An estimated 300 to 500 birds winter in the country, generally in the marshes of Wadi el Natrun and the Nile Delta and Valley. The largest winter count noted in recent years is 40 at El Abbassa on 12 January 1976 (MSm). Spring migration proceeds across most of the country from mid-March to mid-May. Relatively large concentrations occur, e.g. 80 at Suez on 12 March 1982 (Wimpfheimer *et al.* 1983), 40 at Baltim on 29 March 1981 (WS), and 100 in Wadi el Natrun on 28 April 1984 (MBe, SM). Single Wood Sandpipers noted at Sidi Abd el Rahman between 25 May and 3 June 1966 (JT), at El Tor on 26 May 1938 (GZM), and at Hurghada on 21 June 1939 (Marchant 1941) may have been summering locally.

Ringing A bird ringed in Sweden on 14 July 1968 was recovered at El Sinaniya on 6 April 1970 (Watson 1971) and another ringed in Zaire on 20 November 1971 was recovered in Egypt on 15 September 1972 (EDB).

Terek Sandpiper *Xenus cinereus* PV

Local name T̲iyt̲awaa nakaat طيطوى نكات.

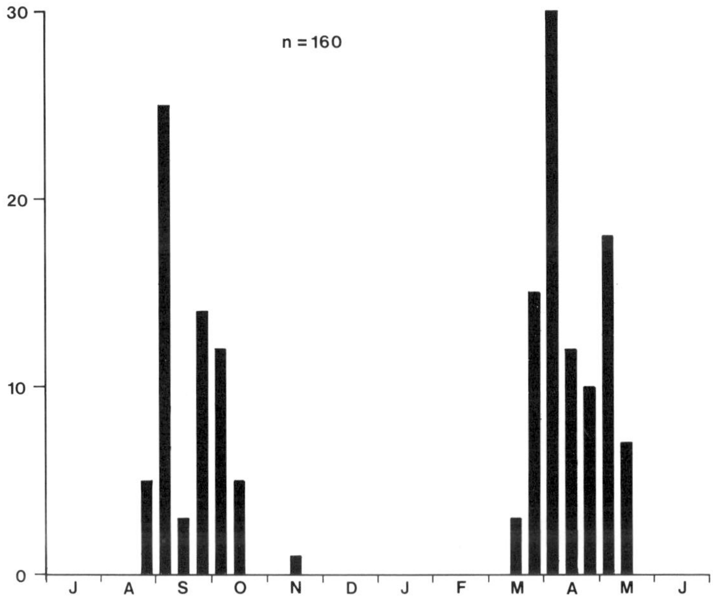

FIG. 6.20. Histogram of the numbers of Terek Sandpipers recorded during 10-day periods throughout the year.

Rare passage visitor, mainly along the Red Sea coast, in autumn from late August to mid-October (mid-November) and in spring from mid-March to mid-May (Fig. 6.20). Monotypic.

Four Terek Sandpipers were observed near Alexandria on 4 May 1902 (Parrot 1903) and one at Rosetta on 15 October 1917 (Nicoll 1919). Meinertzhagen (1930) did not accept these records and considered the species hypothetical in Egypt. In recent years it has become clear that this bird is a rare but apparently regular autumn and spring migrant in the eastern half of the country. Autumn records from Zaranikh include: one on 8 September 1980 (Petersen and Sørensen 1981*a*); at least 23 between 22 August and 10 September 1981, with a maximum of eight on 5 September (IP, UGS); and five on 3 September and two on 20 September 1982 (Baha el Din and Salama 1984). A single bird was observed along the Suez Canal on 8 September 1947 (Elliott and Monk 1952) and a maximum of eight at Suez between 28 September and 18 October 1981 (CAB *et al.*). In late September 1984 up to six were noted on islands in the northern Red Sea (SMB). The latest autumn record is one at Lake Qarun on 15 November 1983 (Bezzel 1987*a*).

Between 15 March and 19 May small flocks have been observed at several localities along the Red Sea coast and near Suez. A flock of 15 was noted on 6 April 1985 at Shalatein (SMG, PLM) and another of 13 on 2 May 1983 at Suez Bay (Wimpfheimer *et al.* 1983). One bird was found south of Port Said on 7 April 1982 (De Bièvre 1983).

The only acceptable inland spring record is one on 21 March 1983 at Aswan (Welch 1983). Rudolf of Austria's (1889) comment that it was common at Lake Qarun is unsubstantiated and there is no evidence of passage in Egypt west of the Nile Valley.

Common Sandpiper *Actitis hypoleucos* PV WV

Local name T̲iyt̲awaa طيطوى.

Fairly common passage visitor and scarce in winter from mid-July to mid- (late) May. Monotypic.

In autumn the first Common Sandpipers arrive in mid-July and the bulk of passage moves across Egypt in a broad front between August and September. At Zaranikh this species is a regular autumn migrant with total counts of 228 between 13 August and 29 September 1973 (YP, UP); 660 between 16 August and 24 September 1981, with 140 on 17 August (Petersen and Sørensen 1981*b*); and 143 between 2 and 24 September 1982 (Baha el Din and Salama 1984).

An estimated 300 to 500 birds winter in the country, mainly in the Nile Delta and Valley, and smaller numbers in Western Desert oases and along the Red Sea coast. Spring migration takes place from mid-March to mid- (late) May, again sweeping over the country along a broad pathway. The largest local spring concentration is 40 at Aswan on 29 March 1982 (JW).

Turnstone *Arenaria interpres* PV WV

Local name Qunbura al-maa- قنبرة الماء.

Scarce passage and winter visitor, mainly along the coasts of the Mediterranean and Red Sea and Suez Canal area, from (mid-) late August to mid-May. All Egyptian material examined is referable to *A. i. interpres*.

Turnstones begin to arrive in mid-August and by late August flocks of up to 40 are commonly observed along the Mediterranean and Red Sea coasts. Counts at Zaranikh include: 196 between 13 August and 29 September 1973 and 132 between 15 August and 5 October 1978 (YP, UP); 117 between 16 August and 24 September 1981 (Petersen and Sørensen 1981*b*); and 84 between 2 and 24 September 1982 (Baha el Din and Salama 1984). Inland records are known mainly from Lake Qarun, where up to five have been recorded in recent years between early September and early October, and single birds on 31 January 1977 (Short and Horne 1981) and 8 April 1985 (JvT). One was noted at Aswan on 25 April 1984 (MBe, SM). Along the Red Sea coast numbers show a slight peak in September–October and in April. The bulk of migrants leave the country by mid-May; although a few summer along the Red Sea coast.

Red-necked Phalarope *Phalaropus lobatus* PV

Local name Falaaruub aẖmar al-'unuq فلاروب أحمر العنق.

Rare passage visitor with most records confined to the autumn months. Monotypic.

During counts of migrating water-birds at Zaranikh, Red-necked Phalaropes were observed in small numbers each year. Records include: a total of six between 7 and 30 September 1980, with a maximum of two on 24 September (Petersen and Sørensen 1981*a*); about 30 between 16 August and 24 September 1981, with a maximum of five on 15 September (Petersen and Sørensen 1981*b*); and single birds or pairs on 12, 15, 16, and 21 September 1982 (Baha el Din and Salama 1984). Other recent autumn records include two at Suez on 6 September 1984 (RvD *et al.*), and five at El Taloul on 5 October 1985 and eight the following day (SMB, DM).

There are only a few spring records. One was observed at Lake Maryut on 11 April 1966 (JWW) and three on Lake Nasser, near Aswan, on 24 April 1983 (SC, PLM, IP). A single bird was noted at Suez between 1 and 7 May 1982 (Wimpfheimer *et al.* 1983). At Elat, Israel, the Red-necked Phalarope is a rare migrant in autumn between August and October and in spring between March and May; the maximum local daily count is 34 birds between 16 and 19 May 1981 (Paz 1987).

Grey Phalarope *Phalaropus fulicarius* (PV) (WV)

Local name Falaaruub ramaadii فلاروب رمادي.

This species is an irregular passage and winter visitor. All records are from the Nile Delta or north Sinai coast. Monotypic.

During counts of water-birds at Zaranikh, single Grey Phalaropes have been observed on 9, 16, and 19 September 1981 (Petersen and Sørensen 1981*b*), and 20 September 1982 (Baha el Din and Salama 1984). Other Egyptian records include four at El Malaha on 10 January 1980 (Meininger and Mullié 1981*a*) and one at Lake Manzala, north-west of El Matariya, on 5 May 1983 (PLM, UGS). There is a sighting of one bird in breeding plumage in May 1983 at Elat, Israel (Paz 1987; Hovel 1987).

Family Stercorariidae

Pomarine Skua *Stercorarius pomarinus* PV WV

Local name Karkar كركر.

Rare passage and winter visitor from mid-August to late May (early June). Small numbers summer in Egyptian waters. Monotypic.

Meinertzhagen (1930) mentioned only one record of the Pomarine Skua in Egypt. In recent years it has become clear that this species is a regular autumn and spring migrant through the country. The first autumn Pomarine Skuas arrive in mid-August, when small numbers are observed almost exclusively along the north coast of Sinai and near Port Said. In September and October there is a distinct influx of migrants and they are regularly seen flying westwards along the north coast of Sinai. It seems plausible that these birds move overland across the Sinai, presumably along a broad front and at high altitudes, to the Red Sea (Meininger and Sørensen 1986). The only known winter records are three on 6 January and one on 7 January 1980 along the southern shore of Lake Bardawil (Meininger and Mullié 1981*a*).

From February to June this species is frequently observed in small numbers near Elat, Israel (UP). Here between late March and early June, mostly adult light-phased birds are seen leaving the Gulf of Aqaba and continuing overland. Some concentration occurs in May, when the maximum number recorded in a single day is 39 (Krabbe 1980). A few are noted between mid-March and late May in the northern portion of the Red Sea, including the Gulf of Suez, and near Port Said. There are several records from June and July, generally of immatures, along the Mediterranean and Red Sea coasts (Meininger and Sørensen 1986).

Arctic Skua *Stercorarius parasiticus* PV WV

Local name Karkar qutbii كركر قطبي.

Scarce passage and winter visitor from mid-August to late May (early June). A few summer in Egyptian waters. Monotypic.

Although there were a few late nineteenth- and early twentieth-century observations of the Arctic Skua in Egypt, Meinertzhagen (1930) considered these records insufficient to admit this species to the Egyptian list. In recent years it has become clear that it regularly occurs along the Mediterranean and Red Sea coasts.

Between August and November the Artic Skua is frequently noted along the north coast of Sinai and near Port Said. Most, if not all, of these birds presumably cross overland or along the Suez Canal to the Red Sea. Autumn observations from the Suez Canal which support this presumption include a group of four on 29 August 1944 near Kasfarit and one adult near Suez on 4 November 1981. The maximum autumn daily count is 34 on 13 October 1980 at Port Said (Meininger and Sørensen 1986). There are several autumn records from the Red Sea.

During the winter a few are noted along the Mediterranean coast. Between late February and early June this species is fairly common in the northern Gulf of Aqaba. At Elat, Israel, migration peaks in April, when up to 48 have been noted in one day (Krabbe 1980). Most birds, mainly light-phase adults, are seen migrating in a northerly direction. The maximum number of Arctic Skuas recorded at this locality during a single spring season is 293 between 20 February and 2 June 1979. Smaller numbers have been recorded in the spring and summer along the coasts of the entire Egyptian Red Sea, and between February and April near Port Said. On 28 May 1978,

28 were seen in the Gulf of Suez (Meininger and Sørensen 1986). The only Egyptian Nile Valley record is one collected at El Balyana on 10 April 1897 (Koenig 1932; ZFMK). The recovery of a Shetland ringed bird along the Nile near Omdurman, Sudan, in January 1985 (Mead and Hudson 1986) is suggestive that more inland records might be expected.

Ringing A bird ringed in Finland on 5 July 1969 was recovered at Sidi Salim on 15 October 1969 (Hildén 1971).

Long-tailed Skua *Stercorarius longicaudus* (PV)

Local name Karkar ṯawiil adh-dhanab كركر طويل الذنب.

Rare and irregular passage visitor. Monotypic.

There are several recent records of Long-tailed Skuas in Egypt, suggesting that this species is probably an irregular migrant to the country in small numbers. Most autumn records are from the north coast of Sinai and all spring records from the northern Red Sea.

Autumn A single bird was observed at Rumana on 20 August 1981 (SMB). Records from Zaranikh include: one on 15 September, two on 17 September, and at least three on 19 September 1981, and one on 16 September 1982 (Meininger and Sørensen 1986).

Spring and summer Single birds were noted on 23 April 1984 at Sharm el Sheikh (Siering 1986), on 20 May 1980 on Gezira Tiran (Hovel 1987), and on 23 May 1984 near Gezira Abu Minqar (Goodman and Storer 1987). A dried mummy was found in Wadi Feiran on 25 June 1981 (TAUM).

At Elat, Israel, between 1970 and 1984 at least 19 birds were noted from late March to 20 May (Meininger and Sørensen 1986). In the late summer of 1980, seven were recorded at this locality between 24 July and 8 August (HSh).

[**Great Skua** *Stercorarius skua* AV?

One Great Skua was seen on 13 September 1978 at Lake Bardawil (Paz 1987). Either this species or the South Polar Skua (*S. (skua) maccormicki*) was observed off Port Said on 27 April 1961, 23 May 1961, 23 May 1959 (two), and 5 July 1964, and in the Gulf of Suez at 28°N., 33°E. on 17 August 1957 (Mörzer Bruyns and Voous 1965). A skua noted on 22 March 1964 off Port Said may have been of this species complex (Bourne 1967). A bird observed on 3 June 1983 at Elat, Israel, was identified as a South Polar Skua (HSh; cf. Paz 1987).]

Family Laridae

Generic local names for gull Nuuras, nawras.

Sooty Gull (Hemprich's Gull) *Larus hemprichii* RB Plate 6

Local name Nuuras as<u>h</u>am نورس أسحم.

Scarce but locally common breeding resident on islands in the Red Sea. Scarce winter visitor to the northern portion of the Red Sea. It nests between July and September. Monotypic.

In 1983 and 1984 Sooty Gulls were documented or suspected to breed on several

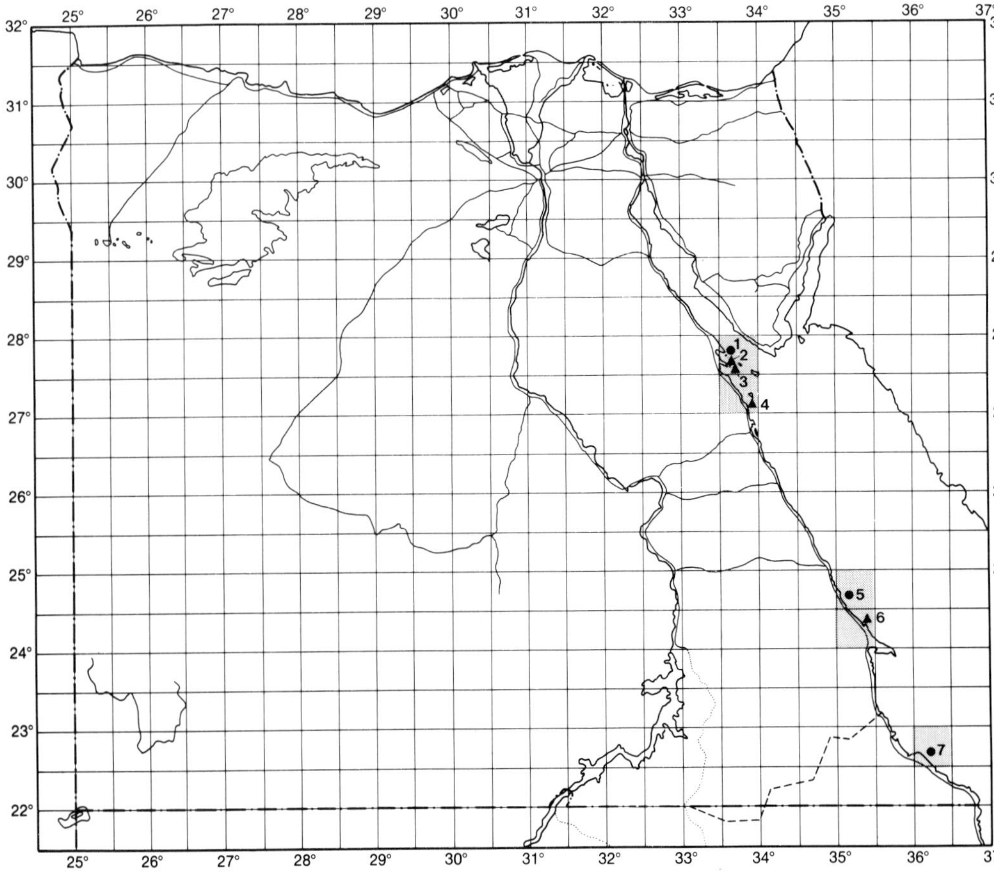

Map B37. Breeding distribution of Sooty Gull *Larus hemprichii* between 1980–87.
Breeding localities (see text for details): (1) Gezira Ashrafi. (2) Gezira Umm el Heimat. (3) Gezira Tawila. (4) Gezira Umm Agawish el Kabir. (5) Gezira Wadi Gimal. (6) Gezira Mahabis. (7) Gezira Siyal group.

islands in the northern Red Sea. Fifty pairs were estimated to nest on Gezira Ashrafi, Gezira Umm el Heimat, and Gezira Tawila. In the autumn of 1984, 233 birds, including fledglings, were noted on these islands (Jennings *et al.* 1985). In mid-May 1984 pairs were found on Gezira Tawila and Gezira Umm Agawish el Kabir in nuptial plumage and performing courtship displays (Goodman and Storer 1987). On Gezira Wadi Gimal 30 juveniles were noted on 10 October 1982 (Barratt 1982), one recent fledgling on 12 November 1982 (Baha el Din and Saleh 1983), and isolated pairs defending territories on 25 June 1984 (Goodman and Storer 1987). In mid-June 1984 and late April 1985 territorial pairs were found on Gezira Mahabis, off Hamata (Goodman and Storer 1987) and the Siyal island group, off Adal Deib (SMG, PLM), respectively. On the latter islands fishermen were familiar with the eggs and plumages of the Sooty and White-eyed Gull and confirmed that both species bred locally.

Small flocks of up to 50 birds, often mixed with White-eyed Gulls, are regularly observed throughout the year on Red Sea islands and along the coast from Hurghada southwards. Small numbers enter the Gulfs of Suez and Aqaba. The species is noted somewhat regularly at Suez but rarely at Elat, Israel (Krabbe 1980; Ben Dov and Golan 1983). At least three Sooty Gulls were reported on 6 February 1925 near Port Said and off Damietta in mixed flocks of *Larus* spp. (Chapman 1930); however, these birds were more likely White-eyed Gulls.

White-eyed Gull *Larus leucophthalmus* RB Plate 6

Local names Nuuras ’ajama نورس عجمة, ’ajama, zummaj al-maaya.

Fairly common breeding resident on islands in the Red Sea. Uncommon summer visitor to the Suez Canal area and rarely to the Mediterranean coast. It nests between July and September. Monotypic.

In 1983 and 1984 the breeding population on islands off Hurghada and at the mouth of the Gulf of Suez was estimated at 1500–2000 pairs (Jennings *et al.* 1985). Quantitative information is lacking on the number of birds breeding on islands further south (see map for details). On 27 June 1945 about 30 White-eyed Gulls in adult plumage were observed near Brothers Reef and it is possible that they nest on this island (Phillips 1947).

Throughout the year flocks of up to several hundred are observed along the coast from Hurghada to the Sudanese border. After the breeding season White-eyed Gulls disperse widely. A few enter the Gulf of Aqaba but the species is scarce at Elat, Israel (Krabbe 1980). A considerable number are found throughout the year in the Gulf of Suez, with the largest numbers present in the spring and summer. Groups of several hundred were noted at Suez in late June 1945 (Phillips 1947).

In spring and summer this species is regularly noted in the Suez Canal. In 1944 it was recorded at the Bitter Lakes between 16 May and 9 September (LB), with a flock of 250 on 17 May (Bodenham 1945*b*). Regular northward movements were noted at

Fayid between 20 April and 9 May 1955, with about 25 birds passing per hour during three morning counts (RN). Some birds reach the Mediterranean coast via the Suez Canal. At Port Said one was seen on 2 April 1960 (Bourne and Radford 1961) and three on 28 June 1977 (Meininger and Dielissen 1979). The record of Searight (1952*a*) of one in the Faiyum is not acceptable.

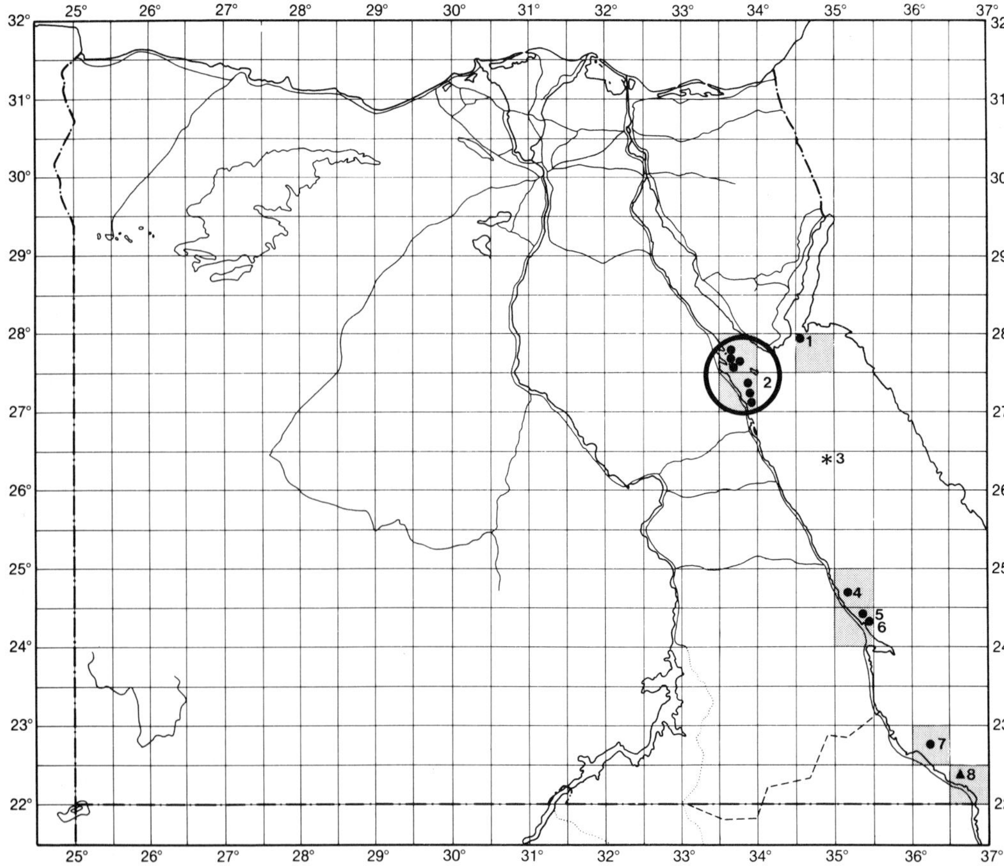

∗ possible breeding ▲ probable breeding ● definite breeding ▒ half degree grid plots with probable or definite breeding

Map B38. Breeding distribution of White-eyed Gull *Larus leucophthalmus* between 1945–87. *Breeding localities* with most recent estimate of breeding pairs: (1) Gezira Tiran (1973–78: 50). (2). Islands mouth Gulf of Suez (1983–84: 1500–2000). (3) Brothers Reef (1945: breeding?). (4) Gezira Wadi Gimal (1982: at least 50). (5) Gezira Showarit (1982: present). (6) Gezira Mahabis (1982: present). (7) Gezira Siyal group (1985: present). (8) Gezira Rawabel el Kabir (1985: present).

Great Black-headed Gull *Larus ichthyaetus* PV WV

Local names Nuuras as-samak نورس السمك, abuu gahsh (F).

Rare passage migrant and winter visitor from early October to mid-May, mainly along the Red Sea coast. Monotypic.

According to Shelley (1872) and Meinertzhagen (1930) the Great Black-headed Gull was a not uncommon winter visitor to Egyptian waters, including the Faiyum, Upper Nile Valley, the Suez Canal area, and the Gulfs of Suez and Aqaba. Nicoll (1919) commented that this species was 'apparently far less common' than it had been 50 years earlier. In the past few decades only small numbers have been recorded in the country. Due to a typesetting error beyond the control of Meininger and Baha el Din (1986) the information presented therein under this species actually concerns the Mediterranean Gull.

Nile Delta In January 1908 Nicoll (1908) saw several and collected one Great Black-headed Gull at Lake Manzala. The only relatively recent records from the Nile Delta are one adult near Alexandria on 19 December 1972 (Haensel 1980), four at Port Said on 5 January 1985 (WL), and one immature at Lake Manzala, near Port Said, on 21 March 1982 (ES, HWu). In the winters of 1978–9 and 1979–80 the Nile Delta lakes, as well as other Egyptian wetlands, were intensively surveyed and this species was not recorded (Meininger and Mullié 1981*b*). It formerly occurred in 'large numbers' during the winter in the Suez Canal area (Meinertzhagen 1930). The maximum recorded in subsequent years is five at Deversoir on 29 April 1953 and five at Fanara the next day (AP), and two at El Kabrit on 12 November 1984 (RvD *et al.*).

Nile Valley Meinertzhagen (1930) mentioned that as many as 50 were seen on Lake Qarun on 20 January [1920], and one was collected (BMNH). Koenig (1932) observed several and collected four in the 1890s. Von Besserer (1908) also mentioned it as common in this area. The only more recent record from the Faiyum is two on 19 March 1986 (JW). There are several recent records from Wadi el Rayan: at least five between 6 and 9 December 1985 (DM) and six between 19 and 31 December 1984 (SMB). There are only a few reports from the Nile Valley proper. Shelley (1872) collected one near El Kab (BMNH) and Meinertzhagen (1930) mentioned its occurrence in 'Upper-Egypt'. The only known recent records from the Nile Valley are one near Luxor on 30 November 1982 (JDG), one between El Kab and Idfu on 4 February 1978 (Short and Horne 1981), and one between Qena and Dishna on 9 February 1986 (FG).

Red Sea The Red Sea coast is the only area in Egypt where this species is still observed with some regularity. The majority of records are from the Gulfs of Aqaba and Suez, where there are distinct influxes of birds in November and between mid-February and late March, suggesting passage to and from areas further south. This species was recently discovered wintering on inland Ethiopian lakes (Ash and Ashford 1977) and these birds may well reach this area via the Red Sea. The earliest autumn record is one on 1 October 1984 at Suez (RvD *et al.*). There are a few winter records—two on 25 December 1987 near Hurghada (ADB, AJA), and one on 1 January 1979 on Gezira Fara'un and one the following day between Nuweiba and Dahab (BCD). Three were seen at Suez on 29 March 1982 (Wimpfheimer *et al.* 1983). From mid-February to late March 1985 small numbers were recorded along the entire Red Sea coast (SMG, PLM, RWS). The latest spring records are single birds on 28 May 1916 at El Kubri (Boyd 1917) and 23 May 1984 on Gezira Abu Minqar (Goodman and Storer 1987). At Elat, Israel, this species is uncommon between January and early May (Krabbe 1980).

North Sinai This species was not recorded during the numerous systematic autumn bird counts near Zaranikh. The only record appears to be one adult on 7 February 1979 near Misfaq (EB).

Mediterranean Gull *Larus melanocephalus* WV

Local name Nuuras al-bahr al-mutawassat نورس البحر المتوسّط.

Scarce winter visitor from (late August) mid-September to late March (mid-April), mainly along the Mediterranean coast. Rare in summer. Monotypic.

North Sinai and Nile Delta Small numbers have been regularly recorded between mid-November and mid-March along the Mediterranean coast of the Nile Delta and north Sinai. The largest concentration noted in recent years is 230 near Port Said on 26 and 27 December 1979, composed mainly of first-year birds (PLM). Surprisingly, no Mediterranean Gulls have been observed during the September and October systematic bird counts at Zaranikh. Thus, it is presumed that winter visitors arrive in the late autumn. The only summer record appears to be one adult in summer plumage and four sub-adults near Port Said on 28 June 1977 (PLM).

Nile Valley The Mediterranean Gull is rare on inland waters and recent records include: one on 24 August 1984 west of Cairo (SH, DM); one immature on 24 December 1983 at Aswan (PE); one on 11 January 1979 on Lake Qarun (Meininger *et al.* 1979); and three adults on 8 March 1982 at Aswan (ES, HWu).

Red Sea There are several reports from the northern Red Sea and the species is probably a regular winter visitor in small numbers to the Gulfs of Aqaba and Suez. The earliest local autumn record is 5 September 1981 at Suez (RGB *et al.*). On 1 and 2 January 1979, seven were seen along the northern Gulf of Aqaba coast of Sinai (BCD). We are not aware of any records from the southern Egyptian Red Sea coast, although the species has recently been observed in spring along the Sudanese coast (Nikolaus 1984). The latest spring record for the country is three immatures in Suez Bay on 16 April 1958 (Fischer 1959). This species is a scarce spring migrant at Elat, Israel, from late April to early June (Krabbe 1980).

Ringing There are 13 recoveries from Egypt of birds ringed in the northern Black Sea, USSR (Shevareva 1955; GZF).

Little Gull *Larus minutus* WV

Local name Nuuras saghiir نورس صغير.

Scarce winter visitor from (late August) mid-September to early May mainly along the Mediterranean coast. Monotypic.

Mediterranean Coast and Nile Delta and Valley Little Gulls are regularly recorded between October and late April on coastal lakes and along the Mediterranean coast of the Nile Delta. An exceptionally early autumn record is a flock in Alexandria harbour on 25 August 1966 (Kattinger 1970). There are several records from September of small flocks of Little Gulls migrating along the north coast of Sinai (YP, UP; Petersen and Sørensen 1981*a*, 1981*b*). The largest concentration noted in recent years is 130 (40 per cent immature) at Lake Idku on 28 January 1979 (Meininger *et al.* 1979); 260 at Lake Burullus on 7 February 1986 (Meininger and Baha el Din 1986); and 30 in the Alexandria harbour on 9 February 1982 (GM). One was collected at Salum in March 1921 (GZM). The latest spring records are an adult and immature on 5 May and one immature on 6 May 1983 on Lake Manzala (PLM, UGS). This species is rarely recorded inland and the following records are known: an immature on 29 September 1985 between El Minya and Beni Hassan (Hume 1985); two (one immature) on 15 November 1977 on Lake Qarun (Short and Horne 1981); one immature on 13 December 1980 on Dahshur Lake (RW); one (collected) in January 1904 on Lake Qarun (Russell 1905); one immature on 12 January 1979 in Wadi el Natrun (Meininger *et al.* 1979); and one on 4 February 1977 at Asyut (Short and Horne 1981).

Red Sea and southern Sinai One was reported at Sharm el Sheikh on 15 September 1981 (MW) and three on 18 September 1980 (DJM, DES). One adult was noted at Suez on 11 October 1981 (CAB *et al.*). At Suez, during virtually continuous observations, single birds or pairs were seen at irregular intervals between 4 April and 9 May 1982 (Wimpfheimer *et al.* 1983). The only record from the Egyptian Red Sea proper is three near Hurghada on 28 April 1982 (Ferguson 1984). The species has also been recorded along the Sudanese Red Sea coast (Madden 1927). At Elat, Israel, this species is an uncommon migrant and winter visitor from November to late May (Krabbe 1980).

[**Sabine's Gull** *Larus sabini* AV?

There are two recent sight records of the Sabine's Gull at Zaranikh; one was observed on 29 September 1973 (Hovel 1987) and an adult on 17 September 1980 (Petersen and Sørensen 1981*a*). A specimen was taken at Tel Aviv, Israel, on 15 January 1939 (TAUM).]

Black-headed Gull *Larus ridibundus* PV WV

Local name Nuuras aswad ar-ra-s نورس أسود الرأس.

Abundant passage and winter visitor from (late August) mid-September to early May. Scarce non-breeding summer visitor. Monotypic.

In mid-September migrant Black-headed Gulls begin to arrive in the Nile Delta and Valley, and along the Red Sea coast, but substantial numbers are not present until November. Spring migration peaks in March and continues into May. In the summer months small numbers, mainly immatures, are observed on Lake Qarun, in the Nile Valley south to Lake Nasser, along the Red Sea coast, and somewhat larger numbers on the Delta lakes. Presently, this species appears to be more common than in the late nineteenth century. This is in accordance with the recent increase in the European breeding populations and the corresponding number wintering in north-eastern Africa (Cramp and Simmons 1983; Nikolaus and Hamed 1984).

Western Desert Small groups or single birds have been recorded in most of the

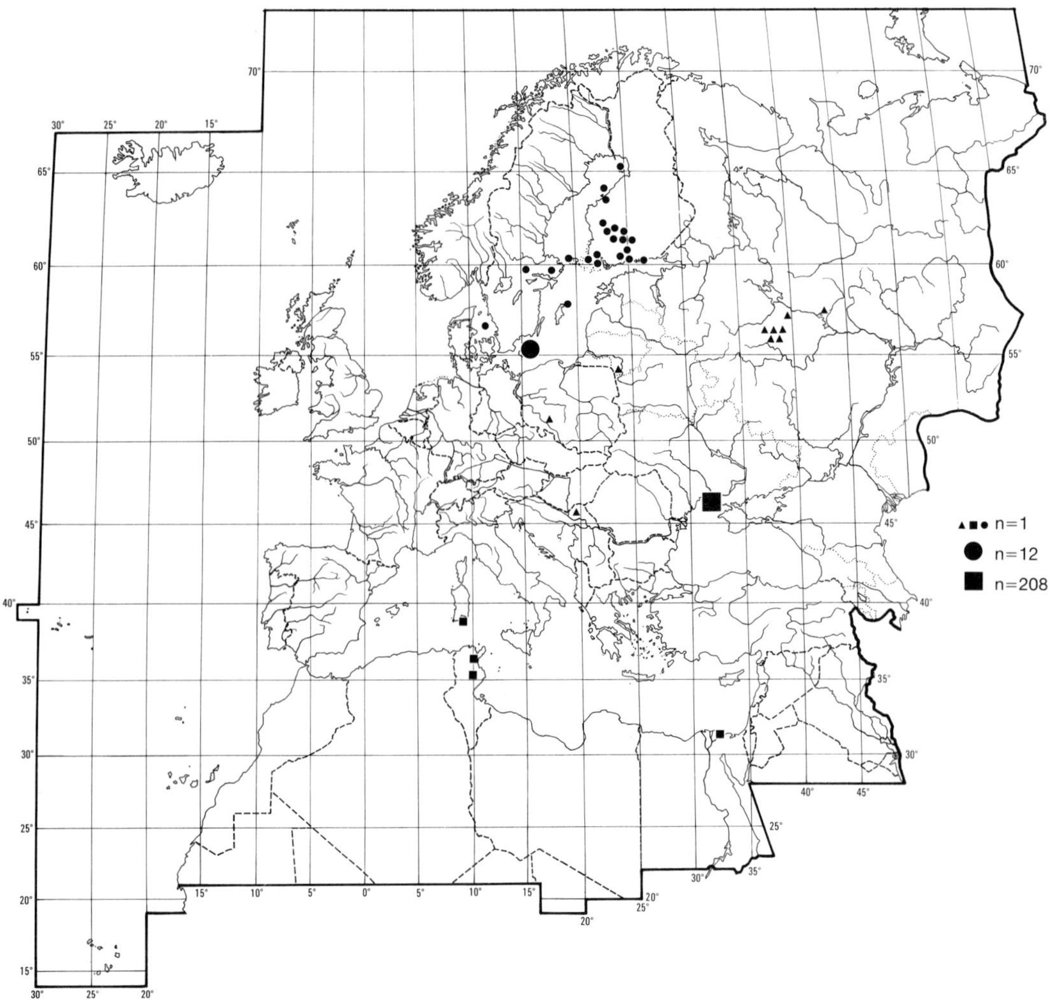

FIG. 6.21. Origin of Black-headed Gulls (▲, n=10), Slender-billed Gulls (■ n=212), and Lesser Black-backed Gulls (●, n=34) recovered in Egypt. Most birds ringed as nestlings.

larger Western Desert oases between late September and late April (Goodman *et al.* 1986).

Nile Delta and Valley The largest concentrations of this species in Egypt are found on the Nile Delta lakes during the winter months. Counts include over 36 000 on Lake Burullus in January 1979 and 24 000 on Lake Manzala in January 1980 (Meininger and Mullié 1981*b*). There may well be several hundred thousand wintering in the Nile Delta. Thousands are also present in winter and spring in the Suez Canal area, on Lake Qarun, and in the Nile Valley south to Lake Nasser, but the numbers vary from year to year. Between 12 and 23 March 1983, 10 000 were seen between Cairo and Aswan (Welch 1983). About 120 were noted at Aswan on 2 May 1978 flying north (PLM, BCD).

Red Sea This species is observed throughout the winter along the entire Red Sea coast. The largest concentrations are found in the Gulfs of Suez (up to 1000) and Aqaba, and numbers generally decrease towards the south. About 200–300 were present at Suez in early May 1982 (Wimpfheimer *et al.* 1983).

Sinai Relatively small numbers have been observed migrating along the north coast of Sinai in September (Petersen and Sørensen 1981*a*, 1981*b*; Baha el Din and Salama 1984) and October (Petersen and Sørensen 1980).

Ringing There are 21 recoveries from the Nile Delta and one from Upper Egypt of birds ringed in Lithuania; near Moscow, USSR; in Poland; and Yugoslavia (Fig. 6.21; Mackintosh 1941; Ptusjenko 1948; Stromar 1971; Watson 1971; EDB; GZF; RP).

Slender-billed Gull *Larus genei* RB PV WV

Local names Nuuras qarqaṯii نورس قرقطي, aruug (ND).

Scarce breeding resident in north-western Sinai, common passage and winter visitor, and fairly common non-breeding summer visitor. Monotypic.

El Negumi *et al.* (1950) mentioned a breeding colony of Slender-billed Gulls on Gezira Dib in El Malaha, east of Port Said (Abdel Megid 1945). On 4 July 1938 El Negumi and Abdel Megid ringed 32 nestlings at this locality (GZF). In 1979 eggs were examined that had been taken from 200–400 nests in El Malaha (Meininger and Mullié 1981*a*). During an aerial survey of El Malaha on 4 July 1986 up to 2000 adults were observed, including concentrations of several hundred birds on islands (DM, WCM), and it seems likely that this species still breeds in the area. At Lake Qarun over 250 adults were present on 31 May 1986 (DM, WCM) and up to 30 in June and July 1984 (SH, DM). In 1986 fishermen from the Faiyum accurately described the eggs of this species (DM, WCM) and local breeding cannot be excluded.

The principal habitat of this species appears to be saline and coastal waters, e.g. the Mediterranean and Red Seas, and Lake Qarun. It is occasionally observed on the Nile; there are no records from the Western Desert.

Between May and August considerable numbers are present on the lakes of the Nile

Delta. In July and August 1979 at least 2000 were seen on Lake Manzala (EM). Small numbers, including adults, summer along the Suez Canal and in the Gulfs of Aqaba and Suez, but it is rare further south along the Red Sea. On 29 July 1981, three were noted at Hurghada (CB, GT).

Autumn migrants begin to arrive in August. By September this species is one of the most abundant gulls along the northern coast of Sinai. During systematic counts at Zaranikh 944 were noted between 13 August and 29 September 1973 and 1123 between 15 August and 5 October 1978 (YP, UP); 1638 between 7 and 30 September 1980 (Petersen and Sørensen 1981*a*); 2011 between 16 August and 24 September 1981 (Petersen and Sørensen 1981*b*); and 515 between 2 and 24 September 1982 (Baha el Din and Salama 1984). There is a clear influx of migrants in September and October along the Red Sea coast and the numbers in the Nile Delta and on Lake Qarun also show a marked increase.

In midwinter the largest numbers are present on the Nile Delta lakes and El Malaha, with probably over 10 000 occurring locally. Considerable numbers winter on Lake Qarun, in the Suez Canal area, and in the Gulf of Suez. The largest midwinter counts are 500 on 13 December 1952 at El Kabrit (AP), 5300 on 9 January 1980 at El Malaha, 640 in early January 1980 on Lake Manzala (Meininger and Mullié 1981*b*), 500 on 10 January 1976 on Lake Burullus (MSm), 1240 on 10 January 1979 on Lake Qarun (Meininger and Mullié 1981*b*), and 410 on 11 December 1980 at Suez (SMB).

Spring movements, which are most notable along the Red Sea coast, begin in February, climax in March and April, and continue into early May. During peak migration up to several hundred birds are noted per day at Suez (e.g. Wimpfheimer *et al.* 1983).

The first known record from Lake Qarun appears to be one taken on 27 January 1938 (HGD). This lake has become saline only during the past few decades and the substantial increase in the number occurring there appears to reflect colonization of the new habitat. There are only a few records from the Nile Valley, particularly along the upper section.

Ringing Egypt is an important wintering area for Slender-billed Gulls breeding in the Black Sea (Fig. 6.21). Over 200 birds ringed at Orlov, northern Black Sea, have been recovered in Egypt (Semenov and Sabinevskii 1957; Erard 1958; Watson 1971). There are also recoveries from Egypt of birds ringed in Tunisia: a chick ringed on 13 June 1972 was recovered on 2 May 1973 and a first-year bird ringed on 25 October 1971 was recovered on 15 March 1973 (EDB). One bird ringed on Sardinia, Italy, on 1 June 1978 was recovered at Port Said on 6 January 1984 (MSp). One of the nestlings ringed at El Malaha on 4 July 1938 was recovered in February 1940 at Lake Manzala (GZF).

Audouin's Gull *Larus audouinii* (PV)

Local name Nuuras udwin نورس أودن.

Rare and irregular passage visitor. Monotypic.

The following records are known from Egypt: Lake Manzala—one adult on 9 November 1981 (CAB *et al.*); Lake Qarun—one shot on an unspecified date (Koenig 1932); Suez Bay—one adult on 6 September 1981 (CAB *et al.*); Zaranikh—one first-year bird on 16 September 1980 (Petersen and Sørensen 1981*a*) and another on 6 and 9 September 1981 (Petersen and Sørensen 1981*b*); and Lake Bardawil—one on 22 March 1976 (SC). Phillips (1915) collected an immature female near Aqaba, Jordan, on 21 April 1914 (MCZ). This is the only known record from the Gulf of Aqaba.

Common Gull *Larus canus* WV

Local names Nuuras shaa-i' نورس شاع, zummaj al-maaya.

Rare winter visitor to the north of the country from mid-September to late April. All Egyptian material examined is referable to *L. c. canus*.

Nicoll (1908) found the Common Gull 'fairly numerous' on Lake Manzala in January 1908. Meinertzhagen (1930) mentioned it as 'not uncommon in winter on the coasts and inland waters'. It cannot be excluded that some of the sight records of immature Common Gulls reported herein are actually of Mediterranean Gulls.

Nile Delta and Valley Nowadays the Common Gull is rare in winter on the Nile Delta lakes and along the Mediterranean coast. It is occasionally noted on Lakes Idku, Manzala, and Burullus. The largest concentration reported in recent years is 10 on 25 December 1980 between Port Said and Damietta (PLM, WCM). The only recent inland records are from Lake Qarun and include a few noted on 14 February 1976 (Short and Horne 1981) and one on 12 January 1979 (Meininger *et al.* 1979). The latest spring record is a single bird on 22 April 1964 near Matruh (HCK).

Red Sea Meinertzhagen (1930) considered the species scarce in the Gulf of Suez. The earliest autumn record for the country is a second-year bird on 4 September 1981 at Suez (CAB *et al.*). On 31 October 1982 an immature was noted 10 km south of Ain Sukhna and on 25 November 1982, three were observed 14 km north of Hurghada (Baha el Din and Saleh 1983). At Elat, Israel, from 1966 to 1979 it has been recorded 25 times between September and March (Krabbe 1980).

Lesser Black-backed Gull *Larus fuscus* PV WV

Local names Nuuras dughba نورس دغبة, guukaa.

Common passage migrant throughout the country and fairly common winter visitor from early September to late May. Rare summer visitor. All Egyptian material examined is referable to *L. f. fuscus*.

Autumn passage of Lesser Black-backed Gulls begins in early September, peaks in October, and continues into November. Flocks of migrants concentrate along the north coast of Sinai. During systematic counts of migrating water-birds at Zaranikh 1156 Lesser Black-backed Gulls were noted between 15 August and 5 October 1978 (YP, UP); 539 between 7 and 30 September 1980 (Petersen and Sørensen 1981*a*); and 634 between 16 August and 24 September 1981 (Petersen and Sørensen 1981*b*). Concentrations of migrants are also noted along the Nile Valley, Suez Canal, and Red Sea.

In midwinter hundreds are present on the lakes of the Nile Delta and along the Mediterranean coast. Small numbers frequent Wadi el Natrun, Lake Qarun, and the entire Red Sea coast. The species is a rare winter visitor to the Nile Valley (south to Aswan).

During spring migration, from late March to early May, the Lesser Black-backed Gull is more abundant than in autumn. Passage takes place along a broad front, including the Western Desert and the mountains of south Sinai. Numerous small flocks come north from the Sudan via Lake Nasser and the Nile Valley. In the Red Sea, migrating flocks do not always follow the coastline. Significant numbers have been recorded at the Daedalus Reef in March and early July (Bulman and Beale 1944). Concentrated movements of spring migrants are also noted at the northern ends of the Gulfs of Suez and Aqaba. Near Suez, 600 birds were observed on 26 March 1982 (Wimpfheimer *et al.* 1983) and at the Great Bitter Lake a flock of about 1000 on 8 April 1945 (LB). This species is a common migrant at Elat, Israel, where an estimated 20 000–50 000 birds pass each spring (Safriel 1968).

During the summer months small numbers, mainly immatures, are found along the Mediterranean coast, on the Nile Delta lakes, in the Suez Canal area, on Lake Qarun (Meininger and Dielissen 1979), and along the Red Sea coast (Goodman and Storer 1987).

Ringing A total of 40 birds has been recovered in Egypt that were ringed in Denmark, Finland, and Sweden (Fig. 6.21). Additionally, 12 birds with Rossitten rings were recovered in Egypt for which the exact ringing localities are unknown (Mackintosh 1941; Watson 1971; EDB; GZF).

Yellow-legged Gull *Larus cachinnans* RB WV

Local names Nuuras asfar al-qadam نورس أصفر القدم, ruunis (NC).

Local breeding resident along the western Mediterranean coast and scarce non-breeding summer visitor to other portions of the country. Fairly common winter visitor, mainly in the north. All Egyptian material examined is referable to *L. c. cachinnans* and presumably birds of this subspecies are winter visitors to the country. However, the populations breeding along the Mediterranean coasts of Europe and north-west Africa are referable to *L. c. michahellis*, and it is presumably this form that breeds in Egypt. *L. c. armenicus* was reported in the springs of 1982 and 1983 from along the Israeli Mediterranean coast (Géroudet 1982; Dubois 1985) but there is no evidence that this form reaches Egyptian waters.

In February 1980 pairs of Yellow-legged Gulls, apparently on territory, were observed in a lagoon east of El Alamein (PI). This site was revisited on 7 May 1985 and a breeding colony of 10–15 pairs was found on an island (SMG, PLM) and again on 17 May 1986 when 25 pairs were counted (MAM). Local inhabitants also reported that they collect gull eggs in a near-by sand-dune area (SMG, PLM). This is the only place in Egypt where this species is known to breed.

The Yellow-legged Gull is a fairly common winter visitor to the Mediterranean coast, Nile Delta lakes, and the Suez Canal area, mainly from mid-August to mid-April. Small numbers remain during the summer months. A few winter on Lake Qarun and occasionally in the Nile Valley south to Aswan.

This species is a regular migrant to the Red Sea, particularly in the north. There are some yearly fluctuations in the number wintering in the Gulf of Suez (Marchant 1941), although it appears to be regular at Suez between mid-September and mid-March. Small numbers are occasionally observed further south between Hurghada and Wadi Gimal (Goodman and Storer 1987). Krabbe (1980) considered it a fairly common migrant and winter visitor to Elat, Israel.

Kittiwake *Rissa tridactyla* (WV)

Local name Nuuras aswad al-qadam نورس أسود القدم.

Rare and irregular winter visitor, mainly to the north of the country. Monotypic.

Boyd (1917) observed a few immature Kittiwakes at Port Said on 15 March 1916. In September 1978, three were seen along the north coast of Sinai (YP). A dead immature bird, in a partially decayed state, was found along Lake Burullus, near Baltim, on 10 May 1985 (skull in UMMZ). The only other known records from Egypt are from the Suez Canal area, and include an adult flying south near Suez on 25 October 1981 (CAB *et al.*) and five near El Shatt on 15 October 1983 (SMB). At Elat, Israel, this species has been recorded 46 times from November to mid-April in five years between 1966 and 1979 (Krabbe 1980).

Family Sternidae

Generic local names for tern Khuttaaf, kharshana, marshak.

Gull-billed Tern *Gelochelidon nilotica* PV WV

Local names Khuttaaf nuurasii al-minqaar خطاف نورسي المنقار, luwayq, uwayq.

Scarce passage visitor in autumn from late August to mid-October and in spring from early March to mid-May (early June). Rare winter visitor to the Nile Delta and Valley. The subspecies *G. n. nilotica* is the one occurring in the country.

Meinertzhagen (1930) considered the Gull-billed Tern a common winter visitor to

Egypt, including Sinai, on both inland and coastal waters. He stated that it was sometimes abundant near Suez and in the Faiyum and probably the most common species of tern in Egypt. These remarks are striking, as in the past few decades the Gull-billed Tern has not been commonly observed in Egypt. This decrease is presumably related to the recent decline of breeding birds in the USSR, eastern Europe, and Turkey (Møller 1975).

Nile Delta and Valley In the Nile Delta, including the Suez Canal area, this species is generally recorded from early September to mid-May. Ninety-two migrants were noted between 28 April and 1 May 1952 at Abu Sultan (AP). Exceptionally late records include 18 on 1 June 1952 at Deversoir (AP) and three on 9 June 1984 at Lake Timsah (BH *et al.*). The only relatively recent winter records are one on 9 December 1985 at Aswan (HD), 10 on 16 December 1952 at Abu Sultan (AP), and nine on 10 January 1980 at El Malaha (Meininger and Mullié 1981*a*).

In the Nile Valley, including the Faiyum, the only recent observations are small numbers during migration. Autumn records range from six on 29 September 1981 at Aswan (FdR) to 10 on 10 October 1983 at Kom Ombo (Hume 1983), and spring records from three on 5 March 1979 at Kom Ombo (Short and Horne 1981) to one on 6 May 1978 on Lake Qarun (Meininger and Dielissen 1979).

Red Sea The Gull-billed Tern is a scarce migrant in the northern Red Sea. There are a few records in September from Suez and along the Sinai coast of the Gulf of Aqaba. One was taken at El Tor on 23 August 1979 (Hovel 1987). Further south along the Egyptian Red Sea there are only a few recent reports, e.g. one on 15 May 1982 at Hurghada (Wimpfheimer *et al.* 1983) and two on 6 April 1985 near Shalatein (SMB, SMG, PLM). At Elat, Israel, small numbers are regularly observed from early March to late May (Krabbe 1980).

North Sinai This species is a regular autumn migrant along the north coast. Counts at Zaranikh include: 211 from 13 August to 29 September 1973 and 410 from 15 August to 5 October 1978 (YP, UP); 91 from 7 to 30 September 1980, with 24 on 8 September (Petersen and Sørensen 1981*a*); 198 from 16 August to 24 September 1981, with 33 on 22 September (Petersen and Sørensen 1981*b*); and 54 from 2 to 24 September 1982, with 40 on 2 September (Baha el Din and Salama 1984).

Generic local names for *Sterna* spp. H̱adma (Ab), haliish (NC), simnaaf (ND).

Caspian Tern *Sterna caspia* RB PV WV

Local names Khuṯṯaaf abuu balaẖa خطاف ابو بلحة, abuu balẖa, abuu jirra, daghaab (Ab).

Locally common breeding resident on islands and along the coast of the Red Sea. Scarce passage migrant and winter visitor to the Mediterranean coast, the Nile Delta, and the Red Sea coast from mid-August to mid-March (late May). Rare passage visitor to inland waters. Monotypic.

Nile Delta and Valley and North Sinai Along the Mediterranean coast and in the Nile Delta (including the Suez Canal area) the Caspian Tern is a regular migrant and winter visitor. Counts of birds passing Zaranikh include: 57 between 13 August and 29 September 1973 and 85 from 15 August to 5 October 1978 (YP, UP); 57 between 7 and 30 September 1980 (Petersen and Sørensen 1981*a*); 48 between 16 August and 24 September 1981 (Petersen and Sørensen 1981*b*); and 14 between 2 and 24 September 1982 (Baha el Din and Salama 1984). The largest winter count in the Delta is 32 on Lake Manzala between 26 and 29 December 1980 (PLM, WCM).

Inland this species is a rare migrant. According to Meinertzhagen (1930), 40 were seen during March in the Faiyum and there were October and March records from

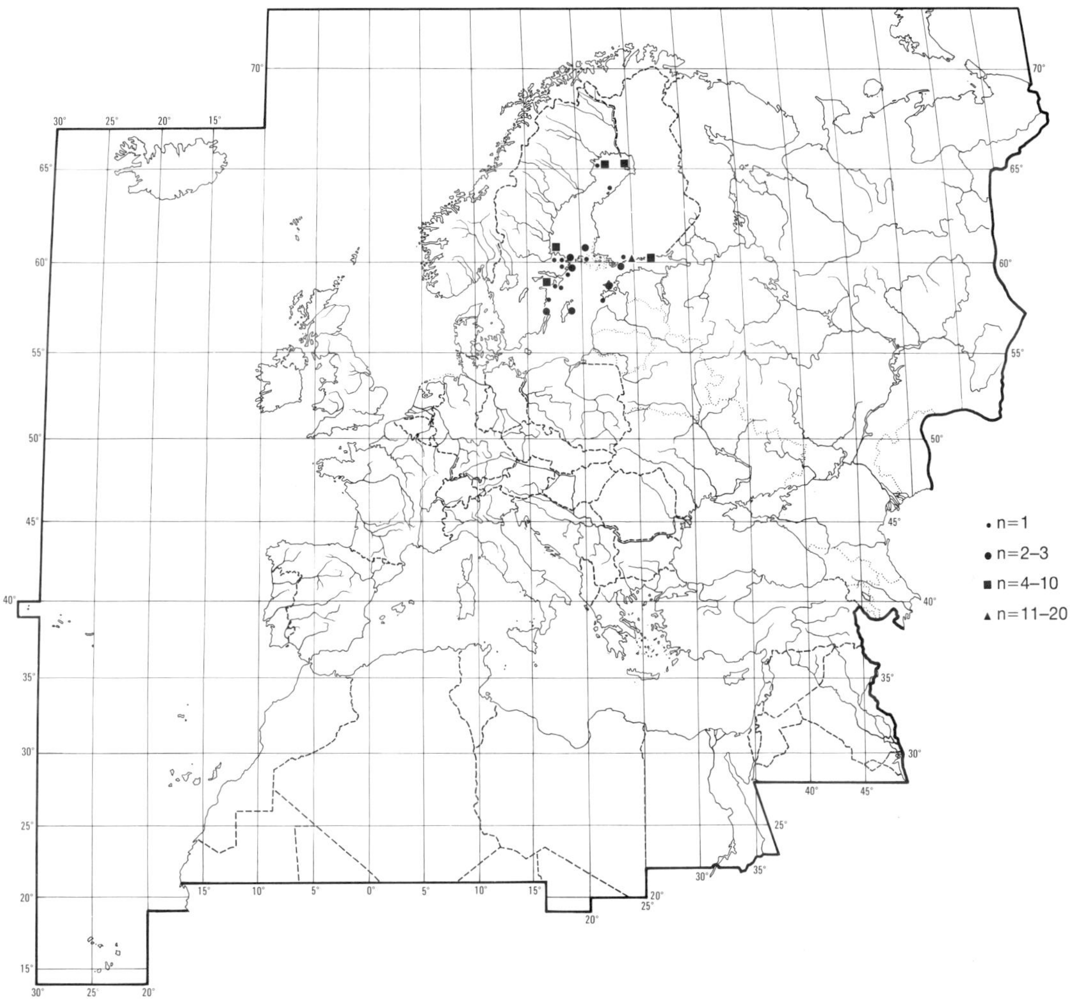

FIG. 6.22. Origin of Caspian Terns (n=75) recovered in Egypt. Most birds ringed as nestlings.

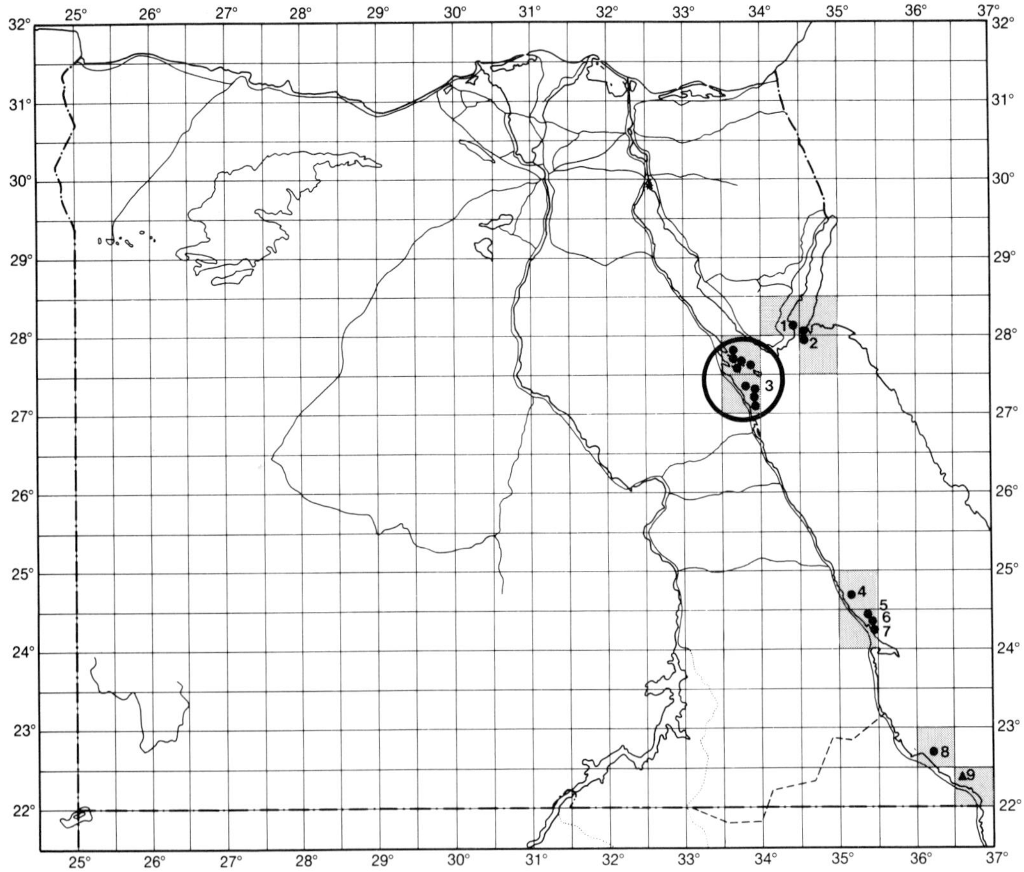

Map B39. Breeding distribution of Caspian Tern *Sterna caspia* between 1970–87.
Breeding localities with most recent estimate of breeding pairs: (1) Nabq (1979: 2–3). (2) Gezira Tiran (1973–79: 25–30). (3) Islands mouth Gulf of Suez (1983: 165). (4) Gezira Wadi Gimal (1982: at least 1). (5) Gezira Siyul (1983: 2). (6) Gezira Showarit (1983: 2). (7) Gezira Mahabis (1983: 10–15). (8) Gezira Siyal group (1985: at least 2). (9) Gezira Rawabel el Kabir (1985: at least 1).

Cairo. Koenig (1932) shot one at Kom Ombo on 29 March 1897. The majority of post-Meinertzhagen (1930) records from the Nile Valley are from the Faiyum, although four were noted on 19 March 1985 near Sohag (Grieve 1985) and three on 21 August 1980 at Aswan (JB). El Negumi *et al.* (1950) mentioned that this species bred in the Faiyum; however, the basis for this statement is unknown. There are old specimens taken in May 1875 at Alexandria and May 1878 at Damietta (BMNH).

Red Sea Caspian Terns can be observed throughout the year on islands and along the coast of the Red Sea. This species generally breeds in small colonies or singly, mainly on islands (see map for details). In spring 1983 an estimated 165 pairs were found breeding on the islands off Hurghada and at the mouth of the Gulf of Suez, including a colony of 110 on Gezira Ashrafi and 23 on Gezira el Hamra (Jennings *et al.* 1985). The total Egyptian breeding population is estimated to be between 250 and 350 pairs. Nests containing fresh eggs have been found between mid-February and late May.

Post-breeding long-distance dispersal of the Egyptian Red Sea breeding population appears to be minimal. Concentrations are noted mainly in the non-breeding season and in the general vicinity of nesting areas. For example, at Hurghada 51 were noted on 18 June 1980 and 74 on 30 January 1981 (SMB). Some birds appear to move into deeper waters as the species was noted at the Daedalus Reef on 13 July 1936 (Bulman and Beale 1944). There is no evidence of significant migratory movements of Eurasian breeding populations into the Red Sea, although there are recoveries of Scandinavian ringed birds at Suez.

Ringing There are about 76 recoveries in the Nile Delta and Valley, along the Mediterranean coast, and at Suez of birds ringed in Finland, Estonia, and Sweden (Fig. 6.22; Mackintosh 1941; Österlöf 1969, 1970; Watson 1971; Anon. 1973, 1976; Staav 1977; Kastepold and Kabal 1985; BEP; EDB; GZF; RSt).

Swift Tern (Crested Tern) *Sterna bergii* RB? WV Plate 6

Local names Khuṯṯaaf sarii' خطاف سريع, ẖiqṯ.

Presumed rare migrant or resident breeding bird on islands in the Red Sea. The subspecies occurring in Egyptian waters is *S. b. velox*.

There is no definitive evidence of the Swift Tern breeding in Egyptian waters. Abdel Megid (1945) found 'Crested Terns' nesting on Gezira Umm Agawish el Saghir in July 1943. This record may indeed be of the Swift Tern but cannot be accepted without further details. Marchant (1941) suspected that it bred near Ras Gemsa or on Gezira Gemsa. In late April 1983 Jennings *et al.* (1985) observed 30 Swift Terns on Gezira Ghanim and four on Gezira Ashrafi. They noted that 'small numbers are no doubt breeding in the area, but only circumstantial evidence was obtained'. Taylor (1867) reported wild bird eggs in the Damietta market that he believed to be of Swift Terns. This record is enigmatic and the possibility that the eggs were actually of another Sternid or Larid, e.g. Slender-billed Gull, cannot be excluded. The Swift Tern is known to nest on islands off the Sudanese Red Sea coast (Moore and Balzarotti 1983).

Small numbers have been recorded throughout the year along the complete Egyptian Red Sea coast, including the Gulfs of Suez and Aqaba. Concentrations of birds include 20 at Hurghada on 30 January 1981 (SMB), up to 20 in Gemsa Bay in January and February 1940 (Marchant 1941), 14 at Marsa Sha'b on 14 February 1984 (Goodman and Storer 1987), 15 at Gezira Ghanim on 14 March 1984 (PDH, EMN), 39 at Gezira Ashrafi on 27 September 1984 (Jennings *et al.* 1985), and 20 at Gezira Abu Minqar in early July 1951 (Searight 1952*c*). Meinertzhagen (1930) mentioned that one was shot at Port Said on 19 March, which is the only known record from the Mediterranean.

Lesser Crested Tern *Sterna bengalensis* MB PV Plate 6

Local names Khuttaaf mutawwaj خطاف متوّج, abuu qushash.

Fairly common but local migrant breeder on islands in the Egyptian Red Sea. Rare passage migrant along the Mediterranean coast, in the Nile Delta and Suez Canal area. Absent from Egyptian waters between late November and early March. There is a breeding population of *S. b. torressii* along the Mediterranean coast of Libya (Bundy 1976). One specimen collected in the Gulf of Suez has been identified as this subspecies (Cramp 1985). The birds nesting in the Red Sea are referable to *S. b. par*.

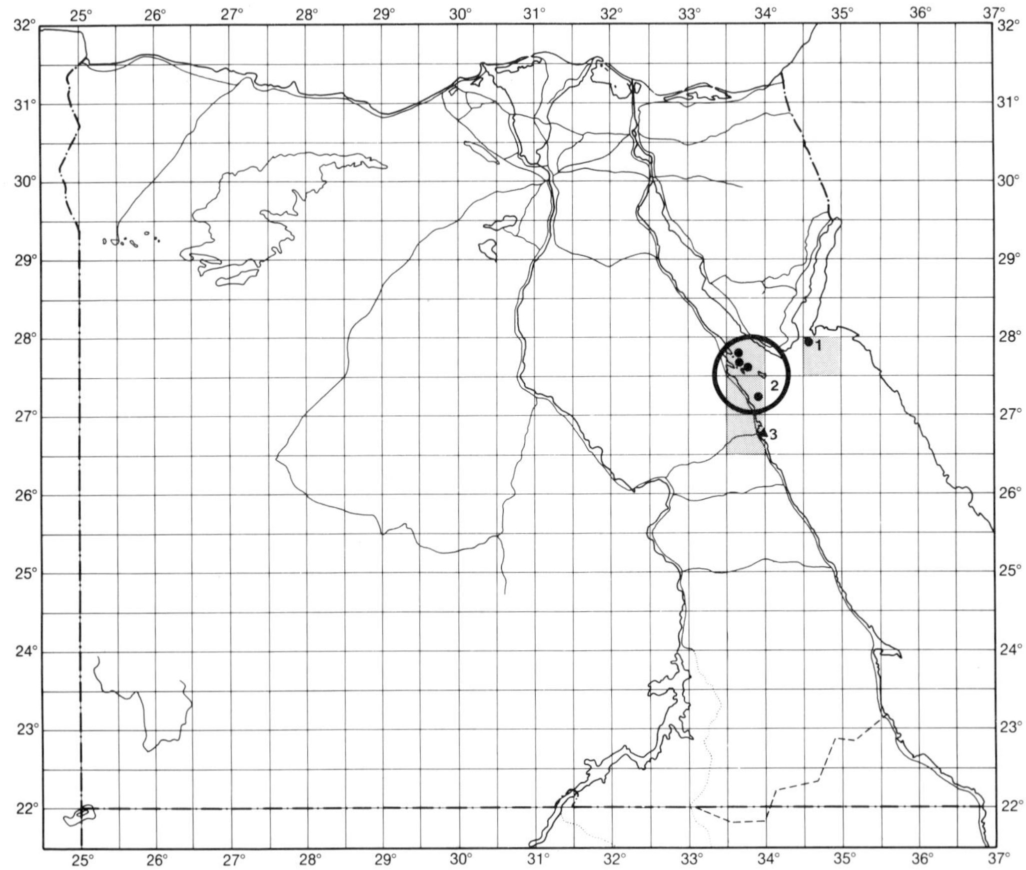

▲ probable breeding ● definite breeding ▒ half degree grid plots with probable or definite breeding

Map B40. Breeding distribution of Lesser Crested Tern *Sterna bengalensis* between 1940–87. *Breeding localities* (see text for details): (1) Gezira Tiran. (2) Islands mouth Gulf of Suez. (3) Gezira Safaga.

Nile Delta Meinertzhagen (1930) mentioned that specimens of the Lesser Crested Tern were collected at Port Said in May and at Alexandria in June. A single bird was seen at Lake Manzala on 10 October 1943 (Meiklejohn 1944) and one off north Sinai on 23 September 1981 (Petersen and Sørensen 1981*b*). Six were noted in the Suez Canal north of El Kubri on 9 June 1984 (BH *et al.*), which seems to indicate that some movements occur between the Red Sea and Mediterranean via the Suez Canal. However, the subspecies involved in these movements is unknown. One exceptional inland record is a bird captured on 26 June 1982 at Khartoum, Sudan (Nikolaus 1984).

Red Sea The Lesser Crested Tern breeds in colonies on islands in the Red Sea from July to October. Between 1973 and 1979 the population breeding on Gezira Tiran varied from 60 to 100 pairs (IBIC; SS). In late September 1984 an estimated 1000–1250 pairs nested on islands in the mouth of the Gulf of Suez and off Hurghada. The largest concentrations included 1800 adults on Gezira Ashrafi and 3000 adults on Gezira Umm el Heimat (Jennings *et al.* 1985). Bulman (1944) thought that this species was 'most probably breeding' on Gezira Safaga in early July 1942. The earliest spring record is two on 3 March 1984 at Ras Abu Darag (SG). Small numbers are noted throughout the summer and autumn along the entire Red Sea coast, including the Gulf of Suez. The species is rarely noted in the northern Gulf of Aqaba. An exceptional record is a flock of about 200 on 16 October 1976 off Sharm el Sheikh (LNA). The latest autumn record is two at Suez on 24 November 1984 (RvD *et al.*).

Sandwich Tern *Sterna sandvicensis* PV WV

Local name Khuttaaf kharshana خطاف خرشنة.

Fairly common passage visitor to the Mediterranean and Red Sea coasts, and fairly common winter visitor and scarce non-breeding summer visitor along the Mediterranean. *S. s. sandvicensis* is the form occurring in Egypt.

Mediterranean coast and Nile Delta From mid-August to mid-May Sandwich Terns are regularly noted along the Mediterranean coast and adjacent inland lakes. A total of 316 was counted at Zaranikh between 2 and 24 September 1982 (Baha el Din and Salama 1984). In December 1979 and January 1980 about 830 were recorded on El Malaha and Lake Manzala (Meininger and Mullié 1981*a*). Small numbers have been observed in late June and early July 1977 near Port Said and Alexandria (Meininger and Dielissen 1979) and on 25 June 1984 near El Arish (SMB).

Nile Valley The only known record is one observed (at close range) south of Nag Hammadi on 2 October 1985 (Hume 1985).

Red Sea The Sandwich Tern is generally observed in small numbers in autumn from early October to late November and in spring from late March to late May. Spring concentrations have been noted in the northern ends of the Gulfs of Suez and Aqaba. For example, a flock of several hundred was recorded in Suez Bay on 19 April (Meinertzhagen 1930).

Ringing A bird ringed on Orlov Island, Black Sea, on 21 June 1952 was recovered at Lake Manzala on 18 November 1952 (GZF).

[**Roseate Tern** *Sterna dougallii* AV?

Five Roseate Terns were observed on 6 May 1980 at Suez (SMB). The observation of two near Beni Suef on 16 March 1985 (Grieve 1985) is not acceptable. Sight records from Elat, Israel, include two on 7 April 1981 (Heinze 1981), three in mid-April 1971 (Krabbe 1980), and one immature between 4 and 20 November 1982 (Ben Dov and Golan 1983). There is no evidence that this species breeds in the Egyptian Red Sea (*pace* Etchécopar and Hüe 1967; Gochfeld 1983).]

Common Tern *Sterna hirundo* PV

Local name Khuttaaf al-bahr خطاف البحر.

Fairly common passage visitor along the Mediterranean and Red Sea coasts and the Nile in autumn from early August to mid-October (early November) and in spring from mid-March to mid-June. Scarce non-breeding summer visitor to the Mediterranean coast, the Gulf of Suez, and northern Red Sea. *S. h. hirundo* is the form occurring in the country.

The Common Tern is a regular autumn migrant along the Mediterranean coast from early August to mid-October. At Zaranikh autumn counts of this species include: 3794 between 13 August and 29 September 1973 and 2769 between 15 August and 5 October 1978 (YP, UP); 3964 between 7 and 30 September 1980 (Petersen and Sørensen 1981*a*); 12 433 between 16 August and 24 September 1981 (Petersen and Sørensen 1981*b*); and 2898 between 2 and 24 September 1982 (Baha el Din Salama 1984). In September and October small numbers pass through the Suez Canal, Nile Valley, and along the Red Sea. The only known record from the Western Desert is 20 in Dakhla on 7 September 1984 (Goodman *et al.* 1986). The latest known autumn report is one south of Quseir on 7 November 1982 (Baha el Din and Saleh 1983).

This species appears to be virtually absent from Egyptian waters between early November and mid-March. The only recent winter record is one on 10 February 1977 at Kom Ombo (Short and Horne 1981). The earliest spring records are one on 14 March 1983 near Asyut (Welch 1983) and several on 15 March 1981 at Port Said (SMB, BB). From late March to mid-June considerable numbers migrate through the Red Sea and concentrations of up to several hundred regularly appear at the northern ends of the Gulf of Suez and Aqaba, and along the Suez Canal. Up to 1000 were noted at Hurghada between 20 and 22 April 1984 (MBe, SM). Small numbers are observed in spring along the Nile Valley and the Mediterranean, and in summer along the northern Red Sea coasts.

White-cheeked Tern *Sterna repressa* MB Plate 6

Local name Khuttaaf abuu batn خطاف ابو بطن.

Fairly common migrant breeding bird on islands in the Red Sea. Occasionally observed in the Suez Canal. Between late November and late March this species seems to be completely absent from Egyptian waters. Monotypic.

The White-cheeked Tern is regularly observed along the entire Red Sea coast between mid-April and early November, and breeding takes place between early July and mid-September. Between 1973 and 1979 the breeding population on Gezira Tiran was generally about 50 pairs, although 150 pairs bred in 1978 and it was completely absent in a few years (IBIC; SS). The total population breeding in 1984 on the islands

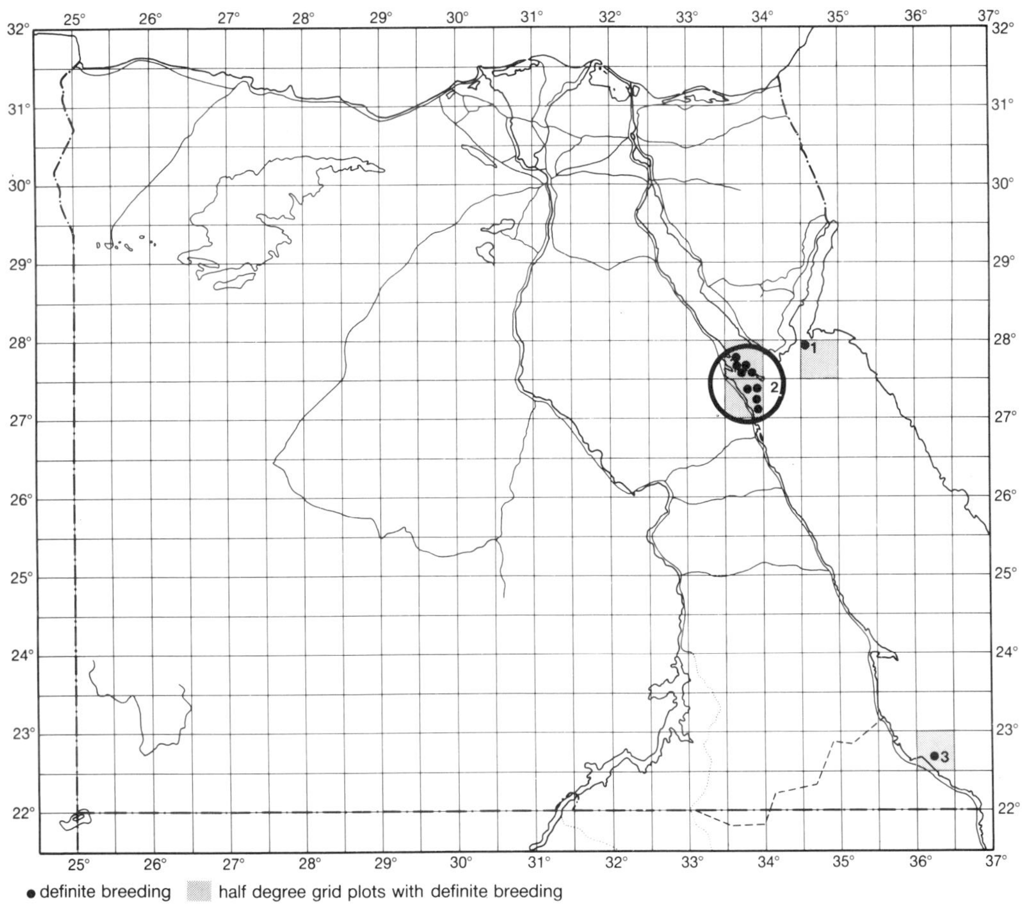

Map B41. Breeding distribution of White-cheeked Tern *Sterna repressa* between 1970–87. *Breeding localities* (see text for details): (1) Gezira Tiran. (2) Islands mouth Gulf of Suez. (3) Gezira Siyal group.

in the mouth of the Gulf of Suez and off Hurghada was tentatively estimated at between 1000 and 1500 pairs (Jennings *et al.* 1986). The largest colonies found in 1984 were about 100 adults on Gezira Siyul el Kabir and 900 on Gezira Fanadir. Hundreds of White-cheeked Terns were present on islands in the Gezira Siyal group, off Adal Deib, on 25 April 1985 and local fishermen accurately described their nest and stated that hundreds of pairs bred there annually (SMG, PLM).

The earliest spring record from the area is two on 21 March 1983 off Shalatein (Goodman and Atta 1987). They occasionally occur in large numbers north of known breeding sites e.g. 150 at Ras Muhammad on 1 April 1979 (Madge 1982) and up to 250 in Suez Bay in late April 1978 (RJM, LRP).

Concentrations of up to several hundred occur in autumn, as late as mid-November, along the Red Sea coast south of Hurghada. Small numbers have been recorded in the Gulfs of Suez and Aqaba in summer and autumn, with a maximum of 40 at Suez on 10 November 1984 (RvD *et al.*). There are a few records from the Suez Canal area. These include single birds on 3 April 1983 at Suez flying north in a flock of Black-headed Gulls (SC) and on 9 June 1984 at Ismailiya flying north (BH *et al.*).

Bridled Tern *Sterna anaethetus* MB Plate 6

Local name Khuṯṯaaf as̱ham خطاف أسحم.

Fairly common migrant breeding bird on islands in the Red Sea. Generally present in Egyptian waters between early May and late September. All Egyptian material examined is referable to *S. a. anaethetus*.

Bridled Terns breed in colonies on islands in the Egyptian Red Sea. Eggs appear to be laid between May and July. In 1984 an estimated 200–300 pairs bred on several islands in the mouth of the Gulf of Suez and off Hurghada. The largest colony of about 100 pairs was on Gezira Ashrafi, and most nests were placed in rock crevices and under bushes (Jennings *et al.* 1985). Between 12 and 20 pairs bred on Gezira Umm Agawish el Kabir in 1951 (JCC). Ticehurst (1924) reported that 'thousands' bred on the Brothers Reef; there is no recent information from this island.

From early May to September this species is regularly recorded in small numbers along the Red Sea coast. It is rare in the Gulfs of Suez and Aqaba. Four were observed near Suez on 9 June 1984 (BH *et al.*). The species may winter in the southern Egyptian Red Sea, as about 20 were noted on Gezira Rawabel el Kabir on 24 February 1985 and one at Shalatein on 21 March 1983 (SMG, RWS). Between 1966 and 1979 it was not recorded at Elat, Israel (Krabbe 1980). In more recent years it has been observed at this locality between late July and late August with up to 29 birds on 12 August 1980 (Ben Dov and Golan 1983).

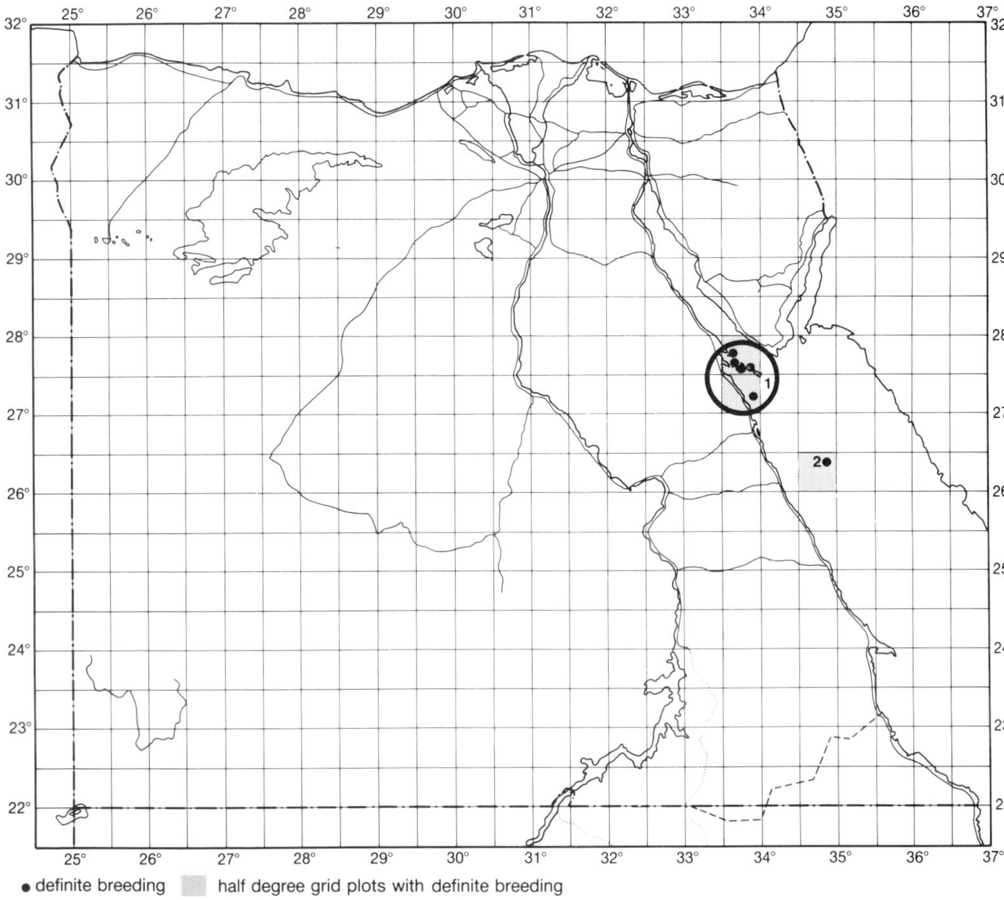

Map B42. Breeding distribution of Bridled Tern *Sterna anaethetus* between 1920–87.
Breeding localities (see text for details): (1) Islands mouth Gulf of Suez. (2) Brothers Reef.

Little Tern *Sterna albifrons* MB PV

Local names Khuttaaf saghiir خطاف صغير, abuu dagga (NC), dughbaz, marshak.

Fairly common migrant breeder along the Mediterranean coast and in the Faiyum. Fairly common passage migrant along the Mediterranean and Red Sea coasts and rare on inland waters. This species is absent from the country between early November and mid-March (*pace* Muselet 1985). All Egyptian material examined is referable to *S. a. albifrons*. Eggs are generally laid between late March and June.

The Little Tern is a fairly common migrant breeder along the Mediterranean coast. The total Egyptian breeding population is tentatively estimated at 2000 pairs. In the Nile Delta, colonies are concentrated along the coastal plain and on islands in the large lakes; the local breeding population may exceed 1000 pairs (Meininger *et al.* 1986).

Between 800 and 1200 pairs are thought to nest in northern Sinai (Dunnet *et al.* 1986; Meininger and Baha el Din 1986). During a 1986 (incomplete) survey of Lake Bardawil about 700 pairs were located (Atta 1986). A few pairs also nest (irregularly?) along the Great Bitter Lake (DM; RN). On 2 August 1985, 100 Little Terns, including recent fledglings, were observed on the eastern shore of Lake Qarun (SMB). Eggs were also found at this locality in June (year unspecified) (El Negumi *et al.* 1950). The breeding of this species along Lake Qarun seems to be variable and in some years apparently does not take place.

During autumn migration, from mid-August to late October, the Little Tern is particularly common along the north coast of Sinai. Counts of migrating Little Terns at Zaranikh include: 659 between 13 August and 29 September 1973 and 1810 between 15 August and 5 October 1978 (YP, UP); 721 between 7 and 30 September 1980 (Petersen and Sørensen 1981*a*); 1347 between 16 August and 24 September

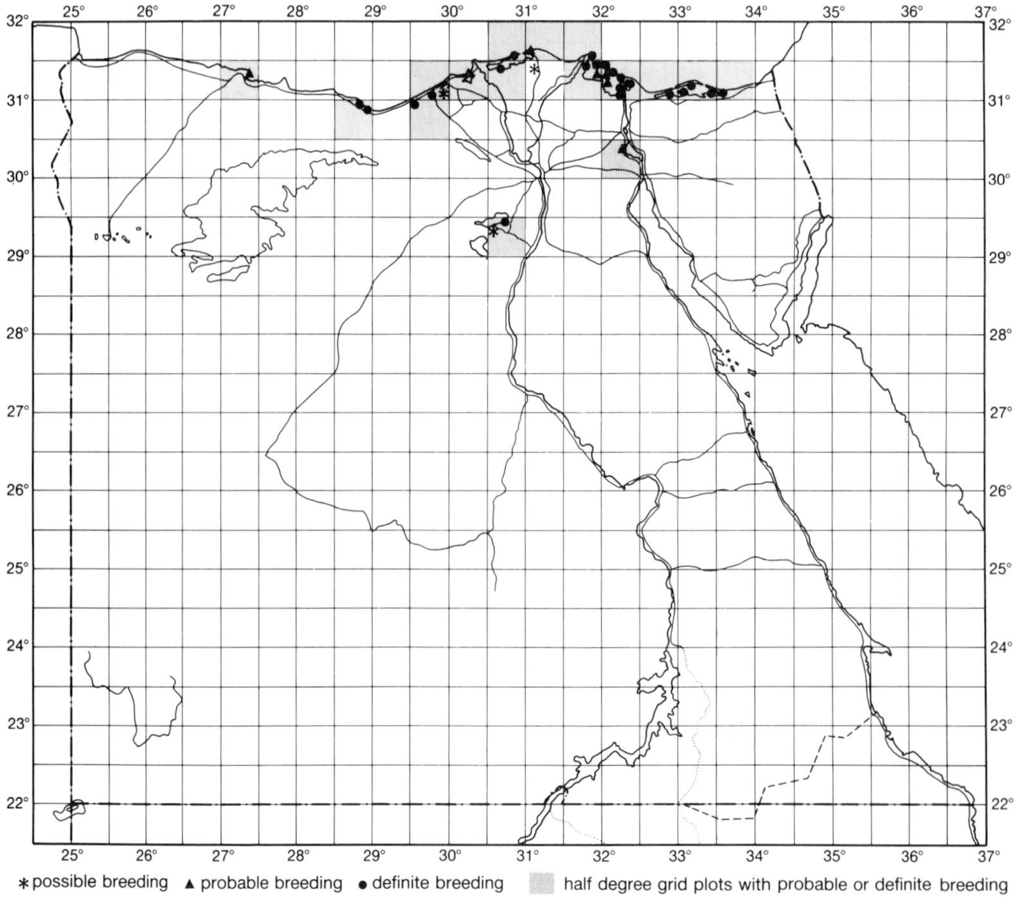

Map B43. Breeding distribution of Little Tern *Sterna albifrons* between 1945–87.

1981 (Petersen and Sørensen 1981*b*); and 872 between 2 and 24 September 1982 (Baha el Din and Salama 1984). In autumn small numbers are regularly observed along the Suez Canal and Red Sea coast. By mid-October the majority of Little Terns have left Egyptian waters. Five observed at Lake Qarun on 20 October 1980 (SMB) represent the latest inland autumn record.

In spring the first Little Terns arrive in late March and passage is most notable in the Gulfs of Suez and Aqaba. This species was common at Suez between 29 March and 13 May 1982, with a maximum daily total of 140 on 17 April (Wimpfheimer *et al.* 1983). Some migrants follow the Suez Canal, e.g. on 30 April 1955, 76 were noted within a 1.5-hour period flying north near Ismailiya (RN). On 22 April 1979, 60 were seen off Ras Muhammad (HW). Two Little Terns were noted on 10 June 1984 in the southern Gulf of Suez flying in a north-western direction (BH *et al.*). At Elat, Israel, this species is fairly common from (early) mid-April to late May (Krabbe 1980).

The Little Tern is a rare spring migrant in the Nile Valley. Koenig (1932) collected two near 'Tachta' (= Tahta?) on 22 April 1899. Thirteen were noted flying north on 13 April and 25 on 21 April 1983 at Lake Nasser near Wadi Halfa, Sudan (PLM).

Saunder's Little Tern *Sterna saundersi* AV

Local name Khut̲taaf al-bah̲r al-ah̲mar خطاف البحر الأحمر.

Accidental visitor or perhaps irregular migrant to the Egyptian Red Sea coast. Monotypic.

A mummified carcass found on 21 May 1982 in an Osprey's nest on Gezira Mahabis (Frazier *et al.* 1984; USNM) was identified as a Saunder's Little Tern (Goodman and Storer 1987). A skull, still with some skin and feathers, picked up on 20 November 1982 on Gezira Umm Agawish (Baha el Din and Saleh 1983) was originally referred to this species (Baha el Din in Frazier *et al.* 1984) but after comparison with a series of known *S. saundersi* it was reidentified as *S. albifrons* (Goodman and Storer 1987).

There are no other records of this species from Egyptian waters. It is conceivable that it breeds on islands in the extreme southern Egyptian Red Sea. On 27 May 1982, six little terns (*S. albifrons* or *saundersi*) were noted calling and displaying mobbing behaviour on an island off Marsa Suma (Frazier *et al.* 1984). This species has been collected at Port Sudan, Sudan (Goodman and Storer 1987), and is known from the Saudi Arabian Red Sea coast as far north as Yanbu (Gallagher *et al.* 1984).

Whiskered Tern *Chlidonias hybridus* PV WV

Local name Khut̲taaf abyad̲ al-khadd خطاف أبيض الخدّ.

Common passage migrant throughout the country and winter visitor, mainly in the

Nile Delta, from (early) late August to early (mid-) June. The form *C. h. hybridus* is the one occurring in the country.

Ticehurst (1912) reported that several Whiskered Terns were observed at Lake Maryut on 15 May 1909 and according to local 'natives . . . many arrive there towards the end of the month and breed in the reed-beds'. However, we are not aware of any definitive evidence that this species breeds or bred in Egypt.

The first autumn Whiskered Terns generally arrive in late August. Exceptionally early records include one on 2 August 1981 at Suez (CB, GT), one on 3 August 1983 in Bahariya (HB *et al.*), and over 50 in July and August 1979 on Lake Manzala (EM). Relatively large numbers migrate along the north coast of Sinai. Counts at Zaranikh include: 370 between 13 August and 29 September 1973 and 896 between 15 August and 5 October 1978 (YP, UP); 802 between 7 and 30 September 1980 (Petersen and Sørensen 1981*a*); 950 between 16 August and 24 September 1981 (Petersen and Sørensen 1981*b*); and 96 between 2 and 24 September 1982 (Baha el Din and Salama 1984). From late August to early November small numbers migrate southwards through the Nile Valley, including the Faiyum, and along the Red Sea coast. The only other autumn record from the Western Desert is three in Dakhla between 2 and 4 October 1962 (Scharlau 1963).

Meinertzhagen (1930) mentioned this species only as a passage visitor. However, in recent years it has become clear that it winters locally in large numbers. In mid-November 1981, 1940 were noted on Lake Manzala, 2970 on Lake Burullus, and 837 on Lake Nasser (CAB *et al.*). Those observed during this period in the Nile Delta were predominantly adults (adults = 295, juveniles = 3) and those in Lake Nasser juveniles (adults = 2, juveniles = 128) (CAB *et al.*), suggesting that within Egypt there is differential migration between the age classes. Winter counts include: Lake Burullus—17 380 on 31 January 1979 and 17 500 from 14 to 16 February 1980; and Lake Manzala—7080 from 6 to 8 February 1979 and 6400 from 28 to 29 December 1979 (Meininger and Mullié 1981*b*). A recent estimate of the total European Whiskered Tern breeding population was 18 000 pairs (Mees 1979). Thus, it is clear that the Nile Delta is an extremely important wintering area for the western Palearctic breeding populations of this species.

This species has also been noted during the winter months in other portions of Egypt. A freshly dead bird was found in Farafra in late January 1958 (Al Hussaini 1959). Observations include two on 2 January 1976 at Lake Qarun (MSm), four on 16 January 1985 in Wadi el Natrun (Goodman *et al.* 1986), and one on 18 January 1981 at Cairo (SV).

In spring small flocks migrate northwards through the Nile Valley, including the Faiyum, between late March and early June. They are also observed with some frequency in Western Desert oases (Goodman *et al.* 1986) and in the northern portion of the Red Sea. A maximum of 61 were recorded near Suez on 22 April 1982 (Wimpfheimer *et al.* 1983). Hundreds are present on the Nile Delta lakes until early May. The latest spring record is one on 15 June 1985 at Lake Qarun (DM).

Black Tern *Chlidonias niger* PV WV

Local names Khuttaaf aswad خطاف أسود, shakiitii, shaniifa.

Scarce passage migrant and rare winter visitor from early August to mid- (late) May. There are a few summer records. The subspecies *C. n. niger* is the one occurring in Egypt.

The Black Tern is the least common of the *Chlidonias* terns in Egypt. Between early August and mid-November small numbers are observed migrating along the north coast of Sinai, in the Nile Delta and Valley, and along the Red Sea coast. The majority pass in September. The only autumn records from the Western Desert are an immature collected in Wadi el Natrun on 24 August 1911 (Nicoll 1912*b*) and one bird in Bahariya from 7 to 12 August 1983 (HB *et al.*).

There are several records of small numbers in late November and December and a few reports from January and February. At least four, nearly in summer plumage, were observed at Lake Burullus on 31 January 1979 (Meininger and Mullié 1981*b*). Counts of 118 Black Terns at Lake Nasser on 22 and 23 January 1980 (Meininger and Mullié 1981*b*) were presumably of winter visitors rather than early spring migrants. Two were seen on 7 February between Luxor and Dandara and one on 11 February 1978 between El Amarna and Beni Hassan (Short and Horne 1981).

Spring migration generally takes place from mid-March to mid-May when small numbers pass through the Nile Delta and Valley and along the Red Sea coast. On 22 May 1984, one was noted at Hurghada (Goodman and Storer 1987). The only spring record from the Western Desert is one observed on 8 May 1985 in Siwa (Goodman *et al.* 1986). An exceptional summer record is an immature on 9 July 1952 at the Gebel Maryam lagoon (AP).

White-winged Black Tern *Chlidonias leucopterus* PV (WV)

Local names Khuttaaf abyad al-jinaah خطاف أبيض الجناح, abuu daqqa.

Common passage migrant in autumn from (late July) early August to early November and in spring from mid-March to late May. Scarce in summer and occasional in winter. Monotypic.

Between July and August 1979 over 1000 White-winged Black Terns were noted on Lake Manzala (EM). Large numbers also pass along the north coast of Sinai. Counts of this species at Zaranikh include: 7657 between 13 August and 29 September 1973 and 8835 between 15 August and 5 October 1978 (YP, UP); 3220 between 7 and 30 September 1980 (Petersen and Sørensen 1981*a*); 18 436 between 16 August and 24 September 1981, with 2562 on 1 September (Petersen and Sørensen 1981*b*); and 3478 between 2 and 24 September 1982 (Baha el Din and Salama 1984). From early August to early November flocks of several hundred are regularly observed throughout the Nile Delta and Valley, including Lakes Qarun and Nasser. In autumn this species is rare in the Western Desert and along the Red Sea coast.

There are a few winter records (Meinertzhagen 1930). Recent observations include one on 4 February 1978 between El Kab and Isna (Short and Horne 1981) and another on 26 February 1980 between Abu Tig and El Balyana (SK). During spring migration, from mid-March to late May, the species is less common in the Nile Delta and Valley than in autumn. It has been recorded in Western Desert oases mainly between mid-March and mid-May (JW; Goodman *et al.* 1986). It is regularly observed along the Red Sea coast, with notable concentrations near Suez of 130 birds on 30 April 1982 (Wimpfheimer *et al.* 1983) and near Elat, Israel, of up to 800 per day in early May 1979 (Krabbe 1980; EB). Summer records include three on 15 June 1985 at Lake Qarun (DM) and several on 9 July 1952 at the Gebel Maryam lagoon (AP). The latter record may have been of early autumn migrants.

Family Rynchopidae

African Skimmer *Rynchops flavirostris* FB? PV

Local names Abuu miqass afriiqii ابو مقص افريقي, 'ujhuum.

Possibly former breeding resident in the Upper Nile Valley. Presently a rare non-breeding summer visitor to the southern Nile Valley and rarely to the Red Sea. Monotypic.

There are numerous reports from the mid- and late nineteenth century of small flocks of African Skimmers in the Nile Valley south of El Balyana between early April and September (Allen 1862; Shelley 1872; Gurney 1876; Flower 1933). A specimen was taken near Damietta in 1878 (BMNH). In April 1897, Koenig (1932) observed several pairs and shot birds with enlarged gonads in Upper Egypt. In late April Shelley (1872) found them evidently preparing to breed near Armant. However, there is no definitive evidence of nesting in the country.

We are not aware of any records of the African Skimmer in Egypt between 1900 and 1979. However, since 1979 there have been several reported observations from the Upper Nile Valley. These include: Aswan—two on 19 July 1981 (CB, GT); El Zeiniya Qibli—nine on 3 October 1985 (Hume 1985); near Kom Ombo—40 on 7 October 1985 (Hume 1985), 43 on 8 October 1984 (Miles 1984), 16 on 10 October 1983 (Hume 1983), and 40 on 17 October 1979 (Short and Horne 1981); and near Luxor—two on 23 and 25 March 1986 (UA). These records are presumably of Sudanese breeding birds dispersing northwards. The only known recent report from the Egyptian Red Sea is one on 15 May 1982 at Hurghada (Wimpfheimer *et al.* 1983).

[Family Alcidae

Razorbill *Alca torda* AV?

A Razorbill was found dead on the beach at Ramleh, near Alexandria, in the winter of 1908–9 (Ticehurst 1912). It is impossible to know if the bird reached Egyptian waters alive.]

ORDER PTEROCLIDIFORMES

Family Pteroclididae

Generic local names for sandgrouse Qaṯaa, baluul aay (Bi), gaṯaa (Ma).

Many *Pterocles* spp. are known to drink only once every three to five days, and thus the birds visiting a watering spot on any one day represent only a portion of the local population (Thomas 1984).

Lichtenstein's Sandgrouse *Pterocles lichtensteinii* RB Plate 4

Local names Qaṯaa nuubii قطا نوبي, anuy (Bi).

Local breeding resident in eastern Sinai and south-eastern Egypt. It visits watering places at dusk. It tends to occur in rocky or mountainous areas. *P. l. lichtensteinii* is the form found in the country.

Eastern Desert The Lichtenstein's Sandgrouse is a locally common resident of the south-eastern region. Several hundred were counted drinking at wells near Gebel Abraq between late March and early April 1985 (SMG, PLM). There are numerous specimens and sight records from the wadis immediately surrounding Gebel Elba and to the east along the coastal plain.

Sinai Lichtenstein's Sandgrouse have been recorded in small numbers near waterholes in eastern Sinai. They presumably nest in desert areas surrounding these drinking places. This species has been recorded at the following localities: Taba—regularly observed between March and April 1978–80 (Olsen 1983); Nuweiba—two on 11 March 1978 (RSS); Dahab—one collected in February (Meinertzhagen 1930) and a pair seen on 27 March 1980 (JO); and Sharm el Sheikh—regularly up to 12 birds between early March and late May (e.g. Olsen 1983) and one on 18 September 1980 (DJM, DES). This species was regularly recorded at Elat, Israel, from January to June and in October with a maximum of 53 on any single day (Krabbe 1980).

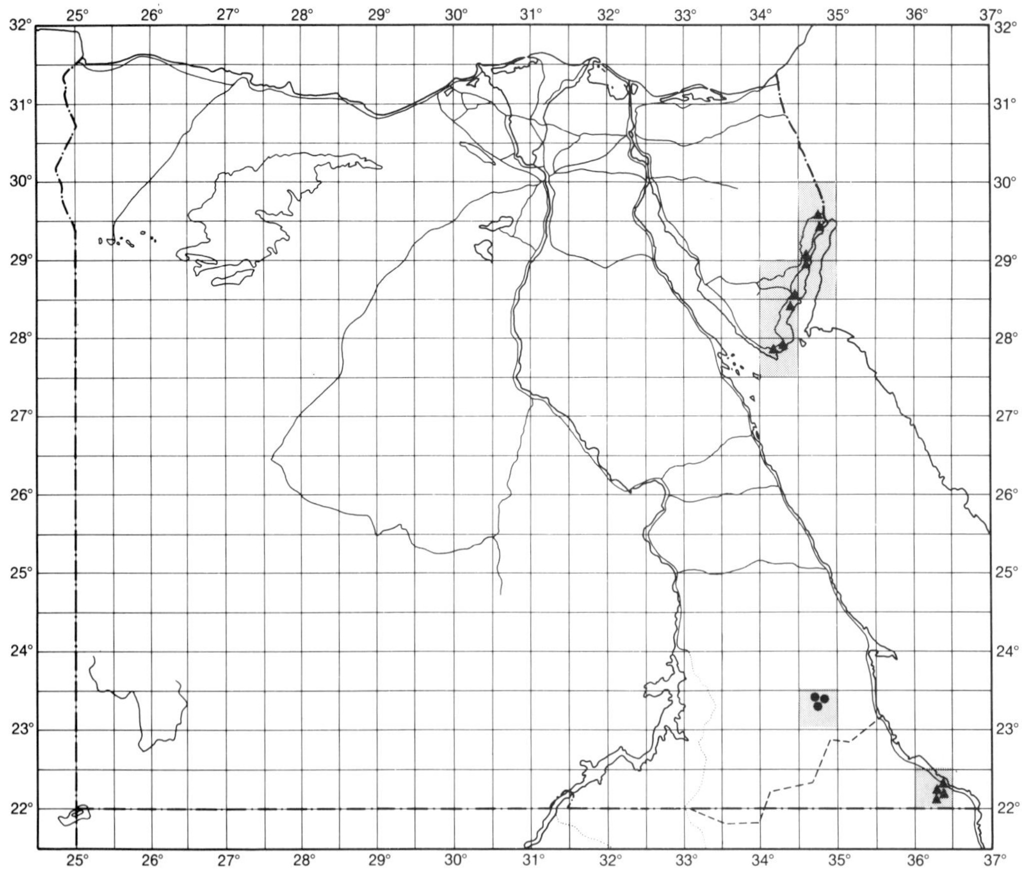

▲ probable breeding ● definite breeding ▒ half degree grid plots with probable or definite breeding

Map B44. Breeding distribution of Lichtenstein's Sandgrouse *Pterocles lichtensteinii* between 1977–87.

Crowned Sandgrouse (Coronetted Sandgrouse) *Pterocles coronatus* RB

Local name Qaṯaa mutawwaj قطا متوّج.

Fairly common, but local, breeding resident in the Eastern Desert and Sinai. Generally observed in flocks at drinking places after sunrise. This species seems to occur mainly in stony habitat with scrub and generally avoids open sandy deserts. It apparently travels considerable distances for water. The breeding subspecies of Sinai is *P. c. vastitas* and in the balance of the country *P. c. coronatus*.

North Coast and Western Desert The Crowned Sandgrouse is a presumed local resident in these areas but there is no definitive proof of breeding. On 6 March 1976 a male with small gonads was taken from a group of approximately 12 at Qara, Qattara Depression (Goodman and Ames 1983). One specimen was collected on 31 January 1920 at Matruh (BMNH), an area from which there is no recent record. This species is widely distributed in the Cyrenaican desert of Libya (Bundy 1976). Two males and two females were obtained in Wadi el Natrun on 22 May 1910 (Goodman *et al.* 1986).

Nile Valley There are a few recent records from the edge of the Nile Valley. Six were observed on 23 February 1984 near the Hauwara Pyramid, Faiyum (SG), 10 on 17 March 1983 south of Qena (Welch 1983), and one on 25 March 1985, 30 km east of Nag Hammadi (JW).

Eastern Desert The Crowned Sandgrouse has been observed on the eastern edge of the Nile Delta, where several were shot in late August 1917 (Raw 1921). Recent records from the northern Eastern Desert include: one on 23 June 1980 (SMB) and five on 9 May 1985 (DM) along the Cairo–Suez road and 17 on 30 October 1981 at Port Taufiq (RGB *et al.*). On 19 March 1982, 46 were noted along the coastal road between Ras Gemsa and Ras Gharib and 27 between Ras Gharib and Ras Abu Darag

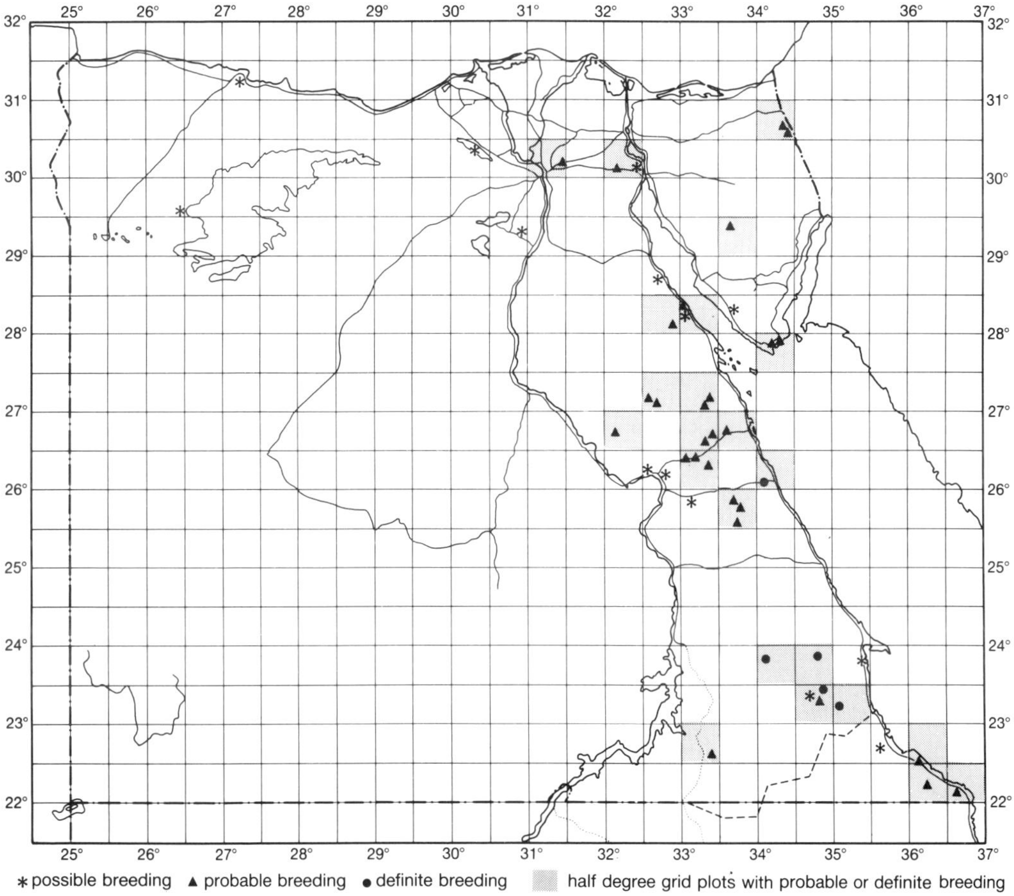

Map B45 Breeding distribution of Crowned Sandgrouse *Pterocles coronatus* between 1930–87.

(Meilstrup and Sørensen 1983). This species is regularly observed in the central Egyptian Red Sea Mountains. Flocks congregate along the Qift–Quseir and Qena–Safaga roads to feed on grain spilt by passing lorries (SMG; JJH). An important watering spot for this species is at Bir Beida, just west of Quseir on the road to Qift. This site is visited daily throughout the year by several hundred Crowned Sandgrouse (SMG, PLM), with a maximum number of 500 on 13 April 1984 (PDH, EMN).

This species is relatively common in the south-eastern desert as far south as Halaib and Gebel Elba (Goodman 1984*b*). Several specimens and three eggs were collected on 22 March 1943 in Wadi Garara (KFMP). In mid-June 1984 adults with young were observed in Wadi Naam on the east side of Gebel Abraq and to the north in Wadi Hodein (SMG). Pairs or small groups of up to six have been recorded along the coastal plain in areas of relatively heavy tamarisk scrub, e.g. near Adal Deib (SMG).

Sinai Flocks have been observed at Ain el Gedeirat, near Qusaima, on 13 March 1928 (Meinertzhagen 1930) and 15 May 1939 (Galal el Din and Fouad 1939–40). In recent years up to 15 have been noted at this locality. Meinertzhagen (1930) mentioned that it had been found in Wadi Feiran and near El Tor, but the only relatively recent record from this area is two birds taken on 2 January 1957 at El Tor (TAUM). Between Sharm el Sheikh and Na'ama there is a watering hole that is visited daily by Crowned Sandgrouse; the numbers recorded vary from 524 on 18 September 1980 (DJM, DES) to 80 in late December 1980 (BCD). At Elat, Israel, this species has been observed in numbers of up to 85 and it is probably present in the area throughout the year (Krabbe 1980).

Spotted Sandgrouse *Pterocles senegallus* RB Plate 3

Local name Qaṯaa arqaṯ قطا ارقط.

Fairly common breeding resident of open sandy desert and generally avoids rocky and mountainous areas. Breeding is usually between March and mid-June. Monotypic.

North Coast and Western Desert Two male Spotted Sandgrouse were taken on 12 August 1964, 11 km south of Bahig (FMNH). The species has not been observed in this immediate area in more recent years. Wings of both males and females were collected at Bir Victoria, near Wadi el Natrun, in April 1922 (Goodman *et al.* 1986) and in May 1985 fresh sandgrouse tracks were found in a plantation near Sadat City (SMG, PLM). In 1986 a hunter from Wadi el Natrun reported that this species bred locally (WCM). Up to 25 birds have been regularly noted in recent years along the Cairo–Faiyum road. This species has also been recorded in small flocks in Bahariya, Farafra, Dakhla, and Kharga and in the desert between these oases (Goodman *et al.* 1986). On 2 February 1983 about 200 were observed in small flocks between Asyut and Kharga (HR).

Nile Delta and Valley and Eastern Desert A specimen was taken at Alexandria in April 1870 (BMNH). Raw (1921) mentioned summer and autumn records from El

Marg, north-east of Cairo. Several specimens were taken at El Qassasin on 6 September 1922 (GZM). Throughout the year flocks of up to 15 birds have been regularly observed on the edge of the Nile Valley near El Amarna, El Fashn, Idfu, Isna, and Daraw. In the Eastern Desert proper the species has been noted along the Cairo–Ismailiya road, Cairo–Suez road, north of Suez, along the El Koriamat–Zafarana road, the coastal plain between Suez and Safaga, and the western portions of the Qena–Safaga and Qift–Quseir roads. Large concentrations include: 250 east of Qift on 11 October 1986 (EH, PK), 200 between El Koriamat and Zafarana on 19 November 1982 (Baha el Din and Saleh 1983), 210 east of Qift on 31 January 1983 (Goodman and Atta 1987), and 250 about 100 km west of Safaga on 14 February 1983 (JJH).

Sinai Records from the Qusaima area include: flocks on 15 May 1939 (Galal el Din and Fouad 1939–40), 185 on 25 March 1975 (SC), and 500–1000 on 18

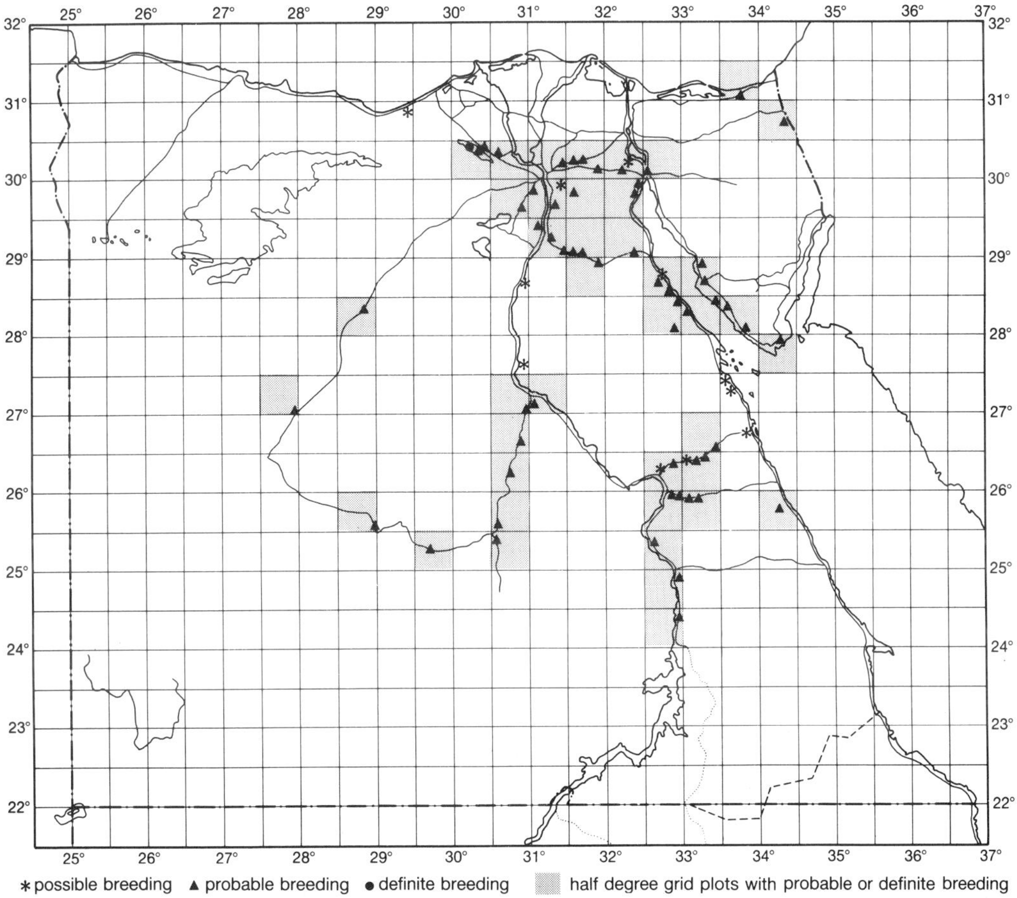

Map B46. Breeding distribution of Spotted Sandgrouse *Pterocles senegallus* between 1930–87.

September 1980 (DJM, DES). The only known record from the Lake Bardawil area is one on 30 April 1985 (Fry *et al.* 1985). This species is frequently observed in small numbers in the southern portion of the Peninsula, along the coastal plain between Sharm el Sheikh and Ras Abu Rudeis. It is absent from the mountainous parts of Sinai. There are several records from Elat, Israel (HSh).

Ringing A bird ringed at Nizzana, Israel, on 7 December 1983 was recovered in Egypt, presumably in Sinai, before April 1985 (GZF).

Chestnut-bellied Sandgrouse *Pterocles exustus* FB (RB)

Local names Qaṯaa maṣrii قطا مصري, ghaṯaaṯ.

Formerly a fairly common breeding resident of the Nile Valley. *P. e. floweri* is the form found in the Egyptian Nile basin and is presently rare or locally extirpated.

Between November 1853 and January 1854 Chestnut-bellied Sandgrouse were reported as abundant in the Nile Valley and in bordering desert (Taylor 1859). The statement of Von Heuglin (1861) that 'zahlreichen Ketten' were found at Suez is rather exceptional. Adams (1864) mentioned this species as common in Upper Egypt and breeding in June. One was taken at Kom Ombo on 2 January 1891 (FMNH). Two eggs were collected on 9 May 1894 on an island in the Nile near El Aiyat (Sparrow in Raw 1921). In 1899 the species was found to be not common in areas of *halfa* (*Imperata cylindrica*) between Beni Suef and Qulusna (Koenig 1926). In the early twentieth century it was reported as common in the Faiyum near Tamiya and at El Idwa, and occasionally noted on the eastern edge of the Nile Delta near El Marg (Raw 1921). By 1929 this species was considered 'certainly scarce' in Egypt (Meinertzhagen 1930). Several were collected at El Hammam, between Beni Suef and the Faiyum, in the first half of June 1938 (HGD). The only recent record is about 10 on 4 March 1979 between Isna and Idfu (Short and Horne 1981).

Black-bellied Sandgrouse *Pterocles orientalis* (WV)

Local names Qaṯaa aswad al-baṯn قطا أسود البطن, ghaḏaf, juunii.

Occasional winter visitor to northern Sinai, the Eastern Desert, and the Nile Valley. All Egyptian material examined is referable to *P. o. orientalis*.

Meinertzhagen (1930) mentioned that at Qusaima a small flock of Black-bellied Sandgrouse was seen on 13 March 1928, and that in most winters only a pair or a single bird could be found locally. Recent records from this area of Sinai and west of Nizzana, near the Israeli border, include 13 on 20 September 1980 (DJM, DES) and one on 29 March 1975 and 89 on 3 April 1973 (SC). In October 1979, five birds were reported near Elat, Israel (Krabbe 1980).

The only known record from the Eastern Desert is a single bird collected in December 1982 from a flock of about 20 Spotted Sandgrouse 70 km east of El

Koriamat on the road to Zafarana (HS). Fourteen were observed along Lake Qarun on 23 November 1982 (PB).

Pin-tailed Sandgrouse *Pterocles alchata* AV

Local name Qaṯaa ṯawiil adh-dhanab قطا طويل الذنب.

Accidental visitor to north-eastern Sinai. The single specimen examined from the area is referable to *P. a. caudacutus*.

One Pin-tailed Sandgrouse was collected at Rafa on 14 November 1917 and in the same year several were taken across the present Israeli border near Gaza (Meinertzhagen 1930; BMNH). Kaiser (1892, and in le Roi 1923) recorded this species on the western side of the Peninsula in January, March, and April but he may have confused it with another sandgrouse species. Meinertzhagen (1930) mentioned that Koenig obtained one at El Tor in April; however, we cannot trace a citation in one of Koenig's papers nor the specimen and this record is unacceptable. This species breeds in the western Negev Desert of Israel (Paz 1987).

ORDER COLUMBIFORMES

Family Columbidae

Generic local name for pigeon and dove H̲amaam.

Rock Pigeon (Rock Dove) *Columba livia* RB

Local names H̲amaam jabalii حمام جبلي, abaal hamaam (Bi), h̲amaam azraq, ṯaraanii.

Common breeding resident in mountainous areas, on cliffs, in caves, and in deep wells. Four relatively distinct subspecies breed in the country. *C. l. livia* occurs along the Mediterranean coast from Salum to at least Matruh, *C. l. gaddi* in the Sinai Peninsula, *C. l. schimperi* in the Nile Delta and Valley, and *C. l. dakhlae* in Western Desert oases. The breeding birds of the Eastern Desert tend to be intermediate in plumage coloration between *schimperi* and *gaddi*.

Domestic Pigeons are commonly raised throughout the country for meat and fertilizer. Interbreeding between wild, feral, and domestic pigeons occurs in many areas. Thus, based on plumage coloration it is sometimes impossible to separate wild birds from domestic pigeons, particularly in areas near human settlements.

North Coast Rock Pigeons are widely distributed in the coastal strip between Alexandria and the Libyan border. In this region they commonly breed on cliffs, ruins, and in wells. In January 1942 the species was common along the coast from Buqbuq to the Libyan border (Rowntree 1943). At Bir el Ghallaz, south of Matruh on the road to Siwa, several were found nesting in a well on 8 May 1985 (SMG, PLM).

Western Desert It is not clear if this species naturally occurs in Bahariya and Farafra oases. On 26 April 1981, one pigeon was seen resting along a road in barren desert midway between Dakhla and Farafra. Four specimens collected in Dakhla in January 1972 are typical of *C. l. dakhlae* and interbreeding in this oasis with local domestic pigeons appears to be minimal (Goodman *et al.* 1986; USNM).

Nile Delta and Valley This species is presumed to be widely distributed along the edges of the Nile Delta and Valley but details are imperfectly known. Interbreeding with domestic pigeons appears to be relatively frequent.

Eastern Desert The species is a locally common resident on cliffs and in wells of the Red Sea Mountains and along the coastal plain. It tends to nest in small colonies, e.g. 10 pairs on cliffs at Bir Beida, west of Quseir, and 15 pairs on the upper cliffs of

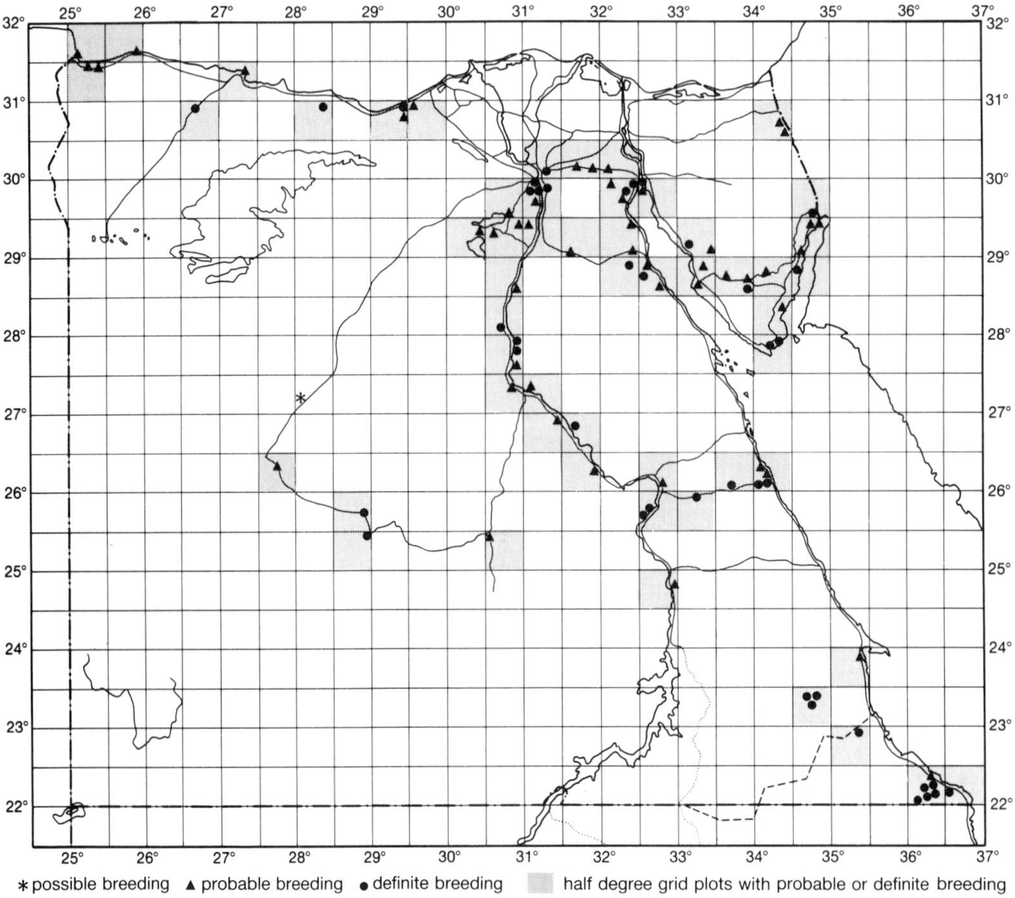

Map B47. Breeding distribution of Rock Pigeon *Columba livia* between 1930–87.

Wadi Akwamtra, Gebel Elba (SMG). Solitary nesting pairs are occasionally noted. In mid-April 1983 this species was probably breeding in the hulk of an old ship 1 km offshore from Adabiya (MCJ *et al.*).

Sinai Common breeding resident in the mountains of southern Sinai and along the coastal plain of the Gulf of Aqaba. Flocks of up to 200 are regularly recorded.

Stock Pigeon (Stock Dove) *Columba oenas* WV

Local names H̱amaam barrii حمام برّي, al-waraqaa-.

Rare winter visitor between early September and mid-March (late April). *C. o. oenas* is the subspecies occurring in Egypt.

Western Desert Stock Pigeons have been collected in Wadi el Natrun on 24 and 26 November 1911 and on 4 October 1914 (Goodman *et al.* 1986), and observed on 19 March 1910 (Nicoll 1912*a*).

Nile Delta and Valley Occasionally large numbers visit the Nile Delta but in most years it is rare or completely absent. A flock of several thousand pigeons noted on 17 October 1917 flying over the Suez Canal at El Qantara was thought to be this species (Meinertzhagen 1930). There are only a few post-1930 records; the largest number being a flock of 76 over Suez on 26 September 1981 (RGB *et al.*).

Sinai According to Meinertzhagen (1930) 'immense flocks' of Stock Pigeons visit northern Sinai in some winters; however, we are unaware of any recent large concentrations. A specimen was obtained at El Tor on 4 November 1911 (Zedlitz 1912). One or two birds were noted at Lake Bardawil on 22 April 1985 (Fry *et al.* 1985). The species is considered a rare winter visitor to Elat, Israel (Krabbe 1980).

Generic local names for *Streptopelia* spp. Yamaam, qimrii, timlii (Si), waṯwaat (Si).

Pink-headed Dove (African Collared Dove) *Streptopelia roseogrisea* RB

Local names Yamaam wardii ar-ra-s يمام وردي الرأس, dibsa, faakhita, kuubaalii (Bi), ṣalṣala.

Local breeding resident near Gebel Elba. The form occurring in this region is *S. r. arabica.*

The Pink-headed Dove is a locally common breeding resident in vegetated areas of Gebel Elba, particularly in the north-eastern wadis such as Wadi Aideib (SMG, PLM). In January 1896 it was observed at Gebel Shellal (Cholmley 1897). Marchant (1941) reported that in 1933 one was found in captivity at Hurghada. The details of this record are ambiguous and it is impossible to determine if the bird 'reached the area by natural means'.

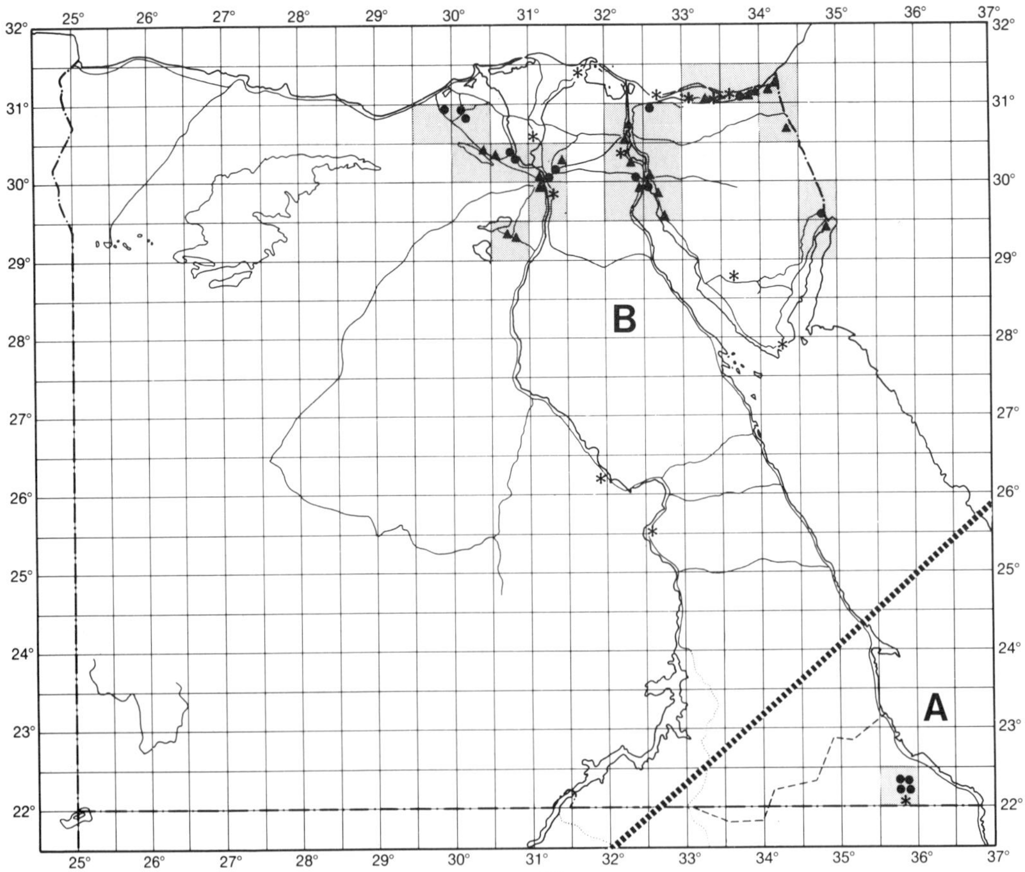

∗ possible breeding ▲ probable breeding ● definite breeding ▒ half degree grid plots with probable or definite breeding

Map B48. Breeding distribution of: Pink-headed Dove *Streptopelia roseogrisea* (A) and Collared Dove *S. decaocto* (B) between 1977–87.

Collared Dove *Streptopelia decaocto* RB

Local names Yamaam muṯawwaq يمام مطوّق, faakhita, qumriyya, yaa kariim.

Locally common breeding resident in Wadi el Natrun, portions of the Nile Delta and northern Valley, the Suez Canal area, and Sinai. It is rapidly colonizing new portions of the country. All of the Egyptian material examined is referable to *S. d. decaocto*.

Sinai The Collared Dove apparently commenced its colonization of Egypt in the mid-1970s. The first Egyptian record that we are aware of is 25 at Ain el Gedeirat on 21 March 1976 (SC). A specimen was collected at Nahal Yam on 17 October 1978 (TAUM). On 27 March 1979, 200 were observed near Rafa (EKr). Other early records include: birds breeding in palm groves in 1979, 10 km west-south-west of Rumana (OH), and a total of 21 at El Arish on 5 and 6 January 1980 (Meininger and

Mullié 1981*c*). In recent years this species has been found nesting in Wadi Taba (HSh) but as of 1987 there is no evidence of breeding at other localities along the Gulf of Aqaba; it has been recorded as far south as Sharm el Sheikh.

Nile Valley and Delta On 8 December 1978, six Collared Doves were observed between Suez and Ismailiya (HW). Beginning in 1979 this species was regularly recorded in numbers of up to 50 at Suez, where it has since been found breeding. On 8 February 1979, three were observed in downtown Cairo and the next day one was collected (Goodman and Houlihan 1981). Although captive Collared Doves are known to have escaped from pet shops in Cairo (Goodman and Houlihan 1981), it seems likely that the greater portion of the original colonizing stock were of wild origin. Other records from the environs of Cairo include: two on 2 April 1980 at Heliopolis (SMB), and single birds on 8 September 1981 in the city centre at Midan Tahrir, and on 16 September 1981 at El Dokki (RGB *et al.*). On 11 December 1980, two were observed along the Cairo–Suez road, 50 km west of Suez (SMB). At least 12 birds and several nests, including one with a nestling, were found on 18 April 1983 in a wadi with acacia 24 km north-west of Suez (MCJ *et al.*).

Between 1980 and 1985 this species appeared in numerous portions of the Nile Delta and breeding has been proven at several localities. Considerable local concentrations have been observed. At El Khatatba one was shot on 7 May 1982, 100 were noted on 11 February 1983, and several hundred were present by 1984 (BO). At least 14 were noted on 7 May 1985 at Sadat City, between Wadi el Natrun and the Nile Delta (SMG, PLM). On 10 March 1983 about 30 were observed at Inchas (BO). A survey of Cairo pet shops in 1984 and 1985 found only '*risoria*' Collared (Barbary) Doves, a domestic type that has not been noted free-living in Egypt (SMG).

This species apparently spread quickly up the Nile Valley. In the Faiyum, two were noted on 14 March 1982 at El Faiyum (Meilstrup and Sørensen 1983) and two on 23 February 1984 near the Hauwara Pyramid (SG). The most southern Nile Valley records to date are single birds at El Balyana on 24 February 1984 (SG) and between Isna and Luxor on 23 March 1986 (UA).

Western Desert The only area this species has been recorded is near Wadi el Natrun, the earliest record being one on 1 May 1981 (Goodman *et al.* 1986).

Turtle Dove *Streptopelia turtur* MB PV (WV)

Local names Qumrii قمري, qumriyya, ṣalṣal, turghul, yimaam (Ma).

Common migrant breeder in Western Desert oases, throughout the Nile Delta and Valley, and in north-east Sinai. It has recently been proposed that the breeding forms of the Western Desert, *S. t. rufescens*, and the Nile Valley, *S. t. isabellina*, were synonymous (Cramp 1985). The differentiation of these forms was confounded by the lack of specimen material from the two areas of the same sex and the possibility that the recognized differences between them were simply sexual. We have been able to compare four adult male specimens taken in the spring in Kharga and Dakhla with adult males from the adjacent Nile Valley. The Western Desert material has consistently darker and redder coloration than the Nile Valley. The forehead is grey in the Western Desert specimens compared with a completely isabelline-brown crown in the Nile Valley material. Based on these characters we follow Vaurie (1965) in recognizing the breeding form of the Western Desert (Kharga and Dakhla) as *S. t. rufescens* and that of the Nile Valley as *S. t. isabellina*. Eurasian migrants, including the forms *S. t. turtur* and *S. t. arenicola*, are common throughout the country in autumn from early August to mid-October and in spring from early April to late May.

Western Desert Several Turtle Doves, referred to *S. t. isabellina*, were observed in Wadi el Natrun in late May 1918 but no evidence of breeding was found (Raw 1921). On 7 May 1985 distinctly pale Turtle Doves were found breeding in new plantations at Sadat City, south-east of Wadi el Natrun (SMG, PLM). This species commonly breeds in Siwa, Bahariya, Farafra, Dakhla, and Kharga, and these local populations

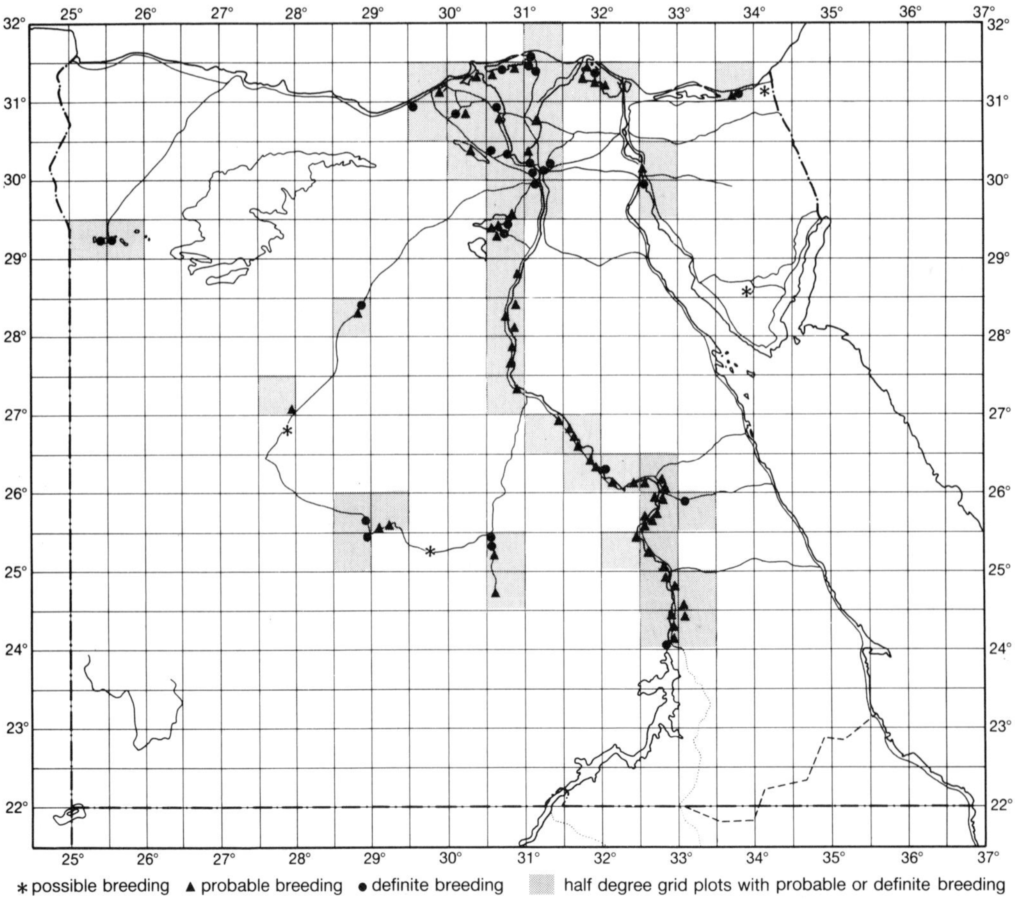

Map B49. Breeding distribution of Turtle Dove *Streptopelia turtur* between 1977–87.

are presumably referable to *S. t. rufescens*. The breeding birds in these oases are in local residence between late February and mid-October. Apparently they are only partially migratory as a specimen referable to *rufescens* was taken in Dakhla on 16 January 1972 (USNM). Eurasian breeding birds also winter in the area—one identified as *S. t. turtur* was collected in Kharga on 14 January 1972 (USNM).

Nile Delta and Valley This species is a common migrant breeder in the Nile Delta and Valley, the Suez Canal area, and the Faiyum. It has been observed in small numbers during the winter months in the Upper Nile Valley. Singing birds were noted at Aswan on 20 January 1980 (PLM, WCM) and a specimen referable to *isabellina* was collected at Kom Ombo on 27 December 1971 (USNM). Birds identified as *isabellina* have been taken in midwinter and during the breeding season in the Nile Valley above Aswan, in the area now inundated by Lake Nasser (YPM). Adults were noted at El Balyana on 8 April 1983 carrying nest material into reed-beds (PLM). This species has also been found nesting in reed-beds in the Nile Delta (PLM, UGS).

Eastern Desert and Sinai Common passage migrant in autumn and spring. Excluding a few sites, this species is not known to nest in these two regions. On 30 April 1983 (PLM) and on 4 June 1984 (SMG) nests were found at El Laqeita, a small oasis in the Eastern Desert, 38 km east of Qift. The species may also nest near Gebel Elba (SMG). Possible breeding records from the Sinai include two singing in Wadi Tarfa on 11 May 1985 and two at St Katherine on 15 May 1985 (DM). A few pairs breed at El Arish (SMB).

Ringing A bird ringed at El Maadi on 11 September 1968 was recovered near Alexandria on 19 September 1969 (Watson 1971) and another ringed at Bahig on 24 April 1970 was recovered at Alexandria on 23 September 1972 (GZF). Both records show migration route fidelity.

Palm Dove (Laughing Dove) *Streptopelia senegalensis* RB

Local names Yamaam baladii يمام بلدي, dibsiya, gimriyya (Ma), qimrii baladii, qubarriyat (Bi).

Abundant breeding resident in numerous portions of the country. The subspecies breeding in the vicinity of Gebel Elba and in extreme Upper Egypt is *S. s. senegalensis*, and in the rest of the country *S. s. aegyptiaca*. There is a broad zone of intergradation in the Upper Nile Valley between *aegyptiaca* and *senegalensis*. We treat *S. s. dakhlae* as a synonym of *S. s. aegyptiaca* and *S. s. sudanensis* as a synonym of *S. s. senegalensis*. We have not examined any Egyptian material referable to *S. s. phoenicophila* (*contra* Hovel 1987). Breeds from early February to mid-June.

Western Desert Palm Doves are locally common breeding birds in all major Western Desert oases, except Farafra. This species has recently colonized Wadi el Natrun, Siwa, and Kharga (Goodman *et al.* 1986).

Nile Delta and Valley It is abundant in cultivated areas, villages, and towns throughout the Nile Delta and Valley (south to Aswan), the Suez Canal area, and the Faiyum. At Abu Simbil, two were observed on 4 May 1978 (Meininger and Dielissen 1979, at least two on 29 March 1982 (JW), and several on 27 April 1984 (MBe, SM).

Eastern Desert In recent years this species has apparently colonized human settlements along the Red Sea. It has been recorded at the following localities: Ain Sukhna—about 30 on 29 April 1965 (HCK); Ras Gharib—found breeding on 4 February 1981 (SMB) and regularly since; Hurghada—about 25 on 15 April 1965 (HCK), five on 21 September 1981 (FdR), and regularly in small numbers since. A maximum of 60 was noted on 1 November 1984 (RvD *et al.*); Safaga—irregularly in small numbers since September 1942 (Bulman 1944); Hamrawein—four on 25 March 1985 (SMG, PLM); Quseir—15 on 28 April 1983 (PLM) and regularly since; Marsa el Alam—two on 27 March 1985 (SMG, PLM). In the vicinity of Gebel Elba it is a common breeding resident in wadis containing acacias.

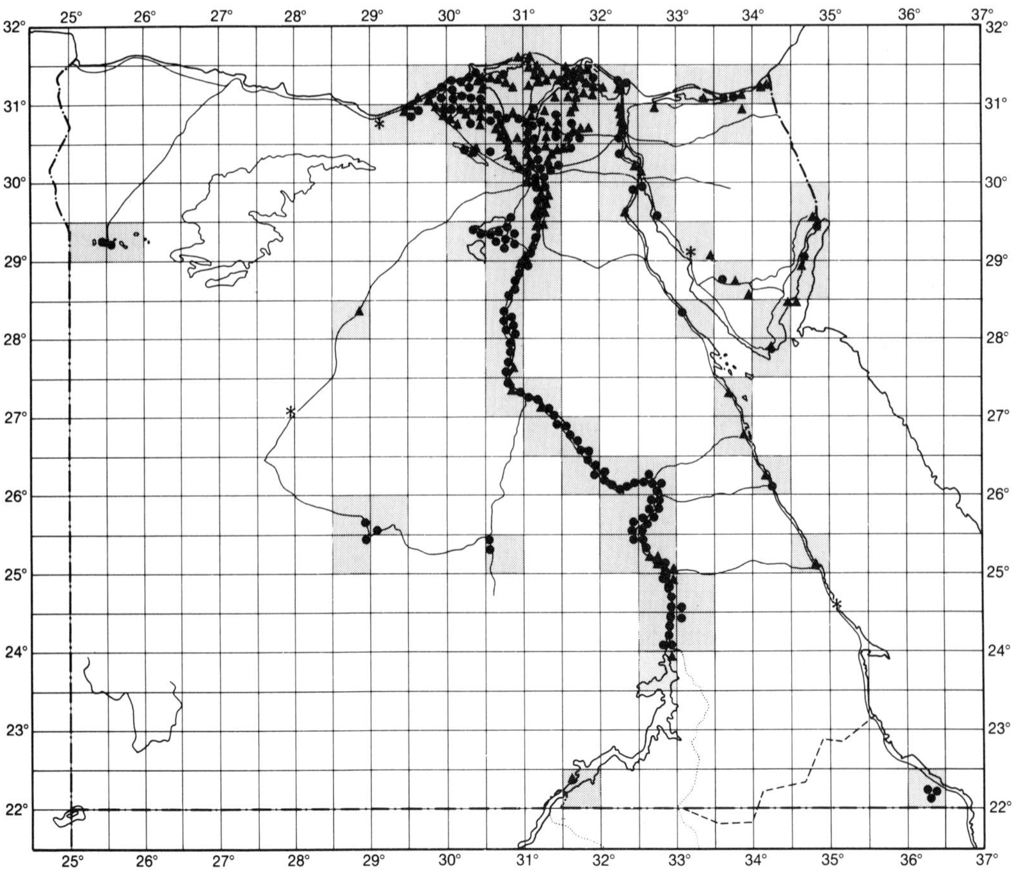

∗ possible breeding ▲ probable breeding ● definite breeding ▒ half degree grid plots with probable or definite breeding

Map B50. Breeding distribution of Palm Dove *Streptopelia senegalensis* between 1977–87.

Sinai The Palm Dove is a local resident in the El Arish–Rafa area and breeds sporadically in palm groves west to Rumana. Small numbers are regularly observed along the coast of the Gulf of Aqaba. Several pairs were found breeding at Nuweiba on 3 April 1983 and by early March 1987, 40–60 pairs were present (PLM). On 1 April 1983, one was observed in Wadi Feiran (PLM) and since then this species has been regularly noted in the area. A fledgling was observed on 26 September 1983 in Wadi el Sheikh (SMB). On 6 March 1987 at least two pairs were seen in the St Katherine Monastery garden displaying and copulating (PLM). Several were observed on 21 March 1986 at Gebel Sarabit el Khadim (AS). Records from the Gulf of Suez coast include: Ras el Sudr—birds collected on 2 June 1969 and 7 January 1970 (HH), on 11 July 1972 (Hovel 1987), and recorded throughout the period 1983–4 (SMB); and 12 km north of Abu Zenima—three on 25 January 1985 (SMG).

Namaqua Dove (Long-tailed Dove) *Oena capensis* RB? WV

Local names Yamaam baaluum يمام بالوم, bayluma, h̠umh̠uma, umm baliima, yah̠muu, yamaam al-kaab.

Rare breeding resident in the Gebel Elba area. A vagrant to other portions of the country. The subspecies occurring in Egypt is *O. c. capensis*.

Nile Valley Two Namaqua Doves were collected on 27 December 1971 at Kom Ombo (Goodman and Watson 1983). Recent observations include two on 10 January 1982 at Aswan (YL), six males and three females on 18 February 1982 at Dandara (SQ), and one male on 22 February 1984 near Lake Qarun and the following day at Dandara (SG). About 40 were for sale in a Cairo pet shop in late January 1985 (SMG) and tens in spring 1986 (SMB, WCM).

Eastern Desert While exploring the African Red Sea coast, Von Heuglin (1871) found the Namaqua Dove as far north as Berenice. This species is a resident of the Gebel Elba region, generally only in wadis with heavy acacia growth. In this area it is apparently scarcer in years with little or no rain compared with years with considerable precipitation. It is plausible that this species is a migrant breeding bird. One was observed on 28 April 1983 at Marsa el Alam (PLM).

Sinai A female was observed on 17 April 1981 south of Dahab (DMe) and a male in March 1982 at Nuweiba (HSh). The first nest found at Elat, Israel, was in 1980 and by 1986 about 15 pairs bred locally (HSh).

ORDER PSITTACIFORMES

Family Psittacidae

Ring-necked Parakeet (Rose-ringed Parakeet) *Psittacula krameri* IB

Local names Durra mut̠awwaqa درّه مطوقة, babbaghaan, babghaa, duuduu, t̠uut̠ii.

Locally common breeding resident in Cairo, local environs, and probably Suez. Occasional at other localities. All specimens examined from Egypt are referable to *P. k. manillensis*.

Meinertzhagen (1930) stated that the Ring-necked Parakeet was introduced into the Nile Delta around 1912 and quickly established itself at Giza and other localities near Cairo. According to Flower (1933) the introduction of this species into Egypt was the result of escaped cage-birds. He further mentioned that they were common in Egypt long before 1901. In the years 1916 to 1919, 127 feral birds were collected on the grounds of the Giza Zoological Gardens (Flower 1933). The earliest record of this species in Egypt appears to be a male collected at Helwan on 19 December 1895 (Goodman 1982).

Nowadays, this species can regularly be seen in gardens and parks in Cairo proper, Giza, El Dokki, Zamalek, Heliopolis, and Abu Rauwash. The total local feral population is estimated at several hundred. They nest in holes in buildings, under roof-eaves, and in tree hollows. A specimen was taken at Bahig on 14 April 1971 (Goodman 1982). In late June 1986 a flock of 10 birds was observed feeding on fruits in a garden near Alexandria (JH). Up to four birds were noted between 23 February and 19 April 1982 at Suez and Port Taufiq (Wimpfheimer *et al.* 1983). Other records of this species from Suez include: two on 27 March 1983 (PLM, HR), two on 4 April 1983 (SC, IP, HR), and five on 2 and 18 November 1984 (RvD *et al.*).

ORDER CUCULIFORMES

Family Cuculidae

Generic local names for cuckoo Waqwaaq, waquuq, waqwaq.

Great Spotted Cuckoo *Clamator glandarius* MB PV

Local names Waqwaaq arqaṯ وقواق أرقط, abuu burbur, huuhuu, kabar, qamiiẖa, qayqab, sagaaw.

Rare migrant breeder in the Nile Valley. Uncommon migrant in the Western Desert, throughout the Nile Delta and Valley, and Sinai. Monotypic.

In Egypt the Great Spotted Cuckoo is a brood parasite of the Hooded Crow. At the turn of the twentieth century, this cuckoo was relatively common south of Asyut, particularly between Luxor and Aswan, but not found south of Aswan (Koenig 1920). Eggs have been collected between January and February at Sohag, Qena, Naqada, Luxor, Armant, Idfu, and Nag Hammadi (Koenig 1920), and in Lower Egypt near Saqqara (Von Heuglin 1871), Abu Zabal (Raw 1921), and the Delta Barrage in late May (Meinertzhagen 1930). A young bird was noted as early as 11 January 1945 in a Hooded Crow's nest on Kitchener's Island (RHG).

Nowadays, this species is much rarer than it was at the beginning of the twentieth century. Small numbers are regularly observed between early February and mid-August in the Nile Valley south of Nag Hammadi and in the Faiyum. Breeding in these areas seems likely but definitive evidence from the past few decades is lacking. The only recent breeding records from Lower Egypt are nestlings taken on 1 April 1970 at Abu Rauwash and 16 April 1970 at El Baragil (USNM).

Meinertzhagen (1930) mentioned a distinct northward-bound spring passage in Upper Egypt, with parties of 10 to 20 birds noted in February. He stated that the species is absent from Egypt between August and December, except for a few that might winter in Upper Egypt. However, this is not always the case, for one was taken at Dahshur on 19 December 1937 (HGD) and in more recent years it has been obtained on 27 August 1969 at Bahig (USNM) and observed in December near 'Edfour' [= Idfu?] (Etchécopar and Hüe 1967).

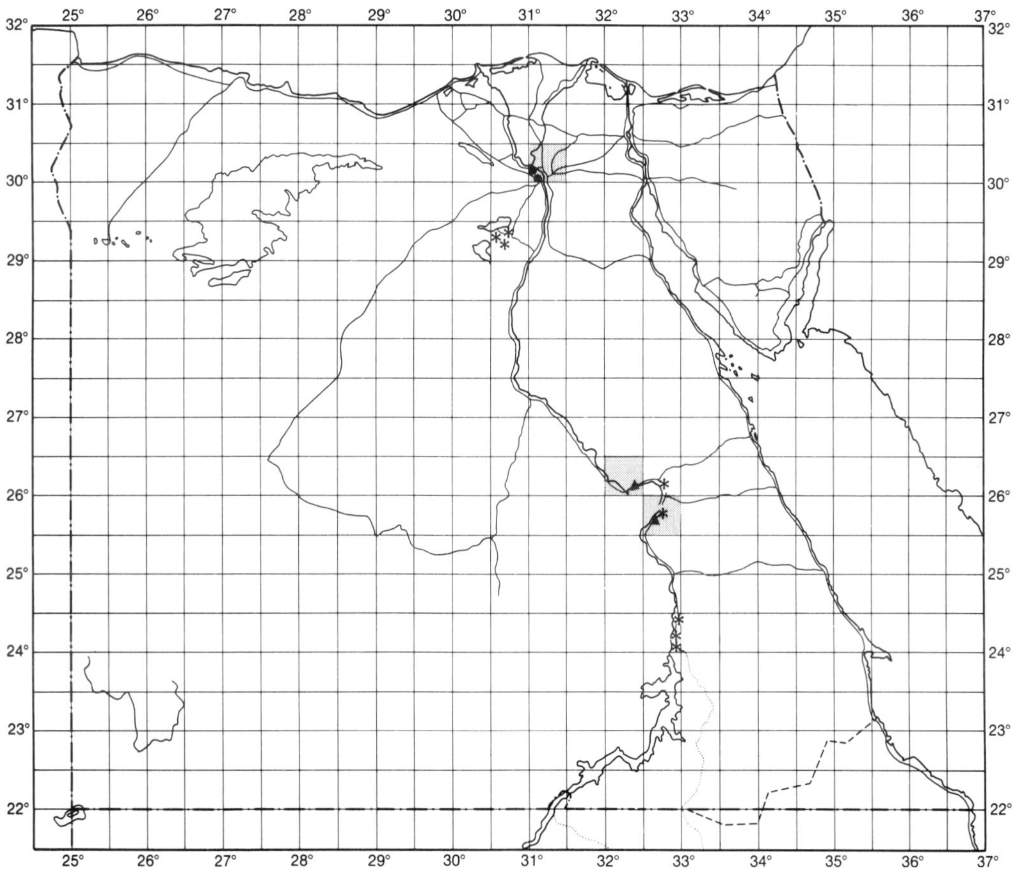

* possible breeding ▲ probable breeding ● definite breeding ░ half degree grid plots with probable or definite breeding

Map B51. Breeding distribution of Great Spotted Cuckoo *Clamator glandarius* between 1970–87.

Between mid-March and mid-April there is a distinct increase in the number of birds observed in the Nile system. Some of these are presumed to be Eurasian migrants and the balance local breeding birds. There are several spring records from Sinai and Bahig. One was seen at Suez on 13 May 1982 (Wimpfheimer *et al.* 1983) and two at Fayid on 8 April 1952 (AP). These extralimital records are presumed to be of migrants. One was found on a ship in the Mediterranean, off Sinai, at 33°09′N., 34°45′E. on 24 August 1964 (Tuck 1964). The only known records from the Western Desert are single migrants in Bahariya on five days between 4 August and 3 September 1983 (HB *et al.*). This species is a scarce migrant at Elat, Israel, between mid-February and mid-April, and has occasionally been observed in July (Krabbe 1980).

Cuckoo *Cuculus canorus* PV

Local names Waqwaaq وقواق, huuhuu, jlima (NC), shakhfuut, taguug.

Scarce passage visitor throughout the country in autumn from mid-August to early October and in spring from (early) mid-March to mid- (late) May. The form occurring in Egypt is *C. c. canorus*. *C. c. telephonus* is treated herein as a synonym of *C. c. canorus* (Vaurie 1965).

The autumn passage of Cuckoos is most notable between mid-August and late September, when small numbers are regularly observed throughout the country. During autumn Cuckoos are distinctly less common than in spring. The latest autumn record is 4 October 1981 at El Faiyum (FdR).

The first spring Cuckoos are usually observed in mid-March. Exceptionally early records are one shot between Aswan and Wadi Halfa in early February 1913 (Koenig 1920) and one at Fanara on 4 March 1953 (AP). During spring passage, which peaks between mid-April and mid-May, this species is regularly recorded in small numbers throughout the country, including the Western and Eastern Deserts, and Sinai. The latest spring records are one on 20 May 1984 at Bir Abu Sha'r (UMMZ) and on 21 May 1917 at Abu Zabal (Raw 1921). A bird noted on 19 July at an undisclosed locality (Meinertzhagen 1930) may have been summering locally or an early autumn migrant.

Ringing One ringed in Estonia on 2 September 1971 was recovered at Bahig on 1 September 1972 (Kastepold 1974).

Senegal Coucal *Centropus senegalensis* RB Plate 2

Local names Mukk مك, burbur.

Fairly common breeding resident in the Nile Delta, the Suez Canal area, the Faiyum, and locally in the Nile Valley. The Egyptian breeding population is referable to *C. s. aegyptius*. It nests between mid-March and early August.

At the beginning of the twentieth century all known breeding records of the Senegal

Coucal were restricted to the Faiyum, the environs of Cairo, and the lower portion of the Rosetta Nile branch (Meinertzhagen 1930; Flower 1933). In recent years this species has apparently extended its range in the Nile Delta and the Lower Valley (Meininger *et al.* 1986). Presently, it is a fairly common and widely distributed breeding resident of orchards, palm groves, and reed-beds in scattered localities throughout the Nile Delta, the Faiyum, and the Nile Valley south to Beni Suef. Nestlings have been taken at Abu Rauwash on 18 June 1969 and 11 August 1969 (USNM). The species has also recently colonized the Suez Canal area. One was observed near the Little Bitter Lake on 6 May 1982 (Wimpfheimer *et al.* 1983) and a pair was seen carrying food just west of Ismailiya on 16 April 1984 (Siering 1986).

Recent records from the Nile Valley south of Beni Suef include: north of El Minya—one on 21 March 1985 (JW); Beni Hassan—one calling on 22 May 1981 (SMG); Hawatka—a few heard on 15 April 1978 (RJM); Luxor—one on 19 January 1979

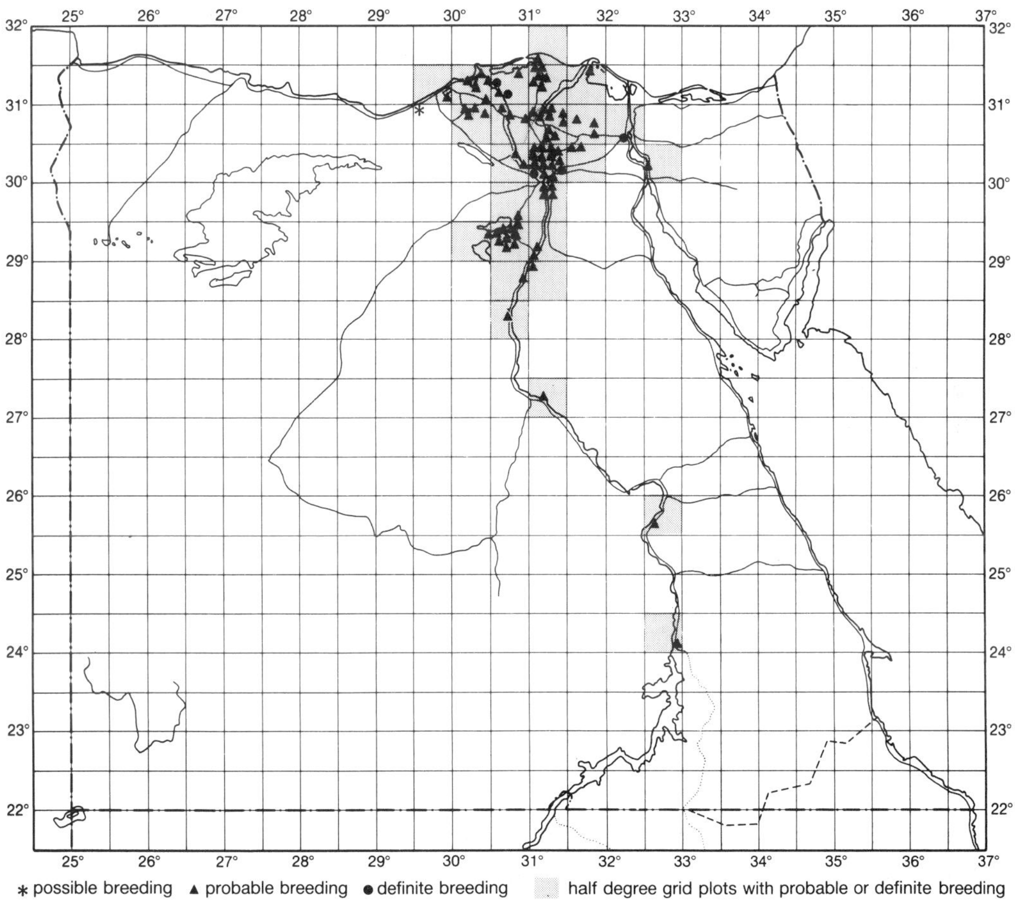

Map B52. Breeding distribution of Senegal Coucal *Centropus senegalenis* between 1977–87.

(Meininger *et al.* 1979) and five on 2 March 1982 (ES, HWu); and Aswan—one calling on 4 March 1980 (Short and Horne 1981).

ORDER STRIGIFORMES

Generic local name for owl: Buuma.

Family Tytonidae

Barn Owl *Tyto alba* RB

Local names Buuma ma<u>ss</u>aa<u>s</u>a بومة مصاصة, buuh, buuma bay<u>d</u>aa, haama, ma<u>ss</u>aa<u>s</u>a, umm a<u>s</u>-<u>s</u>akhr.

Fairly common local breeding resident in Wadi el Natrun, the Nile Delta and Valley, and probably also in Sinai. Specimens examined from Sinai appear to be of *T. a. erlangeri* and those from other portions of the country of *T. a. alba*. None of the Egyptian material we have studied is comparable in size or colour to specimens of *T. a. affinis* from central Sudan.

North Coast According to Meinertzhagen (1930) the Barn Owl occurs on the coast between Alexandria and Salum. The only recent records from this area include a bird taken on 2 August 1938 at Burg el Arab (GZM) and several collected between 1967 and 1971 at Bahig (USNM).

Western Desert This species is a resident of Wadi el Natrun. Its status in other oases is unclear (Goodman *et al.* 1986). Meinertzhagen (1930) mentioned it as a resident in Siwa, but Moreau (1940–1) did not accept this record. One was heard in Bahariya between 20 and 24 March 1925 (Moreau 1927). In Dakhla it was considered a local resident by Meinertzhagen (1930) but no details were given. A specimen collected on 29 March 1928 at Kharga by Meinertzhagen (BMNH) was not mentioned in his 1930 book.

Nile Delta and Valley This species breeds near human habitations of all sizes (including Cairo) and in ruins throughout the Nile Delta and Valley. On 10 July 1984, one was seen during the day in Wadi el Rayan (SMB). It is probably a resident in the Suez Canal area. One was collected at Ismailiya in March 1974 (ZM-HUJ). Since December 1980 the species has been frequently seen at Suez (SMB; PW). In March 1952 a large cache of pellets was located at Gebel el Goza el Hamra, where two pairs were suspected of breeding (AP).

Sinai A Barn Owl was heard at El Arish in March and remains of this species were picked up at Qusaima in the same month (Meinertzhagen 1930). In the summers of 1978 and 1979 it probably bred in a palm grove 10 km west-south-west of Rumana (OH). Recent specimens include a female collected at Nahal Yam in October 1976 (TAUM) and a male at Sharm el Sheikh on 24 July 1975 (TAUM).

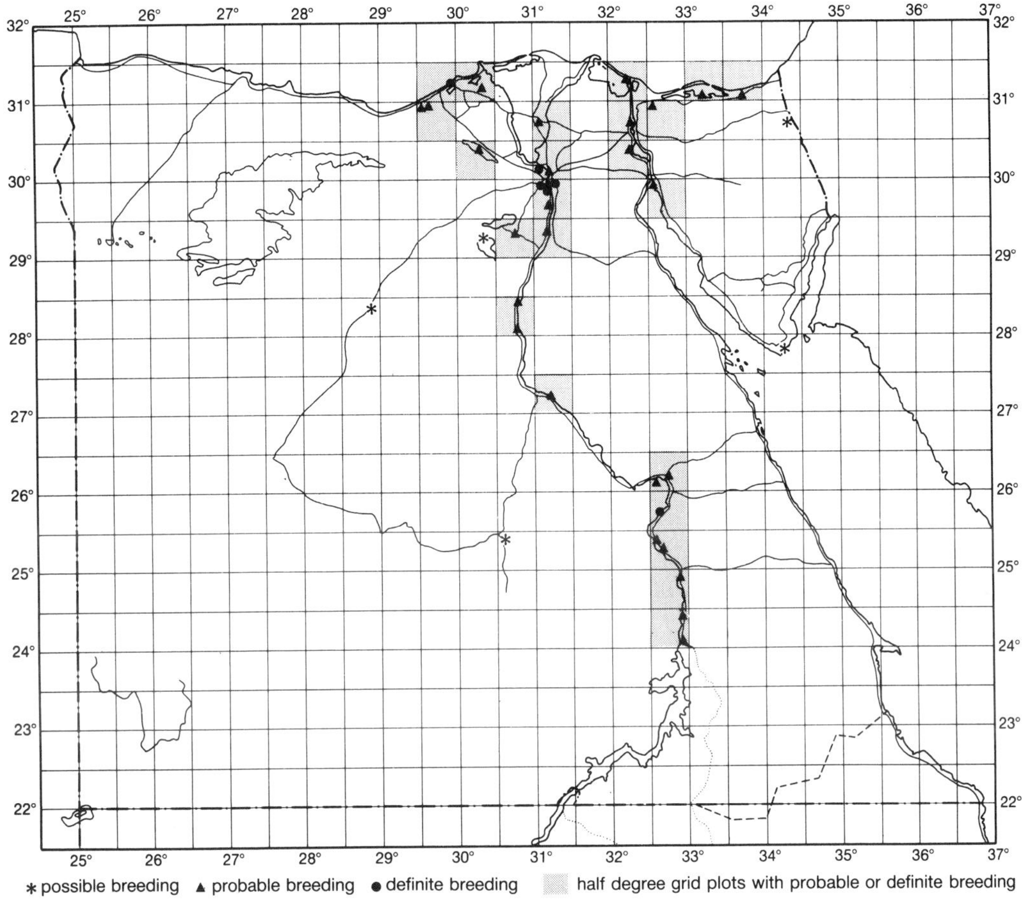

Map B53. Breeding distribution of Barn Owl *Tyto alba* between 1930–87.

Family Strigidae

Striated Scops Owl *Otus brucei* AV

Local name Thiij mukhaṯṯaṯ ثيج مخطط.

The only known Egyptian records of the monotypic Striated Scops Owl are two specimens taken in Sinai. One was collected at El Tor in November 1823 by Ehrenberg and the other near 'Um Heshim' on 1 December 1891 (Meinertzhagen 1930). There are a few recent records from Elat, Israel (Paz 1987).

Scops Owl *Otus scops* PV

Local names Thiij ثيج, buuh, haama, huuma (NC), sabaj, thabaj.

Scarce passage migrant in autumn from mid-August to mid-October (early November) and fairly common in spring from (late February) early March to late April (early May). This species is perhaps often overlooked because of its retiring habits during daylight hours. All Egyptian material examined is referable to *O. s. scops*. The only exception is one bird taken on 13 April 1973 at Matruh that approaches the plumage coloration of *O. s. cycladum* (USNM).

Western Desert Autumn records of the Scops Owl include three captured in Bahariya on 19 September and one on 20 September 1983 (HB *et al.*). There are several spring records from the area (Goodman *et al.* 1986). On 15 April 1977, one was captured at El Maghra, Qattara Depression (Goodman and Ames 1983); one was collected on an unknown date in Farafra (Goodman *et al.* 1986); and one was obtained on 11 March 1963 in Kurkur Oasis (YPM). As many as 14 were observed sitting in a tree in Wadi el Natrun on 14 April 1923 (Meinertzhagen 1930).

Nile Delta and Valley The majority of Egyptian records are from the Nile Delta and Valley, and the Faiyum. A bird collected at Port Said on 28 September 1906 had '2 crab claw ends' in its stomach (BMNH). Other autumn records include a bird collected at Damietta in September 1875 (Gurney 1876) and one observed on 26 August 1904 at Giza (Nicoll 1912*a*). Twenty-four specimens were taken at Bahig in the 1960s between 10 September and 13 October, and one on 13 August (FMNH; USNM; YPM). Kattinger (1970) claimed to have heard one calling in the Faiyum on the night of 6 August 1966. In six spring field seasons at Bahig between 1967 and 1973, the number of Scops Owls caught per season varied from 18 to 70, and 255 were captured in total (Horner 1977).

Red Sea Rare spring migrant along the coast. Birds have been observed at Safaga in April 1932 and on 27 March 1933 (Marchant 1941). An exceptional record is one observed in December 1985 along the beach of Gezira Zabargad, central Egyptian Red Sea (DP).

Sinai There are relatively few records from this area. Five were caught in Quail nets near Zaranikh between 18 and 23 September 1982 (Baha el Din and Salama 1984). One was heard calling at El Arish on the night of 10 October 1979 (Petersen and Sørensen 1980) and one male was collected in Wadi Kid on 24 October 1969 (ZM-HUJ).

Ringing A bird ringed on 9 April 1967 at Bahig was recaptured at the same locality on 10 April, 11 April, and in September 1967 (Watson 1971).

Eagle Owl *Bubo bubo* RB

Local names Ba'afa بعفة, buuh, buuma bi-quruun, buuma ghaamiqa, hid/hid (Ma), kuutif (Bi).

Breeding resident, probably fairly common, in desert areas throughout the country. Its distribution is imperfectly known. We agree with Vaurie (1960) that the form *B. b.*

desertorum is synonymous with *B. b. ascalaphus*. A 'hybrid' specimen intermediate between *B. b. ascalaphus* and *interpositus* (often referred to as *B. b. aharonii*) was collected at El Arish on 30 September 1918 (GZM). The rest of the Egyptian material examined is referable to *B. b. ascalaphus*.

North Coast According to inhabitants of El Omayed the Eagle Owl bred in the local environs in 1986 (WCM).

Western Desert Nicoll (1912*a*) stated that the Eagle Owl occurred in Wadi el Natrun; its present status is uncertain. Meinertzhagen (1930) mentioned it was recorded in Siwa, where one was taken on 7 March 1964 (Goodman and Ames 1983). In May 1985 this species was known to the local Siwans (Goodman *et al.* 1986). Almond (1937) found one in mid-March 1936 midway between Siwa and Matruh. Moreau (1927) saw one on 25 March 1925 at Befi, 120 km south of Burg el Arab on the track to Bahariya. In late March or early April 1979 it was noted in Farafra (Abdel Rahim *et al.* 1980) and birds have been taken locally (Goodman *et al.* 1986). Specimens were collected in Dakhla on 25 and 27 March 1928 by Meinertzhagen (BMNH), and according to local inhabitants it bred in 1981 (SMG, PLM, WCM). Moreau (1934) reported that Clayton found this species inhabiting a cairn at 27°N., 25°10′E., or about 190 km west of Dakhla, in the Great Sand Sea. Moreau (1927) saw one at Kharga in April 1925, and fledglings were taken in May on the outskirts of Dush, 15 km south of Baris (Al Hussaini 1959). In 1979, one was observed and photographed at Ain Shurafa, Kharga (CVH). One was noted between 8 November and 5 December 1968 at Karkur Talh and Wadi Hanzul, Gebel Uweinat (Misonne 1974).

Nile Delta and Valley Recent records from the general environs of Cairo include regular observations at the pyramids of Abu Rauwash, Giza, Saqqara, Dahshur, and Abu Sir, and also at Gebel el Muqqatam and Helwan. Records from Abu Zabal, on the eastern edge of the Nile Delta, include birds obtained on 23 June, 26 August, 8 October, and 16 October 1916 (Raw 1921; AMNH). There are several pre-Meinertzhagen (1930) reports, including specimens, from the Faiyum. The only recent record is a nest with young in 1986 at Qasr Qarun (IH). Short and Horne (1981) mentioned an observation at the Maidum Pyramid, and a cache of pellets thought to be from this species near Tuna el Gebel. The only recent record from the Luxor area, where it was seen in the Karnak Temple in November 1918 (Bartlett 1942), is one calling at Thebes on 2 March 1985 (RV). Specimens have been taken at Gebel Abu Foda on 28 December 1890 (FMNH), at Kom Ombo on 25 December 1971 (USNM), and at Ineiba on 27 February 1965 (YPM). Recent observations from Egyptian Nubia include: one on 24 January 1980 on Gezira Kalabsha, near Aswan (PLM, WCM), one on 16 April 1985 at Aswan (DKT), and one on 27 April 1984 at Abu Simbil (MBe, SM). A bird obtained at Abu Rauwash on 10 January 1910 had a Little Owl in its gizzard (GZM).

Eastern Desert According to Meinertzhagen (1930) Eagle Owl feathers were picked up along the Qena–Quseir road, and in recent years they have been found at Bir Beida, just west of Quseir (SMG). Tregenza (1951) mentioned summer records from several sites in the central Egyptian Red Sea Mountains, including Gebel Shayib el Banat, Wadi Shawak, and Gebel Qattar. A single bird was noted on 13 July 1985 perched on a telephone pole 65 km south of Safaga (SMB). These records are presumably of permanent breeding residents. In Wadi el Allaqi, five to six nests were found in 1964 (IH) and a specimen was collected on 24 March 1965 (FMNH). One was heard calling near Bir Abraq, Gebel Abraq, on 2 April 1985 (SMG, PLM). The Eagle Owl is apparently a breeding resident in the Gebel Elba area, where a bird was collected near Bir Kansisrob on 8 March 1954 (Goodman 1984*a*) and birds have been heard calling on several occasions in recent years (SMG, PLM).

Sinai Meinertzhagen (1930, 1954) reported the species as breeding near St Katherine and noted a bird at El Arish on 30 September 1918 (mentioned above). This

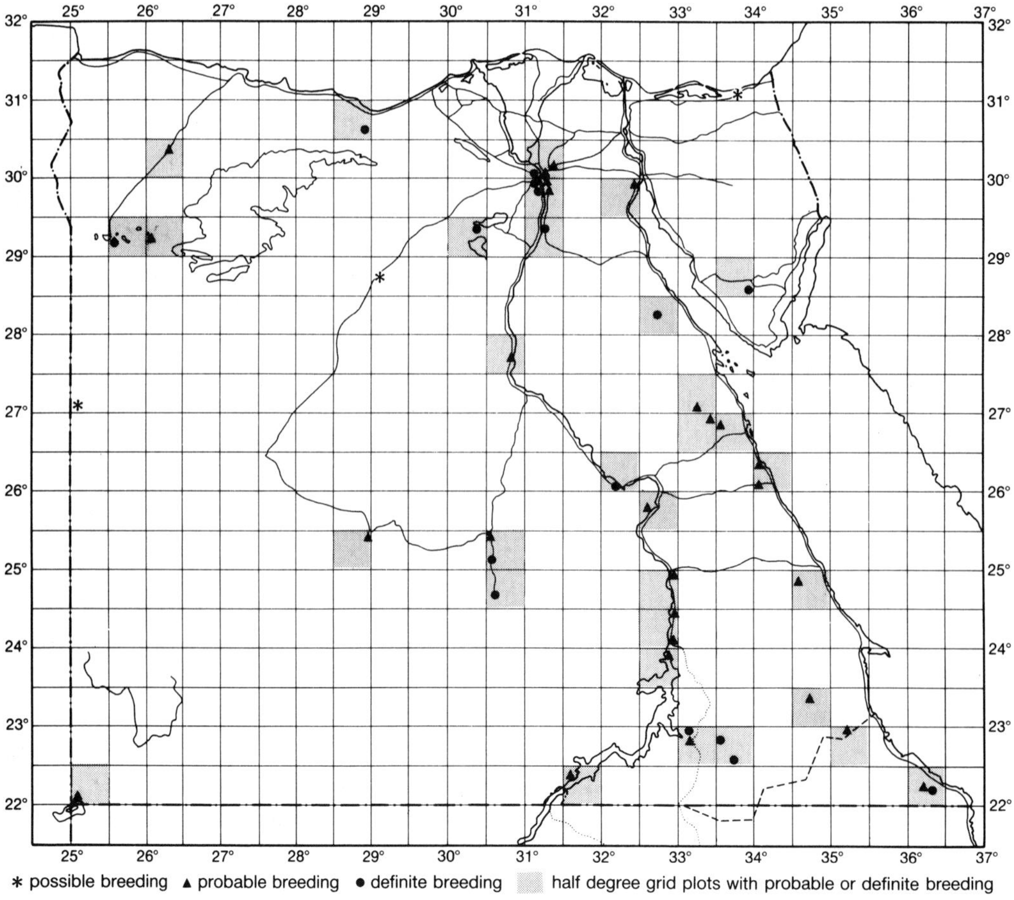

Map B54. Breeding distribution of Eagle Owl *Bubo bubo* between 1930–87.

species probably breeds in the mountains between Qusaima and Elat, Israel (IRIC). An adult was captured on 1 August 1976 at Refidim, just across the border in Israel (TAUM).

Little Owl *Athene noctua* RB

Local names Umm quwayq أم قويق, amjuuj (Bi), buuma (Ma), s̱adan, umm as-sahar.

Fairly common breeding resident of cultivated areas, along the desert fringe, and locally in rocky areas. Specimens from the Western Desert are highly variable in plumage coloration. Within an oasis, the colour variation sometimes spans the range used to differentiate the two subspecies *A. n. glaux* and *A. n. saharae*, and some populations appear to be polymorphic (Goodman *et al.* 1986). The pattern of geographical variation of the Egyptian Little Owl populations is in need of further study. A bird taken at Gebel Elba on 8 March 1954 is exceptionally small and based on size approaches specimens of *A. n. spilogaster* from coastal Sudan and Eritrea (Goodman 1984*a*). The subspecific identity of Little Owls inhabiting Sinai is not clear, although Vaurie (1965) referred this population to *A. n. lilith*.

North Coast Little Owls breed at many localities in the coastal desert between Matruh and Alexandria. Specimens have been taken near Salum in April 1921 (GZM) and on 23 October 1953 (FMNH), and between Matruh and Sidi Barrani on 11 August 1941 (UMZC). There are recent observations and specimens along the Matruh–Siwa road, up to 60 km from the coast. One was noted on 16 May 1986 at El Omayed and according to inhabitants it breeds locally (WCM). Fledglings have been taken at Bahig on 22 May 1975 (FMNH).

Western Desert There are several old specimens and breeding records of this species in Wadi el Natrun. The only recent records from this oasis are one on 18 January entering a burrow (SMG) and another on 9 August 1985 (DM); according to local people it is a common inhabitant (WCM). Presently, it is known to breed in Siwa, the Qattara Depression, Bahariya, Farafra, Dakhla, and Kharga (Goodman *et al.* 1986).

Nile Delta and Valley Meinertzhagen (1930) mentioned the Little Owl as abundant in the Nile Delta and Valley (south to Aswan), and the Faiyum. In January 1980, one was seen along the rocky shore of Lake Nasser, 60 km south of Aswan (PLM). This species can still be found throughout this region but is not as abundant as in the past. In recent years Little Owls have apparently colonized portions of the north-western Delta. Flower (1933) stated that the species was common in the central portion of the Delta but that he was unaware of records near Alexandria or Damanhur. Presently, it is fairly common at these two localities. Flower (1933) never saw or heard it in the Suez Canal area but there are several recent records from this area.

Eastern Desert This species appears to be an uncommon resident in the southern Red Sea Mountains. It may well breed in the Gebel Elba area (Goodman 1984*a*). One bird was heard calling on 21 June 1984 near Bir Kansisrob and another observed at Gebel Hamra Dom on 15 February 1984 (SMG).

Sinai Meinertzhagen (1930) stated that the Little Owl is fairly evenly distributed but scarce, throughout the Peninsula, and more common in the north near Rafa. There are a few recent records from Sinai, the southern-most being one at Dahab on 11 December 1984 (RvD *et al.*). One was taken in Wadi Gerady on 13 October 1975 (TAUM). This species may breed in the mountains between Qusaima and Elat, Israel (IRIC).

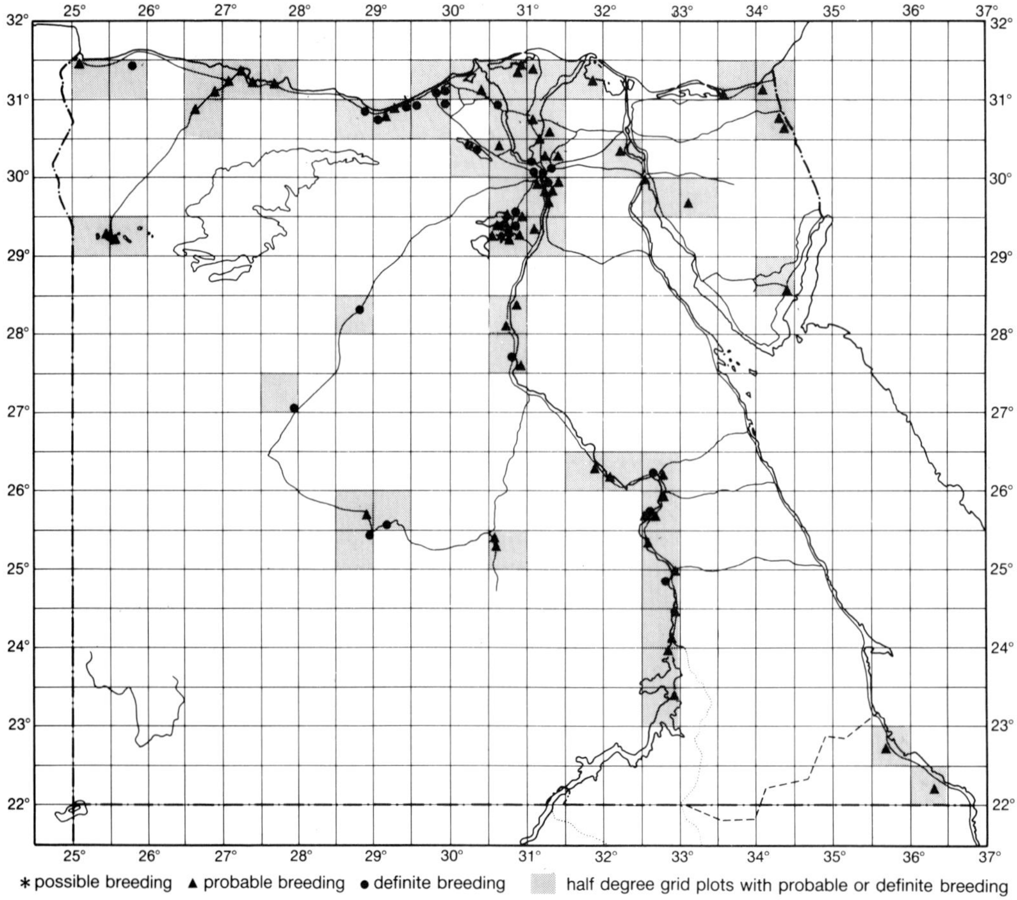

Map B55. Breeding distribution of Little Owl *Athene noctua* between 1930–87.

Hume's Tawny Owl *Strix butleri* RB Plate 5

Local names Buuma butlir بومة بتلر, buuma khabl, khabl.

Rare and local breeding resident in a few wadis and oases of southern Sinai. There is one record from the Red Sea Mountains. Monotypic.

Eastern Desert On 16 February 1982 a Hume's Tawny Owl was collected in Wadi Nugrus, about 54 km west of the Red Sea tarmac road. This is the only known record of this species on the African continent (Goodman and Sabry 1984).

Sinai While conducting a biological survey of Sinai in 1864, Wyatt (1870) heard an owl in Wadi Hibran and 'perhaps half a dozen times' elsewhere on the Peninsula, and he collected a Hume's Tawny Owl in Wadi Feiran (in Liverpool Museum). Meinertzhagen (1930) heard two birds and saw one at this latter locality in March 1928. Other Wadi Feiran specimens include: a male taken on 23 September 1911 (le Roi 1923; ZFMK), a male on 31 March 1914 (Phillips 1915; MCZ), and a female on 15 June 1939 (GZM). Tristram (1879) reported that he had a specimen of this species in his collection obtained at Gebel Musa. Other Sinai specimens include a female

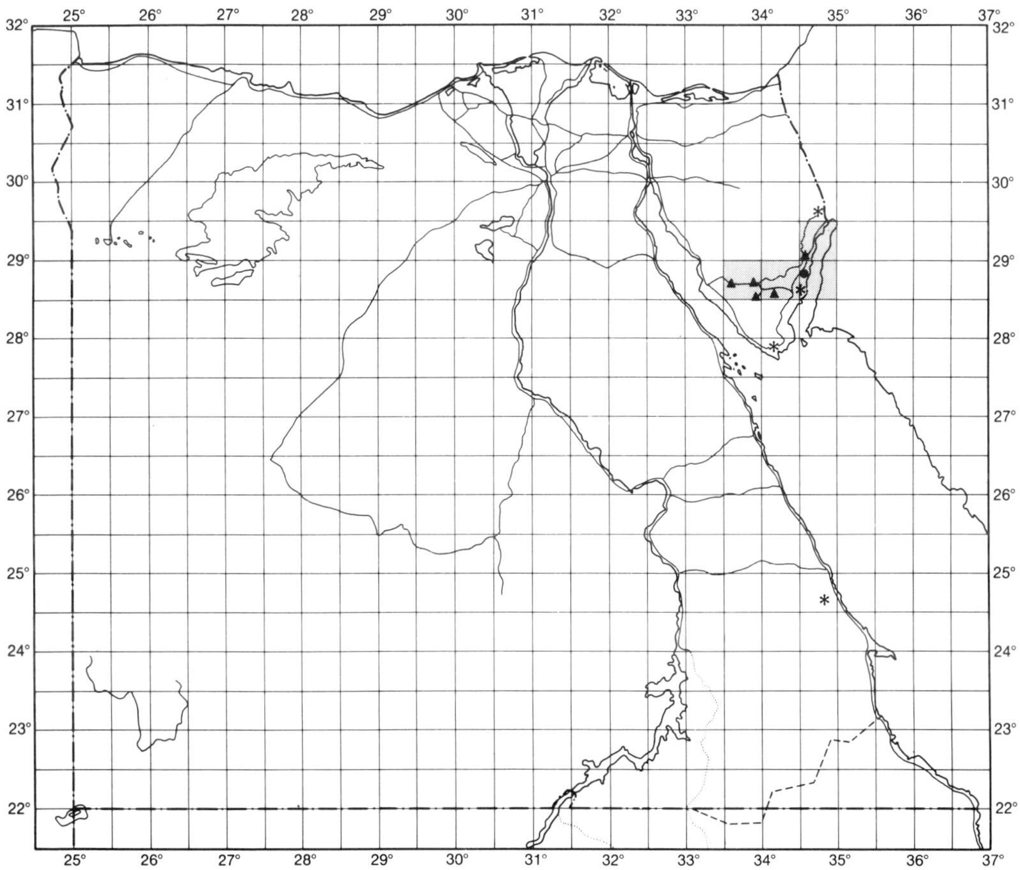

Map B56. Breeding distribution of Hume's Tawny Owl *Strix butleri* between 1970–87.

collected on 17 May 1970 at St Katherine (ZM-HUJ) and an unsexed bird found on 5 September 1976 in Wadi Watir, 7 km west of Nuweiba (TAUM).

In recent years this species has been recorded in several other localities in southern Sinai, mostly in the general area near St Katherine and also in coastal wadis between Elat, Israel, and Sharm el Sheikh (YL; IRIC). It was noted several times in Wadi Feiran in March and April 1986 (DB). Three were heard calling at this locality on 22 August 1986 (EE *et al.*). Feathers were found in Wadi Solaf in November 1983 and in Wadi Nasb in August 1984 (SMB).

Long-eared Owl *Asio otus* PV WV

Local name Buuma ṯawiila al-udhaan بومة طويلة الأذان.

Winter visitor in small numbers between late September and late March, almost exclusively to the Nile Delta. *A. o. otus* is the form occurring in the country.

Western Desert Long-eared Owls have been taken in Wadi el Natrun on 25 November 1911 (Nicoll 1912*b*) and on 18 October 1914 (Meinertzhagen 1930).

Nile Delta and Valley Small numbers have been infrequently noted in these areas, including the environs of Cairo. A roost composed of 11 birds was found on 16 November 1985 at Dahshur (SMB, DM). An exceptionally southern record is one near Abu Simbil on 3 February 1981 (SV). The only known summer record is a recently dead bird on 3 July 1983 near Ismailiya (SMB).

Sinai Records from Zaranikh include single birds on 19 October 1979 and 13 October 1979 (Petersen and Sørensen 1981*a*), and two on 29 September 1980 (Petersen and Sørensen 1981*b*). One was collected in February 1974 at Ras el Sudr (TAUM).

Short-eared Owl *Asio flammeus* PV WV

Local names Buuma qaṣiira al-udhaan بومة قصيرة الأذان, haama.

Scarce passage and winter visitor between mid-August and early May (early June). *A. f. flammeus* is the subspecies occurring in Egypt.

North Coast Short-eared Owls have been recorded in winter at Matruh (Meinertzhagen 1930). One was collected at this locality on 30 January 1920 (BMNH).

Western Desert There are several winter and spring records from Wadi el Natrun (Nicoll 1912*b*; Goodman *et al.* 1986). A single bird was noted at Hatiyet el Saqr, south of El Maghra, on 3 March 1984 (SMB). In Siwa, one was observed on 25 May and another on 4 June 1935 (Moreau 1940–1). Specimens have been collected in this oasis on 25 April and 4 May 1935 and also in Farafra (Goodman *et al.* 1986). An owl foot identified as belonging to this species was picked up in early May 1933 at 27°30′N., 25°35′E., or approximately 190 km south of Farafra (Moreau 1934;

Wingate 1934). Between 9 and 11 April 1984, two were seen at East Uweinat (SMB).

Nile Delta and Valley Winter observations are almost exclusively confined to the Nile Delta, the Faiyum, and Suez Canal area. During autumn and spring the species is also recorded further south in the Nile Valley, e.g. a group of five near Kom Ombo in mid-March (Cecil 1904), two between Beni Suef and El Minya on 17 March 1984 (Durdin 1984), and one flying north at El Balyana on 8 April 1983 (SC, PLM, IP).

Red Sea There are several records from the Red Sea coast, all during the migration periods and of single birds: between 4 and 24 August 1938 at Hurghada (Al Hussaini 1939); 1 November 1957 at Marsa el Alam (Horváth 1959); 13 November 1982 off Gezira Mahabis and another off Gezira Umm el Showra (Baha el Din and Saleh 1983); 24 February 1938 at Halaib (GZM); 6 March 1954 at Bir Kansisrob, Gebel Elba (FMNH); and 5 April and 18 April 1939 at Hurghada (Marchant 1941). There is some passage directly across the northern Red Sea (e.g. Tuck 1964; Casement 1977–8).

Sinai This species has been observed in northern Sinai in March and April, and a flock of 18 was noted near Nekhl in mid-March 1928 (Meinertzhagen 1930). Between 11 and 20 October 1979, 13 migrating Short-eared Owls were seen at Zaranikh (Petersen and Sørensen 1980). One was captured near Sharm el Sheikh on 18 October 1978 (TAUM).

ORDER CAPRIMULGIFORMES

Family Caprimulgidae

Generic local names for nightjar Subad, abuu an-nawm, ibb layy ir (Bi), lubbaada, qirra, ṯayr al-mawt.

Nubian Nightjar *Caprimulgus nubicus* MB? Plate 4

Local name Subad nuubii سبد نوبي.

Only known in Egypt from the Gebel Elba region. A single specimen collected at this locality is of *C. n. nubicus* (UMMZ). Nubian Nightjars breeding near Elat, Israel, are referable to *C. n. tamaricis* (e.g. USNM).

There are only a few records of the Nubian Nightjar in Egypt, all from Gebel Elba. On 18 February 1985, one was observed in Wadi Akwamtra, and three days later two were noted (SMG, DM). A single bird was observed on 25 February 1985 in a tributary of Wadi Serimtai leading to Wadi Rabdeit (SMG). Two to eight pairs of Nubian Nightjars were thought to have bred near Elat, Israel (Krabbe 1980) but there are no records of it at this locality since 1985 (HSh).

Nightjar *Caprimulgus europaeus* PV

Local name Subad uuruubii سبد أوروبي.

Scarce passage visitor throughout the country in autumn from late August to mid-(late) November and in spring from (mid-March) early April to early (late) May. Most migrants passing through Egypt are of *C. e. europaeus*. There are a few recent Egyptian specimens referable to *C. e. meridionalis*, e.g. one taken on 7 October 1966 at Bahig (USNM). Meinertzhagen (1930) identified a bird taken on a ship off Port Said on 30 May 1910 as *C. e. unwini*.

In both autumn and spring the majority of Nightjar records are from the Mediterranean coast and the Nile Delta. The species has also been noted several times in the Western Desert, the Faiyum, and the Suez Canal area.

The bulk of autumn migrants pass through Egypt in September. In some years considerable diurnal migration has been noted at Zaranikh: 11 between 11 and 20 October 1979 (Petersen and Sørensen 1980); 472 between 7 and 30 September 1980, with 148 on 8 September 1980 (Petersen and Sørensen 1981*a*); 224 between 16 August and 24 September 1981 (IP, UGS); and 113 between 2 and 24 September 1982 (Baha el Din and Salama 1984). Excluding observations along the north Sinai coast, migrant Nightjars are usually noted singly. An exception is 10 on 11 September 1942 at Abu Qir (Bodenham 1945*a*).

In spring Nightjars are regularly recorded in small numbers in various portions of the country. The earliest spring record is one heard calling over Suez between 15 and 19 March 1983 (ES, HWu), and the latest are one at St Katherine on 17 May 1970 (ZM-HUJ) and three calling at Fayid on 24 May 1952 (AP).

Ringing A bird ringed at Matruh on 28 April 1970 was recovered at Buqbuq on 30 April 1970 (Watson 1971).

Egyptian Nightjar *Caprimulgus aegyptius* RB PV

Local names Subad masrii سبد مصري, abuu al-'aman, al-bakhaakh, raggaada (Ma).

Scarce and local breeding resident throughout portions of the country. The patterns of geographical variation, distribution, and movements of this species in Egypt are poorly known. As it is currently recognized, *C. a. aegyptius* occurs in 'Egypt south of at least Cairo' and the distinctly paler *C. a. saharae* 'Egypt west of the Nile (breeding in Wadi el Natrun and El Fayum . . .)' (Vaurie 1965).

Western Desert Specimens of both *C. a. aegyptius* and *C. a. saharae* were taken in Wadi el Natrun between 18 March and 13 June 1910. A female (referable to *saharae*) taken on 8 June 1910 had 'much enlarged ovaries' (Nicoll 1912*b*; Goodman *et al.* 1986). On 20 March 1935 a group of 30–40 Egyptian Nightjars was chased up at El Beida (Jourdain and Lynes 1936). There is no recent evidence of this species breeding in Wadi el Natrun. An immature was collected in Bahariya on 13 June 1974 (FMNH). In Dakhla, about 60 were seen between 2 and 4 October 1962 (Scharlau 1963), single birds on 18 October 1978 and in late January 1984 (Goodman *et al.* 1986), and one on 26 February 1987 (AH).

Nile Delta and Valley Between 1915 and 1919 this species was occasionally observed at Abu Zabal, particularly during the period from mid-August to late September, when flocks of up to 50 were noted (Raw 1921). Specimens of both *saharae* and *aegyptius* have been collected at this locality, including a female on 20 May 1916 with a well-defined brood patch and referable to *saharae* (Raw 1921). Specimens have also been taken along the eastern edge of the Nile Delta between March and November (Meinertzhagen 1930). A pair with enlarged reproductive organs was collected at El Qatta on 25 May 1910 (Sparrow in Raw 1921). On 21 May 1986 the species was found breeding on a dry mud-flat along Lake Manzala, half-way between Damietta and Port Said (WCM). Two clutches of eggs were found and one bird collected (referable to *aegyptius*) at Sidi Salim on 29 May (Raw 1921). A pair nested in the summer of 1954 on a roof-top in Sirs el Liyyan (HCK). A clutch of two eggs was collected at El Qassasin on 1 June 1950 (JCC) and a nestling at Abu Rauwash on 28 July 1959 (FMNH). There are a few records from the Suez Canal area. One was collected on 31 May 1910 at Suez (BMNH) and another found dead on 3 April 1955 at Abu Sultan (RN). In 1952 a pair was suspected of breeding near Fayid (AP). One was seen in early July 1986 near El Qantara (DM). Specimens referable to *saharae* have been collected in the greater Cairo area in October, January, and from February to late May (Meinertzhagen 1930). In recent years this species has been recorded during the summer months in the deserts near Cairo, Heliopolis, and Giza, and it is thought to be a relatively common breeding bird on roof-tops of old Cairo (MuS).

Records from the Nile Valley include: Faiyum—eggs collected on 7 June 1902 at 'Wadi de Deir el Beda' (Koenig 1919), a specimen referred to *aegyptius* taken in March (Meinertzhagen 1930), and one observed on 30 May 1980 (SMB, BB); Beni Hassan—two on 2 September 1983 (Hume 1983) and single birds on 17 April 1984 (MBe, SM) and 16 May 1985 (JS *et al.*); El Amarna—three birds on 15 February 1982 (SQ); Girga—nest with two eggs on 29 March 1913 (Meinertzhagen 1930); Luxor—one calling between 24 and 27 March 1982 (JW) and one observed on 5 November 1984 (RvD *et al.*); Aswan—four on 22 March 1988 (JS *et al.*); and Abu Simbil—one calling on 29 March 1982 (JW).

Eastern Desert and Sinai Dead Egyptian Nightjars have been found in the Eastern Desert on 29 October 1957 at Bir Abbad (Horváth 1959) and 29 March 1985 at Hurghada (MCJ *et al.*). One was observed on 6 November 1982, 30 km south of Hurghada (Baha el Din and Saleh 1983). Records from the south include specimens taken on 23 February 1938 (GZM) and 25 December 1938 (HGD) at Gebel Elba, and in mid-January 1896 near Sawakin el Qadim (Cholmley 1897). The only known Sinai records are single birds at Sharm el Sheikh on 7 March 1984 (SG) and 29 March 1985

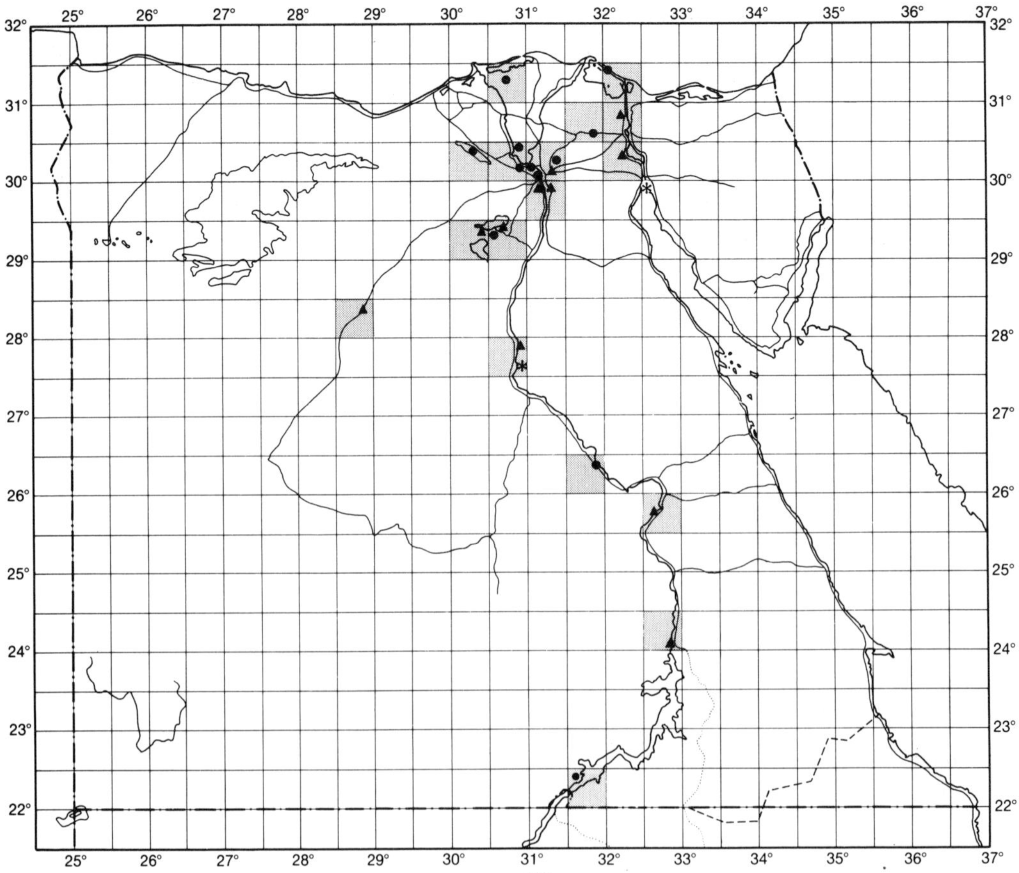

∗ possible breeding ▴ probable breeding • definite breeding ▒ half degree grid plots with probable or definite breeding

Map B57. Breeding distribution of Egyptian Nightjar *Caprimulgus aegyptius* between 1930–87.

(UGS). At Elat, Israel, this species is considered an uncommon spring migrant between late March and mid-April (Krabbe 1980).

ORDER APODIFORMES

Family Apodidae

Generic local names for swift Samaama, khuttaaf, salwaa (NC), samiima.

Swift *Apus apus* PV

Local name Samaama سمامة.

Fairly common passage visitor throughout the country in autumn from mid-August to mid-October and in spring from late February to mid-June. Records of this species in

Egypt are based almost entirely on observations. It is often difficult to separate *A. apus* from *A. pallidus* in the field, particularly the dark forms *A. p. brehmorum* and *A. p. illyricus*. Thus, it is assumed that some of the observations reported in the literature as *A. apus* are actually *A. pallidus*. The only Egyptian specimen examined is a bird taken on 27 May 1984 in Wadi el Natrun and referable to *A. a. apus* (UMMZ).

Swifts apparently pass over Egypt along a broad front. The latest autumn Swift record is a few observed in a mixed flock with Pallid Swifts on 21 October 1979 between El Balyana and Luxor (FV). An exhausted bird was captured on 10 September 1947 aboard a ship in the Red Sea off Daedalus Reef (Elliott and Monk 1952). Madden (1930) recorded this species in late January and early February 1925 near Gebel Elba and Halaib; it is impossible to know if these birds were wintering locally or early spring migrants.

In spring Swifts have been observed throughout Egypt. The bulk of passage is between late March and mid-May. A large number was noted moving through Bahariya on 22 March 1925 (Moreau 1927). The latest spring records are two over the Red Sea, off Brothers Reef, on 10 June 1984 (BH *et al.*) and one at Salum on 11 June 1942 (Rowntree 1943).

Pallid Swift *Apus pallidus* RB MB PV

Local name Samaama baahita سمامة باهتة.

Common, but local, breeding bird in towns and on cliffs. *A. p. pallidus* is the form nesting in the Nile Delta and Valley. Specimens of the distinctly darker forms *A. p. brehmorum* and *A. p. illyricus* have also been taken in the country, the latter being poorly differentiated from the former. A portion of these darker birds are certainly Eurasian migrants. *A. p. brehmorum* is known to nest in caves along the Cyrenaican coast of Libya (BMNH), just across the Egyptian border, and may well be the form breeding near Matruh and in Western Desert oases. This is in need of further study.

North Coast One dark Pallid Swift was shot from a flock on 21 January 1920 at Salum (BMNH). On 8 May 1985, at least five pairs were observed displaying and copulating at Matruh, and the species is presumed to breed locally (SMG, PLM). It was also noted at Matruh on 1 and 2 July 1953, possibly visiting nest sites (ACK). In 1942 it was found breeding in quarries near Bahig (Maclaren 1944). In recent years specimens referable to *illyricus*, *brehmorum*, *pallidus*, and intermediates have been taken at Bahig (Goodman and Watson 1983; FMNH; USNM).

Western Desert Six Pallid Swifts were observed in Bahariya between 14 and 22 February 1938 (Al Hussaini 1938*b*) and four on 28 April 1981 (Goodman *et al.* 1986). A male with enlarged gonads and intermediate in plumage coloration between *brehmorum* and *illyricus* was collected in Bahariya from a flock of 30 on 17 March 1984 (UMMZ). Meinertzhagen (1930) found this species breeding in earth cliffs at Dakhla, an oasis where it has been recorded throughout the year (Goodman *et al.*

1986). A female taken at Dakhla on 16 January 1972 appears to be of *brehmorum* (Goodman and Watson 1983). In Kharga this species was observed on 15 and 16 April 1925 (Moreau 1927) but there is no evidence of local breeding. A male collected in this oasis on 13 January 1972 approaches the plumage coloration of *A. p. brehmorum* (Goodman and Watson 1983).

Nile Delta and Valley The Pallid Swift breeds locally in towns and villages. It should be noted that nesting has been proven in relatively few localities and not south of Luxor. The species is particularly common in Alexandria, Cairo, and on the cliffs above Deir el Bahari, near Thebes. At some known breeding sites Pallid Swifts are absent between October and January, while at others the numbers are reduced during the winter months. A specimen taken in the Faiyum at El Lahun on 4 December 1959 appears to be of *brehmorum* (FMNH). Spring passage appears to begin as early as late January. About 900 were seen migrating north over Lake Nasser between 22 and 24

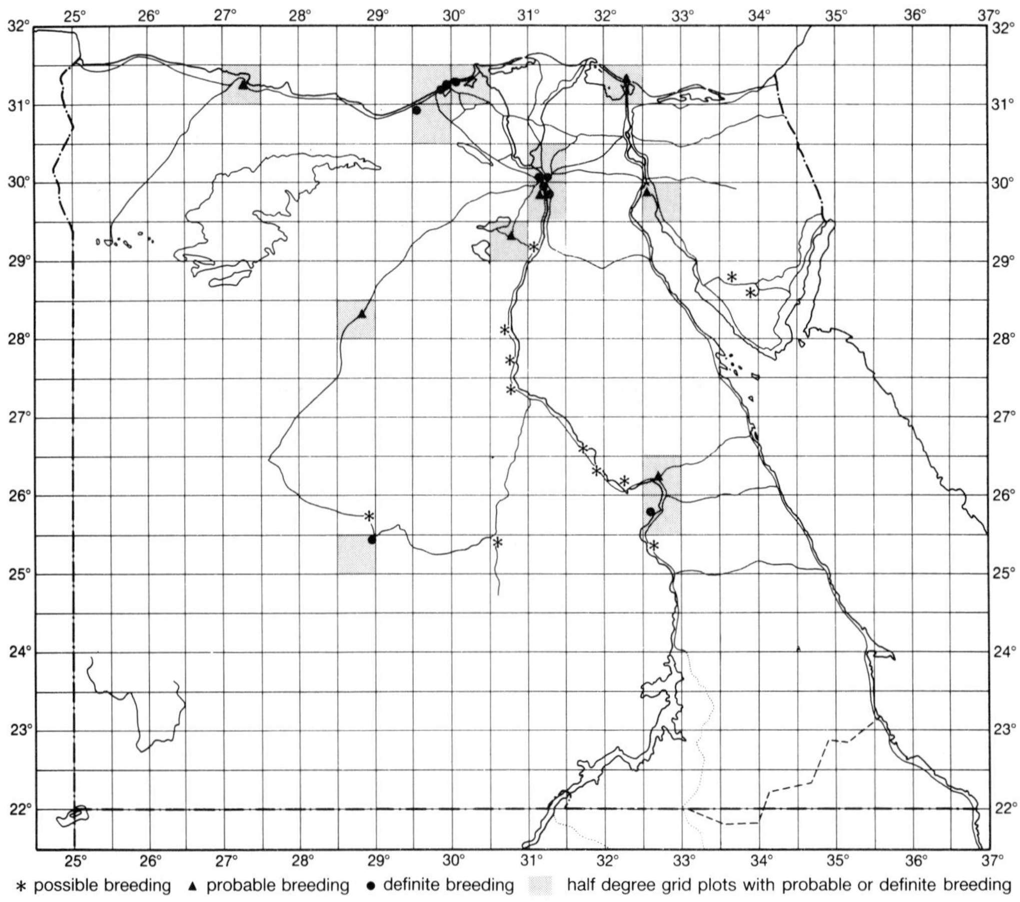

Map B58. Breeding distribution of Pallid Swift *Apus pallidus* between 1970–87.

January 1980 (SMB, PLM, WCM). On 16 March 1982 about 4000 were noted between Qena and Nag Hammadi (Welch 1983), and on 2 May 1978 about 120 were counted flying north between Luxor and Isna (Meininger and Dielissen 1979). This species has been noted during the breeding season at Port Said and Suez, where it is presumed to nest.

Eastern Desert and Sinai The species is apparently an uncommon migrant in the Eastern Desert and along the coastal plain. Relatively recent records include one bird in Gemsa Bay on 16 March 1983 (NM), and two over Gezira Abu Minqar on 14 April 1965 and 22 at Ain Sukhna on 19 April 1965 (HCK). Pallid Swifts have been observed in the southern Sinai mountains throughout the year, and the species possibly breeds in Wadi Feiran and near St Katherine (Meinertzhagen 1930). A specimen taken in Wadi Feiran on 9 May 1939 is referable to nominate *pallidus* (GZM), as is one collected in Wadi Tarfa on 27 April 1911 (RMS). Between mid-February and late April small numbers of migrating Pallid Swifts are observed in southern Sinai. At Elat, Israel, the species is occasionally observed in autumn between November and February and is abundant in spring between mid-February and early May, with up to 15 000 noted on a single day (Krabbe 1980).

Ringing There are several recaptures or recoveries at or near the original ringing sites of Bahig and Ikingi Maryut. These include nine birds ringed in spring and recovered the subsequent spring, two birds ringed in spring and recovered two springs later, and one bird ringed in spring and recovered the following autumn (GZF).

Alpine Swift *Apus melba* PV (WV)

Local names Samaama aṣ-ṣuruud سمامة الصرود, 'uwwaam.

Rare passage and winter visitor. There is no evidence of this species breeding in Sinai, as suggested by Etchécopar and Hüe (1967) and Hovel (1987). A specimen taken on 31 October 1957 at El Dokki is referable to *A. m. melba* (GZM) and one collected on 4 October 1917 at Rafa to *A. m. tuneti* (BMNH).

North Coast The only known Alpine Swift records are one on 18 March 1942 at Sidi Barrani (Tomlinson 1943) and three on 4 April 1980 near Burg el Arab (SV).

Nile Delta and Valley The Alpine Swift is infrequently noted in small numbers in the Nile system. Most records are from September, March, and April. Winter records include: eight at Karnak on 14 January 1954 (HCK); single birds between Beni Suef and El Minya on 18 January 1979 (Meininger and Mullié 1981*c*); and at Thebes on 7 February 1977 (Short and Horne 1981) and 27 February 1979 (SMG). Records from Suez include one on 27 September 1981 (RGB *et al.*), and one on 19 March and three on 30 March 1982 (Wimpfheimer *et al.* 1983). One was seen on 6 May 1985 at Giza (SMG, PLM), and eight on 26 July and one on 27 July 1983 in the centre of Cairo (TF). The only known records west of the Nile Valley (excluding the North Coast) are a 'small party' in March 1923 in the Faiyum (Meinertzhagen 1930) and one on 18 April 1986 in Wadi el Rayan (SMB, DM, WCM).

Eastern Desert and Sinai On 29 April 1965, four were noted at Ain Sukhna (HCK). One specimen was collected on a ship in the Red Sea, off Halaib, on 11 June 1984 (RNHL). This species was observed between 14 and 30 March 1979 on Gezira Tiran (RF). A few were also noted at Ain el Gedeirat on 15 May 1939 (Galal el Din and Fouad 1939–40). It is considered an uncommon migrant at Elat, Israel, in early October and between mid-February and mid-May (Krabbe 1980).

Little Swift *Apus affinis* (PV)

Local name Samaama s̱aghiira سمامة صغيرة.

Rare passage migrant. No Egyptian material of this species is known. On geographical grounds the subspecies occurring in the country are either *A. a. affinis* and/or *A. a. galilejensis*.

We are aware of the following observations of the Little Swift in Egypt. One was noted at Alexandria on 5 September 1957 (Horváth 1959; LH), seven over Suez on 2 November 1981 (RGB *et al.*), and one at Luxor on 12 April 1982 (GdB). At least 50–100 Little Swifts were observed on 14 August 1984 in a mixed flock of larger swifts (*A. apus/pallidus*) at Cairo (DM). The species was also observed on Gezira Tiran between 14 and 30 March 1982 (RF). At Elat, Israel, this species is considered a scarce spring migrant between late February and early May (Krabbe 1980).

[**Palm Swift** *Cypsiurus parvus* AV?

The only twentieth-century record of the Palm Swift in Egyptian territory is a few birds observed at Abu Simbil in February 1928 (Meinertzhagen 1930). Von Heuglin (1869*b*) stated that it was a resident in southern Egypt and that he found nests between May and August near Wadi Halfa, Sudan. Koenig (1919) spent a considerable amount of time in Nubia around the turn of this century and did not observe it north of 18°N. Since the filling of Lake Nasser in the mid-1960s potential breeding habitat of this species in Egyptian Nubia has been virtually destroyed.]

ORDER CORACIIFORMES

Family Alcedinidae

Generic local names for kingfisher S̱ayyaad as-samak, 'aruusa al-mans̱aba (ND).

White-breasted Kingfisher *Halcyon smyrnensis* RB? WV

Local names Qaawanad قاوند, diik al-baẖr, muhalhal.

Irregular winter visitor to the north of the country. Possibly breeding in northern Sinai and since 1985 probably a few pairs in the southern Nile Delta. The subspecies *H. s. smyrnensis* is the one occurring in the country.

Nile Delta Flower examined a White-breasted Kingfisher said to have been collected near Zagazig in 1899 (Meinertzhagen 1930). One was noted by Mackintosh at El Tell el Kebir in early December 1944 (RHG) and another was taken at El Mansuriya on 15 January 1949 (GZM). At Birket el Hagg, north of Cairo, one was seen on 24 November 1984, and up to two or three pairs have been regularly observed between May 1985 and March 1988 (Merrie *et al.* 1986; WCM; JS *et al.*). The species presumably breeds at this locality. Single birds have been found at Suez on 5 October 1981 (JV) and 23 and 24 February 1982 (Wimpfheimer *et al.* 1983).

Sinai This species breeds along the Gaza strip, just across the present Egyptian political border (Meinertzhagen 1930; EKr). It may nest in the Rafa–El Arish area. Pairs were suspected to have bred near Yamit in the 1970s (YYT) and near Rumana in July and August 1979 (OH). Other Sinai records include: Rafa—one in November 1917 (Meinertzhagen 1930); El Arish—one on 25 September 1980 and 26 September 1981 (UGS, IP); Zaranikh—two between 15 August and 5 October 1978 (YP, UP), one ringed on 14 October 1979 (Petersen and Sørensen 1980), three to four between 16 August and 24 September 1981 (Petersen and Sørensen 1981*b*), and one on 17 September 1982 (Baha el Din and Salama 1984); Nuweiba—one on 23 and 24 February 1987 (PdG); and El Tor—recorded at this locality by Von Heuglin (1869*b*). Five to six pairs have bred in recent years at Elat, Israel (HSh).

Kingfisher *Alcedo atthis* CB? WV

Local names S̱ayyaad as-samak صياد السمك, abuu mumbalad (NC), abuu ar-raqs̱, maazuur, rafraaf.

Perhaps occasionally breeds in the country. Fairly common passage and winter visitor between (late July) early August and late April (mid-May). All Egyptian material examined is referable to *A. a. atthis*.

North Coast Small numbers of Kingfishers are regularly recorded, mainly in autumn, along the Mediterranean coast between Alexandria and Salum.

Western Desert This species has been found on a few occasions in the oases of Wadi el Natrun, Siwa, Farafra (Goodman *et al.* 1986), and Qattara Depression (SMB). Single birds were captured in Bahariya on 13 and 28 September 1983 (HB *et al.*) and one was noted on 22 March 1984 (CC).

Nile Delta and Valley Between mid-August and late April this species is a regular visitor throughout the Nile Delta and Valley (south to Aswan), the Suez Canal area, and the Faiyum. It is particularly common near the Nile Delta lakes, where several hundred winter. Smaller numbers are regularly observed along the river and irrigation canals, although it is scarce in the Upper Nile Valley.

Most Kingfishers have left the country by mid-April. On 21 April 1902 a bird was observed with a fish in its bill near El Cap, along the Suez Canal (Parrot 1903). This record has been cited as possible evidence of local breeding (le Roi 1923). A single bird was noted near Suez on 12 June (Parrot 1903). The species was also reported to have bred in the Nile Delta in 1916 but no details are known (Meinertzhagen 1930). In the summer of 1917 Kingfishers were observed near Alexandria (Bartlett 1942) and during the summer months of 1923 and 1924 a pair was noted daily along the seashore near Ramleh (Brooksbank 1925).

Red Sea Kingfishers are regularly observed between mid-August and late April along the entire Egyptian Red Sea coast. The species is generally noted frequenting rocky shores, harbours, and mangroves. The latest spring records appear to be single birds at Safaga on 9 May 1983 (MCJ *et al.*) and at Hurghada on 21 May 1985 (JS *et al.*). One or two pairs were reported to have bred in Marsa Morakh ('The Fjord'), northern Gulf of Aqaba (Krabbe 1980); no details are known and this record is unacceptable (HSh).

North Sinai The earliest autumn record from this area is one bird on 21 July 1984 at El Arish (DM). Considerable numbers migrate along the north coast of Sinai. Systematic observations at Zaranikh have yielded the following counts: 1195 between 15 August and 5 October 1978 (YP, UP); 117 between 7 and 30 September 1980 (Petersen and Sørensen 1981*a*); 825 between 16 August and 24 September 1981 (Petersen and Sørensen 1981*b*); and 330 between 2 and 24 September 1982 (Baha el Din and Salama 1984). A few winter along El Malaha and Lake Bardawil (PLM, WCM). The latest spring record from the area is one at El Arish on 27 April 1985 (DM). Hovel (1987) reported that this species bred in 1982 in Sinai; however, details are lacking and this record is not acceptable.

Pied Kingfisher *Ceryle rudis* RB (WV) Plate 2

Local names S̲ayyaad as-samak al-abqa' صياد السمك الأبقع, abuu ar-raqs̲, 'ajiib, 'arussa al-bah̲r, mulaa'ib z̲illu, qirillaa, shirsh̲aar.

Fairly common breeding resident in the Nile Delta and Valley, and Faiyum. All Egyptian material examined is referable to *C. r. rudis*.

The present distribution of the Pied Kingfisher in the Nile Delta is virtually confined to the coastal lakes and their immediate surroundings. Here they breed in small colonies (up to 25 burrows) in steep banks bordering the lakes. In a May 1983 survey of the area 300 birds were noted (Meininger *et al.* 1986). This species is rare at inland areas of the Nile Delta. Five were observed at San el Hagar on 25 May 1978 (RJM, LRP), one near Bilbeis on 4 January 1985 (DM), and three near El Abbassa on 25 April 1986 (SMB, WCM). The Pied Kingfisher was unknown from Kasfarit before October 1944; thereafter it locally increased in numbers. Between 1944 and 1946 this species was 'noted as being extremely numerous on Lake Timsah' (LB). Small numbers currently breed along the Suez Canal between Ismailiya and Port Said, near Lake Timsah, and on the western shore of the Great Bitter Lake. The only known records from Suez are two on 2 March 1982 (Wimpfheimer *et al.* 1983), one on 20 August 1986 (EE *et al.*), and one on 14 October 1986 (EH, PK).

Until the late 1950s this species was a widely distributed resident throughout the Nile Delta and Valley, and the Faiyum (Raw 1921; Meinertzhagen 1930; Flower 1933). For example, on 30 October 1916, 73 were counted at Giza (Flower 1933) and between 7 and 12 October 1957, 42 were recorded along Lake Qarun and at Kom Aushim, Faiyum (Horváth 1959). In the 1960s and 1970s breeding Pied Kingfishers seem to have almost completely disappeared from the southern Nile Delta, the Faiyum, and the Nile Valley between Cairo and Sohag. The species was also rather uncommon during winter in the Nile Valley between Sohag and El Minya.

On nine Nile boat trips in autumn and winter (1976–81) this species was never

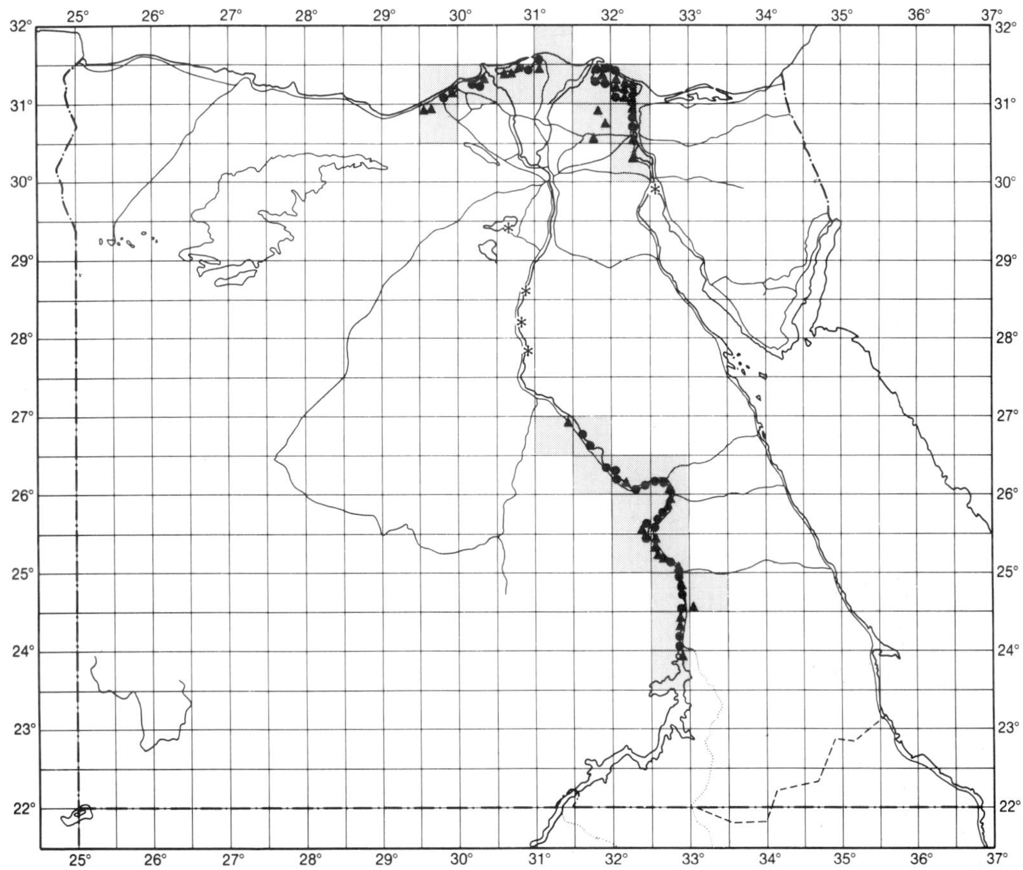

∗ possible breeding ▲ probable breeding ● definite breeding ░ half degree grid plots with probable or definite breeding

Map B59. Breeding distribution of Pied Kingfisher *Ceryle rudis* between 1970–87.

observed north of Tahta (Short and Horne 1981). The reason for this decline is not clear but may be related to water pollution in the Lower Nile Valley and an increase in the water salinity of Lake Qarun. In the Nile Valley between Sohag and Aswan this species has remained common. Pied Kingfishers were known to occur in the Nile Valley above Aswan before the completion of the High Dam; however, their current breeding status in Lake Nasser is unknown.

Since 1982 small numbers have been observed with increasing frequency between Maghagha and Sohag, mainly between September and March, and the species has apparently recolonized this area. Twelve nest burrows were found at El Maragha on 1 March 1985 (AS). No records are known from Lake Qarun between 1957 and the early 1980s but there are several recent reports from the lake. These include: 13 on 13 January 1982 (YL), and single birds on 28 January 1984 (SH, DM), 18 May 1985 (DM), and 26 September 1985 (Hume 1985). In the Nile Valley south of Sohag, the local population seems to have increased slightly in recent years and this section of the river contains a breeding population of several thousand pairs.

Sinai Pied Kingfishers have been recorded in small numbers in El Malaha, just across the Suez Canal from Port Said (PLM, WCM). One was observed at Nabq on 17 October 1976 (LNA). At Elat, Israel, the species is an occasional winter visitor between October and March (Krabbe 1980).

Family Meropidae

Generic local names for bee-eater Warwaar, abuu wahwaa (NC), abuu wah̲waah̲ (NC), daliil as-summaan (ND), ghuraab yabaanii (Su), khud̲d̲aar, khud̲d̲iir.

Little Green Bee-eater *Merops orientalis* RB Plate 1

Local names Khud̲d̲iir خضير, khud̲d̲iir mas̲rii, 'us̲fuur al-janna, warwaar mas̲rii.

Fairly common breeding resident in the Nile Delta and Valley (south to Aswan), and the Faiyum. The subspecies occurring throughout the Nile Delta and Valley is *M. o. cleopatra*. Blue-headed birds, presumably referable to *M. o. cyanophrys* which is known to occur in neighbouring areas of Israel (USNM; Madge 1981), have been observed near Taba. Meinertzhagen (1930) reported that a bird collected near Gebel Elba was referable to *M. o. viridissimus*. Several expeditions to Gebel Elba in the past few years have not turned up this species and the origin of Meinertzhagen's specimen is enigmatic.

Western Desert The Little Green Bee-eater is only known in this region from Wadi el Natrun, where at least eight were noted on 7 May 1981 and two on 10 April 1983 (Goodman *et al.* 1986).

Nile Delta and Valley In the early twentieth century the Little Green Bee-eater extended its range northwards into the Lower Nile Valley and the southern Nile Delta (Borman 1928; Flower 1933). Meinertzhagen (1930) stated that its distribution in the

Nile Delta was unclear and that it was not known to breed north of Benha. However, a clutch of eggs was reported to have been taken at Damietta on 16 April (Sparrow in Raw 1921) and a pair was observed at Alexandria on 1 June 1919 (Stoneham 1930).

Presently, it is a common breeding bird in the Nile Delta north of Cairo but appears to be rather local in other portions of the area. It probably bred near Alexandria in the 1970s (EK). On 11 May 1985 a nest was found near Zawyet Sidi Ghazi and several pairs were observed between Abu el Matamir and Hosh Isa (SMG, PLM). On 24 May 1980 this species was observed near Rosetta (SV). At least three pairs were found breeding on 16 April 1984 in a cultivated area between Cairo and Ismailiya (Siering 1986). Since 1980 it has been recorded several times in the Suez Canal area, north of Ismailiya. The only record from Suez is one flying south on 26 September 1981 (RGB *et al.*).

This species is a common resident in the Faiyum and the entire Nile Valley.

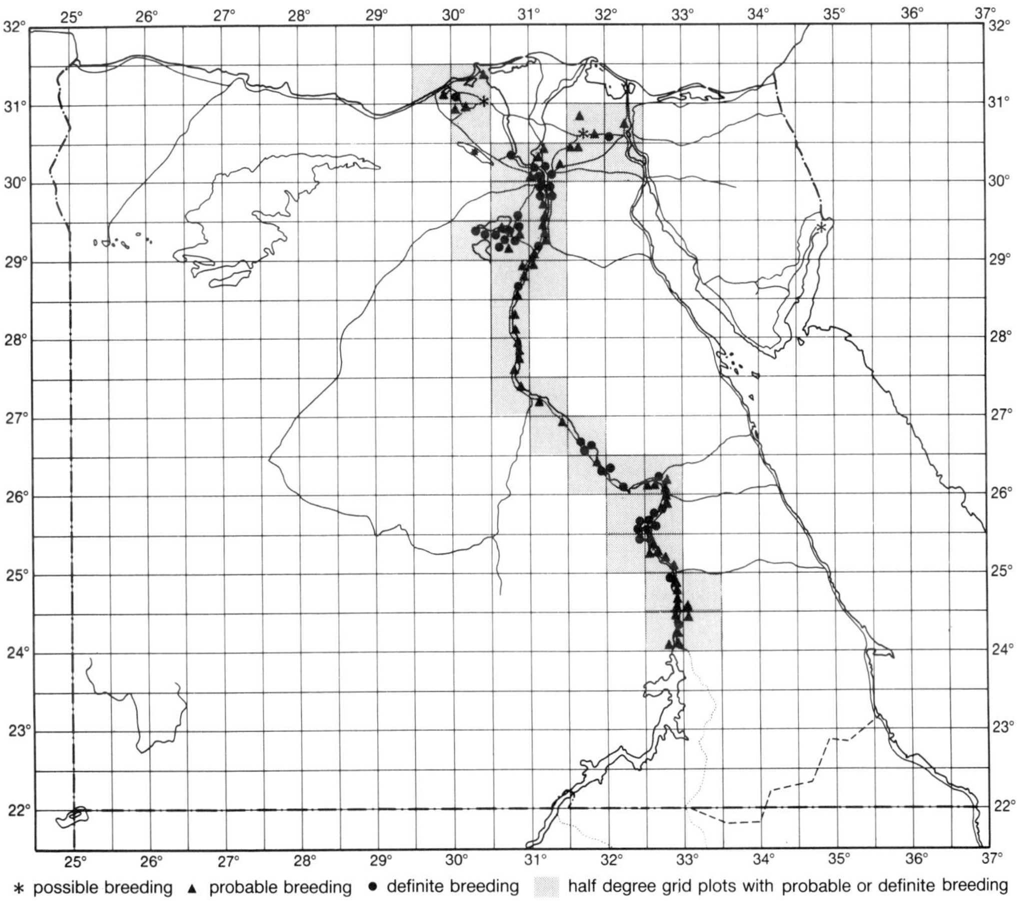

Map B60. Breeding distribution of Little Green Bee-eater *Merops orientalis* between 1970–87.

Meinertzhagen (1930) mentioned that it bred in the Nile Valley south to Luxor. At present it commonly nests at least to Aswan. It is not clear if this is a real range extension, since there are several records from the turn of the twentieth century between Luxor and Aswan (e.g. Gurney 1876; Koenig 1920).

Sinai In early and late March 1987 up to two blue-headed Little Green Bee-eaters were observed at Taba (PLM; RGB). The birds were singing but no sign of breeding was found.

Blue-cheeked Bee-eater *Merops superciliosus* MB PV

Local names Warwaar azraq al-khadd ورروار أزرق الخدّ, shigaag.

Fairly common migrant breeder in the north. Fairly common passage visitor throughout the country in autumn from late August to late October and in spring from (early) mid-March to mid-May. *M. s. persicus* is the subspecies occurring in Egypt.

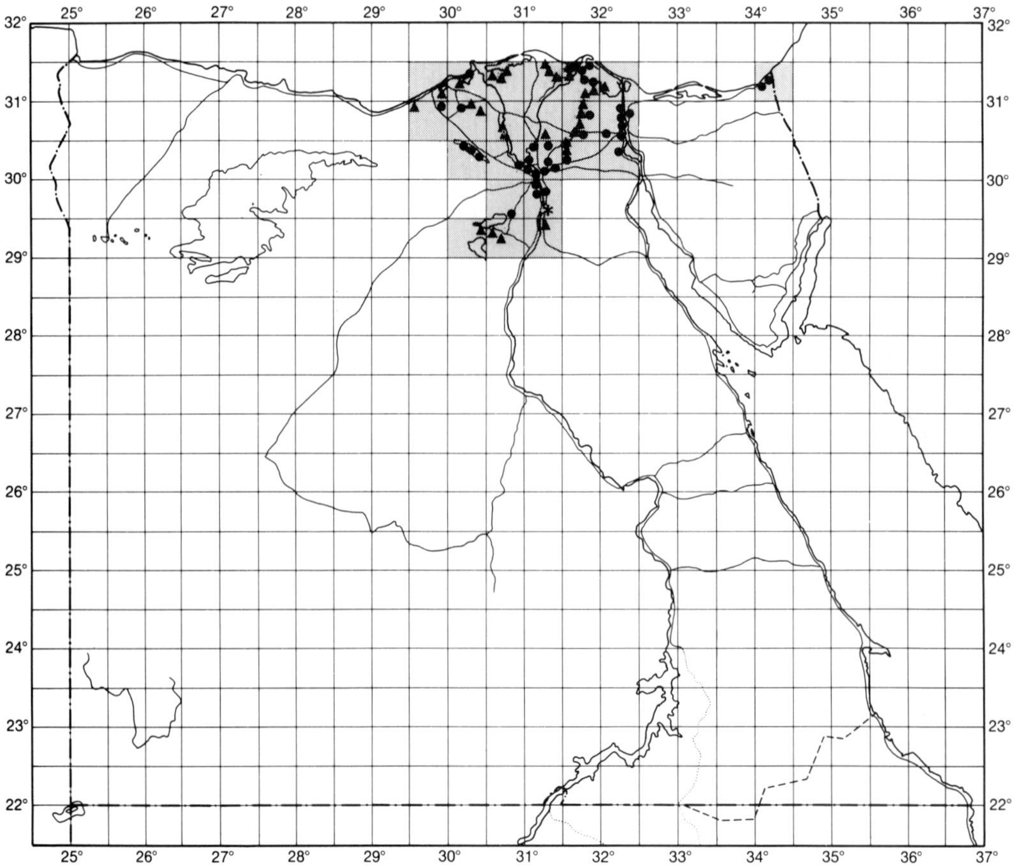

Map B61. Breeding distribution of Blue-cheeked Bee-eater *Merops superciliosus* between 1970–87.

The Blue-cheeked Bee-eater breeds in colonies in Wadi el Natrun, throughout the Nile Delta, the Suez Canal area, Lower Nile Valley south to at least El Badrshein and Helwan, and rarely in the Faiyum and north-eastern Sinai. This species has been observed several times in Wadi el Natrun during the breeding season. Local nesting was documented in June 1985 when at least 10 occupied burrows were found (SMB). One bird taken at Bahig on 6 July 1953 (FMNH) may have been breeding locally. Generally, breeding colonies are composed of up to 25 burrows; although one extended colony found on 7 May 1983 along a few kilometres stretch of dike near Abu Gridi, Lake Manzala, contained about 700 burrows (Meininger *et al.* 1986). Koenig (1920) found colonies in 1904 of over 100 pairs between Cairo and El Minya; however, there is no recent evidence of breeding south of Helwan. The first breeding record for Sinai was in 1979 when it was found nesting near Rafa (Paz 1987). Since then it has nested several times in this area, including 10 pairs in June 1986 (WCM).

The latest autumn record is 15 near El Wasta on 27 October 1979 (FV). There are a few autumn records from the Western Desert (Goodman *et al.* 1986). This species seems to be less common in autumn than spring. Haensel's (1980) record of three near Saqqara on 7 December 1972 is not acceptable.

The earliest spring record is over 100 between 10 and 11 March 1986 at Dakhla (AH). Between mid-March and mid-May flocks migrate along a broad front across the whole country. The latest record of spring migration is one on 17 May 1982 flying north along the Red Sea at Sharm el Bahari (Frazier *et al.* 1984).

Bee-eater *Merops apiaster* MB PV

Local names Warwaar uuruubii وروار أوروبي, kalb as-sahayl (Ma), shigaag, shihaag, 'usfuur al-janna.

Rare migrant breeder in north-eastern Sinai and common passage migrant throughout the country in autumn from late August to late October (late November) and in spring from (mid-March) mid-April to mid-May (late June). Monotypic.

On 26 June 1984 about 20 adult Bee-eaters were noted 5 km west of Rafa in a mixed colony with Blue-cheeked Bee-eaters. The Bee-eaters were observed entering nest holes and feeding fledglings (SMB). This appears to be the only documented breeding record from the country. Brewster (1966) mentioned that this species bred along the coastal belt of Sinai; however, the exact locality and details are ambiguous.

On both autumn and spring passage Bee-eaters migrate across Egypt along a broad front, including the Western and Eastern Deserts. Birds pass during the day and night. Migrants tend to concentrate in areas with vegetation and large numbers can be observed in oases and vegetated wadis and throughout the Nile Delta and Valley. The bulk of autumn migration takes place between late August and late October, with a distinct peak in September. Single birds were observed as late as 16 November 1983 at Abydos (Bezzel 1987) and 27 November 1965 at Alexandria (JWW).

The first spring migrants are usually observed in early April, although there are several mid- and late March records. Between mid-April and mid-May the species is common throughout the country and flocks of up to several hundred or larger are regularly noted. An exceptionally large migratory movement was noted at Fayid on 29 April 1952, when 5500 Bee-eaters were counted in three hours (AP). The species has been recorded as late at 19 June 1984 at Shalatein (SMG) and 22 June 1981 at Hurghada (SMB). There are also several summer records from the Nile Delta and Valley, the North Coast, and the Eastern Desert.

Family Coraciidae

Generic local names for roller Shiqirraaq, shirquuq.

Roller *Coracias garrulus* PV

Local names Ghuraab zaytuunii غراب زيتوني, bakaraadsh, bint barri, d̲u-d̲u-.

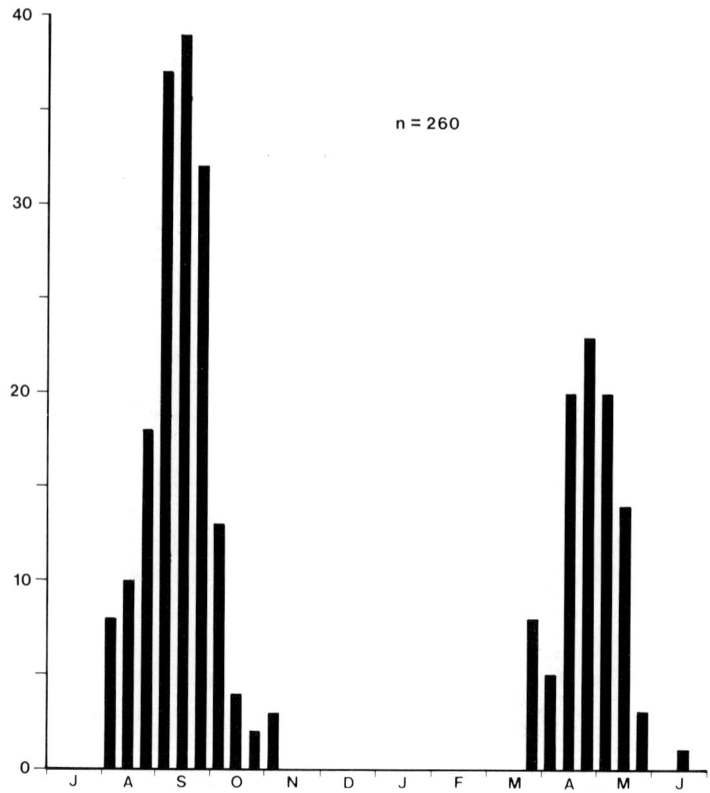

FIG. 6.23. Histogram of the numbers of Rollers recorded during 10-day periods throughout the year.

Scarce passage visitor throughout the country in autumn from (late July) early August to early October (early November) and in spring from late March to mid-May (mid-June) (Fig. 6.23). The subspecies *C. g. garrulus* is the one occurring in the country.

Meinertzhagen (1930) stated that the Roller was common on autumn migration between 30 July and 30 September, with the bulk passing between the end of August and mid-September. He also mentioned that it is rare on spring migration. Recent information shows a different pattern. Both in autumn and spring equally small numbers of Rollers pass through the country, including the Western and Eastern Deserts.

The earliest autumn records are birds on 30 July (Meinertzhagen 1930) at an undisclosed locality and 3 August 1942 at El Hammam, along the North Coast (Maclaren 1944). Passage is concentrated in the period between late August and early October. There is some migration across the Red Sea. At the Daedalus Reef three were noted on 7 August 1936 (Bulman and Beale 1944) and one on 19 September 1913 (Ticehurst 1924). The latest autumn record is two at Suez on 1 November 1981 (RGB *et al.*).

The earliest spring records are one at El Minya on 22 March 1985 (JW) and one at Asyut on 24 March 1984 (CC). The majority of Rollers pass between mid-April and mid-May. The latest spring records are one on 29 May 1966, 75 km from Cairo on the desert road to Alexandria (HCK), and another on 14 June 1984 at Marsa el Alam (SMG).

Ringing A total of four Rollers ringed in Latvia and Poland have been recovered at Damietta (Mackintosh 1941).

Abyssinian Roller *Coracias abyssinicus* AV

Local name Ghuraab zaytuunii h̲abashii غراب زيتوني حبشي.

There are three records of the monotypic Abyssinian Roller in Egypt. One was obtained at Abu Simbil on 14 February 1928 (Meinertzhagen 1930). Misonne (1974) found a wing and tail of this species on 22 November 1968 at Karkur Ibrahim, Gebel Uweinat. One specimen collected in 1874 is labelled 'Egypt' (FMNH).

Family Upupidae

Hoopoe *Upupa epops* RB PV

Local names Hudhud هدهد, abuu 'ab'aab (NC), abuulaal indiir (Bi), buu 'ab'ab (Si), dolo indiil (Bi), fiqiih bilaa dawaaya, gubbaar abuu fi-gafaa (Ma), ibn sakaatuu (Bi).

The subspecies *U. e. major* is a common breeding resident in several Western Desert oases and throughout the Nile Delta and Valley. It breeds between mid-February and

June. *U. e. epops* is a regular passage visitor in autumn between mid-August and early November and in the spring between early March and early (late) May. Birds breeding in north-eastern Sinai may be referable to the nominate subspecies.

Meinertzhagen (1930) stated that the Hoopoe did not breed in any Western Desert oasis. However, in recent years breeding has been proven or strongly suspected in Wadi el Natrun, Bahariya, Farafra, Dakhla, and Kharga (Goodman *et al.* 1986). A nest with young was discovered at El Omayed, south of El Alamein, on 17 May 1986 (WCM). In the Nile Delta and Valley (south to Aswan), the Suez Canal area, and the Faiyum this species commonly breeds in holes in trees, buildings, walls, and banks. There are specimens referable to *major* taken in the Nile Valley above Aswan before the formation of Lake Nasser. This species is a common breeding resident of north-eastern Sinai, particularly in the area between El Arish and Rafa where numerous fledglings were found in June 1984 (SMB).

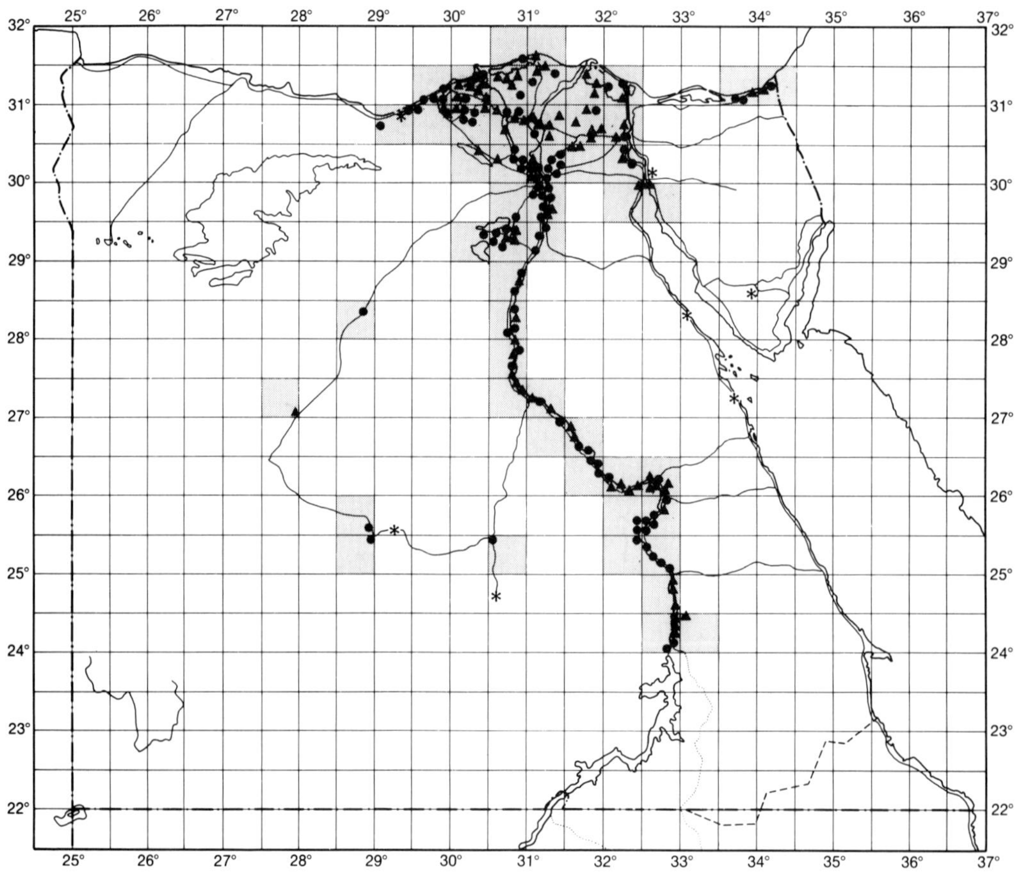

Map B62. Breeding distribution of Hoopoe *Upupa epops* between 1970–87.

During autumn and spring migration Hoopoes have been recorded in all portions of the country, including desert areas, along the Red Sea coast, and on islands. There appears to be considerable autumn passage along the north coast of Sinai, e.g. 438 at Zaranikh between 16 August and 24 September 1981 (Petersen and Sørensen 1981*b*). From 12 to 15 March 1982, 25 Hoopoes were noted at Hurghada leaving the mainland and heading out to sea in a northerly direction (ES, HWu). There are several records of this species crossing the northern Red Sea (e.g. Tuck 1964). In April and May 1983 Hoopoes were noted at Hurghada calling and displaying, and local breeding seems possible (MCJ *et al.*).

Ringing A bird ringed at Bahig on 28 September 1971 was recovered at the same locality on 6 September 1972 (GZF).

ORDER PICIFORMES

Family Picidae

Wryneck *Jynx torquilla* PV (WV)

Local names Lawwaa- لواء, abuu jamiid (NC), abuu lawwaa, umm al-waa.

Scarce passage visitor throughout the country in autumn from late August to mid-(late) October and in spring from early March to late May (Fig. 6.24), including desert areas, the Nile Delta and Valley, and Sinai. Occasionally noted in winter. The majority of Egyptian material examined is referable to *J. t. torquilla*. Two specimens taken at Bahig on 10 March and 18 April 1971 are of *J. t. tschusii* (Goodman and Watson 1983; USNM).

On autumn migration the Wryneck is less common in Egypt than in spring. Extreme dates are from 30 August 1971 at St Katherine (ZM-HUJ) to 11 October 1981 at Suez (RGB *et al.*). Winter records include two noted in December at an undisclosed locality (Meinertzhagen 1930), one collected on 6 January 1908 near Damietta (Nicoll 1908), one taken on 31 January 1984 at Barsiq (AAUC), and one seen on 15 January 1979 at Suez (Meininger *et al.* 1979).

Wrynecks are regularly recorded in small numbers on spring migration, and most birds pass in April. Extreme spring dates are from 4 March 1970 and 1971 at Bahig (Horner 1977) to 21 May 1980 at El Alamein (SV). The number captured at Bahig during the six PMS spring seasons, between 1967 and 1973, varied from 67 to 161 (Horner 1977).

Syrian Woodpecker *Dendrocopos syriacus* RB

Local name Naqqaar al-khashab as-suurii نقار الخشب السوري.

Rare breeding resident in north-eastern Sinai. Monotypic.

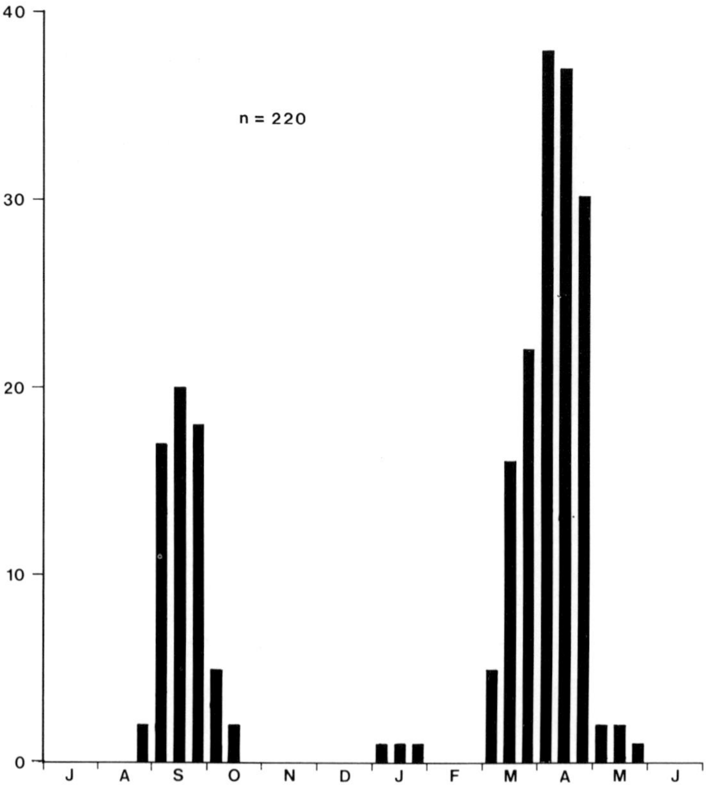

FIG. 6.24. Histogram of the numbers of Wrynecks recorded during 10-day periods throughout the year.

The Syrian Woodpecker reputedly established itself as a breeding bird in the Rafa area in the early 1970s (Paz 1987; OH) but no precise details are known. On 26 June 1984, two adults and one fledgling were noted in an orchard about 5 km west of Rafa. One of the adults was observed feeding the fledgling several times and it is presumed that the pair bred locally (SMB).

ORDER PASSERIFORMES

Family Alaudidae

Generic local names for lark Qunburra, qubbara, qumbar.

Black-crowned Finch Lark *Eremopterix nigriceps* RB Plate 4

Local name Qunburra suudaa- mutawwaja قنبرة سوداء متوجة.

Common local breeding resident in the south-east and occasionally observed in other areas. All Egyptian material examined is referable to *E. n. melanauchen*.

Western Desert In Wadi el Natrun three Black-crowned Finch Larks were observed during the winter of 1912–13 and an adult male was collected on 10 September 1911 (Nicoll 1912*b*; GZM). Moreau (1934) mentioned that one was taken on 3 October 1932 in the desert at 23°50′N., 27°40′E., between Dakhla and Gebel Uweinat.

Eastern Desert A single bird was observed at Berenice on 6 January 1896 (Cholmley 1897). On 3 May 1983, three were observed at Bir Abraq (Saleh 1984). This is the only record from the Gebel Abraq area, which was extensively surveyed in 1984 and 1985 (SMG, PLM). One male was observed on 31 January 1986 in Wadi el Allaqi, near Gebel Murra (SMB).

This species is a locally common breeding resident along the coastal plain of the

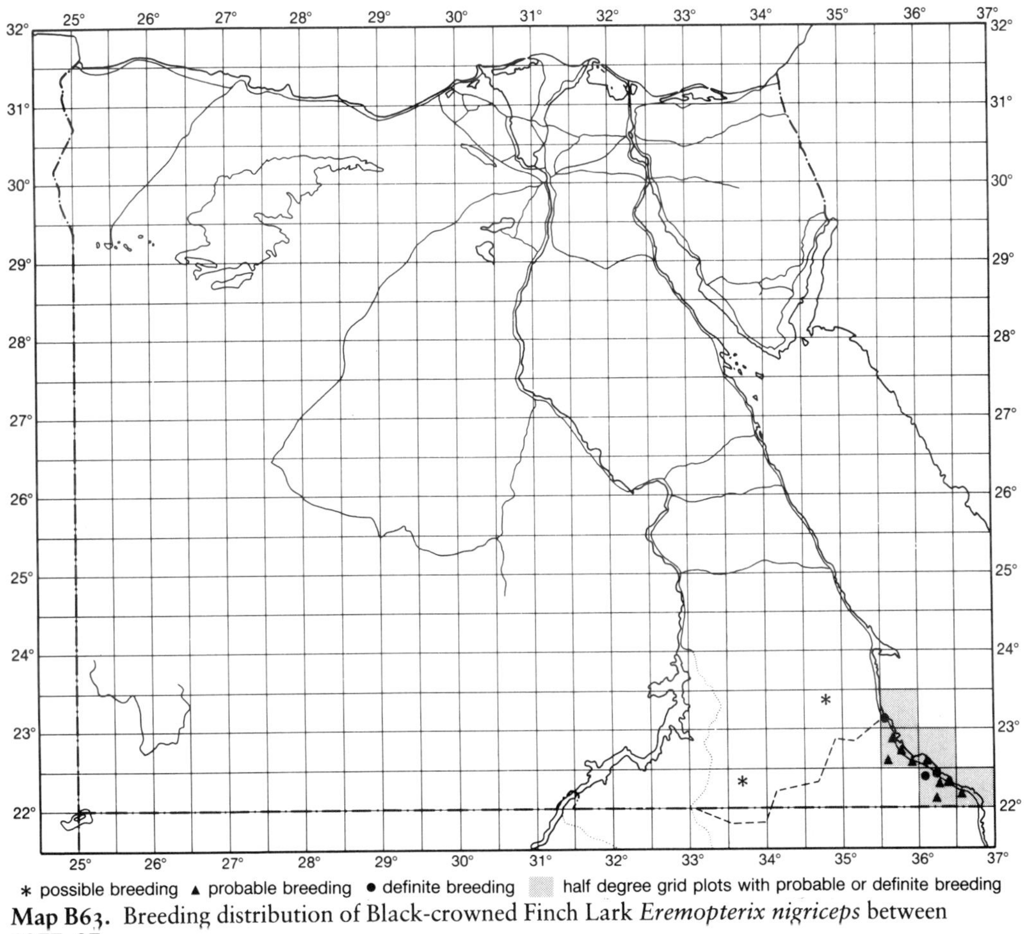

Map B63. Breeding distribution of Black-crowned Finch Lark *Eremopterix nigriceps* between 1977–87.

Sudan Government Administration Area. At Shalatein and Adal Deib it is a relatively common commensal in Bedouin villages. In the winter of 1984, after exceptionally heavy rains, a concentration of singing males was noted in Wadi Abu Fatma, north of Abu Ramad. Nests with eggs were found at this locality on 18 February 1984 (SMG). On 15 February 1984, one male was noted in the rather barren desert of Wadi Meisa, near Gebel Hamra Dom (SMG).

Sinai A flock of about 30 was observed on 20 March 1976 on Gezira Tiran (Ilani 1976), three in April 1976 near Gebel Areif el Naqa (Hovel 1987), a pair on 27 March 1979 (DJF *et al.*), and one bird on 2 November 1978 near Sharm el Sheikh (Hovel 1987).

[Dunn's Lark *Eremalauda dunni* AV?

On 30 March 1978, four or five Dunn's Larks (including two singing) were observed 10 km south-west of Nizzana along the Egyptian–Israeli border. When this site was revisited a year later, the bird was not found (Round and Walsh 1981); although three were noted at this locality on 22 October 1983 (Hovel 1987). Another bird was observed north of Nekhl on 17 March 1981 (SMB).]

Bar-tailed Desert Lark *Ammomanes cincturus* RB Plate 3

Local names Qunburra as̱-s̱aẖraa- muwashamma adh-dhanab قنبرة الصحراء موشمة الذنب, ẖammaada, ẖummara.

Locally common breeding resident, mainly in sandy desert areas with sparse vegetation and generally avoids rocky or mountainous areas. All Egyptian specimens examined are referable to *A. c. arenicolor*.

North Coast and Western Desert Meinertzhagen (1930) mentioned the Bar-tailed Desert Lark as 'common in the desert behind Salum, south to Siwa' and Bulman (1942) observed small numbers along the coast between Salum and Matruh. At El Hammam birds have been recorded on 23 May 1975 (FMNH) and a fledgling on 26 August 1972 (USNM). A series of specimens taken at Bahig in late May had enlarged gonads (USNM). There are several records from the desert between the Matruh–Siwa road and the Nile Delta, including a female at El Maghra on 11 May 1975 (Goodman and Ames 1983) and one at Hatiyet Badr el Din on 4 March 1984 (Baha el Din in press). In 1904, 1910, and 1981 the species was found breeding in Wadi el Natrun (Raw 1921; Goodman *et al.* 1986). The occurrence of this species in Bahariya and Farafra needs further confirmation (cf. Abdel Rahim *et al.* 1980). It was noted in the desert between Dakhla and Kharga (Meinertzhagen 1930) and on 23 April 1981 near Kharga (Goodman *et al.* 1986).

Nile Delta and Valley According to Meinertzhagen (1930) this species occurs in the desert near the Faiyum, where specimens were collected before 1918 (GZM) but there are no more recent records. It apparently occurs locally on the edge of the Nile

Valley and has been observed at Beni Hassan (Short and Horne 1981), on several occasions near Luxor (January to April), and on 8 October 1986 near Aswan (EH, PK).

Eastern Desert This species is common in the northern portion of the Eastern Desert, particularly along the Cairo–Suez road and in the broad wadis between the Nile Valley and Zafarana. Tregenza (1951) stressed its absence during the summer months in the region between Qena and Quseir, and north to Ras Gemsa. There are no recent records from the Red Sea coastal plain between Zafarana and Quseir, although two were observed on 16 January 1986 in Wadi Safaga east of Bir Abu Mureiwat (SMB). From Quseir south to the Sudanese political border it is locally common along the coastal plain and inland sandy wadis. A female collected out of a pair on 12 February 1983 in Wadi Sharm el Qibliya, 36 km south of Quseir, had an enlarged ovary (Goodman and Atta 1987).

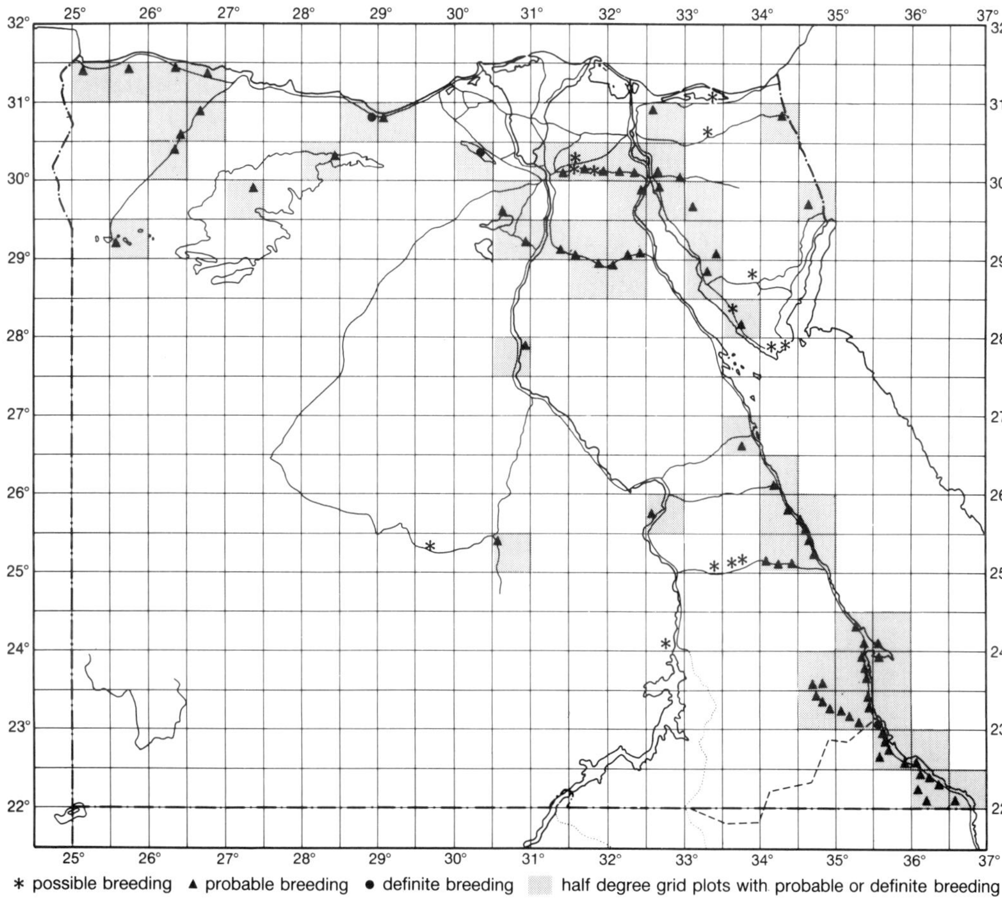

Map B64. Breeding distribution of Bar-tailed Desert Lark *Ammomanes cincturus* between 1930–87.

Sinai The Bar-tailed Desert Lark is locally common in the desert plains of northern, western, and southern Sinai. There are a few winter records from St Katherine, which may concern local non-breeding birds.

Desert Lark *Ammomanes deserti* RB Plate 3

Local names Qunburra as̱-s̱aẖraa- قنبرة الصحراء, ẖammaada, ẖummara al-baadiya, qunburra al-baadiya, rahaydin (Ma).

Common breeding resident in rocky desert areas, including low foothills and mountains, and generally avoids sandy wadis. There is considerable variation in the plumage coloration of Egyptian Desert Larks. Two subspecies have been previously recognized as occurring in the country: *A. d. deserti* from the edge of the Nile Valley and *A. d. isabellinus* from the remaining desert areas, including Sinai. *A. d. katharinae*

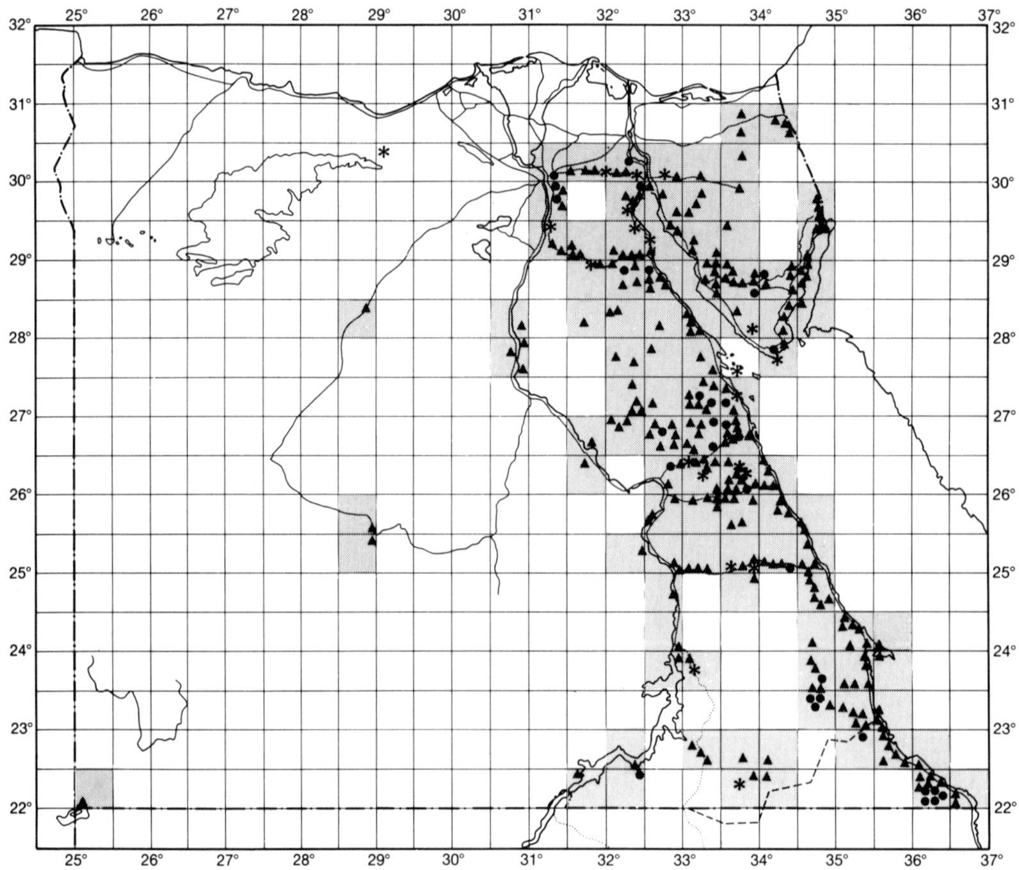

* possible breeding ▲ probable breeding ● definite breeding ▒ half degree grid plots with probable or definite breeding

Map B65. Breeding distribution of Desert Lark *Ammomanes deserti* between 1930–87.

is considered synonymous with *A. d. isabellinus* (Vaurie 1959). Horváth (1958) described a new subspecies, *A. d. borosi*, from Bir Abbad, 15 km east of Idfu along the road to Marsa el Alam. Since we have not been able to examine and compare the type of *borosi*, the validity of this form cannot be assessed. In late February 1985, two specimens were collected in Wadi Aideib, Gebel Elba, that are referable to the distinctive form *A. d. samharensis* (UMMZ).

Western Desert According to Meinertzhagen (1930) the Desert Lark was common at Dakhla, an oasis from where there are only a few recent records. The first known report for Bahariya was one on 29 April 1981 (Goodman *et al.* 1986). One was noted 20 km north-east of El Maghra on 2 March 1984 (Baha el Din in press). In October and November 1968 it was observed at Gebel Uweinat (Misonne 1974). We are unaware of other positive records from the remaining Western Desert oases.

Nile Delta and Valley This species has been noted at various localities along the desert edge of the Nile Delta and Valley, including the shores of Lake Nasser.

Eastern Desert The breeding range of this species appears to be from the Cairo–Suez road and the Fayid hills south to the Sudanese political border. It is common in rocky wadis and mountainous areas with sparse vegetation. Along the coastal plain, Desert Larks occur locally in areas almost completely devoid of vegetation. One was observed on Gezira Tawila, off Hurghada, on 13 May 1984 (SMG, RWS).

Sinai This species is common in rocky wadis and on mountain slopes but apparently not in the northern sandy desert. At Gebel Musa it has been recorded at altitudes of up to 2280 m. Locally high densities have been observed; for example, in April 1984 at least 50 pairs were present around St Katherine (Siering 1986).

Hoopoe Lark *Alaemon alaudipes* RB Plate 3

Local names Mukkaa- مكاء , abuu khamiira (NC), akhraj, cru-aan (NC), giigi (Ma), taawa.

Common breeding resident of open, sandy desert plains, and generally avoids mountainous regions. Several authors have referred material collected in the Eastern Desert and along the Red Sea coastal plain to *A. a. desertorum* (e.g. Meinertzhagen 1930; Horváth 1959). We have examined several specimens from this region and based on size and plumage coloration none is clearly referable to *desertorum*, although a few birds are intermediate between *A. a. alaudipes* and *desertorum*. Hoopoe Larks from other portions of the country are typical of *A. a. alaudipes*. Breeding is generally from March to late June but has also been recorded in December (e.g. Hachisuka 1924).

North Coast and Western Desert The Hoopoe Lark is a characteristic bird of the Mediterranean coastal desert, the Qattara Depression, and the desert west of the Nile Delta (near Wadi el Natrun, the Faiyum, and Wadi el Rayan). There are no recent records from Siwa, although Al Hussaini (1938–9) saw one on 16 February 1939,

20 km north of the oasis. The species is common in the environs of Bahariya, Farafra, Dakhla, Kharga (Goodman *et al.* 1986), and apparently Gebel Uweinat (Misonne 1974). One bird was found dead on 9 April 1984 at East Uweinat (Baha el Din in press). This species is presumably a breeding resident in other isolated areas of the Western Desert.

Nile Valley This species has been recorded at a few places on the desert edge of the Nile Valley. Small numbers are regularly noted near El Amarna, where it probably breeds. Hoopoe Larks have also been recorded in February 1980 near Luxor (SV) and on 16 January 1963 at the Amada Temple (YPM). Apparently it also occurred in the Nile Valley proper—one was taken on 5 January 1875 'on an island below Golosaneh [= Qulusna] on the Nile' (FMNH).

Eastern Desert Locally common resident in the desert between Cairo and Ismailiya south to Ras Gharib, and between Hamata and Halaib. It appears to be

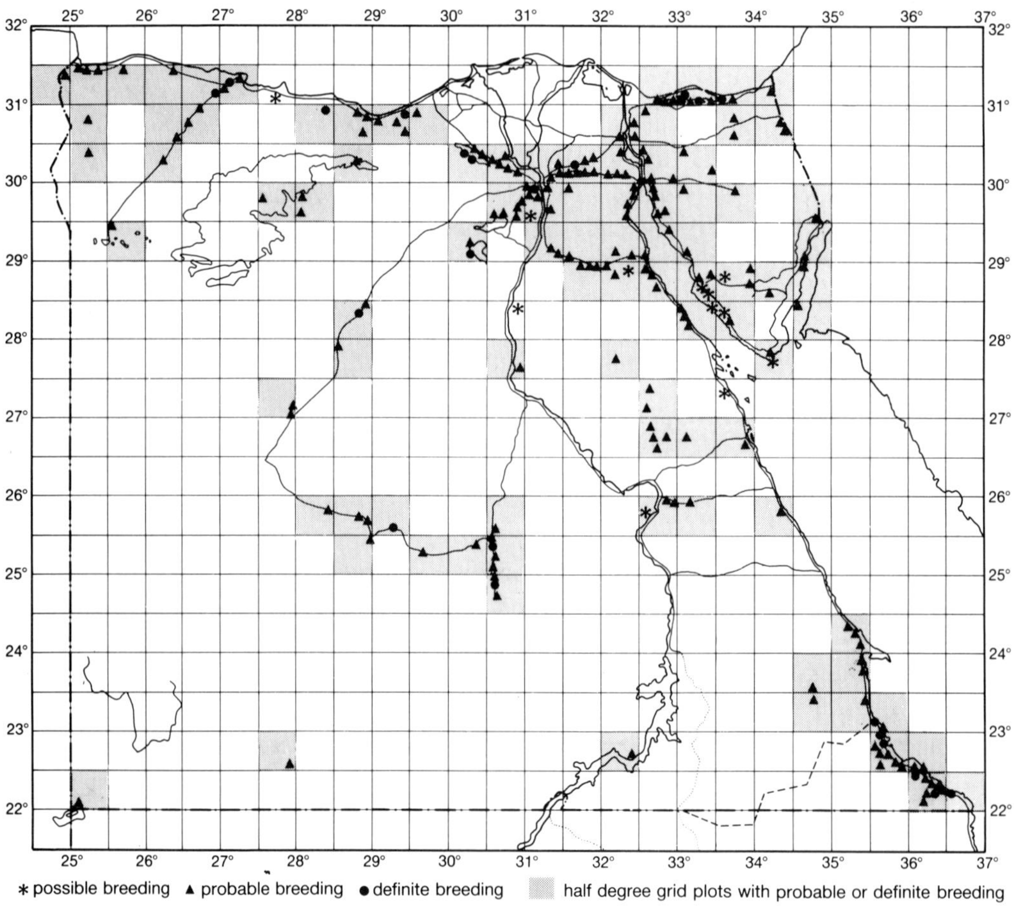

Map B66. Breeding distribution of Hoopoe Lark *Alaemon alaudipes* between 1930–87.

scarce or rather local along the coastal plain between Ras Gharib and Hamata. There is only one record from the Hurghada area—two on the beach near Myos Hormos on 3 August 1949 (Tregenza 1955). The Hoopoe Lark occurs locally in the generally more open plains of the central portions of the Eastern Desert, such as the wide lower reaches of Wadi Qena (Tregenza 1951), near El Laqeita on the Qift–Quseir road, Gebel Hamra Dom, and Gebel Abraq. It is common along the coastal plain of the Sudan Government Administration Area, particularly between Abu Ramad and Gebel Elba.

Sinai Common resident of the open desert plains and dunes in the northern and western portions of the Peninsula but uncommon and rather local along the Gulf of Aqaba coastal plain. It has been noted locally in south-central Sinai at altitudes of up to 1200 m.

Dupont's Lark *Chersophilus duponti* RB

Local name Qunburra ṯawiila al-minqaar قنبرة طويلة المنقار.

Scarce breeding resident along the North Coast from Salum to Bahig. All Egyptian specimens examined are referable to *C. d. margaritae*.

North Coast According to Meinertzhagen (1930) the Dupont's Lark was a resident of the Mediterranean coastal desert from Salum to Matruh but not penetrating further inland than 40 km. Since the greater portion of this zone has not been visited for several decades, this species' current status in certain areas is unknown. Recent records include a fledgling on 23 May 1975 at El Hammam (FMNH) and an adult male on 2 September 1971 at Bahig (USNM).

Thick-billed Lark *Rhamphocorys clotbey* (PV) (WV)

Local names Qunburra maghrabiyya قنبرة مغربية, ẖummara.

Rare and irregular passage and winter visitor. Monotypic. The comment of Paz (1987) that Israeli Aharoni discovered this species breeding in the Sinai Desert in 1930 was wrongly translated from Paz (1986); the nest was found in the Syrian Desert.

Most Egyptian records of Thick-billed Lark are from along the Cairo–Suez road and are presumably of migrants or vagrants from the Middle East breeding population, rather than from the Maghreb. The following is a list of all known records from the general vicinity of the Cairo–Suez road—one on 21 January 1928 (Meinertzhagen 1930; BMNH); at least 12, 12–13 km from Cairo on 15 March 1946 (PDH); a flock of 50 on 7 April 1946 (LB); a female with incubation patch on 2 July 1922 (Meinertzhagen 1930; GZM); two on 21 July 1922 (Jourdain and Lynes 1936); and one in August 1923 (Meinertzhagen 1930). There is a single record west of the Nile Valley—one was taken on 16 September 1941 at Matruh (Goodman and Watson 1983). This species was noted in May 1969 in Wadi el Heisi, eastern Sinai (Hovel

1987). Small flocks are occasionally observed in early April and late May near Elat, Israel (Krabbe 1980; Paz 1987; HSh).

Calandra Lark *Melanocorypha calandra* WV

Local names Qunburra al-gharb al-kabiira قنبرة الغرب الكبيرة, 'ul'aal.

Scarce winter visitor, mainly to the north of the country, between (early September) mid-November and mid- (late) April. Rare migrant along the Red Sea coast. All of the Egyptian material examined is of *M. c. calandra* and none is referable to *M. c. psammochroa* (*pace* Etchécopar and Hüe 1967).

According to Meinertzhagen (1930) the Calandra Lark is a winter visitor along the Mediterranean coast from Salum to Port Said, and rarely along the northern coast of Sinai. More recent information confirms this. The earliest autumn record appears to be on 9 September 1942, when a small flock was noted near Abu Qir (Bodenham 1945*a*). Apparently, this species generally does not penetrate far inland and there are only a few records from the northern Nile Valley. On 15 November 1985 over 20 were noted at Gebel Asfar (SMB), on 22 February 1909 one was 'purchased in Cairo Market' (GZM), and on 9 April 1978 one was observed at Giza (RJM, LRP). There are a few spring records from the Red Sea coast, suggestive of some passage to and from sub-Saharan Africa. These include at least 30, with several Bimaculated Larks, at St Paul Monastery on 26 March 1986 (AS), and single birds in Wadi Aideib, Gebel Elba, on 22 February 1985 (UMMZ), at Taba on 1 April 1978 (MAl), at Safaga on 3 April 1985 (UGS), and at Hurghada on 9 April 1982 (De Bièvre 1983). The Gebel Elba record is exceptionally far south.

Bimaculated Lark *Melanocorypha bimaculata* PV

Local names Qunburra ash-sharq al-kabiira قنبرة الشرق الكبيرة, 'ul'aal.

Accidental migrant in autumn and apparently regular in spring between late February and mid-April (early May) in the eastern half of the country. All Egyptian material examined is referable to *M. b. rufescens*.

The only record of the Bimaculated Lark mentioned by Meinertzhagen (1930) was a pair obtained along the Qena–Quseir road on 20 February 1928 (BMNH 1965 M.7075 is labelled Luxor). There are numerous recent spring records of small groups along the Red Sea coast from Suez to Gebel Elba, in the central Eastern Desert, and Sinai. In the spring of 1985, passage was noted along the Red Sea coastal plain from 24 March to 18 April, with hundreds of birds passing in late March (SMG, PLM). Most of these observations were of single birds or small groups (up to five) generally in the company of Short-toed Larks but a flock of at least 45 was seen near Hurghada on 25 March 1985. The latest spring record is a single bird at Hurghada on 9 May 1983

(MCJ *et al.*). The only known autumn report is one at St Katherine on 12 November 1984 (SH, DM).

Short-toed Lark *Calandrella cinerea* PV (WV) CB

Local names Qunburra qaṣiirat al-asaabi' قنبرة قصيرة الأصابع, 'ulay'ala.

Abundant migrant through the country in autumn from (late August) early September to mid- (late) October and in spring from late February to mid- (late) May. Scarce winter visitor and apparently an irregular breeding resident. *C. c. brachydactyla* is the most common form in Egypt. *C. c. longipennis* has been taken in northern Egypt between 13 September and 10 April. *C. c. hermonensis* has been collected at various localities in northern Egypt, between Dakhla and Kharga, and near Gebel Elba (Meinertzhagen 1930; Goodman 1984*a*). Two birds obtained at Bahig on 20 March 1971 (USNM) have exceptionally rufous crowns and approach the coloration of *C. c. rubiginosa* specimens taken in the Maghreb (see below). We have not examined any Egyptian birds referable to *C. c. artemisiana* (*contra* Hovel 1987).

The only evidence for Short-toed Larks breeding in Egypt is a female taken on 28 March 1910 in Wadi el Natrun and the inscription on the data tag reads 'large incubation patch—eggs have recently been laid' (GZM). The plumage coloration of this specimen, particularly the crown, is rather rufous and the bird appears to be of *C. c. rubiginosa*. According to Meinertzhagen (1930) a clutch of four eggs was collected in Wadi el Natrun on 28 May but further details are lacking.

In autumn this species can be observed throughout the country but groups of over 100 are uncommon. One exceptionally large flock of approximately 1800 was noted on 20 September 1980 south of Nizzana (DJM, DES). During a study of migratory birds passing Zaranikh a total of 5623 was recorded between 3 and 24 September 1982, with 3127 on 18 September (Baha el Din and Salama 1984). After late October the species is rare and only a few winter records are known. These include birds at Abu Zabal on unspecified dates (Raw 1921), El Dab'a and Buqbuq in January 1942 (Rowntree 1943), and Giza in December 1980 (SV).

Spring passage seems to be concentrated in the eastern half of the country and large flocks are regularly observed. For example, flocks of approximately 3000 birds on 6 March 1984 between Hurghada and Quseir (PDH, EMN), 2350 on 19 March 1982 between Suez and Hurghada (Meilstrup and Sørensen 1983), 3600 on 24 and 25 March 1985 between Zafarana and Quseir (SMB, SMG, PLM), and 'thousands' on 24 April 1982 between Safaga and Quseir (JW). During the PMS survey at Bahig, Short-toed Larks were one of the most frequently captured spring migrants, passing between 26 February and 27 April (Horner 1977; USNM).

Ringing Two birds ringed at Matruh in spring 1970 were recovered at the same locality in autumn 1970. Two others ringed near Bahig in spring 1969 and 1971 were recovered the subsequent spring at Alexandria and Bahig, respectively (Watson 1971; GZF). These records clearly show some migration route fidelity.

Lesser Short-toed Lark *Calandrella rufescens* RB WV

Local name Qunburra qaṣiirat al-asaabi' ṣaghiira قنبرة قصيرة الأصابع صغيرة.

Two subspecies nest in the country. *C. r. nicolli* is a fairly common breeding bird of Nile Delta salt-marshes and perhaps vegetated desert areas of northern Sinai. *C. r. minor* breeds along the North Coast south to Siwa and the Qattara Depression. We have examined the specimens taken at Port Said on 21 October 1906 (BMNH) and 27 December 1922 (GZM) which were identified by Meinertzhagen (1930) as *C. r. heinei* but they appear to be of *nicolli*. Recent specimens referable to *minor* have been taken along the North Coast at El Hamman (type locality) in late May 1975 (FMNH), at Bahig between 20 February and 21 May (series in USNM), and Salum in March 1931 (HGD). The species nests from mid-April to late June.

North Coast and Western Desert Along the North Coast, Lesser Short-toed Larks

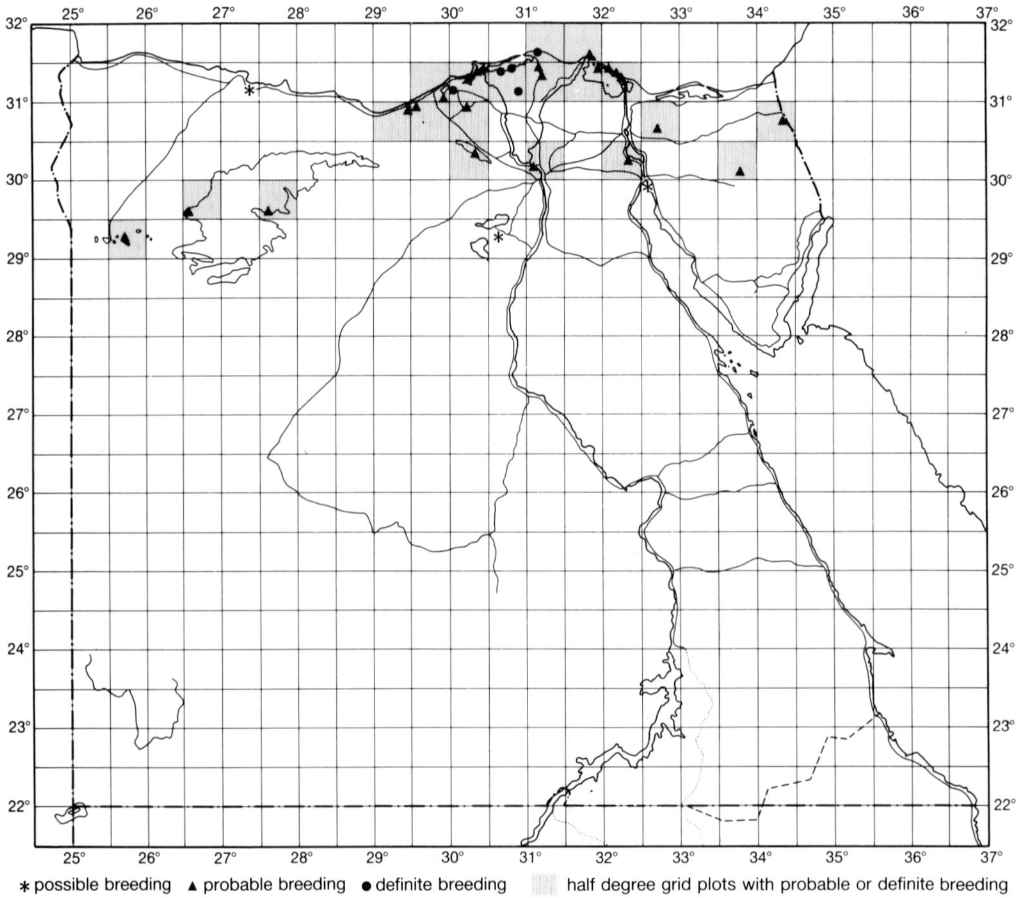

Map B67. Breeding distribution of Lesser Short-toed Lark *Calandrella rufescens* between 1970–87.

have been noted between Salum and Bahig. Two were seen 33 km east of Matruh on 6 October 1980 (HB *et al.*). Records from the Western Desert include an observation in Wadi el Natrun on 3 April 1981 and specimens taken in Siwa and the Qattara Depression in early March (Goodman *et al.* 1986). The species presumably breeds at these localities.

Nile Delta and Valley This species is a locally common resident of the northern Nile Delta and the Suez Canal area (Meininger *et al.* 1986). About 75 birds were noted singing and displaying on 8 February 1986 south of Baltim (SMB, DM). There are a few winter observations from the Nile Valley (which may concern at least in part misidentified Short-toed Larks). Ten were observed near Dandara on 20 February 1982 (GM) and two collected at Shakshuk on 8 February 1939 (HGD).

Red Sea coast This species is occasionally observed along the Red Sea coast, generally in the spring. Bulman (1944) observed 20 on 17 September and four on 25 September 1942 at Safaga, and he identified one collected bird as *C. r. nicolli.*

Sinai Lesser Short-toed Larks probably breed in northern Sinai but are apparently absent from the immediate Lake Bardawil area. In March and April 1973 to 1976 small numbers were observed south of Nizzana along the Egyptian–Israeli border (SC). Several were noted north of Nekhl in March 1986 singing and defending territories (DB). One specimen of *C. r. nicolli* was collected at El Tassa on 24 April 1972 (TAUM). A presumed migrant was seen at St Katherine on 1 October 1983 (SMB).

Crested Lark *Galerida cristata* RB

Local names Qunburra mutawwaja قنبرة متوجة, abuu qubba, qunburra umm shuusha, qunburra bishuusha.

Abundant breeding resident of vegetated desert regions, the Nile Delta and Valley, the Faiyum, along Lake Nasser, and locally in Sinai. Three subspecies are currently recognized as occurring in the country. *G. c. nigricans* is the exceptionally dark form of the Nile Delta, *G. c. altirostris* is the pale form of Wadi el Natrun and the North Coast, and *G. c. maculata* is the form intermediate in coloration between *nigricans* and *altirostris* of the Nile Valley (Vaurie 1959). Birds from northern Sinai appear to be referable to *G. c. maculata*. It should be noted that there is considerable variation in the plumage coloration of adults from the same population, which is not explainable solely by the presently recognized patterns of geographical variation. *G. c. brachyura* is considered a synonym of *G. c. altirostris* (Vaurie 1959). This species breeds from late February to mid-June.

North Coast and Western Desert The Crested Lark is common along the Mediterranean coastal belt between Salum and Alexandria. Near Wadi el Natrun this species is common but there are only a few records from other Western Desert oases. It was observed on 4 March 1976 near Qara, Qattara Depression (Goodman and Ames 1983) and on 7 September 1984 in Dakhla (SH, DM).

Nile Delta and Valley and Eastern Desert This species is a characteristic bird of the Nile Delta and Valley, although it is less common and sometimes locally absent in the central portion of the Delta. Before the completion of the Aswan High Dam it was common in the area now inundated by Lake Nasser. It is presently common along the shores of Lake Nasser but does not appear to occur out of the Nile basin into the Eastern Desert. The only known exceptions are observations at Bir Abbad, 20 km east of the Nile Valley at Idfu (Horváth 1959), a few recent records just south of Suez, and an unverified observation at Ain Sukhna (Searight 1952*b*).

Sinai This species is particularly common along the shores, dunes, and islands of Lake Bardawil and the Gulf of Aqaba coastal plain. There appear to be some local winter movements. A few birds were noted at the St Katherine airport, where the species is not known to breed, between 10 October 1983 and 12 March 1984 (SMB). Several were heard singing near Nuweiba on 9 March 1987 (PLM).

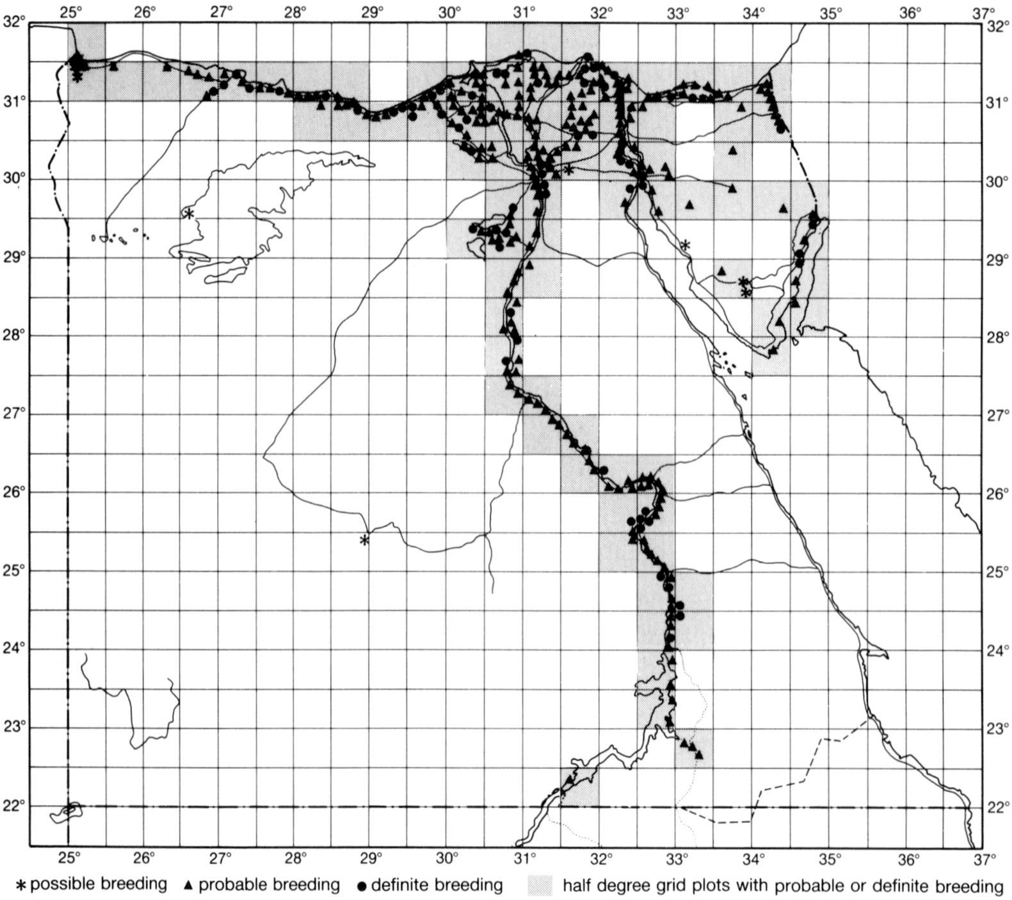

Map B68. Breeding distribution of Crested Lark *Galerida cristata* between 1930–87.

Ringing Single birds ringed at Ikingi Maryut in spring 1969 were recovered at Bahig in spring 1970 and 1971 (Watson 1971; GZF).

Thekla Lark *Galerida theklae* RB?

Local name Qunburra mutawwaja saghiira al-minqaar قنبرة متوجة صغيرة المنقار.

Possible breeding resident in the extreme north-western corner of the country. All Egyptian specimens examined are referable to *G. t. superflua*.

Twelve Thekla Larks were collected at Salum in late January 1920 (Meinertzhagen 1921; three in GZM) and two in April 1931 (HGD); however, the species was not noted at this locality in 1928 (Meinertzhagen 1930). Al Hussaini (1938–9) mentioned that both *G. cristata* and *G. theklae* were common in the streets of Matruh in February 1939 and that four specimens of the latter species were collected. Since Thekla Larks normally avoid human habitation and the specimens cannot be traced, this record must be called into question. Further, the observations of this species in April 1982 at Alexandria, Cairo, and the Faiyum (De Bièvre 1983) cannot be accepted. There are no documented recent Egyptian records. However, the Salum region, which is adjacent to known breeding areas in Cyrenaica, Libya, has not been visited by ornithologists in recent years.

Wood Lark *Lullula arborea* WV

Local name Qunburra al-ghiyaad قنبرة الغياض.

Rare winter visitor to the northern portion of the country between mid-October and late March. Most Egyptian specimens are referable to *L. a. arborea*. Records of *L. a. pallida* include two taken in Wadi el Natrun on 24 November 1911 (Nicoll 1912*b*; GZM); one collected from a group of six at Matruh on 30 January 1928 (Meinertzhagen 1930); one at Ikingi Maryut on 23 November 1969 (USNM); and specimens taken at Bahig on 24 (two) and 28 February 1971 and 25 March 1970 (USNM).

Wood Larks are recorded in small numbers along the North Coast and in the Nile Delta. The only known record south of Cairo is two heard and seen at Aswan on 21 March 1986 (UA). Observations of migrating birds in 1981 at Suez include: three on 15 October, six on 30 October, and one on 2 November (RGB *et al.*). One was observed at St Katherine on 12 October 1983 (SMB), another at El Kabrit on 23 October 1952 (AP), and eight at Heliopolis in November 1979 (SMB). The records near Suez seem to indicate that Wood Larks occur further south along the Red Sea coast, an area for which there are no autumn records. During spring migration small flocks have been noted in March 1983 at Ras Gemsa (NM) and between 14 and 30 March 1979 on Gezira Tiran (RF).

Skylark *Alauda arvensis* WV

Local names Qunburra al-ghayṯ قنبرة الغيط, abuu qawii'.

Fairly common winter visitor to the north of the country between (early September) mid-October and early April (mid-May). All Egyptian material examined is referable to *A. a. cantarella*.

The Skylark is regularly observed in vegetated areas along the complete North Coast, the Nile Delta, the Suez Canal area, and northern Sinai. In the Nile Valley it apparently has not been observed south of the Faiyum. There are several records from Wadi el Natrun. The only other Western Desert record appears to be three birds on 7 March 1980 at Bahariya (Goodman *et al.* 1986). There is one observation from southern Sinai—two on 11 March 1979 at Sharm el Sheikh (HW). The latest spring record is a single bird at Matruh on 12 May 1956 (FMNH).

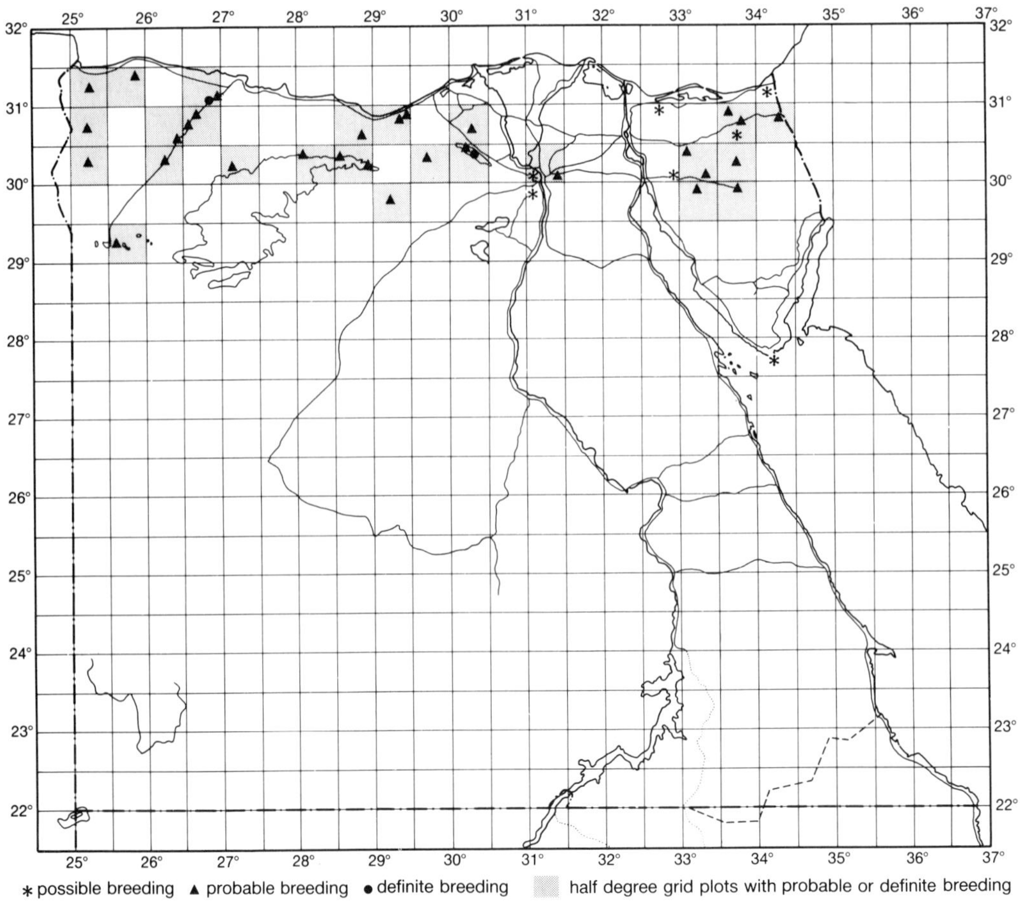

Map B69. Breeding distribution of Temminck's Horned Lark *Eremophila bilopha* between 1930–87.

Temminck's Horned Lark *Eremophila bilopha* RB

Local name Qunburra as-sahraa- al-muqarrana قنبرة الصحراء المقرنة.

Fairly common breeding resident in the northern portion of the country. Monotypic.

North Coast and Western Desert The Temminck's Horned Lark is locally common along the North Coast from the Libyan border to the Nile Delta. It generally occurs in sparsely vegetated areas but is found along the Matruh–Siwa road, an area almost devoid of plant cover. This species is known from Siwa, Wadi el Natrun, and several areas in the Qattara Depression, but there is no positive evidence from any other Western Desert oasis (Goodman *et al.* 1986). Outside the breeding season it is sometimes gregarious. Bulman (1942) reported large flocks 40 km south of Sidi Barrani in November 1940. Birds in breeding condition have been collected between mid-May and mid-June in Wadi el Natrun (Goodman *et al.* 1986).

Nile Delta and Valley There are a few records from the Nile basin. Specimens have been taken at Abu Rauwash on 21 January 1910 (GZM) and along the Cairo–Faiyum road on 24 February 1977 (HS).

Sinai This species is a local breeding resident in the north. During the non-breeding season some local movements occur. The only known record from the southern portion of the Peninsula is one at Ras Muhammad on 23 April 1979 (HW).

Family Hirundinidae

Generic local names for martin and swallow 'Usfuur al-janna, hujayjiyya, irgay'a (Ma), khuttaaf, philiilist (Si), salluu (NC), salwaa (NC), sunuuniyya, sunuunuu, ziwwaar.

Sand Martin *Riparia riparia* MB PV WV

Local name Sunuunuu ash-shawaati- سنونو الشواطىء.

The subspecies *R. r. riparia* and *R. r. diluta* are Eurasian passage migrants to Egypt and *R. r. shelleyi* is the local breeding form. Eurasian birds migrate through Egypt, including desert areas, in autumn from late July to mid-November and in spring from late January to late May (mid-June). The species is a scarce winter visitor in the Nile Delta and Valley. It is occasionally noted throughout the summer months in areas where it is not known to breed, e.g. along the Red Sea coast.

The form *R. r. shelleyi* is an abundant migrant breeder, sometimes in colonies composed of several thousand active burrows, throughout the Nile Delta and Valley (relatively scarce between Luxor and Aswan), the Faiyum, Suez Canal area, and Wadi el Natrun. Breeding birds return to colonies in the Upper Nile Valley in early February and in the Delta in mid-March. In early March 1984, during a non-comprehensive survey, over 40 000 burrows were counted along the river between Cairo and Luxor

(AS), and the total Egyptian breeding population certainly exceeds several hundred thousand pairs.

After the nesting season large concentrations are again noted. For example, over 20 000 were observed on 27 July 1980 between El Minya and Asyut (RD). In the Nile Valley numbers sharply decline in August and September and the species is generally scarce between November and December, although small numbers winter in the Nile Delta and Valley. An exceptional autumn record of hundreds of thousands on 10 November 1981 at Lake Manzala presumably involved at least some Eurasian migrants (RGB *et al.*).

In the southern Nile Valley spring passage starts in mid-January and sometimes the number of migrants is considerable—about 50 000 were noted on 8 February 1978 near Nag Hammadi (Short and Horne 1981) and over 150 000 from 14 to 21 March 1985 between Cairo and Aswan (Grieve 1985). Out of the Nile basin the first spring

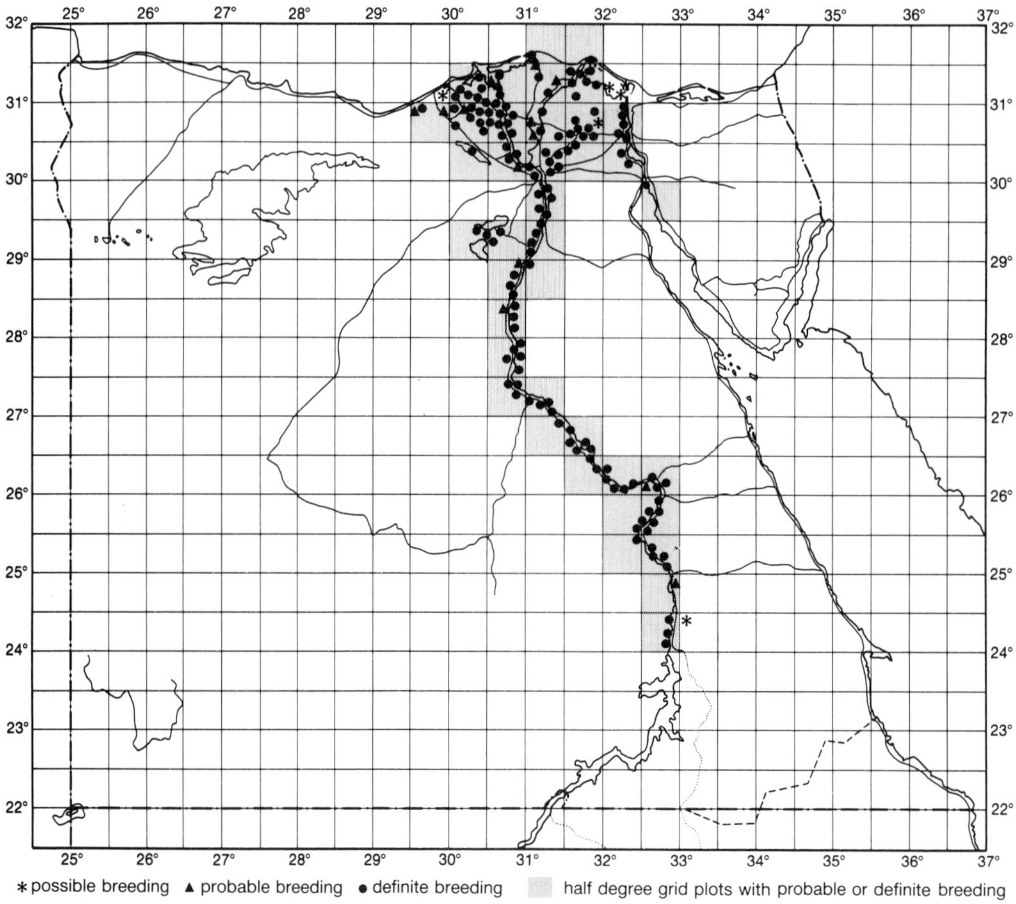

Map B70. Breeding distribution of Sand Martin *Riparia riparia* between 1970–87.

migrants are observed in late January and the species is common from March to late May. Single birds or small flocks are regularly observed until mid-June.

Ringing A bird ringed at Ikingi Maryut on 14 May 1969 was recovered in Czechoslovakia on 12 July 1969 (GZF).

Pale Crag Martin (Rock Martin) *Ptyonoprogne obsoleta* RB

Local name Sunuunu aṣ-ṣakhr al-baahit سنونو الصخر الباهت.

Common breeding resident of rocky habitats throughout the country, particularly in the Eastern Desert. All Egyptian material examined is referable to *P. o. obsoleta*. The only exception is a single bird collected at Gebel Elba which is intermediate in size and coloration between *P. o. obsoleta* and *P. o. arabica* (BMNH). The subspecific identity of birds occurring at Gebel Elba is in need of further study. This species breeds from

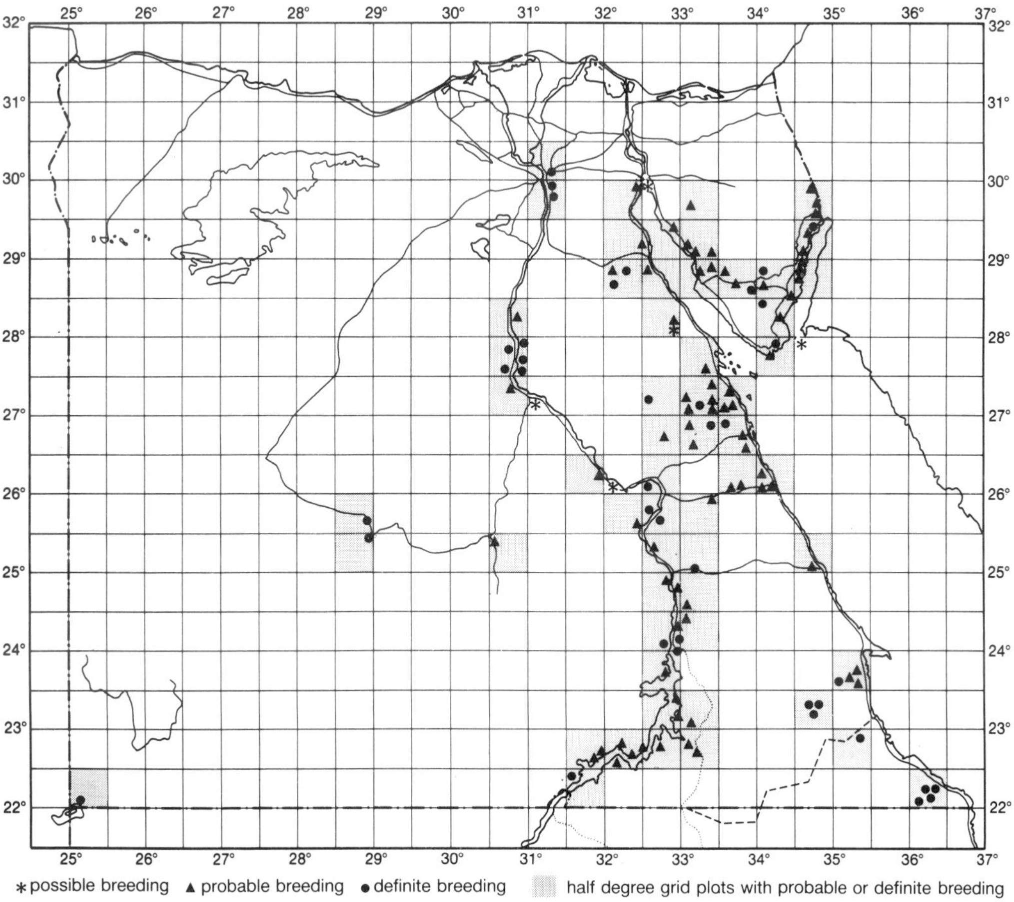

Map B71. Breeding distribution of Pale Crag Martin *Ptyonoprogne obsoleta* between 1945–87.

mid-January to late June. It can be observed in breeding areas throughout the year but some migratory movements seem to occur. At Suez small numbers were seen flying south between 14 September and 15 October 1981 (RGB *et al.*) and about 490 on 9 December 1980 (SMB).

Western Desert There appear to be no positive Pale Crag Martin breeding records from Wadi el Natrun, Siwa, Farafra, or Bahariya (Goodman *et al.* 1986). At Dakhla the species was found nesting near Mut and El Qasr in April 1925 (Moreau 1927) and there are several recent records from this oasis. Meinertzhagen (1930) mentioned a nest at Kharga in a native dwelling. In this oasis birds have been noted on 13 January 1972 and 23 April 1981 (Goodman *et al.* 1986). A nest was located at Gebel Uweinat in November 1968 (Misonne 1974).

Nile Valley Pale Crag Martins are locally common breeding birds along the edge of the Nile Valley on cliffs and in buildings, particularly ancient tombs and temples. They are common along the shores of Lake Nasser. There are no recent records from the Faiyum (cf. Meinertzhagen 1930).

Eastern Desert This species is a locally common breeding bird of mountainous areas. It is regularly observed from Suez south to the Sudanese political border. The only record north of Suez appears to be one on 18 January 1980 at the Bitter Lakes (PLM, WCM). It occasionally breeds in buildings in the Red Sea Mountains, such as the monasteries of St Anthony and St Paul and the mosque of Sheikh Shadli.

Sinai This species is a common breeding resident in the mountains of the south and the complete Gulf of Aqaba coastal plain. At Gebel Musa it has been found breeding at altitudes of up 2000 m.

Crag Martin *Ptyonoprogne rupestris* (PV) (WV)

Local name Sunuunuu aṣ-ṣakhr سنونو الصخر.

Rare and irregular migrant and winter visitor, mainly in the north of the country, between late October and late April. Some sight records of this species are presumably misidentified Pale Crag Martins. Monotypic.

Meinertzhagen (1930) mentioned Crag Martin specimens collected in January and February at Helwan and in January at Giza, and observations in March at Luxor and in February at Quseir. More recent sight records from the Nile Valley include: one on 29 October 1982 at Luxor (Davies 1982), eight on 27 December 1985 at El Kab (AS), five on 14 January 1954 at Karnak (HCK), three on 26 and 27 March 1982 at Aswan (Meilstrup and Sørensen 1983), one on 5 April 1985 at the Giza Pyramids, and four on 26 April 1983 at Aswan (UGS). Single birds have been observed in Wadi el Rayan on 23 January 1985 (SMB) and at Suez on 17 and 24 April 1982 (Wimpfheimer *et al.* 1983).

Swallow *Hirundo rustica* RB PV WV

Local name 'Usfuur al-janna عصفور الجنة.

The Eurasian breeding forms, *H. r. rustica* and *H. r. transitiva*, are passage visitors to Egypt. Both have relatively light-coloured breasts and the nominate subspecies is the more common of the two. The distinctly dark-breasted form *H. r. savignii* is an abundant breeding resident in towns and villages throughout the Nile Delta and Valley, the Faiyum, and the Suez Canal area. Breeding is from mid-February to late May. Light-breasted adult birds have been observed attending nests in Port Said and their subspecific identity is unknown. *H. r. transitiva* nests in Israel not far across the border from Rafa. Eurasian breeding birds are abundant migrants through the country in autumn from (early) late July to early December and in spring from (late February) late March to mid-May (late June), and rare winter visitors.

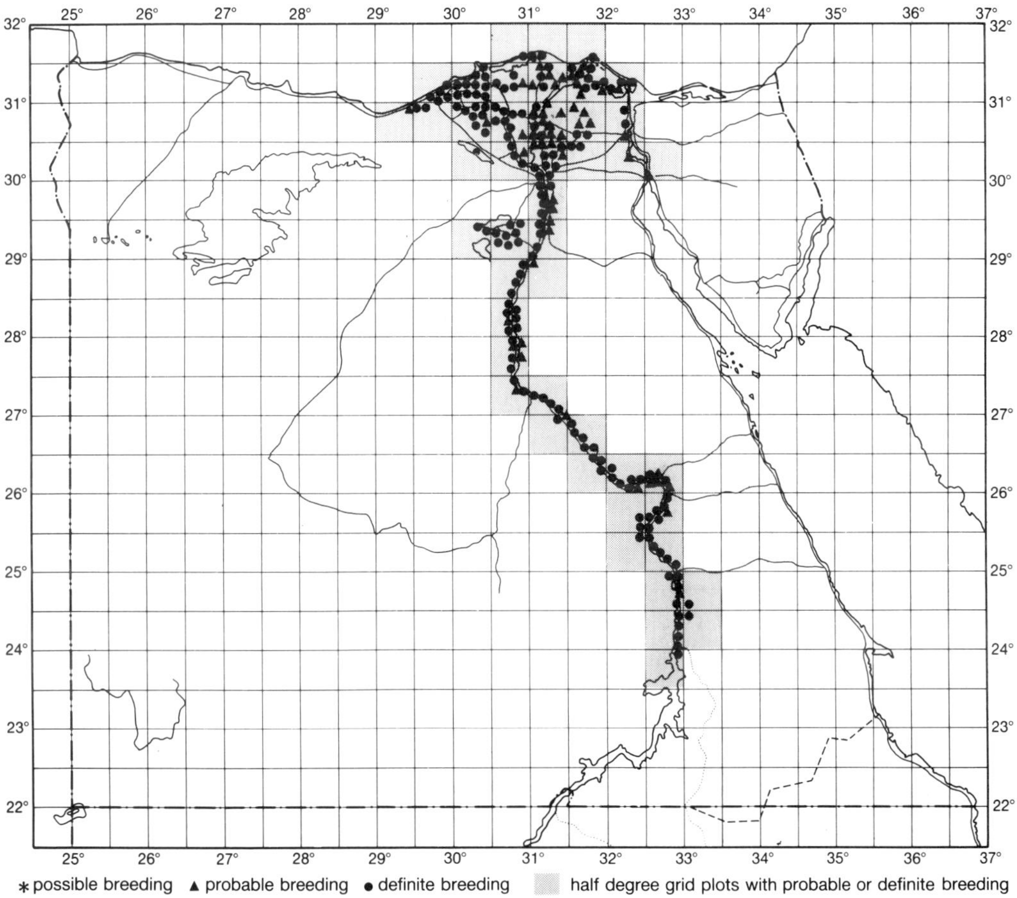

Map B72. Breeding distribution of Swallow *Hirundo rustica* between 1970–87.

According to Meinertzhagen (1922; 1930) *H. r. savignii* did not occur south of Luxor or as of 1920 west of Alexandria. Nowadays it is common as far south as the Aswan High Dam and occurs along the North Coast west to at least El Hammam. Apparently Swallows do not breed in Sinai or along some portions of the Suez Canal, such as at Suez. They are known to nest in city centres, such as Cairo, Alexandria, and Port Said. Dark-breasted birds, presumably *H. r. savignii*, have been recorded several times in Wadi el Natrun but as yet have not been found breeding there (Goodman *et al.* 1986). Two dark-breasted Swallows were observed on 1 November 1984 at Hurghada and one on 30 October 1984 at Ain Musa (RvD *et al.*).

In winter local concentrations occur, for example, 8500 at Lake Idku on 28 January 1979 (Meininger *et al.* 1979). These flocks are presumably composed mostly of local breeding birds.

Eurasian breeding birds pass through the country on both migrations along a broad front. Autumn passage begins in late July, peaks in September and October, and continues into December. Small numbers of adult light-breasted birds are observed with some regularity in late December and early January in Western Desert oases, the Nile Delta and Valley, and southern Sinai. North-directed movements start in January and the bulk of spring passage is between late March and mid-May. Small numbers summer along the Mediterranean and Red Sea coasts.

Ringing An adult ringed in Czechoslovakia in August was recovered at Port Said in October of the same year (Zink 1969).

Red-rumped Swallow *Hirundo daurica* PV (WV)

Local name 'Us̱fuur al-janna aẖmar al-'ajaz عصفور الجنة أحمر العجز.

Fairly common migrant throughout the country in autumn from mid-September to early (late) November and more common in spring from early March to early (late) May. Occasional in summer and winter. All Egyptian specimens examined are referable to *H. d. rufula.*

Winter records of Red-rumped Swallows include two observed on 25 December 1980 at Sharm el Sheikh (BCD), two on 15 January in the Faiyum (Meinertzhagen 1930), and one at Fayid on 18 February 1950 (HGB). There are two exceptional summer records of single birds on 9 June 1984 at Lake Timsah (BH *et al.*) and on 4 and 5 July 1984 at Hurghada (SH, DM).

House Martin *Delichon urbica* PV (WV)

Local name Sunuunuu abyad̲ al-bat̲n سنونو أبيض البطن.

Common migrant throughout the country in autumn from (late August) early September to mid-October (early November) and more common in spring from (early) late February to late May (mid-June). Rare summer visitor and occasional in winter. All Egyptian material examined is referable to *D. u. urbica*.

The only recent midwinter record of the House Martin appears to be one on 20 December 1980 at Idfu (SV). In addition to the early February observations presented by Short and Horne (1981) other records include single birds on 8 February 1976 at Sharm el Sheikh (HW) and 8 February 1979 at Lake Manzala (Meininger *et al.* 1979); it is presumed that these birds were early spring migrants. There are several summer records from the coastal areas of Sinai and the Red Sea. This species apparently crosses the northern Red Sea on migration (e.g. Tuck 1964).

Family Motacillidae

Generic local names for pipit Abuu fussiya, khushna, ṯayr al-rabii'a (Ma), 'uzayzaa.

Richard's Pipit *Anthus novaeseelandiae* PV WV

Local name Abuu fussiya kabiir ابو فصية كبير.

Rare migrant and winter visitor from (mid-September) early October to mid-May. According to Meinertzhagen (1930) it is an uncommon winter visitor to the Faiyum and Nile Delta from October to March but had not been recorded in the Sinai. All Egyptian material examined is referable to *A. n. richardi*.

Western Desert The only Richard's Pipit report from this region is one observed on 7 April 1925 at Kharga (Moreau 1927).

Nile Delta and Valley All known records from the Nile Delta and Lower Valley include: Damietta—several in January 1908 (Nicoll 1908) and one collected on 6 January (GZM); Birket Accrashi—one collected on 26 January 1917 (Raw 1921); Giza—one collected on 17 November 1908 (Nicoll 1909; GZM); Abu Zabal—one collected on 27 January 1917 (GZM); Fayid—one noted between 5 and 12 October 1952 (AEG, AP); Fanara—one on 4 March 1953 (AP); Ismailiya—one on 6 April 1966 (LGG); Suez—single birds on 6 October 1980 (Meininger *et al.* 1981), 18 and 19 October 1981 (Bijlsma 1982), 28 March 1965 and two on 1 April 1966 (HCK); and Faiyum—single birds on 12 March and 12 November 1909 (GZM). The species was noted two or three times in March and April 1903 at Lake Manzala (Loat 1906). The only known record from the Upper Nile Valley is one collected on 5 February 1913 at Abu Simbil (Koenig 1926).

Eastern Desert and Sinai There are few records from these regions. Marchant (1941) noted single birds at Hurghada on five days between 15 March and 14 April 1939. One was observed on 17 March 1981 at Lake Bardawil (SMB, BB) and another on 16 September 1981 at Sharm el Sheikh (MW). Between 1 and 13 October 1983 up to two were present at the St Katherine airport (SMB). At Elat, Israel, the species is a scarce migrant and winter visitor from late October to late March (Krabbe 1980).

Tawny Pipit *Anthus campestris* PV WV

Local name Abuu fus̲siya as̲-s̲ah̲raa- ابو فصية الصحراء.

Fairly common passage migrant in autumn from late August to early November and in spring from (mid-) late February to late April (mid-May). It is a scarce winter visitor. Egyptian specimens have been examined of both *A. c. campestris* and *A. c. griseus*.

During both migration periods Tawny Pipits have been recorded throughout the country. The earliest autumn record is two at Nuweiba on 26 August 1985 (DM). At Suez, the autumn migration of 1981 peaked in mid-September, with a maximum of 50 on 18 September and 124 on 20 September (Bijlsma 1982). This species was noted at Bahariya between 29 August and 26 September 1983, with a maximum of five on 14 September (HB *et al.*). Midwinter records are known from various localities in the Nile Delta and Valley, the Faiyum, and southern Sinai.

The bulk of spring migrants pass from March to April, and some apparently cross the Red Sea (Ticehurst 1924). The earliest known record from a non-wintering area appears to be one on 13 February 1984 in Wadi Ibib, near Gebel Hamra Dom (SMG), and the latest records are one on 9 May 1981 at Suez (SMB) and 14 May 1970 at Matruh (Horner 1977). There is no evidence to support Meinertzhagen's (1954) claim that it 'has certainly bred at Helwan . . .'.

Ringing A bird ringed at Bahig on 16 March 1971 was recovered in Bulgaria on 20 March 1972 (GZF).

Tree Pipit *Anthus trivialis* PV

Local name Abuu fus̲siya ash-shajar ابو فصية الشجر.

Common passage migrant throughout the country in autumn from late August to early (late) November and in spring from early March to mid- (late) May. All Egyptian material examined is referable to *A. t. trivialis*.

At Suez, Tree Pipit autumn migration commences in late August, peaks in mid-September (Bijlsma 1982), and a few linger through October and November. The latest autumn record is on 27 November 1965 at St Katherine (HCK).

In spring this species is more common than in autumn. The first migrants are usually noted in mid-March, although the species has been noted as early as 2 March 1973 at Bahig (Horner 1977) and 7 March 1987 at Nuweiba (PLM). Spring passage peaks between early and mid-April and continues into mid-May. The latest spring record is one on 25 May 1955 at Fayid (RN).

Ringing Birds ringed at Matruh in March 1970 and at Bahig on 10 April 1973 were recovered at the same locality in autumn 1970 and in Italy on 19 October 1973, respectively (GZF; MSp).

Meadow Pipit *Anthus pratensis* PV WV

Local name Abuu fus̲s̲iya al-ghayt̲ ابو فصية الغيط.

Fairly common winter visitor from (mid-) late September to mid-April, mainly in the north. A few migrants pass through the country. All Egyptian material examined is referable to *A. p. pratensis*.

The first autumn Meadow Pipits generally arrive in late September. Bijlsma (1982) studied bird migration at Suez between 4 September and 5 November 1981 and only recorded this species once—seven on 5 November. It appears that some birds pass through the country to points further south.

They remain fairly numerous throughout the winter in the Nile Delta, the Faiyum, Wadi el Natrun, the Suez Canal area, and northern Sinai. Some recent Western Desert records include: Bahariya—single birds captured on 22 and 28 October 1983 (HB *et al.*), and one seen on 12 February 1972 and a few on 15 March 1981 (Goodman *et al.* 1986); and Dakhla—two observed on 15 January 1982 (Goodman *et al.* 1986) and one on 18 March 1988 (AH).

Spring migration extends from mid-February to mid-April, peaking at Bahig between early and mid-March (Horner 1977). At this time of the year the species is rare in the Lower Nile Valley and the southern-most record appears to be at Asyut on 14 March 1983 (Welch 1983). The latest records are 16 April 1966 at Lake Maryut (JWW) and 19 April 1984 at St Katherine (Siering 1986).

Ringing Two birds ringed at Bahig in early 1971 were retrapped at the same locality in early 1972 (Horner 1977; GZF).

Red-throated Pipit *Anthus cervinus* PV WV

Local name Abuu fus̲s̲iya ah̲mar az-zawr ابو فصية أحمر الزور.

Abundant passage and winter visitor throughout the country from (late August) early October to early May (early June). Monotypic.

The earliest autumn Red-throated Pipit record is one on 31 August 1984 at Birket el Hagg (DM). The species is common on migration throughout the country from early October to mid-November. Red-throated Pipits are commonly observed throughout most of Egypt during the winter months in suitable habitat, such as cultivation, marshes, and along sea-shores.

In spring the species appears to be more common in the eastern portion of the country, and movements begin in late February and peak between late March and mid-April. The number of migrants passing through Bahig is not constant from spring to spring, and the species is occasionally completely absent (Horner 1977). The latest known spring records are single birds on 17 May 1984 at Gezira Abu Minqar (UMMZ) and on 7 June 1984 at Bir Beida (SMG).

Water Pipit *Anthus spinoletta* PV WV

Local name Abuu fus̲siya al-maa- ابو فصية الماء.

Common winter visitor mainly to the northern part of the country from (late September) early October to mid-April (mid-May). Some migrants apparently pass through Egypt to and from points further south. All Egyptian material examined is referable to *A. s. coutellii*. We follow Vaurie (1959) in treating *A. s. blakistoni* as a synonym of *A. s. coutellii*.

In autumn Water Pipits generally reach Egypt in early October but the earliest record appears to be four on 20 September 1981 at Hurghada (FdR). During the autumn of 1981 the species was recorded at Suez from 21 October to 5 November (when observations ceased), with up to 950 gathering on 29 October at a pre-roosting site (Bijlsma 1982). Autumn movements have been noted at Aswan on 1 October 1981 (FdR) and in spring along the Red Sea coast and in the Western Desert. This would indicate that a small number of migrants winter in areas further south. However, there appears to be no mention of such a passage in the standard works on Sudanese birds (Cave and Macdonald 1955; Nikolaus 1987) or Palearctic–African bird migration (Moreau 1972; Curry-Lindahl 1981).

In winter the Water Pipit is common in wet areas, particularly in the Nile Delta, Wadi el Natrun, the Faiyum, the Suez Canal area, and northern Sinai. It is uncommon in the Nile Valley, along the Red Sea coast, and in most Western Desert oases. During March the numbers decrease and by mid-April the species is scarce. The latest spring record is 14 May 1981 at Suez (SMG).

Generic local names for wagtail Abuu fas̲aada, ruwayii (Ma), umm sayyis (NC).

Yellow Wagtail *Motacilla flava* RB PV WV

Local names Abuu fas̲aada as̲far ابو فصادة أصفر, bad badiid (Bi), dha'ra, umm sak'ak', zat̲zaat̲a, zayt̲a.

M. f. pygmaea is a common breeding resident in various portions of the country. It is unknown which form breeds in northern Sinai. The remaining subspecies, *M. f. flava*, *cinereocapilla*, *thunbergi*, and *feldegg*, are passage migrants throughout the country in autumn from (early) mid-August to mid-October (mid-November) and in spring from (early) mid-March to mid-May (mid-June). Yellow Wagtails of subspecies other than *pygmaea* are uncommon winter visitors to southern Egypt. We are unaware of any specimens of *M. f. leucocephala* taken in Egypt (cf. Grote 1930).

M. f. pygmaea is a common breeding resident in marshes and cultivation near Bahig, Wadi el Natrun, the Nile Delta and Valley (south to Aswan), the Faiyum, and Suez Canal area. It is particularly common in the Delta marshes, with densities of up to 15–20 breeding pairs/ha (Meininger *et al.* 1986). In the south-central Nile Delta and

Nile Valley the densities are generally lower and the species is sometimes absent from seemingly suitable habitat. A female of *M. f. pygmaea* collected at Matruh on 10 April 1970 may have been a local breeding bird (USNM). *M. f. pygmaea* can be observed throughout the year in the breeding areas and nesting takes place between late March and May. This species was found breeding on 21 April 1977 in north-eastern Sinai at Kurum el Sheikh (Sabkhet e-Sheikh) (Hovel 1987; Paz 1987).

Systematic observations of motacillid migration were carried out at Suez from 4 September to 5 November 1981 and passage of Eurasian Yellow Wagtails was noted from 5 September to 17 October, peaking in late September and early October (Bijlsma 1982). This species was observed daily in Bahariya between 6 August and 28 October 1983, with the bulk of migrants passing between late August and late September (HB *et al.*). On 11 September 1981 about 1490 were counted at Zaranikh (Petersen and Sørensen 1981*b*) and on 30 September 1983 more than 1000 Yellow

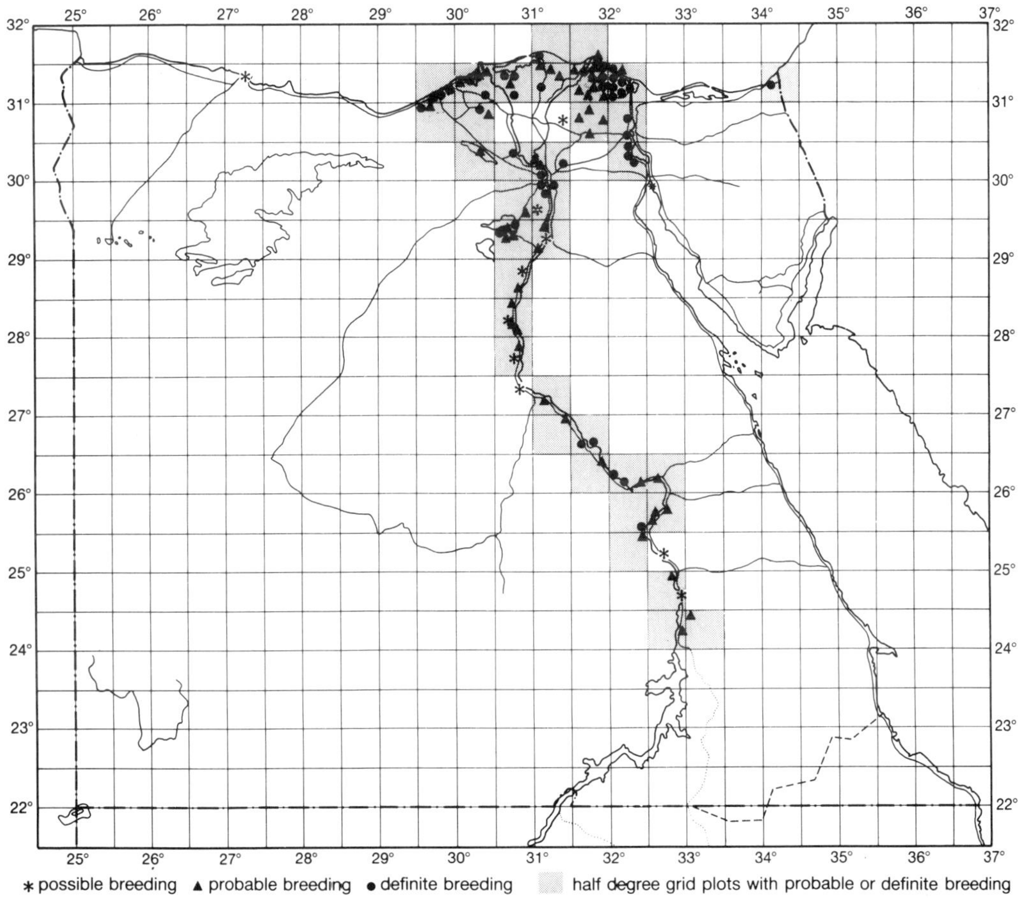

Map B73. Breeding distribution of Yellow Wagtail *Motacilla flava* between 1970–87.

Wagtails were found at a roost between El Wasta and Beni Suef (Hume 1983). *M. f. flava* seems to be the most common subspecies, followed by *M. f. feldegg*. There are winter records of these two forms near Ismailiya, in Upper Egypt, southern Sinai, Gebel Elba, and Gebel Uweinat. In spring both are common on passage. An exceptionally late migrant was observed on 10 June 1984 near the Brothers Reef, central Red Sea (BH *et al.*).

Several other subspecies of Yellow Wagtail have been recorded in Egypt, many of which are difficult to distinguish in the field. The following summaries are based primarily on specimens. *M. f. thunbergi* is apparently fairly common on autumn and spring passage (+20 recent specimens) and has been found during late December in Egyptian Nubia (USNM). *M. f. cinereocapilla* is rare in autumn with records on 27 September 1969 at Matruh (USNM) and 12 October at Rosetta (Meinertzhagen 1930), and in spring (nine post-Meinertzhagen specimens examined) between 23 March and 11 April. Recent observations of male *M. f. beema* include single birds on 21 April 1984 along the east coast of Sinai (Siering 1986) and on the same date at Hurghada (MBe, SM); however, without further documentation these records are not acceptable. A number of *M. f. 'dombrowskii'* specimens have been taken in Egypt (e.g. USNM) that are best considered secondary intergrades between *M. flava* subspecies.

Ringing A bird ringed in Finland on 10 September 1968 was recovered at Damietta on 25 October 1971 (EDB). One ringed at Bahig on 25 April 1972 was recovered in Czechoslovakia on 14 April 1974 (GZF).

[**Citrine Wagtail** *Motacilla citreola* AV?

Nicoll (1909) identified an immature bird taken at Giza on 18 September 1907 as *M. citreola*. Meinertzhagen (1930) examined this specimen and one other referred to this species and concluded that both were immature *M. flava feldegg*. More recently one Citrine Wagtail was observed near Bir el Abd, Lake Bardawil, on 19 April 1985 (CHF). This species is a regular autumn and spring migrant in small numbers at Elat, Israel (Inskipp 1979; Krabbe 1980; HSh).]

Grey Wagtail *Motacilla cinerea* PV WV

Local name Abuu fas̱aada ramaadii ابو فصادة رمادي.

Fairly common passage migrant and winter visitor throughout the country from (mid-August) mid-September to mid-April (mid-May). All Egyptian material examined is referable to *M. c. cinerea*.

Along the north coast of Sinai autumn migration of Grey Wagtails commences in mid-August (Shariv 1970). During a study of bird migration at Suez between 4 September and 5 November 1981, this species was scarce from 14 September to 16 October (Bijlsma 1982). In winter Grey Wagtails can be observed along the North Coast, in the Western Desert, the Nile Delta and Valley, the Suez Canal area, the

Faiyum, and near wells throughout the mountains of the Eastern Desert and Sinai. Most migrants leave the country in March and early April. The latest spring records appear to be one on 2 May 1978 between Kom Ombo and Aswan (Meininger and Dielissen 1979) and on 16 May 1985 in Wadi el Natrun (JS *et al.*).

White Wagtail *Motacilla alba* PV WV

Local names Abuu fas̱aada abyaḏ ابو فصادة أبيض, umm brayma (NC).

Abundant passage migrant and winter visitor throughout the country from (late September) early October to late April (mid-May). All Egyptian material examined is referable to *M. a. alba*.

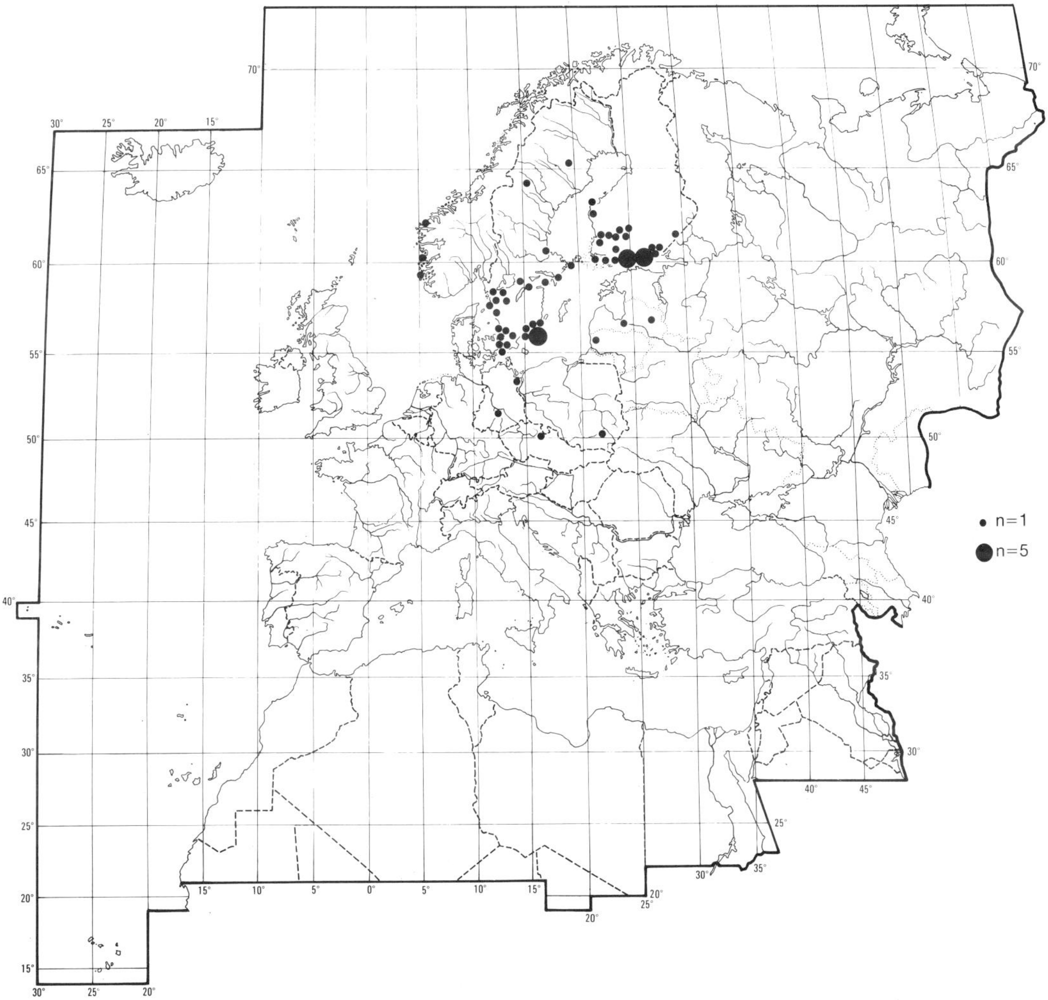

FIG. 6.25. Ringing or recovery sites of White Wagtails (n=67) occurring in Egypt.

Apart from exceptional observations of single White Wagtails on 2 August 1985 at Shakshuk (SMB) and on 19 August 1986 at Suez (EE *et al.*), the earliest autumn record in Egypt is 24 September 1944 at Kasfarit (LB). During a systematic study of bird migration at Suez between 4 September and 5 November 1981, this species was recorded from 29 September to 5 November, with the heaviest passage during the last 10 days of October (Bijlsma 1982). It is particularly common in cultivated areas of the Nile Delta and Valley, with as many as 100 000 noted on 21 October 1979 at a night roost in Heliopolis (SMB).

In winter this species is common throughout the country, with small numbers present in remote desert areas with sparse vegetation or near human settlements. In the Eastern Desert small flocks are often found associated with domestic animal herds belonging to local Bedouins. Several Bedouin informants mentioned that White Wagtails can also be found feeding near dorcas gazelle and Nubian ibex (SMG, JJH).

Spring migration begins in February and continues to mid-April. Large flocks are occasionally noted. One exceptional observation, made from a ship off the Egyptian Mediterranean coast, was an estimated 'one million' migrating White Wagtails which passed on an evening in the last week of March 1962. The next morning the ship's deck was littered with thousands of White Wagtail corpses (Larmuth 1973). By late April this species is scarce and the latest spring record is on 17 May 1935 at Siwa (Moreau 1940–1).

Ringing There are over 60 recoveries in Egypt of birds ringed in Denmark (1), Finland (26), Norway (2), Poland (2), Sweden (28), and western USSR (3) (Fig. 6.25; Mackintosh 1941; Tolstoy 1960; Österlöf 1970, 1971; Watson 1971; Anon. 1972, 1973, 1976; Holgersen 1974; Zink 1985; BEP; EDB; GZF; JG; RSt), and at least seven birds ringed in Egypt have been recovered in Finland, the German Democratic Republic, Norway, Poland, and Sweden (GZF).

African Pied Wagtail *Motacilla aguimp* RB

Local name Abuu fas̱aada abqa' ابو فصادة أبقع.

The form *M. a. vidua* is a rare breeding resident in the Lake Nasser area. The type locality of this subspecies is Aswan.

Until the beginning of the twentieth century the African Pied Wagtail was regularly recorded in small numbers along the Nile, north to the First Cataract near Aswan (e.g. Adams 1864; Gurney 1876; Russell 1905). Shelley (1872) found it common in the vicinity of boat moorings near Aswan and about to breed in April. Koenig collected 11 specimens in the general vicinity of El Shallal between January and March 1897–1910 (ZFMK). Meinertzhagen (1930) observed several pairs near Aswan in February 1928 and collected a male at Abu Simbil on 11 February 1928 that was apparently approaching breeding condition (BMNH).

The only recent report from Aswan is one in early March 1980 (SK). This species probably breeds along the shores of Lake Nasser. There are several recent records in autumn, winter, and spring between Aswan and Wadi Halfa (Sudan). These include single birds at Abu Simbil on 4 January 1978 (AS) and 26 March 1985 (Grieve 1985), and a pair 15 km south of Abu Simbil on 23 April 1983 (SC, PLM, IP).

Family Pycnonotidae

Black-capped Bulbul (Yellow-vented Bulbul) *Pycnonotus xanthopygos* RB Plate 5

Local name Bulbul 'arabii بلبل عربي.

Locally common breeding resident in Sinai and rarely noted west of the Suez Canal. Monotypic.

The Black-capped Bulbul is a breeding bird of oases and cultivated areas in Sinai. Along the north coast it frequents gardens in the El Arish–Rafa area and small date palm groves as far west as Rumana. In the southern portion of the Peninsula it is regularly found in high mountain areas (up to 2000 m) with suitable habitat, such as St Katherine and Wadi Feiran, and east to the Gulf of Aqaba coastal plain (between Taba and Sharm el Sheikh). A recent fledgling was noted at Taba on 9 March 1987 (PLM). The species is not known from the coastal areas along the Gulf of Suez or the south-western shores of Sinai. Adult specimens taken near Nuweiba have exceptionally pale vents (ZM-HUJ).

A male Black-capped Bulbul collected north of Cairo on 11 April 1920 was 'apparently paired with *P. arsinoe* [= *barbatus*]' and had enlarged testes (GZM; see also Meinertzhagen 1930). The record of Wimpfheimer *et al.* (1983) of 'Yellow Vented Bulbul' at Suez is in error and these birds were actually white-vented Common Bulbuls (SMB). No clear *P. xanthopygos* × *P. barbatus* hybrid specimens have been examined from Egypt.

Common Bulbul *Pycnonotus barbatus* RB

Local names Bulbul shaai' بلبل شاع, aṯ-ṯayr al-mughanii (NC), jumayl, kam lawiin (Bi), ku'ayt, shanzuur ash-shaami (F).

Common breeding resident in Wadi el Natrun, the Nile Delta and Valley, the Faiyum, the Suez Canal area, and the Gebel Elba area. All Egyptian material examined is referable to *P. b. arsinoe*. Throughout the Egyptian range densities of breeding birds are variable and in some places the species is locally scarce.

The Common Bulbul has extended its Egyptian breeding range in the Nile Delta and Valley during this century. Adams (1864) made no mention of this conspicuous bird north of Wadi Halfa, Sudan. Specimens were taken in late January 1891 at 'Belauyeh, Nubia' (FMNH). By the beginning of the twentieth century the species was common

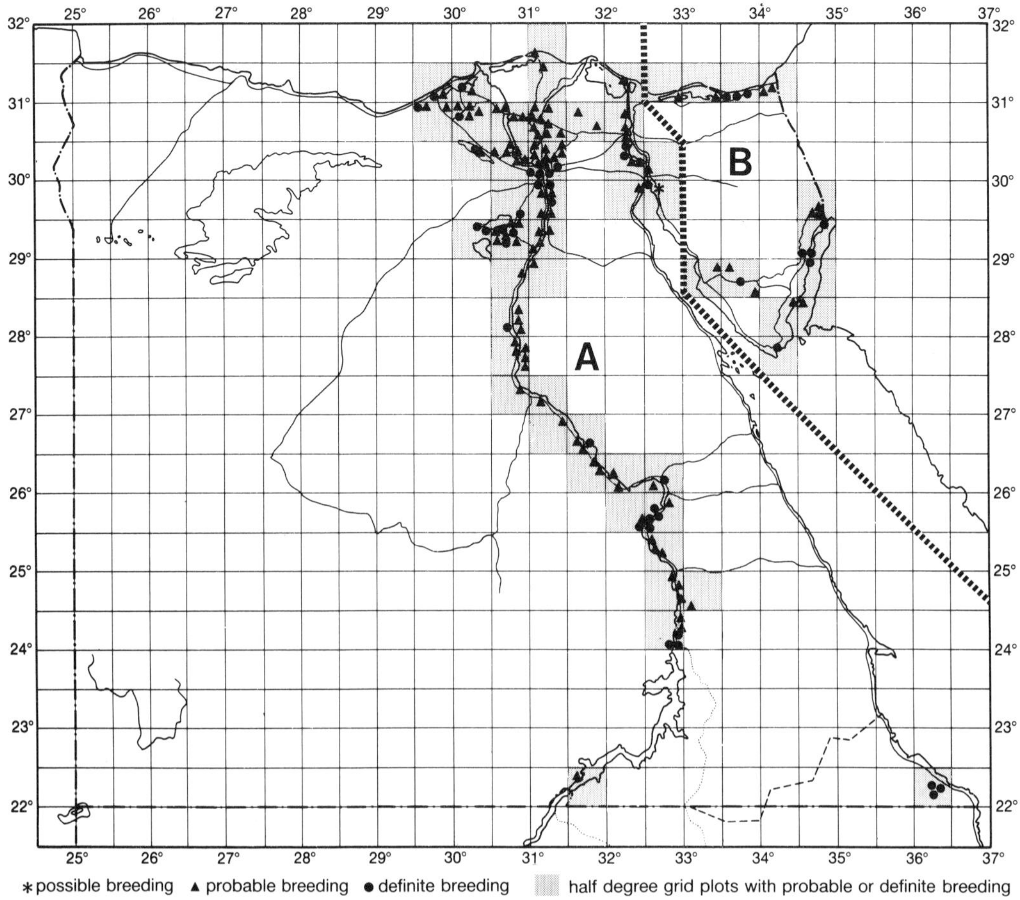

Map B74. Breeding distribution of: Common Bulbul *Pycnonotus barbatus* (A); and Black-capped Bulbul *P. xanthopygos* (B) between 1970–87.

in the Faiyum but rare near Cairo (Koenig 1920; Raw 1921). According to Meinertzhagen (1930) it first appeared in the Nile Delta about 1909 and quickly established itself in the environs of Cairo. In the 1920s it moved northwards in the Nile Delta and became more common near Cairo (Borman 1928). Bodenham (1945*a*) observed it sporadically near Alexandria between 1942 and 1944. The earliest record at Bahig appears to be May 1967 (USNM).

The first known report from the Suez Canal area was in 1944 at Ismailiya (LB). Presently this species breeds in most portions of the Nile Delta west to Bahig, the Suez Canal area, the Faiyum, and the Nile Valley south to Aswan. In the mid-1960s it was found in the region of Egyptian Nubia now inundated by Lake Nasser (COM; YPM). It apparently still occurs in this area: at Abu Simbil a singing bird was noted on 30 March 1982 (JW) and several on 27 April 1984 (MBe, SM). Common Bulbuls have been observed in Wadi el Natrun since 1980 and are apparently local breeding residents (Goodman *et al.* 1986). Two were observed in Sinai at Ain Musa, just across the canal from Suez, on 6 and 17 October 1984 (RvD *et al.*).

This species is a common resident in vegetated wadis and on mountain slopes of Gebel Elba. Here birds have been observed at about 1100 m singing and defending territories in the local *Dracaena ombet* forest. Concentrations are often noted near fruiting plants; for example, over 10 birds were seen on 22 February 1985 feeding on the fruits of a single *Balanites aegyptiaca* tree (SMG). Although the Gebel Elba and nearest Egyptian Nile Valley populations are separated by at least 500 km of desert, there appears to be no discernible morphological differences between them.

Family Bombycillidae

Grey Hypocolius *Hypocolius ampelinus* AV

Local name Khanaaq ramaadii خناق رمادي.

The only verifiable record of the Grey Hypocolius in Egypt is a female collected on 22 December 1938 in Wadi Shallal, south of Gebel Elba (Goodman 1984*a*). The basis for the 1943 records cited by Mackworth-Praed and Grant (1980) and Etchécopar and Hüe (1967) is unknown.

Family Prunellidae

Dunnock *Prunella modularis* (WV)

Local names 'U<u>s</u>fuur ash-shawk عصفور الشوك, shuwiikii.

Rare and irregular winter visitor to the Mediterranean coast and a few other localities between late October and mid-March. Egyptian material contains birds referable to both *P. m. modularis* and *P. m. obscura* (see below).

Three specimens referable to *P. m. obscura* were collected at Bahig between 4 and 23 November (Goodman and Watson 1983). *P. m. modularis* is represented by birds taken near Giza on 26 February (Meinertzhagen 1930) and at Bahig and Matruh in the second half of November and mid-March (Goodman and Watson 1983). This species has also been found at Ikingi Maryut on 22 October 1975 (DMNH), St Katherine on 26 February 1973 (ZM-HUJ), Giza on 12 February 1937 (HGD), and Suez on 12 March 1982 (Wimpfheimer *et al.* 1983).

Family Turdidae

Rufous Bush Robin *Cercotrichas galactotes* MB PV (WV)

Local names Du<u>kh</u>ala <u>h</u>umraa- دخلة حمراء, adabs (NC), du<u>kh</u>aala, tanjuu gist (Si), wagdii (Si).

Locally common migrant breeding bird in various vegetated portions of Egypt. Fairly common passage migrant in autumn from mid-August to mid-October and in spring from (early) mid-March to mid-May. It is rare during the winter. The Egyptian breeding populations are referable to *C. g. galactotes*. On the basis of wing measurements specimens from Egyptian Nubia are closer to *C. g. galactotes* than to *C. g. minor* of Sudan. The subspecies *C. g. syriacus* is an uncommon migrant to the country.

The Rufous Bush Robin is a migrant breeding resident in all of the major Western Desert oases (Goodman *et al.* 1986) and is locally common along the North Coast, in the Nile Delta and Valley, the Faiyum, and Suez Canal area. In the central Delta it is not common and generally restricted to orchards. In the Valley between Cairo and Qena it tends to be rather localized in suitable habitat. In June 1984 pairs were observed along the Qift–Quseir road at El Laqeita and Bir Beida (SMG). One

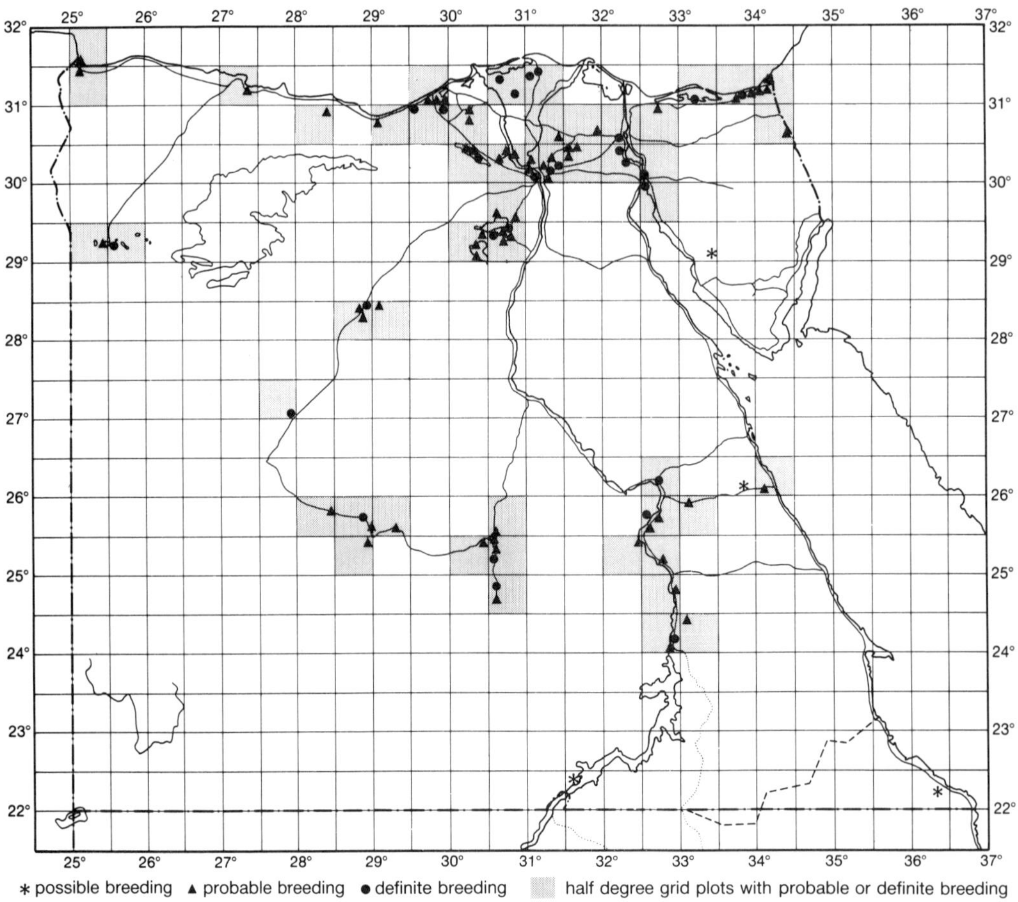

Map B75. Breeding distribution of Rufous Bush Robin *Cercotrichas galactotes* between 1970–87.

observed on 20 June 1984 in Wadi Akwamtra, Gebel Elba, may have been part of a local breeding population (SMG). In Sinai, this species commonly breeds near El Arish, Rafa, Quseima, and locally along the southern shore of Lake Bardawil.

During August and September most local breeding birds leave the nesting grounds and migrate to sub-Saharan Africa (Moreau 1972). At this same time northern migrants also pass through Egypt. Records of single birds at Suez on 28 October 1984 (RvD *et al.*); in Karkur Talh, Gebel Uweinat, on 4 November 1968 (Misonne 1974); and in Dakhla on 16 January 1982 (Goodman *et al.* 1986) are exceptional.

Spring migration generally commences in mid-March and returning Egyptian birds quickly establish their breeding territories. During March and April migrants are regularly observed at localities where this species is not known to nest, such as sparsely vegetated wadis in the Eastern Desert.

Robin *Erithacus rubecula* WV

Local names Abuu al-h̲innaa- ابو الحناء, abuu al-h̲usnaa-, abuu s̲adr, al-h̲usn, h̲abnaa (NC).

Fairly common winter visitor to the north of the country from (late September) mid-October to late March (mid-April). All Egyptian material examined is typical of *E. r. rubecula*.

The only known record of the Robin in the Western Desert is one on 24 October 1983 in Bahariya (HB *et al.*). It is sometimes a common winter visitor along the entire Mediterranean coast but the numbers vary from year to year. Robins generally do not penetrate far south of the coast, although there are a few records from the Suez Canal area, near Cairo, and the Faiyum. Two females were found in Wadi Qiseib, near Ain Sukhna, on 20 and 21 December 1966 (USNM). One was observed at St Katherine on 27 January 1985 (SMG) and another on 7 March 1987 (PLM). The latest spring record is 18 April at Bahig (Horner 1977).

Thrush Nightingale *Luscinia luscinia* PV

Local name 'Andaliib عندليب.

Fairly common passage migrant in autumn from mid-August to mid-October and scarce in spring from (mid-) late March to early May. Monotypic.

In autumn the Thrush Nightingale is sometimes locally common throughout the country. During three autumn field seasons in Bahariya between 1982 and 1985, 39 Thrush Nightingales were captured (extreme dates 22 August to 3 October), compared with six Nightingales (extreme dates 14 September to 4 October) (HB *et al.*). Between 16 August and 24 September 1981 a total of 52 was observed at Zaranikh (IP, UGS). Maclaren (1944) found the species to be widespread near El Amiriya between 26 August and 18 September 1942.

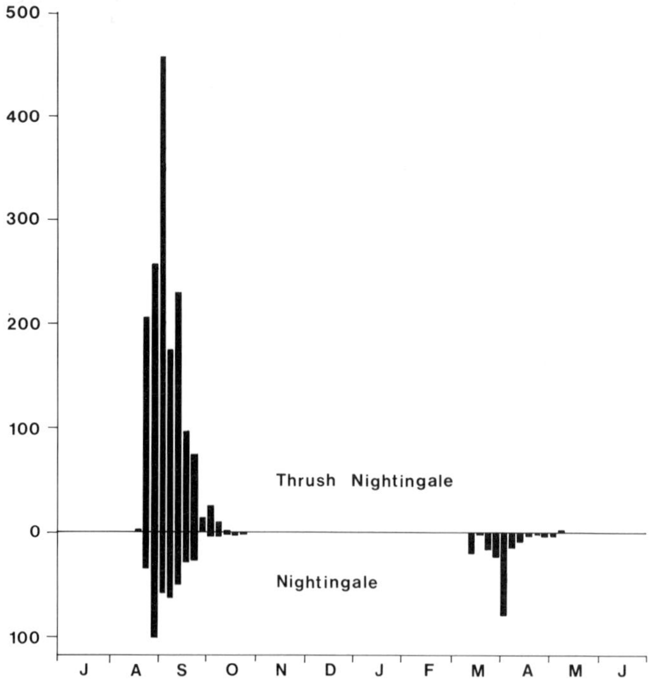

FIG. 6.26. Migration patterns of Thrush Nightingales and Nightingales along the North Coast (after Watson 1971).

In spring it is generally scarce throughout the country but apparently more common in the east. The earliest record is 13 March 1983 near El Amarna (Welch 1983) and the latest is 9 May along the North Coast (Horner 1977). During the six PMS spring study seasons at Bahig between 1967 and 1973 about five Thrush Nightingales were captured in total, compared with 5968 Nightingales (Fig. 6.26; Horner 1977; USNM). Conversely, 30 Thrush Nightingales were ringed at El Arish between 22 April and 3 May 1985 and only four Nightingales (Fry *et al.* 1985).

Ringing There are 10 recoveries in autumn and one in spring of Danish, Finnish, and Swedish ringed birds in Egypt (Österlöf 1969; Watson 1971; Anon. 1972; Zink 1973; EDB; RSt). A bird ringed at Bahig in September 1972 was recovered in the Federal Republic of Germany in May 1975 (GZF).

Nightingale *Luscinia megarhynchos* PV

Local names Hazaar هزار, abuu haaruun, al-mughannii al-asmar, atbaz (NC), muzqa.

Scarce passage migrant in autumn from early August to mid-October and common in spring from mid-March to early May (early June). All Egyptian material examined is referable to *L. m. megarhynchos*.

Small numbers of Nightingales are found in autumn throughout the country. The earliest record is on 2 August (year not stated) at an undisclosed locality (Meinertzhagen 1930) and the latest is on 13 October 1981 at Suez (RGB *et al.*).

On spring passage this species is regularly noted throughout the country, and singing birds are frequently heard. The first Nightingales usually appear in the second half of March. The earliest date at Bahig is 12 March 1971 (Horner 1977). Spring passage generally peaks in April, and small numbers can be observed into late May (Fig. 6.26). One exceptionally late record is a singing bird at Fayid from 4 to 9 June 1955 (RN).

Ringing Birds ringed at Bahig in the spring have been recaptured at the same locality in the autumn of that year and the following spring (Watson 1971; GZF). Another ringed at Bahig on 7 April 1967 was recovered in Bulgaria (GZF).

Bluethroat *Luscinia svecica* PV WV

Local names H̱usaynii حسيني, mus/har, muzqa.

Common passage migrant and winter visitor from (mid-September) early October to mid-April (early May). We have found it difficult to place subspecific identifications on males in winter plumage, and two types are recognized herein, birds intermediate between *L. s. svecica* and *volgae*, and those referable to *L. s. cyanecula*.

The earliest recent autumn record of the Bluethroat in Egypt is on 18 September 1982 at Bahariya (HB *et al.*). Kaiser (1892) recorded *L. s. cyanecula* some time during September in Sinai. Autumn passage normally begins in early October and continues into late November.

In winter the species can often be found along the North Coast, in Western Desert oases, the Nile Delta and Valley, Faiyum, Suez Canal area, Sinai, and along the Red Sea coast. Generally, it is restricted to marshes, reed-beds, and wet cultivated areas.

Spring migration commences in February and is particularly notable in March when Bluethroats are observed in areas where they do not winter, for instance sparsely vegetated areas of the Eastern Desert. Generally Bluethroats leave the country by mid-April but have been observed as late as 'May' at Abu Zabal (Raw 1921) and on 4 May at Bahig (Horner 1977).

White-throated Robin *Irania gutturalis* AV

Local name Abuu al-ẖinnaa- abyaḏ az-zawr ابو الحناء أبيض الزور.

The only documented record of the monotypic White-throated Robin in Egypt is one on 30 May 1984, 8 km north of Abu Zenima (Goodman and Storer 1985).

Generic local names for *Phoenicurus* spp. Abuu ẖumra (NC), ẖumayraa-, ẖumayraa dabsaa.

Black Redstart *Phoenicurus ochruros* PV WV

Local name Ḫumayraa- suudaa- حميراء سوداء.

Fairly common passage migrant and winter visitor to the northern portion of the country from (late August) late October to mid- (late) March. Most Egyptian specimens are typical of *P. o. gibraltariensis*. A bird collected at Giza on 16 February 1901 is referable to *P. o. phoenicuroides* (Meinertzhagen 1930). The type specimen of *P. o. semirufus* is not from Egypt but Syria (Stresemann 1954; cf. Meinertzhagen 1930) and there is no definitive evidence that this form occurs in Egypt.

In autumn Black Redstarts generally arrive in late October or early November. Exceptionally early records include birds east of Helwan on 25 August 1984 (DM) and at Suez on 4 September 1943 (Meiklejohn 1944).

The species is not uncommon during the winter along the North Coast, in the Nile Delta and Lower Valley, the Suez Canal area, and the Sinai. There are relatively few records from the Nile Valley south of Cairo and none further than Luxor. In the Western Desert it has been recorded on several occasions in Siwa and Wadi el Natrun, and a few times in Bahariya and Dakhla (Goodman *et al.* 1986). The only known records from the Eastern Desert are from Gebel Elba and include one or two on 24 February 1927 (Madden 1930) and a male and a female in late February 1985 (DM). The number of Black Redstarts decreases during February and most have left the country by mid-March. The latest spring record is 27 March 1967 at Bahig (Horner 1977).

Ringing A bird ringed in the Federal Republic of Germany as a nestling was recovered at Cairo in February of the following calendar year (Zink 1981) and another ringed at Ikingi Maryut on 13 March 1969 was recaptured at Bahig on 3 March 1970 (Watson 1971).

Redstart *Phoenicurus phoenicurus* PV (WV)

Local name Humayraa- حميراء.

Common migrant throughout the country in autumn from (late August) mid-September to early (late) November and in spring from (early February) mid-March to mid-May (mid-June). This species is an irregular winter visitor. *P. p. phoenicurus* is the most common form passing through the country. A number of specimens taken on the North Coast have broadly white-edged flight feathers and are referable to *P. p. samamisicus* (USNM).

The earliest autumn Redstart records in Egypt are single birds on 27 August 1983 in Bahariya (HB *et al.*) and on 30 August 1966 at Alexandria (JWW). From late September to early November the species is commonly observed throughout the country, including desert regions. The latest autumn records are a female on 17 November 1946 at sea, 95 km north-east of Alexandria (Elliott and Monk 1952) and one on 25 November 1984 at Suez (RvD *et al.*). The only midwinter report appears to be one observed in the Faiyum on 31 December 1963 (HCK).

There are several observations from early February, presumably of early spring migrants. These include Redstarts in Wadi el Natrun on 2 and 9 February 1903 (Loat 1905), in the Nile Valley in early February (Short and Horne 1981), and near Gebel Hamra Dom on 16 February 1984 (SMG). Spring migrants pass through Egypt on a broad front and numbers peak between late March and April. Some birds fly directly across the northern Red Sea (Tuck 1961). An exceptionally late record is one on 18 June 1984 on Gezira Siyul, off Hamata (SMG).

Ringing Birds ringed at Bahig on 6 April 1967 and 19 October 1969 were recovered in Libya before early February 1968 and on 5 May 1971, respectively (Hubbard and Horner 1969; GZF). Another ringed at Bahig on 20 October 1971 was recovered in Lebanon on 20 September 1972 (GZF). One ringed at Bahig on 5 April 1967 was recaptured at the same locality exactly three years later (Watson 1971).

Blackstart *Cercomela melanura* RB

Local name Qulay'ii aswad adh-dhanab قليعى أسود الذنب.

Locally common breeding resident in Sinai (*C. m. melanura*) and near Gebel Elba (*C. m. lypura*). It is also thought to nest near Gebel Uweinat, where the local form is presumably closer to Saharan *C. m. airensis* than to *C. m. lypura*. It is not clear if records outside the known residential areas are of migrants or representatives from undiscovered breeding populations.

Western Desert One Blackstart was observed in December 1968 at Gebel Uweinat (Misonne 1974), where it may breed (Goodman *et al.* 1986).

Nile Valley There are a number of Blackstart sight records in this area, which is outside their known breeding range. These include single birds at Giza on 9 October 1907 (Nicoll 1912*b*) and on 23 December 1985 (AS), at Philae on 16 October 1979 and at Saqqara on 22 February 1980 (Short and Horne 1981), at Aswan on 8 March 1985 (AS), and a pair at Abu Simbil on 22 March 1983 (Welch 1983). It is conceivable that the species breeds in vegetated wadis that drain into Lake Nasser, which would account for the Upper Nile Valley records.

Eastern Desert There are two well-documented sight records from the northern portion of this area; one was observed 6 km south of Suez on 14 October and two on 16 October 1986 (EH, PK). This species is a common breeding resident in the Gebel Elba region. In late June 1984 groups of presumed males were observed displaying in what appeared to be leks (SMG).

Sinai Blackstarts are locally common breeding residents in rocky wadis of southern and central Sinai, to altitudes of about 1500 m, and along the Gulf of Aqaba coastal plain. Generally they are found in wadis with acacia trees but several were noted on 30 May 1984 in a tamarisk thicket 8 km north of Abu Zenima (SMG, RWS).

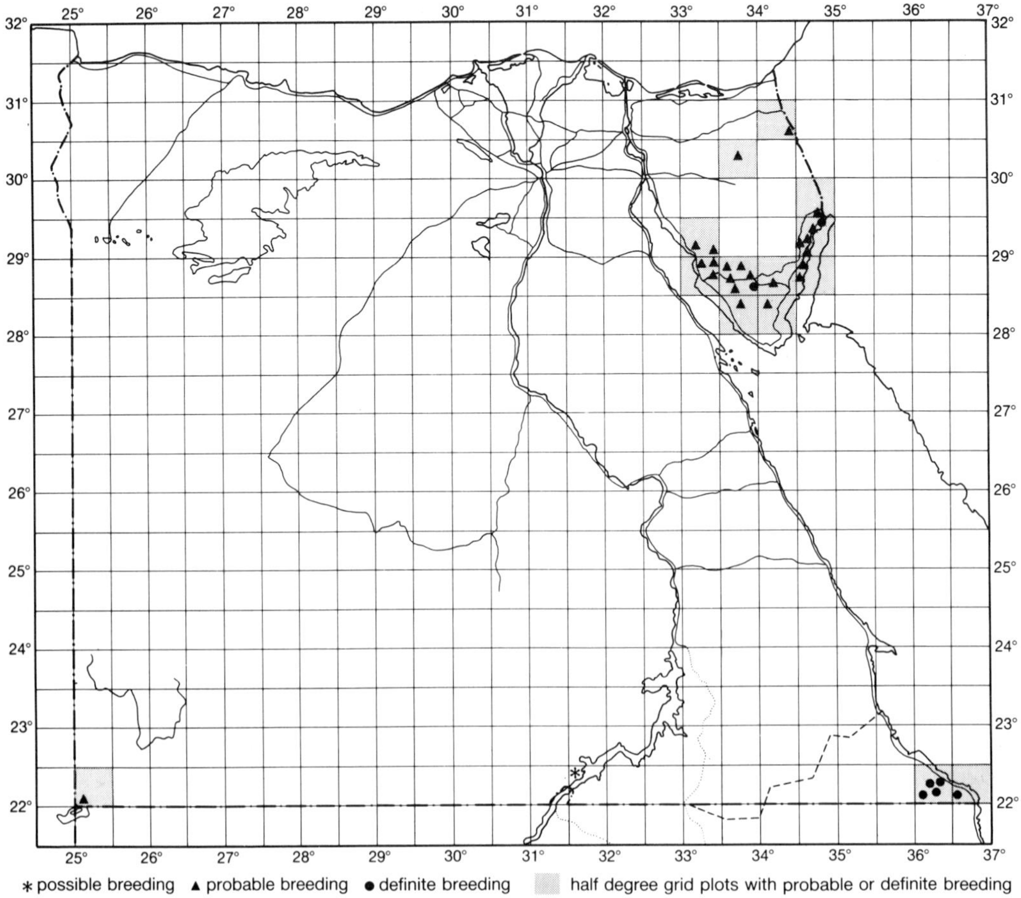

Map B76. Breeding distribution of Blackstart *Cercomela melanura* between 1970–87.

Whinchat *Saxicola rubetra* PV

Local names Qulay'ii a<u>h</u>mar قليعى أحمر, fustuqa, qulay'ii.

Fairly common migrant throughout the country in autumn from (late August) early September to late October (late November) and common in spring from (early February) early March to late May. Monotypic.

The earliest autumn records of Whinchats are 27 August 1983 at Bahariya (HB *et al.*) and 25 on 29 August 1982 at Zaranikh (IP, UGS). Between early September and late October the species is regularly observed and passage peaks between mid-September and mid-October. There are numerous November records from various parts of the country.

Spring passage appears to commence in February. Migrants have been observed as early as 2 February 1983 near Quseir (SMG) and 11 February 1913 in Wadi el Natrun

(GZM). Since there are no December or January records from the country it is presumed that these birds are early migrants rather than winter visitors. Between early March and late May the species is regularly noted throughout the country and passage peaks between early April and early May. The latest spring record is on 23 May 1971 at Bahig (Horner 1977).

Stonechat *Saxicola torquata* PV WV

Local names Qulay'ii muṯawwaq قليعى مطوق, fustuqa, qulay'ii.

Fairly common migrant and winter visitor from (mid-September) mid-October to mid-March (late April), mainly in the north of the country. On the basis of museum specimens *S. t. rubicola* seems to be the most common subspecies in Egypt. *S. t. variegata* and *S. t. armenica* have also been ṯaken in the country. We are unaware of any firm evidence to support the occurrence of *S. t. maura* in Egypt.

The earliest autumn Stonechat record in Egypt is 12 September 1965 at Lake Maryut (JWW). There are several late September and early October records. Between mid-October and mid-March the Stonechat is locally common along the North Coast, in Western Desert oases, the Nile Delta and Lower Valley, the Faiyum, the Suez Canal area, and throughout Sinai. Although this species is apparently not uncommon in Dakhla and Kharga (Goodman *et al.* 1986), it is rarely recorded in the Nile Valley south of Mallawi. Records from Upper Egypt include a male on 25 December 1983 at Aswan (PE) and a female on 20 March 1965 at El Dakka (YPM). This species is uncommon along the Red Sea coast and the southern-most record is one at Abu Ramad on 18 February 1985 (DM). By the end of March most Stonechats have left the country. The latest spring records are one south of Ismailiya on 17 April 1984 (MS) and one at Hurghada on 27 April 1984 (SH, DM).

Generic local names for wheatear Ablaq, abuu ghamirra (NC), afqaaq (Su), slaygaw (Ma).

Isabelline Wheatear *Oenanthe isabellina* PV WV

Local name Ablaq ash/hab أبلق أشهب.

Common passage migrant and fairly common winter visitor throughout the country between (mid-July) mid-August and mid-April (late May). Monotypic.

In autumn the earliest Isabelline Wheatears arrive in mid-July (Meinertzhagen 1930) and the bulk of passage takes place between mid-August and mid-October. Throughout the winter this species can be observed in most of the country, mainly in open areas such as coastal plains and inland deserts.

Most spring migration is from mid-March to mid-April. The latest record is one on 23 May 1984 on Gezira Giftun el Saghir (SMG, RWS). One bird collected on 26 June 1969 in Wadi Gabria, Sinai (ZM-HUJ), was presumably summering.

Wheatear *Oenanthe oenanthe* PV (WV)

Local names Ablaq abuu bulayq أبلق ابو بليق, abuu bulayq.

Common migrant throughout the country in autumn from (mid-) late August to mid-(late) November and in spring from (mid-) late February to mid-May (mid-June). It is an irregular winter visitor to Egypt. *O. o. oenanthe* is the form found in the country and we are unaware of any evidence for the occurrence of *O. o. seebohmi*.

The earliest autumn Wheatear records are 15 August along the Mediterranean coast (Meinertzhagen 1930) and 24 August 1986 in Wadi el Natrun (EE *et al.*). There is some passage across the northern Red Sea (Casement 1976–7). From late August to late October the species is recorded throughout the country and the vast majority of birds have left by late November. There are scattered midwinter records from Sinai, near Gebel Elba, and in the Upper Nile Valley. In mid-February an influx of migrants is noticeable, and from late February to mid-May the species is common. During the six PMS spring field seasons at Bahig, passage was recorded from early March to mid-May, peaking in mid-March and mid-April (Horner 1977). The latest spring record is one at Hurghada on 11 June 1981 (SMB).

Ringing A bird ringed on 31 March 1970 at Matruh was recaptured on 16 April 1970 near Salum (Watson 1971).

Pied Wheatear *Oenanthe pleschanka* PV WV

Local names Ablaq abqa' أبلق أبقع, ablaq muraqqaṯ.

Rare migrant and winter visitor from (late September) early October to mid-April (mid-May). Sluys and van den Berg (1982) presented evidence that the Pied Wheatear breeding on Cyprus should be recognized as a distinct species, *O. cypriaca*; with this change *O. pleschanka* becomes monotypic. It is likely that some Egyptian records reported in the literature as *O. pleschanka* are actually of *O. cypriaca*, which is also known to occur in the country (see next species account).

In late October and early November there is a distinct influx of Pied Wheatears to Egypt. The majority of observations and specimens are from the Nile Valley, Suez Canal area, Eastern Desert, and Sinai. Between 1 and 10 October 1984 several birds were observed on Red Sea islands off Hurghada (MCJ *et al.*), which would indicate trans-sea migration from the Sinai and/or Arabian Peninsulas. The only known records from the Western Desert are one in Bahariya on 23 September 1982 (HB *et al.*) and three in Wadi el Natrun on 1 October 1965 (HCK). Most migrants continue to areas further south where they spend the winter; although a few remain in Egypt. Returning migrants begin to arrive in mid-March and by early to mid-April the majority have left the country. The latest spring record appears to be one on 14 May 1984 in Wadi Bali (SMG).

Cyprus Wheatear *Oenanthe cypriaca* PV WV

Local name Ablaq qubruṣii أبلق قبرصي.

Uncommon, but apparently regular, passage migrant and winter visitor from mid-October to mid-April. Monotypic.

Raw (1921) mentioned that Cyprus Wheatears were collected at Abu Zabal on 1 November 1918 and 1919, and several were seen between 1 and 7 November 1918. Male specimens in the BMNH were collected at Abu Zabal on 25 October 1918, 30 October 1918 (two), and 30 October 1919 (two). Another was taken at El Tor on 20 October 1911 (ZFMK). Three were noted at Zaranikh on 19 and 20 October 1979 (Petersen and Sørensen 1980). A single bird was noted at the Aswan High Dam on 28 January 1986 (SMB). Hoogstraal *et al.* (1961) reported that in the general vicinity of Cairo and/or the Faiyum 13 were trapped between 18 March and 12 April 1960, and two on 15 April 1959 and 16 April 1958.

Black-eared Wheatear *Oenanthe hispanica* PV (WV) MB?

Local names Ablaq aswad al-udhun أبلق أسود الأذن, abuu bulayq.

Fairly common migrant in autumn from (mid-August) early September to early October (mid-November) and common in spring from (early) late February to early May (late June). Possible migrant breeder in north-east Sinai. It appears to be an irregular winter visitor to the country. All Egyptian material examined is referable to *O. h. melanoleuca.*

The earliest autumn Black-eared Wheatear record is 17 August 1984 at Suez (SH, DM). This species is fairly common throughout the country from early September to early October. Late autumn records include birds at Suez on 27 October 1981 (RGB *et al.*) and on 15 November 1984 (RvD *et al.*).

Black-eared Wheatears have been noted in early January 1980 at Sharm el Sheikh (HSh), on 28 January 1985 midway between Sharm el Sheikh and El Tor (SMG), on 31 January 1981 at Hurghada (SMB), and on 4 February 1978 at El Kab (Short and Horne 1981). Many were observed in early February near Wadi Halfa and Aswan (Meinertzhagen 1930). It is not clear if the late January and early February records are of wintering birds or early spring migrants. Between late February and early May there is a distinct influx of migrants throughout the country. The latest spring record is one at Abu Rauwash on 30 May 1956 (FMNH). An adult was noted near El Arish on 25 and 27 June 1984 (SMB), and as suggested by Meinertzhagen (1920) this species may breed in northern Sinai.

Ringing Some birds show migration route fidelity. One ringed at Bahig on 6 April 1967 was retrapped at the same locality in September 1967 (Watson 1971).

Desert Wheatear *Oenanthe deserti* RB PV WV

Local names Ablaq as-sahraa- أبلق الصحراء, abuu bulayq, abuu zaara.

Breeding resident of desert regions in northern Egypt. The breeding populations inhabiting the North Coast and the Western Desert are referable to *O. d. homochroa*, and those in Sinai (and possibly the Eastern Desert) to *O. d. deserti*. These populations are augmented between mid-August and early May by migrants, which include birds referable to the above two subspecies and also to *O. d. atrogularis*. *O. d. homochroa* is a poorly differentiated subspecies and is often difficult to distinguish from *O. d. deserti*, particularly for birds taken in the Western Desert. Migrants and winter visitors are frequently observed throughout the country. To our knowledge no nest has been found in Egypt and the breeding records mentioned herein are generally based on observations of fledglings or inferred by dates.

North Coast and Western Desert Bulman (1942) regularly observed Desert Wheatears between Salum and Matruh from June to December 1940, and Maclaren (1944) noted birds on territory near El Hammam in July. Two fledglings, apparently capable of sustained flight, were found on 19 and 21 July 1922 at Matruh (GZM). Nicoll (1912*a*) stated that Desert Wheatears collected during the summer of 1910 in Wadi el Natrun 'had evidently been breeding'. Several specimens, all referable to *homochroa*, were collected in June 1910 at this locality. There are numerous other records of presumed non-breeding birds from various portions of the Western Desert (Goodman *et al.* 1986).

Nile Delta and Valley Raw (1921) occasionally saw this species throughout the year at Abu Zabal and stated that 'a brood of young birds usually appeared during August, but I never discovered the nest'. A fledgling was collected at this locality on 22 July 1916 (GZM). A pair with recently fledged young was observed 10 km west of El Khatatba on 14 June 1985 (SMB, DM) and a single bird at the Giza Pyramids on 24 June 1978 (PLM, BCD). In the mid-1960s specimens intermediate between *O. d. deserti* and *homochroa* were collected in the area of the Upper Nile Valley now inundated by Lake Nasser (YPM). Several of the specimens had enlarged gonads and may have been local breeding birds. One specimen, referable to *O. d. atrogularis*, was collected at El Malki on 11 February 1963 (YPM).

Eastern Desert In their extensive travels across the northern and central Eastern Desert neither Tregenza (1951) nor JJH found evidence of this species breeding. Meinertzhagen (1922) stated that it bred between Cairo and Suez but did not present any documentation, and no subsequent information is known. There appear to be migratory movements across the Red Sea; one was found at Daedalus Reef on 1 September 1913 (Ticehurst 1924).

Sinai Specimens in reproductive condition were taken on 16 April 1927 in Wadi el Arish at Bir Lahfan and at El Magdaba (GZM). One found in Wadi Rahaya, south of El Kuntilla, on 14 August 1969 may have been a local breeding bird (TAUM). It nests

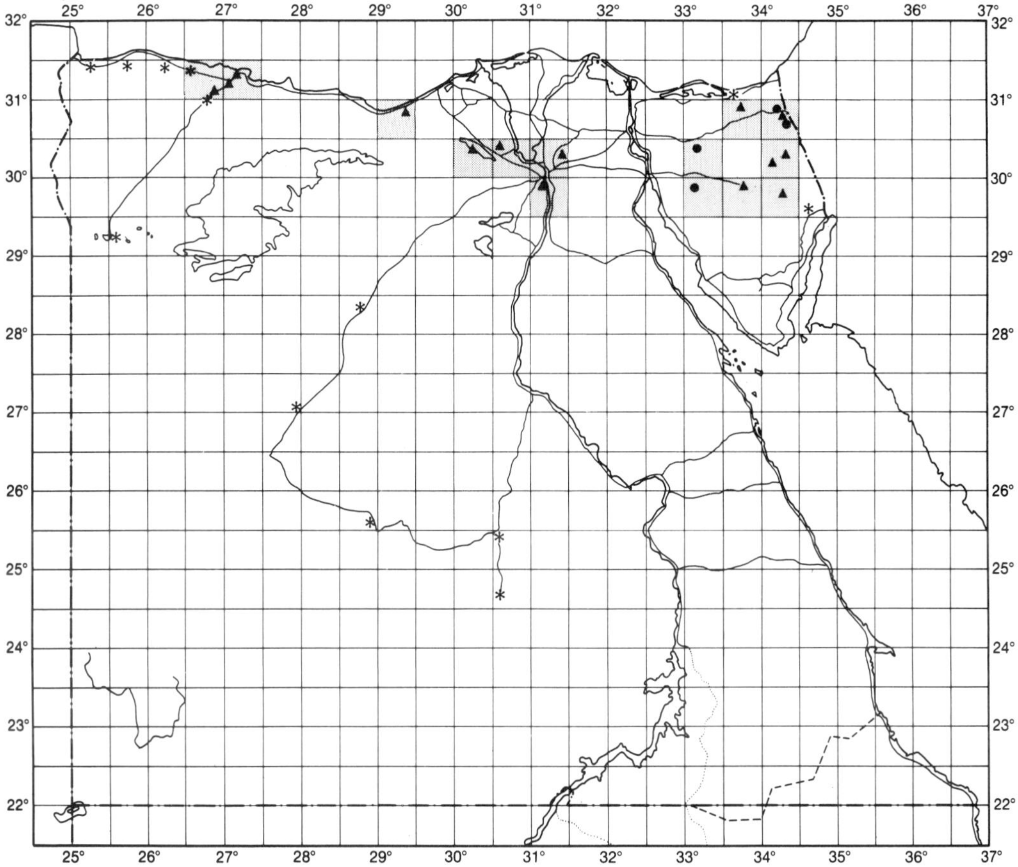

∗ possible breeding ▲ probable breeding ● definite breeding ░ half degree grid plots with probable or definite breeding

Map B77. Breeding distribution of Desert Wheatear *Oenanthe deserti* between 1930–87.

near Qusaima (SC) and in other areas of north-eastern Sinai (UP). There are several older records of birds observed or collected during the breeding season near El Tor (Kaiser 1892) and Nekhl (le Roi 1923).

Finsch's Wheatear *Oenanthe finschii* WV

Local name Ablaq 'arabii أبلق عربي.

Rare winter visitor to eastern Egypt between (early October) early November and late February (mid-May), and a vagrant to the Western Desert. All Egyptian specimens examined are referable to *O. f. finschii.*

The Finsch's Wheatear is an uncommon winter visitor to the northern Nile Valley, Suez Canal area, and Sinai. The earliest autumn record is 2 October 1971 at St Katherine (ZM-HUJ). It has also been noted on 25 October 1978 west of El Arish

(TAUM). By late February this species has generally left the country and the latest known spring dates are single birds at Aswan on 17 April 1984 (DKT) and at El Maghra on 11 May 1975 (Goodman and Ames 1983). The latter record is the only one known from the Western Desert (Goodman *et al.* 1986).

Red-rumped Wheatear *Oenanthe moesta* RB

Local names Ablaq ahmar al-'ajuz أبلق أحمر العجز, buu jihaar shayb.

Fairly common local breeding resident along the North Coast (*O. m. moesta*). The present status of the north Sinai population (*O. m. brooksbanki*) is uncertain.

North Coast According to Meinertzhagen (1930) the Red-rumped Wheatear did not occur further east along the coast than 20 km east of El Dab'a. Bulman (1942) found one every 3 km or so in suitable habitat within 40 km of the coast between

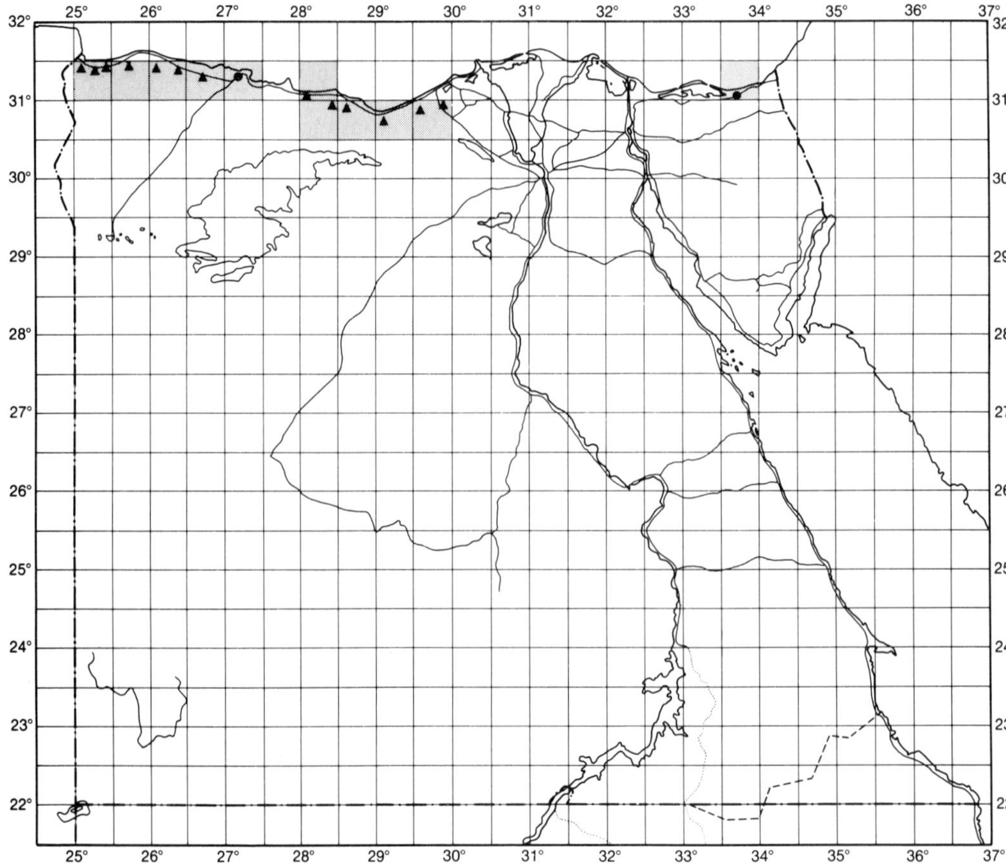

Map B78. Breeding distribution of Red-rumped Wheatear *Oenanthe moesta* between 1928–87.

Salum and Matruh, from June to December 1940. Maclaren (1944) noted that this species was abundant after 25 August 1942 at Marcopolis, near Alexandria, with 'twelve males within half a mile ..., each with a strict territory, and definite singing post'. There are numerous post-Meinertzhagen (1930) records from near Burg el Arab, Matruh, Sidi Barrani, and El Hammam.

Nile Delta and Valley There is no clear evidence that Red-rumped Wheatears migrate during the winter months to areas outside their breeding range. The record of Short and Horne (1981) of five birds at Giza on 31 January 1977 is rather irregular. A bird observed at Giza and reported as this species (Meininger and Mullié 1981*c*; Cramp 1988) was actually *O. xanthoprymna*. Whymper (1909) also reported this species twice near the 'edge of cultivation ... in the stubble just by the river-side'. One of the birds he observed at Deir el Bahari on 1 November 1907 is illustrated in his book and is clearly a Red-tailed Wheatear. Thus, his records are also not acceptable.

Sinai Meinertzhagen (1930) mentioned *O. m. brooksbanki* as a 'not uncommon' resident in northern Sinai. On 12 March 1928 he found a nest 24 km south-east of El Arish. The only more recent records from Sinai appear to be a singing male, with well-developed testes and paired with a female, collected in 'northern Sinai' in April 1936 (TAUM) and two seen within close range at Nuweiba on 27 August 1985 (DM). A record of this species at Dahab mentioned by Hovel (1987) based on *Dutch Birding*, 4, 4, is in error. (There is no mention of this species in the first ten volumes of this journal.) It appears that the Sinai breeding population of *brooksbanki* is either extremely small or locally extirpated.

Red-tailed Wheatear *Oenanthe xanthoprymna* WV

Local names Ablaq aḫmar adh-dhanab أبلق أحمر الذنب, umm al-hinnaa.

Scarce but locally common winter visitor between late October and early (late) May. All Egyptian material examined is referable to *O. x. xanthoprymna*. Meinertzhagen (1930) mentioned only a few records of this species in Egypt. The number of records has increased in recent years and it is now clear that this species is a regular winter visitor to eastern portions of the country.

Western Desert Misonne (1974) saw two Red-tailed Wheatears at Gebel Uweinat on 2 December 1968, one of which was collected (IRSNB). This appears to be the only record from the Western Desert (Goodman *et al.* 1986).

Nile Delta and Valley The earliest autumn record from Egypt is one at Saqqara on 22 October 1982 (Davies 1982). Small numbers are observed with some frequency throughout the winter in the Nile Valley and adjacent desert edge. A specimen was taken at Abu Rauwash on 30 December 1970 (USNM). A late spring record is one bird on 4 May 1977 at Suez (MB).

Eastern Desert There are several recent records of this species from the Red Sea coast and inland areas. One male was noted at Safaga on 21 March 1982 (Meilstrup and Sørensen 1983) and a female on Gezira Giftun el Kabir, off Hurghada, on 28 December 1983 (PE). Tregenza (1951) saw two in the late 1940s near Qena on 27 November and 28 May. Single birds were observed at Bir Abu Sha'r on 25 October 1986 (SMG, JJH) and at Quseir on 17 February 1985 (DM). A male was noted at Gebel Misikat el Gukh on 17 January 1986 (SMB). In February and March 1985 it was common in the vegetated wadis at the base of Gebel Elba but absent by April (SMG, PLM).

Sinai One was collected in 'west Sinai' on 5 March 1928 (BMNH). Between 27 and 30 December 1980 a single bird was present at Sharm el Sheikh (BCD). This species was observed on Gezira Tiran in February 1975 (Hovel 1987). There is no acceptable record from Elat, Israel (HSh; *contra* Krabbe 1980).

Mourning Wheatear *Oenanthe lugens* RB (WV)

Local name Ablaq h̲aziin أبلق حزين.

Common breeding resident of mountainous regions of the Eastern Desert and Sinai, perhaps also locally on the west bank of the Nile (*O. l. lugens*). The form *O. l. halophila*, which is known to breed in Tunisia, Algeria, and Libya, has been collected during the winter near Giza, Abu Rauwash, and Thebes (Meinertzhagen 1930). A more recent specimen of *halophila* was taken on 17 December 1971 at Dandara (USNM). A bird collected on 21 February 1928 at Quseir is referable to *O. l. persica* (Meinertzhagen 1930; BMNH).

North Coast and Western Desert Bulman (1942) reported observations of single Mourning Wheatears in May and October 1940 in the general Matruh area and in March 1941, 80 km south of Matruh on the road to Siwa. On 4 October 1980 single birds were noted 20 km east of and 20 km west of Matruh (HB *et al.*). A juvenile specimen, referred to *O. l. halophila* by Steinbacher (1965), was taken 125 km south-west of Matruh on 15 September 1963.

Nile Delta and Valley The Mourning Wheatear may breed on the west bank of the Nile at Thebes, where small numbers have been regularly noted in recent years. In the non-breeding season some birds from the Eastern Desert migrate to the Nile Valley (Hartley 1949). Thus, it is not possible to separate winter visitors to the region from potential local breeding birds. The species was reported to occur on the edge of the Faiyum (Meinertzhagen 1930), an area for which there are no recent records. In Wadi el Rayan one specimen, referable to *O. l. lugens*, was collected in June 1984 (AAUC), others observed throughout the year, and it probably breeds locally (SMB). There are also numerous records from Egyptian Nubia, e.g. one collected at Madiq on 10 February 1963 (YPM) which is referable to *O. l. lugens*. In 1952 it probably bred in the hills near Fayid, west of the Bitter Lakes (AEG) and on 27 April 1944 one was seen at Ismailiya (CJG). A bird collected in the centre of the Nile Delta at Benha on 18 July 1963 was identified as this species (Steinbacher 1965); however, this record is enigmatic.

Eastern Desert and Sinai *O. l. lugens* is locally common in wadi beds and on rocky slopes of the Eastern Desert (south to Gebel Elba) and Sinai. Tregenza (1951, 1955) mentioned this species as the most common wheatear in the central Red Sea mountains from autumn to spring, except in high mountain areas, and noted a few birds in late summer. It probably has a continuous distribution in the relatively low areas of mixed rock and sand east of the Nile. Mourning Wheatears are generally rare at Zaranikh but in late September and early October 1985 this species was the most common local wheatear (SMB); this increase in numbers is almost certainly an influx of immigrants. There are also considerable dispersal and altitudinal movements of local populations during the non-breeding season.

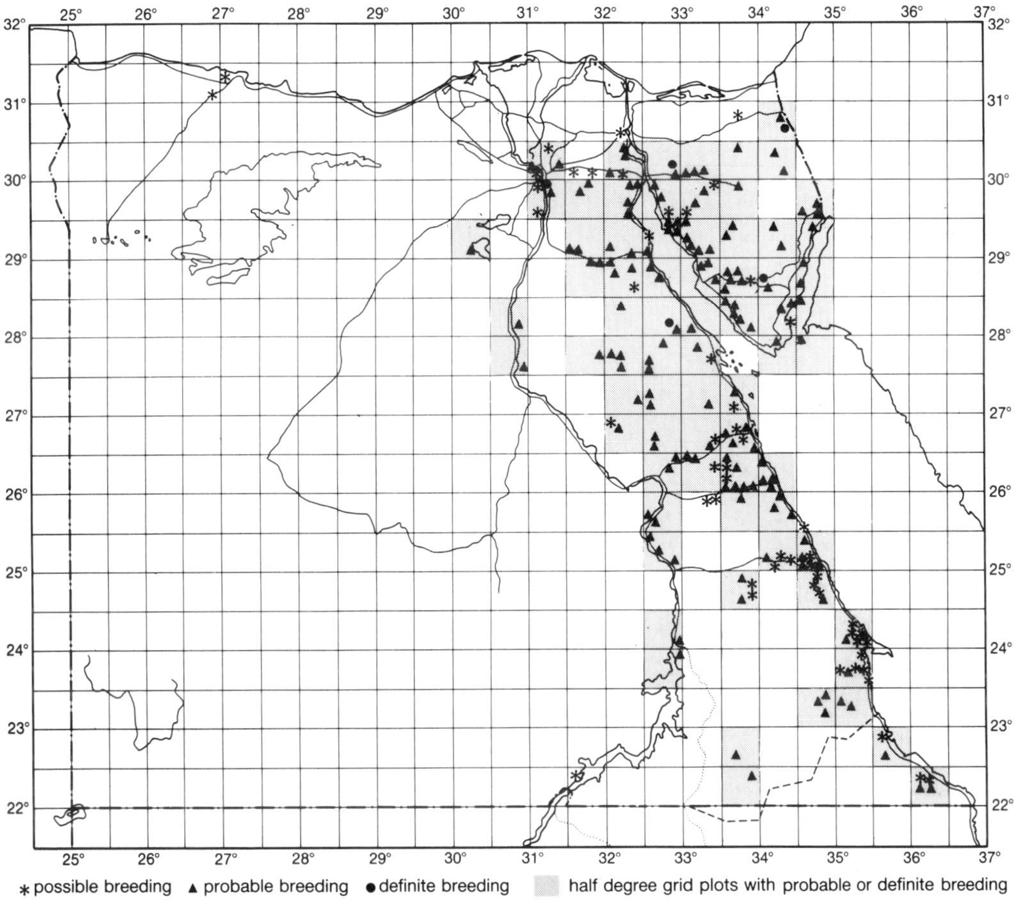

Map B79. Breeding distribution of Mourning Wheatear *Oenanthe lugens* between 1940–87.

Hooded Wheatear *Oenanthe monacha* RB

Local names Ablaq abuu ṯaaqiya أبلق ابو طاقية, ablaq mutaqallas.

Fairly common breeding resident in the mountainous regions of the Eastern Desert and southern half of Sinai. Monotypic.

North Coast Bulman (1942) observed one Hooded Wheatear between Salum and Sidi Barrani on 14 September 1940. This appears to be the only record west of the Nile basin.

Eastern Desert and Nile Valley In the summer months this species tends to be restricted to the higher portions of the central Red Sea Mountains (Tregenza 1951). In the winter it is occasionally observed on the banks of the Nile and it probably breeds locally in some rocky areas along the Valley edge. Recent reports from the west bank include several February records from Thebes and near Abydos (Short and Horne 1981; SG), and birds on 4 January 1978 at Abu Simbil (AS) and on 4 March 1985 at Giza (UGS). This species appears to be absent from the Sudan Government Administration Area.

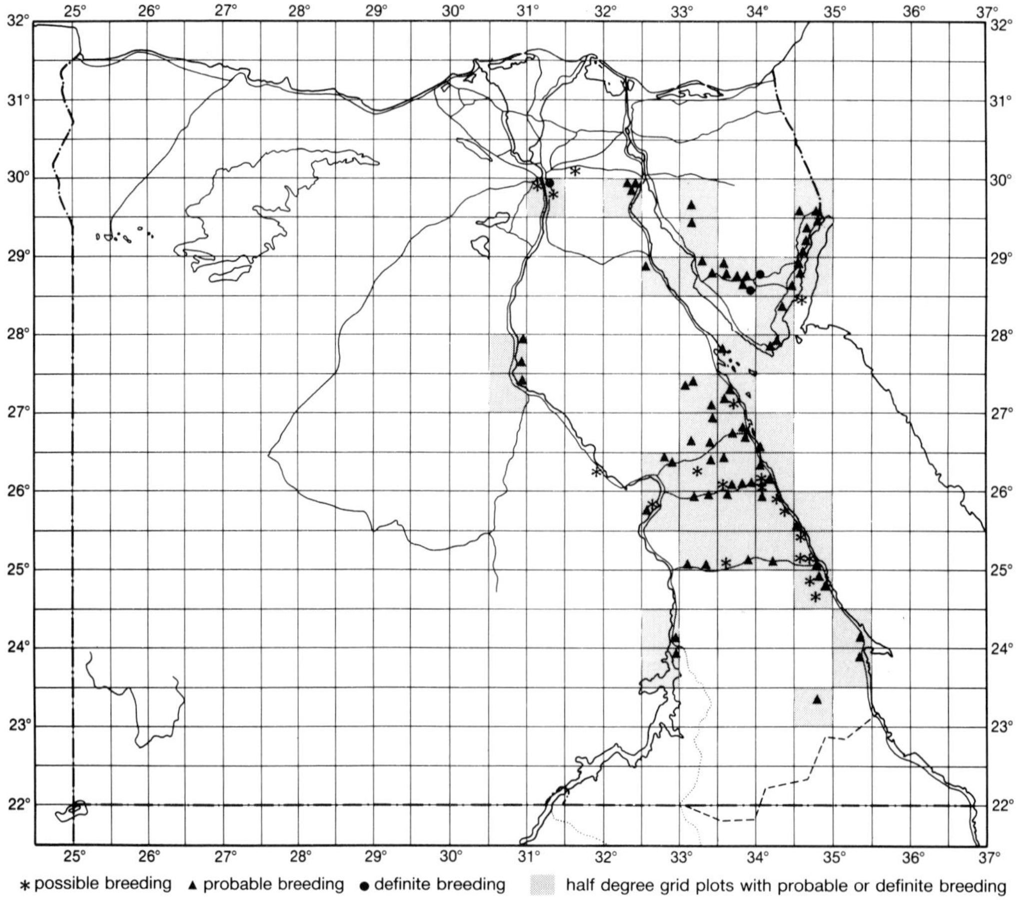

Map B80. Breeding distribution of Hooded Wheatear *Oenanthe monacha* between 1940–87.

Sinai In the southern portion of the Peninsula this species is a fairly common resident in the mountains and at lower altitudes in rocky areas.

White-crowned Black Wheatear *Oenanthe leucopyga* RB Plate 3

Local names Ablaq aswad abyaḏ al-ra-s أبلق أسود أبيض الرأس, abuu futfuut, abuu slimaan, am salaman (Bi), bugwa (SC), hajj mulayn (Si), ṯaa-irat ash-shaykh.

Common breeding resident in Western Desert oases, along the shores of Lake Nasser, in the Eastern Desert, the mountains of Sinai, and locally along the Nile Valley plateau. Two forms occur in the country—*O. l. ernesti* in the mountains of southern Sinai and *O. l. leucopyga* in the rest of the Egyptian breeding range. The subspecific identity of the north Sinai population is unknown.

Western Desert The White-crowned Black Wheatear is a common breeding bird

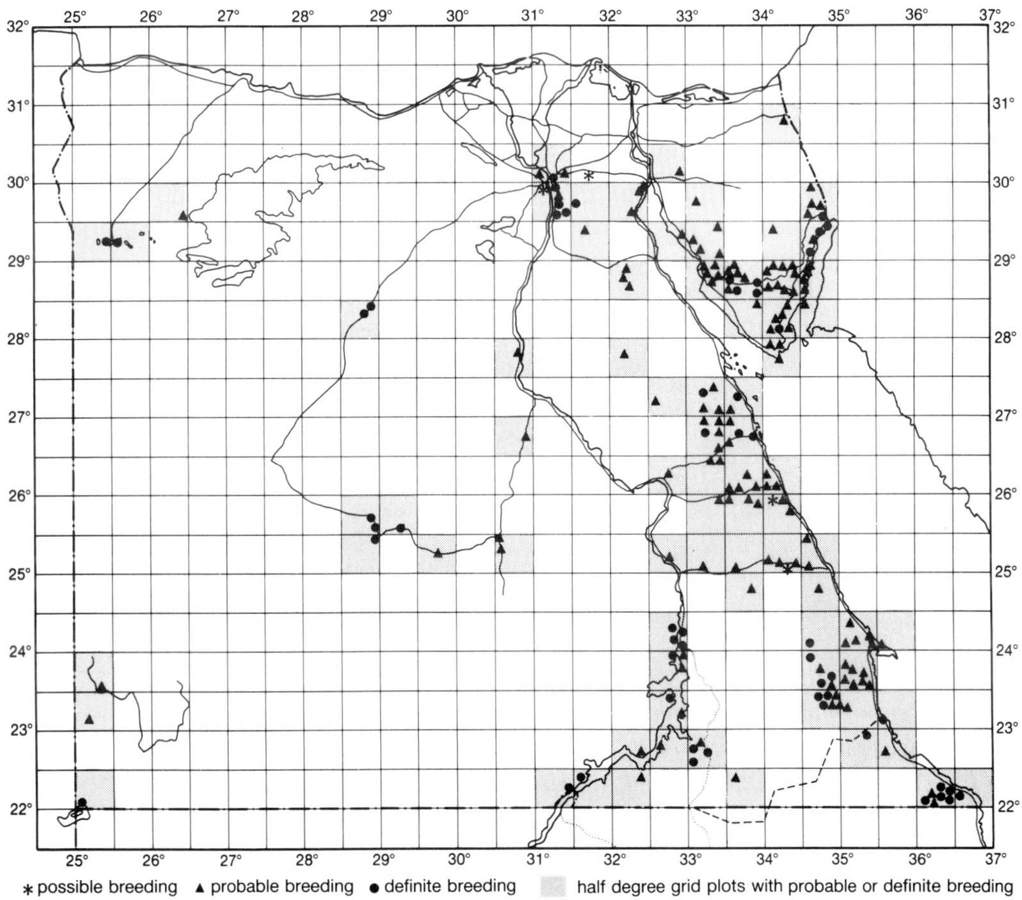

Map B81. Breeding distribution of White-crowned Black Wheatear *Oenanthe leucopyga* between 1940–87.

in most oases of this region, particularly Siwa, portions of the Qattara Depression, Bahariya, and Dakhla. In recent years this species has been found to be uncommon in Kharga and absent from Farafra (Goodman *et al.* 1986). It is a resident at Gilf Kebir and Gebel Uweinat (Jany 1963; Misonne 1974).

Nile Delta and Valley This species is a breeding resident along the edge of the Upper Nile Valley. It is rather local in its distribution south of Luxor, including the shores of Lake Nasser. Birds are occasionally noted at other localities, e.g. the Giza Pyramids and Saqqara. An exceptional record is one collected in the centre of the Nile Delta at Benha on 15 July 1963 (Steinbacher 1965).

Eastern Desert and Sinai This species is one of the characteristic birds of the mountainous and rocky areas of these two regions. It is sometimes found in relatively high densities, particularly in rugged granitic country.

Rock Thrush *Monticola saxatilis* PV (WV)

Local names Sakala سكلة, abuu shalk (NC), abuu suukla.

Scarce migrant in autumn between late August and mid-November and fairly common in spring between (late February) mid-March and late April (mid-May). Apparently an irregular winter visitor. Monotypic.

There are relatively few autumn records of Rock Thrushes in Egypt, all between 27 August and 16 November. The only known autumn records from the Western Desert are single males on 11 September 1983 in Bahariya and on 20 September 1910 in Wadi el Natrun, and several birds in early November 1968 at Gebel Uweinat (HB *et al.*; Goodman *et al.* 1986). Four were taken at Bahig between 27 August and 22 September (USNM). The only known recent winter record is a male in the Faiyum on 31 January 1977 (Short and Horne 1981). During spring passage this species is decidedly more common than in autumn, and the bulk of migrants move through in the first half of April. Late records include birds on 11 May (year not stated) at Bahig (Horner 1977), 11 May 1972 at St Katherine (ZM-HUJ), and 13 May 1983 at Hurghada (MCJ *et al.*).

Blue Rock Thrush *Monticola solitarius* CB? PV WV

Local names Sakala zarqaa- سكلة زرقاء, ḫamaama zarqaa-.

Perhaps occasional breeding resident. It is a fairly common passage and winter visitor from early September to late April (mid-May) (Fig. 6.27). The majority of specimens taken in Egypt are referable to *M. s. solitarius*. According to Meinertzhagen (1930), *M. s. longirostris* is probably the more abundant form on spring migration. However, based on the material we have examined this is not the case, at least in part because *longirostris* is not easily differentiated from nominate *solitarius*.

Meinertzhagen (1930) reported that the Blue Rock Thrush was seen throughout the

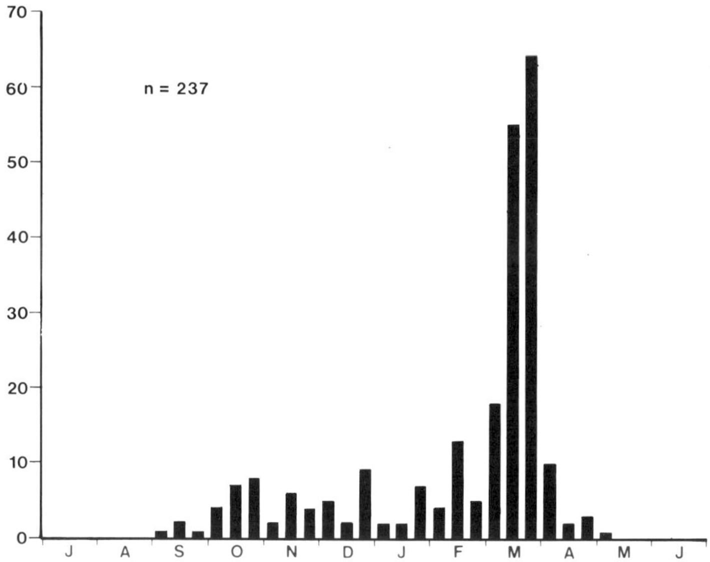

FIG. 6.27. Histogram of the numbers of Blue Rock Thrushes recorded during 10-day periods throughout the year.

year near the Giza Pyramids and Salum but no definite proof of breeding was found. On 5 April 1981 a female was seen at the Citadel, Cairo, flying under the roof-eaves with food in its bill (WS).

The Blue Rock Thrush is a fairly common autumn migrant throughout the country; the earliest record is 10 September 1966 at Suez (JT). This species is a winter visitor in the Nile Valley, Sinai, the Eastern Desert, and several Western Desert oases. It is common on spring migration in March and early April, and numbers peak in the second half of March. The latest post-Meinertzhagen (1930) spring record is on 10 May 1984 at Hurghada (MCJ *et al.*).

Generic local names for *Turdus* spp. Dujj, sumna.

Ring Ouzel *Turdus torquatus* (WV)

Local name Dujj muṯawwaq دج مطوق.

Irregular winter visitor, mainly along the North Coast, in the Nile Delta, Suez Canal area, and Sinai, between late October and late March. All recent Egyptian material examined is of *T. t. alpestris*. Two specimens collected at St Katherine on 3 March 1928 are referable to *T. t. amicorum* (Meinertzhagen 1930; BMNH).

Meinertzhagen (1930) considered the Ring Ouzel a straggler to Egypt. More recent information indicates that it is an irregular migrant to the country. The earliest autumn records are birds at Bahig on 25 October 1972 (Goodman and Watson 1984) and in Wadi Kid on 26 October 1969 (ZM-HUJ). Other autumn records include two at Abu Rauwash on 16 November 1936 (HGD) and at Bahig on 16 November 1969 (Goodman and Watson 1984). We are unaware of any records from the Egyptian Nile Valley, although it has been recorded at Dongola and Khartoum, Sudan (Butler 1905; Nicoll 1922). Winter records include a flock at St Katherine from late February to early March 1928, one bird at Giza on 9 December 1906, and one at Abu Rauwash on 8 February 1936 (Meinertzhagen 1930; HGD). More recently this species has been noted at St Katherine in early March 1970 and 1987 (ZM-HUJ; HH; PLM) and 19 March 1987 (RGB). The latest spring record appears to be one at Bahig on 20 March 1972 (Horner 1977).

Blackbird *Turdus merula* RB WV

Local name Shahruur شحرور.

Locally common breeding resident of the Nile Delta and north-eastern Sinai. It is a fairly common winter visitor to the country, mostly in the north, from (mid-September) late October to mid-April. Two subspecies, *T. m. merula* and *T. m. syriacus*, have been taken in Egypt, although numerous specimens are intermediate between these two forms (Goodman and Watson 1984). The subspecific identity of the Nile Valley and Sinai breeding populations is unknown but presumably *T. m. syriacus* is the form in the latter area.

Before the mid-1970s the Blackbird was unknown to breed in Egypt. In 1975 it was found nesting at El Arish (UP; HM) and by 1984 it was a locally common resident of orchards in the El Arish and Rafa area (SMB). On 13 March 1976 a nest with three eggs was located at Ain el Gedeirat (UP) and in early May 1982 a single bird was noted at El Gorah (Wimpfheimer *et al.* 1983).

In the mid-1980s the species was found to have also colonized the southern and central Nile Delta. In late May 1985 numerous males were observed near El Khatatba singing and defending territories (SMG, PLM). In mid-June 1985 it was found to be a common breeding resident of mango and guava plantations in this area (DM), and in May 1986 nesting in similar habitat north of Abu Rauwash, near Benha, and Tanta (WCM). A pair was noted several times in March 1987 on Zamalek, Cairo, once collecting nest material (DJE).

The earliest autumn migrants arrive in mid-September and it is a fairly common winter visitor. The majority of records are restricted to the North Coast, the Nile Delta, Suez Canal area, and northern Sinai. There appear to be no reports from the Egyptian Nile Valley south of Cairo. Records from the Western Desert include birds in Wadi el Natrun in mid-November and March 1912, in Siwa on 9 March 1976, and at El Maghra on 10 February 1974 (Goodman *et al.* 1986). In some areas of the Nile Delta the species is apparently a regular winter visitor. Raw (1921) reported Blackbirds 'every' winter in agricultural areas of Lower Egypt. El Negumi *et al.* (1950)

mentioned that this species was abundant in winter near El Arish and often poisoned, presumably because of damage to crops. There are several recent winter records from the southern half of Sinai, generally in oases or gardens. During the six PMS spring seasons at Bahig, between 1967 and 1973, the maximum number handled in any year was nine, and passage was noted between 25 February and 17 April (Horner 1977).

Ringing One ringed at Bahig on 12 November 1969 was recovered in Cyprus on 5 November 1972 (Flint and Stewart 1983).

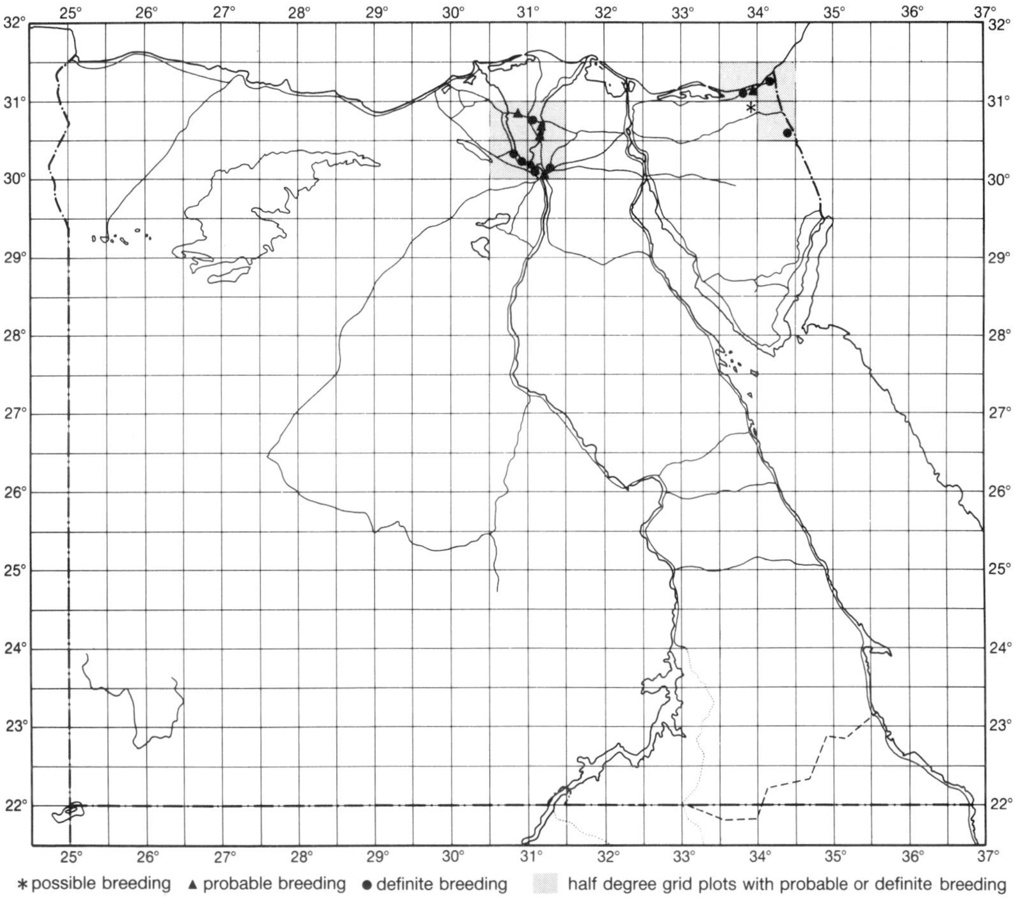

Map B82. Breeding distribution of Blackbird *Turdus merula* between 1970–87.

Black-throated Thrush *Turdus ruficollis* AV

Local name Dujj aswad az-zawr دج أسود الزور.

Accidental visitor. The single Egyptian specimen examined is referable to *T. r. atrogularis*.

The only specimen of the Black-throated Thrush from Egypt is a male collected in the autumn of 1833 in Sinai (RMS). On 3 January 1982 a male with a black throat was observed at El Arish (Baha el Din 1984*a*).

Fieldfare *Turdus pilaris* (WV)

Local name Dujj al-ghayṯ دج الغيط.

Irregular winter visitor, mainly to the north of the country, from late October to mid-March. Monotypic.

In some years the Fieldfare is fairly common and in others completely absent. The earliest autumn record is 18 September 1917 at Rafa (Meinertzhagen 1930). The vast majority of records are from the North Coast and the Nile Delta. However, the species has been noted in Wadi el Natrun (Goodman *et al.* 1986), Bahariya (HB *et al.*), the Faiyum (Goodman and Watson 1984; Bezzel 1987*a*), near Suez (RGB *et al.*), and in the Nile Valley as far south as between Luxor and Dandara (Short and Horne 1981). Sinai records include birds at El Arish on 6 January 1980 (PLM, WCM), at Ras el Sudr on 19 February 1974 (HH), and at St Katherine on 6 February 1976 (HW) and 3 March 1971 (ZM-HUJ). The latest spring record is one on 14 March 1957 at Abu Rauwash (Goodman and Watson 1984).

Song Thrush *Turdus philomelos* WV

Local names Sumna muṯriba سمنة مطربة, ghirriida, shaẖruur.

Fairly common winter visitor to the country between (early) late October and late March (late April). All Egyptian material examined is referable to *T. p. philomelos.*

The earliest autumn record of the Song Thrush in Egypt is 3 October 1980 near Matruh (HB *et al.*). In most winters it can be met with in Egypt, particularly along the North Coast, in the Nile Delta, and Suez Canal area. Records of large winter flocks include 250 flying west on 7 January 1980 at Lake Bardawil (PLM, WCM) and 320 on 9 January 1980 at Burg el Arab (PI). Small numbers have also been noted in Wadi el Natrun, Farafra, the Nile Valley south to Aswan, the Faiyum, along the Red Sea coast, and at numerous places throughout Sinai. One was seen at Gebel Elba on 11 April 1985 (SMG, PLM). During the six PMS spring surveys at Bahig, between 1969 and 1973, this species was recorded from 19 February to 23 April, with a maximum of 65 handled in 1970 (Horner 1977).

Ringing One bird ringed at Bahig on 26 March 1967 was recovered at the same locality on 10 March 1970 (Watson 1971).

Redwing *Turdus iliacus* (WV)

Local name Sumna ẖumraa- al-janaaẖ سمنة حمراء الجناح.

Scarce and irregular winter visitor to the northern portion of Egypt between late October and late April. All Egyptian material examined is referable to *T. i. iliacus*.

Meinertzhagen (1930) noted that in some years the Redwing was not an uncommon winter visitor to Egypt. The earliest autumn records appear to be on 30 October at Bahig in both 1968 and 1971. The majority of records are from the Nile Delta and northern Sinai, although it has been noted along the North Coast as far west as Burg el Arab, in Wadi el Natrun (Goodman and Watson 1984), and at St Katherine (Hovel 1987). The latest spring record is 23 April 1939, 40 km south of Suez (HGD).

Mistle Thrush *Turdus viscivorus* (WV)

Local names Sumna kabiira سمنة كبيرة, dujj ad-dibaq, sumna ad-dibaq.

Rare and irregular winter visitor from late October to early March. All Egyptian material examined is referable to *T. v. viscivorus*.

The majority of Mistle Thrush records in Egypt are from the Nile Delta and near Bahig. The earliest autumn and latest spring reports are from Matruh on 21 October 1969 and 9 March 1970, respectively (Goodman and Watson 1984). It was observed by Rüppell near Suez in April 1831 (Meinertzhagen 1930). Sinai records include single birds near Rumana on 29 October 1916 (Boyd 1917) and at St Katherine on 4 March 1970 (ZM-HUJ).

Family Sylviidae

Generic local name for warbler 'Ifrii (NC).

[**Cetti's Warbler** *Cettia cetti* AV?

Meinertzhagen (1930) did not know of any authenticated records of the Cetti's Warbler in Egypt. Since then single birds have been observed on 11 March 1979 at Sharm el Sheikh (HW) and on 30 March 1982 between Suez and Adabiya (Wimpfheimer *et al.* 1983). At Elat, Israel, it is considered an uncommon winter visitor, departing in spring by mid- to late March (Krabbe 1980).]

Fan-tailed Warbler *Cisticola juncidis* RB

Local names Fus̱siya marwaẖiyya adh-dhanab فصية مروحية الذنب, umm al-fus̱sii.

Abundant breeding resident in the Nile Delta and Valley south to Aswan, the Suez Canal area, the Faiyum, Wadi el Natrun, and Kharga, and possibly eastern Sinai. All Egyptian material examined is referable to *C. j. juncidis*. If indeed a breeding population exists in eastern Sinai it might represent a westward range extension of *C. j. neurotica* from the Negev Desert of Israel. This species inhabits cultivated fields, open *Salicornia* and tamarisk scrub, and less commonly reed-beds. It is often sympatric with the Graceful Warbler. Breeding takes place from mid-February to early June (Meinertzhagen 1930). In recent years the breeding range has significantly expanded.

North Coast Meinertzhagen (1930) noted that the Fan-tailed Warbler did not occur west of Alexandria. Hubbard and Seymour (1968) caught a fledgling near Bahig on 17 March 1967 and saw a few adults singing in local grain fields. In March 1983 it was fairly common in fields south of Burg el Arab and near Bahig (PLM).

Western Desert According to Meinertzhagen (1930), Wadi el Natrun was the only oasis where the Fan-tailed Warbler occurred. In the past two decades or so this species has colonized new portions of the Western Desert. Kiepenheuer and Linsenmair (1965) observed it in Siwa in September 1963. Several were seen in Kharga on 18

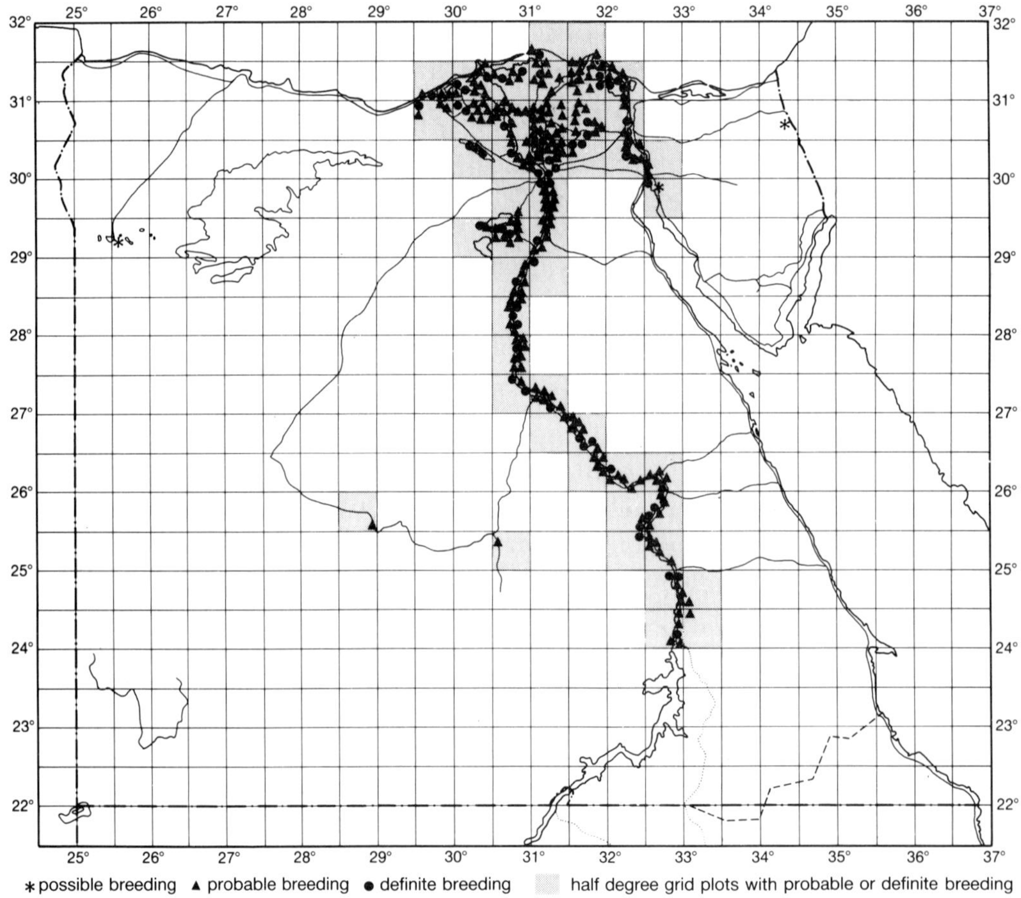

Map B83. Breeding distribution of Fan-tailed Warbler *Cisticola juncidis* between 1970–87.

October 1980 and at least four on 23 April 1981 (Goodman *et al.* 1986). A singing bird was heard in Dakhla on 17 March 1986 (AS).

Nile Delta and Valley Meinertzhagen (1930) considered this species a breeding resident of the Nile Delta and Valley south to at least Luxor. Presently, it is found throughout the Valley at least to Aswan, although it is distinctly less common in the Upper portion. Specimens were taken in mid-January 1963 at Arminna, a region now inundated by Lake Nasser (YPM).

Sinai Singing birds were heard on 25 January 1985 in a marsh 35 km south of the Suez Tunnel (SMG) and one was found on 27 January 1974 on the east side of the Suez Canal, opposite Ismailiya (ZM-HUJ). The only known record from eastern Sinai is one heard singing at Ain el Gedeirat on 13 March 1976 (UP).

Graceful Warbler *Prinia gracilis* RB Plate I

Local names Fussiya فصية, haazija.

Abundant breeding resident in Wadi el Natrun (*P. g. natronensis*); the Nile Delta (*P. g. deltae*); the Suez Canal area, the complete Nile Valley, the Faiyum, Wadi el Rayan (*P. g. gracilis*); and parts of Sinai (*P. g. palaestinae*). This species is a characteristic bird of agricultural areas, scrub, marshes, and less commonly reed-beds. It is also found in gardens and parks of highly urbanized areas such as Cairo. Breeding is from mid-February to mid-September.

Western Desert The Graceful Warbler has not been recorded in any Western Desert oasis except Wadi el Natrun, where it is common (Goodman *et al.* 1986). This population is subspecifically distinct (Goodman 1984*b*).

Nile Delta and Valley Graceful Warblers are locally abundant. Specimens taken along the Suez Canal appear intermediate in plumage coloration between *P. g. deltae* and *P. g. palaestinae*. Meinertzhagen (1930) noted that this species occurs in the Nile Valley as far south as Luxor. In 1957, Horváth (1959) observed it at Aswan. It was known from Egyptian Nubia before the filling of Lake Nasser. Recent Lake Nasser records include birds in low pioneering vegetation along the shore near the Aswan High Dam in April 1983 (PLM), and up to six birds at Abu Simbil on 26 April 1984 (MBe, SM) and several on 8 October 1986 (EH, PK). In January 1986 this species was found to be common in thick tamarisk scrub of Wadi el Allaqi, east of Lake Nasser (SMB). It has also colonized Wadi el Rayan, apparently after the creation of lakes in the early 1980s (SMB).

Eastern Desert In March 1928, two specimens referable to *P. g. carlo* were reputedly collected at Gebel Elba (Meinertzhagen 1930; BMNH). The Gebel Elba region has been extensively surveyed in recent years and no evidence of this conspicuous species was found. These specimen records are enigmatic. The species is not known to breed along the Egyptian Red Sea coast (*pace* Etchécopar and Hüe 1967).

Sinai Meinertzhagen (1930) stressed the lack of Sinai records, particularly in suitable country 'around El Arish and other localities'. Nowadays it is fairly common in the region between El Arish and Rafa (SMB). Single birds have been seen and heard on 28 March 1979 at Nuweiba (DJF *et al.*), on 1 April 1980 near Nabq (KO) and on 4 March 1984 at Ain Musa (SG). The Graceful Warbler was a common breeding bird at Elat, Israel, before 1973 (HSh; cf. Krabbe 1980).

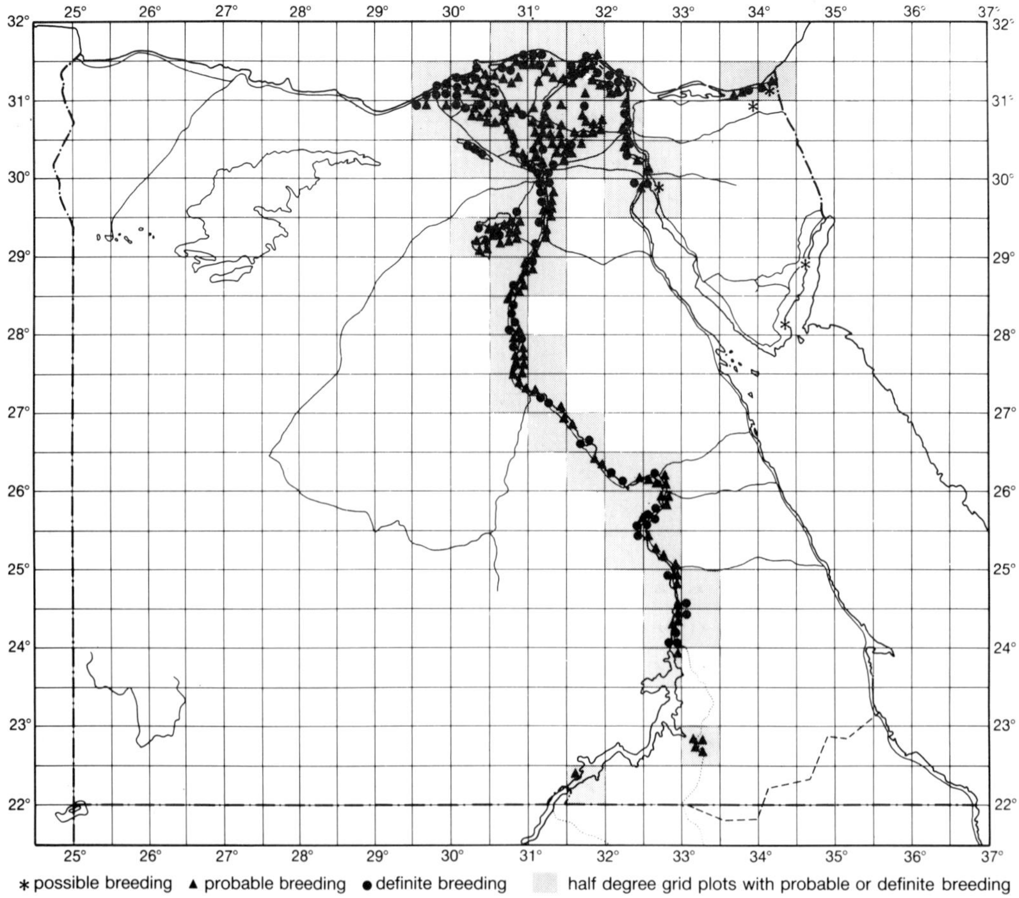

Map B84. Breeding distribution of Graceful Warbler *Prinia gracilis* between 1970–87.

Scrub Warbler *Scotocerca inquieta* RB Plate 5

Local names Namnamat ash-shajar نمنمة الشجر, du<u>kh</u>alat ad-daghal, haazija ad-daghal, rawa'iya (Ma), shuwaala.

Local breeding resident in the northern Eastern Desert and in southern Sinai. It tends to occur in areas with relatively heavy vegetation. Breeding commences in early March and is completed by mid-June. The Egyptian breeding form is *S. i. inquieta*. *S. i. saharae* is known from Cyrenaica and it is plausible that it also occurs across the Libyan border in Egypt, particularly near Salum.

Nile Valley There are no documented records of the Scrub Warbler in the Nile Valley proper. A specimen taken in the Faiyum on 9 August 1963 and reported as this species (Steinbacher 1965) was actually a Graceful Warbler (DSP). Meinertzhagen (1930) mentioned its occurrence near Luxor and Aswan but presented no basis for these records.

Eastern Desert This species occurs in the mountains east of Cairo (including Gebel el Muqattam), in wadis along the Cairo–Suez road, near the Gebel el Galala plateaux, and in other parts of the northern Red Sea Mountains. Tregenza (1951) considered it rare in the desert between Qena and the Red Sea. The only known record from this area is one observed in Wadi Semna on 7 February (Tregenza 1958).

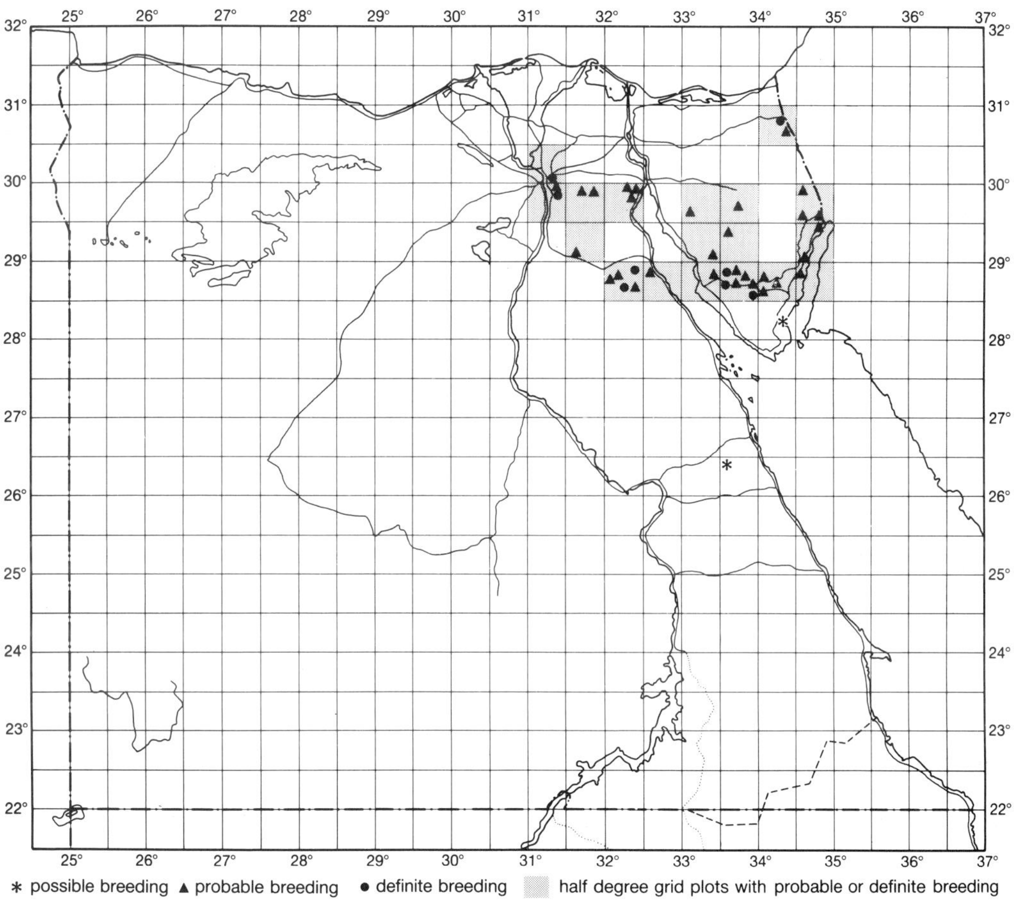

Map B85. Breeding distribution of Scrub Warbler *Scotocerca inquieta* between 1940–87.

Sinai This species is common in the mountains of southern Sinai. It also occurs at lower elevations in wadis draining towards the Gulf of Aqaba, e.g. at El Kuntilla (Galal el Din and Fouad 1939–40), west of Nizzana, and at Ain el Gedeirat. Marchant (1939–40) found it common and widespread at Gebel Somar. Fledglings were observed in late May 1984, 20 km north of Wadi Feiran (SMG, RWS).

Generic local names for *Locustella* and *Acrocephalus* spp. Haazija, fisaysi (Ma), khansa', khashiir (Su).

[**Grasshopper Warbler** *Locustella naevia* AV?

Nicoll (1909) listed a Grasshopper Warbler observation at Giza on 28 September 1907. This record was not accepted by Meinertzhagen (1930). Both Grote (1930) and Etchécopar and Hüe (1967) stated that it occurs in Egypt but presented no details. There are a few recent observations: one on 5 September and two on 20 September 1981 at Suez (RGB) and one on 26 April 1983 at Ineiba, near Kom Ombo (SC, IP). Four were ringed at Elat, Israel, in autumn 1986 (HSh).]

River Warbler *Locustella fluviatilis* PV

Local name Haazija an-nahar هازجة النهر.

Rare migrant in autumn between mid-August and early (late) October and in spring between early March and mid-May. It is apparently more regular in autumn than spring. Monotypic.

Six River Warblers were captured in two field seasons at Bahariya between 12 August and 27 October 1982–3 (HB *et al.*) and eight at Bahig between 28 August and 3 October (Goodman and Watson 1983; USNM). A single bird was observed on 19 August 1966 at Lake Maryut (Kattinger 1970) and a male collected at Rafa on 29 September 1917 (BMNH). Birds were captured at St Katherine on 25 August and 29 September 1971 (DL). Two were ringed at Zaranikh between 18 and 27 September 1978 (YP, SS).

Spring records include: one on 2 March 1962 at El Maadi (HCK), an 'apparent River Warbler' at Luxor on 3 March 1979 (Short and Horne 1981), and single birds at Bahig on 13 April 1970 (Horner 1977) and 10 May 1970 (USNM). One was ringed at El Arish on 23 April and two on 3 May 1985 (Fry *et al.* 1985). At Elat, Israel, the River Warbler is an uncommon migrant between late April and mid- (late) May, with 26 spring records from 1966 to 1979 (Krabbe 1980).

Savi's Warbler *Locustella luscinioides* PV WV?

Local names Haazija al-mustanqa' هازجة المستنقع, bulbul al-ghaab.

Scarce passage visitor in autumn from mid-August to late October and in spring from (early) mid-March to mid-April (early May). It appears to migrate across the whole country and is presumably an irregular winter visitor. All recent Egyptian material examined is referable to *L. l. luscinioides*. One specimen collected in the Faiyum on 12 March 1909 is the only known record of *L. l. fusca* from Egypt (Meinertzhagen 1930). The vast majority of Egyptian records are specimens or netted birds and presumably this skulking species is frequently overlooked, except in spring when it often sings.

The Savi's Warbler is scarce on autumn passage. Thirty-six were captured at Bahariya between 18 August and 24 October 1983 (HB *et al.*). There is a total of 14 specimens in the collections of the FMNH and USNM taken between 2 September and 28 October along the North Coast. A single bird was ringed at Zaranikh between 3 and 7 October 1978 (YP, SS) and one was taken at St Katherine on 20 August 1971 (ZM-HUJ).

Single birds have been taken in the Faiyum on 4 February 1871, 10 February 1909, and 11 February 1909 (BMNH); in Wadi el Natrun on 9 February 1903 (Loat 1905); and at Port Said on 14 January 1918 (Meinertzhagen 1930). More recently this species has been found at Tushka on 16 February 1963 (YPM) and two others at Ras El Sudr on 19 and 27 February 1974 (HH). It is not clear if the February records are of wintering birds or early spring migrants.

There are substantially more sight records from Egypt in spring, when this species is regularly heard singing in reed-beds and sometimes farm lands, than in autumn. At Elat, Israel, it is considered a common spring migrant (Krabbe 1980).

Moustached Warbler *Acrocephalus melanopogon* PV WV?

Local name Haazija umm shaarib هازجة ام شارب.

Scarce but regular passage visitor in autumn from mid-August to late November and in spring from (early) late February to mid-May. Records tend to be concentrated along the Sinai and Suez Canal corridors. It is not clear if the February records are of wintering birds or early spring migrants. All Egyptian material examined is referable to *A. m. melanopogon*. Generally noted in reed-beds.

Nile Delta and Valley Meinertzhagen (1930) stated that the Moustached Warbler was formerly a common winter visitor to the Nile Delta but mentioned only one record between 1910 and 1930. More recently it has been noted during autumn at Fayid on 21, 22, and 27 August 1949 (AGH, KLS) and at Port Said on 4 October 1980 (IP, UGS). There are several records from Suez between late February and mid-May (Wimpfheimer *et al.* 1983). Single birds were observed on 19 March 1965 at Lake Maryut (JWW) and on 1 April 1950 at Fayid (HGB). The only record we are aware of from the Nile system south of the Delta is 10 on 22 February 1984 near Lake Qarun (SG).

Sinai One was noted on 7 September and three on 9 September 1981 at Zaranikh

(IP, UGS), and one on 4 October 1985 at El Taloul (DM). Specimens from the Meinertzhagen collection were taken at Rafa on 5 October and 10 and 25 November 1917 (BMNH). Late winter and spring records include one at Misfaq on 7 February 1979 (EB), three at Sharm el Sheikh on 13 March 1979 (HW), and one at Ras el Sudr on 14 April 1983 (SFN). At Elat, Israel, it is a scarce winter visitor between early November and early April (Krabbe 1980).

Sedge Warbler *Acrocephalus schoenobaenus* PV (WV)

Local names Haazija as-sa'd هازجة السعد, wish ad-diiba.

Common passage migrant in autumn from (late July) mid-August to mid-October (mid-November) and in spring from (late February) early March to early (late) May. Passage proceeds along a broad front and birds have been noted in the Western and Eastern Deserts and Sinai. It is an irregular winter visitor to the country. Monotypic.

The earliest autumn record of a Sedge Warbler in Egypt is one at Lake Bardawil in late July (Shariv 1970). During a study of autumn migration at Bahariya, this species was recorded between 3 August and 26 October 1983 and the heaviest passage was from mid-August to early September (HB *et al.*). Records from midwinter include: one at Port Said on 9 December 1906 (Lynes 1912), one at Luxor on 21 December and two at Aswan on 25 December 1983 (PE), and one at Lake Manzala on 12 January 1980 (PI). During spring migration this species is often heard singing in reed-beds or other types of vegetation, such as desert scrub.

Ringing Two birds ringed in Finland in August 1982 were recovered in Egypt in September 1982 (EDB).

Marsh Warbler *Acrocephalus palustris* PV

Local name Haazija al-bataa-ih هازجة البطائح.

Scarce passage migrant in autumn from (mid-August) early September to late October (early November) and apparently rare in the spring from mid-March to early May. The abundance of this species on passage in Egypt is unclear because of the difficulty in separating it in the field from the Reed Warbler. Monotypic.

The main migration route of the Marsh Warbler through Egypt appears to be via the Nile corridor. Thirty-one birds were captured at Bahariya between 12 August and 13 October 1983, with heaviest passage between mid-August and early September (HB *et al.*). There are only six definite records of this species from the six PMS spring surveys at Bahig (USNM), compared with a total of 512 Reed Warblers (Horner 1977). There is one definite spring record from the Western Desert; a single bird was collected at Farafra on 19 March 1984 from a flock of Reed Warblers (Goodman *et al.* 1986). An exceptional record is one near Cairo on 7 July (Meinertzhagen 1930).

Reed Warbler *Acrocephalus scirpaceus* MB? PV (WV)

Local names Haazija al-ghaaba هازجة الغاب, abuu dukhna, afrii (ND), dakhnaan, dukhnaa.

Presumed uncommon migrant breeder in the Nile Delta. It is a common migrant in autumn between (early) mid-August and late October (late November) and in spring between (mid-February) mid-March and mid- (late) May. It occasionally winters in Egypt. The subspecies *A. s. fuscus* is a more common passage migrant than *A. s. scirpaceus*, and is presumably the form breeding in Egypt.

In May 1983 Reed Warblers were found breeding for the first time in Egypt at Lake Manzala; a few pairs were seen carrying nest material and food near El Matariya (Meininger *et al.* 1986). There are several May records of singing birds at other Nile Delta lakes and presumably it breeds at some of these localities as well. It is unknown if local nesting populations are migratory.

This species passes throughout the country on both migrations, including movements directly across the Red Sea (e.g. Ticehurst 1924). In 1983 autumn passage at Bahariya was noted between 3 August and 25 October, peaking from mid-August to early September (HB *et al.*). There are several winter records, which are possibly in part early migrants. Birds have been observed from December to early February in the Faiyum and the Nile Delta lakes (e.g. Meinertzhagen 1930; Meininger *et al.* 1979).

Ringing There are eight records of this species ringed in Austria and Hungary and recovered in Egypt during autumn, mainly in September (Watson 1971; Reiser 1973; Schmidt 1983; BEP; EDB).

Clamorous Reed Warbler *Acrocephalus stentoreus* RB Plate 2

Local names Haazija al-qaṣab aṣ-ṣayyaaḥa هازجة القصب الصياحة, dukhnaa-, khansha' ṣayyaaḥa.

Abundant breeding bird in the Nile Delta and Valley south to Aswan, the Suez Canal area, the Faiyum, Wadi el Rayan, and a few Western Desert oases. All Egyptian material examined is referable to *A. s. stentoreus*. This species is regularly found singing in sugar-cane fields and other types of crop land. It nests in reed-beds from March to late June. Outside the breeding season there appears to be some dispersal and birds are seen in exceptional places, such as the centre of Cairo.

Western Desert The Clamorous Reed Warbler is a fairly common breeding resident of Wadi el Natrun. In recent years it has been found or suspected of breeding in Dakhla and Bahariya (Goodman *et al.* 1986). There are a few records of unidentified *Acrocephalus* warblers in the area. In April 1925 Moreau (1927) found either Clamorous Reed Warblers or Great Reed Warblers at Dinaria, Dakhla. On 15 June 1935 'a reed warbler with fledged young' was noted at Sitra (Moreau 1940–1).

Thus, it seems likely that some *Acrocephalus* species colonized the area earlier than presently known.

Nile Delta This species is abundant in reed-beds of the Delta lakes. Densities of 10 to 20 singing birds per/ha are typical, which would equal a local breeding population of tens of thousand pairs (Meininger *et al.* 1986).

Nile Valley Meinertzhagen (1930) considered this species a breeding resident of Lower Egypt but made no mention of its occurrence south of Cairo. At present, it is common in the entire Egyptian Nile Valley, including the Faiyum, south to Aswan. It seems likely that the change in the Nile's flood pattern, after the completion of the Aswan High Dam in the late 1960s, allowed reed-beds to colonize new parts of the Nile Valley and is directly related to this species' range expansion. There appears to be a similar pattern in Wadi el Rayan. In the early 1980s this area was partially flooded, reed-beds became established, and the Clamorous Reed Warbler colonized the region; by 1986 it was abundant in this area (SMB).

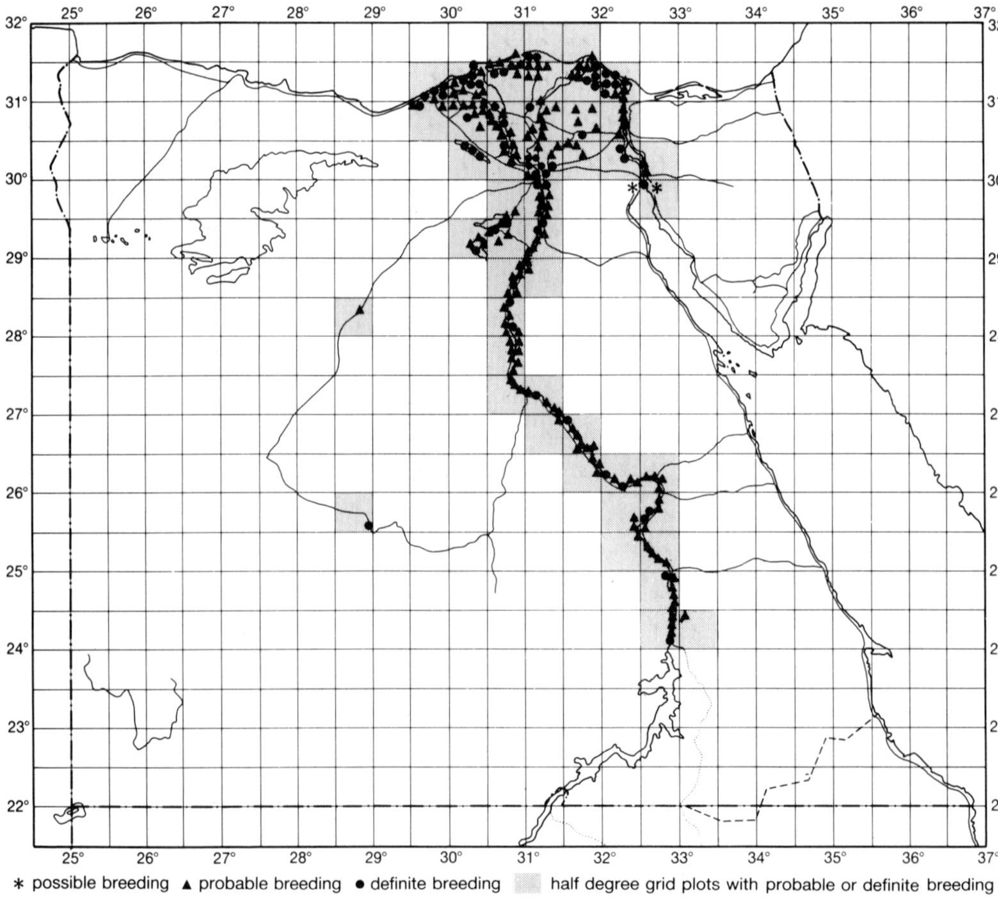

Map B86. Breeding distribution of Clamorous Reed Warbler *Acrocephalus stentoreus* between 1970–87.

Sinai Records include one on 26 December 1980 at Sharm el Sheikh (BCD) and singing birds on 25 January 1985 in a marsh 35 km south of the Suez Tunnel (SMG). This species is a breeding resident at Elat, Israel (Krabbe 1980).

Great Reed Warbler *Acrocephalus arundinaceus* PV

Local names Haazija al-qas̱ab al-kabiira هازجة القصب الكبيرة, umm nuuẖ.

Apparently a scarce migrant in autumn from (mid-) late August to late October (late November) and distinctly more common in spring from mid-March to late May. However, its abundance is somewhat unclear because of potential confusion in the field with the similar looking Clamorous Reed Warbler. All Egyptian material examined is referable to *A.a. arundinaceus*. *A.a. zarudnyi* has also been reported from the country (Vaurie 1959).

The Great Reed Warbler passes along a broad front throughout the country. The earliest autumn record is 18 August 1983 and the latest appears to be 25 November (year not stated) (Meinertzhagen 1930; HB *et al.*). Five specimens were taken in relatively recent years at Bahig between 2 and 23 September (FMNH; USNM). Between 18 August and 18 September 1983, three were captured in Bahariya (HB *et al.*). During the six PMS spring field seasons at Bahig between 1967 and 1973, passage was noted (on the basis of captured birds) from 13 March to 25 May and peaked in April (Horner 1977).

Generic local name for *Hippolais* spp. Khansha'.

Olivaceous Warbler *Hippolais pallida* MB PV (WV?)

Local names Khansha' zaytuunii خنشع زيتوني, zaqq.

H.p. pallida is an abundant migrant breeding bird in Western Desert oases, parts of the Nile Delta and Valley, the Suez Canal area, the Faiyum, and Wadi el Rayan. It breeds from mid-April to late July (Simmons 1952) and is thought for the most part to winter in sub-Saharan Africa (Vaurie 1959), although there are a few winter records from the Egyptian Nile Valley. *H.p. elaeica* is probably a common passage migrant and perhaps the form breeding in the El Arish–Rafa area. *H.p. opaca* is a rare migrant to the country, and one specimen taken on 13 May 1970 at Matruh is referable to this form (USNM).

Meinertzhagen (1930) mentioned that the Olivaceous Warbler bred in the Nile Valley as far south as Beni Suef; nowadays it occurs south to at least Aswan. Specimens taken at Bahig are referable to *H.p. pallida* (USNM) and this form may breed along the eastern North Coast. It possibly breeds at isolated places in the Eastern Desert such as El Laqeita. In early July 1951 and May 1984 Olivaceous Warblers were heard singing in mangrove swamps on Gezira Abu Minqar and may nest on the island (Searight 1952*c*; SMG, RWS). A bird taken at Gebel Elba on 10 April 1985 is referable to *H. p. pallida* (UMMZ).

Olivaceous Warblers pass through Egypt in autumn from mid-August to late October (early November) and in spring from (late February) mid-March to mid-(late) May. Singing birds were noted near Cairo on 24 December 1952 and 12 December 1954 (HCK). Birds observed in early February at Aswan (Short and Horne 1981) and one taken at Tushka on 16 February 1963 (YPM) may have been local wintering birds. Males arrive at the northern Egyptian breeding grounds in the last week of March or the first week of April (Simmons 1952). At Elat, Israel, spring movements of *H. p. pallida* peak before *H. p. elaeica* and the converse is true in autumn (Krabbe 1980).

Ringing There are at least 12 records of nestlings or breeding birds ringed at Fayid which returned to the area in subsequent nesting seasons (Brownlow 1960; HGB).

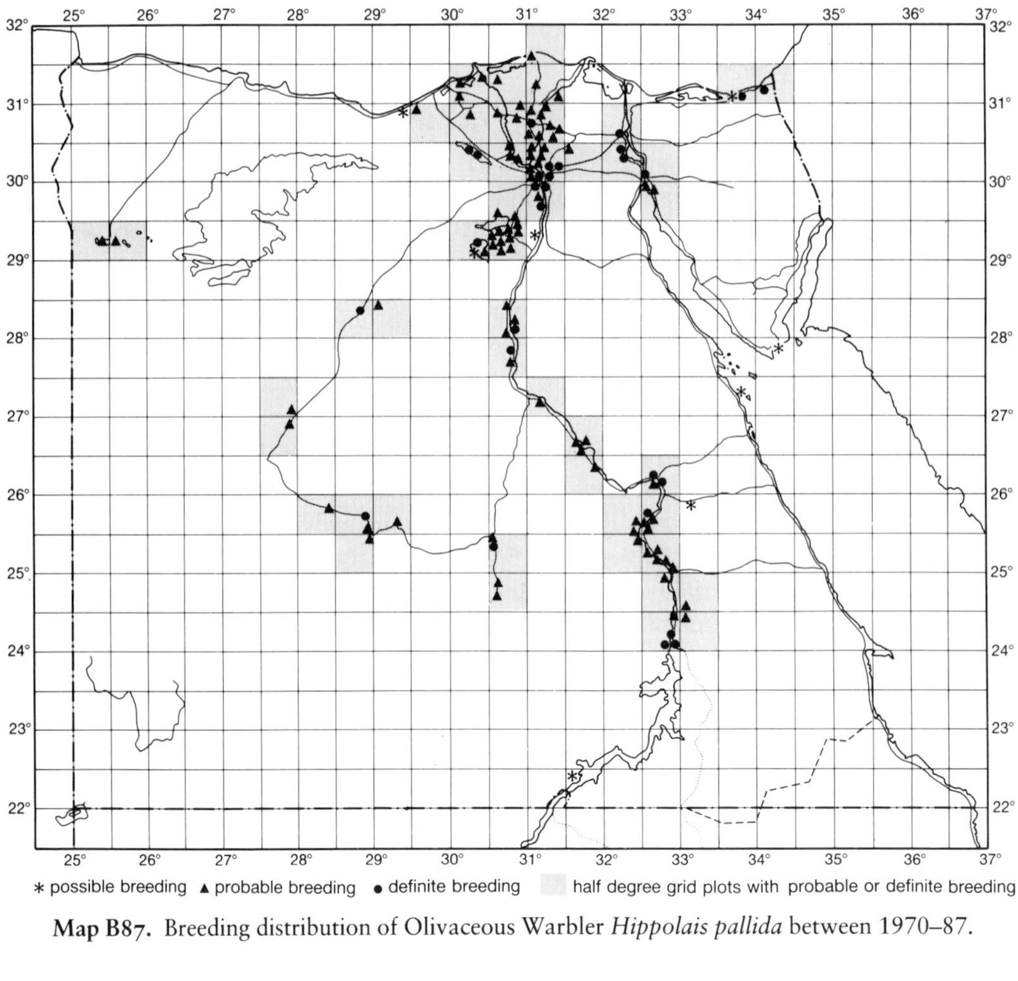

Map B87. Breeding distribution of Olivaceous Warbler *Hippolais pallida* between 1970–87.

Olive-tree Warbler *Hippolais olivetorum* PV

Local name Khansha' az-zaytuun al-kabiir خنشع الزيتون الكبير.

Probably a rare passage migrant in autumn from early August to late November and in spring from late February to mid-April. There is one summer record. Monotypic.

The following is a summary of all known dated records of the Olive-tree Warbler in Egypt: one observed at Cairo on 10 August 1980 (JB); one found dead in Wadi Madsus on 17 August 1982 (Baha el Din 1984*a*); several seen and one collected in Wadi Sa'l on 19 August 1984 (SMB); three collected at Port Said in September 1960 (Moreau 1961); single birds observed on 1 October 1984 at El Amarna and on 10 October 1984 at Aswan (Miles 1984); one noted in Wadi Feiran on 28 November 1965 (HCK); one taken at Quseir on 21 February 1928 (Meinertzhagen 1930; BMNH); one observed at Aswan on 6 March 1979 (Short and Horne 1981); one seen at Cairo on 17 April 1979 (SMB); and one collected at Gebel Elba on 20 June 1984 (UMMZ). At Elat, Israel, this species is a scarce migrant between late April and mid-May (Krabbe 1980).

Icterine Warbler *Hippolais icterina* PV (WV)

Local name Khansha' liimuunii خنشع ليموني.

Scarce passage visitor throughout the country in autumn from mid-August to mid-October (early November) and in spring from (late March) mid-April to mid- (late) May. Occasional in winter. Monotypic.

Thirty-three Icterine Warblers were captured in 1983 and 1985 at Bahariya between 14 August and 24 September, with the bulk passing between late August and mid-September (HB *et al.*). There are several observations of this species in late October and early November from the Upper Nile Valley (e.g. Short and Horne 1981; Davies 1982). Single birds have been noted on 19 January 1964 at Mit Rahina (HCK) and on 10 February 1977 at Kom Ombo (Short and Horne 1981).

An exceptionally early spring record is one at Dakhla on 26 March 1928 (BMNH). During the PMS survey near Bahig, about 330 Icterine Warblers were captured in six springs between 1967 and 1973, compared with only three Olivaceous Warblers (Horner 1977). At Elat, Israel, this species is a regular autumn and spring migrant in small numbers (HSh).

Generic local names for *Sylvia* spp. Dukhala, fisaysi (Ma), zurayqa.

Marmora's Warbler *Sylvia sarda* AV

Local name Dukhala marmara دخلة مرمرة.

Accidental visitor. The single Egyptian specimen examined is referable to *S. s. sarda*.

There is only one authenticated record of the Marmora's Warbler in Egypt. A male was collected at Salum on 28 January 1928 (Meinertzhagen 1930; BMNH). Wyatt (1870) reported that he shot two in Wadi Hibran but the specimens were lost.

Spectacled Warbler *Sylvia conspicillata* RB? WV

Local name Dukhala umm nazaara دخلة ام نظارة.

Scarce winter visitor, mainly to the northern portion of the country, between early September and mid-April. The breeding status of this species in Egypt is unclear. Meinertzhagen (1922, 1930) mentioned it as a partial migrant and believed that it was a resident in the Eastern Desert near Wadi Hof and in the Sinai. He further stated that it was not uncommon along the desert edge–cultivation ecotone in the Faiyum and Wadi el Natrun. Meinertzhagen found old nests in the desert bordering Lake Qarun and the Nile Delta that he tentatively identified as belonging to this species. There appears to be no recent documentation of it breeding in the country. The following records summarize post-Meinertzhagen (1930) information. All Egyptian specimens examined are referable to *S. c. conspicillata*.

North Coast Spectacled Warblers have been recorded at Bahig on 15 October 1971 and near Burg el Arab on 24 October 1971 (USNM). Other records include five on 9 January 1980 at El Omayed (PI) and one on 15 April 1942 at Sidi Barrani (Tomlinson 1943).

Nile Delta and Valley Single birds were noted on 23 and 26 October 1971 at Giza (Haensel 1975). Over 10 Spectacled Warblers were observed in the El Khanka sand-dunes, north-east of Cairo, on 15 November 1985 and at least eight on 22 November 1985 (SMB). Two were seen at Abu Simbil on 22 November 1981 (RGB *et al.*). There are several winter records of small numbers from the northern Nile Delta and two specimens were taken at Bir Victoria on 28 February 1923 (GZM). Short and Horne (1981) regularly noted this species during winter in the wadis leading to El Amarna and once near Luxor. The only known spring record is one observed at El Amarna on 18 March 1985 (Grieve 1985).

Eastern Desert Tregenza (1951, 1958) saw this species during the winter in the triangular area between Qena, Quseir, and Ras Gemsa, and one on 20 July 1947 in Wadi Umm Anfei. More recent records include single birds on 1 April 1966, 48 km south of Ain Sukhna (HCK), on 10 February 1983 at Safaga (UMMZ), and on 21 February 1985 in Wadi Aideib, Gebel Elba (DM).

Sinai This species has been recorded throughout the year in the desert near Qusaima. In late March 1976 and early April 1973 males were observed singing and defending territories. Nesting in this region seems likely (SC). The species is known to breed in the northern Negev Desert of Israel (Paz 1987). Other recent records include single birds on 30 December 1978 at St Katherine (EB), on 6 September 1981 at Zaranikh (IP, UGS), on 6 March 1973 at Nahal Yam (ZM-HUJ), and on 20 November 1979 near Ain el Gedeirat (TAUM).

Subalpine Warbler *Sylvia cantillans* PV WV

Local name Dukhala as-suruud دخلة الصرود.

Scarce migrant in autumn from late August to late September (late October) and common in spring from (mid-) late February to mid-April (mid-May). It is known to winter in the extreme south-western corner of the country. All Egyptian specimens examined are referable to *S. c. albistriata*.

During the six PMS autumn field seasons at Bahig, 48 Subalpine Warblers were handled. All were captured between 31 August and 22 September, except for single birds on 17 and 23 October (Goodman and Watson 1983). Twenty-one were captured at Bahariya in 1983 and 1985 between 31 August and 17 September, with a maximum of seven on 15 September (HB *et al.*). The only autumn record from the Delta proper appears to be one on 6 September 1966 at Alexandria (JWW). Between 4 and 6 September 1945, one was seen with two Rüppell's Warblers at Kasfarit (LB). Birds have been reported near El Maadi on 31 August and 13 September 1967 (HS). Tregenza (1955) noted it on 11 September at Gebel Naq el Teir in the central Red Sea Mountains.

Misonne (1974) found this species wintering in unfluctuating numbers from 29 October 1968 to mid-January 1969 in the Gebel Uweinat area. He observed about one bird per 30 acacias, or approximately 165 near Karkur Talh and 300 at Gebel Uweinat. The species is not known to winter elsewhere in Egypt.

During spring migration this species is common throughout the country, though less so in the eastern portion. Two noted in Wadi Abu Fatma on 18 February 1984 were presumably early spring migrants (SMG). At Bahig there is a distinct surge of migrants between mid-March and early April (Horner 1977). At Elat, Israel, it is a scarce spring migrant (Krabbe 1980).

Ringing A bird ringed at Bahig on 28 February 1971 was recovered at the same locality on 8 March 1972 (GZF).

Ménétries' Warbler *Sylvia mystacea* (PV) (WV)

Local name Dukhala aasiyawiyya دخلة آسياوية.

Rare and irregular migrant and winter visitor. Most records are from the spring. Monotypic.

The following summarizes all known dated records of the Ménétries' Warbler in Egypt. It has been collected on 21 February 1985 at Gebel Elba (UMMZ); on 4 March 1914 at Qena (Meinertzhagen 1930); on 21 March 1913 at Mazghuna (Meinertzhagen 1930); and on 25 March 1985 at Hurghada (UMMZ). Single birds have been observed on 30 December 1984 at Sharm el Sheikh (WL) and on 14 March 1981 east of El Maadi (SMB, BB), and photographed on 30 March 1985 at Na'ama (UGS).

Sardinian Warbler *Sylvia melanocephala* RB PV WV

Local name Dukhala ra-saa- دخلة رأساء.

Breeds in north-eastern Sinai and formerly in the Faiyum. It is a common passage migrant and winter visitor throughout the country from early September to mid- (late) April in a variety of habitats, from the vegetated Nile basin to arid desert. Three forms have been recorded in the country of which *S. m. melanocephala* is the most common migrant and winter visitor. Recent specimens referable to *S. m. momus* have been taken at Arminna on 3 March 1964 (YPM) and at Bahig on 22 March 1971 (USNM). *S. m. norrisae*, which is thought to have been restricted to the Faiyum, formerly bred along the shores and on islands of Lake Qarun (Meinertzhagen 1930). A bird captured at Elat, Israel, on 12 April 1970 was referred to this form (Krabbe 1983); however, the specimen was not compared to Faiyum material and the identification is clearly tentative.

Faiyum Sardinian Warbler nests and nestlings were noted numerous times before 1920 in the second half of March. At the turn of the twentieth century Lake Qarun contained freshwater but it is now saline and the shore-line and islands are virtually devoid of vegetation. Although this species has recently been searched for during the breeding season there are no records and *S. m. norrisae* appears to be extinct. The last verifiable record is a bird collected on 7 February 1939 (HGD). The records of Kattinger (1970), who heard *S. m. 'norrisae'* in the Faiyum on 6 August 1966, are not acceptable. This form appears to have been a migrant breeder since all specimens examined ($n = 18$) were taken near Lake Qarun between early February and late June. It should be noted that other subspecies presently winter in the Faiyum.

Sinai In June 1984 Sardinian Warblers were found to be common between El Arish and Rafa. Birds were observed carrying nesting material and food, and several fledglings noted. On 2 May 1985, four fledglings were seen being fed by adults west of Rafa (SMB). The subspecies breeding in this region is unknown but presumably is *S. m. momus*, which nests in portions of the Levant.

Cyprus Warbler *Sylvia melanothorax* AV

Local name Dukhala qubrusiyya دخلة قبرصية.

Accidental visitor. Monotypic.

Four records of the Cyprus Warbler in Egypt are known: a female in Wadi Umm Rus on 16 December 1938 (HGD); one near Gebel Elba on 22 December 1938 (Goodman and Atta 1987); one female at Sharm el Sheikh on 30 January 1982 (HH); and a male in Wadi Hof on 1 March 1910 (GZM). This species was noted in north-eastern Sudan in November 1980 (Nikolaus 1981) and also winters near Elat, Israel (Krabbe 1980). Thus, it may migrate through and/or winter in the Eastern Desert and Sinai with greater frequency than currently known.

Rüppell's Warbler *Sylvia rueppelli* PV (WV)

Local names Dukhala ruubil دخلة روبل, kahayhayla, qarqafanna, zurayqa qussaabii.

Fairly common migrant in autumn from mid-August to late September (mid-October) and common in spring from (mid-February) early March to late April (early May). Rare in winter. Monotypic.

Rüppell's Warblers pass throughout the country on both migrations. Twenty-five birds were captured in 1983 and 1985 at Bahariya between 29 August and 16 September, with a maximum of eight on 10 September (HB *et al.*). The only known winter record is one observed at Aswan on 5 January 1978 (AS). The bulk of spring migrants at Bahig pass between mid-March and early April (Horner 1977).

Desert Warbler *Sylvia nana* PV WV

Local names Dukhala as-sahraa- دخلة الصحراء, shuwaala.

Rare passage migrant and apparently local winter visitor to the country between (late July) early September and late April, mainly in Sinai and the south-eastern desert. All Egyptian material examined is referable to *S. n. nana*.

North Coast Meinertzhagen (1930) obtained a male Desert Warbler on 28 January 1920 in the desert 16 km west of Salum that is referable to *S. n. deserti* (BMNH). He also saw several Desert Warblers at this locality in January 1928. On the basis of present political boundaries this site is across the Egyptian border in Libyan territory.

Western Desert Only one report is known from this area. A single Desert Warbler was seen on 2 October 1980 along the Bahariya–Cairo road, 174 km north of El Gedida (HB *et al.*).

Nile Valley and Eastern Desert Records from these two areas include single Desert Warblers on 12 December 1938, 88 km east of Qift along the Qift–Quseir road (GZM); 16 December 1938 in Wadi Umm Rus (HGD; GZM); 21 December 1987, 50 km west of Marsa el Alam on the road to Idfu (ADB, AJA); 22 January 1986 at Quseir (SMB); 1 February 1986 at Gebel Abu Brush (SMB); 10 March 1914 at Matana (Meinertzhagen 1930); 12 March 1965 at El Dakka (YPM); and 23 March 1916 at Helwan (Meinertzhagen 1930). Six were observed on 17 February 1984, 1 km east of Aswan (FRL). Tregenza (1951, 1955, 1958) recorded Desert Warblers on numerous occasions between September and March in the central Red Sea Mountains. He noted two near Bir Himreyr on 30 July 1947, an exceptionally early date for autumn migrants. This species is observed with some frequency during the winter months in the Sudan Government Administration Area. The latest spring record appears to be one in Wadi Akwamtra, Gebel Elba, on 20 April 1985 (SMG, PLM).

Sinai The type specimen of *S. n. nana* is from El Tor. Several pre-1930 workers found this species on the Peninsula during November, February, and March (e.g. Koenig 1924; Meinertzhagen 1930). More recent late autumn and winter records include single birds on 26 November 1965 at St Katherine (HCK), on 2 and 3 December 1984 at Sharm el Sheikh (RvD *et al.*), on 5 February 1976 at Nabq (OT), on 6 and 7 February 1976 between Wadi Feiran and St Katherine (OT), and on 18 and 19 February 1980 at Sharm el Sheikh (TAUM).

Arabian Warbler *Sylvia leucomelaena* RB Plate 4

Local name Dukhala al-bahr al-ahmar دخلة البحر الأحمر.

Locally common breeding resident in south-eastern Egypt. A few observations from Sinai. All Egyptian material examined is referable to *S. l. blanfordi* (GZM; UMMZ; Goodman, 1988).

Eastern Desert The Arabian Warbler is a locally common breeding resident in the wadis immediately surrounding and on the slopes of Gebel Elba (Goodman and Meininger in press). It is characteristically found in heavy acacia savanna and at altitudes of up to 550 m. There are no positive records from other areas of the Eastern Desert, although seemingly appropriate habitat exists. On several occasions during the summer of 1947, Tregenza (1951) observed a warbler in the central Red Sea Mountains and on the basis of his description the bird could well have been this species.

Sinai There are two recent observations of Arabian Warblers in this area. Single birds were noted on 17 August 1984 in Wadi Solaf and the following day in Wadi Nasb (SMB). It is unknown if these observations were of local breeding birds or migrants. The species breeds in Wadi Arava, north of Elat, Israel (Afik and Pinshow 1984).

Orphean Warbler *Sylvia hortensis* PV (WV)

Local names Dukhala mughanniyya دخلة مغنية, zurayqa mughanniyya.

Rare passage migrant in autumn from mid-August to mid- (late) October and in spring from late March to late April. It is apparently an irregular winter visitor to the south of the country. All Egyptian material examined is of *S. h. crassirostris*.

Orphean Warbler autumn passage appears to be principally to the east of the country (including Sinai). Between 17 and 21 August 1982 and 1984, 1–10 birds were seen each day in south Sinai (SMB). Three birds were noted at St Katherine on 23 August 1986 (EE *et al.*). Recent Sinai specimens include single birds taken on 14 August 1969 in Wadi Rahaya (TAUM); 19 August 1984 in Wadi Sa'l (AAUC); and 20 August 1971 at St Katherine (ZM-HUJ). One was noted on 15 August 1986 at Bir Beida, just west of Quseir (EE *et al.*) and one trapped on 18 August 1955 at Fayid

(RN). The only known autumn records west of the Nile basin are birds taken at Bahig on 24 August 1969 and 18 September 1969 (Goodman and Watson 1983).

Winter records include one near Gebel Elba on 28 December 1938 (Goodman and Atta 1987), one at Kom Ombo on 10 February 1977 and two at Luxor on 21 February 1976 (Short and Horne 1981), one east of Aswan on 17 February 1984 (FRL), and one in the Faiyum on 28 February 1964 (HCK). The February records might be of early spring migrants. De Meersman (1979) reported a singing Orphean Warbler at an undisclosed locality in Egypt between 19 and 25 January 1979 but details are lacking.

Spring passage is concentrated between late March and late April. At El Maadi birds were taken on 24 and 26 March 1967 (HS). There are six relatively recent spring records from the North Coast between 23 March and 30 April (Horner 1977; Goodman and Watson 1983), and another was noted at Sidi Barrani on 15 April 1942 (Tomlinson 1943). Two males were observed at Abu Simbil on 26 March 1985 (Grieve 1985), a single bird at Sharm el Sheikh on 29 March 1985 (UGS), and four at Suez on 29 April 1978 (RJM, LRP).

Ringing A bird ringed in Rome, Italy, on 29 April 1935 and identified as an Orphean Warbler was recovered in Siwa on 26 September 1935 as a Whitethroat (Mackintosh 1941; GZF). This bird was ringed at the Castel Fusano Centre, near Rome, the files of which were destroyed during World War II (Zink 1973; GZF). Clearly, the bird was misidentified by the person who ringed or recovered it. On the basis of the migration patterns of these two species it is most likely that the bird was a Whitethroat.

Barred Warbler *Sylvia nisoria* PV

Local name Du<u>kh</u>ala mukha<u>tt</u>ata دخلة مخططة.

Apparently a rare passage migrant in autumn from late August to late October (mid-November) and perhaps more common in spring from late April to mid-May. All Egyptian material examined is referable to *S. n. nisoria.* The only exception is a specimen taken at Gebel Elba on 21 April 1985 which is distinctly darker than typical *S. n. nisoria* and approaches the coloration of *S. n. merzbacheri* (UMMZ).

There is a total of 13 relatively recent Barred Warbler records at Bahig between 27 August and 23 October (Goodman and Watson 1983; FMNH). Three were seen at St Katherine on 23 August 1986 (EE *et al.*). Single birds were captured at Bahariya on 14 and 15 September 1982 (HB *et al.*). One immature was observed at Suez on 18 October 1981 (RGB *et al.*) and another bird at Lake Qarun on 15 November 1983 (Bezzel 1987*a*).

Recent spring specimens include one on 24 April at Matruh, four birds between 4 and 9 May at Bahig, and one on 14 May 1984 in Wadi Bali, north of Hurghada (Goodman and Watson 1983; UMMZ). In late April 1985 this species was a relatively common migrant in the area surrounding Gebel Elba with at least 22 birds in Wadi

Akwamtra on 21 April (SMG, PLM). One was noted on 27 April 1983 east of Idfu (MCJ *et al.*). Single birds were ringed at El Arish on 24 and 28 April 1985 (Fry *et al.* 1985). At Elat, Israel, this species is a regular autumn and spring migrant in small numbers (HSh) as it is along the Sudanese Red Sea coast (Nikolaus 1987).

Lesser Whitethroat *Sylvia curruca* PV WV

Local names Du<u>kh</u>ala fiiraanii <u>s</u>aghiira دخلة فيراني صغيرة, zurayqa fiiraanii.

Common passage migrant throughout the country in autumn from (late July) mid-August to mid-October (early November) and abundant in spring from (early) mid-February to mid- (late) May. It is a fairly common winter and rare summer visitor. The vast majority of Egyptian specimens examined are referable to *S. c. curruca*. The Siberian form, *S. c. blythi*, has been collected several times in spring and autumn in the Nile Delta and Faiyum (Meinertzhagen 1930).

Between May and August 1983 the Lesser Whitethroat was seen and heard singing regularly in a garden at Gebel Rahaba, near St Katherine, but no evidence of breeding

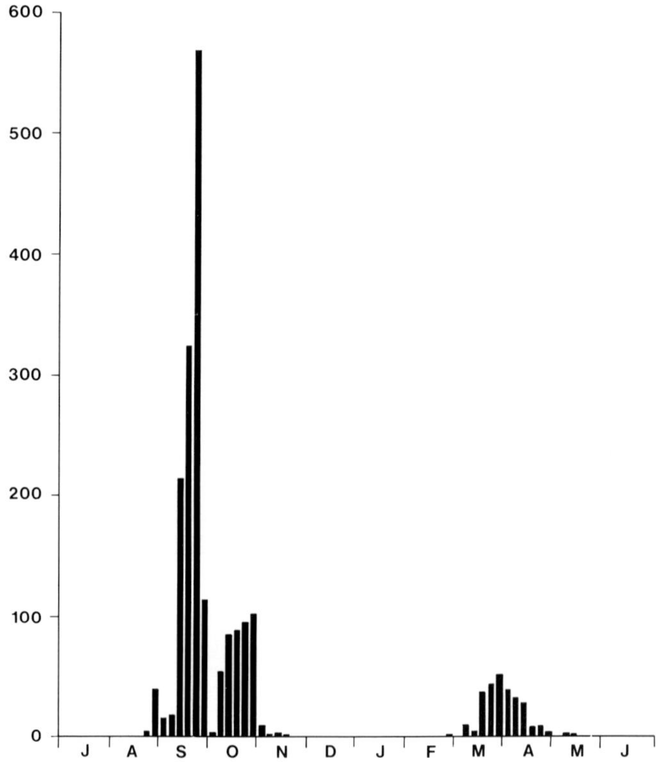

FIG. 6.28. Migration pattern of Lesser Whitethroats along the North Coast (after Watson 1971).

was found (SMB). It has also been noted in late July along the north Sinai coast (Shariv 1970).

Autumn passage peaks between mid-September and early October (Fig. 6.28). The species appears to be a fairly numerous winter visitor (records from December and January) in several Western Desert oases, the Faiyum, and the Nile Valley, but is distinctly more common at this time of the year in Upper Egypt. Spring migration peaks in late March and early April.

Ringing There are 40 recoveries from the autumn and one from the spring in Egypt of birds ringed in Austria, Belgium, Czechoslovakia, Denmark, England, the Federal Republic of Germany, Finland, France, the German Democratic Republic,

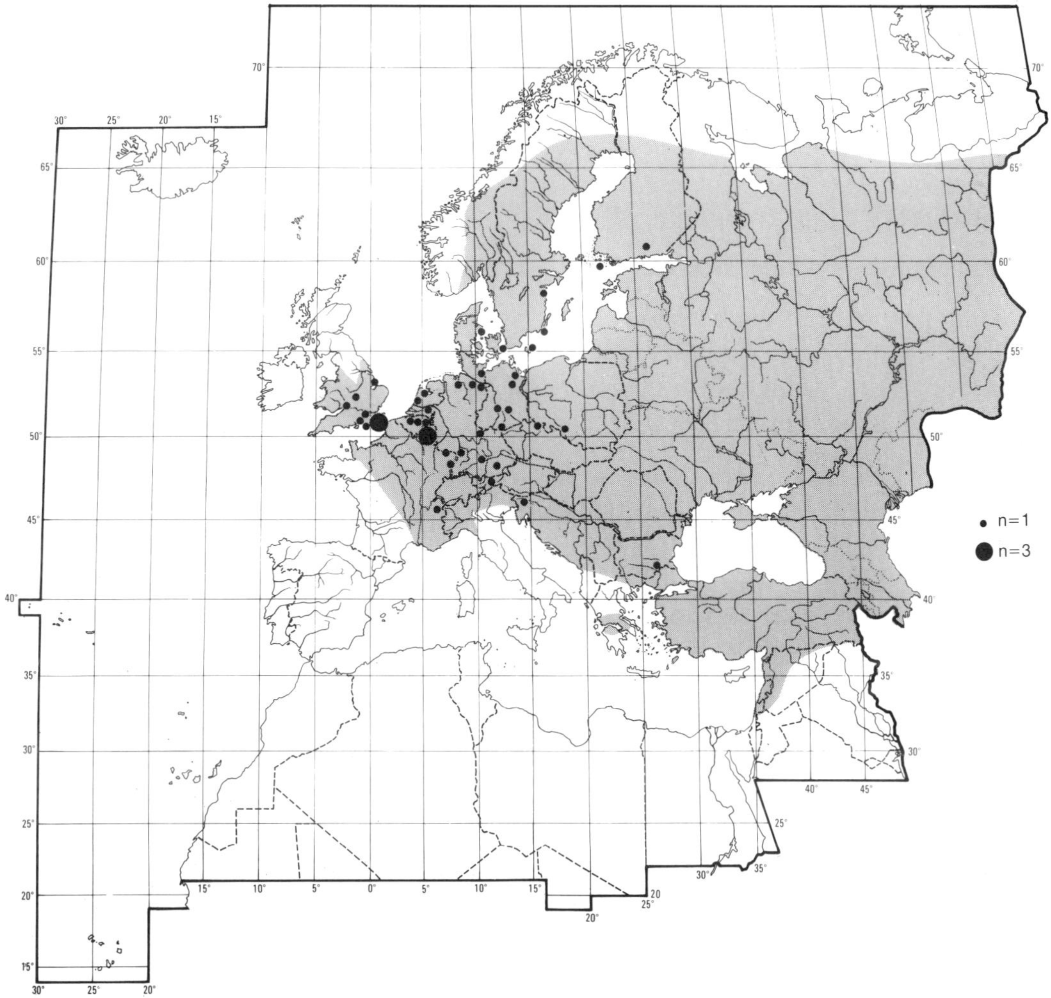

FIG. 6.29. Ringing or recovery sites of Lesser Whitethroats (n=46) occurring in Egypt. Shaded area is breeding range based on Harrison (1982).

The Netherlands, Poland, and Sweden (Fig. 6.29; Watson 1971; Anon. 1972, 1984*b*; Zink 1973; Siefke 1982; EDB; GZF; RSt; WT). This seasonal variation is probably a result of a more easterly route in spring of populations that migrate in autumn over western Egypt (cf. Horner 1980). Two birds ringed at the same site in the Federal Republic of Germany on 17 August 1979 were recovered together along the North Coast on 22 September 1979 (EDB). There are five records in total of birds recovered in Austria, Bulgaria, the Netherlands, and the Federal Republic of Germany which were ringed at Bahig, El Maadi, or Fayid (Brownlow 1952; Watson 1971; GZF; RWa). There appears to be some migration route tenacity. One Lesser Whitethroat ringed in El Maadi on 24 September 1968 was taken the following year at Burg el Arab in September or October and another ringed at El Maadi on 18 October 1968 was recovered at Galal in the autumn of the following year (Watson 1971). One bird ringed at Bahig on 24 March 1971 was recovered at the same place on 6 March 1972 (Horner 1977; GZF). There are no autumn recoveries in Egypt of birds ringed in the country in spring.

Whitethroat *Sylvia communis* PV

Local names Dukhala fiiraanii دخلة فيراني, zurayqa fiiraanii.

Common migrant throughout the country in autumn between (late July) mid-August and late October (early November) and in spring between (mid-February) mid-March and early (late) May. Most Egyptian specimens examined are referable to *S. c. communis*. According to Meinertzhagen (1930), *S. c. icterops* is not as common in Egypt as nominate *communis*.

Fewer Whitethroats move through Egypt on autumn passage than in spring (Fig. 6.30). An exceptionally early migrant collected at Matruh on 26 July is referable to *S. c. icterops* (Meinertzhagen 1930). Whitethroats have been noted in February at Luxor and Aswan (Short and Horne 1981) and taken on 18 February 1964 at Nag Misaw (YPM). Since we are unaware of Egyptian records from midwinter, the February reports are presumably of early spring migrants. Spring passage at Bahig peaks in early April to early May (Horner 1977).

Ringing The migration pattern of the Whitethroat in Egypt is complex. Some birds clearly show loop-migration as exemplified by the following three recoveries. A Whitethroat ringed at Lake Chad, Nigeria, on 13 April 1967 was recaptured at Bahig on 23 September 1968 (Hubbard and Horner 1969; Moreau and Dolp 1970); one ringed at Cape Bon, Tunisia, on 3 May 1968 was recovered at El Dab'a on 17 September 1969 (Bortoli *et al.* 1970); and another ringed on Malta on 1 May 1981 was recovered at Matruh on 10 September 1981 (JS). Six Swedish ringed birds have been recovered in Egypt in both autumn and spring (Österlöf 1969, 1971; Anon. 1972; RSt) and four Finnish birds in autumn (EDB). A bird ringed as a nestling in the German Democratic Republic was recaptured in April in Upper Egypt (Zink 1973). A

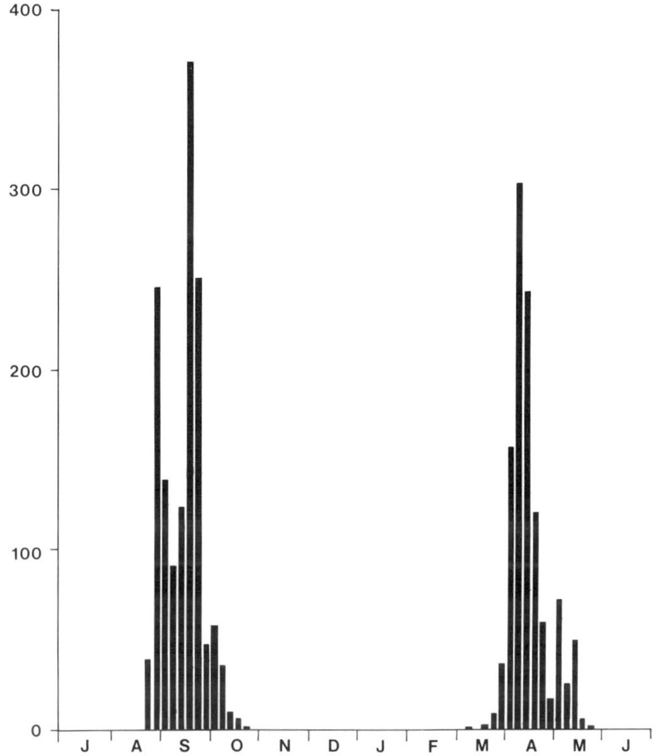

FIG. 6.30. Migration pattern of Whitethroats along the North Coast (after Watson 1971).

bird ringed at Bahig on 23 September 1969 was recovered in Finland on 30 May 1970 (GZF).

Garden Warbler *Sylvia borin* PV WV?

Local names Dukhala al-basaatiin دخلة البساتين, dukhala kahlaa-, qarqafanna.

Fairly common migrant in autumn from mid-August to late September (late October) and in spring from (early March) early April to late May (early June). It appears to pass along a broad front. Perhaps rare winter visitor. We have found it difficult to find consistent differences in size and plumage coloration between *S. b. borin* and *S. b. woodwardi*, both of which have been reported from the country, and all Egyptian material is provisionally assigned to *S. b. borin*.

The only previously published record of the Garden Warbler in the Western Desert was one observed between Kharga and Asyut on 25 March 1984 (Goodman *et al.* 1986). However, on 18 March 1986, two were seen in Kharga (AS) and between 20

August and 13 October 1983, 22 were captured in Bahariya (HB *et al.*). This species is presumably a regular migrant through the Western Desert.

There are a few winter observations of Garden Warblers in Egypt. Short and Horne (1981) listed three sight records from the Upper Nile Valley during the first half of February and a singing male was noted at Abu Simbil on 17 February 1982 (GM). These records may well be in part misidentified birds.

During the six PMS spring study seasons at Bahig, between 1967 and 1973, migration was noted from 11 April to 30 May and perhaps into early June; the number captured each spring varied substantially (Horner 1977). In mid- to late May 1984 a wave of migrants was observed in the barren wadis of the central Red Sea Mountains north of Hurghada (SMG, RWS). The latest spring record is one bird captured on 9 June 1955 at Fayid (RN).

Ringing A bird ringed in Tunisia on 6 May 1968 and another in the Netherlands on 12 July 1971 were recovered in Egypt on 3 May 1972 and 15 September 1971, respectively (EDB). One ringed on 7 May 1950 at Fayid was retrapped on 9 May 1952 at the same site (Brownlow 1960).

Blackcap *Sylvia atricapilla* PV WV

Local names Abuu qalansuwa ابو قلنسوة, dibsiyya (NC).

Scarce migrant in autumn from (late August) mid-September to late October (mid-November) and fairly common in spring from (early) mid-February to mid- (late) May. It migrates across a broad front. Scarce winter visitor. The vast majority of Egyptian material examined is referable to *S. a. atricapilla*. A few specimens taken at Bahig match the plumage coloration of *S. a. dammholzi* from Iran (USNM).

Kiepenheuer and Linsenmair (1965) found Blackcaps a few times in Siwa between 19 and 27 September 1963. This is the most recent known record of this species in the Western Desert. The earliest autumn record is on 23 August 1986 at St Katherine (EE *et al.*). Small numbers have been observed in December and January in Wadi el Natrun and parts of the Nile Delta and Valley. The bulk of spring migration is between March and April. This species is distinctly more common on spring passage than autumn (Horner 1977). There is some passage directly across the northern Red Sea (Casement 1976–7). The latest spring record is one on 25 May 1939 at Hurghada (Marchant 1941).

Ringing An adult male ringed at Fayid on 8 April 1951 was recovered on Cyprus on 11 October 1951 (Brownlow 1960). A bird ringed in the German Democratic Republic on 24 June 1961 was recovered in Egypt on 7 October 1962 (Zink 1969; EDB).

Generic local names for *Phylloscopus* spp. Niqshaara, fisaysi (Ma), haazija.

Yellow-browed Warbler *Phylloscopus inornatus* (PV)

Local name Niqshaara s̲afraa- al-h̲aajib نقشارة صفراء الحاجب.

Rare and irregular migrant to the country. All Egyptian specimens examined are of *P. i. inornatus*.

Nicoll (1919) observed Yellow-browed Warblers in the Faiyum on 10 November 1909 and at Giza on 19 April 1912. These records were not accepted by Meinertzhagen (1930). The first Egyptian specimen was taken on 17 October 1966 at Bahig (Hubbard and Seymour 1968). Subsequent records include single birds at Bahig on 22 October 1969 (Goodman and Watson 1983); at St Katherine on 27 October 1972 (ZM-HUJ) and on 26 November 1965 (HCK); and at Giza on 28 March 1982 (Meilstrup and Sørensen 1983). Between 1966 and 1979 this species was recorded five times in total at Elat, Israel, in November and from mid-March to early April (Krabbe 1980).

Bonelli's Warbler *Phylloscopus bonelli* PV

Local name Niqshaara s̲afraa- al-ajuz نقشارة صفراء العجز.

Fairly common migrant throughout the country in autumn from (early) mid-August to late September (late November) and in spring from (mid-February) early March to mid- (late) May. Rare summer visitor. Both *P. b. bonelli* and *P. b. orientalis* have been collected in Egypt but the latter form is distinctly more common.

There are a few November records of Bonelli's Warbler in Egypt, the latest being two birds on 22 November 1981 at Abu Simbil (RGB *et al.*). The species has been noted several times in early February (e.g. Short and Horne 1981), and one specimen was collected at Qena on 15 February 1897 (Koenig 1924). Since there are no midwinter reports from Egypt, the February records presumably concern early migrants and/or misidentified birds. The bulk of spring passage is from early March to mid-May. Summer records include single birds observed at El Arish on 25 June 1984 (SMB) and at the base of Gebel Shayib el Banat on 14 July (Tregenza 1958).

Ringing A bird ringed at Ma'agan Michael, Israel, on 21 March 1970 was recovered near Kom Ombo and reported on 15 June 1970 (Watson 1971). The reported date is rather striking and presumably in error.

Wood Warbler *Phylloscopus sibilatrix* PV

Local names Niqshaara ash-shajar نقشارة الشجر, al-mughannii al-as̲far, t̲arghaluus shamaaliyya.

Uncommon migrant in autumn from mid-August to late September (early November) and common in spring from (early March) early April to mid-May (early June). Monotypic.

Between mid-August and late September small numbers of Wood Warblers have been recorded along the North Coast, in the Nile Delta and Valley, the Faiyum, and Suez Canal area. In the autumn of 1949 the earliest Wood Warblers at Fayid were noted on 9 September in a distinct influx of *Phylloscopus* warblers (AGH, KLS). One was trapped at St Katherine on 16 September 1972 (DL). Relatively recent autumn Western Desert records include single birds captured in Bahariya on 3 September and 4 October 1983 (HB *et al.*) and three in Kharga in early October 1962 (Scharlau 1963). Horváth (1959) observed this species at Aswan on 26 October 1957 and there are a few other sight records from late October and early November. Two Wood Warblers collected at Cairo on 17 July may have been early autumn migrants (Meinertzhagen 1930).

The earliest spring record is one on 7 March 1985 at Abu Simbil (UGS). Between early April and mid-May this species is common throughout the country. Along the North Coast spring passage peaks in April (Horner 1977). A specimen taken at Abu Zenima on 7 June 1963 (Steinbacher 1965) was presumably a late spring migrant.

Chiffchaff *Phylloscopus collybita* PV WV

Local names Suksuka سكسكة, al-mughannii al-akhḏar, ghudhayr (NC), shaadiya al-khamaa-il.

Abundant migrant in autumn from (early) late September to early November and in spring from mid-February to late April (mid-May). It is a common winter visitor throughout most of the country. Two subspecies occur in Egypt, *P. c. collybita* and *P. c. abietinus*, the former being distinctly more common. A recent specimen referable to *P. c. abietinus* was taken at Wadi Qiseib, near Ain Sukhna, on 20 December 1966 (USNM). We are unaware of any Egyptian specimens referable to the Siberian breeding form *P. c. tristis*.

Chiffchaff autumn passage is concentrated between late September and late October. Some migration continues into early November, as six birds were found aboard a ship off Alexandria on 3 November 1960 (Tuck 1961). In winter the species is abundant throughout the country, including vegetated wadis and oases in desert areas. Calculated densities vary from 8–10 birds/km of wadi at Gebel Elba in mid-February 1985 to 2–4 birds/km in central Wadi Qena on 29 November 1986 (SMG). On the basis of these figures tens of thousands of Chiffchaffs winter in the Eastern Desert alone. Along the North Coast spring migration commences in mid-February and peaks in March (Horner 1977). The latest known spring record is 18 May 1983 at Hurghada (MCJ *et al.*).

Ringing A bird ringed in the Federal Republic of Germany was recovered in Matruh (Hoogstraal *et al.* 1963). There are also recoveries in Egypt of Hungarian,

Swedish, and Polish ringed birds (Pátkai 1966–7; Osterlöf 1970; Zink 1973). Two birds ringed along the North Coast in October and November returned to the same site in the subsequent spring and two autumns later, respectively (Horner 1977; GZF). Birds ringed at Fayid on 26 November 1949 and on 14 January 1950 were both retrapped at the same locality on 2 February 1951 (Brownlow 1960; GZF).

Willow Warbler *Phylloscopus trochilus* PV (WV)

Local name Niqshaara as̲-s̲afs̲aaf نقشارة الصفصاف.

Common migrant throughout the country in autumn from (mid-) late August to mid-

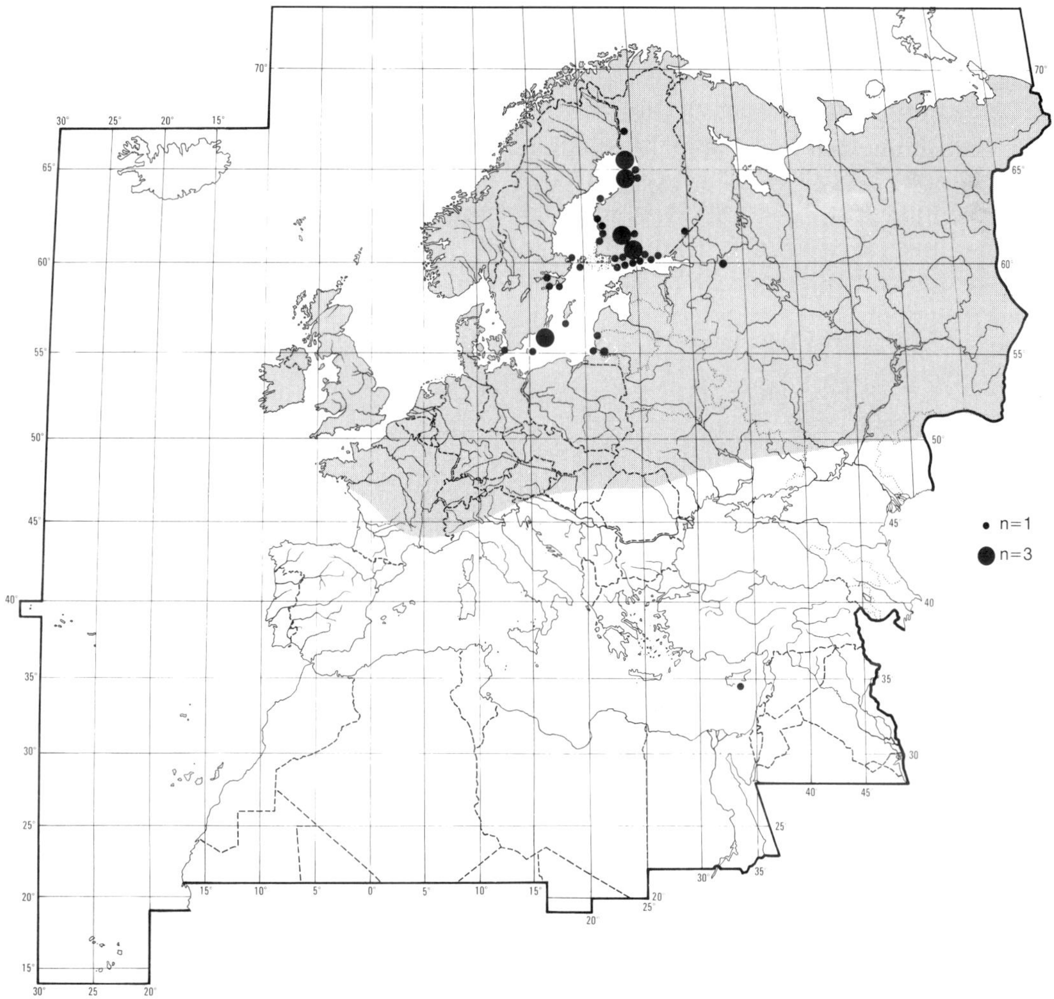

FIG. 6.31. Ringing or recovery sites of Willow Warblers (n=47) occurring in Egypt. Shaded area is breeding range based on Harrison (1982).

October (late November) and less common in spring from (mid-) late March to early (late) May. In winter it has been recorded in the extreme south-western corner near Gebel Uweinat (Misonne 1974) and possibly elsewhere in the country. Two subspecies, *P. t. trochilus* and *P. t. acredula*, have been identified from material collected in Egypt.

During a study of bird migration in Bahariya in 1983, Willow Warblers were recorded between 15 August and 29 October, with an estimated 1000 birds being present on 29 August (HB *et al.*). Late autumn records include several at Marsa el Alam between 1 and 5 November 1957 (Horváth 1959) and one at Sarabit el Khadim on 27 November 1967 (TAUM). Misonne (1974) saw two on 11 December 1968 at Gebel Uweinat, one of which was collected. Winter sight records from the Nile Valley are not acceptable (e.g. Short and Horne 1981) because of possible confusion with the Chiffchaff. Madden (1930) reported a few near Halaib sometime in February 1925. During the six PMS spring surveys at Bahig passage was noted from 19 March to 24 May, with the majority moving through in April (Horner 1977).

Ringing There are over 45 recoveries in Egypt of birds ringed in Denmark, Finland, Sweden, and USSR (Kalingrad and Leningrad regions) (Fig. 6.31; Watson 1971; Zink 1973; EDB; RSt). A bird ringed at Ikingi Maryut on 10 September 1969 was recaptured in Finland on 27 May 1970 (Watson 1971) and one ringed on Cyprus on 15 September 1970 was recovered at El Hauwariya on 20 October 1972 (Flint and Stewart 1983). One bird ringed on 18 March 1949 at Fayid was recaptured on 3 February 1951 at the same locality (Brownlow 1960).

Goldcrest *Regulus regulus* (WV)

Local names S̲a'uu as̲far al-'urf صعو أصفر العرف, nimnima, sa'wa, suksuuka, wasa'.

Rare and irregular winter visitor. All Egyptian material examined is referable to *R. r. regulus*.

Several Goldcrests were seen or collected at Giza between 22 November and 21 December 1910, and one was observed on 22 February 1911 (Nicoll 1912*a*; Meinertzhagen 1930; GZM). The only recent record appears to be one at Bahig on 30 October 1968 (USNM).

Firecrest *Regulus ignicapillus* AV

Local names S̲a'uu ah̲mar al-'urf صعو أحمر العرف, sa'wa, wasa'.

Accidental visitor. The single Egyptian specimen is referable to *R. i. ignicapillus*.

There is one documented record of the Firecrest in the country. A male was taken on 5 March 1973 at Bahig (Goodman and Tewfik 1983). In November 1916, one was observed by E. H. Buxton 'on the north-west frontier' of Egypt (Nicoll 1919).

Family Muscicapidae

Generic local names for flycatcher Khaaṯif adh-dhubaab, shuurib.

Spotted Flycatcher *Muscicapa striata* PV (WV)

Local name Khaaṯif adh-dhubaab al-anqaṯ خاطف الذباب الأنقط.

Common migrant throughout the country in autumn from (early) mid-August to late October (early November) and in spring from (early March) mid-April to late May. It is an irregular winter visitor. Two subspecies have been recorded in the country, *M. s. striata* and *M. s. neumanni*; the former is distinctly more common.

During a study of autumn migration in Bahariya in 1983, Spotted Flycatchers were captured between 20 August and 28 October, with the bulk passing in September (HB *et al.*). Winter records include single birds observed at Gebel Asfar on 3 December 1981 (RGB, SMB), 16 km north of El Tor on 28 January 1985 (SMG), in the Faiyum on 2 February 1965 (HCK), and at Giza on 12 February 1976 (Short and Horne 1981). During six PMS spring study seasons at Bahig, between 1967 and 1973, migrants were captured between 30 March and 21 April, except for in 1971 when the species was recorded as early as 3 March (Horner 1977).

Ringing Three birds ringed in Finland and Norway have been recovered in Egypt during the autumn (Zink 1975; BEP; EDB). One bird ringed in Lithuania on 16 May 1982 was recaptured at Sadat City on 4 September 1983 (RP) and another ringed at Bahig on 3 May 1967 was recovered in Greece on 20 November 1969 (Watson 1971).

Red-breasted Flycatcher *Ficedula parva* PV (WV)

Local name Khaaṯif adh-dhubaab aḫmar aṣ-ṣadr خاطف الذباب أحمر الصدر.

Scarce migrant in autumn between early September and mid-November and rare in spring between mid- and late April. Rare winter visitor. All Egyptian specimens examined are referable to *F. p. parva*.

Meinertzhagen (1930) mentioned six Red-breasted Flycatcher specimens in the GZM collected between 14 October and 18 November, three of which were taken in Wadi el Natrun (Goodman *et al.* 1986). Recent work in the Western Desert has shown that this species is a regular autumn migrant through the area. During a field season in Bahariya from 3 August to 29 October 1983, 21 Red-breasted Flycatchers were captured between 25 September and 28 October. Further, three were noted in the oasis on 1 October 1980 and two on 9 October 1982 (HB *et al.*). Other autumn records from the Western Desert include a few birds in Siwa between 19 and 27 September 1983 (Kiepenheuer and Linsenmair 1965) and one half-way between Bahariya and Cairo on 8 September 1984 (Goodman *et al.* 1986). At least eight were recorded during the six PMS autumn surveys at Bahig and Matruh between 20

September and 30 October (Watson 1971; USNM). Other records from the North Coast include two on 3 October 1980, 131 km west of El Alamein and one on 6 October 1980, 61 km east of Matruh (HB *et al.*); and one on 5 November 1975 at Bahig (DMNH). There are only a few records from the Eastern Desert and Sinai. One bird was found at the base of Gebel el Galala el Bahariya on 10 October 1965 (JT) and another at St Katherine on 16 October 1971 (ZM-HUJ).

The only winter record we are aware of is two observed on 29 December 1984 at St Katherine (WL). There is a paucity of spring reports from the country. Some include: one at Cairo on 12 April 1967 (JWW), two in the Faiyum on 13 April 1962 (HCK), one near Lake Bardawil on 20 April 1985 (Fry *et al.* 1985), one at El Arish on 26 April 1985 (DM), and one at Matruh on 27 April 1970 (USNM). Trans-Saharan passage appears to cross to the west of Egypt. The species is a regular spring migrant in Serir Oasis, eastern Libya (Hogg 1974). There are a few spring records of this bird at Elat, Israel (HSh).

Semi-collared Flycatcher *Ficedula semitorquata* PV

Local name Khaaṯif adh-dhubaab shibh al-muṯawwaq خاطف الذباب شبه المطوق

Rare migrant in autumn and fairly common in spring from mid-March to mid-May. In Egypt it is the least numerous of the three species of 'pied flycatchers'. Monotypic.

The only known autumn records of the Semi-collared Flycatcher in Egypt include one male on 4 September 1981 at Zaranikh (IP, UGS), an immature on 11 September at Port Said (Lynes 1912), one on 10 October 1908 at Giza (GZM), and one on 13 October 1962 at Bahig (FMNH). On spring migration it has been noted throughout the country. Specimens were taken in four of the six spring PMS field seasons at Bahig, between 19 March and 11 April (Horner 1977).

Collared Flycatcher *Ficedula albicollis* PV

Local names Khaaṯif adh-dhubaab al-muṯawwaq خاطف الذباب المطوق, abuu shayquuna.

Rare migrant in autumn from late August to late September (late October) and fairly common in spring from (mid-) late March to early (late) May. Monotypic.

Recent work in Bahariya has shown that the Collared Flycatcher is a rare, but apparently regular, autumn migrant through the Western Desert. During a study season held in this oasis between 3 August and 29 October 1983, eight birds were captured between 22 August and 16 September (HB *et al.*). Other autumn records from the country include single birds on 16 September 1918 at Gheit el Nassara (GZM), on 23 September 1966 at Bahig (USNM), on 24 September 1984 at Hurghada (MCJ *et al.*), and on 21 October 1967 at St Katherine (Ferguson 1970). The earliest spring record is a male at Memphis on 17 March 1986 (UA). In six PMS spring field seasons at Bahig, between 1967 and 1973, the extreme dates this species (presumably including some *F. semitorquata*) was recorded were between 19 March and 21 May (Horner 1977).

Ringing A bird ringed at Gotland, Sweden, on 27 June 1980 was recovered at Matruh on 10 April 1981 (RSt).

Pied Flycatcher *Ficedula hypoleuca* PV

Local name Khaaṯif adh-dhubaab al-abqa' خاطف الذباب الأبقع.

Rare in autumn from late August to early October and fairly common in spring from (mid-) late March to mid- (late) May. All Egyptian specimens examined are referable to *F. h. hypoleuca*. We have not been able to compare any autumn material taken in Egypt.

During a study conducted in Bahariya between 3 August and 29 October 1983, six Pied Flycatchers were captured between 24 August and 5 October. In 1982, at this same locality, one was noted on 22 September and two on 24 September (HB *et al.*). Other dated autumn records from the country include single birds on 1 September 1944 near the Great Bitter Lake (CJG), 6 September 1966 at Alexandria (JWW), 10 September 1966 at Suez (JT), 16 September (year and locality unknown) (Meinertzhagen 1930), and 24 September 1984 at Hurghada (MCJ *et al.*). One first-year male flycatcher ringed at Fayid on 9 November 1952 was identified as *F. hypoleuca* (AP). This is an exceptionally late record for any of the pied-flycatcher species in Egypt.

In spring the species has been recorded in all parts of the country, including Western Desert oases (Goodman *et al.* 1986). Spring passage peaks in the second to third week of April. During six PMS spring field seasons at Bahig, between 1967 and 1973, it was noted from 14 March to 21 May (Horner 1977).

Ringing A bird ringed in the French Pyrenees on 17 August 1968 was captured at Bahig on 13 May 1969 (Watson 1971).

Family Timaliidae

Brown Babbler (Arabian Babbler) *Turdoides squamiceps* RB

Local name Tharthaara 'arabiyya ثرثارة عربية.

Local breeding resident of eastern Sinai in areas with relatively heavy vegetation. *T. s. squamiceps* is the form occurring in the area.

Meinertzhagen (1930) mentioned records of Brown Babblers from Ain el Gedeirat and Wadi el Arish. The only recent records from this latter area, which has not been frequently visited, are single birds just south of El Arish on 17 March 1981 (SMB) and at El Gorah on 7 May 1982 (Wimpfheimer *et al.* 1983). Galal el Din and Fouad (1939–40) noted this species in Wadi Gerafi on 14 May 1939, and one bird was found

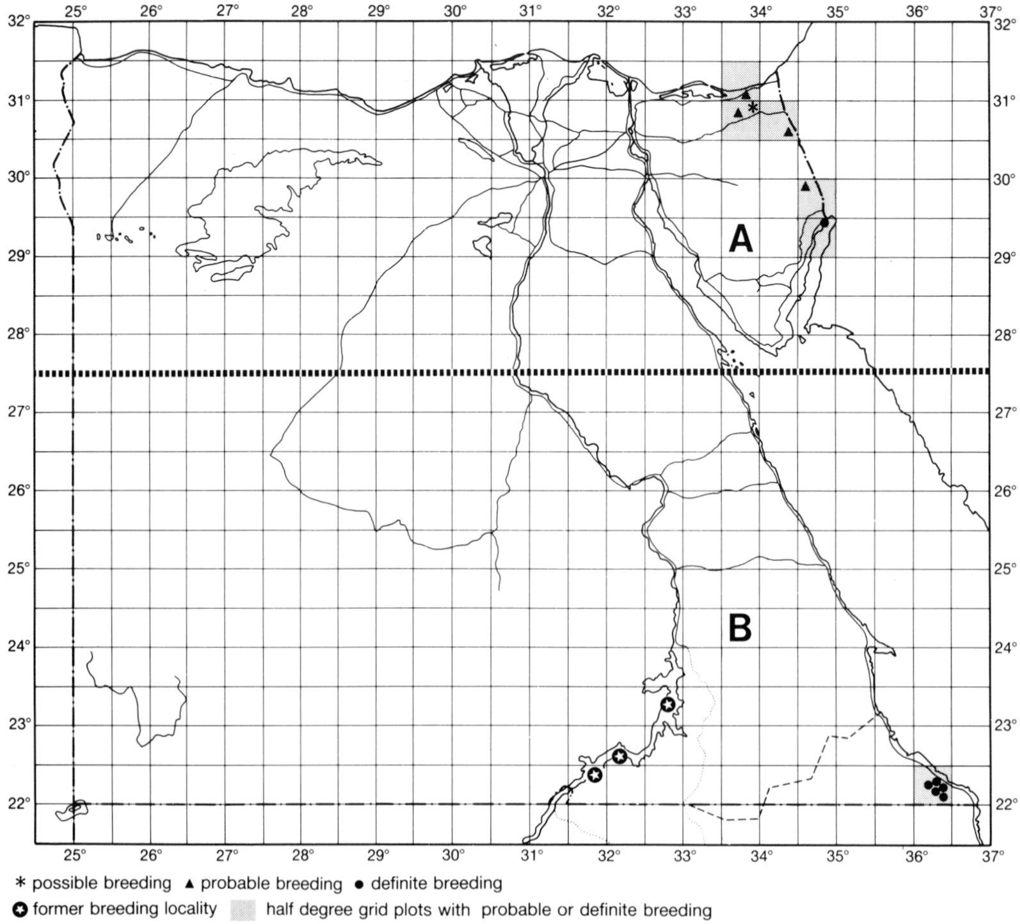

Map B88. Breeding distribution of Brown Babbler *Turdoides squamiceps* (A); and Fulvous Babbler *T. fulvus* (B) between 1970–87.

in this area on 25 June 1969 (ZM-HUJ). This species is known to nest in Wadi Taba (HSh). At Elat, Israel, it is a common breeding bird (HSh).

Fulvous Babbler *Turdoides fulvus* RB Plate 4

Local names Tharthaara ash-shajar ثرثارة الشجر, biishaawaata (Bi), shawataab (Bi).

Formerly a breeding resident of the Upper Nile Valley. It is locally common in the Gebel Elba area. All Egyptian material examined is referable to *T. f. acaciae*.

Nile Valley The Fulvous Babbler was formerly a breeding resident in the Nile Valley from Aswan south to the Sudanese border, an area now inundated by Lake Nasser. Koenig (1924) found it common in this region in the late nineteenth century.

Ripley (1963) noted flocks in early March 1962 composed of four to five birds and numerous pairs near Tushka. He also observed birds copulating and building nests. Specimens were taken in this region in 1963 and 1964 (YPM), several of which were in breeding condition. There are no documented Egyptian Nile Valley reports after the completion of the Aswan High Dam. This area has been extensively visited in recent years and the complete lack of substantiated records is indicative that the bird is locally extirpated. Recolonization of the region from populations in the northern Sudanese Nile Valley is plausible. One was reputedly observed on 6 August 1966 south of Lake Qarun (Kattinger 1970); however, details are lacking and this record cannot be accepted.

Sudan Government Administration Area The only area in Egypt where this species presently occurs is near Gebel Elba. Here it is common in wadis with acacias and breeds communally (Goodman and Atta 1987). This area was visited on four separate occasions between 1983 and 1985; groups of Fulvous Babblers varied in size from 2 to 10 (mean = 4.7, $n = 33$) and only once was a single bird noted. In March 1983, four birds were attending a nest close to a Bischarin camp. Every morning the birds would remove, within 1 m of humans, small pieces of uncooked bread dough from a bowl that was purposely left for them by the Bedouins (SMG).

Family Paridae

Great Tit *Parus major* RB

Local name Qarqaf kabiir قرقف كبير.

Locally common breeding resident of north-eastern Sinai, between El Arish and Rafa. No Egyptian material has been examined; *P. m. terraesanctae* is known from neighbouring Israel and is presumably the form occurring in Egypt.

Meinertzhagen (1930) did not mention the occurrence of the Great Tit in Egypt. It is not clear when this species first colonized north Sinai but it was already common near El Arish in the early 1970s (UP). A nest was found at this locality in April 1979 (OH). On 26 June 1984, two adults were noted feeding four fledglings about 5 km west of Rafa and on 2 May 1985, four adults and four fledglings were observed in the same area (SMB). Two adults, carrying food, were seen at El Arish on 26 April 1985 (DM).

Family Remizidae

Penduline Tit *Remiz pendulinus* WV

Local name Qarqaf al-banduul قرقف البندول.

Rare and local winter visitor to Egypt between late November and late March. It is plausible that this species will be found breeding in the country sometime in the near future. No material from Egypt has been examined. *R. p. pendulinus* and/or *R. p. menzbieri* are the most likely subspecies to occur in the country.

The first Penduline Tit record from Egyptian territory appears to be one on 21 March 1976 at Ain el Gedeirat, north-eastern Sinai (SC). A single bird was noted in a reed-bed west of Rumana in December 1978 (OH). Between 24 February and 31 March 1982 up to 20 birds were seen by various observers in the general Suez area (Wimpfheimer *et al.* 1983). Small numbers were also observed at Suez on 24 January 1984 and 22 March 1985 (SMB, DM, UGS), and on 22 and 26 November 1985 (RvD *et al.*). A single bird was heard on 7 March 1984 at Sharm el Sheikh (SG). The southern-most locality this species has been noted in Egypt is Bir Beida, just west of Quseir, where one was observed and several heard on 24 March 1985 (SMB, PLM). At Elat, Israel, it is considered a fairly common winter visitor from early November to early April (Krabbe 1980).

Family Nectariniidae

Generic local name for sunbird Tumayr.

Nile Valley Sunbird *Anthreptes platurus* RB Plate I

Local names Tumayr waadii an-niil تمير وادي النيل, abuu er-riish.

Fairly common breeding resident in the Upper Nile Valley. There is evidence that this species has recently started to breed near Cairo. In winter it is observed throughout the Nile Valley, the Faiyum, and occasionally in the Suez Canal area. Winter movements to the Nile Delta and Lower Valley are presumably influenced by the sequence and abundance of flowering plants. The twentieth-century northward range expansion of this species is probably related to the planting of flowering trees and shrubs, many of which are exotics, in parks and urban gardens (Greaves and Tregenza 1937). All Egyptian material examined is referable to *A. p. metallicus*.

Nineteenth-century records of Nile Valley Sunbirds in Egypt are generally restricted to the Upper Nile Valley south of Aswan (e.g. Shelley 1872). Meinertzhagen (1930) mentioned this species as a rare resident of Upper Egypt, where a few could be seen in the gardens of Luxor and Aswan. Greaves and Tregenza (1937) noted it breeding in 1915 as far north as Luxor, and by 1936–7 it was a fairly plentiful nesting resident at Qena.

Nowadays it is a common breeding bird of the Nile Valley between Aswan and Qena. This is the region where acacias, particularly *Acacia nilotica* a favourite food source and nesting tree, are abundant. Locally high densities occur. For example, on 16 April 1980, three active nests were found in one acacia tree near Aswan (JV) and on 1 April 1982 between 20 and 30 pairs were present on Kitchener's Island, near Aswan (JW). High densities of birds also occur around single flowering trees—in March 1979 over 30 females and immature males were observed feeding in a flowering *Bombax* tree outside Luxor (SMG). The species was common in the mid-1960s in the region of Egypt now inundated by Lake Nasser (YPM).

Meinertzhagen (1930) mentioned only one record near Cairo. Between 1928 and 1935 a few birds were observed in Cairo every spring; in 1935 it was noted continuously from 16 December to March; and in 1937 as early as 23 October (Greaves and Tregenza 1937). In the greater Cairo area it seems to be mainly a non-breeding visitor from early October to May in gardens and parks (e.g. Greaves 1945*b*). On 9 March 1980 birds were observed in the Giza Zoological Gardens displaying and carrying nesting material (SMB) and in late June 1984 a male was noted in breeding plumage (SMG). The only evidence of this species breeding in the Cairo area is fledglings being fed by adults in late September 1986 in both Giza and Zamalek (SMG).

The majority of records from the Nile Valley between Cairo and Qena are from the winter months. Observations during the known breeding period include: a pair on 8 April 1982 at Mallawi (MU), a male on 26 June 1981 north of El Minya (SMB), and a male on 12 March 1983 between Beni Suef and El Minya (Welch 1983). There are several recent winter records from the Faiyum (e.g. Meininger and Mullié 1981*c*; Short and Horne 1981). This species is a rare winter visitor to the northern Nile Delta and Suez Canal area. One was observed at Alexandria on 14 November 1980 (SV). Simmons and Hurrell (1951) recorded it from 1 November 1949 to 7 January 1950 in the Suez Canal area. A single bird was noted on 23 March 1953 at Fayid (AP).

Shining Sunbird *Nectarinia habessinica* RB Plate 4

Local names Tumayr ḫabashii تمير حبشي, bilay'ir (Bi), haawtiiruub (Bi).

Locally common breeding resident in the Gebel Elba region. The subspecies occurring in this area is *N. h. habessinica* (Williams 1955; Goodman 1984*a*), rather than *N. h. hellmayri* (*contra* Meinertzhagen 1930). Nests are generally placed in acacia trees.

Cholmley (1897) found Shining Sunbirds 'common everywhere' at Gebel Shellal in January 1896. El Negumi (1949) found a nest sometime between December 1938 and January 1939 in Wadi Kansisrob, Gebel Elba, with a pair of six-day-old nestlings. There appear to be local movements in the Gebel Elba area. In February and March 1985 this species was common in the inner portions of Wadi Kansisrob near stands of flowering Malvaceae and a legume tree *Delonix elata*. By April 1985 these plants were no longer in blossom and the majority of Shining Sunbirds had left the immediate area (SMG).

Orange-tufted Sunbird *Nectarinia osea* RB Plate 5

Local name Tumayr 'arabii تمير عربي.

Local breeding resident in the gardens of the El Arish and Rafa area and vegetated wadis in southern Sinai. All Egyptian material examined is referable to *N. o. osea*.

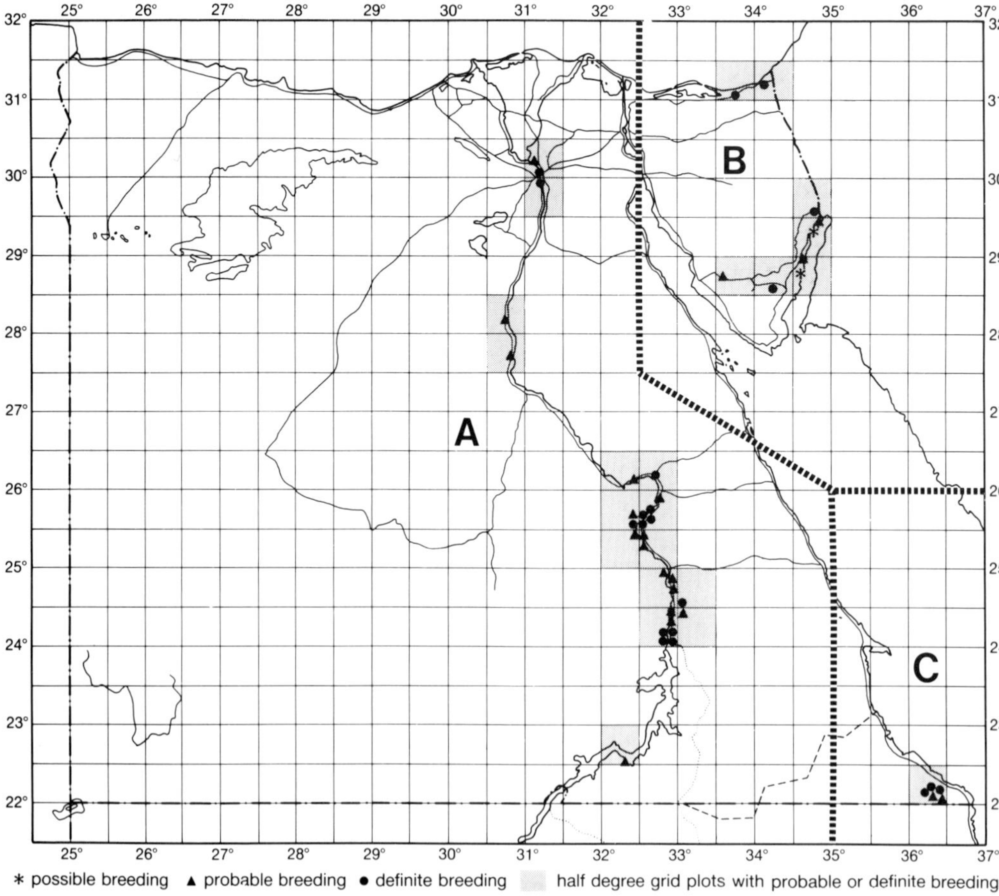

Map B89. Breeding distribution of Nile Valley Sunbird *Anthreptes platurus* (A); Orange-tufted Sunbird *Nectarinia osea* (B); and Shining Sunbird *N. habessinica* (C), between 1970–87.

The first record of the Orange-tufted Sunbird in Egypt appears to be at El Arish on 11 October 1979 (Petersen and Sørensen 1980). More recent information indicates that it is present throughout the year in the area from El Arish to Rafa (Baha el Din 1985), where it breeds in gardens and orchards.

On 18 August 1984 territorial males in breeding plumage, pairs, and juveniles were observed in Wadi Nasb (SMB), and there are several subsequent records from the area. One was observed on 26 March 1985 in the Wadi Feiran Monastery garden (UGS), on 23 August 1986 in the wadi proper (EE *et al.*), and a pair on 17 August 1985 between Nuweiba and Dahab (SMB). Orange-tufted Sunbirds often feed on the flowers of *Loranthus acaciae* (UP), a semi-parasitic plant of *Acacia* and *Zizyphus* trees (Täckholm 1974), and this species' colonization of Sinai should be partially linked to the distribution of this plant. The species is known to breed in Wadi Taba

(HSh; PLM) and is rather common across the border at Elat, Israel, and outlying wadis (HSh).

Family Oriolidae

Golden Oriole *Oriolus oriolus* PV

Local names S̱uffayr صفير, abuu s̱uffayr (NC), s̱ufaariya, 'us̱fuur at-tuut.

Fairly common passage visitor in autumn from (mid-) late August to mid- (late) October and in spring from (late March) mid-April to mid- (late) May. It occurs throughout the country. All Egyptian specimens examined are of the nominate subspecies.

Large concentrations of Golden Orioles are sometimes noted; for example, on 27 April 1982, thousands were seen at El Hauwariya, south-west of Alexandria (BO).

Ringing There are single autumn recoveries along the North Coast and in Bahariya Oasis of birds ringed in the German Democratic Republic and Italy, respectively (Zink 1975; Feige 1986; MSp). Four birds ringed in Egypt have been recovered in Lebanon, Malta, and Greece (Watson 1971; GZF).

Family Laniidae

Generic local names for shrike Daqnaash, abuu al-'ala (Ma), nuhas, qanuush, s̱urad.

Rosy-patched Shrike *Rhodophoneus cruentus* RB Plate 4

Local name Daqnaash wardii as̱-s̱adr دقناش وردي الصدر.

The Rosy-patched Shrike is a locally common breeding resident in wadis surrounding Gebel Elba, particularly those with extensive acacia groves. In the spring and early summer months pairs are often heard duetting. An adult was seen carrying a faecal sac on 28 March 1983 and fledglings were noted on 21 June 1984 (SMG). Specimens taken locally are referable to *R. c. cruentus*.

Isabelline Shrike *Lanius isabellinus* (PV)

Local name Daqnaash ash/hab دقناش اشهب.

Rare and irregular migrant. There are scattered autumn and spring records. It may be a more regular migrant through the area than currently known. Two birds taken in 1928 (see below) are referable to *L. i. isabellinus*. *L. i. phoenicuroides* has been collected at Elat, Israel, on 16 October 1968 (USNM) and it may also pass through Egypt.

Specimens of Isabelline Shrikes taken in Egypt include: a male on 21 February 1928 at Quseir and a male on 28 March 1928 at Gebel Elba (Meinertzhagen 1930; BMNH), and two birds taken from a group of five on 26 October 1957 at Kitchener's Island, Aswan (Horváth 1959). Two were reputed to have been collected in the Western Desert between Dakhla and Gebel Uweinat in early October 1932 (Moreau 1934); however, the specimens cannot be traced or the locality verified. Observations from Egypt include: one male at Suez on 27 September 1981 (JV), two on Gezira el Hamra on 2 October 1984 (SMB), and one at Gebel Elba on 8 March 1984 (PDH, EMN). Between early April and early May 1979, three or four were observed at Elat, Israel (Krabbe 1980).

Red-backed Shrike *Lanius collurio* PV (WV)

Local name Daqnaash ak<u>h</u>al دقناش اكحل.

Common migrant in autumn from (early) mid-August to early (late) November and rare in spring from mid-February to late May. Irregular winter visitor. All Egyptian material examined is referable to *L. c. collurio*.

In autumn the Red-backed Shrike passes through Egypt along a broad front but is apparently more common along the North Coast, in the Western Desert, and in the western portion of the Nile Delta (Maclaren 1944; Molteno 1944; Bodenham 1945*a*; D. Goodwin 1947). The earliest known autumn records are 6 August 1942 at Abu Qir (Bodenham 1945*a*) and 8 August 1952 at Fayid (AP). This species was noted in Bahariya between 9 August and 29 October 1983 with the bulk of passage from late August to mid-September and a maximum of 170–200 birds on 3 September (HB *et al.*). There are several sight records and specimens of immature birds in the country up until late November.

An adult male was obtained at Rafa on 18 December 1979 (TAUM) and one was observed at Sirs el Liyyan on 11 January 1953 (HCK). Misonne (1974) observed several Red-backed Shrikes at Gebel Uweinat on 23 December 1968 and one male on 25 December 1968. This species presumably winters in the general Gebel Uweinat area and perhaps irregularly in other portions of the country.

Spring migration appears to bypass much of Egypt and the majority of records are restricted to the Eastern Desert, the Upper Nile Valley, and Sinai. Short and Horne (1981) noted several birds along various portions of the Valley between 11 February and 5 March, which were presumably early spring migrants. There are a few spring records from the Delta and Lower Valley. These include: single birds at El Fashn on 24 February 1979 (Short and Horne 1981) and at Bahig on 10 May 1969 (USNM), and one ringed at Suez on 12 May 1982 (SMB). A pair was found north of Hurghada on 20 May 1984 (UMMZ). There are numerous records from Sinai between early April and early May. At Elat, Israel, the species is abundant from (mid-) late August to mid-October (mid-November) and is common from (mid-) late April to mid-June (Krabbe 1980).

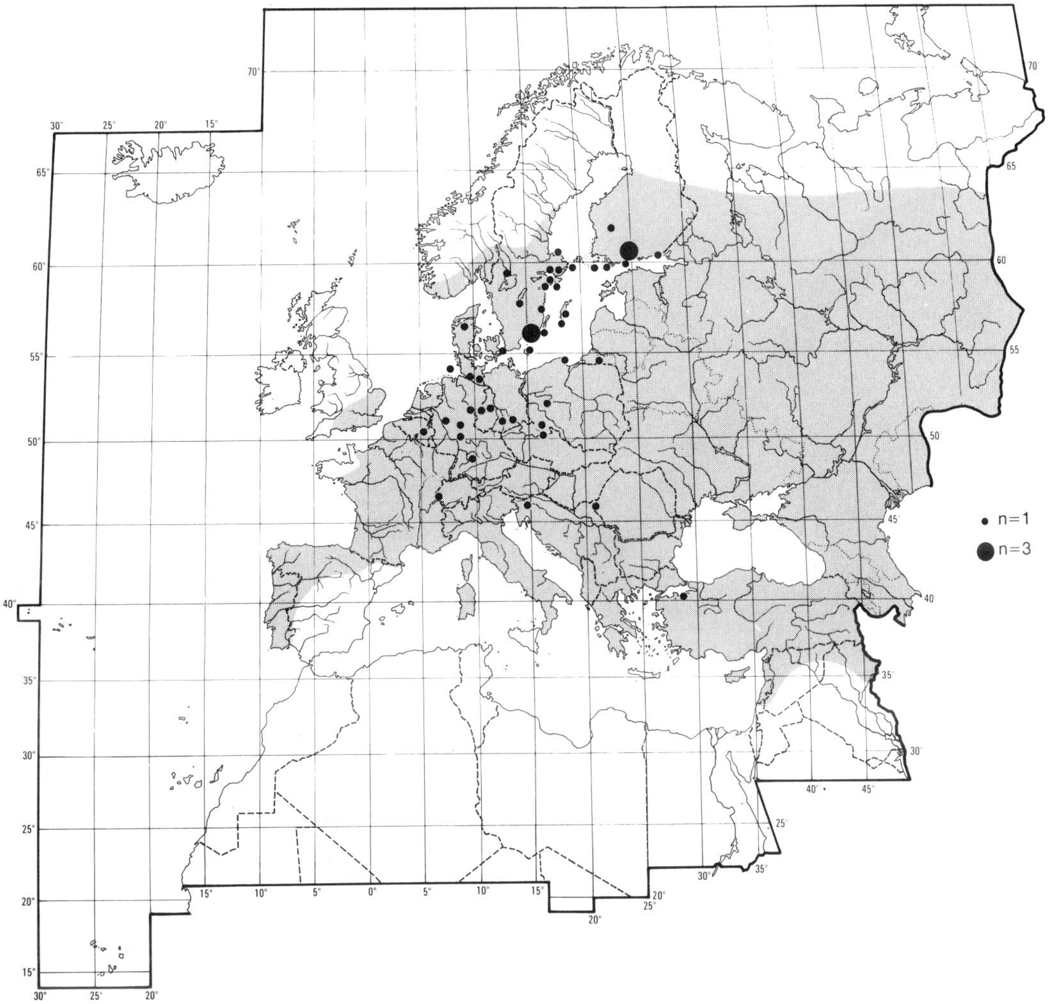

FIG. 6.32. Ringing sites of Red-backed Shrikes (n=49) recovered in Egypt. Shaded area is breeding range based on Harrison (1982).

Ringing There are 59 records of birds ringed in the Baltic area of the USSR (1), Belgium (1), Denmark (3), the Federal Republic of Germany (11), Finland (9), the German Democratic Republic (5), Hungary (1), Poland (5), 'Rossitten' (1), Sweden (19), Switzerland (1), Turkey (1), and Yugoslavia (1) recovered in autumn in Egypt (Fig. 6.32; Mackintosh 1941; Danielsson 1959; Pátkai 1966–7; Österlöf 1968, 1969; Watson 1971; Anon. 1972; Zink 1975; Sere 1986; EDB; GZF; JG; RSt; WT). A bird ringed at Bahig on 7 October 1969 was recovered at El Rekabiya in September 1970 (GZF).

Lesser Grey Shrike *Lanius minor* PV

Local names Daqnaash s̱urdii دقناش صردي, abuu allah.

Fairly common migrant in autumn from early August to early October (late November) and uncommon in spring from late February to late May. All Egyptian specimens examined are of *L. m. minor*.

In autumn the Lesser Grey Shrike is generally noted in small numbers throughout the country. Records of relatively large concentrations include 20 near Sharm el Sheikh on 21 August 1986 and 20 near Baltim on 25 August 1986 (EE *et al.*). The latest autumn records are two on 26 and one on 27 November 1981 in the Faiyum (RGB *et al.*).

The spring migration pattern of this species in Egypt resembles that of the Red-backed Shrike. There are only a few recent records from the Nile Delta and Valley. These include birds on 28 February 1979 at El Balyana (Short and Horne 1981), between 17 and 22 March 1949 at Fayid (HGB), on 4 April 1964 at Alexandria (JWW), and on 9 April 1982 in the Faiyum (PW). The species is regularly noted in small numbers along the Red Sea coast and Suez Canal from early April to mid-May (Marchant 1939–40; Bulman 1944). Several were noted in the Hurghada area in late May 1984 (SMG, RWS) and one at El Kuntilla on 14 May 1939 (GZM).

Great Grey Shrike *Lanius excubitor* RB WV Plate I

Local names Daqnaash al-baadiyya دقناش البادية, abuu allah, abuu s̱uruunid, adhaydha (NC), daqnaash salata, shakalit (Bi).

Fairly common breeding bird in various portions of the country. Local populations are augmented during the winter by northern migrants. *L. e. elegans* breeds locally along the North Coast, in the Nile Delta and Valley, the Suez Canal area, the northern Eastern Desert, and northern Sinai. *L. e. aucheri* breeds in the rest of the Sinai, and along the Red Sea coast south to Gebel Elba. In parts of the Sinai and central Eastern Desert *aucheri* grades into *elegans*. The seasonal status and distribution of the Egyptian breeding populations is not clear. According to Meinertzhagen (1930) both subspecies are thought to be sedentary but there is some evidence that local winter movements occur. A specimen collected at Quseir on 20 February 1928 is referable to *L. e. pallidirostris* (Meinertzhagen 1930; BMNH). Breeding has been noted from mid-January to mid-June.

North Coast Meinertzhagen (1930) stated that 'a few [Great Grey Shrikes] occur on the edge of the desert and along the coast from Sallum to Alexandria', but in his 1921 paper the species was noted as common at Matruh and absent from Salum. Single birds were observed on 3 October 1980, 103 km and 135 km west of El Alamein (HB *et al.*). Except for the immediate Bahig region, there is no recent evidence of breeding from the North Coast.

Western Desert The only record of this species in Wadi el Natrun is two observed on 17 January 1979 (Meininger *et al.* 1979). Moreau (1927) noted one at Dakhla between 10 and 13 April 1925 and Meinertzhagen (1930) considered it rare in Kharga. There is no evidence of breeding or recent records from these two oases (Goodman *et al.* 1986). One was observed at El Maghra on 2 March 1984 (Baha el Din in press). It was found breeding near Gebel Uweinat in mid-April 1961 (Jany 1963).

Nile Delta In the Nile Delta and the Suez Canal area this species is a locally common breeding bird of desert edge and thorn scrub. Flower (1933) mentioned its regular occurrence in several cultivated portions of the Delta but nowadays it appears to be rare in the greater portion of the area. Raw (1921) noted it 'was not uncommon at Abu Zabal during most of the year but disappeared during the breeding-season'. Nestlings have been found at Abu Rauwash on 1 May 1959 (FMNH) and 1 April 1970 (USNM).

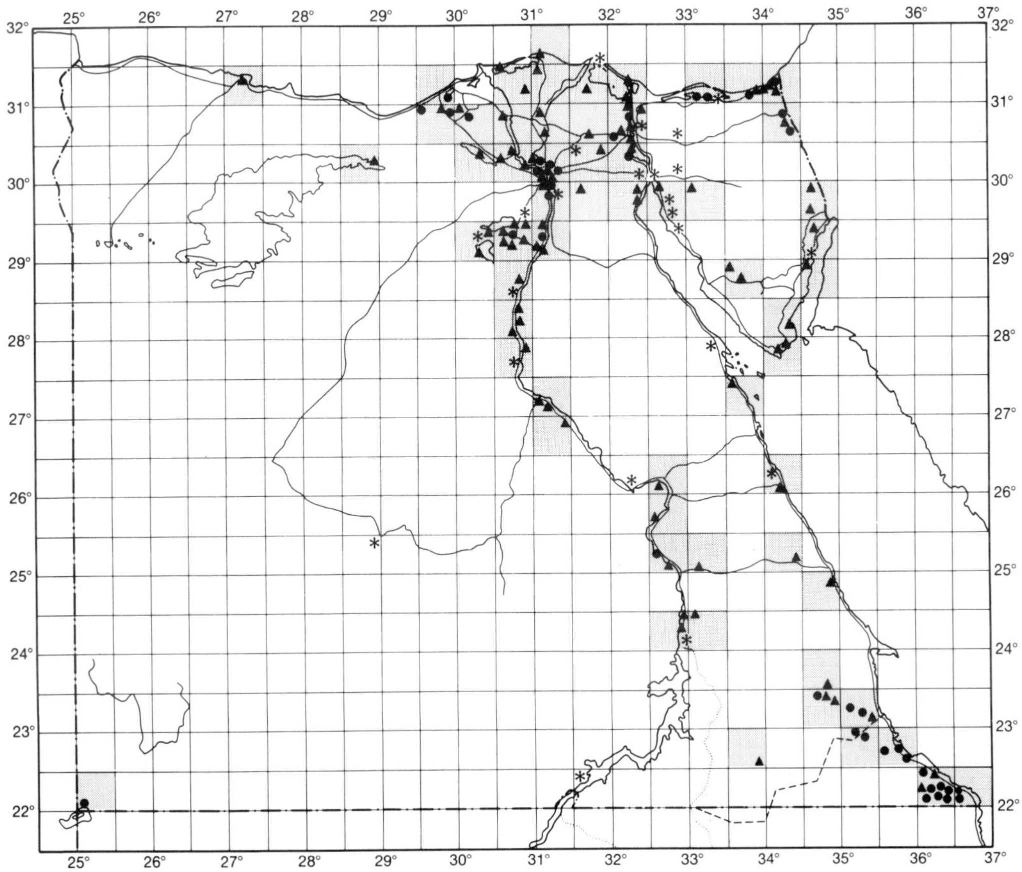

∗ possible breeding ▲ probable breeding ● definite breeding ░ half degree grid plots with probable or definite breeding

Map B90. Breeding distribution of Great Grey Strike *Lanius excubitor* between 1930–87.

Nile Valley This species is a fairly common breeding bird in suitable habitat near Cairo and in the Faiyum and is rather sparsely distributed in the remaining areas of the Valley, including Lake Nasser. Two were seen in Wadi el Rayan on 13 July 1984 (SMB).

Eastern Desert This species is a rather local breeding bird in the vegetated wadis of this region, mainly along the Red Sea coastal plain. Horváth (1959) collected two specimens at Bir Abbad on 29 October 1957 that he identified as *L. e. aucheri*. Other recent observations include birds on 8 November 1982 in Wadi Asal, on 29 November 1982 on Gezira Qeisum (Baha el Din and Saleh 1983), on 1 February 1986 in Wadi Umm Ghalaqa, north of Gebel Abu Brush (SMB), and on 13 March 1984 north of Quseir (PDH, EMN). The species is relatively common near Gebel Abraq (SMB, SMG, PLM).

The Great Grey Shrike is a common breeding bird throughout much of the Sudan Government Administration Area, particularly in vegetated wadis. In 1984 fledglings were noted between 16 February and 8 March (SMG; PDH, EMN). In late June 1984 it was particularly common in Wadi Serimtai, on the eastern side of Gebel Elba (SMG).

Sinai This species is a locally common breeding bird in the north between El Arish and Rafa, and in vegetated wadis in the east and south. Records from the Gulf of Suez coastal plain may be in part of migrants.

Woodchat Shrike *Lanius senator* PV

Local names Daqnaash uuruubii دقناش أوروبي, daqnaash niilii, daqnaash shaamii.

Rare migrant in autumn from (late July) early August to mid-October and common in spring from (early) mid-March to early (late) May (Fig. 6.33). Two subspecies pass through the country, *L. s. senator* and *L. s. niloticus*.

In spring the Woodchat Shrike is observed throughout the country, including the desert regions and Sinai. Records on exceptional dates include single birds on 31 July 1938 at 'Qaar el Laban', Western Desert (GZM); on 1 and 2 August 1983 at Bahariya (HB *et al.*); and on 30 May 1956 at Abu Rauwash (FMNH).

Masked Shrike *Lanius nubicus* PV

Local name Daqnaash qubṯii دقناش قبطي.

Scarce migrant in autumn from (late July) late August to mid- (late) October and common in spring from (late February) mid-March to late April (late June) (Fig. 6.34). Monotypic.

The earliest autumn record of the Masked Shrike appears to be a male on 29 July 1952 between Fayid and Abu Sultan (AP). Autumn passage is concentrated in the Nile Valley and eastern Delta east to Sinai. During a study season in Bahariya from 3

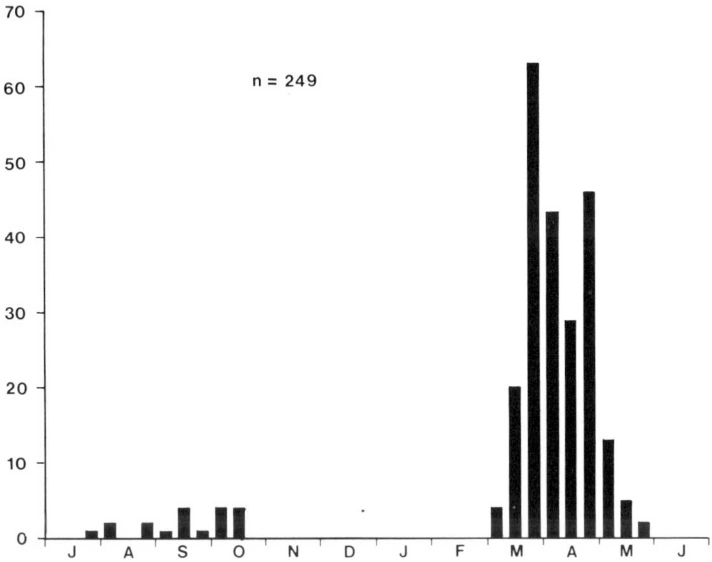

FIG. 6.33. Histogram of the numbers of Woodchat Shrikes recorded during 10-day periods throughout the year.

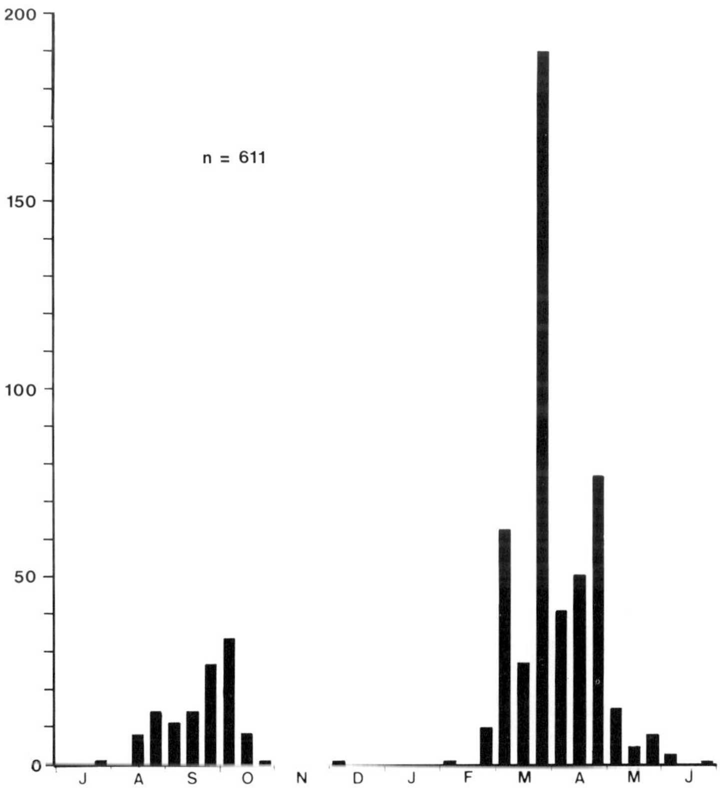

FIG. 6.34. Histogram of the numbers of Masked Shrikes recorded during 10-day periods throughout the year.

August to 29 October 1983, 16 birds were captured or observed between 18 August and 4 October (HB *et al.*). Immatures have been recorded at Kharga on 18 October 1980 (Goodman *et al.* 1986) and at Bahig on 30 August and 16 September 1969 (USNM). Meinertzhagen (1930) stated that a few winter in southern Egypt. The only more recent records falling between November and January are single birds at Ismailiya on 3 December 1982 (PB) and at Fayid and El Kabrit on 18 January 1953 (AP). There are several February records (e.g. Short and Horne 1981), which concern winter visitors and/or early spring migrants.

In spring this species is observed throughout the country but is distinctly more common in the southern Nile Valley, along the Red Sea coast, and in Sinai. There are only two spring records from the Western Desert (Goodman *et al.* 1986) and it is rare at Bahig (Horner 1977). Small numbers are observed until early June and exceptionally to 25 June (Meinertzhagen 1930). One adult was observed at El Arish on 24 June 1984 (SMB).

Family Corvidae

Generic local names for *Corvus* spp. Ghuraab, kiikaay (Bi).

House Crow *Corvus splendens* IB

Local name Ghuraab al-manzil al-hindii غراب المنزل الهندي.

Scarce introduced breeding bird in several towns and villages along the coast of the Gulf of Suez, Red Sea, and Suez Canal. No Egyptian material has been examined but on the basis of notes taken on the plumage coloration of living birds they appear to be referable to *C. s. splendens* (Bijlsma and Meininger 1984).

It is plausible that House Crows established themselves at Suez in or before 1922 where they were originally mistaken for Jackdaws (e.g. Greaves 1939). Since then the species has increased in numbers and can commonly be seen in Port Taufiq and other parts of Suez. In 1981 this local population was estimated to be 800–850 birds (Bijlsma and Meininger 1984). Here they breed in loose colonies in trees of parks and gardens. In April 1967 the species was discovered in Ismailiya; more recent records from this locality include two on 27 June 1977, at least three on 5 August 1981 (Meininger and Dielissen 1979; Bijlsma and Meininger 1984), one on 1 May 1985 (Hovel 1987), and more than 25 (including used nests) on 8 November 1985 (SMB). At Port Said two birds were observed on 6 April 1982 (De Bièvre 1983) and 11 on 8 November 1985 (DM).

House Crows have also been observed at other localities along the Gulf of Suez and northern Egyptian Red Sea coast: Ain Sukhna—two on 1 October 1981 (Bijlsma and Meininger 1984); Ras Gharib—a pair with four recently fledged young on 13 August 1979 (Meininger *et al.* 1980) and regularly observed since then in groups of up to seven; Safaga—up to three regularly seen since late June 1980 (Meininger *et al.* 1980) and a nest in March 1985 (SMG, PLM); and Quseir—two observed on 27 March 1986 (JW).

The only definite record from Sinai is two birds at Nabq on 8 March 1984 (SG). The species is known to have bred in recent years at Elat, Israel, and Aqaba, Jordan (Meininger *et al.* 1980; Paz 1987). In the TAUM collection there is a specimen taken 'from Sinai' in February 1975 but all collecting details are lacking. This may be the bird mentioned by Hovel (1987) which was apparently from El Tor.

The House Crow has successfully established itself as a breeding bird in numerous localities outside its natural range. Transportation by ship may well account for the original introduction of this species into Egypt. That all known localities where it occurs in Egypt are situated on major sea shipping routes supports this idea (Meininger *et al.* 1980; Bijlsma and Meininger 1984).

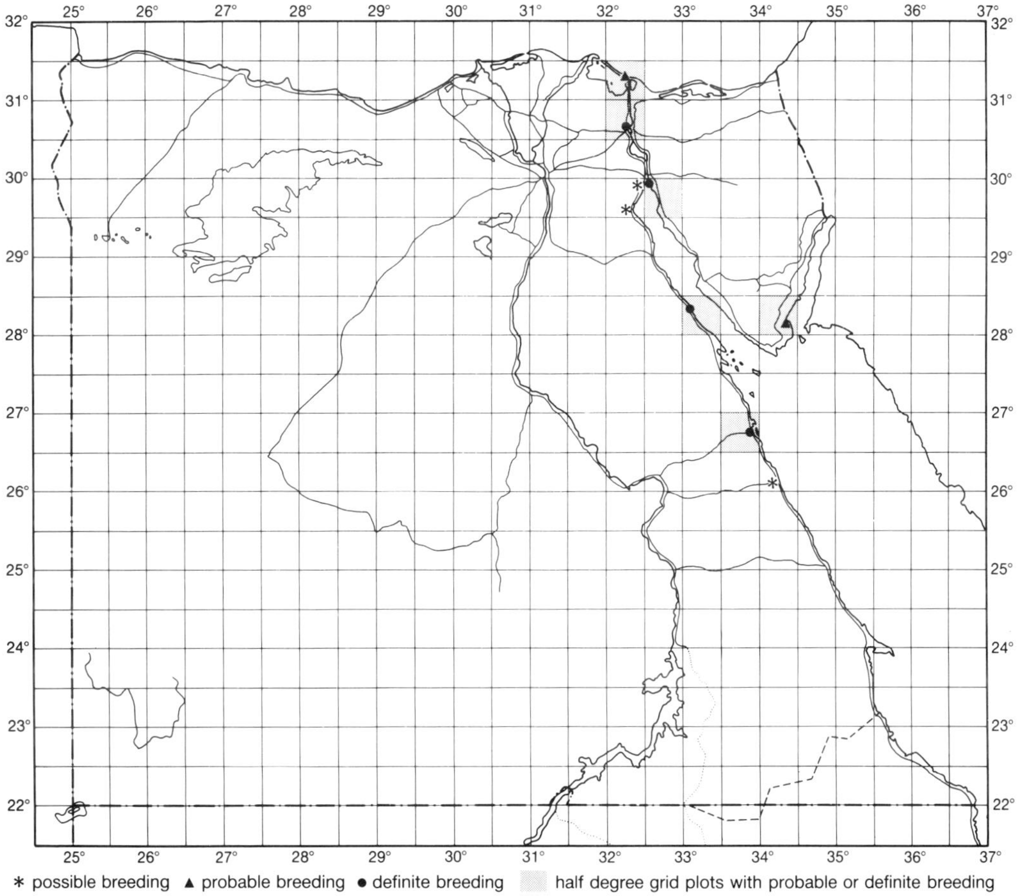

Map B91. Breeding distribution of House Crow *Corvus splendens* between 1970–87.

Rook *Corvus frugilegus* (WV)

Local names Ghadaaf غداف, ghuraab al-ghayṯ.

Rare and irregular winter visitor from late November to mid-March. Generally noted in small flocks or singly in cultivated areas. All Egyptian specimens examined are referable to *C. f. frugilegus*.

A flock of about 20 Rooks was observed at Giza on 29 January 1864 (Taylor 1867). Groups were noted near El Marg on 21 February 1891 (Koenig 1920) and one was collected at El Qantara on 11 December 1915 (GZM). There are several archaic records from near Suez (Von Heuglin 1871; Meinertzhagen 1930). On 8 May 1916 'a flock of Rooks' was recorded in the Egyptian Red Sea crossing over from the African to Arabian shore (Whistler 1916). More recent observations include two near Luxor on 9 January 1975 (HSw), nine at Ismailiya in February 1982 (MAS), and eight flying north near Gezira Fara'un on 12 March 1979 (SC).

Hooded Crow *Corvus corone* RB

Local names Ghuraab baladii غراب بلدي, ghaaq, qaaq, zaagh.

Common resident in most cultivated parts of the Nile Delta and Valley (south to Aswan), along the Suez Canal, in the Faiyum, and north-eastern Sinai. The form breeding in these areas is *C. c. sardonius*. *C. c. cornix*, a European form, has been identified from specimens taken along the Suez Canal in January 1924 and one from Wadi el Natrun in April 1923 (Meinertzhagen 1930). We have examined this material (GZM) and are of the opinion that on the basis of plumage coloration and wing length they are referable to local *C. c. sardonius*. The Hooded Crow is rarely sympatric with the Brown-necked Raven except along the immediate ecotone between the desert and the Nile basin. Nesting is between mid-January and late June.

Western Desert In Wadi el Natrun a small flock of Hooded Crows was observed in March 1911 and many were noted in the autumn of that same year (Nicoll 1912*b*). One bird was taken there in April 1923 (Meinertzhagen 1930; see above).

Nile Delta and Valley In the first decades of the twentieth century this species was unknown from the northern portion of the Nile Delta. In January 1908 it was noted as absent in the area between Tanta and Damietta (Nicoll 1908). Borman (1928) stated that it was plentiful between Cairo and Damanhur, decreased in numbers from Damanhur northwards, and was rarely seen near Alexandria. According to Flower (1933) it extended its range between 1914 and 1924 to the west as far as Kafr el Dauwar, between Damanhur and Alexandria, and to the east as far as Ismailiya. He stressed its absence in the Port Said and Suez areas. Apparently this species continued to expand its range, for in 1944 it was plentiful north of Deversoir, although still rare at Suez (LB), and in September 1957, 35 were noted at Alexandria (Horváth 1959). The species bred in 1953 at Fayid (AP). Nowadays Hooded Crows occur in all portions of the Nile Delta and Suez Canal area, including Port Said and Suez. However, in recent years the number in the Delta has dramatically decreased, possibly due to the extensive use of pesticides (Mullié and Meininger 1985). A juvenile was found at Bahig on 25 April 1967 (USNM) and presumably this species breeds just west of Alexandria. Hooded Crows are still common in gardens and parks of Cairo and the Faiyum. In the Nile Valley the species is widely distributed but is more abundant south of Asyut than to the north. Horváth (1959) observed seven birds at Bir Abbad, 15 km west of the Nile Valley at Idfu, in late October 1957. The species appears to be absent from the Lake Nasser area.

Sinai This species is an uncommon resident in north-east Sinai between El Arish and Rafa. Several were observed near El Mazar on 18 December 1986 (SMG).

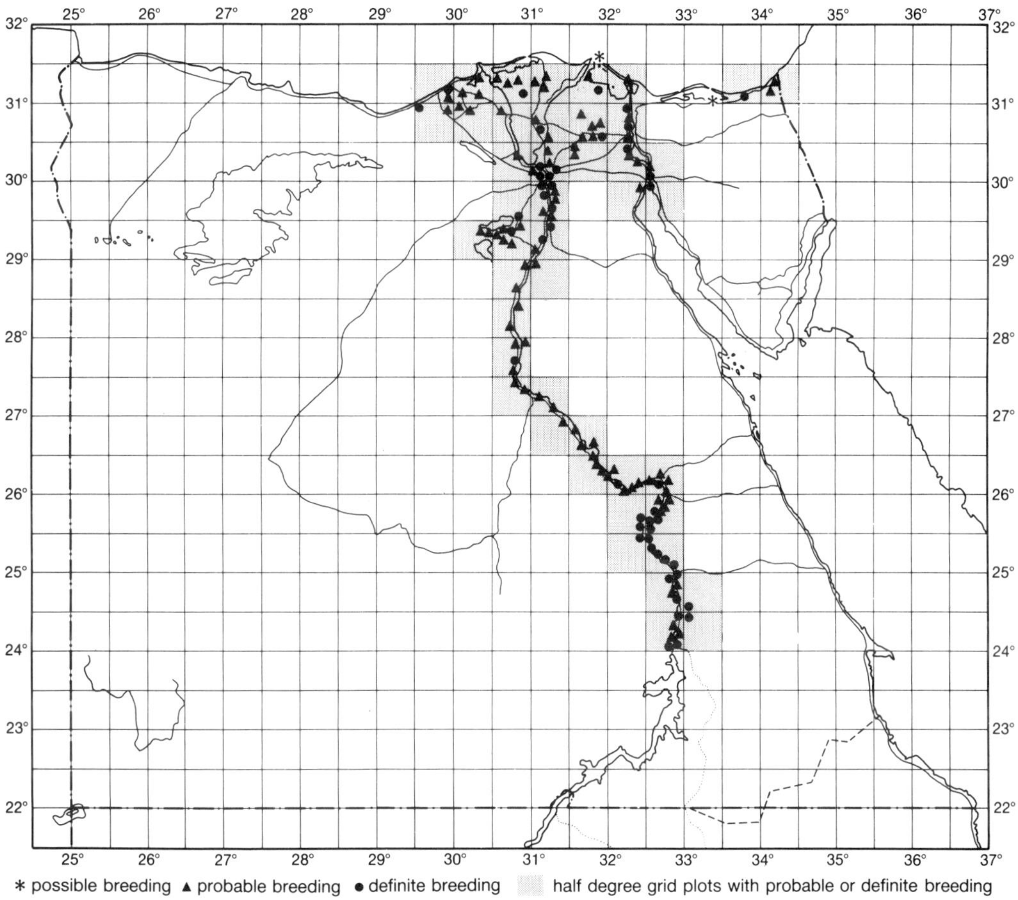

Map B92. Breeding distribution of Hooded Crow *Corvus corone* between 1970–87.

Brown-necked Raven *Corvus ruficollis* RB

Local names Ghuraab nuuhii غراب نوحي, ghuraab (Ma).

Common breeding resident in most desert areas of the country. All Egyptian material is referable to *C. r. ruficollis*. Nestlings have been noted from early March to mid-May.

North Coast Brown-necked Ravens seem to be completely absent from the coastal belt between Salum and Alexandria. In this region the species appears to be replaced by the Raven.

Western Desert The Brown-necked Raven is a fairly common breeding bird of the Qattara Depression, Siwa, Bahariya, Farafra, Dakhla, Kharga, and numerous other scattered localities. There is no evidence of it breeding in Wadi el Natrun as suggested

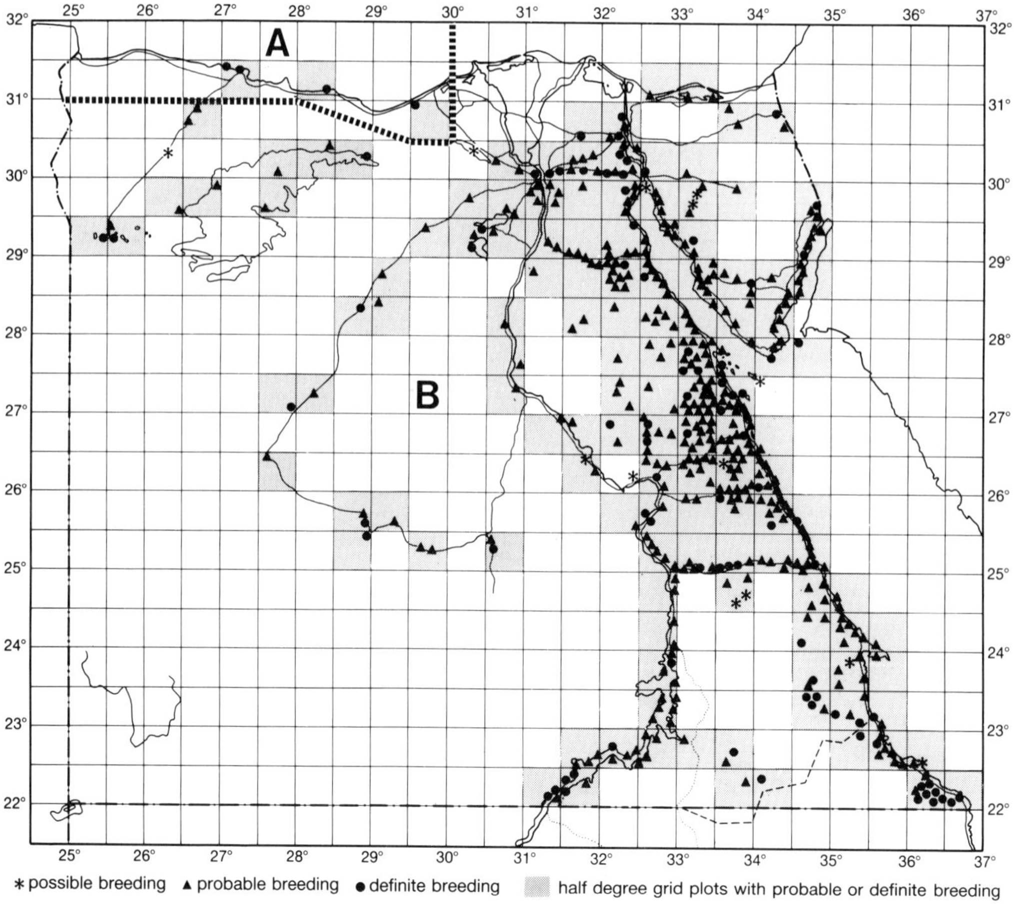

Map B93. Breeding distribution of: Raven *Corvus corax* (A); and Brown-necked Raven *C. ruficollis* (B) between 1930–87.

by Meinertzhagen (1930). It is unknown from the south-west of the Western Desert, including the Great Sand Sea, Gilf Kebir, and Gebel Uweinat (Goodman *et al.* 1986).

Nile Delta and Valley Since this species inhabits desert areas it is virtually unknown from cultivated portions of Egypt. Brown-necked Ravens have been noted along the ecotone between desert and cultivation in the Nile Delta and Valley south to Aswan, the Faiyum, and the Suez Canal area. The species appears to be more common along the Suez Canal south of Ismailiya, where large concentrations (up to 150) are occasionally noted. In the Nile Valley it is regularly seen south of Luxor, particularly where the desert plateau abruptly abuts the flood-plain and to some extent in cultivated areas. This species is abundant along the shores of Lake Nasser.

Eastern Desert and Red Sea Brown-necked Ravens are common throughout the Eastern Desert. Breeding sites have been noted from sea-level up to 2000 m. A pair was observed on 14 October 1983 at the summit (2187 m) of Gebel Shayib el Banat (JJH). Concentrations, mainly of non-breeding birds, occur at refuse dumps, near towns, villages, Bedouin and military camps, and along roads. Near Gebel Elba they are partially sympatric with Fan-tailed Ravens. In this area Brown-necked Ravens tend to inhabit the outlying sandy wadis and coastal plain, while Fan-tailed Ravens occur at higher altitudes.

The species is regularly observed feeding along the shores of the Red Sea and is known to have bred on several islands (e.g. Borman 1929). On 18 May 1984, two birds were observed on Gezira Umm Agawish el Kabir systematically plundering Western Reef Heron nests (Goodman and Storer 1987).

Sinai This species is common in most of the Peninsula except for the sandy desert along the north coast. Concentrations of up to 100 birds often occur near human habitation.

Raven *Corvus corax* RB

Local names Ghuraab as<u>h</u>am غراب أسحم, ghuraab nuu<u>h</u>ii.

Rare breeding resident along the North Coast. The subspecies occurring in Egypt is *C. c. tingitanus*.

Meinertzhagen (1930) found a Raven nest containing three young in May 1920 near Matruh and another with five young in March near El Dab'a. Nestlings have also been noted in the Matruh area on 11 May 1927 (GZM). More recent records include an inactive nest 25 km west of Matruh in September 1979 (SMB) and a pair breeding in the ruins of Abu Sir, near Burg el Arab, in April 1981 (PLM). This species has recently been reported from the Umm Agawish Tourist Village, south of Hurghada (Ferguson 1984); however, Brown-necked Ravens abound in the area and this record cannot be accepted. An observation of two Brown-necked Ravens near Burg el Arab on 24 September 1942 may well have been of Ravens (Maclaren 1944). The basis for the statement by Hovel (1987) that it is 'probably [a] scarce resident, N. Sinai' is unknown.

Fan-tailed Raven *Corvus rhipidurus* RB

Local name Ghuraab marwahii adh-dhanab غراب مروحي الذنب.

Local breeding resident in the mountains of extreme south-eastern Egypt and south-central Sinai. It is occasionally observed outside these areas. Monotypic.

Nile Valley A Fan-tailed Raven was observed by Von Heuglin (1871) near Thebes and in January or February 1904 by Russell (1905) near Kurusku (in the area presently inundated by Lake Nasser). A specimen in the BMNH was taken in 'Upper Egypt' in February 1870. Recent records include birds seen on 24 December 1980 on the desert edge between Kom Ombo and Idfu (RW) and on 25 May 1981 near Luxor (SMG). It is not clear from which breeding population(s) these birds disperse.

Sudan Government Administration Area Fan-tailed Ravens are common breeding residents of Gebel Elba, with a local population of several hundred. The species

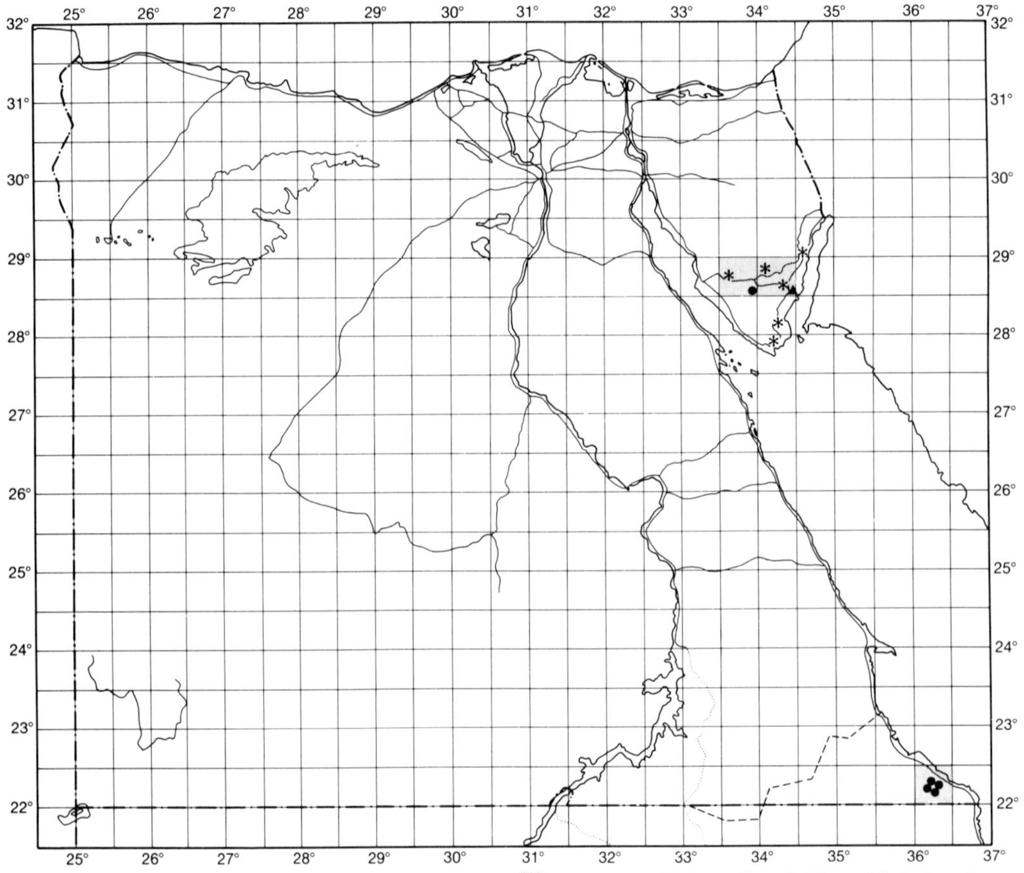

Map B94. Breeding distribution of Fan-tailed Raven *Corvus rhipidurus* between 1970–87.

breeds on cliff faces, often at high altitudes, and frequently descends into wadi bottoms to feed, for instance, on refuse from Bedouin encampments (SMG, PLM). Two nests, one with three eggs, were discovered at approximately 1000 m on 27 March 1983 (SMG).

Sinai Small numbers breed on the cliffs of Gebel Katherina and Gebel Musa. Two were observed near Dahab on 3 April 1981 (OT) and 28 March 1985 (UGS). Small numbers are irregularly seen near Nuweiba and Sharm el Sheikh (UP). A few pairs breed near Elat, Israel, and this population is occasionally augmented during the winter months by migrants (Krabbe 1980).

Family Sturnidae

Tristram's Grackle *Onychognathus tristramii* RB Plate 5

Local names Suwaadiyya سوادية, shah̲ruura (SC).

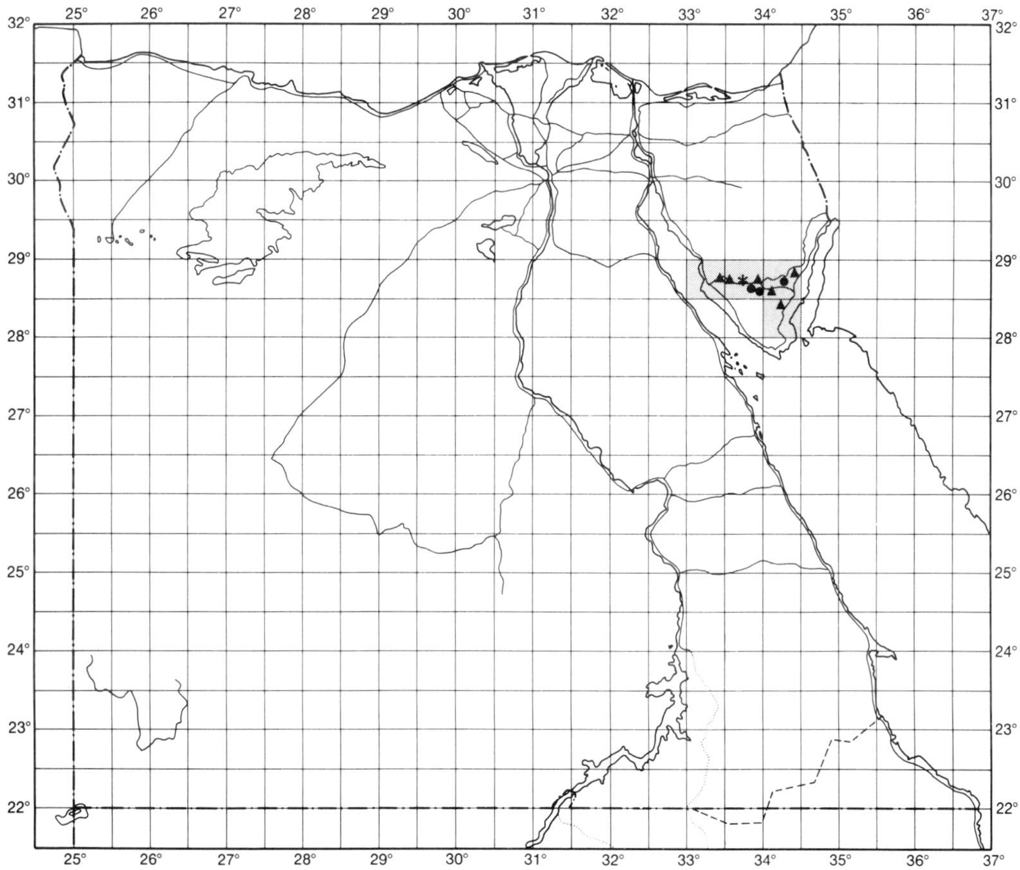

* possible breeding ▲ probable breeding ● definite breeding ▒ half degree grid plots with probable or definite breeding

Map B95. Breeding distribution of Tristram's Grackle *Onychognathus tristramii* between 1970–87.

Locally common breeding resident in the mountains of south-central Sinai. Monotypic.

The Tristram's Grackle is a resident in the environs of St Katherine. Here it breeds in crevices on cliffs surrounding the walls of the monastery. It is also known to occur in Wadi Feiran, Wadi Kid, Wadi Nasb, and Wadi Zaghra, and presumably other parts of southern Sinai. Recent large concentrations include 70 on 7 February 1976 between St Katherine and Wadi Feiran (HW) and at least 35 on 26 March 1985 in Wadi Bosela (UGS). Almost fledged young were observed on 19 April 1984 at St Katherine (MS). The species breeds at Elat, Israel (PLM).

Starling *Sturnus vulgaris* WV

Local names Zarzuur زرزور, khaliish (NC), sursuur.

Winter visitor in varying numbers from (early) late October to early March (mid-April). It mainly occurs along the North Coast, in the Nile Delta south to Cairo, and Suez Canal area. Since the various subspecies are distinguished by differences in plumage coloration, particularly iridescence, and the majority of the Egyptian material is in winter or worn plumage, it is difficult to assign them to subspecies. Most Egyptian specimens we have examined appear closest to *S. v. purpurascens*. The subspecies *S. v. sophiae* mentioned by Meinertzhagen (1930) is a synonym of *S. v. vulgaris* (Vaurie 1959).

In some winters Starlings are found in flocks of several hundred while in other winters they are virtually absent from Egypt. There are several old records from Wadi el Natrun (Goodman *et al.* 1986). Single birds were observed in Bahariya four times between 4 and 14 October 1983 (HB *et al.*). Meinertzhagen (1930) noted the species' occurrence in Siwa but presented no precise information. It is seen with some regularity in the Faiyum but is scarce in the Nile Valley and unknown south of Luxor. There are a few records from western and southern Sinai.

Ringing A bird ringed in the USSR was recovered in Cairo on 12 November 1969 (Watson 1971).

Rose-coloured Starling *Sturnus roseus* (PV)

Local names Zarzuur wardii زرزور وردي, samarmar.

The monotypic Rose-coloured Starling is a rare and irregular migrant to the country. Records include: an immature male at Giza on 25 August 1864 (Von Heuglin 1871); one at Rafa on 10 August 1917 (Meinertzhagen 1930; specimen in AMNH is dated 19 August 1917); one immature at Suez on 12 March 1982 (Wimpfheimer *et al.* 1983); and another at Sharm el Sheikh on 21 August 1982 (Baha el Din 1984*a*). In 1983 single immature birds were captured at Bahariya on 30 and 31 August and two observed, perhaps the same ones, between 31 August and 7 September (HB *et al.*). At

Elat, Israel, the species is a regular autumn and spring migrant in small numbers (HSh).

Family Passeridae

Generic local names for *Passer* spp. H̲uzwii (NC), 'us̲fuur.

House Sparrow *Passer domesticus* RB

Local names 'Us̲fuur duurii عصفور دوري, 'us̲fuur al-ghayt̲, 'us̲fuur 'arabii, 'us̲fuur baladii, 'us̲fuur buyuutii, zarzuur (Ma).

Abundant breeding resident along the North Coast, in portions of the Western Desert, the Nile Delta and Valley, and a few localities along the Red Sea coast and in Sinai. It breeds in buildings and often places hanging nests from trees and poles for overhead wires. Specimens from Egyptian Nubia (YPM) are intermediate in plumage coloration and wing length between typical Egyptian *P. d. niloticus* and northern Sudanese *P. d. rufidorsalis*. Vaurie (1959) believed that the zone of integradation between these two subspecies was near Wadi Halfa but it appears to be further north. Specimens taken near El Arish are referable to *P. d. biblicus*. The inclusion of *P. d. italiae* is based on the specimen reported by Meinertzhagen (1930). We have not examined Egyptian birds referable to *P. d. indicus* (*contra* Hovel 1987). We follow Vaurie (1959) in treating *P. d. halfae* as a synonym of *P. d. niloticus*.

North Coast According to Meinertzhagen (1930), House Sparrows occurred from Alexandria west to El Dab'a and were absent from the area between Matruh and Salum. In February 1939 the western limit of House Sparrows was Burg el Arab (Al Hussaini 1938–9). Presently this species is common near human habitation between Alexandria and Matruh. It is not known if it occurs further west, since this area has not been visited in recent years by ornithologists.

Western Desert In Meinertzhagen's time (1930) the House Sparrow was unknown from the Western Desert. Several nests were found on 1 May 1981 in Wadi el Natrun and the species has also recently colonized Dakhla and Kharga (Goodman *et al.* 1986). It is common in the environs of Sadat City, between Wadi el Natrun and the Nile Delta (SMG, PLM).

Nile Delta and Valley House Sparrows are abundant residents in cultivated areas of the Nile Delta and Valley south to Aswan, the Faiyum, and the Suez Canal area. In January 1980 hundreds of thousands were observed flying to a night roost in Giza (PLM, WCM). Numerous specimens were taken in the mid-1960s in the portion of the Nile Valley now inundated by Lake Nasser (YPM). The species is common at Abu Simbil and Wadi Halfa, Sudan (PLM).

Eastern Desert House Sparrows are common residents at Ain Sukhna and Ras Gharib, scarce at Hurghada and Safaga, and altogether lacking at Quseir and southwards. There are several isolated areas in the Eastern Desert where the species has been found, e.g. at the rest-house midway along the Cairo–Suez desert road (PLM) and at Bir Abbad, 15 km east of Idfu on the road to Marsa el Alam (Horváth 1959).

Sinai House Sparrows were already common breeders in the El Arish and Rafa area in the 1920s (Flower 1933; *contra* Meinertzhagen 1930). Presently, this species is found near human settlements and in date palm groves all along the north coast, and is uncommon at Nekhl, Na'ama, and Taba. According to Meinertzhagen (1930) this species occurred at El Tor and Abu Zenima. There are no recent records from these localities, although five were noted at Ras Abu Rudeis on 26 January 1985 (SMG).

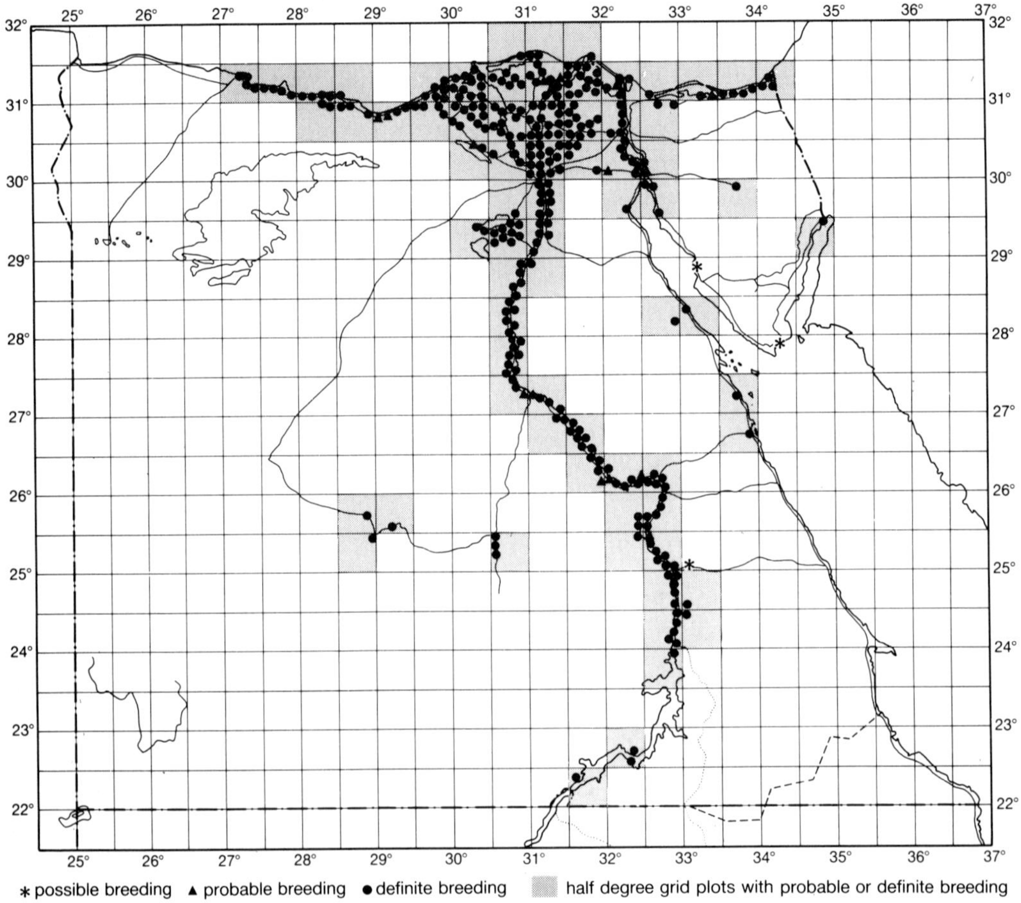

Map B96. Breeding distribution of House Sparrow *Passer domesticus* between 1970–87.

Spanish Sparrow *Passer hispaniolensis* PV WV

Local names 'Usfuur isbaanii عصفور اسباني, zarzuur.

Abundant passage migrant and winter visitor to Egypt from (early) late September to mid-April (early May). Occasional in summer. All of the recent material we have examined (80+ specimens) appears closer in plumage coloration to *P. h. hispaniolensis* than to *P. h. transcaspicus*. Meinertzhagen (1930) and Vaurie (1959) referred some Egyptian material to the latter form.

Spanish Sparrows are winter visitors to the cultivated areas along the North Coast, the Nile Delta and Valley, Faiyum, Suez Canal area, and Sinai. The species sporadically occurs in vegetated areas of the Western and Eastern Deserts. During a study of bird migration in Bahariya from 3 August to 29 October 1983, Spanish Sparrows were first recorded on 7 September, a distinct influx was noted in early October, and hundreds were present for the rest of the month (HB *et al.*). In the Nile Valley the species is scarce south of Luxor. There is a specimen from Kurusku taken on 12 February 1963 (YPM). Most records from the Eastern Desert are during the migration period (October, November, and March), although there are reports from January at Gebel Abraq and Halaib (Goodman and Atta 1987), and February at Wadi Qiseib (USNM) and Bir Beida (SMG). Two birds were taken on 9 May 1968 in Wadi Atalla, approximately midway between Qift and Quseir (FMNH). A flock of about 700 was observed on 7 March 1987 at Nuweiba flying to a night roost (PLM). It was observed near El Arish in July 1984 (DM). This species has recently been found breeding at Elat, Israel (HSh).

Ringing Birds ringed at Bahig on 7 April 1972 and on 13 April 1967 were recovered at the same site on 8 April 1973 and at Samotraci, Greece, sometime before May 1968, respectively (Hubbard and Horner 1969; GZF). One ringed at Elat, Israel, on 25 February 1969 was recovered near Cairo sometime before July 1970 (Watson 1971).

[Dead Sea Sparrow *Passer moabiticus* AV?

The only known record of the Dead Sea Sparrow in Egypt is a flock of about 15 noted in the winter of 1971–2 near Nuweiba (UP). The basis of Hovel's (1987) statement that this species has been recorded in the gardens of St Katherine is unknown. The form *P. m. moabiticus* is a common winter visitor to Elat, Israel, from early October to mid-April (Krabbe 1980) and recently has been found breeding there (HSh). Since Elat is less than 20 km from the Israeli–Egyptian frontier, this species presumably occurs in Egypt.]

Desert Sparrow *Passer simplex* AV

Local name 'Us̱fuur as̱-s̱aẖraa-عصفور الصحراء.

Accidental visitor. All known Egyptian records of Desert Sparrows are from the Gebel Uweinat area, and on geographical grounds they are presumably of birds referable to *P. s. simplex*.

Between 21 October and December 1968, five were seen at Ain Zuweina and between 27 and 28 December 1968, 22 were noted at Karkur Talh (Misonne 1974). This species is not known to breed across the Egyptian border in Libya (Bundy 1976).

Tree Sparrow *Passer montanus* AV

Local name 'Usfuur ash-shajar عصفور الشجر.

Accidental visitor. It is uncertain which subspecies occurs in Egypt but on geographical grounds it is presumably *P. m. montanus* and/or *P. m. transcaucasicus*.

Meinertzhagen (1930) rejected Von Heuglin's (1871) statement that the Tree Sparrow occurs in Lower Egypt. There have since been two observations of this species in the country: one on 29 April 1982 at Suez (Wimpfheimer *et al.* 1983) and one (photographed) at Safaga on 12 and 13 March 1985 (UGS).

Sudan Golden Sparrow *Passer luteus* MB?

Local name 'Usfuur dhahabii عصفور ذهبي.

Presumed migrant breeding bird of the Gebel Elba area. Monotypic.

Two Sudan Golden Sparrows were collected in Wadi Kansisrob on 16 February 1964 (Goodman and Watson 1983) and a flock of over 29 was observed at this same locality on 28 March 1983 (SMG). Two old nests thought to be constructed by this species were discovered on the south side of the mountain in Wadi Rabdeit on 25 February 1985 (SMG). An adult and immature were taken in 'Égypte' sometime before 1876 (IRSNB). This species is known to wander in flocks during the non-breeding season (Mackworth-Praed and Grant 1980).

Pale Rock Sparrow *Petronia brachydactyla* (PV)

Local name 'Usfuur as-sakhr al-baahit عصفور الصخر الباهت.

Rare and irregular migrant to the country. Monotypic.

Meinertzhagen (1930) collected one Pale Rock Sparrow out of a pair at El Arish on 2 June 1917. The bird showed signs of breeding. There are a few recent observations in Egypt: 35 on 25 October 1979 at Abu Qurqus (FV), six or seven on 27 October 1978 at Saqqara (JB), and two on 9 April 1977 at Nuweiba (RSS). The basis for Etchécopar and Hüe's (1967) statement that it 'sometimes comes to Sinai in flocks' is not clear. This species is a rare winter visitor to Elat, Israel (Krabbe 1980) and has also been observed near Wadi Halfa, Sudan (Nikolaus 1987).

Family Ploceidae

Streaked Weaver *Ploceus manyar* IB Plate 2

Local name 'Usfuur nassaaj mukhattat عصفور نساج مخطط.

Introduced breeding bird in the Nile Delta, particularly common in reed-beds. The natural range of this species is the Indian subcontinent and south-east Asia (Ali and Ripley 1974). In 1971 several appear to have escaped from the Alexandria Zoo (BO) and it is assumed that these were the progenitors of the present Egyptian population. The subspecies occurring in the country has not been determined.

The earliest record of wild Streaked Weavers in Egypt appears to be 11 May 1978, when about 20 birds and several nests were found at El Mahmudiya, Lake Idku (RJM, LRP). The species was also noted at Barsiq, Lake Idku, in 1982 (MAS) and in February

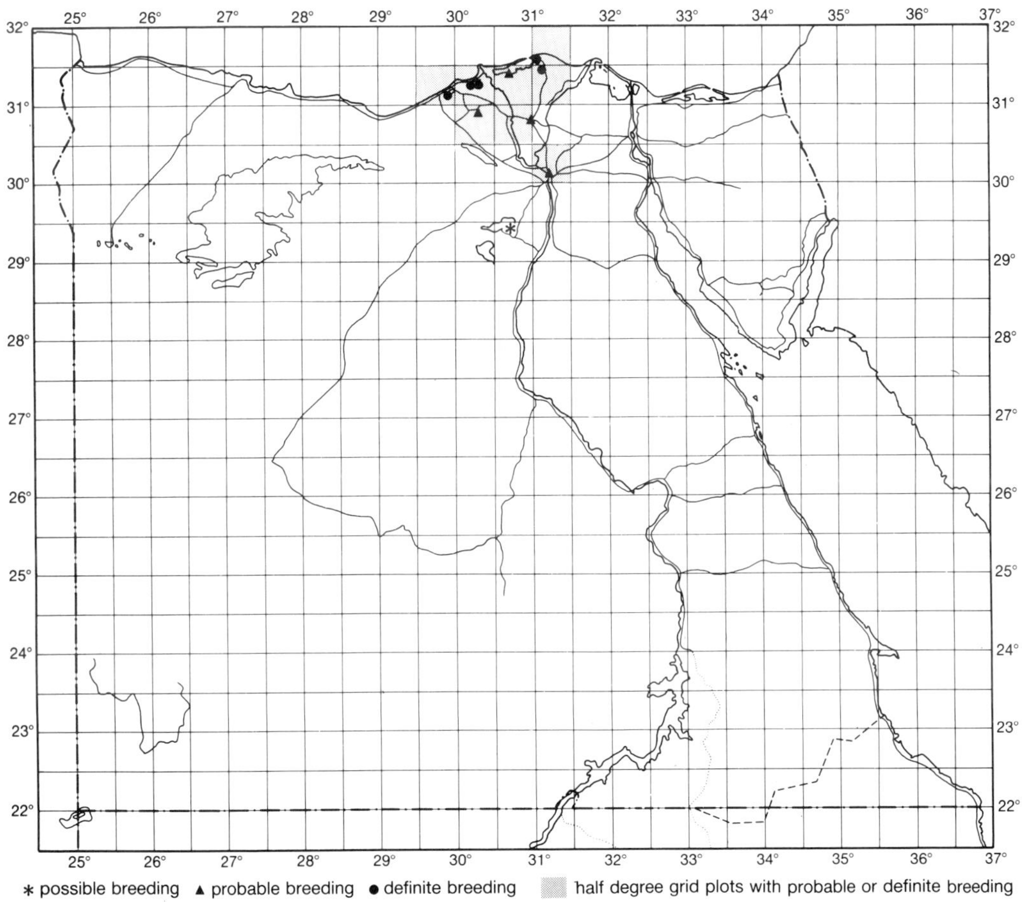

Map B97. Breeding distribution of Streaked Weaver *Ploceus manyar* between 1970–87.

1984 (SMB). Two colonies were discovered in May 1983 near Baltim in various stages of breeding, including several nests with eggs on 10 May, and the total local population was estimated to be at least 25 birds. On 11 May 1983, 15 km south-east of Baltim a flock of at least 16 birds was noted and one nest was under construction. On that same day, on the southern shore of Lake Burullus, 15 km north of El Haddadi, another bird was observed in an area of extensive reed-bed (Meininger and Sørensen 1984). Between late May and July 1986, four separate colonies composed of 8–12 birds were discovered just south-west of Alexandria (JH).

This species quickly spread across the Nile Delta. One male was observed at Tanta on 6 April 1984 (SMG) and three birds were noted near Hosh Isa on 11 May 1985 (SMG, PLM). It also rapidly moved up the Nile; one male was seen on 17 April 1984 south-east of Lake Qarun, Faiyum (MBe, SM) and several males, in breeding plumage, and females were found on 16 June 1985 in a small reed-bed along the Nile opposite Shubra el Kheima (SMB).

Family Estrildidae

Avadavat *Amandava amandava* IB Plate 2

Local name 'Us̱fuur aẖmar hindii عصفور أحمر هندي.

Locally common introduced breeding bird in the Nile Delta and portions of the northern Nile Valley, the Faiyum, and Suez Canal region. The natural range of this species is the Indian subcontinent east to south-eastern Asia (Ali and Ripley 1974). All material examined from Egypt is referable to *A. a. amandava*. The local breeding season is rather extended; nests and fledglings have been noted from August to late November and a laying female was collected in early May.

The first mention of Avadavats in Egypt was by Antinori (1862), who shot several near Alexandria on 17 December 1861. Walter Innes (in Koenig 1926) collected specimens near Cairo on 15 May 1903 and on 13 January 1905. He also mentioned that Avadavats were often absent from the Giza Zoological Gardens for several months and returned in large flocks late in the calendar year to breed.

Meinertzhagen (1930) stated that this species appeared in the Nile Delta in 1905 and that 'it was until quite recently common near Giza, Inshas [= Inchas], Bilbeis and at the Delta Barrage, but now appears to have disappeared'. There is a series of 18 specimens, including juveniles, in the GZM collected in the Giza–Inchas area between 1908 and 1918. This species was noted in February 1928 at Luxor (Meinertzhagen 1930) but there are no more recent records from Upper Egypt. Hachisuka (1924) saw this species in the Faiyum in February 1924. The only known records of this species in Egypt between 1928 and 1974 are one at Abu Rauwash on 30 August 1971 (USNM) and another at El Mansura on 6 October 1970 (HS).

In January and February 1974 flocks of 5–10 Avadavats were frequently observed about 4 km south-west of Ismailiya (Safriel 1975) and there are many recent observations from this area. One specimen was collected near Ismailiya on 8 February 1974 (ZM-HUJ). Since the late 1970s Avadavats have been noted with ever increasing frequency in the Nile Delta and it is clear that this species is widely distributed in this region. In 1978 several were noted at Asher Talaaf, 30 km south-west of Alexandria (EK). The species is presently well established in reed-beds near Suez, where up to 100 birds have been regularly observed since 1980. On 10 April 1982 a female was observed at this locality carrying nest material (MU). Between May and July 1986 up to 30 birds, including immatures, were noted in reed-beds near Alexandria (JH).

This species is common along the river just south of Cairo. Other recent observations in the Nile Valley include 25 on 4 February 1981 at El Saff (SMB), a total of six on 27 September 1985 at a few localities between Mazghuna and El Wasta, and one the following day south of Beni Suef (Hume 1985). Since autumn 1983 this species has

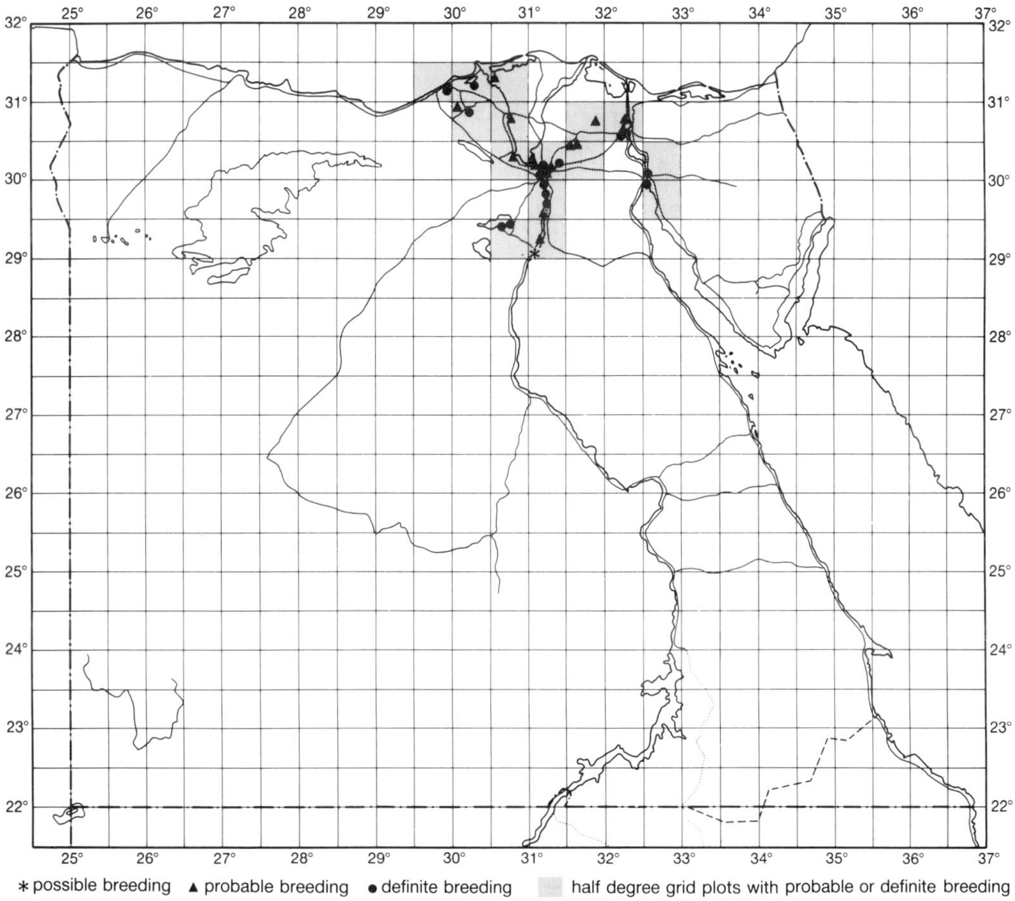

Map B98. Breeding distribution of Avadavat *Amandava amandava* between 1970–87.

been noted along Lake Qarun, where it was found breeding in November 1983 (Bezzel 1987*a*).

African Silverbill *Euodice cantans* RB?

Local name ’Us̱fuur faḏḏii al-minqaar عصفور فضي المنقار.

The African Silverbill is a presumed breeding resident of the Gebel Elba region. It has been taken in Wadi Rabdeit on the south side of Gebel Elba and at Gebel Shellal, the next massif to the south (Goodman 1984*a*). In April 1985 a few were observed in the heavily vegetated tributaries of Wadi Aideib (SMG, PLM). *E. c. orientalis* is the form naturally occurring in the Gebel Elba region. A bird identified as this species (presumably an escaped cage-bird) was trapped on the grounds of the Giza Zoological Gardens on 24 April 1911 (Nicoll 1912*a*).

Family Fringillidae

Chaffinch *Fringilla coelebs* WV

Local names ’Us̱fuur ẕaalim عصفور ظالم, s̱aghnaj, ẕaalim.

Scarce winter visitor to the northern portion of the country from late October to late March (early April). All Egyptian specimens examined are of *F. c. coelebs*.

During the winter Chaffinches regularly occur in small flocks along the North Coast, in the Nile Delta, the Suez Canal area, and northern Sinai. In the Western Desert the species has been occasionally recorded in Wadi el Natrun and Bahariya. Single birds were captured in Bahariya on 22 and 23 October 1983 (HB *et al.*) and three observed on 8 March 1980 (Goodman *et al.* 1986). It is rarely seen in the Nile Valley or the Faiyum. The southern-most record is at Aswan on 24 February 1976 (Short and Horne 1981). There are a few observations from the north of the Eastern Desert at the monasteries in Gebel el Galala el Qibliya but none further south. Invasions are noted in some years. Jarvis (1942) mentioned tens of thousands arriving in October to northern Sinai. Twelve were observed at St Katherine in early March 1987 (PLM). One exceptionally late record is a bird on 1 June 1917 at Giza (GZM).

Ringing A bird ringed in Cairo on 9 November 1937 was recovered in Siberia on 27 April 1939 (Mackintosh 1941).

Brambling *Fringilla montifringilla* WV

Local names ’Us̱fuur zaghaarii عصفور زغاري, shurshuur jabalii.

Rare winter visitor from late October to late March mainly in the northern half of the country. Monotypic.

Meinertzhagen (1930) mentioned 11 Brambling records from the Nile Delta. More recently a male was obtained at Port Said on 8 November 1959 (FMNH) and a female on 20 November 1966 (Hubbard and Seymour 1968). Along the Nile this species has not been recorded south of Cairo. On 21 March 1982, two were seen at Safaga (Meilstrup and Sørensen 1983) and on 5 January 1980, one was observed at El Arish (Meininger and Mullié 1981*c*). Records from St Katherine include: single birds on 29 and 30 October 1972 (ZM-HUJ), one on 1 November 1972 (DL), two on 6 February 1976 (HW), one on 6 March 1987 (PLM), and four on 18 March 1987 (RGB).

Red-fronted Serin *Serinus pusillus* AV

Local name Na''aar ah̠mar al-jabha نعار أحمر الجبهة.

There are only a few records of the monotypic Red-fronted Serin in Egypt. These include a female on 18 March 1967 at Bahig (Hubbard and Seymour 1968) and three on 15 February 1969 at Bir Naba, Wadi el Biyar, Sinai (ZM-HUJ; Hovel 1987).

Serin *Serinus serinus* RB? WV

Local names Na''aar نعار, s̠ayyaah̠, tinjii, 'us̠fuur na''aar.

Winter visitor from mid-September to mid-March (late April), mainly to the northern portion of the country. Probably a local breeding resident in the Nile Delta. Monotypic.

North Coast There are several recent October and November records of Serins from this area. The only spring report is one bird at Matruh on 28 April 1970 (USNM).

Nile Delta and Valley Small flocks are regularly observed during the winter in the cultivated areas of the Nile Delta and Valley (south to Beni Suef), and the Faiyum. The southern-most record is one at Aswan on 22 March 1986 (UA). The first indication of this species breeding in Egypt was a singing male observed in a Cairo garden on 6 June 1953 (ACK). More recently at least three singing and displaying males were noted on 11 May 1985 in citrus orchards near El Khatatba (SMG, PLM). However, no Serins were observed or heard on a visit to this same area in June and July of the same year (SMB). Singing males, presumably on territory, were noted near Shatanuf on 20 May 1986 and north of Abu Rauwash on 20 June 1986 (WCM).

Sinai This species is a regular winter visitor to the area between El Arish and Rafa, and the mountainous region in the south. Records of substantial flocks include about 20 on 27 November 1965 at St Katherine (HCK), 20 on 5 January 1980 at El Arish (PLM, WCM), and 50 on 2 February 1984 in Wadi el Arbain (SMB).

Ringing A bird ringed in Cairo on 9 November 1937 was recovered in Kazakhstan, USSR, on 3 May 1939 (Mackintosh 1941).

Tristram's Serin (Syrian Serin) *Serinus syriacus* WV

Local names Na''aar suurii نعار سوري, s̲ayyaah̲.

Apparently a rare winter visitor to the country, perhaps somewhat regular in south Sinai. Monotypic.

On 27 December 1919 a male Tristram's Serin was collected near Helwan (Meinertzhagen 1930) and on 25 November 1969 another was taken at Bahig (USNM). A singing male and an 'apparent' female were found by Short and Horne (1981) in a garden at Tuna el Gebel on 25 February 1979. This species appears to be an irregular winter visitor to Sinai. During March 1970 it was apparently common at St Katherine and numerous specimens were collected (HH; UMMZ; ZM-HUJ). At least 20 were seen at this locality from 28 to 29 December 1978 (EB), seven on 24 March 1979 (DJF *et al.*), and a few at El Arish in March 1986 (DB).

Greenfinch *Carduelis chloris* RB WV

Local names 'Us̲fuur khud̲ayrii عصفور خضيري, akhd̲ar, khud̲ayr, khud̲ayrii.

Local breeding resident in the Nile Delta, Suez Canal area, and northern Sinai. Fairly common winter visitor to the north of the country from mid-October to early April. Two subspecies, *C. c. chlorotica* and *C. c. aurantiiventris*, have been identified from material taken in Egypt. *C. c. chlorotica* is the distinctly more common of the two and presumably the form breeding in the country. Some specimens are intermediate in plumage coloration between these two subspecies.

According to Meinertzhagen (1930) the Greenfinch possibly bred in north-east Sinai between El Arish and Rafa. It is presently a common breeding bird in this area. Meinertzhagen also mentioned that this species was occasionally observed between November and March in the Delta and predicted that it might colonize this region. This forecast has been realized in recent years. On 11 May 1985 at least two singing males were observed in orchards near El Khatatba (SMG, PLM) and on 14 June 1985 the species was found to be common and breeding in this area (DM). Records from other localities include: three birds on 17 May 1985 at Birket el Hagg and several on 24 May 1985 in Wadi el Natrun; three to four on 30 July 1985 and 20 June 1986 near Abu Rauwash; three on 25 April 1986 near Bilbeis; one on 25 April 1986 near El Abbassa; and one on 4 June 1986 about 75 km south of Port Said (DM, WCM).

Greenfinches are fairly common winter visitors to the North Coast between Alexandria and Matruh, the Nile Delta, Suez Canal area, the Faiyum, Wadi el Natrun, and Sinai. There are only a few records from the Valley above Cairo, the most southerly being one near Isna on 31 October 1982 (Davies 1982). This species is occasionally recorded along the Red Sea coast: eight on 19 January 1980 at Hurghada (Meininger and Mullié 1981*c*) and one on 6 February 1983 at Safaga (UMMZ).

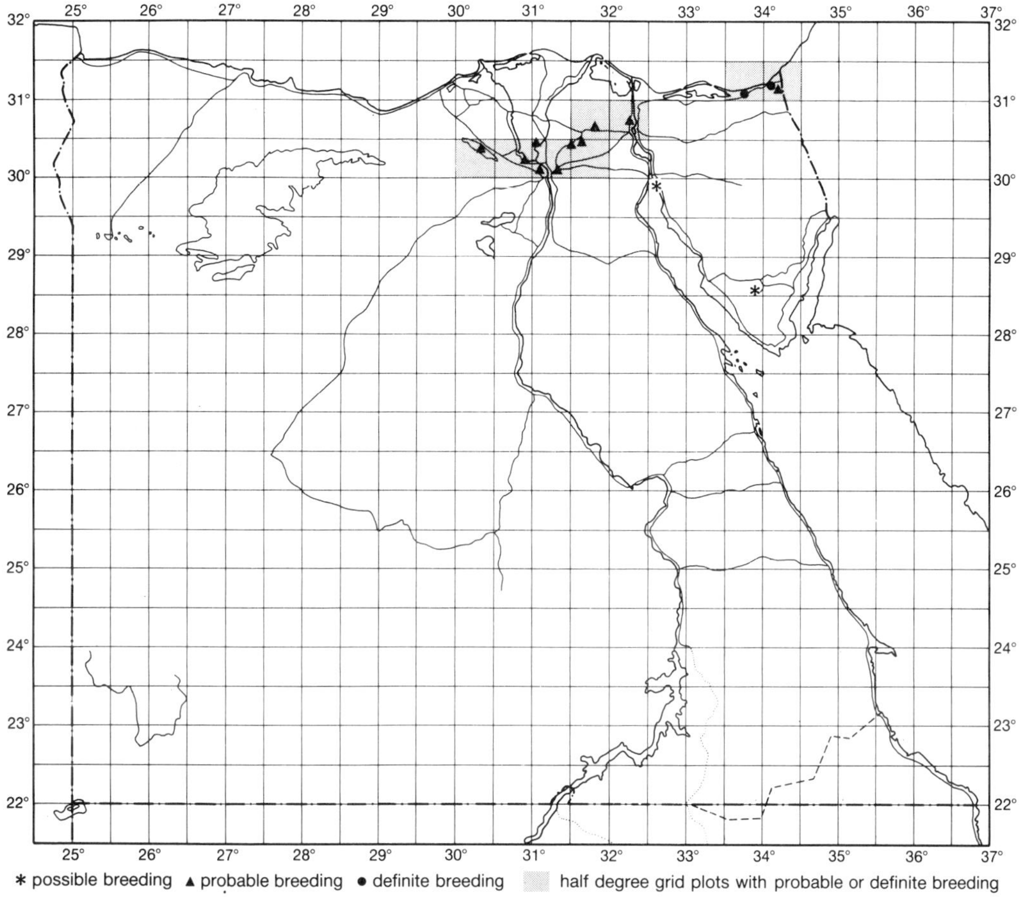

Map B99. Breeding distribution of Greenfinch *Carduelis chloris* between 1977–87.

Goldfinch *Carduelis carduelis* RB WV

Local names 'Us̱fuur ẖassuun عصفور حسون, abuu al-ẖusn, abuu saqaaya, abuu zaqaaya, ẖassuun, shuwaykii.

Common breeding resident in the Nile Delta and Valley, in several Western Desert oases, and north-eastern Sinai. The local breeding populations are presumably augmented by migrants during the winter months. All Egyptian material examined is referable to *C. c. niediecki*. The only exception is a specimen taken on 18 December 1966 at Port Said (USNM) which is similar in plumage coloration to winter birds from continental Europe and appears to be referable to *C. c. carduelis*.

North Coast The first known breeding record of the Goldfinch from this area was in 1922 at Burg el Arab (Moreau 1969). This record was unknown to Hubbard and Seymour (1968) and thus they reported fledglings at Bahig in May 1967 as a significant breeding-range extension. Since 1967 fledglings have been noted at Bahig between 9 April and 25 May (USNM) and it appears that this species is now more common in the area than in the 1920s. Adult specimens have been taken at Matruh on 9 March and 14 March 1970 (USNM).

Western Desert The earliest record of the Goldfinch in Wadi el Natrun seems to be 29 September 1912 (GZM). This species was found breeding in this oasis in 1981 (SV; WS). The first record of Goldfinches in any other Western Desert oasis appears to be a small flock in Kharga on 10 December 1965 (JT). Between 1981 and 1984 evidence of breeding was found in Bahariya, Farafra, Dakhla, and Kharga (Goodman *et al.* 1986).

Nile Delta and Valley According to Meinertzhagen the Goldfinch was apparently a resident in the environs of Cairo. Nests were found near Giza in 1907 (Nicoll 1908), 1910, and 1917 (Raw 1921). D. Goodwin (1947) stated 'the bird is said to have bred at one time regularly near Cairo but later to have become extinct'. Until the mid-1950s

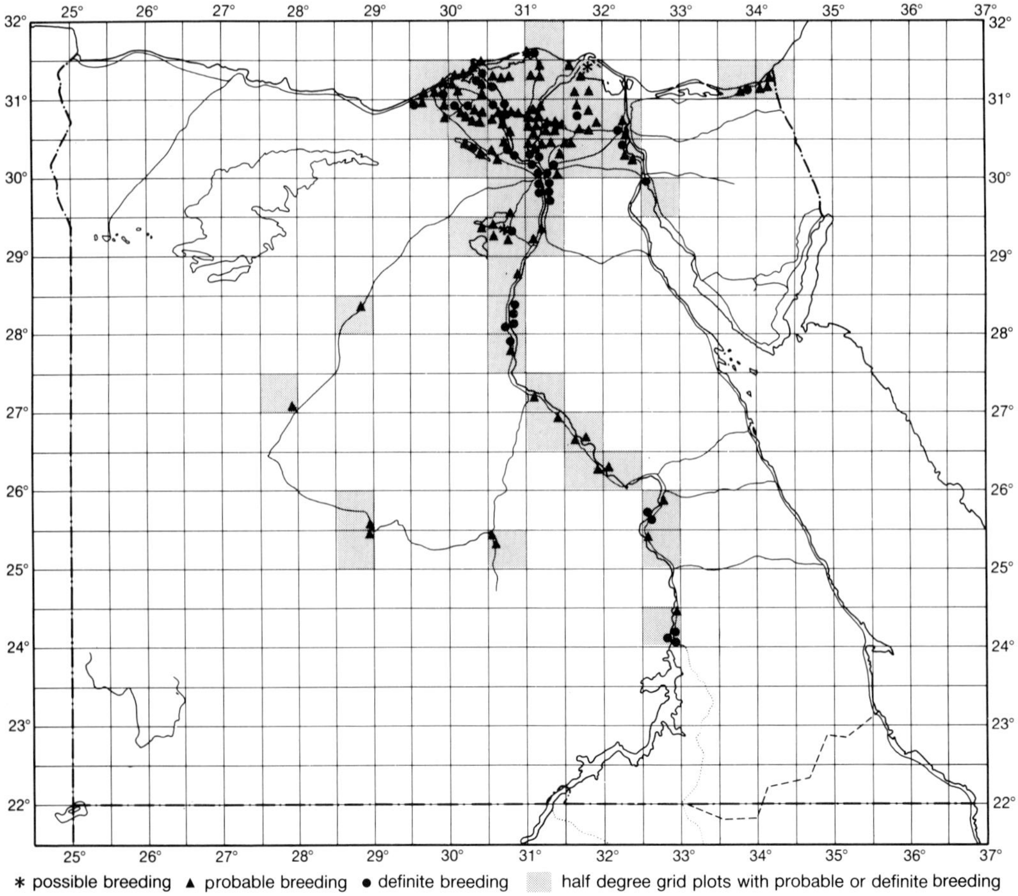

Map B100. Breeding distribution of Goldfinch *Carduelis carduelis* between 1970–87.

Goldfinches were considered rare winter visitors to northern Egypt and the Suez Canal area. The first report of breeding in the latter area was in June 1950 at Fayid, when an empty nest was located (JCC). Nowadays it is a common breeding bird in gardens, orchards, and parks throughout the Nile Delta, the Suez Canal area, the Faiyum, and most parts of the Nile Valley south to Aswan.

Eastern Desert The Goldfinch is an irregular winter visitor along the Red Sea coast south of Suez. Records include: Ain Sukhna—two on 3 March 1984 (SG); Hurghada—two on 27 December 1983 (PE) and one on 10 March 1984 (PDH, EMN); and Safaga—one on 25 February 1985 (DM).

Sinai This species is a common resident in the El Arish and Rafa area. It was first recorded nesting in this region at El Arish in 1924 (Jarvis 1942). Outside the breeding season it has been observed in small numbers in vegetated areas as far south as St Katherine and Sharm el Sheikh. Goldfinches might nest in the oases of southern Sinai, as the species is known to breed at Elat, Israel (Krabbe 1980).

Ringing One ringed at Akrotiri, Cyprus, on 27 April 1968 was recovered in Alexandria on 6 December 1968 (Flint and Stewart 1983).

Siskin *Carduelis spinus* WV

Local name 'Us̱fuur sumaylii عصفور سميلي.

Scarce winter visitor to the northern portion of the country from mid-October to mid-March. Monotypic.

In most winters Siskins are rare in Egypt and records are generally from the Nile Delta and Sinai. In some years, however, large numbers invade the country, e.g. in February and March 1907 (Meinertzhagen 1930). The 1907 invasion apparently resulted in serious agricultural damage, as a substantial portion of the barley crop near Maryut was said to have been destroyed by this species (Nicoll 1909). Between 19 and 21 November 1952 up to 13 birds were observed at Fayid (AP). The winter of 1966–7 seems to have been another invasion year; large numbers were observed near Port Said (Hubbard and Seymour 1968), where numerous birds were obtained in November (USNM), and the species was common in El Maadi between 3 January and 28 March (HS). The most southerly Nile Valley records are from the Faiyum. Here small flocks are occasionally recorded.

There are a few records from the Western Desert—one in Wadi el Natrun on 17 October 1914 (Meinertzhagen 1930; Goodman *et al.* 1986) and single birds in Bahariya between 12 and 29 October 1983 (HB *et al.*). Reports from St Katherine include four in late October 1972 (ZM-HUJ), 35–40 on 6 February 1976 (HW), two on 17 November 1983 (SMB), and five on 6 March 1984 (SG).

Linnet *Carduelis cannabina* WV

Local names 'Us̱fuur tuffaaẖii عصفور تفاحي, zaqayqiya.

Fairly common winter visitor to the northern part of the country from early September to early April. Two subspecies have been recorded in Egypt, *C. c. cannabina* and *C. c. bella*. Nominate *cannabina* is a more common migrant than *bella*. Numerous specimens are intermediate in plumage coloration between these two forms. We follow Vaurie (1959) in treating *C. c. mediterranea* as a synonym of *C. c. cannabina*.

The earliest autumn record of Linnet in Egypt is 6 September 1944 at Kasfarit, and six days later about 200 were seen flying south (LB). Small flocks are noted throughout the winter in cultivated areas along the North Coast, in the Nile Delta and Valley, Suez Canal area, the Faiyum, and Sinai. In December 1951 and January 1952 up to 200 Linnets roosted in a garden at Fayid (AP). The southern-most Nile Valley record appears to be one at 'New Nubia' on 3 December 1970 (USNM). There are several February and March reports from Wadi el Natrun and a few from Bahariya (Goodman *et al.* 1986). A flock of about 50 was noted in Dakhla on 3 March 1987 (AH). The only information from the Red Sea coastal plain south of Suez is a single bird at Safaga on 12 and 13 March 1985 (UGS). The latest spring report is 7 April 1954 at Firdan (RN).

[**Desert Finch** *Rhodospiza obsoleta* AV?

There are a few well-documented observations of Desert Finches in north-eastern Sinai. On 5 February 1979 a female was observed at Yamit (EB) and two sub-adults were noted just south of El Arish on 24 June 1984 (SMB). At Elat, Israel, the species is a regular autumn and winter visitor and has bred locally on a few occasions (HSh).]

Trumpeter Finch *Bucanetes githagineus* RB

Local names Zummayr زميّر, ṯabbaal, 'uṣfuur ẖamra, zarzuur (Ma), zummaar.

Locally common breeding resident of rocky desert areas along the Nile Valley, in the Western and Eastern Deserts, and Sinai. Two specimens collected at Gebel Uweinat (USNM) are similar in plumage coloration to birds taken in southern Algeria, distinctly paler than those from other portions of Egypt, and are referable to *B. g. zedlitzi*. The balance of Egyptian material examined is of *B. g. githagineus*.

Western Desert The Trumpeter Finch has been recorded in Bahariya, Farafra, Dakhla, and Kharga oases (Goodman *et al.* 1986). In these areas it generally breeds along the desert edge and visits the cultivated areas for food and water. It is common in the Gebel Uweinat region near wells (Jany 1963; Misonne 1974).

Nile Valley This species is a locally common breeding bird along the edge of the Valley between Cairo and Aswan. It was common in the mid-1960s in regions of Egyptian Nubia now at least in part inundated by Lake Nasser (YPM). It appears to be more regular on the east bank of the Nile, although it is often observed on the west bank near Sohag, Luxor, and Isna. Birds have been observed in March at the Temple of Hatschepsut, Deir el Bahari, carrying food and faecal sacs to and from holes in the edifice (SMG). Outside the breeding season it congregates in flocks and frequently visits cultivated areas. In the late nineteenth century Koenig (1926) observed tremendous flocks in fields between Aswan and Wadi Halfa. In December 1908 Whymper (1909) noted hundreds in the cultivated fields near Thebes. Nowadays, flocks of less than a hundred are observed in the Upper Nile Valley.

Eastern Desert This species is a locally common resident of rocky wadis. It is often abundant around wells in the northern and central Red Sea Mountains and in the Gebel Abraq area, particularly during the summer months. It is apparently absent from the extreme south-east, for example near Gebel Elba. The southern-most locality in the Egyptian Red Sea Mountains appears to be Gebel Hamra Dom, where it was common in mid-February 1984 (SMG). In late June 1984, 350–500 birds congregated around Bir Abraq (SMG). An old record of this species at Berenice (Von Heuglin 1861) is the only one known from the Red Sea coastal plain.

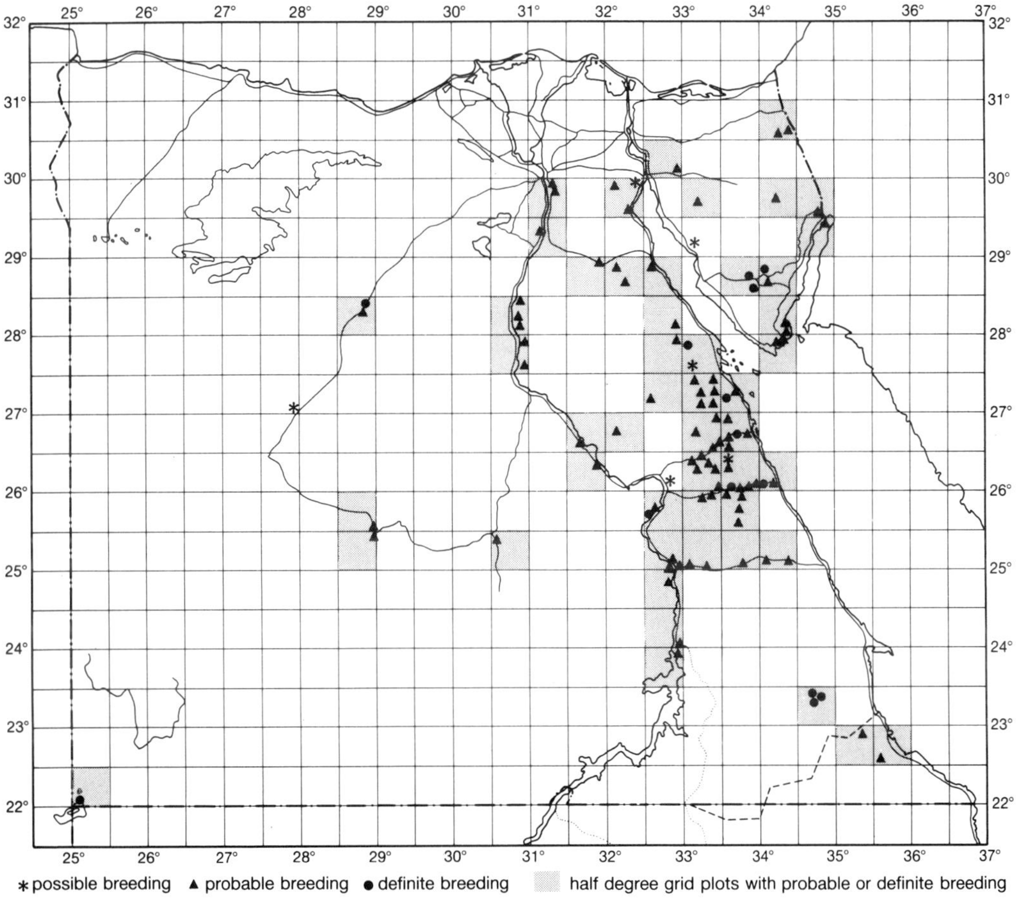

Map B101. Breeding distribution of Trumpeter Finch *Bucanetes githagineus* between 1930–87.

In the past decade there has been a substantial increase in the amount of whole grains shipped from abroad to the Red Sea port at Safaga. The grain is trucked to the Nile Valley via the Safaga–Qena and Quseir–Qift desert roads. Considerable quantities are spilt on the road, particularly around bends (e.g. near Wadi el Hammamat and Umm Fawakhir along the Qift–Quseir road), and in such areas Trumpeter Finches are abundant.

Sinai This species is a widespread, but local resident, in the rocky wadis and mountain areas of the Peninsula. It appears to be completely absent from the sandy desert areas of northern Sinai.

Scarlet Rosefinch (Common Rosefinch) *Carpodacus erythrinus* (PV)

Local name 'Us̲fuur wardii عصفور وردي.

Rare and irregular passage visitor to the country between late September and late October. Accidental in spring. Three specimens were taken at Bahig (Hubbard and Seymour 1968; Goodman and Watson 1983) which appear referable to *C. e. kubanensis*.

Records of Scarlet Rosefinches from Bahig include single birds on 7 October 1968, 21 October 1969, and 22 May 1967 (Goodman and Watson 1983). Eighteen were captured at St Katherine between 28 September and 20 October 1971 (DL). It is unknown if such migratory movements in Sinai are regular. According to Krabbe (1980) this species is rare on autumn passage at Elat, Israel, from late October to mid-November and scarce in the spring from late April to mid-May. Two specimens were taken at Elat on 16 September 1968 (USNM).

Sinai Rosefinch *Carpodacus synoicus* RB Plate 5

Local names 'Us̲fuur wardii siinaa-ii عصفور وردي سينائي, gazama (SC).

Fairly common breeding resident in mountainous areas of southern Sinai. The local breeding form is *C. s. synoicus*.

The Sinai Rosefinch inhabits the high mountains of southern Sinai, including Gebel Musa, Gebel Serbal, and Gebel Umm Shomar. On Gebel Musa it breeds at altitudes of up to at least 2000 m. During the winter it can be seen in flocks, sometimes composed of hundreds, in wadis, near human habitation, and at high altitudes (up to 2240 m). It has also been observed during the winter in wadis draining towards the Gulf of Suez (Inbar 1979). On 26 August 1985, two were noted in the mountains north-west of Nuweiba (DM). Kaiser's (1890) record of this species in the mountains east of Cairo is not acceptable.

Ringing At least eight birds ringed at St Katherine in the spring of 1975 returned to the same site the following spring (Darom 1979).

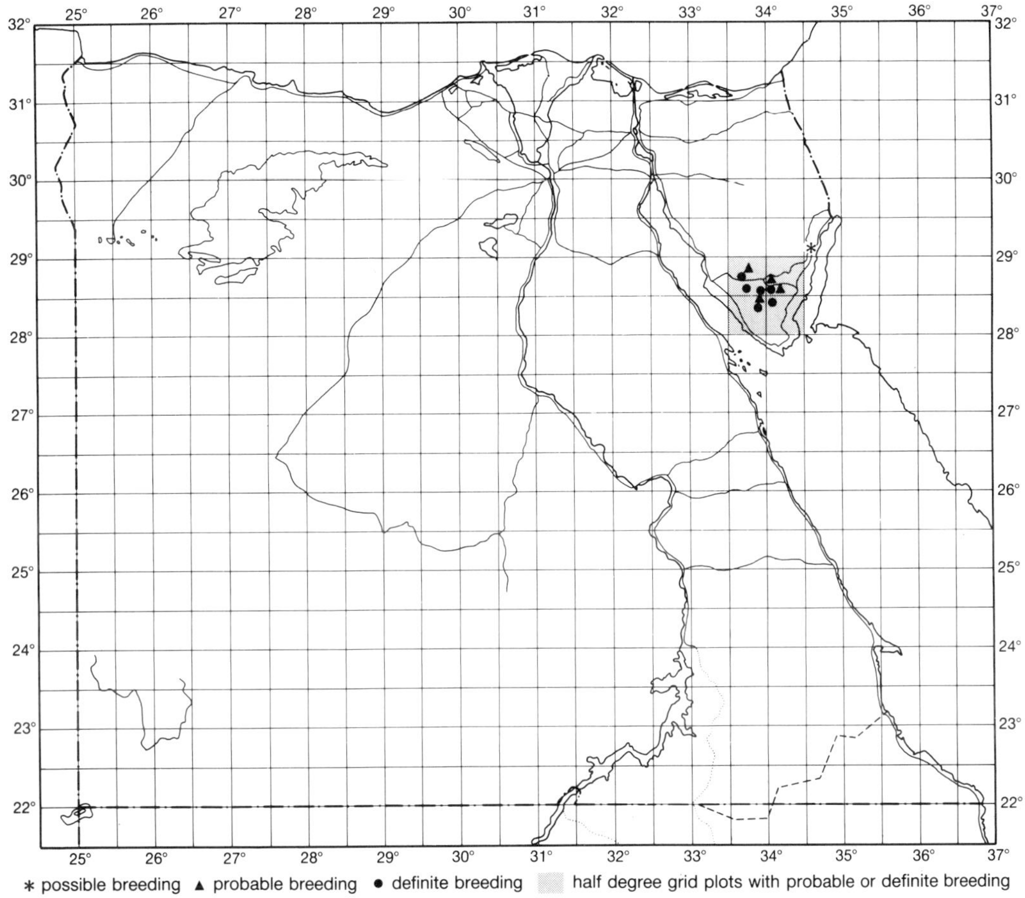

Map B102. Breeding distribution of Sinai Rosefinch *Carpodacus synoicus* between 1970–87.

Hawfinch *Coccothraustes coccothraustes* WV

Local names Bulbul zaytuunii بلبل زيتوني, h̲uzwii (NC).

Rare winter visitor from mid-October to late March in the northern part of the country. All Egyptian material examined is of *C. c. coccothraustes*.

Nile Delta Hawfinches are infrequently recorded in the Nile Delta south to Cairo. Lynes (1912) noted a marked passage along the Mediterranean coast from early November to early December. At Burg el Arab one was reported on 25 February 1938 (HGD) and three on 10 January 1980 (PI). Four were taken at Bahig in the late 1960s and early 1970s between 30 October and 23 November (USNM).

Western Desert According to Meinertzhagen (1930) this species has been collected in Wadi el Natrun in October and December. Three were obtained at this locality on 17 October 1914 (Goodman *et al.* 1986). A fresh specimen was found dead in Siwa on 25 March 1927 (Moreau 1928, 1940–1).

Family Emberizidae

Generic local names for bunting Durrsa, bulbul ash-sha'iir.

[**Yellowhammer** *Emberiza citrinella* AV?

The only record of the monotypic Yellowhammer in Egypt is one bird observed in a flock of 1200 Corn Buntings on 21 February 1980 at Heliopolis (Baha el Din 1984*a*).]

Cirl Bunting *Emberiza cirlus* AV

Local name Durrsa suudaa- az-zawr درسة سوداء الزور.

Accidental visitor. On geographical grounds the form most likely to occur in Egypt is *E. c. cirlus*.

There is a single record of the Cirl Bunting in the country. One was collected at Damietta in 1879 (Schrader 1892). The specimen cannot be traced.

House Bunting *Emberiza striolata* RB Plate 5

Local name Durrsa mukhattata درسة مخططة.

Local and fairly common breeding resident of Sinai and the southern Eastern Desert (*E. s. striolata*) and possibly in the Gebel Uweinat area (presumably *E. s. sahari*).

North Coast Meinertzhagen (1930) collected a male at Salum on 28 January 1928 that he identified as *E. s. sahari*. This is the only known record from this area.

Western Desert Six House Buntings were seen at Gebel Uweinat from 5 November to 4 December 1968 (Misonne 1974). The species may breed locally (Goodman *et al.* 1986).

Eastern Desert This species is common in the Gebel Abraq area, where fledglings have been noted in the first half of April 1985. It is a locally common bird in rocky and vegetated wadis of Gebel Elba (SMG, PLM) and has also been collected at Gebel Shellal (Goodman 1984*a*).

Sinai Hardy (1946) remarked that a small colony of House Buntings occurred in Wadi Feiran; there are several subsequent records from this locality. Marchant (1939–40) saw numerous birds in October and November at Gebel Somar and he assumed that it was a local resident. It has also been noted at St Katherine in early to mid-September 1972 (ZM-HUJ) and at Ain el Furtaga on 4 July 1973 (TAUM). Several were noted on 26 August 1982 in Wadi Nasb and between Nuweiba and St Katherine (SMB).

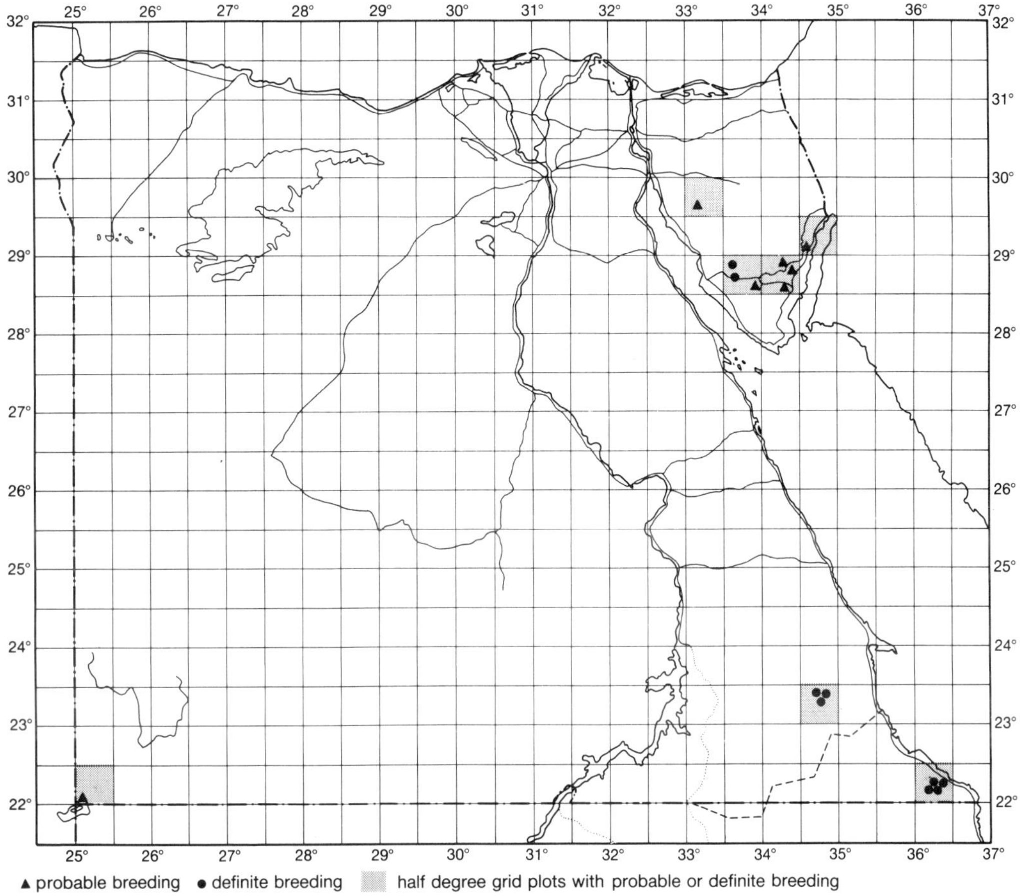

Map B103. Breeding distribution of House Bunting *Emberiza striolata* between 1930–87.

Cinnamon-breasted Rock Bunting *Emberiza tahapisi* AV

Local name Durrsa as-sakhr al-afriiqiyya درسة الصخر الأفريقية.

Accidental visitor. On geographical grounds the form most likely to occur in Egypt is *E. t. septemstriata.*

On 19 and 20 April 1984 an adult male Cinnamon-breasted Rock Bunting was observed near St Katherine (Anon. 1984*a*; Siering 1986). We have examined a photograph taken of the bird (MS) and concur with the identification.

Cinereous Bunting *Emberiza cineracea* AV

Local name Durrsa shaamiyya درسة شامية.

Accidental visitor. Specimens of both *E. c. cineracea* and *E. c. semenowi* have been collected in Egypt.

There are only a few records of the Cinereous Bunting from the country. A specimen of *E. c. cineracea* was taken on 21 March 1972 at St Katherine (ZM-HUJ) and one of *E. c. semenowi* on 12 April 1928 at Gebel Elba (Meinertzhagen 1930; BMNH). On 27 August 1983 a female was observed 20 km north-east of St Katherine (SMB). A male noted on 11 May 1982 at El Gorah was identified as *E. c. cineracea* (Bruun 1984). At Elat, Israel, there are at least 10 records of *E. c. cineracea* and three of *E. c. semenowi* (Krabbe 1980).

Ortolan Bunting *Emberiza hortulana* PV (WV)

Local names Durrsa ash-sha'iir درسة الشعير, s̲a'wa al-h̲at̲ab.

Fairly common passage visitor in autumn from mid-August to mid-October and more common in spring from (late February) mid-March to early (mid-) May. Apparently occasional winter visitor. Monotypic.

In autumn the Ortolan Bunting is regularly observed in small numbers mainly in the Suez Canal area, Eastern Desert, and Sinai. Between 31 August and 24 September 1981 a total of 41 migrants were counted at Zaranikh (IP, UGS). The species is rare along the North Coast and in the Nile Delta and Valley. The only known autumn records from the Western Desert are 15 seen or captured in Bahariya between 30 August and 15 September 1983 (HB *et al.*). The latest autumn reports are one at Port Said on 17 October 1961 (FMNH) and eight at Suez on 18 October 1981 (RGB *et al.*). The single winter record is a bird at Port Said on 3 December 1961 (FMNH).

The earliest spring record is one at Kom Ombo on 23 February 1976 (Short and Horne 1981). Between mid-March and late April the species is observed in small numbers in the Nile Delta and Valley, the Faiyum, Eastern Desert, and Sinai. On 2 May 1985, 34 were noted at Lake Bardawil (Fry *et al.* 1985). Western Desert records include several at Wadi el Natrun in March and April (Meinertzhagen 1930) and a few in Siwa on 27 March 1927 (Moreau 1928). At Bahig this species was recorded during all six PMS spring field seasons between 20 March and 26 April (one late bird on 14 May 1971), with a maximum of 97 in any one year (Horner 1977).

Cretzschmar's Bunting *Emberiza caesia* PV WV?

Local name Durrsa zarqaa- al-ra-s درسة زرقاء الرأس.

Fairly common passage visitor in autumn from mid-August to late September (early November) and in spring from (mid-February) mid-March to late April. According to Meinertzhagen (1930) a few remain in Egypt during some winters; there is no recent information confirming this. Monotypic.

Raw (1921) noted Cretzschmar's Buntings regularly in autumn and spring at Abu Zabal. In more recent years this species has been found to be uncommon on autumn passage in the Nile Delta and Valley. The only autumn records from the Western

Desert are single birds in Bahariya on 4 and 28 September 1983 (HB *et al.*). Between 23 August and 24 September 1981, 39 migrants were observed at Zaranikh (IP, UGS). The species has been taken at Bahig between 30 August and 30 September (USNM). The latest recent autumn reports are on 20 October 1983 at Port Said (Casement 1984) and on 8 November 1983 at Suez (JDG).

This species was noted at Gebel Elba in February 1925 (Madden 1930) and collected on 19 February 1928 (BMNH) and 25 February 1938 (GZM). Presumably these records are of early spring migrants. It has been observed at the Daedalus Reef, Red Sea, between 29 February and 6 April 1936 (Bulman and Beale 1944). From mid-March to mid-April a few are regularly recorded in the Nile Delta and Valley, the Eastern Desert, and Sinai. During the PMS survey at Bahig between 1967 and 1973, this species was captured in small numbers during all six spring seasons, with a maximum of 36 birds in 1969 and 1971 (Horner 1977). In late March 1910 it was reported to be 'exceedingly common' in Wadi el Natrun (Nicoll 1912*b*); the only more recent spring record from the Western Desert is two males in Bahariya on 9 March 1980 (Goodman *et al.* 1986). The latest spring report appears to be one on 26 April 1984 at Hurghada (SH, DM).

Rustic Bunting *Emberiza rustica* AV

Local name Durrsa ṣad-iyya درسة صدئية.

Accidental visitor. The single Egyptian specimen examined is referable to *E. r. rustica*.

The only known record of the Rustic Bunting in Egypt is one collected at Bahig on 22 October 1971 (Goodman and Watson 1983; USNM). At Elat, Israel, it is a regular autumn migrant in small numbers and a few remain for the winter (HSh). Specimens have been taken at this locality on 23 October 1968 (USNM), 11 November 1968 (USNM), and 2 March 1978 (TAUM).

Little Bunting *Emberiza pusilla* AV

Local name Durrsa ṣaghiira درسة صغيرة.

Accidental visitor. Monotypic.

One Little Bunting was noted at Agamy on 2 November 1980 (SV). Sinai records include one taken at St Katherine on 5 March 1970 (ZM-HUJ) and single birds captured on 18 and 19 October 1971 (DL). At Elat, Israel, this species is apparently a regular autumn migrant in small numbers and occasionally winters locally (Paz 1987; HSh). A specimen was taken at this locality on 19 November 1968 (USNM).

Yellow-breasted Bunting *Emberiza aureola* AV

Local name Durrṣa ṣafraa- aṣ-ṣadr درسة صفراء الصدر.

Accidental visitor. The single Egyptian specimen examined is referable to *E. a. aureola*.

There appears to be only one record of the Yellow-breasted Bunting in Egypt. An adult male was collected at St Katherine on 3 September 1972 (ZM-HUJ). This species has been seen on a few occasions in September between 1979 and 1986 at Elat, Israel (Paz 1987; HSh).

Reed Bunting *Emberiza schoeniclus* AV

Local names Durrsa al-ghaab درسة الغاب, s̱a'wa al-qas̱ab.

Accidental visitor to Egypt. All of the material examined, including specimens discussed by Meinertzhagen (1930), is referable to *E. s. intermedia*.

There are only a few Reed Bunting records from Egypt. These include: two birds at Port Said on 6 December 1906, a flock at Salum on 28 January 1928, one female at Damietta on 22 March 1871 (Meinertzhagen 1930), and one in Wadi Feiran on 19 November 1956 (ZM-HUJ). At Elat, Israel, this species is an occasional winter visitor between November and mid-March (Krabbe 1980).

Black-headed Bunting *Emberiza melanocephala* PV

Local name Durrsa suudaa- al-ra-s درسة سوداء الرأس.

Rare passage visitor in autumn from early August to mid-September and in spring from late March to mid-May. Monotypic.

The Black-headed Bunting appears to be a more regular migrant in Sinai than in other portions of the country. Single birds were noted at Zaranikh on 16 August and 2 September 1981 (IP, UGS). In two autumn seasons of systematic bird netting at St Katherine, six Black-headed Buntings were captured from 26 August to 6 September 1971 and 16 from 9 August to 19 September 1972 (DL). Five were seen at this locality on 23 August 1986 (EE *et al.*). Other autumn records from Sinai include a female on 25 August 1911 in Wadi Feiran (Meinertzhagen 1930) and one female on 18 September 1980 at Sharm el Sheikh (DJM, DES).

The earliest known spring record is 30 March 1946 when one flew aboard a ship that had just departed Port Said (Hardy 1946). A male was observed at Suez on 3 May 1982 (Baha el Din 1984*a*). The only known spring records from Sinai are birds on 10 and 20 May 1972 at St Katherine (ZM-HUJ). This is surprising because this species is considered a fairly common migrant between mid-April and late May at Elat, Israel (Krabbe 1980).

Corn Bunting *Miliaria calandra* PV WV

Local names Durrsa درسة, s̱a'wa.

Fairly common passage and winter visitor between (early September) mid-October and early (mid-) April, mostly to the northern portion of the country. All Egyptian material examined is referable to *M. c. calandra* (cf. Vaurie 1959).

Corn Buntings are regular winter visitors in small flocks to the cultivated portions of the Nile Delta and Valley (as far south as Mallawi), Suez Canal area, the Faiyum, and northern Sinai. Recent autumn Corn Bunting records from the Red Sea include two on 13 November 1982 on Gezira Umm el Showra, two on 14 November 1982 at Ras Banas, and one on 15 November 1982 at Hurghada (Baha el Din and Saleh 1983). An exceptionally large flock of 1200 was seen on 21 February 1980 at Heliopolis (SMB). Along the Mediterranean coast this species has been noted at El Dab'a in September 1942 (Molteno 1944), at sea off Alexandria on 26 November 1951 (Zink 1954), and at Salum on 21 January 1920 (Meinertzhagen 1930; AMNH).

This species was captured in two of the six PMS spring field seasons at Bahig between 1967 and 1973 (Horner 1977). In the Western Desert, Corn Buntings have been recorded with some regularity in Wadi el Natrun and small flocks were noted in Bahariya on 4 March 1980 and in Kharga on 7 and 9 April 1925 (Goodman *et al.* 1986). The species was observed between 14 and 30 March 1979 on Gezira Tiran (RF).

APPENDIX 1

SPECIES REJECTED FROM THE EGYPTIAN LIST

In this appendix we have summarized the species which have been reported in the literature as occurring in Egypt but for which this information is considered unacceptable. The basis for rejection is when specimen, photographic, or observational details are lacking, vague, or at fault.

Meinertzhagen (1930) did an excellent job of tabulating the literature and records of species poorly documented in Egypt. To avoid redundancy we have only presented detailed information for those species not treated by Meinertzhagen, reported in the post-1930 literature, or discussed by Meinertzhagen but for which new information has been published.

Black-throated Diver *Gavia arctica*

Koenig (1932) reported observing a Black-throated Diver on 9 March 1898 at Abu Zenima. The details of this record are ambiguous and the species cannot be admitted to the Egyptian list. It has been observed on at least five occasions at Elat, Israel (Krabbe 1980; Paz 1987), where on 7 February 1970, one was found dead on the beach (Goodman and Storer 1987).

Slavonian Grebe *Podiceps auritus*

Rejected by Meinertzhagen (1930). No more recently published details are known for its occurrence in Egypt.

Wedge-tailed Shearwater *Puffinus pacificus*

Winterbottom (1936) mentioned that a shearwater observed on 27 March 1935 in the Gulf of Suez may have been this species. One was observed near Quseir in late November 1983 (Bezzel 1987*b*). Neither of these records is acceptable.

Storm Petrel *Hydrobates pelagicus*

Storm Petrels were reported 80 km north-west of Port Said in March and October but Meinertzhagen (1930) did not accept these observations. We do not know of any verifiable records of Storm Petrels from Egyptian waters (*pace* Etchécopar and Hüe 1967).

Red-footed Booby *Sula sula*

A Red-footed Booby was reported to have been observed near Suez in June 1942 (Maclaren 1946). This record is not acceptable. The wings of a booby that hit the Sanganeb lighthouse, off Port Sudan, Sudan, on 7 November 1920 were identified as this species (Ticehurst 1924).

Masked Booby *Sula dactylatra*

Rejected by Meinertzhagen (1930). No more recently published details are known for its occurrence in Egypt.

Pygmy Cormorant *Phalacrocorax pygmeus*

Rejected by Meinertzhagen (1930). One found in an exhausted state on 2 January 1982 at Sharm el Sheikh (Hovel 1987) is the only recent report from Egypt; however, details are lacking and this record is not acceptable. See Long-tailed Cormorant in species accounts for further details.

Great Frigatebird *Fregata minor*

This species or *Fregata* sp. have been observed a few times off the southern shores of Sinai (Paz 1986; Hovel 1987); however, details are lacking and these records are not acceptable.

Black Egret *Egretta ardesiaca*

On 2 and 3 August 1980, two black egrets believed to be this species were observed near Aswan (Dekker 1981). Some of the details of this record are lacking and the possibility that the birds were inland Western Reef Herons or melanistic Little Egrets cannot be eliminated. Moreover, dark ('dirty') Purple Herons and Cattle Egrets have recently been noted near Aswan.

African Open-billed Stork *Anastomus lamelligerus*

On 1 July 1919 an African Open-billed Stork was observed flying about the Giza Zoological Gardens (Nicoll 1919). Meinertzhagen (1930) did not accept this record. In August 1938 a single bird was noted near Qena (El Negumi *et al.* 1950).

Bewick's Swan (Whistling Swan) *Cygnus columbianus*

In the early 1950s, three Bewick's Swans were reported to have been shot from a group of six (Anon. 1953). The details of the identification are ambiguous and the specimens cannot be traced.

White-faced Whistling Duck *Dendrocygna viduata*

While boating down the Nile, soon after crossing from Sudan into Egypt, Chapman (1930) saw a flock of White-faced Whistling Ducks preening on a sand-bank. He further noted that these birds had been favourite daily companions for months past in the Sudan but he had not expected to find them so far north.

Scaup *Aythya marila*

Rejected by Meinertzhagen (1930). No more recently published details are known for its occurrence in Egypt.

Goldeneye *Bucephala clangula*

Meinertzhagen (1930) rejected the Goldeneye from the Egyptian list. Hardy (1946) reported that this species occasionally occurs in the Gulf of Aqaba. We are unaware of the evidence this statement was based on.

Goosander *Mergus merganser*

Rejected by Meinertzhagen (1930). No other published details are known for its occurrence in Egypt. The only more recent record is five seen near Fayid on 28 January 1950 (HGB). It is interesting to note that bone remains recovered from a late Pleistocene site near Kom Ombo (15 000–10 000 BC) have been identified as this species (Churcher 1972).

Pallas's Fish Eagle *Haliaeetus leucoryphus*

Meinertzhagen (1930) observed 'A large Sea Eagle with pale golden brown head and nape and dark brown back, and with a pure white bar across the centre of the tail, ... in Lake Timsah on the Suez Canal in February 1923, and was almost certainly this species. ... But the species cannot be admitted on a single observation record.' There has been no subsequent Egyptian record. Records from Israel include one observed on 3 April 1981 at Hazeva (Ilani 1981) and an adult on 27 December 1980 at En Yahav (Hovel 1987).

Hooded Vulture *Necrosyrtes monachus*

Rejected by Meinertzhagen (1930). No more recently published details are known for its occurrence in Egypt.

Dark Chanting Goshawk *Melierax metabates*

Two Dark Chanting Goshawk specimens were collected in 'Égypte' in 1851 and 1861 (RNHL). It is impossible to know if the birds were collected within the modern political boundaries of Egypt or south in former colonial territories.

Rough-legged Buzzard *Buteo lagopus*

Meinertzhagen (1930) rejected Nicoll's statement that the Rough-legged Buzzard was not uncommon during autumn and spring migration. This species has also been reported from the country by other authors (e.g. Boyd 1917; Molteno 1944). It is presumed that birds reported as this species were actually unusual colour phases of the Buzzard and/or the Long-legged Buzzard.

Secretary Bird *Sagittarius serpentarius*

This species was reported near Cairo in the winter of 1886–7 (Kaiser 1890). The record is not acceptable.

Black Francolin *Francolinus francolinus*

Rejected by Meinertzhagen (1930). No more recently published details are known for its occurrence in Egypt.

Black Crake *Limnocorax flavirostra*

In the ZMA there are two adult Black Crake specimens from the Sillem–van Marle collection taken at 'Damiette, N.W. Egypte'. The normal range of this sedentary species is sub-Saharan Africa (Snow 1978) and there is no other record of it in Egypt. The procurement of these specimens is enigmatic.

Allen's Gallinule *Porphyrula alleni*

Shelley (1872) reported that S. Allen obtained one Allen's Gallinule in the Alexandria market. This record was accepted by Meinertzhagen (1930). Allen published numerous pre-1872 papers on his observations and collections of Egyptian birds, including those obtained in markets, and never mentioned this record. Since the specimen cannot be traced and this is the only report of this species from the country it cannot be accepted.

Crested Coot *Fulica cristata*

Rejected by Meinertzhagen (1930). No more recently published details are known for its occurrence in Egypt.

Andalusian Hemipode *Turnix sylvatica*

Rejected by Meinertzhagen (1930). No more recently published details are known for its occurrence in Egypt.

Arabian Bustard *Ardeotis arabs*

Rejected by Meinertzhagen (1930). No more recently published details are known for its occurrence in Egypt.

Black-headed Lapwing *Hoplopterus tectus*

Johnsgard (1981) mentioned that the Black-headed Lapwing straggles to Egypt and sometimes further north. An old specimen in the BMNH labelled 'Egypt' was not accepted by Meinertzhagen (1930) as evidence for this species' occurrence in the country. Another bird was obtained in 1869 in Wadi Arava, Israel (Meinertzhagen 1930; Hovel 1987). These specimens are presumably the basis for Johnsgard's statement.

Herring Gull *Larus argentatus*

Rejected by Meinertzhagen (1930). El Negumi *et al.* (1950) reported that two Herring Gulls were collected at Damietta on 12 April 1899. These specimens, which are housed in the GZM, have been examined and are Yellow-legged Gulls. There is no evidence that Herring Gulls of the *argentatus* group reach Egyptian waters. A 'Herring Gull' fitted with a Rossitten ring on 24 June 1934 was recovered at El Arish on 23 June 1936 (Mackintosh 1941). This record is rather problematic. The ringed bird was presumably a chick of *L. fuscus*, a species that breeds sympatrically with Herring Gulls throughout parts of north-western Europe.

Glaucous Gull *Larus hyperboreus*

There is no evidence that the Glaucous Gull occurs in Egyptian waters proper. Phillips (1915) collected a male on 18 April 1914 at Aqaba, Jordan (MCZ). Meinertzhagen (1930) mentioned this specimen as 'the only record from Egyptian waters'.

Great Black-backed Gull *Larus marinus*

There are several older records of this species in Egypt (e.g. Von Heuglin 1873; Schrader 1892; Nicoll 1919). However, this species was not admitted to the national bird list by Meinertzhagen (1930). On 23 December 1924, two gulls were observed off Port Said and based on a written description may well have been this species (Bolam in Chapman 1930). One reputed Great Black-backed Gull was noted at Port Said on 25 June 1945 (Phillips 1947). There were a few unconfirmed sightings from September of 1973 and 1978 in northern Sinai (Paz 1987). To-date positive documentation of this species' occurrence in Egypt is lacking.

Arctic Tern *Sterna paradisaea*

Meinertzhagen (1924) reported that an Arctic Tern hit the Damietta lighthouse on 5 November 1920 but did not admit the species to the Egyptian list in his 1930 book. More recently Fischer (1959) noted a flock of 30 to 35 birds in Suez Bay on 16 April 1958. Without further details these records are not acceptable. At Elat, Israel, six birds were observed on 24 April 1978 (Argyle and Gal 1978) and a single bird on 2 July 1982 (Ben Dov and Golan 1983). An exceptional inland record, which gives some credence to the above observations, is one Arctic Tern captured on 16 May 1982 at Khartoum, Sudan (Nikolaus 1984).

Sooty Tern *Sterna fuscata*

There are no substantiated reports of the Sooty Tern in Egyptian waters proper. The basis of Hovel's (1987) statement that this species is 'most probably a visitor to S.[outh] Sinai coasts' is unknown. At Elat, Israel, it is considered an accidental vagrant. Records from this locality include one noted on 27 July 1980 and three on 31 July and 2 August 1980 (Ben Dov and Golan 1983). Two adults were noted at Muhammad Qol, about 125 km south of the Egyptian–Sudanese border, on 13 August 1982 (Nikolaus 1984).

Common Noddy *Anous stolidus*

An old specimen of the Common Noddy in the IRSNB is labelled 'Égypte'. The bird was received at the museum sometime before 1841. This species has recently been found nesting on islands off the coast of Suakin, Sudan (Moore and Balzarotti 1983) and there are several records from the Saudi Arabian Red Sea coast (Halfawi 1945; King 1978).

Little Auk *Alle alle*

Rejected by Meinertzhagen (1930). No more recently published details are known for its occurrence in Egypt.

Woodpigeon *Columba palumbus*

Yom-Tov (1987) included the Woodpigeon in a list of the birds of Sinai. The basis for this is unknown and its inclusion on the Egyptian list is not acceptable.

Oriental Cuckoo *Cuculus saturatus*

One Oriental Cuckoo was reported to have been seen on Gezira Tiran in the 1970s (Su-Aretz 1976). The details of this observation are lacking and the record cannot be accepted.

Tawny Owl *Strix aluco*

Rejected by Meinertzhagen (1930). No more recently published details are known for its occurrence in Egypt.

Tengmalm's Owl *Aegolius funereus*

Rejected by Meinertzhagen (1930). No more recently published details are known for its occurrence in Egypt.

Red-necked Nightjar *Caprimulgus ruficollis*

Casement (1984) reported that on 24 September 1983 a Red-necked Nightjar flew aboard a ship anchored off Port Said. This record is not acceptable.

Striped Kingfisher *Halcyon chelicuti*

Rejected by Meinertzhagen (1930). No more recently published details are known for its occurrence in Egypt.

Carmine Bee-eater *Merops nubicus*

Rejected by Meinertzhagen (1930). No more recently published details are known for its occurrence in Egypt.

Small Skylark (Oriental Skylark) *Alauda gulgula*

Meinertzhagen (1930) identified a specimen taken on 15 October 1914 in Wadi el Natrun as a Small Skylark. We have examined this specimen (GZM) and it is referable to the Skylark. The occurrence of the Small Skylark at Elat, Israel, has recently been documented (Shirihai 1986a).

Shore Lark *Eremophila alpestris*

Meinertzhagen (1930) reported a bird observed by Mrs Moreau on 9 April 1927 near Cairo as a Shore Lark. This record is not acceptable and the bird was presumably a Temminck's Horned Lark.

African Sand Martin *Riparia paludicola*

De Bièvre (1983) mentioned that African Sand Martins were observed in spring 1982 at Lake Maryut, Lake Manzala, and other places in the Nile Delta, and that one was found dead near Lake Qarun. We have examined a photograph of the dead bird (specimen was not saved) and are of the opinion that the bird was a Sand Martin. Further, De Bièvre's sight records are not acceptable. This species occurs along the Nile in Sudan as far north as Wadi Halfa (Nikolaus 1987).

Dipper *Cinclus cinclus*

Rejected by Meinertzhagen (1930). No more recently published details are known for its occurrence in Egypt.

Wren *Troglodytes troglodytes*

Rejected by Meinertzhagen (1930). No more recently published details are known for its occurrence in Egypt.

Black Bush Robin *Cercotrichas podobe*

One was observed in south-eastern Egypt, near Gebel Elba, on 29 January 1896 (Cholmley 1897). Details of this observation are lacking and it is not acceptable. This species is, however, known to occur in the Red Sea hills of Sudan (Nikolaus 1987). There are several recent records of this species in the general area of Elat, Israel (Eames 1986; Shirihai 1986*b*).

Black Wheatear *Oenanthe leucura*

The type specimen of *O. l. syenitica* was taken in June 1852 'bei El Kab in Oberegypten' (Von Heuglin 1869*a*). El Kab is a considerable distance from any region where this species has been documented and the type locality is presumably in error. Meinertzhagen (1930) collected a pair referable to *syenitica* on 28 January 1928, 20 km west of Salum, just across the border in Libya. Recent observations of 'Black Wheatears' in Egypt (Meininger and Mullié 1981*c*; Baha el Din 1984*d*) may have been this species, Eastern Pied Wheatears (*O. picata ophistoleuca*), or most likely White-crowned Black Wheatears in aberrant plumage.

Aquatic Warbler *Acrocephalus paludicola*

Von Heuglin (1869*b*) stated that the Aquatic Warbler was sometimes common in Lower Egypt. Meinertzhagen (1930) did not admit it to the Egyptian list. More recent sight records include one near Ismailiya on 17 October 1943 (Meiklejohn 1944), single birds at Lake Maryut on 19 April 1964 and 19 February 1966 (JWW), and one in Wadi Kabrit on 8 March 1984 (SH, DM).

Blyth's Reed Warbler *Acrocephalus dumetorum*

One bird identified as this species was captured and ringed at St Katherine in 1970 or 1971 (Hovel 1987). However, details are lacking and the record is not acceptable.

Booted Warbler *Hippolais caligata*

Specimens previously reported as *H. c. rama* were reidentified by Meinertzhagen (1930) as *H. pallida*. The basis for its inclusion as a summering bird in Sinai (Yom-Tov 1987) is unknown and this statement is not acceptable. One was recorded at Elat, Israel, on 3 May 1983 (Paz 1987).

Upcher's Warbler *Hippolais languida*

According to Hovel (1987) this species is a scarce passage visitor to Sinai. However, the basis for this statement is unknown. It has been documented in Israel and Sudan (Hovel 1987; Nikolaus 1987).

Dartford Warbler *Sylvia undata*

Rejected by Meinertzhagen (1930). No more recently published details are known for its occurrence in Egypt.

Nuthatch *Sitta europaea*

Rejected by Meinertzhagen (1930). No more recently published details are known for its occurrence in Egypt.

Wall Creeper *Tichodroma muraria*

Rejected by Meinertzhagen (1930). No more recently published details are known for its occurrence in Egypt.

Tropical Boubou *Lanarius ferrugineus*

Rejected by Meinertzhagen (1930). No more recently published details are known for its occurrence in Egypt.

Magpie *Pica pica*

Rejected by Meinertzhagen (1930). No more recently published details are known for its occurrence in Egypt.

Alpine Chough *Pyrrhocorax graculus*

Rejected by Meinertzhagen (1930). No more recently published details are known for its occurrence in Egypt.

Chough *Pyrrhocorax pyrrhocorax*

Several nineteenth-century ornithologists recorded the Chough in Sinai but to our knowledge no specimens were collected. Meinertzhagen (1930) did not admit this species to the Egyptian list. In more recent years there have been two published records of Chough in Egypt. Horváth (1959) observed two on 13 September 1957 on the cliffs of Gebel el Muqattam near Cairo, and one was noted on 9 May 1982, 35 km north of Hurghada (Frazier *et al.* 1984). Without further details these records are not acceptable.

Jackdaw *Corvus monedula*

Meinertzhagen (1930) mentioned that M. J. Nicoll observed a pair and three young Jackdaws near Suez in May 1922. Further, five Jackdaws were reported at Port Taufiq on 7 August 1939 (Greaves 1939). Bijlsma and Meininger (1984) have presented evidence that these records are almost certainly of House Crows, an introduced species still nesting near Suez.

Red-cheeked Cordon-bleu *Uraeginthus bengalus*

This species has apparently been released or escaped from captivity on several occasions. In 1906, 1908, and 1911 the Red-cheeked Cordon-bleu was reported to be free-living in the Giza Zoological Gardens (Nicoll 1912*a*). It was observed with some frequency in the El Maadi area during the mid-1960s and one was collected on 5 May 1964 (HS). There are no more recent records from the country and it is presumed that a viable breeding population has not become established. In late January 1985 about 100 were for sale in a Cairo pet shop (SMG).

Common Waxbill *Estrilda astrild*

Nicoll (1912*b*) mentioned that Common Waxbills, which had escaped from cages, were free-living in the Giza Zoological Gardens in 1906, 1907, and 1910. On 13 February 1976 a flock of 15, composed of males and females, was observed for 10 minutes north of Saqqara (Short and Horne 1981). This is the only recent record of this sub-Saharan species in Egypt and almost certainly represents escaped domestic birds. It is rather striking that the Saqqara observations are from an area that has been extensively visited in recent years and this species is either relatively rare or the Short and Horne record is actually of Avadavats.

Common Redpoll *Carduelis flammea*

Rejected by Meinertzhagen (1930). No more recently published details are known for its occurrence in Egypt.

APPENDIX 2

GAZETTEER

This gazetteer contains all of the localities mentioned in the text and their latitude and longitude coordinates. Alternative place names are also cross-referenced. Egypt is divided into 27 governorates (Fig. A2.1) and these are mentioned under separate entries.

There is considerable variation in the systems used to transliterate Egyptian place names into English. In an attempt to standardize the spelling of locality names mentioned in the text, we have used those given on the 'Survey of Egypt' maps published by the Egyptian Geological Survey. In a few cases we deviate from this series for places well known by another spelling. Further, several localities mentioned in the text are not identified on the 'Survey of Egypt' maps and for these we have adopted parallel spellings. Place names mentioned in the main text within parentheses are generally not listed in the gazetteer.

Throughout the book we have used geographical or habitat terms from the Arabic. A few are from the Hebrew and designated by '(H)'.

Ain (Ein, Uyun)—spring or well
Bahr—sea, but sometimes used locally for large body of water, river, or canal
Bir (Beer)—well or sometimes spring or cistern
Beheira—lake
Birket—pond or sometimes lake
Bur—port
Darb—camel trail
Ein (Ain)—spring or well
Feddan—land unit measurement, 1 feddan = 1.038 acres or 4200 m^2
Gebel (Jebel)—mountain or large hill
Gezira—island
Ghubbet (Ghubba)—bay
Hamada—rocky desert covered with non-smooth pebbles
Har (H)—mountain, *see* gebel
Hatiyet—area with vegetation, particularly acacia trees, in otherwise barren desert (generally used in the Western Desert)
Hod—basin
Karkur—winding gorge (generally used in the Western Desert and extreme Upper Egypt)
Kef (H)—cape or point, *see* ras

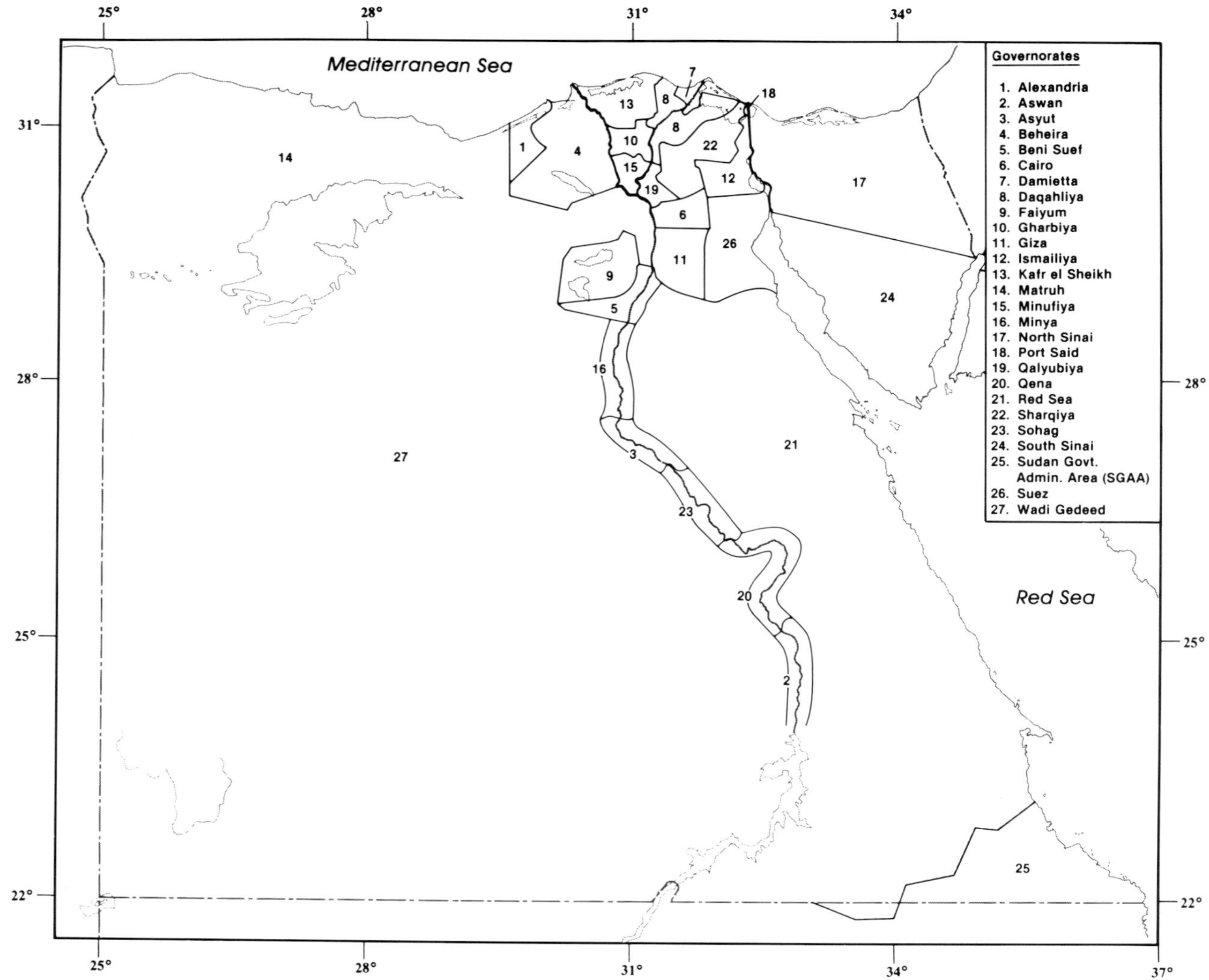

FIG. A2.1. Map of Egyptian governorates.

Khor—dry river course or valley (generally used in Upper Egypt), *see* wadi
Maayan (H)—spring, *see* ain
Marsa (Mersa)—bay, inlet, or cove
Medinet (Medina)—city
Minqar—distinctive cliff or bluff
Nahal (H)—valley, *see* wadi
Neve (H)—oasis, *see* wahat
Oasis—vegetated area surrounded by desert and generally with naturally occurring water
Qara—prominent hill in flat desert
Qasr—literally castle, now used to denote central place such as a square
Ramat (H)—plateau
Ramleh—sand
Ras—cape or point
Sanyet (Sanya)—well
Sebakha—salt-pan
Serir—undulating desert covered with small wind-polished pebbles
Sharm (Sherm)—bay, often used for anchorage
Tell—hill
Wadi—dry river-course or valley
Wahat (Waha)—oasis
Yam (H)—sea or large body of water, *see* bahr

Locality	Governorate, country, or cross-reference[1]	N. Lat.		E. Long.	
		(°)	(′)	(°)	(′)
Abd el Malik (Abdel Malek)	Wadi Gedeed	23	50	25	18
Abu el Matamir	Beheira	30	55	30	11
Abu Ghalib	Minufiya	30	16	30	54
Abu Ghusun (Ghusoon)	Red Sea	24	26	35	12
Abu Girgeh (Girga)	see Girga				
Abu Gridi	Damietta	31	18	31	46
Abu Hammad	Sharqiya	30	32	31	40
Abu Handal	Aswan	22	40	32	15
Abu Himeira	Red Sea	23	30	35	29
Abu Minqar	Wadi Gedeed	26	30	27	38
Abu Qir (Kir)	Alexandria	31	19	30	04
Abu Qurqas	Minya	27	56	30	50
Abu Ramad	SGAA	22	21	36	27
Abu Rauwash (Ruwaysh)	Giza	30	02	31	06
Abu Rudeis	see Ras Abu Rudeis				
Abu Sha'r	Red Sea	27	18	33	45
Abu Simbil (Simbel)	Aswan	22	22	31	38
Abu Sir	Giza	29	53	31	13
Abu Sir	Matruh	30	57	29	31
Abu Sultan	Ismailiya	30	25	32	19

Locality	Governorate, country, or cross-reference[1]		N. Lat. (°)	N. Lat. (′)	E. Long. (°)	E. Long. (′)
Abu Tig	Asyut		27	02	31	19
Abu Zabal	Qalyubiya		30	15	31	21
Abu Zenima	South Sinai		29	03	33	06
Abydos	Qena		26	11	31	55
Adabiya (Adabeya)	Suez		29	53	32	28
Adal Deib	SGAA		22	41	36	05
Aga	Daqahliya		30	56	31	17
Agamy	Alexandria		31	09	29	47
Aidhab	see Sawakin el Qadim					
Ain el Fogeia (Fukia)	South Sinai		29	36	33	11
Ain el Furtaga (Fortage)	South Sinai		29	03	34	33
Ain el Gedeirat (Qudeirat)	North Sinai		30	39	34	26
Ain Hegiya (Chagia)	South Sinai		29	23	33	24
Ain Musa (Ayun Musa)	South Sinai		29	52	32	39
Ain Shurafa	Wadi Gedeed		25	11	30	39
Ain Sukhna (Sokhna)	Suez		29	35	32	20
Ain Zuweina (Zueia, Zuya)	LIBYA		21	53	24	50
Alexandria	Alexandria		31	12	29	54
Almaza	Cairo		30	35	31	20
Amada Temple	Aswan		22	43	32	15
Aqaba	JORDAN		29	32	35	00
Armant	Qena		25	37	32	32
Arminna (Armenna)	Aswan		22	31	31	52
Asher Talaaf	Alexandria	approx.	30	50	29	45
Asirat	Qena		26	22	31	49
Aswan (Assouan)	Aswan		24	05	32	53
Aswan High Dam	Aswan		24	03	32	52
Asyut	Asyut		27	11	31	11
Atfih	Giza		29	24	31	15
Aushim	see Kom Aushim					
Ba'alwa	Sharqiya		30	33	31	51
Bahariya Oasis	Giza	approx.	28	21	28	52
Bahig	Matruh		30	56	29	35
Ballah	Ismailiya		30	46	32	19
Baltim	Kafr el Sheikh		31	33	31	05
Baris (Berys)	Wadi Gedeed		24	40	30	36
Barramiya	Red Sea		25	05	33	48
Barsiq	Beheira		31	07	30	20
Befi	Giza or Matruh	not located				
Beheira Land Co[2]	Kafr el Sheikh		31	21	30	48
Beheirat Murrat el Kubra	see Great Bitter Lake					
Benha	Qalyubiya		30	28	31	11
Beni Hassan	Minya		27	54	30	51
Beni Mazar	Minya		28	30	30	48
Beni Suef	Beni Suef		29	05	31	05
Berenice	Red Sea		23	58	35	22
Bilbeis	Sharqiya		30	25	31	34
Biq'at ha Yareah	see Moon Valley					
Bir Abbad	Red Sea		25	02	33	04
Bir Abraq	Red Sea		23	25	34	48

Locality	Governorate, country, or cross-reference[1]		N. Lat. (°)	N. Lat. (′)	E. Long. (°)	E. Long. (′)
Bir Abu Mureiwat	Red Sea		26	34	33	42
Bir Abu Sha'r (Shaar)	Red Sea		27	22	33	37
Bir Ambagi	see Bir Beida					
Bir Beida	Red Sea		26	05	34	07
Bir el Abd	North Sinai		31	01	33	00
Bir el Ghallaz	Matruh		30	52	26	48
Bir el Nuss	Matruh		30	16	26	17
Bir el Qaraya	see Bir Victoria					
Bir el Shab	Wadi Gedeed		22	20	29	45
Bir el Themada (Tmade)	North Sinai		30	10	33	28
Bir Gahliya	Red Sea		23	31	35	08
Bir Gifgafa (Gafgafa)	North Sinai		30	27	33	11
Bir Gindali	Suez		29	55	31	40
Bir Hasana (el Hasne)	North Sinai		30	28	33	47
Bir Himreyr	Red Sea		27	25	33	25
Bir Kansisrob (Kansathrope)	SGAA		22	15	36	22
Birket Accrashi[3]	Qalyubiya	not located				
Birket el Hagg	Qalyubiya		30	08	31	23
Birket el Qarun	see Lake Qarun					
Bir Kiseiba	Wadi Gedeed		22	42	29	56
Bir Lahfan (Lhfan)	North Sinai		31	01	33	52
Bir Matar	Giza		28	22	28	54
Bir Milaha (Mallaha)	Red Sea		27	34	33	27
Bir Misaha	Wadi Gedeed		22	12	27	57
Bir Naba (Nbea)	South Sinai		29	02	34	09
Bir Rabdeit (Rabdate)	SGAA		22	11	36	24
Bir Semna	Red Sea		26	27	33	42
Bir Sheitun	Red Sea		26	48	32	07
Bir Shellal	ISRAEL		31	17	34	29
Bir Taba	South Sinai		29	30	34	53
Bir Tarfawi (Terfawi)	Wadi Gedeed		22	55	28	53
Bir Umm Fawakhir	Red Sea		26	02	33	35
Bir Victoria	Beheira		30	24	30	37
Bitter Lakes		approx.	30	25	32	25
Brothers Reef			26	20	34	53
Buqbuq (Bugbug)	Matruh		31	31	25	34
Bur Fuad	see Port Fuad					
Burg el Arab	Matruh		30	55	29	32
Bur Said	see Port Said					
Caene	see Qena					
Cairo	Cairo		30	03	31	15
Chirbe	see Nahal Yam					
Clysma	see Suez					
Coral Island	see Gezira Fara'un					
Daedalus Reef			24	55	35	51
Dahab (Dhahab)	South Sinai		28	29	34	32
Dahshur	Giza		29	48	31	12
Dakhla Oasis	Wadi Gedeed	approx.	25	29	28	59
Damanhur	Beheira		31	02	30	28
Damietta	Damietta		31	25	31	48

Locality	Governorate, country, or cross-reference[1]		N. Lat. (°)	N. Lat. (′)	E. Long. (°)	E. Long. (′)
Damietta Lighthouse	Damietta		31	32	31	51
Dandara (Dendera)	Qena		26	10	32	39
Daraw	Aswan		24	25	32	56
Deir el Bahari	Qena		25	43	32	37
Deir el Raya	see Gebel el Teir					
Delta Barrage	Qalyubiya		30	12	31	08
Deversoir	Ismailiya		30	25	32	23
Dinaria (Dinariya)	Wadi Gedeed		25	44	28	54
Dineiqil	Wadi Gedeed		23	25	31	38
Dishna	Qena		26	07	32	28
Di-Zahav	see Dahab					
Dukka	see El Dakka					
Dumyat	see Damietta					
Dunqul	Wadi Gedeed		23	26	31	37
Dush (Doash)	Wadi Gedeed		24	34	30	42
East Uweinat	Wadi Gedeed	approx.	22	35	27	50
Edfu	see Idfu					
Edwa	see El Idwa					
Ein el Foqeia (Fukia, Foqeiya)	see Ain el Fogeia		29	36	33	11
Ein Gedi (Jedi)	ISRAEL		31	27	35	23
El Abbasia	Cairo		30	04	31	17
El Abbassa	Sharqiya		30	32	31	42
El Aiyat (Ayat)	Giza		29	37	31	15
El Akhawein	see Brothers Reef					
El Alamein	Matruh		30	49	28	57
El Amarna	Minya		27	39	30	58
El 'Ambag	see Great Bitter Lake					
El Amia		not located				
El Amiriya	Alexandria		31	05	29	48
El Arish	North Sinai		31	09	33	48
El Ashmunein	Minya		27	47	30	49
Elat (Eilat)	ISRAEL		29	33	34	57
El Badrshein	Giza		29	51	31	16
El Bahrain	Matruh		28	40	26	32
El Balyana	Sohag		26	14	32	00
El Baragil	Giza		30	04	31	09
El Bawiti	Giza		28	21	28	52
El Beida	Beheira		30	27	30	15
El Bersheh	Minya		27	45	30	56
El Burg	Kafr el Sheikh		31	35	30	59
El Cap	Ismailiya		30	57	32	18
El Dab'a (Dabah)	Matruh		31	02	28	26
El Dakka	Aswan		23	12	32	45
El Dilingat	Beheira		30	50	30	32
El Dokki	Cairo		30	04	31	15
Elephantine	Aswan		24	05	32	53
El Faiyum (city)	Faiyum		29	19	30	48
El Fashn	Beni Suef		28	49	30	54
El Ferden	see Firdan					
El Gedida	Giza		28	29	29	02

Locality	Governorate, country, or cross-reference[1]		N. Lat. (°)	N. Lat. (′)	E. Long. (°)	E. Long. (′)
El Gharbaniyat	Matruh		30	53	29	29
El Gorah	North Sinai		30	58	33	57
El Haddadi	Kafr el Sheikh		31	20	30	47
El Hammam	Beni Suef		29	14	31	01
El Hammam	Matruh		30	50	29	23
El Hamul	Kafr el Sheikh		31	19	31	10
El Hauwariya	Matruh		30	58	29	41
El Hawamidiya (Hawamdieh)	Giza		29	54	31	15
El Idwa	Faiyum		29	21	30	55
El Iskandariyah	see Alexandria					
El Kab	Aswan		25	08	32	48
El Kabrit	Ismailiya		30	17	32	23
El Karnak	Qena		25	43	32	38
El Khanka	Qalyubiya		30	13	31	21
El Kharga	Wadi Gedeed		25	26	30	33
El Kharruba	North Sinai		31	11	33	57
El Khatatba	Beheira		30	23	30	50
El Koriamat (Kuraimat)	Giza		29	18	31	13
El Kubri	Suez		30	02	32	33
El Kuntilla	North Sinai		30	00	34	41
El Lahun	Faiyum		29	13	30	59
El Laqeita	Red Sea		25	53	33	07
El Maadi	Cairo		29	58	31	15
El Madiq	Aswan		22	44	32	09
El Magdaba (Magdhaba)	North Sinai		30	53	34	01
El Maghra (Magrah, Moghra)	Matruh		30	15	28	55
El Mahalla el Kubra	Gharbiya		30	58	31	10
El Mahmudiya	Beheira		31	11	29	52
El Malaha	North Sinai		31	13	32	19
El Malki	Aswan		22	39	32	23
El Mansura	Daqahliya		31	03	31	23
El Mansuriya	Giza		30	08	31	05
El Manzala	Daqahliya		31	09	31	54
El Maragha	Sohag		26	42	31	36
El Marg	Qalyubiya		30	09	31	20
El Matariya	Cairo		30	07	31	19
El Matariya	Daqahliya		31	11	32	02
El Mazar	North Sinai		31	02	33	23
El Minya	Minya		28	06	30	45
El Mitla	see Mitla Pass					
El Moghra	see El Maghra					
El Omayed	Matruh		30	45	29	12
El Qa'	South Sinai	approx.	28	10	33	40
El Qahira	see Cairo					
El Qalag	Qalyubiya		30	11	31	21
El Qantara	Ismailiya		30	52	32	18
El Qaraya	see Bir Victoria					
El Qasr	Giza		28	21	28	51
El Qasr	Wadi Gedeed		25	47	28	52
El Qassasin	Sharqiya		30	32	31	51

Locality	Governorate, country, or cross-reference[1]		N. Lat. (°)	N. Lat. (′)	E. Long. (°)	E. Long. (′)
El Qatta	Giza		30	13	30	58
El Rashda (Rashida)	Wadi Gedeed		25	35	28	56
El Rekabiya	Damietta		31	25	31	34
El Saff	Giza		29	34	31	18
El Shallal	Aswan		24	03	32	53
El Shallufa	Suez		30	07	32	34
El Shatt	Suez		29	57	32	35
El Sinaniya	Damietta		31	24	31	47
El Suweiz	see Suez					
El Taloul	North Sinai		31	02	33	08
El Tell el Kebir	Sharqiya		30	33	31	47
El Tih	South Sinai	approx.	29	10	33	35
El Till el Amarna	see El Amarna					
El Tor	South Sinai		28	14	33	37
El Umbarakab	Aswan		23	40	32	53
Elwa el Agramiya	see Ilwa el Agramiya					
El Wasta	Beni Suef		29	20	31	12
El Zeiniya Qibli	Qena		24	45	32	41
En Fortaga	see Ain el Furtaga					
En Fukia	see Ain el Fogeia					
En Yahav	ISRAEL		30	40	35	14
Eskenderiya	see Alexandria					
Esna	see Isna					
Et Tassa	North Sinai		30	38	32	41
Et Tih	see El Tih					
Et Tina (E-Tina)	see El Malaha					
Et Tur	see El Tor					
Ezbet Zayat	Wadi Gedeed		25	14	29	45
Faiyum (region)	Faiyum	approx.	33	25	30	47
Fanara	Ismailiya		30	16	32	20
Farafra Oasis	Wadi Gedeed	approx.	27	00	28	00
Fariskur (Faraskur)	Damietta		31	20	31	43
Fayid	Ismailiya		30	20	32	18
Feiran Oasis	see Wadi Feiran					
Firdan	Ismailiya		30	41	32	20
First Cataract	Aswan		24	04	32	52
The Fjord	see Marsa Morakh					
Galal	Matruh		31	02	28	10
Gamasa	Damietta		31	28	31	34
Gaza	ISRAEL		31	30	34	27
Gebel Abraq	Red Sea		23	23	34	45
Gebel Abu Brush	Red Sea		22	31	33	59
Gebel Abu Dukhan (Dokhan)	Red Sea		27	13	33	16
Gebel Abu el Hasan	Red Sea		26	57	33	21
Gebel Abu Foda (Fawdah, Foeder)	Asyut		27	24	30	58
Gebel Abu Hamamid	Red Sea		24	15	34	48
Gebel Abu Harba	Red Sea		27	17	33	13
Gebel Abu Hareigal (Haragiel)	Red Sea		23	22	33	20
Gebel Adda	Aswan		22	16	31	36
Gebel Areif el Naqa	North Sinai		30	22	34	26

Locality	Governorate, country, or cross-reference[1]		N. Lat. (°)	N. Lat. (′)	E. Long. (°)	E. Long. (′)
Gebel Asfar	Qalyubiya		30	08	31	23
Gebel Ataqa	Suez		29	55	32	20
Gebel Elba	SGAA		22	11	36	21
Gebel el Farayid	Red Sea		23	35	35	22
Gebel el Galala el Bahariya	Red Sea	approx.	29	20	32	00
Gebel el Galala el Qibliya	Red Sea	approx.	28	50	32	30
Gebel el Goza el Hamra	Ismailiya		30	16	32	17
Gebel el Hamma (Hama)	North Sinai		30	44	33	43
Gebel el Igma (Egma)	South Sinai	approx.	29	10	33	45
Gebel el Muqattam (Mokattam)	Cairo		30	02	31	17
Gebel el Safra (Tsafra)	South Sinai		27	52	34	16
Gebel el Teir	Minya		28	12	30	46
Gebel el Thabt	South Sinai		28	16	34	01
Gebel el Zeit	Red Sea		27	56	33	30
Gebel Gharib	Red Sea		28	07	32	54
Gebel Halal	North Sinai		30	37	34	01
Gebel Hamata	Red Sea		24	12	35	00
Gebel Hamra	South Sinai		29	40	34	47
Gebel Hamra Dom	SGAA		22	39	35	39
Gebel Katherina (Catherine)	South Sinai		28	31	33	57
Gebel Libni	North Sinai		30	44	33	50
Gebel Maghara	North Sinai		30	44	33	22
Gebel Maryam	Ismailiya		30	33	32	18
Gebel Migif	Red Sea		24	47	34	27
Gebel Misikat el Gukh	Red Sea		26	28	33	24
Gebel Murra	Red Sea		22	22	33	47
Gebel Musa	South Sinai		28	32	33	59
Gebel Naq el Teir	Red Sea		26	42	32	55
Gebel Nuqrus	Red Sea		24	49	34	36
Gebel Qattar	Red Sea		27	05	33	22
Gebel Rahaba (Raba, Rabbah)	South Sinai		28	26	33	59
Gebel Ruweished (Rueis)	Red Sea		26	59	33	22
Gebel Sabbagh (Sabt)	South Sinai		28	12	34	04
Gebel Safsafa	South Sinai		28	34	33	58
Gebel Serbal (Sirbal)	South Sinai		28	39	33	39
Gebel Shayib el Banat	Red Sea		26	59	33	29
Gebel Shellal (Shallal)	SGAA		22	01	36	31
Gebel Somar	South Sinai		29	39	33	10
Gebel Umm Khisheib	North Sinai		30	15	32	58
Gebel Umm Samiuki	Red Sea		24	15	34	52
Gebel Umm Shomar (Shaumar)	South Sinai		28	22	33	55
Gebel Uweinat	Wadi Gedeed		21	50	25	00
Gebel Yelleq (Yeleg)	North Sinai		30	22	33	31
Gebel Zabara	Red Sea		24	45	34	42
Gedeirat	see Ain el Gedeirat					
Gemsa Bay	see Ghubbet el Gemsa					
Gezira	not located[4]					
Gezira Abu Minqar	Red Sea		27	13	33	52
Gezira Abu Rimathi (Ramada)	Red Sea		27	10	33	59
Gezira Amalawaya	see Gezira Umm Ladid					

Locality	Governorate, country, or cross-reference[1]		N. Lat. (°)	N. Lat. (′)	E. Long. (°)	E. Long. (′)
Gezira Ashrafi	Red Sea		27	46	33	41
Gezira Aswan	see Elephantine					
Gezira Atrun	see Kitchener's Island					
Gezira Baruda	see Gezira Wadi Gimal					
Gezira Dib	North Sinai[5]	approx.	31	13	32	19
Gezira el Akhawein	see Brothers Reef					
Gezira el Dibia	see Gezira Rawabel el Kabir					
Gezira el Hamra	Red Sea		27	33	33	45
Gezira el Makawarat	Red Sea		27	47	33	41
Gezira Fanadir	Red Sea		27	18	33	49
Gezira Fara'un (Fir'awn)	South Sinai		29	28	34	52
Gezira Gemsa	Red Sea		27	39	33	36
Gezira Ghanim	Red Sea		27	46	33	36
Gezira Giftun el Kabir	Red Sea		27	13	33	56
Gezira Giftun el Saghir	Red Sea		27	12	33	59
Gezira Gisum	see Gezira Qeisum					
Gezira Gubal el Kabir	Red Sea		27	40	33	48
Gezira Gubal el Saghir	Red Sea		27	12	33	47
Gezira Halaib	SGAA		22	16	36	39
Gezira Jubal	see Gezira Gubal					
Gezira Kalabsha	Aswan		23	33	32	52
Gezira Magawish	see Gezira Umm Agawish					
Gezira Mahabis	Red Sea		24	19	35	23
Gezira Mikauwa (Mukawwa)	Red Sea		23	50	35	48
Gezira Qeisum (south Geisum)	Red Sea		27	42	33	42
Gezira Qiftun	see Gezira Giftun					
Gezira Rawabel el Kabir (Ruwabel)	SGAA		22	25	36	32
Gezira Safaga	Red Sea		26	45	33	59
Gezira Saluga	Aswan		24	04	32	52
Gezira Shadwan	Red Sea		27	30	33	59
Gezira Showarit (Showra)	Red Sea		24	22	35	23
Gezira Sinafir	SAUDI ARABIA		27	57	34	42
Gezira Siyal group[6]	SGAA	approx.	22	47	36	16
Gezira Siyul (Siyal)	Red Sea		24	23	35	23
Gezira Siyul el Kabir	Red Sea		27	33	33	50
Gezira Tawila	Red Sea		27	35	33	46
Gezira Tiran	South Sinai		27	56	34	33
Gezira Umm Agawish el Kabir	Red Sea		27	09	33	52
Gezira Umm Agawish el Saghir	Red Sea		27	10	33	51
Gezira Umm el Heimat (Heimet)	Red Sea		27	42	33	37
Gezira Umm el Kiman	Red Sea		27	50	33	35
Gezira Umm el Showra	see Gezira Showarit					
Gezira Umm Geshat	SGAA		22	41	36	13
Gezira Umm Ladid	Red Sea		24	19	35	24
Gezira Umm Qamar (Gammar, Qanar)	Red Sea		27	21	33	54
Gezira Wadi Gimal (Gemal)	Red Sea		24	40	35	10
Gezira Zabargad			23	37	36	12

Locality	Governorate, country, or cross-reference[1]		N. Lat. (°)	N. Lat. (′)	E. Long. (°)	E. Long. (′)
Ghardaqa	see Hurghada					
Ghatta		not located[7]				
Gheit el Nassara	Damietta		31	24	31	49
Ghubbet el Gemsa	Red Sea		27	39	33	32
Ghubbet el Zeit	Red Sea		27	53	33	31
Gilf Kebir (Gilf el Kabir)	Wadi Gedeed	approx.	23	27	26	00
Girga	Qena		26	20	31	53
Giza	Giza		30	01	31	13
Golosaneh	see Qulusna					
Great Bitter Lake	Ismailiya		30	20	32	23
Great Sand Sea	Wadi Gedeed	approx.	27	00	26	00
Gulf of Aqaba		approx.	28	45	34	45
Gulf of Suez		approx.	29	00	33	00
Halaib	SGAA		22	13	36	38
Hamata	Red Sea		24	17	35	21
Hamifraz haNe'elan	see Marsa Morakh					
Hamrawein	Red Sea		26	15	34	12
Hash Medibah	Faiyum	not located				
Hatiyet Badr el Din	Matruh	approx.	29	50	27	50
Hatiyet el Saqr	Matruh	approx.	29	25	28	50
Hauwara Pyramid	Faiyum		29	16	30	54
Hauwaret	Faiyum		29	15	30	54
Hawatka	Asyut		27	18	31	00
Hazeva	ISRAEL		30	49	35	15
Heliopolis	Cairo		30	06	31	20
Helwan (Hulwan)	Cairo		29	51	31	20
Homran Barrage	see Nag Hammadi					
Hosh Isa	Beheira		30	55	30	17
Hurghada	Red Sea		27	14	33	50
Ibshawai	Faiyum		29	21	30	41
Idfina	Beheira		31	18	30	31
Idfu	Aswan		24	58	32	52
Idku	Beheira		31	18	30	18
Ikingi Maryut	Matruh		31	00	29	45
Ikyad	Sharqiya		30	45	31	52
Ile de Graye	see Gezira Fara'un					
Ilwa el Agramiya	South Sinai		28	45	33	58
Imbaba	Cairo		30	05	31	13
Inchas (Inshas)	Sharqiya		30	23	31	27
Ineiba	Aswan		22	38	32	20
Iskanderiya	see Alexandria					
Ismailiya	Ismailiya		30	35	32	16
Isna	Qena		25	18	32	33
Ityai el Barud	Beheira		30	53	30	40
Jedda	SAUDI ARABIA		21	28	39	10
Kadesh Barnea	see Ain el Gedeirat					
Kafr Aydub	Sharqiya		30	27	31	35
Kafr el Ayid	Sharqiya		30	41	31	31
Kafr el Dauwar (Dawar)	Beheira		31	10	30	07
Kafr el Sheikh	Kafr el Sheikh		31	07	30	56

Locality	Governorate, country, or cross-reference[1]		N. Lat. (°)	N. Lat. (′)	E. Long. (°)	E. Long. (′)
Kafr Hakim	Giza		30	05	31	07
Kafr Hemaid		not located				
Kalabsha	Aswan		23	33	32	52
Kantara	see El Qantara					
Karanis	see Kom Aushim					
Karkur Hamid	LIBYA		22	01	24	58
Karkur Ibrahim	see Abd el Malik					
Karkur Idris	LIBYA		22	03	24	52
Karkur Murr	SUDAN		21	53	25	06
Karkur Talh (Tahl)	Wadi Gedeed		22	02	25	08
Karnak	see El Karnak					
Kasfarit (Kasfareet)	Ismailiya		30	15	32	22
Kenah (Kina)	see Qena					
Kfar Kasem	ISRAEL		32	07	34	59
Kharga Oasis	Wadi Gedeed	approx.	25	26	30	33
Khartoum	SUDAN		15	39	32	30
Khor el Allaqi	see Wadi el Allaqi					
Khor Musa	SUDAN		21	51	31	16
Kitchener's Island	Aswan		24	05	32	53
Kom Aushim (Oshim)	Faiyum		29	34	30	55
Kom el Wahal	Kafr el Sheikh		31	14	30	57
Kom Ombo	Aswan		24	28	32	57
Kosti	SUDAN		13	10	32	42
Kurkur	Aswan		23	54	32	19
Kurum el Sheikh	North Sinai		31	13	34	07
Kurusku (Korosko)	Aswan		22	36	32	20
Kuseima	see Qusaima					
Kusseir	see Quseir					
Lake Bardawil	North Sinai	approx.	31	09	33	13
Lake Burullus	Kafr el Sheikh	approx.	31	30	30	50
Lake Idku	Beheira	approx.	31	15	30	15
Lake Ikyad	Sharqiya		30	44	31	54
Lake Manzala (Menzaleh)		approx.	31	15	32	05
Lake Maryut (Mareotis, Maeriotis)	Alexandria		31	08	29	56
Lake Moeris	see Lake Qarun					
Lake Nasser	Aswan	approx.	23	20	32	45
Lake Qarun	Faiyum	approx.	29	28	30	40
Lake Sirbonis	see Lake Bardawil					
Lake Timsah	Ismailiya		30	34	32	17
Little Bitter Lake	Suez		30	13	32	31
Lord Kitchener Island	see Kitchener's Island					
Luxor	Qena		25	41	32	39
Ma'agan Michael	ISRAEL		32	33	34	55
Ma'atin el Garawla	Matruh		31	14	27	24
Madiq	Aswan		22	44	32	09
Maghagha	Minya		28	39	30	50
Mahsama	Sharqiya		30	34	32	01
Maidum (Meidum)	Beni Suef		29	22	31	10
Mallahet Maryut	Alexandria		31	05	29	48
Mallawi (Mellawi)	Minya		27	44	30	50

Locality	Governorate, country, or cross-reference[1]	N. Lat. (°)	N. Lat. (′)	E. Long. (°)	E. Long. (′)
Mandara	Beheira	31	16	30	01
Manfalut	Asyut	27	20	30	50
Mareopolis	Beheira	30	59	29	52
Mareotis	see Lake Maryut				
Marsa Abu Suma	see Marsa Suma				
Marsa Baqqush	Matruh	31	11	27	40
Marsa el Alam	Red Sea	25	04	34	54
Marsa Matruh	Matruh	31	21	27	14
Marsa Mirakhi	see Marsa Morakh				
Marsa Morakh (The Fjord)	South Sinai	29	26	34	50
Marsa Sha'b (Shaab)	SGAA	22	50	35	48
Marsa Suma (Soma)	SGAA	22	53	35	44
Maryut (Mariut)	Alexandria	31	01	29	48
Matana	Qena	25	23	32	34
Matruh	Matruh	31	21	27	14
Mazghuna	Giza	29	45	31	15
Medinet Ashara Ramadan	see Tenth of Ramadan City				
Medinet el Sadat	see Sadat City				
Medinet Habu	Qena	25	42	32	38
Memphis	Giza	29	50	31	15
Mifraz Shelomo	see Sharm el Sheikh				
Misfaq	North Sinai	31	01	33	10
Mitla Pass	North Sinai	30	02	32	53
Mit Rahina (Riheina)	Giza	29	51	31	15
Mons Claudianus	Red Sea	26	48	33	29
Mons Pentadactylus	see Gebel el Farayid				
Moon Valley	South Sinai	29	38	34	50
Moses Well	see Ain Musa				
Mount Sinai	see Gebel Musa				
Muhammad Ali Barrage	see Delta Barrage				
Muhammad Qol	SUDAN	20	53	37	09
Mut	Wadi Gedeed	25	29	28	59
Myos Hormos	Red Sea	27	22	33	42
Na'ama	South Sinai	27	52	34	20
Nabq (Nabk, Nabek, Nebk)	South Sinai	28	04	34	25
Nag Hammadi	Qena	26	03	32	15
Nag Misaw	Aswan	22	22	31	42
Nahal Yam	North Sinai	31	04	33	05
Naqada	Qena	25	54	32	42
Nekhl (Nakhl)	North Sinai	29	55	33	45
Neviot	see Nuweiba				
New Kalabsha	Aswan	24	38	32	57
Nile Delta	approx.	31	00	31	00
Nizzana	ISRAEL	30	53	34	20
Nuweiba (Noiba)	South Sinai	28	58	34	39
Ofira	see Sharm el Sheikh				
Old Dam	Aswan	24	04	32	52
Pelusium	North Sinai	31	03	32	31
Philae	Aswan	24	01	32	53

Locality	Governorate, country, or cross-reference[1]		N. Lat. (°)	N. Lat. (′)	E. Long. (°)	E. Long. (′)
Port Fuad	North Sinai		31	15	32	19
Port Said	Port Said		31	16	32	18
Port Sudan	SUDAN		19	38	37	07
Port Taufiq (Tewfik)	Suez		29	57	32	34
Qad el Hamdan	South Sinai		27	48	33	55
Qadesh Barnea	see Ain el Gedeirat					
Qad Idn Haddan	see Qad el Hamdan					
Qara (Gara)	Matruh		29	37	26	31
Qasr Qarun	Faiyum		29	25	30	25
Qattara Depression	Matruh	approx.	29	40	27	10
Qena	Qena		26	10	32	43
Qift	Qena		26	00	32	49
Qulusna	Minya		28	21	30	44
Qus	Qena		25	55	32	45
Qusaima (Kossaima)	North Sinai		30	40	34	22
Quseir (Qosseir)	Red Sea		26	06	34	17
Rafa (Rafah)	North Sinai		31	17	34	14
Ramleh	Alexandria		31	14	29	58
Ras Abu Darag	Suez		29	23	32	33
Ras Abu Fatma	SGAA		22	26	36	26
Ras Abu Rudeis	South Sinai		28	54	33	10
Ras Banas (Benas)	Red Sea		23	54	35	48
Ras Budran	South Sinai		28	56	33	10
Ras Dib	Red Sea		28	03	33	22
Ras el Bahr (Barr)	Damietta		31	32	31	50
Ras el Gindi (Jundi)	South Sinai		29	51	33	08
Ras el Millan	South Sinai		27	54	33	55
Ras el Naqb (Nagb)	South Sinai		29	36	34	51
Ras el Sudr (Suder)	South Sinai		29	36	32	40
Ras Gemsa	Red Sea		27	39	33	35
Ras Gharib	Red Sea		28	21	33	06
Rashid	see Rosetta					
Rashida	see El Rashda					
Ras Muhammad (Mohammed)	South Sinai		27	44	34	15
Ras Nakiv	see Ras el Naqb					
Ras Shukheir	Red Sea		28	08	33	16
Ras Toronbi	Red Sea		25	39	34	35
Ras Umm Sid	South Sinai		27	51	34	19
Ras Zeit	Red Sea		27	56	33	31
Refidim (Revivim)	ISRAEL		31	02	34	43
Romani	see Rumana					
Rosetta	Beheira		31	24	30	25
Rumana (Romani)	North Sinai		31	00	32	40
Sabkhet el Bardawil	see Lake Bardawil					
Sadat City	Beheira		30	20	30	40
Sadd el Ali	see Aswan High Dam					
Sadot	North Sinai		31	13	34	11
Safaga	Red Sea		26	44	33	56
St Anthony Monastery	Red Sea		28	56	32	21
St John's Reef	see Gezira Zabargad					

Locality	Governorate, country, or cross-reference[1]		N. Lat. (°)	N. Lat. (′)	E. Long. (°)	E. Long. (′)
St Katherine Monastery	South Sinai		28	31	33	57
St Paul Monastery	Red Sea		28	51	32	33
St Simeon Monastery	Aswan		24	06	32	51
Salum (Sallum)	Matruh		31	34	25	09
Samalut	Minya		28	18	30	42
Sanabu	Asyut		27	30	30	47
San el Hagar	Sharqiya		30	58	31	52
Sanganeb Lighthouse	SUDAN		19	43	37	26
Saqqara	Giza		29	51	31	13
Sarabit el Khadim (Hadm)	South Sinai		29	01	33	27
Sawakin el Qadim (Suakin Khadim)	SGAA		22	18	36	29
Sede Boqer	ISRAEL		30	52	34	47
Serir (Sarir)	LIBYA		27	39	22	30
Shakshuk (Shakshouk)	Faiyum		29	28	30	42
Shalatein (Shelatin)	SGAA		23	08	35	36
Sharam Ash	see Sharm el Sheikh					
Sharm el Bahari	Red Sea		25	52	34	25
Sharm el Moiya (Moya)	South Sinai		27	51	34	17
Sharm el Sheikh	South Sinai		27	51	34	17
Shatanuf	Qalyubiya		30	13	31	04
Sheikh Shadli	Red Sea		24	12	34	38
Shellal	see Bir Shellal					
Shora el Manqata (Manketa)	South Sinai		28	13	34	26
Shubra el Kheima	Cairo		30	06	31	15
Shurtan el Manketa	see Shora el Manqata					
Sidi Abd el Rahman	Matruh		30	58	28	44
Sidi Barrani	Matruh		31	36	25	55
Sidi Omar	Matruh		31	24	24	52
Sidi Salim	Kafr el Sheikh		31	17	30	48
Sinnuris	Faiyum		29	25	30	52
Sirs el Liyyan	Minufiya		30	26	30	58
Sitra	Matruh		28	42	26	54
Siwa Oasis	Matruh		29	12	25	31
Sohag	Sohag		26	33	31	42
Sollum	see Salum					
Spring of Moses	see Ain Musa					
Suakin	SUDAN		19	05	37	20
Suez	Suez		29	58	32	33
Suez Bay	Suez	approx.	29	50	32	30
Suez Tunnel	Suez		30	03	32	33
Syene	see Aswan					
Taba	South Sinai		29	30	34	53
Tahta	Sohag		26	46	31	30
Tamiya	Faiyum		29	29	30	58
Tanta	Gharbiya		30	47	31	00
Tel el Kebir	see El Tell el Kebir					
Tell el Farame	see Pelusium					
Tenth of Ramadan City	Sharqiya		30	15	31	42
Thebes	Qena		25	41	32	39

Locality	Governorate, country, or cross-reference[1]		N. Lat. (°)	N. Lat. (′)	E. Long. (°)	E. Long. (′)
Tima	Sohag		26	54	31	26
Timna	ISRAEL		29	47	35	00
Tuna el Gebel	Minya		27	46	30	44
Tushka (Toshka)	Aswan		22	31	31	52
Umar Seraya	Wadi Gedeed		25	42	28	54
Umm Agawish Tourist Village	Red Sea		27	09	33	49
Umm Fawakhir	Red Sea		26	01	33	36
Umm Kataf	North Sinai		30	51	34	16
Uyun Musa	see Ain Musa					
Wadi Abu Fatma	SGAA		22	27	36	15
Wadi Abu Khasiba	Red Sea		28	45	32	19
Wadi Abu Qada (Gaada)	South Sinai		29	20	33	12
Wadi Aideib (Adeib)	SGAA		22	15	36	26
Wadi Akwamtra (Agwamtra)	SGAA		22	13	36	18
Wadi Araba	Red Sea	approx.	29	07	32	39
Wadi Arava	ISRAEL/JORDAN	approx.	30	18	35	13
Wadi Asal	Red Sea		25	57	34	23
Wadi Atalla (Atallah)	Red Sea		26	03	33	36
Wadi Aukau (Akau)	SGAA		22	15	36	22
Wadi Bali	Red Sea		27	27	33	39
Wadi Bosela	South Sinai		28	29	33	59
Wadi Buhluk (Buhlug)	Red Sea		26	20	33	30
Wadi Cansathrope	see Wadi Kansisrob					
Wadi Deir el Beda	Faiyum	not located				
Wadi Digla	Cairo		29	58	31	18
Wadi el Abyad	Ismailiya		30	18	32	17
Wadi el Allaqi	Aswan	approx.	23	08	32	46
Wadi el Arbain (Arbaein)	South Sinai		28	32	33	57
Wadi el Arish	North Sinai		31	03	33	50
Wadi el Asyuti	Red Sea		27	25	31	30
Wadi el Atrash	Red Sea	approx.	26	50	29	58
Wadi el Biyar (Biar)	North Sinai		29	55	34	10
Wadi el Gedeirat	see Ain el Gedeirat					
Wadi el Giddi	North Sinai		30	10	32	50
Wadi el Hadhira (Hadirah)	North Sinai		30	19	33	21
Wadi el Hammamat	Red Sea		25	58	33	33
Wadi el Heisi	South Sinai		29	25	34	33
Wadi el Kharit	Red Sea	approx.	24	26	33	03
Wadi el Lega	South Sinai		28	33	33	58
Wadi el Markh (Merkh)	Red Sea		26	21	33	03
Wadi el Mashash	North Sinai		29	40	34	41
Wadi el Natrun	Beheira	approx.	30	25	30	13
Wadi el Qubbaniya (Kubbaniya)	Aswan		24	16	32	48
Wadi el Quffa	Aswan		33	35	33	20
Wadi el Rayan (Raiyan)	Faiyum	approx.	29	05	30	20
Wadi el Sheikh	Red Sea		29	40	31	56
Wadi el Sheikh	South Sinai		28	44	33	50
Wadi el Tarfa	Red Sea	approx.	28	25	30	50
Wadi Feiran	South Sinai	approx.	28	45	33	25
Wadi Gabria	'Sinai'	not located				

Locality	Governorate, country, or cross-reference[1]		N. Lat. (°)	N. Lat. (′)	E. Long. (°)	E. Long. (′)
Wadi Gallaba	Red Sea		26	56	33	37
Wadi Garara	Red Sea	approx.	23	47	34	05
Wadi Garn el Kabsh	Red Sea		28	43	32	17
Wadi Gerady	North Sinai		31	02	33	32
Wadi Gerafi (Girafi)	North Sinai		29	56	34	40
Wadi Gharandal	South Sinai		29	20	33	00
Wadi Gharba	South Sinai		28	57	33	36
Wadi Gimal (Gemal)	Red Sea	approx.	24	40	35	06
Wadi Gurdi	Red Sea		26	40	32	40
Wadi Hagul	Suez		29	42	32	22
Wadi Halfa	SUDAN		21	56	31	20
Wadi Hamid	Aswan	approx.	22	10	32	10
Wadi Hanzul (Hamrul)	Wadi Gedeed		22	03	25	06
Wadi Hareitra	SGAA		22	05	36	13
Wadi Hegiya (Chagia)	South Sinai		29	23	33	34
Wadi Hibran (Hebron)	South Sinai		28	31	33	42
Wadi Himreyr	Red Sea		27	26	33	25
Wadi Hodein	Red Sea	approx.	23	04	35	20
Wadi Hof	Cairo		29	53	31	18
Wadi Ibib	SGAA		22	42	35	47
Wadi Kabrit	Red Sea		27	58	33	28
Wadi Kansisrob (Kansathrope)	SGAA		22	15	36	22
Wadi Kid	South Sinai		28	07	34	30
Wadi Kubbaniya	see Wadi el Qubbaniya					
Wadi Kurkur	Aswan		23	55	32	30
Wadi Kurusku (Korosko)	Aswan		22	36	32	20
Wadi Lagah (Legah)	see Wadi el Lega					
Wadi Lahami	Red Sea		24	13	35	25
Wadi Madsus	South Sinai		27	50	34	16
Wadi Meisa	SGAA		22	35	35	52
Wadi Mukattab	South Sinai		28	52	33	25
Wadi Muweih (Muwayh)	Red Sea		26	04	33	22
Wadi Naam	Red Sea	approx.	23	18	34	59
Wadi Nasb (Nassab)	South Sinai		28	31	34	14
Wadi Nugrus	Red Sea		24	37	34	47
Wadi Or	Aswan		22	15	31	45
Wadi Qattar	Red Sea		27	05	33	17
Wadi Qena	Red Sea	approx.	26	12	32	44
Wadi Qiseib (Quseib)	Suez		29	24	32	39
Wadi Rabdeit	SGAA		22	11	36	24
Wadi Rahaba	South Sinai		28	25	33	59
Wadi Rahaya	South Sinai		29	32	34	37
Wadi Rishrash	Red Sea		29	20	31	40
Wadi Safaga	Red Sea		26	38	33	59
Wadi Sa'l (Saal)	South Sinai		28	45	34	16
Wadi Sannur	Red Sea		28	59	31	03
Wadi Semna	Red Sea		26	27	33	35
Wadi Serimtai	SGAA		22	12	36	28
Wadi Shait	Red Sea		24	45	34	24
Wadi Shallal (Shellel)	SGAA		22	12	36	40

Locality	Governorate, country, or cross-reference[1]		N. Lat. (°)	N. Lat. (′)	E. Long. (°)	E. Long. (′)
Wadi Sharm el Qibliya	Red Sea		25	50	34	22
Wadi Shawak (Showak)	Red Sea		26	57	33	29
Wadi Solaf (Selaf, Silaf)	South Sinai		28	43	33	45
Wadi Sudr	South Sinai		29	39	32	41
Wadi Sukkari	Red Sea		25	03	34	49
Wadi Taba	South Sinai		29	32	34	52
Wadi Tarfa	South Sinai		28	44	33	43
Wadi Umm Anfei	Red Sea		27	15	33	23
Wadi Umm Ghalaqa	Red Sea		22	33	34	05
Wadi Umm Khisheib (Hashiba)	North Sinai		30	17	32	58
Wadi Umm Rus	Red Sea		25	28	34	35
Wadi Umm Taghir	Red Sea		26	41	33	46
Wadi Umm Yasar	Red Sea		27	03	33	11
Wadi Wadden (Waddan)	SUDAN		21	51	25	02
Wadi Watir (Wattir)	South Sinai		29	01	34	40
Wadi Yoider	SGAA		22	17	36	18
Wadi Zaghra	South Sinai		28	40	34	20
Wadi Zeidun	Red Sea	approx.	25	54	33	03
Wadi Zerara	South Sinai		28	49	34	34
Yamit	North Sinai		31	17	34	16
Yanbu	SAUDI ARABIA		24	07	38	02
Yotvata	ISRAEL		29	53	35	03
Zafarana	Red Sea		29	07	32	39
Zagazig	Sharqiya		30	35	31	31
Zamalek	Cairo		30	03	31	15
Zaranikh (Zarneech)	North Sinai		31	08	33	28
Zawyet Sidi Ghazi	Beheira		31	03	30	05
Zeit Bay	see Ghubbet el Zeit					

[1] SGAA = Sudan Government Administration Area
[2] Presumably the same locality as Meinertzhagen's Behera Land Coy
[3] Within a few hours walking distance of Abu Zabal (Raw 1921)
[4] Thought to be near El Qassasin
[5] In El Malaha (Abdel Megid 1945)
[6] Includes Siyal Kabir, Siyal Mutawaset, and Siyal Saghir
[7] Presumably El Qatta

REFERENCES

Abdel Aziz Ali, A. Q. (1982). *Atlas of the climate of Egypt by the computer.* Al-Hayita al-Masriyya lil-Kutub, Cairo [in Arabic].

Abdel Megid, I. M. (1945). Report on an expedition to Hurghada and the northern Red Sea islands. *Bull. Zool. Soc. Egypt*, 7, 48–53 [partially in Arabic].

Abdel Rahim, M. E., Fadil, A. E. M., Ali, A. Maher, and Selim, A. (1980). Technical report on field trip to Baharria, Farafra Oasis and Rayan Valley during the period 29/3/1979–10/4/1979. *Egypt. J. Wildl. Nat. Res.*, **4**, 127–42.

Abdel Rahman, K. (1955). A report about the society's camp at Marsa Matruh. *Bull. Zool. Soc. Egypt*, **12**, 14 [in Arabic].

Abu al Izz, M. S. (1971). *Landforms of Egypt.* The American University in Cairo Press, Cairo.

Adams, A. L. (1864). Notes and observations on the birds of Egypt and Nubia. *Ibis*, **6**, 1–36.

Admiralty of Great Britain. (1946). *Western Arabia and the Red Sea.* Geographical Handbook Series BR 527.

Afik, D. and Pinshow, B. (1984). Notes on the breeding biology of the Arabian Warbler *Sylvia leucomelaena* in the Arava (Rift Valley), Israel. *Ibis*, **126**, 82–9.

Alexander, H. G. (1929). Some birds seen in the Indian Ocean and the Mediterranean. *Ibis*, Ser. 12, 5, 41–53.

Al Hussaini, A. H. (1938*a*). Some birds observed in Ghardaqa (Hurghada) Red Sea coast. *Ibis*, Ser. 14, **2**, 541–4.

Al Hussaini, A. H. (1938*b*). The avifauna of the Bahariya Oasis in winter. *Ibis*, Ser. 14, **2**, 544–7.

Al Hussaini, A. H. (1938–9). An ornithological trip to Siwa Oasis. *Bull. de l'Inst. d'Égypte*, **21**, 145–52.

Al Hussaini, A. H. (1939). Further notes on the birds of Ghardaqa (Hurghada), Red Sea. *Ibis*, Ser. 14, **3**, 343–7.

Al Hussaini, A. H. (1954*a*). *Egyptian birds.* Anglo-Egyptian Bookshop, Cairo [in Arabic].

Al Hussaini, A. H. (1954*b*). Bird protection in Egypt. *Bull. Zool. Soc. Egypt*, **12**, 2–6.

Al Hussaini, A. H. (1959). The avifauna of Al-Wadi Al-Gadid in the Libyan Desert. *Bull. Zool. Soc. Egypt*, **14**, 1–14.

Ali, S. and Ripley, S. D. (1974). *Handbook of the birds of India and Pakistan: Flowerpeckers to Buntings*, Vol. 10. Oxford University Press, Bombay.

Allen, S. S. (1862). Notes on the birds of Egypt. *Ibis*, **4**, 357–61.

Allen, S. S. (1863). Notes on the birds of Egypt. *Ibis*, 5, 32–4, 156–7.

Allen, S. S. (1864). Remarks on Dr. A. Leith Adams's 'Notes and observations on the birds of Egypt and Nubia'. *Ibis*, **6**, 233–43.

Almond, W. E. (1937). Some observations on the bird life of Siwa, the oasis of Jupiter Ammon. *Vasculum*, **23**, 92–4.

Amr, B. and Burgstaller, H. (1984). Estimation of rodent damage in Egypt. *News about the Field Rat Control Project*, **2**, 30–44.

Anon. (1898). La capture des cailles en Égypte. *Naturaliste*, 20, 143.

Anon. (1912). Varia. *Jaarbericht Club Ned. Vogelk.*, 2.

Anon. (1941). Birds of Egypt. *Bull. Zool. Soc. Egypt*, 3 (Suppl.), 1–38 [partially in Arabic].

Anon. (1953). A new record for the Egyptian birds—*Cygnus bewicki* [*sic*]. *Bull. Zool. Soc. Egypt*, 11, 18 [in Arabic].

Anon. (1961). *Ringing and marking of animals*, Information Bulletin, No. 1. Ringing Centre Moscow, Moscow.

Anon. (1972). *Report on Swedish bird-ringing for 1972*. Bird-ringing Centre, Swedish Museum of Natural History, Stockholm.

Anon. (1973). *Report on Swedish bird-ringing for 1973*. Bird-ringing Centre, Swedish Museum of Natural History, Stockholm.

Anon. (1976). *Report on Swedish bird-ringing for 1976*. Bird-ringing Centre, Swedish Museum of Natural History, Stockholm.

Anon. (1977). *Suez Canal region feasibility and design studies*, Terms of reference Lake Manzala study EGY/76/001–07. ARE/UNDP Office for Projects Execution, Cairo.

Anon. (1981). Some breeding birds of the Red Sea and the Persian Gulf. *Bull. Jourdain Soc.*, 10-2, 21–6.

Anon. (1982). *Strategies for accelerating agricultural development. A report of the presidential mission on agricultural development in Egypt.* Ministry of Agriculture of the Arab Republic of Egypt and the US Agency for International Development.

Anon. (1984*a*). A new species for Egypt [*Emberiza tahapsi*]. *Bull. Orn. Soc. Middle East*, 13, 13–14.

Anon. (1984*b*). Beringungen und Wiederfunde 1981. *Ber. Vogelwarte Hiddensee*, 5, 98–116.

Antinori, M. O. (1862). Ueber eine kleine Vogelart aus der Familie der Ploceiden. *J. Orn.*, 10, 462.

Argyle, F. and Gal, B. (1978). *Birds in Eilat. Spring 1978*. Society for the Protection of Nature in Israel, Tel Aviv.

Arid Lands Information Center, University of Arizona. (1980). *Draft environmental report on Arab Republic of Egypt*. University of Arizona Office of Arid Lands Studies, Tucson.

Ash, J. S. and Ashford, O. M. (1977). Great Black-headed Gulls *Larus ichtyaetus* and Red-necked Phalaropes *Phalaropus lobatus* inland in Ethiopia. *J. East Afr. Nat. Hist. Soc. Nat. Mus.*, 31, 1–3.

Ash, V. S. and Pomeroy, D. E. (1981). Mapping schemes in the Afrotropical Region. *Ibis*, 123, 552–3.

Atta, G. A. M. (1986). Status of the breeding population of Little Tern (*Sterna albifrons*) in the Bardawil Nature Reserve (Egypt). Unpublished M.Sc. thesis. University of Aberdeen.

Attiah, H. H. (1977). Ecological assessment of pesticide management on terrestrial ecosystem in Egypt. *Proceedings University of California/AID–University of Alexandria, ARE seminar/workshop in pesticide management. Alexandria, 5–10 March 1977*, pp. 61–74.

Ayyad, M. A. (1980). A preliminary assessment of the effect of protection on the vegetation of the Mediterranean ecosystem. *Taeckholmia*, 9, 85–101.

Ayyad, M. A. and Ghabbour, S. I. (1977). Systems analysis of Mediterranean desert ecosystems of northern Egypt (SAMDENE). *Environ. Conserv.*, 4, 91–101.

Ayyad, M. A. and Ghabbour, S. I. (1986). Hot deserts of Egypt and the Sudan. In *Hot deserts and arid shrublands* (ed. M. Evenari, I. Noy-Meir, and D. W. Goodall), pp. 149–202. Elsevier, Amsterdam.

Baddar, T., Pease, M., Turtiainen, T., Silbiger, S., and Robinson, J. (1985). *Staff appraisal report, A.R.E. second agricultural development project*, World Bank Report, No. 5342a-EGT.

Bagnold, R. A. (1939). An expedition to the Gilf Kebir and 'Uweinat, 1938. *Geogr. J.*, **93**, 281–7.

Baha el Din, S. (1984*a*). New and interesting records for Egypt—1980–82. *Courser*, **1**, 47–51.

Baha el Din, S. (1984*b*). Early raptor migration at Suez. *Bull. Orn. Soc. Middle East*, **12**, 8.

Baha el Din, S. (1984*c*). Notes on the breeding of the Sooty Falcon *Falco concolor* on islands in the Red Sea, and their prey. *Bull. Orn. Soc. Middle East*, **12**, 2–3.

Baha el Din, S. (1984*d*). The Black Wheatear *Oenanthe leucura* in Egypt. *Courser*, **1**, 5–7.

Baha el Din, S. (1984*e*). Birding spots in Egypt. Island of Wadi Gimal. *Courser*, **1**, 52–3.

Baha el Din, S. (1985). The occurrence of the Palestine Sunbird (*Nectarina* [*sic*] *osea*) in Egypt. *Bull. Orn. Soc. Middle East*, **14**, 1–2.

Baha el Din, S. (in press). A contribution to the ornithology of the Western Desert. *Courser.*

Baha el Din, S. and Bruun, B. (1981). Raptor migration survey over Suez, spring 1981. *Torgos*, **1**, 22–4 [in Hebrew].

Baha el Din, S. and Salama, W. (1984). Waterbird migration study at Zaranik, Sinai. 2–24 September 1982. *Courser*, **1**, 28–43.

Baha el Din, S. and Saleh, M. A. (1983). *Report on the ornithological results of the Egyptian Red Sea pollution expedition October–December 1982*. Ornithological Society of Egypt, Cairo.

Bailey, R. S. (1966). Migrant land-birds in the Mediterranean and Red Seas and the Indian Ocean. *Ibis*, **108**, 421–2.

Ballmann, P. (1980). Report on the avian remains from sites in Egyptian Nubia, Upper Egypt and the Fayum. In *Loaves and fishes: the prehistory of Wadi Kubbaniya* (ed. F. Wendorf, R. Schild, and A. E. Close), pp. 307–10. Southern Methodist University, Dallas.

Barnes, R. S. K. (1980). *Coastal lagoons. The natural history of a neglected habitat*. Cambridge University Press, Cambridge.

Baron, H. (1966). Verfolgung von Störchen bei Sues. *Vogelwarte*, **23**, 219.

Barratt, L. (1982). *Scientific report of the Joint Services expedition to the Egyptian Red Sea. 1982*. Coral Reef and Tropical Marine Research Unit, University of York, York.

Bartlett, T. L. (1942). Egyptian bird tour, 1917–18. *Bird Notes and News*, **20**, 41–3.

Beadnell, H. J. L. (1927). *The wilderness of Sinai. A record of two years' recent exploration.* Edward Arnold, London.

Beadnell, H. J. L. (1931). Zerzura. *Geogr. J.*, **77**, 245–50.

Beets, C. (1953). Notes on the dredging in the Great Bitter Lake of the Suez Canal. *Zool. Meded. Rijksmus. Nat. Hist. Leiden*, **32**, 97–106.

Beltagy, A. I. (1976). Egypt: marine parks and reserves on the Red Sea coast. In *Promotion of the establishment of marine parks and reserves in the northern Indian Ocean including the Red Sea and Persian Gulf*, IUCN Publ. New Ser., No. 35, pp. 43–4.

Bender, B. (1975). *Farming in prehistory: from hunter-gatherer to food producer*. St Martin's, New York.

Ben Dov, A. and Golan, Y. (1983). New birds for Israel. *Tzufit* (*Sunbird*), **1**, 13–35 [in Hebrew].

Berney, F. L. (1919). An ornithologist with the A.I.F. in Egypt and Palestine. *Emu*, **19**, 43–8.

Bezzel, E. (1987*a*). Ornithologische Beobachtungen in Ägypten: November 1983. *Garmischer Vogelkundliche Berichte*, **16**, 36–51.

Bezzel, E. (1987*b*). Keilschwanzsturmtaucher (*Puffinus pacificus*) in Ägypten: 'Neue' Art für die Westpaläarktis? *Die Vogelwelt*, **108**, 71–72.

Biebach, H. (1985). Sahara stopover in migratory flycatchers: fat and food affect the time program. *Experientia*, **41**, 695–7.

Biebach, H., Friedrich, W., and Heine, G. (1986). Interaction of bodymass, fat, foraging and stopover period in trans-Sahara migrating passerine birds. *Oecologia (Berlin)*, **69**, 370–9.

Bijlsma, R. G. (1982). The visible migration of pipits (*Anthus*) and wagtails (*Motacilla*) near Suez (Egypt), autumn 1981. *Vogelwarte*, **31**, 423–7.

Bijlsma, R. G. (1983). The migration of raptors near Suez, Egypt, autumn 1981. *Sandgrouse*, **5**, 19–44.

Bijlsma, R. G. and de Roder, F. E. (1982). Goliath Herons in Egypt in 1981. *Dutch Birding*, **4**, 82–4.

Bijlsma, R. G. and Meininger, P. L. (1984). Behaviour of the House Crow, *Corvus splendens*, and additional notes on its distribution. *Gerfaut*, **74**, 3–13.

Bishai, H. M. and Kirollus, S. I. (1980). The water budget of Lake Qarûn and its physicochemical characteristics. *Water Supply and Management*, **4**, 93–7.

Bodenham, K. L. (1945*a*). Some notes on birds at R.A.F. Station Aboukir. *Bull. Zool. Soc. Egypt*, **7**, 14–20.

Bodenham, K. L. (1945*b*). Notes on migrant and resident birds in the northern Suez Canal area from March 1st to December 5th 1944. *Bull. Zool. Soc. Egypt*, **7**, 21–47.

Boessneck, J. (1985). Zum Vorkommen des Schreiseeadlers, *Haliaeetus vocifer* (Daudin, 1800) im alten Ägypten. *Spixiana*, **8**, 17–23.

Bonhote, J. L. (1918). *Reestablishment of the Buff-backed Heron* (Ardea ibis) *in Egypt, 1912–1918*. Government Press, Cairo.

Bonhote, J. L. (1919). *Bird liming in Lower Egypt*. Government Press, Cairo.

Booth, B. D. McD. (1961). Breeding of the Sooty Falcon in the Libyan Desert. *Ibis*, **103a**, 129–30.

Boraey, F. A. (1980). Studies on the changes of some ecological factors affecting fish life in Lake Qarun, Faiyum, Egypt. *Water Supply and Management*, **4**, 99–102.

Borman, F. W. (1928). Birds in Egyptian gardens. *Bird Notes and News*, **13**, 1–4, 27–30.

Borman, F. W. (1929). An ornithological trip in the Gulf of Suez and Red Sea. *Ibis*, Ser. 12, **5**, 639–50.

Bortoli, L., Cantoni, J., Lachaux, M., and Smart, M. (1970). *Liste provisoire des reprises et contrôles intéressant la Tunisie*. Société des sciences naturelles de Tunisie, Tunis.

Botje, H. (1985). *Egyptische stedeling kijkt vreemd aan tegen revolutie op het platteland*, NRC Handelsblad 8–2–1985.

Boulos, L. (1966). A natural history of Kurkur Oasis, Libyan Desert, Egypt. IV. The vegetation. *Postilla*, **100**, 1–22.

Boulos, L. (1982). Flora of Sinai. In *Environment in Sinai*, pp. 45–58. Arab Office of Youth and Environment, Al-Ahram, Cairo [in Arabic].

Boulos, L. (1985*a*). The Middle East. In *Plant resources of arid and semiarid lands* (ed. J. R. Goodin and D. K. Northington), pp. 129–85. Academic Press, Orlando.

Boulos, L. (1985*b*). The arid eastern and south-eastern Mediterranean region. In *Plant* *conservation in the Mediterranean area* (ed. C. Gómez-Campo), pp. 123–40. Junk, Dordrecht.

Boulos, L. and El Hadidi, M. N. (1984). *The weed flora of Egypt*. The American University in Cairo Press, Cairo.

Boulos, L. and Hobbs, J. (1986). Three arboreal species new to the Eastern Desert of Egypt. *Candollea*, **41**, 183–91.

Bourne, W. R. P. (1967). Observations of seabirds and review of literature. *Sea Swallow*, **19**, 51–76.

Bourne, W. R. P. and Radford, M. C. (1961). Notes on observations of sea-birds received during 1960. *Sea Swallow*, **14**, 7–27.

Bourne, W. R. P. and Radford, M. C. (1962). Observations of seabirds. *Sea Swallow*, **15**, 7–27.

Boyd, A. W. (1917). Birds of the Suez Canal Zone and Sinai Peninsula. *Ibis*, Ser. 10, **5**, 539–57.

Brewer, D. J. (1986). Cultural and environmental change in the Fayum, Egypt: an investigation based on faunal remains. Unpublished Ph.D. thesis. University of Tennessee.

Brewster, R. W. (1966). Some birds of the Sinai coastal belt. *Birds Illust.*, **12**, 44–9.

Broggi, M. F. (1985). *L'Europe et l'Afrique défendent la nature: la Convention de Berne*, Comité européen pour la conservation de la nature et des ressources naturelles. Conseil de l'Europe, Strasbourg.

Brooksbank, F. H. (1925). *Egyptian birds*. MacMillan, London.

Brown, L. H., Urban, E. K., and Newman, K. (1982). *The birds of Africa*, Vol. 1. Academic Press, London.

Brownlow, H. G. (1952). A ringing station in Egypt. *Ibis*, **94**, 128–32.

Brownlow, H. G. (1960). Migrational fidelity in Egypt. *Ibis*, **102**, 126.

Bruun, B. (1982). Egyptian institutions and organizations. *Sinai Newsletter*, **1**(1), 4–5.

Bruun, B. (1983). Wildlife offices established in all governorates. *Sinai Newsletter*, **1**(3), 3.

Bruun, B. (1984). Cinereous Bunting *Emberiza cineracea* in Egypt. *Courser*, **1**, 44–6.

Bruun, B. (1986). A law for financing nature conservation in Egypt. *Sinai Newsletter*, **4**(1), 2.

Bruun, B., Mendelssohn, H., and Bull, J. (1981). A new subspecies of Lappet-faced Vulture *Torgos tracheliotus* from the Negev Desert, Israel. *Bull. Brit. Orn. Club*, **101**, 244–7.

Bulman, J. F. H. (1942). Notes on the birds of the Libyan Desert. *Bull. Zool. Soc. Egypt*, **4**, 5–12.

Bulman, J. F. H. (1944). Notes on the birds of Safaga. *Ibis*, **86**, 480–92.

Bulman, J. F. H. and Beale, S. J. (1944). Bird notes from the Daedalus Lighthouse. *Bull. Zool. Soc. Egypt*, **6**, 32–9.

Bundy, G. (1976). *The birds of Libya. An annotated checklist*. British Ornithologists' Union, London.

Burckhardt, J. L. (1822). *Travels in Syria and the Holy Land*. John Murray, London.

Butler, A. L. (1905). A contribution to the ornithology of the Egyptian Soudan. *Ibis*, Ser. 8, **5**, 301–401.

Butzer, K. (1959). Environment and human ecology in Egypt during predynastic and dynastic times. *Bull. Soc. Géogr. d'Égypte*, **32**, 46–55.

Butzer, K. (1976). *Early hydraulic civilization in Egypt*. University of Chicago Press, Chicago.

Butzer, K. (1977). Geographie. In *Lexikon der Ägyptologie* (ed. W. Helck and W. Westendorf), pp. 526–30. Otto Harrassowitz, Wiesbaden.

Casement, M. B. (1976–7). Landbird summary. *Sea Swallow*, **27**, 18–43.

Casement, M. B. (1977–8). Landbird summary. *Sea Swallow*, 28, 21–42.
Casement, M. B. (1984). Landbirds from ships at sea 1983. *Sea Swallow*, 33, 22–35.
Cave, F. O. and Macdonald, J. D. (1955). *Birds of the Sudan*. Oliver & Boyd, London.
Cecil, W. (1904). *Bird notes from the Nile*. Constable, Westminster.
Central Agency for Public Mobilization and Statistics, Arab Republic of Egypt. (1984). *Statistical year book, Arab Republic of Egypt, 1952–1983*. Cairo.
Chapman, A. (1921). *Savage Sudan*. Gurney & Jackson, London.
Chapman, A. (1930). *Memories of fourscore years less two 1851–1929*. Gurney & Jackson, London.
Chapman, S. E. (1982). Notes on seabird reports received 1980–81. *Sea Swallow*, 31, 5–24.
Chapman, S. E. (1984). Notes on seabird reports received 1983. *Sea Swallow*, 33, 12–21.
Cheke, A. S. (1967). Further notes on land birds over the Mediterranean and Red Seas. *Ibis*, 109, 442–4.
Cholmley, A. J. (1897). Notes on the birds of the western coast of the Red Sea. *Ibis*, Ser. 7, 3, 196–209.
Christensen, S., Lou, O., Müller, M., and Wohlmuth, H. (1981). The spring migration of raptors in southern Israel and Sinai. *Sandgrouse*, 3, 1–42.
Churcher, C. S. (1972). Late Pleistocene vertebrates from archaeological sites in the plain of Kom Ombo, Upper Egypt. *Life Sci. Contr., Royal Ont. Mus.*, No. 82.
Clark, G. (1948). Fowling in prehistoric Europe. *Antiquity*, 22, 116–30.
Clawson, M., Landsberg, H. H., and Alexander, L. T. (1971). *The agricultural potential of the Middle East*. Elsevier, New York.
Cooper, J., Williams, A. J., and Britton, P. L. (1984). Distribution, population sizes and conservation of breeding seabirds in the Afrotropical region. *ICBP Tech. Publ.* No. 2, 403–19.
Cowley, J. C. (1953). 'Red Sea' Ospreys (*Pandion h. haliaetus*). *Oologist's Record*, 27, 23–4, 41–5, 55–9.
Cramp, S. (ed.) (1985). *The birds of the Western Palearctic*, Vol. 4. Oxford University Press, Oxford.
Cramp, S. (ed.) (1988). *The birds of the Western Palearctic*, Vol. 5. Oxford University Press, Oxford.
Cramp, S. and Simmons, K. E. L. (ed.) (1977). *The birds of the Western Palearctic*, Vol. 1. Oxford University Press, Oxford.
Cramp, S. and Simmons, K. E. L. (ed.) (1980). *The birds of the Western Palearctic*, Vol. 2. Oxford University Press, Oxford.
Cramp, S. and Simmons, K. E. L. (ed.) (1983). *The birds of the Western Palearctic*, Vol. 3. Oxford University Press, Oxford.
Crivelli, A. and Vizi, O. (1981). The Dalmatian Pelican, *Pelecanus crispus*, Bruch 1832, a recently world-endangered bird species. *Biol. Conserv.*, 20, 297–310.
Curry-Lindahl, K. (1974). The conservation story in Africa during the 1960's. *Biol. Conserv.*, 6, 170–8.
Curry-Lindahl, K. (1981). *Bird migration in Africa. Movements between six continents*. Academic Press, London.
Danielsson, B. (1959). Verksamheten vid Ottenby fågelstation 1958. *Vår Fågelvärld*, 18, 249–79.
Danin, A. (1973). Contributions to the flora of Sinai. II. New records. *Isr. J. Bot.*, 22, 18–32.

Danin, A. (1979). The flora. In *Sinai* (ed. B. Rothenberg), pp. 77–92. Kummerly & Frey, Berne.

Danin, A. (1983). *Desert vegetation of Israel and Sinai*. Cana, Jerusalem.

Darby, W. J., Ghalioungui, P., and Grivetti, L. (1977). *Food: the gift of Osiris*. Academic Press, London.

Darom, L. (1979). More about Sinai Rose Finches. *Israel Land and Nature*, 5(1), 23.

Davies, S. (1982). *Birds along the Egyptian Nile. 20 October–4 November 1982*. Royal Society for the Protection of Birds, Exeter.

De Bièvre, G. (1983). Ornithologisch verslag van de reis naar Egypte in 1982. *Wielewaal*, **49**, 108–17.

Dekker, R. W. R. J. (1981). Black Heron in Egypt in August 1980. *Dutch Birding*, **3**, 80.

De Meersman, L. (1979). Waarnemingen van enkele Euraziatische trekvogels in Soedan en Egypte. *Ornis Brabant*, 80, 4–5.

De Openshaw, R. D. (1961). Red Sea—Red Kites and Sparrow Hawks. *Sea Swallow*, **14**, 43.

Dovrat, E. (1982). Summary of five years survey of raptor migration at Kfar Kasem. *Torgos*, 2(1), 63–115 [in Hebrew].

Dubois, P. (1985). Considérations sur le Goéland d'Arménie *Larus armenicus* Buturlin en Israel. *Alauda*, 53, 226–8.

Dunnet, G. M., Crick, H. Q. P., and Baha el Din, S. M. (1986). Bardawil Lagoon baseline environmental study and vulnerability to oil-pollution. In *Mediterranean marine avifauna* (ed. Medmaravis and X. Monbaillu), pp. 335–58. Springer-Verlag, Berlin.

Durdin, C. (1984). *Swan Hellenic/RSPB Nile cruise 14–30 March 1984, ornithological report*. Royal Society for the Protection of Birds, Norwich.

Durrell, L. (1957). *Justine*. Faber & Faber, London.

Duse, A. (1935). Contributo allo studio della migrazione della Quaglia (*Coturnix coturnix coturnix* L.) nella Valle Padana. *Ric. Zool. Appl. Caccia*, **9**, 3–32.

Eames, J. C. (1986). A record of the Black Bush Chat in southern Israel. *Sandgrouse*, **7**, 60–1.

Edwards, S., Hind, S., and Rosenthal, D. (1981). *Red Sea reef study 1981*. General report of the Cambridge University Underwater Exploration Group, Cambridge.

Eidsvik, H. K. (1980). An introduction to international conservation programs in N.E. Africa. Pt 1. *Egypt. J. Wildl. Nat. Res.*, **3**, 1–6.

Ekins, G. and Hendriks, H. (1983). Bateleur in Israel in April 1982. *Dutch Birding*, **5**, 86–7.

El Badrawy, S. I. (1986). *Report of the bird migration expeditions in north Sinai in August and September 1986*. Ministry of Agriculture, Egyptian Wildlife Service, Cairo.

El Bakri, A. M. (1927–8). The Cattle Egret. A useful bird—with drawbacks. Proposed method of control. *Bull. Zool. Soc. Egypt*, **1**, 19–23.

El Hadidi, M. N. (n.d.). Flora of Sinai. In *International training course on wildlife, Sharm el Sheikh*, Background Paper, No. 4. Al-Ahram, Cairo.

El Hadidi, M. N. (1967). Observations on the flora of the Sinai mountain region. *Bull. Soc. Géogr. d'Égypte*, **40**, 123–55.

El Hadidi, M. N. (1971). Distribution of *Cyperus papyrus* L. and *Nymphaea lotus* L. in inland waters of Egypt. *Mitt. Bot. Staatssamml.*, **10**, 470–5.

El Hadidi, M. N. (1976). The riverain flora in Nubia. In *The Nile, biology of an ancient river* (ed. J. Rzóska), pp. 87–91. Junk, The Hague.

El Hadidi, M. N. and Ayyad, M. (1975). Floristic and ecological features of Wadi Hibis (Egypt). *Colloques Internationaux du CNRS*, **235**, 247–58.

El Hadidi, M. N. and Boulos, L. (1979). *Street trees in Egypt*. Cairo.

El Hadidi, M. N., Kosinová, J., and Chrtek, J. (1970). Weed flora of southern Sinai. *Acta Univ. Carolinae-Biologica*, 1969, 367–81.

El Hinnawi, E. E. (1980). The state of the Nile environment: an overview. *Water Supply and Management*, **4**, 1–11.

Elkington, J. (1975). Beware of the wrath of Osiris. *New Scientist*, **68**, 626–8.

Elliott, H. F. I. and Monk, J. F. (1952). Land-bird migration over the Suez route to East Africa. *Ibis*, **94**, 528–30.

El Negumi, A. (1949). List of, desert animals seen or collected during the periods shown. *Bull. Zool. Soc. Egypt*, **8**, 20–1.

El Negumi, A., Zain el Din, H. F., El Monery, M. A., and Fayed, M. K. (1950). *Birds of Egypt* (2nd edn). Daar el Fikr el Arabi, Cairo [in Arabic].

Elster, H. J. and Vollenweider, R. (1961). Beiträge zur Limnologie Ägyptens. *Arch. Hydrobiol.*, **57**, 241–343.

El Tom, M. A. (1975). *The rains of the Sudan*. Khartoum University Press, Khartoum.

El Zarka, S. E. D. (1963). Acclimatization of *Solea vulgaris* (Linn.) in Lake Quarun, Egypt. *J. Cons. Int. Explor. Mer.*, **28**, 126–36.

Entz, B. (1976). Lake Nasser and Lake Nubia. In *The Nile, biology of an ancient river* (ed. J. Rzóska), pp. 271–98. Junk, The Hague.

Entz, B. (1980). Ecological aspects of Lake Nasser–Nubia. *Water Supply and Management*, **4**, 67–72.

Erard, C. (1958). Sur les zones de reproduction et d'hivernage et les migrations du goéland railleur *Larus genei* Brême. *Alauda*, **26**, 86–104.

Etchécopar, R. D. and Hüe, F. (1964). *Les oiseaux du nord de l'Afrique*. N. Boubée & Cie, Paris.

Etchécopar, R. D. and Hüe, F. (1967). *The Birds of North Africa*. Oliver & Boyd, Edinburgh.

Etherington, J. R. (1983). *Wetland ecology*. Edward Arnold, London.

Euroconsult. (1984). *Land master plan, working paper no. 10. A note on the current five-year plan in light of the preliminary findings of the LMP-project*. Arab Republic of Egypt, Ministry of Development, General Authority for Rehabilitation Projects and Agricultural Development.

Ezzat, I. (1981). Eight nuclear power plants for Egypt by year 2000. *Cairo Today*, **2**(4), 9.

Fahmy, I. R. (1936). *Report on Gabel Elba*. University Press, Cairo.

Fakhry, A. (1974). *The oases of Egypt, Vol. II. Bahariya and Farafra Oases*. The American University in Cairo Press, Cairo.

Feige, K.-D. (1986). *Der Pirol*. A. Ziemsen Verlag, Wittenberg Lutherstadt.

Ferguson, D. A. (1984). Ornithological observation in the Hurghada area, April 22–May 5, 1982. *Courser*, **1**, 8–16.

Ferguson, W. (1970). Wildlife of Sinai. *Audubon*, **72**(2), 32–41.

Fischer, W. (1959). Seevogelbeobachtungen während einer Schiffsreise vom Indischen Ozean bis zur Ostsee. *Anz. Orn. Ges. Bayern*, **5**, 356–62.

Fischer, W. (1963). Vogelbeobachtungen in tropischen Häfen und am Suezkanal. *Orn. Mitt.*, **15**, 99–102.

Fishelson, L. (1980). Marine reserves along the Sinai Peninsula (northern Red Sea). *Helgolander Meeresunters*, **33**, 624–40.

Flint, P. R. and Stewart, P. F. (1983). *The birds of Cyprus*. British Ornithologists' Union, London.

Flower, S. S. (1933). Notes on some birds in Egypt. *Ibis*, Ser. 13, 3, 34–46.

Flower, S. S. and Nicoll, M. J. (1918). *The principal species of birds protected by law in Egypt.* Government Press, Cairo.

Frazier, J. and Salas, S. (1984). The status of marine turtles in the Egyptian Red Sea. *Biol. Conserv.*, 30, 41–67.

Frazier, J. G., Salas, S. S., and Saleh, M. A. (1984). Ornithological observations along the Egyptian Red Sea coast, spring 1982: with notes on migratory and breeding species. *Courser*, 1, 17–27.

Frumkin, R. (1986). The status of breeding raptors in the Israeli deserts, 1980–1985. *Sandgrouse*, 8, 42–57.

Frumkin, R. and Pinshow, B. (1983). Notes on the breeding ecology and distribution of the Sooty Falcon *Falco concolor* in Israel. *Ibis*, 125, 251–9.

Fry, C. H., Atta, G. A. M., Baha el Din, S., and McKay, C. R. (1985). *Bardawil Lagoon baseline environmental studies: 1985 spring ornithological survey. A report to BP Petroleum Development Ltd., Egypt.* Department of Geography, University of Aberdeen, Aberdeen.

Galal el Din, H. (1959). The Sooty Falcon, *Falco concolor* Temm. *Bull. Zool. Soc. Egypt*, 14, 52–3.

Galal el Din, S. and Fouad, S. H. (1939–40). A trip to Sinai Peninsula. *Bull. Zool. Soc. Egypt*, 2, 16–19.

Gallagher, M. D., Scott, D. A., Ormond, R. F. G., Connor, R. J., and Jennings, M. C. (1984). The distribution and conservation of seabirds breeding on the coasts and islands of Iran and Arabia. *ICBP Tech. Publ.*, 2, 421–56.

Gautier, A. (1976). Animal remains from archeological sites of terminal Paleolithic to Old Kingdom Age in the Faiyum. In *Prehistory of the Nile Valley* (ed. F. Wendorf and R. Schild), pp. 369–82. Academic Press, New York.

Gautier, A. (1984). Archaeozoology of the Bir Kiseiba region, eastern Sahara. In *Cattle-keepers of the eastern Sahara: the Neolithic of Bir Kiseiba* (ed. F. Wendorf, R. Schild, and A. Close), pp. 49–72. Pauls Press, New Delhi.

George, C. J. (1972). The role of the Aswan High Dam in changing the fisheries of the southeastern Mediterranean. In *The careless technology. Ecology and international development* (ed. M. T. Farvar and J. P. Milton), pp. 159–78. The Natural History Press, New York.

Géroudet, P. (1982). Le Goéland d'Arménie, *Larus* (*cachinnans*) *armenicus*, en Israël. *Alauda*, 50, 310–11.

Ghabbour, S. I. (1971). Some aspects of conservation in Egypt. *Biol. Conserv.*, 4, 66–7.

Ghabbour, S. I. (1972). Birds as agricultural pests in Africa. *Cairo Univ. Afr. Stud. Rev.*, 1, 101–23 [in Arabic].

Ghabbour, S. I. (1974). Insecticides and cotton in Egypt. *Biol. Conserv.*, 6, 62–3.

Ghabbour, S. I. (1975). National parks in Arab countries. *Environ. Conserv.*, 2, 45–6.

Ghabbour, S. I. (1976). The ecology and pest status of sparrows *Passer* in Egypt. *Int. Studies on Sparrows*, 9, 17–29.

Girgis, W. A., Zahran, M. A., Reda, K., and Shams, H. (1971). Ecological notes on Moghra Oasis (Western Desert, Egypt). *UAR J. Bot.*, 14, 147–55.

Glutz von Blotzheim, U. N., Bauer, K. M., and Bezzel, E. (ed.) (1973). *Handbuch der Vögel Mitteleuropas*, Vol. 5. Akademische Verlagsgesellschaft, Frankfurt am Main.

Gochfeld, M. (1983). The Roseate Tern: world distribution and status of a threatened species. *Biol. Conserv.*, 25, 103–25.

Goodman, S. M. (1982). The introduction and subspecies of the Rose-ringed Parakeet *Psittacula krameri* in Egypt. *Bull. Brit. Orn. Club*, 102, 16–18.

Goodman, S. M. (1984*a*). Report on two small bird collections from the Gebel Elba region, southeastern Egypt. *Bonn. zool. Beitr.*, 35, 39–56.

Goodman, S. M. (1984*b*). The validity and relationships of *Prinia gracilis natronensis* (Aves: Sylviidae). *Proc. Biol. Soc. Wash.*, 97, 1–11.

Goodman, S. M. (1985). *Natural resources and management considerations, Gebel Elba Conservation Area, Egypt/Sudan.* World Wildlife Fund/International Union for the Conservation of Nature and Natural Resources Project, No. 3612.

Goodman, S. M. (1986). The prey of Barn Owls (*Tyto alba*) inhabiting the ancient temple complex of Karnak, Egypt. *Ostrich*, 57, 109–12.

Goodman, S. M. (1988). Patterns of geographic variation in the Arabian Warbler *Sylvia leucomelaena* (Aves: Sylviidae). *Proc. Biol. Soc. Wash.*, 101, 898–911.

Goodman, S. M. and Ames, P. L. (1983). A contribution to the ornithology of the Siwa Oasis and Qattara Depression, Egypt. *Sandgrouse*, 5, 82–96.

Goodman, S. M. and Atta, G. A. M. (1987). The birds of southeastern Egypt. *Gerfaut*, 77, 3–41.

Goodman, S. M. and Helmy, I. (1986). The sand cat *Felis margarita* Loche, 1858, in Egypt. *Mammalia*, 50, 120–3.

Goodman, S. M. and Houlihan, P. F. (1981). The Collared Turtle Dove *Streptopelia decaocto* in Egypt. *Bull. Brit. Orn. Club*, 101, 334–6.

Goodman, S. M. and Meininger, P. L. (in press). The discovery of the Arabian Warbler *Sylvia leucomelaena* in southeastern Egypt. *Courser.*

Goodman, S. M. and Sabry, H. (1984). A specimen record of Hume's Tawny Owl *Strix butleri* from Egypt. *Bull. Brit. Orn. Club*, 104, 79–84.

Goodman, S. M. and Storer, R. W. (1985). A specimen of the White-throated Robin *Irania gutturalis* from the Sinai Peninsula. *Bull. Brit. Orn. Club*, 105, 84–5.

Goodman, S. M. and Storer, R. W. (1987). The seabirds of the Egyptian Red Sea and adjacent waters, with notes on selected Ciconiiformes. *Gerfaut*, 77, 109–45.

Goodman, S. M. and Tewfik, S. A. (1983). A specimen record of the Firecrest, *Regulus ignicapillus*, from north-western Egypt. *Gerfaut*, 73, 201–4.

Goodman, S. M. and Watson, G. E. (1983). Bird specimen records of some uncommon or previously unrecorded forms in Egypt. *Bull. Brit. Orn. Club*, 103, 101–6.

Goodman, S. M. and Watson, G. E. (1984). Records of Palearctic thrushes (*Turdus* spp.) in Egypt and northeastern Africa. *Gerfaut*, 74, 145–61.

Goodman, S. M., Meininger, P. L., and Mullié, W. C. (1983). The atlas of breeding birds in Egypt. *Dutch Birding*, 5, 22–3.

Goodman, S. M., Houlihan, P. F., and Helmy, I. (1984). Recent records of the Ostrich *Struthio camelus* in Egypt. *Bull. Brit. Orn. Club*, 104, 39–44.

Goodman, S. M., Meininger, P. L., and Mullié, W. C. (1986). The birds of the Egyptian Western Desert. *Misc. Publ. Univ. Mich. Mus. Zool.*, No. 172.

Goodman, S. M., Kraus, F., and Baha el Din, S. M. (in press). Records of terrestrial reptiles from Egyptian Red Sea islands. *Bull. Zool. Soc. Egypt.*

Goodwin, D. (1947). Notes on winter and passage migrants in eastern Libya, Egypt, and Malta. *Avicult. Mag.*, 53, 174–85.

Goodwin, D. (1949). Notes on the migration of birds of prey over Suez. *Ibis*, **91**, 59–63.

Goodwin, E. G. M. (1947*a*). A Middle East migration. New facts about a little-known movement of birds of prey. *Field*, **189**, 182.

Goodwin, E. G. M. (1947*b*). Radar and the speed of birds. *Illust. London News*, 15 November 1947, 552.

Greaves, R. H. (1936). *Sixty common birds of the Nile Delta: an introduction for beginners.* British War Fund for Welfare of Forces, Cairo.

Greaves, R. H. (1939). Notes from Egypt. *Oologist's Record*, **19**, 49–51.

Greaves, R. H. (1945*a*). Notes on the future study of birds migration in Egypt. *Bull. Zool. Soc. Egypt*, **7**, 58–63.

Greaves, R. H. (1945*b*). The Nile Valley Sunbird in Egypt. *Ibis*, **87**, 570–1.

Greaves, R. H. and Tregenza, L. A. (1937). The Nile Valley Sunbird (*Nectarinia metallica*) in Egypt. *Oologist's Record*, **17**, 79–83.

Grieve, A. (1985). *Swan Hellenic–RSPB Nile cruise 13–29 March 1985, ornithological report.* Royal Society for the Protection of Birds, Sandy.

Groombridge, B. (1982). *The IUCN amphibia–reptilia red data book: Testudines, Crocodylia, Rhynchocephalia*, Pt 1. International Union for the Conservation of Nature and Natural Resources, Gland.

Grote, H. (1930). Wanderungen und Winterquartiere der Paläarktischen Zugvögel in Afrika. *Mitt. Zool. Mus. Berlin*, **16**, 1–116.

Gurney, J. H. (1876). *Rambles of a naturalist in Egypt and other countries.* Jarrold, London.

Hachisuka, M. U. (1924). Notes on some birds from Egypt. *Ibis*, Ser. 11, **6**, 771–3.

Haensel, J. (1975). Ornithologische Eindrücke während eines Studien-aufenthalts in Ägypten im Herbst 1971. *Beitr. Vogelkunde*, **21**, 312–22.

Haensel, J. (1980). Ornithologische Eindrücke während eines weiteren Studien-aufenthalts in Ägypten im Herbst und Frühwinter 1972. *Beitr. Vogelkunde*, **26**, 19–29.

Haim, A. and Tchernov, E. (1974). The distribution of myomorph rodents in the Sinai Peninsula. *Mammalia*, **38**, 201–23.

Hale, W. G. (1973). The distribution of the Redshank *Tringa totanus* in the winter range. *Zool. J. Linn. Soc.*, **53**, 177–236.

Halfawi, S. (1945). A collection of birds from western Saudi-Arabia January–June 1944. *Bull. Zool. Soc. Egypt*, **7** (Suppl.), 1–11.

Hamdan, G. (1980). *The personality of Egypt.* 'Alam al-Kutub, Cairo [in Arabic].

Hardy, E. (1946). *A handlist of the birds of Palestine.* Office of the Command Education Officer, British Troops, North Levant.

Harrison, C. (1982). *An atlas of the birds of the Western Palaearctic.* Collins, London.

Hartley, P. H. T. (1949). The biology of the Mourning Chat in winter quarters. *Ibis*, **91**, 393–413.

Hassan, F. A. (1986). Desert environment and origins of agriculture in Egypt. *Norw. Arch. Rev.*, **19**, 63–76.

Hasselquist, F. (1766). *Voyages and travels in the Levant; in the years 1749, 50, 51, 52.* Davis & Reymers, London.

Hassib, M. (1951). Distribution of plant communities in Egypt. *Bull. Fac. Sci. Fouad I Univ.*, **29**, 59–261.

Hayes, W. C. (1965). *Most ancient Egypt*. University of Chicago Press, Chicago.

Haynes, C. V. (1982). Great Sand Sea and Selima Sand Sheet, eastern Sahara: geochronology of desertification. *Science*, **217**, 629–33.

Hefny, K. (1982). Land-use and management problems in the Nile Delta. *Nature and Resources*, 18(2), 22–7.

Heinze, J. (1981). Contributo all'Avifauna di Israele. *Gli Uccelli d'Italia*, **6**, 238–53.

Hildén, O. (1971). Occurrence, migration and colour phases of the Arctic Skua (*Stercorarius parasiticus*) in Finland. *Ann. Zool. Fennici*, **8**, 223–30.

Hindmarsh, H. (1949). *The annals of Tel el Kebir Shooting Club 1923–1943*.

Hiraldo, F., Delibes, M., and Calderón, J. (1984). Comments on the taxonomy of the Bearded Vulture *Gypaetus barbatus* (Linnaeus, 1758). *Bonn. zool. Beitr.*, **35**, 91–5.

Hobbs, J. J. (1980). *Sands of the shore: animals in Egyptian life*. Unpublished M.A. thesis. University of Texas, Austin.

Hobbs, J. J. (1986). *Bedouin reconciliation with the Egyptian desert*. Unpublished Ph.D. thesis. University of Texas, Austin.

Hogg, P. (1974). Trans-Saharan migration through Sarir, 1969–70. *Ibis*, **116**, 466–76.

Holgersen, H. (1974). Stavanger museums Gjenfunn 1971–1973. Del 2. Passeriformes. *Sterna*, **13**, 279–307.

Holz, R. K. (1969). Man-made landforms in the Nile Delta. *Geogr. Review*, **59**, 253–69.

Hoogstraal, H. and Kaiser, M. N. (1961). Ticks from European–Asiatic birds migrating through Egypt into Africa. *Science*, **133**, 277–8.

Hoogstraal, H., Wassif, K., and Kaiser, M. N. (1957). Results of the NAMRU-3 southeastern Egypt expedition, 1954, I: Introduction, itinerary and environmental conditions. *Bull. Zool. Soc. Egypt*, **13**, 1–15.

Hoogstraal, H., Kaiser, M. N., Traylor, M. A., Gaber, S., and Guindy, E. (1961). Ticks (Ixodoidea) on birds migrating from Africa to Europe and Asia. *Bull. Wld. Hlth. Org.*, **24**, 197–212.

Hoogstraal, H., Kaiser, M. N., Traylor, M. A., Guindy, E., and Gaber, S. (1963). Ticks (Ixodidae) on birds migrating from Europe and Asia to Africa, 1959–61. *Bull. Wld. Hlth. Org.*, **28**, 235–62.

Hoogstraal, H., Traylor, M. A., Gaber, S., Malakatis, G., Guindy, E., and Helmy, I. (1964). Ticks (Ixodidae) on migrating birds in Egypt, spring and fall 1962. *Bull. Wld. Hlth. Org.*, **30**, 355–67.

Hornell, J. (1937). Traps and snares from Upper Egypt. *Ethnos*, **2**, 64–73.

Horner, K. O. (1977). *A statistical analysis of spring bird migration at Bahij, Arab Republic of Egypt*. Unpublished Ph.D. thesis. Virginia Polytechnic Institute and State University.

Horner, K. O. (1980). Spring migration of *Sylvia* spp. on the north coast of the Arab Republic of Egypt. *Proc. IV Pan-African Orn. Cong.*, 215–26.

Horner, K. O., Nufer, W. L., and Baker, P. S. (1982). Density and laying dates of Blackshouldered Kites in Egypt. *Ostrich*, **53**, 249–51.

Horváth, L. (1958). A new race of the Desert Lark from Egypt. *Bull. Brit. Orn. Club*, **78**, 124–5.

Horváth, L. (1959). The results of the zoological collecting trip to Egypt in 1957, of the Natural History Museum, Budapest. *Ann. Hist.-Natur. Mus. Nat. Hung.*, **51**, 451–81.

Houlihan, P. F. and Goodman, S. M. (1986). *The birds of ancient Egypt*. Aris & Phillips, Warminster.

Hovel, H. (1987). *Check-list of the birds of Israel*. Society for the Protection of Nature in Israel, Tel Aviv.

Hubbard, J. P. (1967*a*). Ringing program in the U.A.R. (Egypt). *Ring*, **50–1**, 15–16.

Hubbard, J. P. (1967*b*). Birds ringed in Egypt 1966/67. *Ring*, **52–3**, 62–3.

Hubbard, J. P. and Horner, K. O. (1969). Recent progress of the Egyptian ringing program. *Ring*, **59**, 216–17.

Hubbard, J. P. and Seymour, C. (1968). Some notable bird records from Egypt. *Ibis*, **110**, 575–8.

Hume, R. (1983). *Swan Hellenic/RSPB Nile cruise 28 September–14 October 1983, ornithological report*. Royal Society for the Protection of Birds, Sandy.

Hume, R. (1985). *Swan Hellenic/RSPB Nile cruise 25 September–11 October 1985, ornithological report*. Royal Society for the Protection of Birds, Sandy.

Hurst, H. E. (1952). *The Nile*. Constable, London.

Ilani, G. (1974). Nature news from Israel. *Israel Land and Nature*, **16**(3), 150 [in Hebrew].

Ilani, G. (1976). Nature news from Israel. *Israel Land and Nature*, **18**(4), 181–4 [in Hebrew].

Ilani, G. (1979–80). Rare birds. *Israel Land and Nature*, **5**(2), 82–3.

Ilani, G. (1980). Nature news from Israel. *Israel Land and Nature*, **4**(22), 168–71 [in Hebrew].

Ilani, G. (1981). Wildlife news. *Israel Land and Nature*, **7**(1), 36–8.

Impekoven, M. (1964). Zugwege und Verbreitung der Knäkente, *Anas querquedula*; eine Analyse der europäischen Beringungsresultate. *Orn. Beobach.*, **61**, 1–34.

Inbar, R. (1979). The secret life of the Sinai Rose Finch. *Israel Land and Nature*, **5**(1), 20–2.

Innes, W. (1902). Mesures à prendre pour la protection des oiseaux en Égypte. *Bull. de l'Inst. d'Égypte*, Ser. 8, **4**(3), 169–77.

Innes, W. (1909). La protection des oiseaux migrateurs en Égypte. *Revue Franç. d'Ornith.*, **1**, 212–14.

Inskipp, T. (1979). Recent west Palearctic records of Citrine Wagtail. *Brit. Birds*, **72**, 44.

Isa, A. L. (1980). Pesticide regulations in Egypt to avoid hazards to the environment and human health. *J. Environ. Sci. Health*, **B15**(6), 615–24.

IUCN. (1980). *World conservation strategy: living resource conservation for sustainable development*. International Union for the Conservation of Nature and Natural Resources, Gland.

Jany, E. (1963). Salma Kabir–Kufra–Djabal al-Uwenat. Ein Reisebericht aus der östlichen Sahara. *Erde*, **94**, 334–62.

Jarvis, C. S. (1932). *Yesterday and today in Sinai*. Houghton Mifflin, Boston.

Jarvis, C. S. (1942). *Scattered shots*. John Murray, London.

Jennings, M. C. (1985). Atlas of the breeding birds of Arabia. *Phoenix*, No. 1 (plus inserts).

Jennings, M. C., Fryer, R. N., and Stagg, A. J. (1982). First breeding record of the Pink-backed Pelican, *Pelecanus rufescens* Gmelin, from Arabia. *Fauna of Saudi Arabia*, **4**, 478–82.

Jennings, M. C., Heathcote, P. C., Parr, D., and Baha el Din, S. M. (1985). *Ornithological survey of the Ras Dib area and the islands at the mouth of the Gulf of Suez, Egypt*. Oil Pollution Research Unit, Pembroke.

Johansen, H. (1962). Resultater af ringmærkningerne i 1957 og 1958. *Vidensk. Medd. Dansk Naturh. Foren.*, **124**, 217–303.

Johnsgard, P. A. (1981). *The plovers, sandpipers, and snipes of the world.* University of Nebraska Press, Lincoln.

Jones, M. E. (1961). Extracts from an interesting voyage: Gannets off Port Said. *Sea Swallow*, **14**, 44–5.

Jourdain, F. C. R. (1934). The Bearded Vulture (*Gypaëtus barbatus*) in Egypt. *Oologist's Record*, **14**, 73.

Jourdain, F. C. R. (1935). Notes on a recent visit to Egypt and Palestine. *Bull. Brit. Orn. Club*, **55**, 168–70.

Jourdain, F. C. R. and Lynes, H. (1936). Notes on Egyptian birds, 1935. *Ibis*, Ser. 13, **6**, 39–47.

Kadry, I. (1942). The economic importance of the Buff-backed Egret (*Ardea ibis*, L.) to Egyptian agriculture. *Bull. Zool. Soc. Egypt*, **4**, 20–6.

Kaiser, A. (1890). Beiträge zur Ornithologie von Aegypten. *Ornis*, **6**, 455–546.

Kaiser, A. (1892). Zur Ornis der Sinaihalbinsel. *Orn. Jahrbuch. Organ für das palaearktische Faunengebiet*, **3**, 207–48.

Kania, W. (1985). Results of bird ringing in Poland. Migrations of White Stork *Ciconia ciconia. Acta Orn.*, **21**, 1–41 [in Polish].

Kashef, A. I. (1981). Technical and ecological impacts of the Aswan High Dam. *J. Hydrol.*, **53**, 73–84.

Kassas, M. (1953). Habitat and plant communities in the Egyptian desert, II. The features of a desert community. *J. Ecol.*, **41**, 248–56.

Kassas, M. (1955). Rainfall and vegetation belts in arid north-east Africa. In *Plant Ecology*. Proceedings of the Montpellier Symposium, pp. 49–57. UNESCO Arid Zone Research.

Kassas, M. (1972). A brief history of land-use in Mareotis region, Egypt. *Minerva Biologica*, **1**, 167–74.

Kassas, M. (1981). Egypt. In *Handbook of contemporary developments in world ecology* (ed. E. J. Kormondy and J. F. McCormick), pp. 447–55. Greenwood Press, Westport, Connecticut.

Kassas, M. and Girgis, W. A. (1969). Plant life in the Nubian Desert east of the Nile, Egypt. *Bull. de l'Inst. d'Égypte*, **51**, 47–71.

Kassas, M. and Girgis, W. A. (1970). Habitat and plant communities in the Egyptian desert, VII. Geographical facies of plant communities. *J. Ecol.*, **58**, 335–50.

Kassas, M. and Girgis, W. A. (1972). Studies on the ecology of the Eastern Desert of Egypt, I. The region between latitude 27°30′ and latitude 25°30′N. *Bull. Soc. Géogr. d'Égypte*, **41–2**, 43–72.

Kassas, M. and Imam, M. (1959). Habitat and plant communities in the Egyptian desert, IV. The gravel desert. *J. Ecol.*, **47**, 289–310.

Kassas, M. and Zahran, M. A. (1962). Studies on the ecology of the Red Sea coastal land, I. The district of Gebel Ataqa and El-Galala El-Bahariya. *Bull. Soc. Géogr. d'Égypte*, **35**, 129–75.

Kassas, M. and Zahran, M. A. (1965). Studies on the ecology of the Red Sea coastal land, II. The district from El-Galala El-Qibliya to Hurghada. *Bull. Soc. Géogr. d'Égypte*, **38**, 155–93.

Kassas, M. and Zahran, M. A. (1967). On the ecology of the Red Sea littoral salt marsh, Egypt. *Ecol. Monogr.*, **37**, 297–316.

Kassas, M. and Zahran, M. A. (1971). Plant life on the coastal mountains of the Red Sea, Egypt. *J. Indian Bot. Soc.*, **50a**, 571–89.

Kastepold, K. E. (1974). *Estonian region 1973: bird ringing report no. 4*. Valgus, Tallin [in Estonian].

Kastepold, T. and Kabal, R. (1985). *Estonian region in 1983: ringing report no. 14*. Valgus, Tallin [in Estonian].

Kattinger, E. (1970). Vogelbeobachtungen im vorderen Orient: Syrien und unteres Niltal. *Naturforschende Ges. Bamberg*, **45**, 57–79.

Kees, H. (1961). *Ancient Egypt: a cultural topography*. University of Chicago Press, Chicago.

Kennedy, T. N. (1918). Birds of Egypt. *Ibis*, Ser. 10, **6**, 174.

Keve, A. and Pátkai, E. (1959). Hungarian ringed-birds in Africa. *Ostrich*, (Suppl.) **3**, 321–30.

Kiepenheuer, J. and Linsenmair, K. E. (1965). Vogelzug an der nordafrikanischen Küste von Tunesien bis Rotes Meer nach Tag- und Nachtbeobachtungen 1963 und 1964. *Vogelwarte*, **23**, 80–94.

King, B. (1978). April bird observations in Saudi Arabia. *J. Saudi Arabian Nat. Hist. Soc.*, **21**, 3–24.

Kinzelbach, R. K. (1986). New records of Goliath Heron, Yellow-billed Stork, Blyth's Reed Warbler and Clamorous Reed Warbler in the Middle East. *Bull. Orn. Soc. Middle East*, **17**, 13–17.

Kistchinski, A. A. (ed.) (1978). *Migration of birds of Eastern Europe and Northern Asia. Gaviiformes–Ciconiiformes*. Nauka, Moscow [in Russian].

Kistchinski, A. A. (ed.) (1979). *Migration of birds of Eastern Europe and Northern Asia. Ciconiiformes–Anseriformes*. Nauka, Moscow [in Russian].

Koch, A., Magnus, D., Seilkopf, H., and Baron, H. (1966). Der Weissstorch-Zug im Raum Sinai bis Kena in Landschaftsmorphologischer Sicht. *Vogelwarte*, **23**, 209–20.

Koenig, A. (1907). Die Falconiden Aegyptens. *J. Orn.*, **55**, 391–469, 549–82.

Koenig, A. (1919). Die Sperrschnäbler (Fissirostres) Aegyptens. *J. Orn.*, **67**, 431–85.

Koenig, A. (1920). Die Sitzfüssler (Insessores), die Klettervögel (Scansores) und die Rabenartigen Vögel (Coraces) Aegyptens. *J. Orn.*, **68** (Sonderheft), 1–148.

Koenig, A. (1924). Die Sänger (Cantores) Aegyptens. *J. Orn.*, **72** (Sonderheft), 1–247.

Koenig, A. (1926). Ein weiterer Teilbeitrag zur Avifauna Aegyptiaca, enthaltend die Ordnungen der Kegelschnäbler (Conirostres), der Tauben (Columbae), der Scharr- oder Hühnervögel (Rasores) und drei Vertreter aus der Ordnung der Wat- oder Sumpfvögel (Grallatores). *J. Orn.*, **74** (Sonderheft), 1–152.

Koenig, A. (1928). Fortsetzung und Schluss der Watvögel (Grallatores) Aegyptens. *J. Orn.*, **76** (Sonderheft), 1–311.

Koenig, A. (1932). Die Schwimmvögel (Natatores) Aegyptens, die Ruderfüsser (Steganopodes) Aegyptens und die Flügeltaucher (Urinatores) Aegyptens. *J. Orn.*, **80** (Sonderheft), 1–237.

Koenig, A. (1936). *Die Vögel am Nil. Band II. Die Raubvögel*. Private printing, Bonn.

Krabbe, N. (1980). *Checklist of the birds of Elat*. Private printing, Copenhagen.

Krabbe, N. (1983). A puzzling *Sylvia* from Eilat, Israel. *Bull. Orn. Soc. Middle East*, **10**, 13–14.

Kumerloeve, H. (1962). Basstölpel (*Sula bassana*) und Zwergmöwen (*Larus minutus*) vor der libanesischen Küste. *Vogelwarte*, **21**, 221–2.

Kumerloeve, H. (1983). Zur kenntnis altägyptischer Ibis-Darstellungen unter besonderer

Berücksichtigung des Waldrapps, *Geronticus eremita* (Linnaeus, 1758). *Bonn. zool. Beitr.*, **34**, 197–234.

Kurylas, H. (1985). National campaign promoted by the Field Rat Control Project. *News about the Field Rat Control Project*, **3**, 34–8.

Kyllingstad, H. C. (1986). A record of Bald Ibis from the Sinai Mountains. *Bull. Orn. Soc. Middle East*, **17**, 1–2.

Larmuth, J. (1973). Migration of *Motacilla alba alba*. *Bull. Brit. Orn. Club*, **93**, 97–8.

Lasserre, P. (1979). Coastal lagoons: sanctuary ecosystems, cradles of culture, targets for economic growth. *Nature and Resources*, **15**(4), 2–21.

Latif, A. F. A. (1984). Lake Nasser—the new man-made lake in Egypt (with reference to Nubia). In *Lakes and reservoirs* (ed. F. B. Taub), pp. 385–410. Elsevier Science Publishers, Amsterdam.

Lavee, D. and Safriel, U. (1974). Utilization of an oasis by desert-crossing migrants. *Isr. J. Zool.*, **23**, 219.

Lipkin, Y. (1972). Vegetation of the Bitter Lakes in the Suez Canal water system. *Isr. J. Zool.*, **21**, 447–57.

Loat, W. L. S. (1905). On a small collection of birds from the Wadi-en-Natrûn, Egypt. *Ibis*, Ser. 8, **5**, 453–61.

Loat, W. L. S. (1906). On a small collection of birds from the vicinity of Lake Menzalah, in the delta of Egypt. *Ibis*, Ser. 8, **6**, 113–24.

Long, R. C. (1961). Some land birds noted at sea between Mombasa and the English Channel. *Ibis*, **103a**, 131–3.

Lortet, L. and Gaillard, C. (1903). La faune momifiée de l'ancienne Égypte. *Arc. Mus. d'Hist. nat. Lyon*, **8**, 1–205.

Luther, H. and Rzóska, J. (1971). *Project Aqua: a source book of inland waters proposed for conservation.* Blackwell Scientific Publications, Oxford.

Lynes, H. (1912). Field-notes on a collection of birds from the Mediterranean. With systematic notes by H. F. Witherby. *Ibis*, Ser. 9, **6**, 121–87.

Mackintosh, D. R. (1941). Bird ringing recoveries. *Bull. Zool. Soc. Egypt*, **3**, 7–29.

Mackworth-Praed, C. W. and Grant, C. H. B. (1980). *Birds of eastern and north eastern Africa.* Longmans, London.

Maclaren, P. I. R. (1944). Bird notes from the neighbourhood of Amriya, Egypt. July–October 1942. *Bull. Zool. Soc. Egypt*, **6**, 19–25.

Maclaren, P. I. R. (1946). Bird notes of the Arabian and Red Seas. *J. Bombay Nat. Hist. Soc.*, **46**, 543–5.

Madden, J. F. (1927). Notes on some shore birds of the Red Sea Province. *Sudan Notes and Records*, **10**, 135–46.

Madden, J. F. (1929). Additional notes on the shore birds of the Red Sea Province. *Sudan Notes and Records*, **12**, 104–6.

Madden, J. F. (1930). Bird migration in the Red Sea Province. *Sudan Notes and Records*, **13**, 123–35.

Madge, S. (1981). The subspecific identity of the Little Green Bee-eater in southern Israel. *Sandgrouse*, **2**, 107.

Madge, S. (1982). April records of White-cheeked Tern in Sinai. *Dutch Birding*, **4**, 104–5.

Magaud d'Aubusson, M. (1891). Les canards d'Égypte et leur chasse. *Rev. Sc. Nat. Appl.*, No. 18.

Magaud d'Aubusson, M. (1892). Les échassiers d'Égypte. *Bull. de la Soc. Zool. d'Acclimatation*, 39, 49–62, 108–26.

Maratos, E. N. (1938). Quail shooting and netting. In *Sport in Egypt* (ed. J. W. Day), pp. 113–18. Country Life, London.

Marchant, S. (1939–40). Some notes on the birds of the Gulf of Suez. *Bull. Zool. Soc. Egypt*, 2, 6–14.

Marchant, S. (1941). Notes on the birds of the Gulf of Suez. *Ibis*, Ser. 14, 5, 265–95, 378–96.

Marx, H. (1968). *Checklist of the reptiles and amphibians of Egypt.* United States Naval Medical Research Unit Number Three, Cairo.

Mead, C. J. and Hudson, R. (1986). Report on bird-ringing for 1985. *Ringing & Migration*, 7, 139–88.

Mees, G. F. (1979). Verspreiding en getalssterkte van de Witwangstern, *Chlidonias hybridus* (Pallas), in Europa en Noord-Afrika. *Zool. Bijdr. Rijksmus. Nat. Hist. Leiden*, No. 26.

Meiklejohn, M. F. M. (1944). Bird notes from Egypt, autumn 1943. *Bull. Zool. Soc. Egypt*, 6, 16–18.

Meilstrup, H. and Sørensen, U. G. (1983). *DOF–Storturen Aegypten 1982 13.–28. marts.* Dansk Ornithologisk Forening, Copenhagen.

Meinertzhagen, R. (1920). Notes on the birds of southern Palestine. *Ibis*, Ser. 11, 2, 195–259.

Meinertzhagen, R. (1921). Notes on some birds from the Near East and from tropical East Africa. *Ibis*, Ser. 11, 3, 621–71.

Meinertzhagen, R. (1922). Notes on some birds from the Near East and from tropical East Africa. *Ibis*, Ser. 11, 4, 1–74.

Meinertzhagen, R. (1924). Birds at Egyptian lighthouses. *Ibis*, Ser. 11, 6, 643–4.

Meinertzhagen, R. (1930). *Nicoll's birds of Egypt.* Hugh Rees, London.

Meinertzhagen, R. (1954). *Birds of Arabia.* Oliver & Boyd, Edinburgh.

Meininger, P. L. and Baha el Din, S. M. (1986). Seabirds along the Mediterranean coast of Egypt. In *Mediterranean marine avifauna* (ed. Medmaravis and X. Monbailliu), pp. 107–21. Springer-Verlag, Berlin.

Meininger, P. L. and Dielissen, B. (1979). Ornitologische waarnemingen in Egypte in 1977 en 1978. *Veldorn. Tijdschrift*, 2, 78–86.

Meininger, P. L. and Mullié, W. C. (1981*a*). Egyptian wetlands as threatened wintering areas for waterbirds. *Sandgrouse*, 3, 62–77.

Meininger, P. L. and Mullié, W. C. (1981*b*). *The significance of Egyptian wetlands for wintering waterbirds.* The Holy Land Conservation Fund, New York.

Meininger, P. L. and Mullié, W. C. (1981*c*). Some interesting ornithological records from Egypt. *Bull. Orn. Soc. Middle East*, 6, 2–5.

Meininger, P. L. and Sørensen, U. G. (1984). Streaked Weaver *Ploceus manyar* breeding in Egypt. *Bull. Brit. Orn. Club*, 104, 54–7.

Meininger, P. L. and Sørensen, U. G. (1986). The occurrence of skuas (*Stercoraridae* [*sic*]) in the Middle East, with special reference to Egypt and the northern Red Sea. *Vogelwarte*, 33, 281–94.

Meininger, P. L., Mullié, W. C., van der Kamp, J., and Spaans, J. (1979). *Report of the Netherlands Ornithological Expedition to Egypt in January and February 1979.* Private printing, Middelburg.

Meininger, P. L., Mullié, W. C., and Bruun, B. (1980). The spread of the House Crow, *Corvus splendens*, with special reference to the occurrence in Egypt. *Gerfaut*, 70, 245–50.

Meininger, P. L., Baha el Din, S. M., and Mullié, W. C. (1981). Some notes on bird migration in the area of the Gulf of Suez (Egypt), autumn 1980. *Sandgrouse*, 3, 84–6.

Meininger, P. L., Sørensen, U. G., and Atta, G. A. M. (1986). Breeding birds of the lakes in the Nile Delta, Egypt. *Sandgrouse*, 7, 1–20.

Mendelssohn, H. (1972). The impact of pesticides on bird life in Israel. *Bull. Int. Council for Bird Pres.*, 11, 75–104.

Mendelssohn, H. (1975). Report on the status of some bird species in Israel in 1974. *Bull. Int. Council for Bird Pres.*, 12, 265–70.

Mendelssohn, H. and Leshem, Y. (1983). The status and conservation of vultures in Israel. In *Vulture biology and management* (ed. S. R. Wilbur and J. A. Jackson), pp. 86–98. University of California Press, Berkeley.

Mendelssohn, H., Marder, U., and Yom-Tov, Y. (1969). On the decline of migrant Quail (*Coturnix c. coturnix*) populations in Israel and Sinai. *Isr. J. Zool.*, 18, 317–23.

Merrie, D., Baha el Din, S., Howe, S., and Riad, A. (1986). Probable breeding of White-breasted Kingfisher in Egypt. *Bull. Orn. Soc. Middle East*, 16, 11–13.

Meshal, A. H. (1977). The problem of the salinity increase in Lake Qarun (Egypt) and a proposed solution. *J. Cons. Int. Explor. Mer.*, 37, 137–43.

Michelson, C. A. (ed.) (1985). *Migrations of birds of Eastern Europe and Northern Asia. Gruiformes–Charadriiformes.* Nauka, Moscow [in Russian].

Migahid, A. M., El Shafei Ali, M., Abd el Rahman, A. A., and Hammouda, M. A. (1960). An ecological study of Kharga and Dakhla Oases. *Bull. Soc. Géogr. d'Égypte*, 33, 279–310.

Miles, J. (1984). *Swan Hellenic–RSPB Nile cruise 26th September–12th October 1984, ornithological report.* Royal Society for the Protection of Birds, Sandy.

Misonne, X. (1970). Animals of the Libyan Desert. *Bull. Int. Union Conserv. Nat.*, 2, 131–2.

Misonne, X. (1974). Les oiseaux de Kufra et du Jebel Uweinat. *Gerfaut*, 64, 41–73.

Moheyi el Din, Y. (1986). *Agricultural economic statistics bulletin for 1979–1983.* Ministry of Agriculture, Central Administration for Agricultural Economics, Cairo [in Arabic].

Møller, A. P. (1975). The breeding population of Gull-billed Terns *Gelochelidon n. nilotica* Gmel. in 1972 in Europe, Africa and western Asia, with a review of fluctuations during the present century. *Dansk Orn. Foren. Tidsskr.*, 69, 1–8.

Molteno, D. I. (1944). List of migrants seen in Egypt, the Western Desert, Tripolitania. *Bull. Zool. Soc. Egypt*, 6, 27–31.

Montasir, A. H. (1954). Protection of flora and conservation of nature in Egypt. *Bull. Soc. Géogr. d'Égypte*, 27, 145–52.

Moore, R. J. and Balzarotti, M. A. (1983). Observations of sea birds nesting on islands of the Sudanese Red Sea. *Bull. Brit. Orn. Club*, 103, 65–71.

Moreau, R. E. (1927). Some notes from the Egyptian oases. *Ibis*, Ser. 12, 3, 210–45.

Moreau, R. E. (1927–8). Quail. *Bull. Zool. Soc. Egypt*, 1, 6–13.

Moreau, R. E. (1928). Some further notes from the Egyptian deserts. *Ibis*, Ser. 12, 4, 453–75.

Moreau, R. E. (1934). A contribution to the ornithology of the Libyan Desert. *Ibis*, Ser. 13, 4, 595–632.

Moreau, R. E. (1940–1). The ornithology of Siwa Oasis, with particular reference to the results of the Armstrong College expedition, 1935. *Bull. de l'Inst. d'Égypte*, 23, 247–61.

Moreau, R. E. (1951). The British status of the Quail and some problems of its biology. *Brit. Birds*, 44, 257–76.

Moreau, R. E. (1961). Problems of Mediterranean–Saharan migration. *Ibis*, **103a**, 373–427, 580–623.

Moreau, R. E. (1967). Water-birds over the Sahara. *Ibis*, **109**, 232–59.

Moreau, R. E. (1969). The Sooty Falcon *Falco concolor* Temminck. *Bull. Brit. Orn. Club*, **89**, 62–7.

Moreau, R. E. (1972). *The Palaearctic–African bird migration systems*. Academic Press, London.

Moreau, R. E. and Dolp, R. M. (1970). Fat, water, weights and wing-lengths of autumn migrants in transit on the north-west coast of Egypt. *Ibis*, **112**, 209–28.

Moreau, W. M. (1969). Hubbard and Seymour on recent Egyptian records. *Ibis*, **111**, 257.

Mörzer Bruyns, W. F. J. and Voous, K. H. (1965). Great Skuas *Stercorarius skua* in northern Indian Ocean. *Ardea*, **53**, 80–1.

Mullié, W. C. (1985). Bedreigingen van (trek) vogels in NO-Afrika. *Limosa*, **58**, 125–6.

Mullié, W. C. (1988). Catching of Common Terns by children in Egypt, *Bull. Orn. Soc. Middle East*, **20**, 22–23.

Mullié, W. C. and Meininger, P. L. (1981*a*). Numbers, measurements and stomach contents of Dunlins, Little Stints and Kentish Plovers from Lake Manzala, Egypt. *Wader Study Group Bull.*, **32**, 25–9.

Mullié, W. C. and Meininger, P. L. (1981*b*). Egypt's disappearing wetlands. *Birds*, **8**(8), 28–30.

Mullié, W. C. and Meininger, P. L. (1983). Waterbird trapping and hunting in Lake Manzala, Egypt, with an outline of its economic significance. *Biol. Conserv.*, **27**, 23–43.

Mullié, W. C. and Meininger, P. L. (1985). The decline of bird of prey populations in Egypt. *ICBP Tech. Publ.*, No. 5, 61–82.

Murphy, G. W. and Redman, N. J. (1983). An inland record of the Red-billed Tropicbird (*Phaethon aethereus*) in Israel. *Bull. Orn. Soc. Middle East*, **10**, 10–11.

Murray, G. W. (1951). The Egyptian climate: an historical outline. *Geogr. J.*, **117**, 422–34.

Muselet, D. (1985). Les quartiers d'hivernage des Sternes naines européennes *Sterna albifrons albifrons*. *Oiseau*, **55**, 183–193.

Muzinic, J. (1984–5). Recoveries of banded birds in the years 1976, 1977 and 1978. *Larus*, **36–7**, 23–48 [in Croato-Serbian].

Nicoll, M. J. (1908). Contributions to the ornithology of Egypt. No. I. Lake Menzaleh. *Ibis*, Ser. 9, **2**, 490–510.

Nicoll, M. J. (1909). Contributions to the ornithology of Egypt. No. II. Birds of the Province of Giza. *Ibis*, Ser. 9, **3**, 285–302, 471–84, 623–49.

Nicoll, M. J. (1912*a*). *Wild birds of the Giza Gardens 1898–1911*. Public Works Department, Cairo.

Nicoll, M. J. (1912*b*). Contributions to the ornithology of Egypt. No. III. The birds of the Wadi Natron. *Ibis*, Ser. 9, **6**, 405–53.

Nicoll, M. J. (1919). *Handlist of the birds of Egypt*. Government Press, Cairo.

Nicoll, M. J. (1921). Exhibition and descriptions of new subspecies (*Galerida cristata halfae*) and (*Charadrius varius allenbyi*) from Egypt. *Bull. Brit. Orn. Club*, **42**, 7.

Nicoll, M. J. (1922). On a collection of birds made in the Sudan by Major S. S. Flower. *Ibis*, Ser. 11, **4**, 688–701.

Nikolaus, G. (1981). Palaearctic migrants new to the north Sudan. *Scopus*, **5**, 121–4.

Nikolaus, G. (1984). Further notes on birds new or little known in the Sudan. *Scopus*, 8, 38–42.

Nikolaus, G. (1987). Distribution atlas of Sudan's birds with notes on habitat and status. *Bonn. zool. Monog.*, No. 25.

Nikolaus, G. and Hamed, D. H. (1984). Distinct status changes of certain Palaearctic migrants in the Sudan. *Scopus*, 8, 36–8.

O'Brien, R. B. (1982). Egypt's ambitious settlement plan for the Sinai. *San Francisco Examiner*, 3 January 1982, A18.

Olsen, K. M. (1983). Lichtenstein's Sandgrouse in Israel and Sinai. *Dutch Birding*, 5, 75–6.

Omer-Cooper, J. (1937). Zoological expedition to the Oasis of Siwa, Egyptian Libya. *Nature*, 140, 919–20.

Ormond, R. F. G. (1981). *Report on the need for management and marine parks in the Egyptian Red Sea.* Institute of Oceanography and Fisheries, Cairo.

Osborn, D. J. and Helmy, I. (1980). The contemporary land mammals of Egypt (including Sinai). *Fieldiana: Zoology*, New Ser., No. 5.

Osborn, D. J. and Krombein, K. V. (1969). Habitats, flora, mammals and wasps of Gebel 'Uweinat, Libyan Desert. *Smiths. Contr. Zool.*, 11, 1–18.

Österlöf, S. (1968). *Report for 1968 of the bird ringing office.* Swedish Museum of Natural History, Stockholm.

Österlöf, S. (1969). *Report for 1969 of the bird ringing office.* Swedish Museum of Natural History, Stockholm.

Österlöf, S. (1970). *Report for 1970 of the bird ringing office.* Swedish Museum of Natural History, Stockholm.

Österlöf, S. (1971). *Report for 1971 of the bird ringing office.* Swedish Museum of Natural History, Stockholm.

Österlöf, S. (1977). Migration, wintering areas, and site tenacity of the European Osprey *Pandion h. haliaetus* (L.). *Ornis Scand.*, 8, 61–78.

Otto, J. M. (1981). *Egypte.* Koninklijk Instituut voor de Tropen. Landendocumentatie 1980, No. 6. Staatsuitgeverij, 's-Gravenhage.

Paran, Y. (1979). Sooty Shearwaters (*Puffinus griseus*) in the eastern Mediterranean. *Bull. Orn. Soc. Middle East*, 2, 5–6.

Paran, Y. (1980). Some notes on waterbirds observed in Egypt and north Sinai—1978/79. *Bull. Orn. Soc. Middle East*, 4, 2–5.

Paran, Y. and Shlüter, P. (1981). The diurnal mass migration of the Little Bittern. *Sandgrouse*, 2, 108–10.

Parrot, C. (1903). Ornithologische Wahrnehmungen auf einer Fahrt nach Aegypten. *Jahresbericht Orn. Vereins München*, 3, 1–50.

Pátkai, I. (1966–7). Bird-banding of the Hungarian Ornithological Institute in the years 1960–66, 25th report on bird-banding. *Aquila*, 73–4, 81–107.

Payne, R. B. (1979). Family Ardeidae. In *Check-list of birds of the world*, Vol. 1, 2nd edn (ed. E. Mayr and G. W. Cottrell), pp. 193–244. Museum of Comparative Zoology, Cambridge, Massachusetts.

Paz, U. (1986). *The birds of Israel.* Ministry of Defense and Society for Protection of Nature, Tel Aviv [in Hebrew].

Paz, U. (1987). *The birds of Israel.* Christopher Helm, London.

Petersen, I. and Sørensen, U. G. (1980). *Ornithological observations from the Bardawil-lagoon, north Sinai, Egypt, October 1979*. Private printing, Humble/Copenhagen.

Petersen, I. and Sørensen, U. G. (1981*a*). *Migration studies from the eastern part of the lagoon 'Sabkhet el Bardawil' on the north coast of the Sinai Peninsula. 7–30 September 1980*. Private printing, Copenhagen.

Petersen, I. and Sørensen, U. G. (1981*b*). *Third Danish ornithological expedition to Egypt, autumn 1981. Preliminary report*. Private printing, Copenhagen.

Phillips, J. C. (1915). Some birds from Sinai and Palestine. *Auk*, **32**, 273–89.

Phillips, W. W. A. (1947). The ornithological diary of a voyage of S.S. *Samluzon*. *J. Bombay Nat. Hist. Soc.*, **46**, 593–613.

Phillips, W. W. A. and Rosenberg, H. (1959). Zum Zug des Weissstorchs im Gebiet von Rotem Meer, Suesgolf und Kanalzone. *Vogelwarte*, **20**, 116–21.

Prendergast, H. D. V. (1975–6). Up the Nile. *Adjutant*, **8**, 22–8.

Preuss, N. O. (1963). Bird-ringing in Denmark 1961: Report No. 30. *Dansk Orn. Foren. Tidsskr.*, **57**, 39–70 [in Danish].

Ptusjenko, E. S. (1948). Recoveries of *Larus r. ridibundus* L. ringed in the U.S.S.R. In *Trudy Tsentralnogho Bjuro Kol'tsevanija*, Vypusk VII (ed. Anon.), pp. 195–269. Ringing Centre Moscow, Moscow [in Russian].

Rabinowitz, D. (1980). *St. Catherine, nomination for World Heritage List*. International Union for the Conservation of Nature and Natural Resources, Gent.

Rabinowitz, D. (1982). *Environmental planning in the Santa Caterina area, Sinai, Egypt*. B.Sc. Final Year Project, Kings College, University of London.

Raw, W. (1921). Field notes on the birds of Lower Egypt. *Ibis*, Ser. 11, **3**, 238–64, 359–87.

Reed, C. A. (1964). A natural history of Kurkur Oasis, Libyan Desert, Western Governorate, Egypt. *Postilla*, **84**, 1–20.

Reed, C. A. and Lovejoy, T. E. (1969). The migration of the White Stork in Egypt and adjacent areas. *Condor*, **71**, 146–54.

Reiser, K.-H. (1973). Ringfunde des Teichrohrsängers (*Acrocephalus scirpaceus*). *Auspicium*, **5**, 47–58.

Ripley, S. D. (1963). Brief bird observations in Nubia. *Ibis*, **105**, 108–9.

le Roi, O. (1923). Die Ornis der Sinai-Halbinsel. *J. Orn.*, **71**, 28–95, 196–252.

Ross, J. P. and Barwani, M. A. (1981). Review of sea turtles in the Arabian area. In *The biology and conservation of sea turtles* (ed. K. Bjornal), pp. 373–83. Smithsonian Institution Press, Washington, DC.

Round, P. D. and Walsh, T. A. (1981). The field identification of Dunn's Lark. *Sandgrouse*, **3**, 78–83.

Rowntree, M. H. (1943). Some notes on Libyan and Egyptian birds. *Bull. Zool. Soc. Egypt*, **5**, 18–32.

Rudolf of Austria. (1889). *Notes on sport and ornithology*. Gurney & Jackson, London.

Rüger, A., Prentice, C., and Owen, M. (1986). *Results of the IWRB international waterfowl census 1967–1983*. International Waterfowl Research Bureau, Slimbridge.

Russell, F. (1905). Birds seen in Egypt in December, January and February. *Zoologist*, Ser. 4, **9**, 208–12.

Saad, M. A. H. (1980). A limnological study on Lake Nasser and the Nile in Egypt. *Water Supply and Management*, **4**, 81–92.

Safriel, U. (1968). Bird migration at Elat, Israel. *Ibis*, **110**, 283–320.

Safriel, U. N. (1975). Reoccurrence of the Red Avadavat *Amandava amandava* (L.) (Aves: Estrildidae) in Egypt. *Isr. J. Zool.*, **24**, 79.

Safriel, U. N., Ben-Hur, Y., and Ben-Tuvia, A. (1985). The diet of the Osprey on Tiran Island: management implications for populations on the northern Red Sea Islands. *ICBP Tech. Publ.*, No. 3, 181–93.

Sagi, Y., Boneh, M., Gidalison, E., and Porat, G. (1979). *Proposed nature conservation plan in Sinai.* Society for the Protection of Nature in Israel, Tel Aviv.

Saleh, M. A. (1984). Birds of the Eastern Desert of Egypt, the Bir Abraq area. *Courser*, **1**, 58–61.

Samman, A. A. (1974). Primary production of Lake Edku. *Bull. Inst. Ocean and Fish*, **4**, 259–317.

Sanford, K. S. and Arkell, W. J. (1933). *Paleolithic man and the Nile Valley in Nubia and Upper Egypt.* University of Chicago Press, Chicago.

Sauer, E. G. F. (1972). Ratite eggshells and phylogenetic questions. *Bonn. zool. Beitr.*, **23**, 3–48.

Savigny, J. C. (1805). *Histoire naturelle et mythologique de l'Ibis.* Paris.

Schäfer, H. (1919). Ägyptischer Vogelfang. *Amtliche Berichte aus den Preusz. Kunstsammlungen*, **40**, 163–84.

Scharff, A. (1928). Storchjagd in Nubien. *Orientalistische Literaturzeitung*, **11**, 928–36.

Scharlau, W. (1963). Ornithologische Beobachtungen in der ägyptischen Oase El-Dachla. *Orn. Mitt.*, **15**, 246–8.

Schlegel, H. (1862). *Revue méthodique et critique de la collection des oiseaux.* E. J. Brill, Leiden.

Schmidt, E. (1983). Bird-banding of the Hungarian Institute for Ornithology—34th report on bird-banding. *Aquila*, **90**, 139–48.

Schrader, G. (1892). Ornithologische Beobachtungen auf meinen Sammelreisen. *Orn. Jahrb.*, **3**, 41–54.

Schüz, E. (1959). Problems about the White Stork *Ciconia ciconia* in Africa seen from a European viewpoint. *Ostrich*, **3** (Suppl.), 333–41.

Sclater, P. L. (1895). Remarks upon the zoological institutions which he had recently visited in Egypt. *Proc. Zool. Soc. Lond.*, **1895**, 400–1.

Searight, R. G. (1952*a*). Notes on birds. (A). A trip to Fayoum. *Bull. Zool. Soc. Egypt*, **10**, 55–7.

Searight, R. G. (1952*b*). Notes on birds. (B). A trip from Suez to Ain Sukhana. *Bull. Zool. Soc. Egypt*, **10**, 57–9.

Searight, R. G. (1952*c*). Notes on birds. (C). Hurghada and Dishit el Daba—Gemsa. *Bull. Zool. Soc. Egypt*, **10**, 60–3.

Semenov, S. M. and Sabinevskii, B. V. (1957). Recoveries of *Larus genei* ringed in the U.S.S.R. In *Trudy Bjuro Kol'tsevanija*, Vypusk IX (ed. T. P. Shevareva), pp. 86–132. Ringing Centre Moscow, Moscow [in Russian].

Sere, D. (1986). Recoveries: Red-backed Shrike *Lanius collurio*. *Acrocephalus*, **29**, 25–7 [in Croato-Serbian].

Shariv, T. (1970). Breeding and migrating birds of the Bardawil area (northern Sinai). Unpublished M.Sc. thesis, The Hebrew University in Jerusalem [in Hebrew].

Sharrock, J. T. R. (1976). *The atlas of breeding birds in Britain and Ireland.* T. & A. D. Poyser, Berkhamsted.

Shelley, G. E. (1872). *A handbook to the birds of Egypt.* John van Voorst, London.

Sherman, A. R. (1915). Birds by the wayside in Egypt and Nubia. *Wilson Bull.*, **27**, 369–93.

Shevareva, T. P. (1955). Recoveries of *Larus melanocephalus* ringed in the U.S.S.R. In *Trudy Bjuro Kol'tsevanija*, Vypusk VIII (ed. V. F. Ryabova), pp. 46–90. Ringing Centre Moscow, Moscow [in Russian].

Shirihai, H. (1982). The autumn migration of Steppe Eagles at Eilat, Israel, 1980. *Sandgrouse*, **4**, 108–10.

Shirihai, H. (1986*a*). The Small Skylark, a species new to Israel and the Middle East. *Sandgrouse*, **7**, 47–54.

Shirihai, H. (1986*b*). Black Bush Robin in Israel. *Dutch Birding*, **8**, 85–8.

Shirihai, H. (1987). Shearwaters and other tubenoses at Eilat. *Dutch Birding*, **9**, 152–157.

Shirihai, H. and van den Berg, A. B. (1987). Influx of Kittlitz's Sand Plover in Israel in 1986–87. *Dutch Birding*, **9**, 85–8.

Short, L. L. and Horne, J. F. M. (1981). Bird observations along the Egyptian Nile. *Sandgrouse*, **3**, 43–61.

Siefke, A. (1982). Beringungen und Wiederfunde 1978. *Ber. Vogelwarte Hiddensee*, **2**, 85–112.

Siering, M. (1986). Verlauf und Ergebnisse einer Frühjahrsreise nach Unterägypten und in den Sinai vom 15. bis zum 29. April 1984. *Verhand. Orn. Gesell. Bayern*, **24**, 319–32.

Simmons, K. E. L. (1951). Raptor migration in the Suez area: autumn 1949–spring 1950. *Ibis*, **93**, 402–6.

Simmons, K. E. L. (1952). Some observations on the Olivaceous Warbler *Hippolais pallida* in Egypt. *Ibis*, **94**, 203–9.

Simmons, K. E. L. (1954*a*). The behaviour and general biology of the Graceful Warbler *Prinia gracilis. Ibis*, **96**, 262–92.

Simmons, K. E. L. (1954*b*). Field-notes on the behaviour of some passerines migrating through Egypt. *Ardea*, **42**, 140–51.

Simmons, K. E. L. and Hurrell, A. G. (1951). Nile Valley Sunbirds in the Suez Canal Zone. *Ibis*, **93**, 468–9.

Sluys, R. and van den Berg, M. (1982). On the specific status of the Cyprus Pied Wheatear *Oenanthe cypriaca. Ornis Scand.*, **13**, 123–8.

Smith, H. S. (1974). *A visit to ancient Egypt. Life at Memphis and Saqqara (c.500–30 BC).* Aris & Phillips, Warminster.

Snow, D. W. (ed.) (1978). *An atlas of speciation in African non-passerine birds.* British Museum (Natural History), London.

Sonnini, C. S. (1800). *Travels in Upper and Lower Egypt.* J. Debrett, London.

Sørensen, U. G. (1983). Raptor migration over the Red Sea Mountains near Port Safaga, Egypt. *Bull. Orn. Soc. Middle East*, **10**, 7–9.

Staav, R. (1977). Etude du passage de la Sterne caspienne *Hydroprogne caspia* en Méditerranée à partir des reprises d'oiseaux bagués en Suède. *Alauda*, **45**, 265–70.

Stamation, E. A. (1951). Notes on the White Stork on the 5th September 1951. *Bull. Zool. Soc. Egypt*, **9**, 28.

Steinbacher, J. (1965). Zur Vogelfauna nordafrikanischer Küstengebiete. *Senckenbergiana Biologica*, **46**, 429–59.

Stejneger, L. (1882). Outlines of a monograph of the Cygninae. *Proc. U.S. Natl. Mus.*, **5**, 174–221.

Stoneham, H. F. (1930). Notes on Egyptian birds. Pt I. *Bateleur*, **2**, 73–87.

Stresemann, E. (1954). Hemprich und Ehrenberg. Reisen zweier naturforschender Freunde im Orient geschildert in ihren Briefen aus den Jahren 1819–1826. *Abh. Akad. Wiss. Berlin. Math.- natur-wissen.*, No. 1.

Stromar, L. (1970). Bird-banding in 1966. *Larus*, **21–2**, 5–30 [in Croato-Serbian].

Stromar, L. (1971). Bird-banding in 1967 and 1968. *Larus*, **23**, 5–37 [in Croato-Serbian].

Su-Aretz, S. (1976). Desert island—a haven for birds. *Israel Land and Nature*, **18**(6), 263–7 [in Hebrew].

Su-Aretz, S. (1979). Spoonbills on Tiran Island. *Israel Land and Nature*, **4**(4), 170.

Su-Aretz, S. (1983). The Ospreys of Tiran—the bird island. *Hawk Trust Ann.*, **13**, 24–6.

Täckholm, V. (1974). *Student's flora of Egypt*, 2nd edn. Cairo University, Cairo.

Täckholm, V. and Drar, M. (1950). Flora of Egypt, Vol. II. *Bull. Fac. Sci. Cairo Univ.*, No. 28.

Täckholm, V. and Drar, M. (1969). Flora of Egypt, Vol. IV. *Bull. Fac. Sci. Cairo Univ.*, No. 36.

Tadros, T. M. (1953). A phytosociological study of halophilus communities from Mareotis (Egypt). *Vegetatio Acta Geobot.*, **8**, 137–60.

Taylor, E. C. (1859). Ornithological reminiscences of Egypt. *Ibis*, **1**, 41–55.

Taylor, E. C. (1867). Egypt revisited. *Ibis*, New Ser., **3**, 48–73.

Taylor, E. C. (1896). A few notes on birds of Egypt, from observations made at Cairo in the months of January and February, 1896. *Ibis*, Ser. 7, **2**, 477–82.

Tchernov, E. (1968). Peregrine Falcon and Purple Gallinule of late Pleistocene age in the Sudanese Aswan Reservoir area. *Auk*, **85**, 133.

Thomas, D. H. (1984). Adaptations of desert birds: Sandgrouse (Pteroclididae) as highly successful inhabitants of Afro–Asian arid lands. *J. Arid Environ.*, **7**, 157–81.

Ticehurst, C. B. (1912). The birds of Lower Egypt. *The Zoologist*, Ser. 4, **16**, 41–59.

Ticehurst, C. B.(1924). Birds from the Red Sea lights. *Ibis*, Ser. 11, **6**, 282–3.

Ticehurst, C. B. (1931). Notes on Egyptian birds. *Ibis*, Ser. 13, **1**, 575–8.

Tolstoy, A. (1960). Verksamheten vid Ottenby Fågelstation 1959. *Vår Fågelvärld*, **19**, 292–330.

Tomlinson, W. (1943). Some notes on Cyrenaican migrants. *Bull. Zool. Soc. Egypt*, **5**, 8–17.

Toschi, A. (1956). Esperienze sul comportamento di Quaglie (*Corturnix c. coturnix* (L.)) a migrazione interrotta. *Ric. Zool. Appl. Caccia*, **27**, 1–275.

Tree, A. J. (1987). Ringing recoveries and migration of Greenshank between Europe and Africa. *Safring News*, **16**, 51–66.

Tregenza, L. A. (1951). Observations on the birds of the S.E. desert of Egypt. *Bull. Zool. Soc. Egypt*, **9**, 1–18.

Tregenza, L. A. (1955). *The Red Sea Mountains of Egypt*. Oxford University Press, London.

Tregenza, L. A. (1958). *Egyptian Years*. Oxford University Press, London.

Treous, V. D. (1957). Recoveries of *Anas clypeata* ringed in the U.S.S.R. In *Trudy Bjuro Kol'tsevanija*, Vypusk IX (ed. T. P. Shevareva), pp. 187–207. Ringing Centre Moscow, Moscow [in Russian].

Tristram, H. B. (1879). [Letter to the editor.] *Stray Feathers*, **8**, 416–17.

Tuck, G. S. (1961). Reports of land birds at sea—analysis. *Sea Swallow*, **14**, 31–40.

Tuck, G. S. (1963). Reports on land birds at sea. *Sea Swallow*, **16**, 40–59.

Tuck, G. S. (1964). Reports of land birds at sea. *Sea Swallow*, **17**, 40–51.

Tuck, G. S. (1966). Reports of land birds at sea. *Sea Swallow*, **18**, 44–7.

Undeland, R. E. (1964). Verreaux's Eagle *Aquila verreauxii* in Sinai. *Ibis*, **106**, 258.

UNEP. (1985). *Environmental law in the United Nations Environment Programme*. UNEP, Nairobi.

Urban, E. K., Fry, C. H., and Keith, S. (1986). *The birds of Africa*, Vol. 2. Academic Press, London.

van Noten, F. (1978). *Rock art of the Jebel Uweinat*. Akademische Druck-Verlagsanstalt, Graz.

Vaurie, C. (1959). *The birds of the Palearctic fauna: Passeriformes*. H. F. & G. Witherby, London.

Vaurie, C. (1960). Systematic notes on Palearctic birds. No. 41 Strigidae: the genus *Bubo*. *Amer. Mus. Novit.*, **2000**, 1–31.

Vaurie, C. (1965). *The birds of the Palearctic fauna: non-Passeriformes*. H. F. & G. Witherby, London.

Von Besserer, Freiherr. (1908). Ornithologische Beobachtungen aus Ägypten. *Verhand. Orn. Gesell. Bayern*, **9**, 53–67.

Von Heuglin, T. (1856). Systematische Übersicht der Vögel Nord-Ost-Afrika's, mit Einschluss der arabischen Küste des rothen Meeres und der Nil-Quellen-Länder südwärts bis zum 4. Grade nördl. Breite. *Sitzung. Akad. Wiss. Math.-Naturw. classe, Vienna*, **19**, 255–324.

Von Heuglin, T. (1861). Forschungen über die Fauna des Rothen Meeres und der Somáliküste. *Mitt. aus Justus Perthes' Geografischen Anstalt*, **7**, 11–32.

Von Heuglin, T. (1869*a*). Synopsis der Vögel Nord-Ost-Afrikas, des Nilquellengebietes und der Küstenländer des Rothen Meeres. *J. Orn.*, **17**, 145–68.

Von Heuglin, T. (1869*b*–75). *Ornithologie Nordost-Afrika's der Nilquellen- und Küsten-Gebiete des Rothen Meeres und des nördlichen Somal-Landes*. Theodor Fischer, Cassel.

Voous, K. H. (1973). List of recent Holarctic bird species: non-passerines. *Ibis*, **115**, 612–38.

Voous, K. H. (1977). List of recent Holarctic bird species: passerines. *Ibis*, **119**, 223–50, 376–406.

Walter, H. (1979). *Eleonora's Falcon*. University of Chicago Press, Chicago.

Walters, M. (1982). A small Ostrich egg from Egypt. *Sandgrouse*, **4**, 116–17.

Wassif, K. (1982). Animals and wildlife in Sinai. In *Environment in Sinai*, pp. 37–44. Arab Office of Youth and Environment, Al-Ahram, Cairo [in Arabic].

Waterbury, J. (1979). *Hydropolitics of the Nile Valley*. Syracuse University Press, Syracuse.

Watson, G. E. (1971). *A serological and ectoparasite survey of migratory birds in northeast Africa. Final Report*. Smithsonian Institution, National Museum of Natural History, Washington, DC.

Wehr, H. (1976). *A dictionary of modern written Arabic* (ed. J. M. Cowan). Spoken Language Services, Ithaca.

Welch, G. (1983). *Swan Hellenic/RSPB Nile cruise ornithological report 9–25 March 1983*. Royal Society for the Protection of Birds, Fairburn.

Wells, S. (in prep.). Egypt. In *A directory of coral reefs of international importance*, pp. 85–116. International Union for the Conservation of Nature and Natural Resources Conservation Monitoring Unit, Cambridge.

Werner, Y. L. (1973). *The reptiles of Sinai Peninsula*. The Hebrew University in Jerusalem, Jerusalem [in Hebrew].

Whistler, H. (1916). Migration notes from a passage steamer. *Zoologist*, **20**, 453–8.

White, C. A. (1949). Bird notes. *Bull. Zool. Soc. Egypt*, **8**, 22–41.

Whymper, C. (1909). *Egyptian birds for the most part seen in the Nile Valley.* Adam & Charles Black, London.

Williams, J. G. (1955). A systematic revision and natural history of the Shining Sunbird of Africa. *Condor*, 57, 249–62.

Williams, S. M. (1981). Egypt: duck gunner's oasis. *Sports Afield.* January, 55–7, 103–4.

Wilson, R. T. (1982). Environmental changes in western Darfur, Sudan, over half a century and their effects on selected bird species. *Malimbus*, 4, 15–26.

Wimpfheimer, D., Bruun, B., Baha el Din, S. M., and Jennings, M. C. (1983). *The migration of birds of prey in the northern Red Sea area.* The Holy Land Conservation Fund, New York.

Wingate, O. (1934). In search of Zerzura. *Geogr. J.*, 83, 281–308.

Winterbottom, J. M. (1936). Five African voyages. *Ostrich*, 7, 90–8.

Woldhek, S. (1979). *Bird killing in the Mediterranean.* European Committee for the Prevention of Mass Destruction of Migratory Birds, Zeist.

World Development Report, 1986. (1986). Oxford University Press, Oxford.

Wuczeticz, W. and Tugarinov, A. (1937). *Seasonal distribution and migration of ducks (subfam.* Anatinae*) on the base of bird ringing in the U.S.S.R. II. The Pintail* Anas acuta L. Ringing Centre Moscow, Moscow.

Wyatt, C. W. (1870). Notes on the birds of the Peninsula of Sinai. *Ibis*, New Ser., 6, 1–18.

Yohe, J. (1982). Crop production. Annex F. In *Strategies for accelerating agricultural development* (ed. Anon.). Ministry of Agriculture of the Arab Republic of Egypt and the US Agency for International Development.

Yom-Tov, Y. (1987). The vertebrates of the Sinai Peninsula. In *Sinai* (ed. G. Gvirtzman, A. Shmueli, Y. Gradus, I. Bet-Arie, and M. Harel). Ministry of Defense, Tel Aviv [in Hebrew].

Zahran, M. A. (1965). Distribution of mangrove vegetation in Egypt. *Bull. Inst. Désert d'Égypte*, 15, 7–10.

Zahran, M. A. (1966). Ecological study of Wadi Dunkul. *Bull. Inst. Désert d'Égypte*, 16, 127–43.

Zahran, M. A. (1970–1). Wadi El-Rayan: a natural water reservoir (Western Desert, Egypt). *Bull. Soc. Géogr. d'Égypte*, 43–4, 83–99.

Zahran, M. A. (1972). On the ecology of Siwa Oasis, Egypt. *Egypt. J. Bot.*, 15, 223–42.

Zahran, M. A. and Girgis, W. A. (1972). On the ecology of Wadi el-Natrun. *Bull. Inst. d'Égypte*, 20, 229–67.

Zain el Din, H. F. (1951). The Darter. *Bull. Zool. Soc. Egypt*, 9, 5–6 [in Arabic].

Zedlitz, O. (1912). Von Suez zum Sankt Katharinen-Kloster. (Ein ornithologischer Streifzug.) *J. Orn.*, 60, 325–64, 529–69.

Zimmerman, D. (1984). Sanctuary on a powder keg. *Intern. Wildl.*, 14, 28–34.

Zink, G. (1954). Zugvogelbeobachtungen zwischen Venedig und Aden 22. November bis 4. December 1951. *Vogelwarte*, 17, 205–7.

Zink, G. (1969). Ringfunde der Vogelwarte Radolfzell 1947–68: Aufgliederung nach Fundgebieten und Fundmonaten. 1. Teil: Passeres. *Auspicium*, 3, 195–291.

Zink, G. (1973). *Der Zug europäischer Singvögel. Ein Atlas der Wiederfunde beringter Vögel.* 1. Lieferung. Vogelwarte Radolfzell am Max-Planck-Institut für Verhaltensphysiologie, Möggingen.

Zink, G. (1975). *Der Zug europäischer Singvögel. Ein Atlas der Wiederfunde beringter Vögel.* 2. Lieferung. Vogelwarte Radolfzell am Max-Planck-Institut für Verhaltensphysiologie, Möggingen.

Zink, G. (1981). *Der Zug europäischer Singvögel. Ein Atlas der Wiederfunde beringter Vögel.* 3. Lieferung. Vogelwarte Radolfzell am Max-Planck-Institut für Verhaltensphysiologie, Möggingen.

Zink, G. (1985). *Der Zug europäischer Singvögel. Ein Atlas der Wiederfunde beringter Vögel.* 4. Lieferung. Vogelwarte Radolfzell am Max-Planck-Institut für Verhaltensphysiologie, Möggingen.

Zohary, M. (1935). Die phytogeographische Gliederung der Flora der Halbinsel Sinai. *Beih. Bot. Centralblatt*, **52**, 549–621.

Zohary, M. (1944). Vegetational transects through the desert of Sinai. *Palest. J. Bot.*, Jerusalem Ser. 3, 57–78.

Zottola, G. (1969). Operation 017. *Ceres*, **2**, 34–40.

Zuckerbrot, Y. D., Safriel, U. N., and Paz, U. (1980). Autumn migration of Quail *Coturnix coturnix* at the north coast of the Sinai Peninsula. *Ibis*, **122**, 1–14.

INDEX OF SCIENTIFIC NAMES

INDEX OF ENGLISH NAMES

Figures in **bold type** refer to species accounts.

INDEX OF 'OFFICIAL' EGYPTIAN ARABIC NAMES (TRANSLITERATED)

INDEX OF 'OFFICIAL' EGYPTIAN ARABIC NAMES (ARABIC SCRIPT)

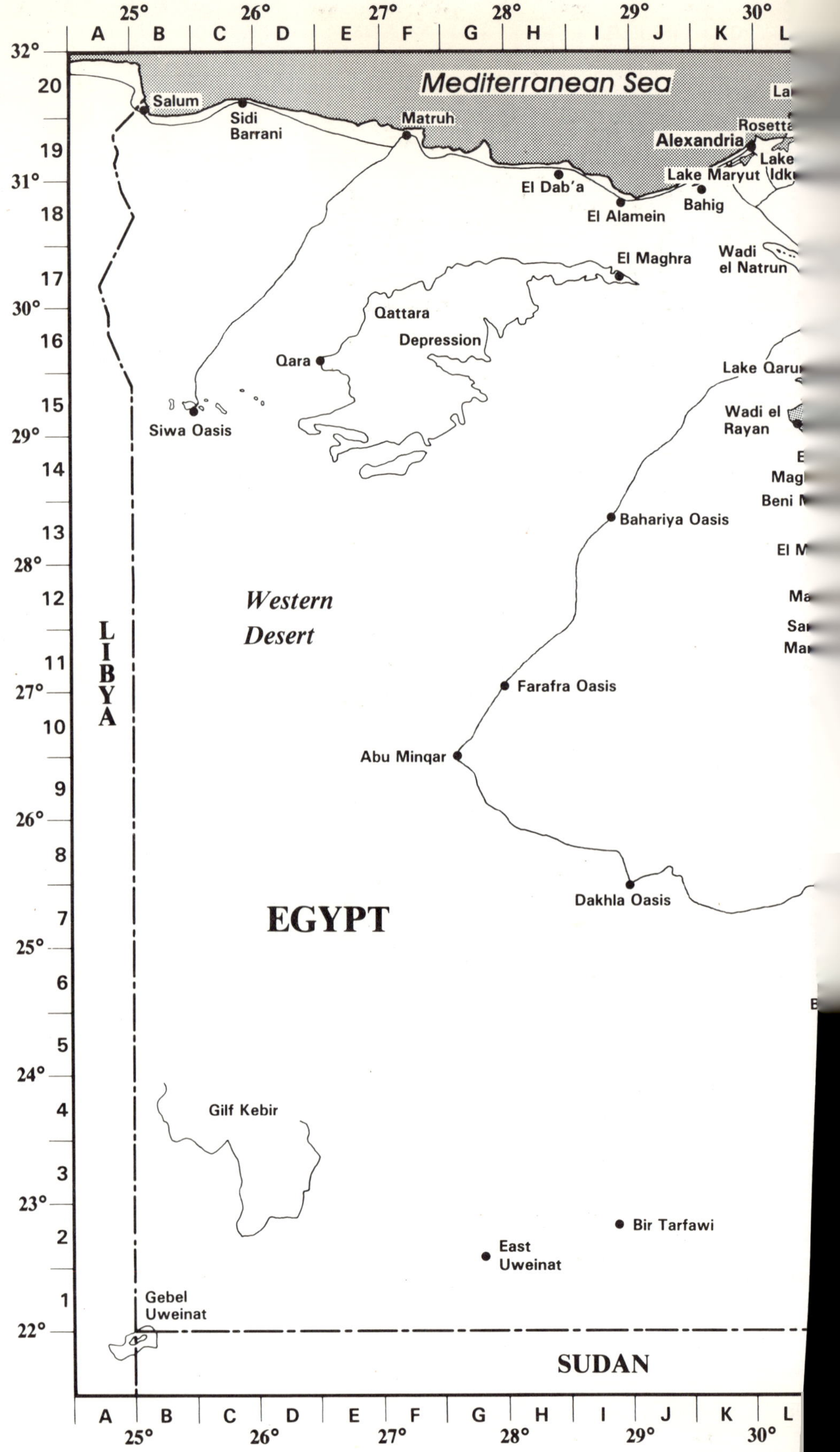

Mediterranean Sea
Salum
Sidi Barrani
Matruh
Alexandria
Lake Maryut
El Dab'a
El Alamein
Bahig
El Maghra
Wadi el Natrun
Qattara Depression
Qara
Siwa Oasis
Lake Qarun
Wadi el Rayan
Bahariya Oasis
Western Desert
LIBYA
Farafra Oasis
Abu Minqar
Dakhla Oasis
EGYPT
Gilf Kebir
Bir Tarfawi
East Uweinat
Gebel Uweinat
SUDAN
A B C D E F G H I J K L
25° 26° 27° 28° 29° 30°
32° 31° 30° 29° 28° 27° 26° 25° 24° 23° 22°
20 19 18 17 16 15 14 13 12 11 10 9 8 7 6 5 4 3 2 1